NAYFEH—Perturbation Methods
NAYFEH and MOOK—Nonlinear Oscillations
ODEN and REDDY—An Introduction to the Mathematical Theory of Finite Elements
PASSMAN—The Algebraic Structure of Group Rings
PETRICH—Inverse Semigroups
PIER—Amenable Locally Compact Groups
PRENTER—Splines and Variational Methods
RIBENBOIM—Algebraic Numbers
RICHTMYER and MORTON—Difference Methods for Initial-Value Problems, 2nd Edition
RIVLIN—The Chebyshev Polynomials
ROCKAFELLAR—Network Flows and Monotropic Optimization
RUDIN—Fourier Analysis on Groups
SAMELSON—An Introduction to Linear Algebra
SCHUMAKER—Spline Functions: Basic Theory
SHAPIRO—Introduction to the Theory of Numbers
SIEGEL—Topics in Complex Function Theory
Volume 1—Elliptic Functions and Uniformization Theory
Volume 2—Automorphic Functions and Abelian Integrals
Volume 3—Abelian Functions and Modular Functions of Several Variables
STAKGOLD—Green's Functions and Boundary Value Problems
STOKER—Differential Geometry
STOKER—Nonlinear Vibrations in Mechanical and Electrical Systems
STOKER—Water Waves
TURÁN—On A New Method of Analysis and Its Applications
WHITHAM—Linear and Nonlinear Waves
WOUK—A Course of Applied Functional Analysis
ZAUDERER—Partial Differential Equations of Applied Mathematics

APPLIED AND COMPUTATIONAL COMPLEX ANALYSIS

VOLUME 3

APPLIED AND COMPUTATIONAL COMPLEX ANALYSIS

VOLUME 3

Discrete Fourier Analysis—Cauchy Integrals—Construction of Conformal Maps—Univalent Functions

PETER HENRICI

Professor of Mathematics
Eidgenössische Technische Hochschule, Zürich

William Rand Kenan, jr., Professor of Mathematics
University of North Carolina, Chapel Hill, N.C.

A WILEY-INTERSCIENCE PUBLICATION

JOHN WILEY & SONS

New York • London • Sydney • Toronto

Library of Congress Cataloging in Publication Data

Henrici, Peter, 1923-

Applied and computational complex analysis.

(Pure and applied mathematics)

"A Wiley-Interscience publication."

Includes bibliographies and indexes.

Contents: v. 1. Power series—integration—conformal mapping—location of zeros.—v. 2. Special functions—integral transforms—asymptotics—continued fractions.—v. 3. Discrete Fourier analysis—Cauchy integrals—Construction of conformal maps—univalent functions.

1. Analytic functions. 2. Functions of complex variables. 3. Mathematical analysis. I. Title. II. Series.

QA331.H453 1974 515'.9 73-19723

ISBN 0-471-37244-7 (v. 1)

ISBN 0-471-08703-3 (v. 3)

Printed in the United States of America

10 9 8 7 6 5 4 3 2 1

To the Memory of
STEFAN BERGMAN

PREFACE

This is the third volume of a work aiming at presenting computational aspects and applications of complex analysis, the earlier volumes of which appeared in 1974 and 1977. The general goals of the project have remained unchanged: I try to present, on a level that is mathematically precise but nevertheless accessible to nonspecialists, complex analysis not merely as a logical structure of great beauty and coherence, but also as a tool for modeling phenomena of the physical world, and as a source of algorithms for the efficient use of these models. It is still my basic premise not to consider a problem solved unless we are able to provide an algorithm for constructing the solution. Moreover, this algorithm should not merely make the construction possible "in principle," but it should be efficient, and it should be implementable on computing equipment that is available today.

In the preface of Volume 1 I gave an outline of the contents of the whole work. Of the topics envisaged there for Volume 3 I here am able to present two, namely, two-dimensional potential theory and the construction of conformal maps for simply and multiply connected regions. These topics represent the core of the book. In addition I offer, under the heading "Cauchy integrals," an introduction to the boundary value problems of complex analysis, which will be found to have numerous applications, both to physical models and to topics presented subsequently. Much of the algorithmic underpinning for these theories is provided by the discrete Fourier transform, a thorough account of which is given in Chapter 13. Chapter 19 is a short essay on the theory of univalent functions, some aspects of which are required elsewhere in this series. The chapter culminates in an elementary account of the proof of the Bieberbach conjecture that was discovered early in 1984 by L. de Branges. Although it is, of course, not applied mathematics in the usual sense, this proof offers the opportunity to apply, in a context of undoubted significance, much of the "computational" material that had been presented earlier, especially in the areas of power series and ordinary differential equations.

The whole field of computational complex analysis has grown so rapidly during the past decade that it would have been difficult to do justice to the topics presented here in the space of less than a full volume. Thus, to my regret, I found myself unable to deal with the other topics originally planned for this volume, namely, approximation theory, several complex variables, and elliptic partial differential equations. I hope to present some applied and computational aspects of these topics in a fourth unit of this series, or perhaps in a different context.

Some readers will find the present volume lacking in quite a different direction. While discussing, for instance, a great many numerical methods for the construction of conformal maps, I have only rarely given practical advice concerning the implementation of any such method, nor did I render many value judgments. To the extent to which they are not based on mathematical analysis, such matters could be dealt with competently only on the basis of practical experience that considerably exceeds my own. Practical matters frequently are dealt with in the references provided; in addition, a forthcoming special issue of the *Journal of Computational and Applied Mathematics* under the editorship of L. N. Trefethen will be devoted exclusively to numerical conformal mapping, frequently from a practical point of view.

While the general thrust of the work has remained unchanged, the style of the present volume differs from that of its predecessors in some details. First, because the topics dealt with owe so much to research that was done only recently, I found it necessary to expand the "Notes" greatly, indicating my sources and providing references to further work. Such notes are now to be found at the end of each section instead of each chapter. Second and more important, the bias toward the Weierstrassian point of view that dominated the earlier volumes now has given way to a more balanced presentation in which the geometric point of view receives its due share. The Riemann mapping theorem, with all its ramifications, indeed is one of the central themes of this volume. The reader undoubtedly will appreciate the new freshness that emanates from this more visual approach to complex analysis.

What has not been changed, on the other hand, is the general mode of presentation. Authors who primarily write for professional mathematicians may cultivate a style where a large number of facts are presented as concisely and economically as possible. However, the present work is not directed exclusively, and perhaps not even primarily, toward such mathematicians. The readers whom I try to reach include the potential users of mathematics. A lifelong career in teaching this kind of reader has convinced me that, however great their appreciation for the logical coherence of the subject, their even greater concern is why they should be interested in it. Thus, time and again, I have allotted valuable space to the task of motivating what is ahead. Moreover, whenever facts or "theorems" are stated—and there are plenty of these—I have endeavored to find formulations that in their essence are intelligible also to readers who did not memorize all the preceding definitions. If I am accused of wordiness and of being, on occasion, repetitive, this is the price that I must pay for attempting to reach a larger audience.

Once again it is my humble duty to express my thanks to a number of individuals and organizations who in various ways helped me shape this volume. The list is even longer than on previous occasions because so many of the topics treated are close to current research.

First I wish to name some organizations that provided a forum for the exchange of some of the ideas presented here. The Department of Mathematics of the Swiss Federal Institute of Technology repeatedly permitted me to teach courses based on selected chapters of this work. The Mathematische Forschungsinstitut Oberwolfach under the directorship of M. Barner organized meetings on computational complex analysis in the summers of 1978, 1980, and 1983. G. H. Golub and L. N. Trefethen at the Computer Science Department of Stanford University organized a workshop on the same topic in 1981. H. Werner and L. Wuytack were instrumental in bringing about a NATO-sponsored conference on computational aspects of complex analysis in 1982. In the summer of 1983, J. Marti and M. Gutknecht organized a symposium at the Swiss Federal Institute of Technology where some 80 speakers presented papers on the same topic. The Department of Mathematics at the University of North Carolina (Chapel Hill) under the chairmanship of J. A. Pfaltzgraff in the academic year 1983–1984 offered hospitality during the final phase of the completion of this work.

It is futile to try to remember all the individuals who shaped my outlook on the topics presented herein. However, I wish to record my indebtedness to D. Gaier, W. B. Gragg, J. Hersch, J. A. Pfaltzgraff, and S. Warschawski for teaching me various aspects of function theory, and to J. W. Cooley and A. Schönhage for clarifying my ideas on discrete Fourier analysis. At the Swiss Federal Institute of Technology, H. Brauchli, A. Moser, P. Leuthold, and N. Rott opened my eyes to remarkable applications of complex analysis in engineering.

Further, I wish to thank the following younger collaborators at the Seminar für angewandte Mathematik for their contributions which enrich this volume: J.-P. Berrut, H. Daeppen, P. Geiger, H.-P. Hoidn, and M. Trummer. L. Oscity has drawn most of the figures.

Martin Gutknecht has been with me almost from the beginnings of this project, and has become a trusted advisor and friend. He not only contributed a large number of results, but also read portions of the manuscript and patiently corrected errors and suggested improvements, including numerous pertinent references.

John A. Pfaltzgraff likewise has read parts of the manuscript and rectified a number of errors.

Finally, L. N. Trefethen, from the time when I knew him as a graduate student at Stanford University, has taken a persevering interest in this work. His contribution goes far beyond his brilliant solution of the Schwarz-Christoffel parameter problem which is recorded in Chapter 16. On behalf of the publisher he read the entire manuscript, and he compiled many pages of suggestions for improvements in both presentation and substance. All of these suggestions had merit, and most of them I have been able to follow. As a result, this volume has a decidedly more contemporary and practical outlook than would otherwise be the case.

The combined efforts of these individuals have made this a better book. But all these efforts would have been to no avail if Beatrice Shube had not, on behalf of John Wiley & Sons, accepted this volume for publication and seen it through the production process in her usual competent manner. My heartfelt thanks go to her.

I dedicate this volume to the memory of Stefan Bergman who, long before the advent of modern computing machinery, had the vision to see the fruitfulness of computational methods in complex analysis, and who inspired my own first attempts in this field.

Zürich, Switzerland PETER HENRICI
September, 1985

CONTENTS

APPLIED AND COMPUTATIONAL COMPLEX ANALYSIS

VOLUME 3

13

DISCRETE FOURIER ANALYSIS

Fourier analysis is one of the most pervasive tools in applied analysis. Among other places, it occurs

(a) In the modeling of time-dependent phenomena that are exactly or approximately periodic.
(b) In the study of problems that involve a circular, spherical, or rectangular geometry.

Examples of (a) include the theory of alternating currents in electrical engineering; the digital processing of information such as speech, electrocardiograms, and electroencephalograms; and the analysis of geophysical phenomena such as earthquakes and tides. Examples for (b) occur not only in classical mathematical physics, notably in the study of vibrations of circular, spherical, or rectangular structures, but also in the transmission and in the processing of pictures, such as satellite pictures of the weather or the pictures of remote planets taken by space probes. As a limiting case of (a), Fourier analysis is also able to deal with nonperiodic processes, provided there is no action at times $\pm\infty$.

For functions defined on the real line, Fourier analysis was discussed briefly in § 4.5 and § 10.7. Frequently in the modeling or analysis of real situations, this theory cannot be applied directly, because the functions to be discussed are known only on a discrete set of "sampling points." It then becomes necessary to replace continuous Fourier analysis by a discretized version of it. Rather surprisingly, the mathematical theory of discrete Fourier analysis enjoys an even greater symmetry than the continuous theory. It is true that the original impact of discrete Fourier analysis was limited by the very large computational demands made by the theory in its naive form.

This was changed in 1965 by the introduction by Cooley and Tukey of a new family of algorithms, called fast Fourier transforms (FFT), which reduced the computational work required to carry out a discrete Fourier transform by orders of magnitude.

Since the FFT became available, discrete Fourier transform techniques have become so effective that they are being applied to problems and mathematical models that are not a priori periodic, such as the analysis of time series, the processing of speech and pictures and of digital data and signals in general. In the realm of mathematics, the FFT is used in efficient methods of computation with polynomials and with power series.

In our account of discrete Fourier analysis we shall emphasize its algorithmic and computational aspects as well as its applications to constructive complex analysis.

NOTES

Our presentation owes much to Cooley, Lewis, and Welch [1967a]. For accounts of discrete Fourier analysis specifically oriented toward electrical engineering and signal processing, see Rabiner and Gold [1975] and Oppenheim and Schafer [1975]. For the algebraist's view, see Auslander and Tolmieri [1979].

§ 13.1. THE DISCRETE FOURIER OPERATOR

I. Existence and Basic Properties

Let n be a positive integer. We denote by Π_n the space of bilaterally infinite sequences

$$\mathbf{x} = \{x_k\}_{k=-\infty}^{\infty}, \qquad x_k \in \mathbb{C},$$

that are periodic with period n,

$$x_{k+n} = x_k, \qquad \forall k.$$

Important examples of elements of Π_n are the sequences

$$\mathbf{f} = \left\{ f\left(\frac{k}{n}\right) \right\}$$

derived from functions $f: \mathbb{R} \to \mathbb{C}$ that are periodic with period 1.

A sequence $\mathbf{x} \in \Pi_n$ is fully determined by the elements of a period segment, for instance, by $x_0, x_1, \ldots, x_{n-1}$. If it is desired to exhibit the elements of $\mathbf{x}$ explicitly, we shall on occasion write

$$\mathbf{x} := \|: x_0, x_1, \ldots, x_{n-1} :\|,$$

the symbol $\|: \quad :\|$ indicating periodic repetition.

Defining addition of two sequences $\mathbf{x} = \{x_k\}$ and $\mathbf{y} = \{y_k\}$ in Π_n by

$$(\mathbf{x} + \mathbf{y})_k := x_k + y_k$$

and multiplication by a scalar $c \in \mathbb{C}$ by

$$(c\mathbf{x})_k := cx_k,$$

it is clear that Π_n becomes a linear space. The zero element of the space is the sequence $\mathbf{0} = \{0\}$, all elements of which are zero. Any sequence $\mathbf{x} \in \Pi_n$ can be expressed as a linear combination of the n sequences

$$\mathbf{e}^{(m)} = \{e_k^{(m)}\}, \qquad m = 0, 1, \ldots, n-1,$$

where

$$e_k^{(m)} = \begin{cases} 1, & k \equiv m \pmod{n} \\ 0, & k \not\equiv m \pmod{n}. \end{cases}$$

Because the sequences $\mathbf{e}^{(m)}$ are linearly independent, Π_n has dimension n.

If we define the *norm* of an element $\mathbf{x} = \{x_k\} \in \Pi_n$ as

$$\|\mathbf{x}\| := (|x_0|^2 + |x_1|^2 + \cdots + |x_{n-1}|^2)^{1/2},$$

it is clear that Π_n becomes a *normed linear space* (see § 2.1). It is obvious that this space is complete, and thus that it is a Banach space. In fact, Π_n is isomorphic to $\mathbb{C}^n$.

We take for granted the notion of a *linear operator* from one linear space to another. Some important examples of linear operators from Π_n to Π_n are the **reversion operator** R, which maps the sequence $\mathbf{x} = \{x_k\} \in \Pi_n$ onto

$$R\mathbf{x} := \{x_{-k}\}, \tag{13.1-1}$$

and the **shift operator** E, which maps $\mathbf{x} = \{x_k\}$ onto

$$E\mathbf{x} = \{x_{k+1}\} = \|\!: x_1, x_2, \ldots, x_{n-1}, x_0 :\!\|. \tag{13.1-2}$$

That is, E shifts a sequence one step to the *left*. Using the shift operator, any sequence $\mathbf{x} = \{x_k\} \in \Pi_n$ can be represented in terms of the **unit sequence**

$$\mathbf{e} := \mathbf{e}^{(0)} = \|\!: 1, 0, \ldots, 0 :\!\|. \tag{13.1-3}$$

We have, for instance,

$$\|\!: 0, 1, 0, \ldots, 0 :\!\| = E^{-1}\mathbf{e},$$

and therefore

$$\mathbf{x} = x_0\mathbf{e} + x_1 E^{-1}\mathbf{e} + \cdots + x_{n-1} E^{-n+1}\mathbf{e}. \tag{13.1-4}$$

Proceeding axiomatically, we now study the existence of a linear operator $\mathscr{F}$ from Π_n to Π_n satisfying the following three properties:

(F$_1$) For all $\mathbf{x} \in \Pi_n$,

$$\|\mathscr{F}\mathbf{x}\| = \|\mathbf{x}\|. \tag{13.1-5}$$

(F_2) For any $\mathbf{x}\in\Pi_n$, if $\mathscr{F}\mathbf{x}=\mathbf{y}=\{y_k\}$ and if E denotes the shift operator defined by (13.1-2), then

$$\mathscr{F}E\mathbf{x}=\{w^k y_k\} \tag{13.1-6}$$

for some complex constant w. Since the sequence $\mathscr{F}E\mathbf{x}$ should belong to Π_n and hence is to be periodic with period n, we must have $w^{k+n}y_k=w^k y_k$ for all k. It follows that w must be an nth root of unity. We demand that (13.1-6) hold for

$$w:=\exp\left(\frac{2\pi i}{n}\right). \tag{13.1-7}$$

(F_3) If $\mathbf{e}$ denotes the unit sequence defined by (13.1-3), there holds

$$(\mathscr{F}\mathbf{e})_k\geqq 0, \qquad \forall k. \tag{13.1-8}$$

Let $\mathscr{F}$ be an operator with the required properties, and let

$$\mathscr{F}\mathbf{e}=\{f_k\},$$

where all $f_k\geqq 0$. If $\mathbf{x}=\{x_k\}\in\Pi_n$, then by (13.1-4) we may write

$$\begin{aligned}\mathscr{F}\mathbf{x}&=\mathscr{F}(x_0\mathbf{e}+x_1E^{-1}\mathbf{e}+\cdots+x_{n-1}E^{-n+1}\mathbf{e})\\&=x_0\mathscr{F}\mathbf{e}+x_1\mathscr{F}E^{-1}\mathbf{e}+\cdots+x_{n-1}\mathscr{F}E^{-n+1}\mathbf{e}.\end{aligned}$$

Using (F_2), we have $\mathscr{F}\mathbf{x}=\{y_k\}$, where

$$y_k=(x_0+w^{-k}x_1+\cdots+w^{-(n-1)k}x_{n-1})f_k.$$

Property (F_1) now requires

$$\sum_{k=0}^{n-1}|x_k|^2=\sum_{k=0}^{n-1}|y_k|^2, \tag{13.1-9}$$

where, using (F_3),

$$|y_k|^2=y_k\bar{y}_k=\sum_{i,j=0}^{n-1}x_i\bar{x}_j w^{-k(i-j)}f_k^2.$$

We thus have

$$\begin{aligned}\|\mathbf{y}\|^2&=\sum_{k=0}^{n-1}f_k^2\sum_{i,j=0}^{n-1}x_i\bar{x}_j w^{-k(i-j)}\\&=\sum_{i,j=0}^{n-1}x_i\bar{x}_j\sum_{k=0}^{n-1}w^{-k(i-j)}f_k^2.\end{aligned}$$

This equals $\|\mathbf{x}\|^2$ for all $\mathbf{x}$ if and only if

$$\sum_{k=0}^{n-1}w^{-k(i-j)}f_k^2=\begin{cases}1, & i=j\\0, & i\neq j.\end{cases} \tag{13.1-10}$$

These relations may be regarded as a system of linear equations for the n quantities $f_0^2, f_1^2, \ldots, f_{n-1}^2$. The system will be satisfied for all $i, j \in \{0, 1, \ldots, n-1\}$ if and only if it is satisfied for $i-j=0, 1, \ldots, n-1$. The system then has the unique solution

$$f_k^2 = \frac{1}{n}, \qquad k = 0, 1, \ldots, n-1,$$

and it follows that

$$f_k = \frac{1}{\sqrt{n}}, \qquad k = 0, 1, \ldots, n-1.$$

Thus the operator $\mathcal{F}$, if it exists, is uniquely determined and is given by $\mathcal{F}\mathbf{x} = \mathbf{y}$ where, for all k,

$$y_k = (\mathcal{F}\mathbf{x})_k = \frac{1}{\sqrt{n}} \sum_{j=0}^{n-1} x_j w^{-jk}, \tag{13.1-11}$$

$$w := \exp\left(\frac{2\pi i}{n}\right). \tag{13.1-12}$$

Conversely, it is easily verified that the operator defined by (13.1-11) indeed has the properties (F_1), (F_2), and (F_3). This operator is called the **discrete Fourier operator** of order n. It will be denoted by $\mathcal{F}_n$ if it is necessary to indicate its dependence on n; likewise, the root of unity defined by (13.1-12) will on occasion be denoted by w_n.

Property (F_1) expresses the fact that the operator $\mathcal{F}$ is *isometric.* It is known from linear algebra that for any isometric operator from a finite-dimensional normed linear space to itself the inverse operator exists and is equal to the *adjoint* operator, which in the case of $\mathcal{F}$ is given by

$$(\bar{\mathcal{F}}\mathbf{y})_k = \frac{1}{\sqrt{n}} \sum_{j=0}^{n-1} y_j w^{jk}. \tag{13.1-13}$$

In this notation we thus have

$$\mathcal{F}^{-1} = \bar{\mathcal{F}}. \tag{13.1-14}$$

In relations such as (13.1-11) and (13.1-13) the summation with respect to j may be extended over any set of n integers j that are distinct modulo n. We thus have, for instance,

$$(\bar{\mathcal{F}}\mathbf{y})_k = \frac{1}{\sqrt{n}} \sum_{j=0}^{n-1} y_{-j} w^{-jk}, \qquad \forall k,$$

which is just the result of applying $\mathcal{F}$ to the sequence $R\mathbf{y}$. Thus there also holds

$$\mathcal{F}^{-1} = \mathcal{F}R. \tag{13.1-15}$$

Multiplying from the right by $\mathcal{F}$ and from the left by $\mathcal{F}^{-1}$ we also get

$$\mathcal{F}^{-1} = R\mathcal{F}. \tag{13.1-16}$$

All together we have established:

THEOREM 13.1a. *The only operator from Π_n to Π_n enjoying the properties (F_1), (F_2), and (F_3) is the discrete Fourier operator $\mathcal{F}$ defined by*

$$(\mathcal{F}\mathbf{x})_k = \frac{1}{\sqrt{n}} \sum_{j=0}^{n-1} x_j w^{-jk} \tag{13.1-17}$$

where $w := \exp(2\pi i/n)$. The inverse operator $\mathcal{F}^{-1}$ satisfies

$$\mathcal{F}^{-1} = \bar{\mathcal{F}} = \mathcal{F}R = R\mathcal{F}. \tag{13.1-18}$$

Attaching the name of Fourier to $\mathcal{F}$ is justified by the many connections that exist between $\mathcal{F}$ and the operators of continuous Fourier analysis; see the following sections. The sign of the exponent in w^{-jk} is likewise justified by this analogy. The choice of the scalar factor $1/\sqrt{n}$, on the other hand, is less important. Our choice is dictated by the mathematical fact that with this choice of factor $\mathcal{F}$ becomes a strict isometry. Some formal relations in § 13.2 would become simpler if a different factor, such as 1 or $1/n$, where chosen. From a numerical point of view the choice of factor is largely irrelevant, because in extended linear computations involving $\mathcal{F}$ the factor 1 is usually chosen and the correct factor is inserted only at the end, depending on the particular application on hand.

II. The Existence of a "Fast" Discrete Fourier Transform for Composite *n*

Here we consider the algorithmic problem of computing the sequence $\mathbf{y} := \mathcal{F}_n\mathbf{x}$ for a given $\mathbf{x} \in \Pi_n$. Because $\mathbf{y}$ is periodic, it suffices to calculate the elements of one full period of $\mathbf{y}$, for instance, the elements $y_0, y_1, \ldots, y_{n-1}$. To compute one such element directly from the formula (13.1-11), assuming that the required powers of $w = w_n$ have already been formed, clearly requires $n-1$ complex multiplications (μ). Modern applications of discrete Fourier analysis, for instance in the analysis of time series (see § 13.7), require values of n as large as 2^{14}. Then $n(n-1) \doteq 2.68 \times 10^8$, and the time required for forming even a single Fourier transform would appear to prohibit large-scale applications of the Fourier method.

It is therefore a fundamental fact of practical Fourier analysis that by a clever arrangement of the arithmetic operations the work required to form $\mathscr{F}_n\mathbf{x}$ can be drastically reduced. This reduction takes place for all values of n, but it is especially easy to implement if the integer n is highly composite. Here the basis for the reduction is a reduction formula that can be traced to the precomputer era, but whose importance for large-scale applications was recognized for the first time by Cooley and Tukey [1965]. If $n=2^l$, the reduction formula results in an algorithm that permits the evaluation of $\mathscr{F}_n\mathbf{x}$ in only $\frac{1}{2}ln=\frac{1}{2}n \operatorname{Log}_2 n$ complex multiplications.

Let $n=pq$, and for a given $\mathbf{x}=\{x_k\}\in\Pi_n$ let

$$\mathbf{x}^{(j)}=\{x_k^{(j)}\}:=\{x_{j+pk}\}, \qquad j=0,1,\ldots,p-1.$$

($\mathbf{x}^{(j)}$ is the subsequence of those elements of $\mathbf{x}$ whose index is $\equiv j \bmod p$.) Evidently

$$\mathbf{x}^{(j)}\in\Pi_q, \qquad j=0,1,\ldots,p-1.$$

We assume that the p sequences

$$\mathbf{y}^{(j)}=\{y_m^{(j)}\}:=\mathscr{F}_q\mathbf{x}^{(j)}, \qquad j=0,1,\ldots,p-1,$$

are known, and we try to express $\mathbf{y}:=\mathscr{F}_n\mathbf{x}$ in terms of the elements of the sequences $\mathbf{y}^{(j)}$. For all integers m,

$$y_m=\frac{1}{\sqrt{n}}\sum_{k=0}^{n-1} w_n^{-mk}x_k=\frac{1}{\sqrt{p}}\sum_{j=0}^{p-1}\frac{1}{\sqrt{q}}\sum_{h=0}^{q-1} w_n^{-m(j+ph)}x_{j+ph},$$

and in view of $w_n^p=w_q$,

$$y_m=\frac{1}{\sqrt{p}}\sum_{j=0}^{p-1} w_n^{-mj}\frac{1}{\sqrt{q}}\sum_{h=0}^{q-1} w_q^{-mh}x_{j+ph}.$$

Considering that the quantities

$$\frac{1}{\sqrt{q}}\sum_{h=0}^{q-1} w_q^{-mh}x_{j+ph}=y_m^{(j)}$$

are known by hypothesis, this yields

$$y_m=\frac{1}{\sqrt{p}}\sum_{j=0}^{p-1} w_n^{-mj}y_m^{(j)}, \tag{13.1-19}$$

which already is a representation of the desired sort. To construct the sequence $\mathbf{y}$ from the sequences $\mathbf{y}^{(j)}$ by means of (13.1-19) evidently requires $(p-1)\mu$ for each value of m ($\mu:=$ unit of cost of one multiplication), which results in a total of $(p-1)m\mu$. The number of additions is approximately the same. We ignore the divisions by $\sqrt{p}$, which in reality need not be carried out, as well as the multiplications that could be saved for $m=0$.

Now let the period n be factored into l factors,

$$n = n_1 n_2 \cdots n_l. \tag{13.1-20}$$

It is not required that the n_i be prime factors. Formula (13.1-19) may then be used recursively. To compute $\mathscr{F}_n\mathbf{x}$ we require n_1 transforms of period $q = n_2 n_3 \cdots n_l$; to compute these, we need $n_1 n_2$ transforms of period $q = n_3 \cdots n_l$, and so on, until we arrive at n transforms $\mathscr{F}_1\mathbf{x}$, which are trivial since $\mathscr{F}_1\mathbf{x} = \mathbf{x}$. The total number of multiplications will then be

$$n(n_1-1) + n_1 \frac{n}{n_1}(n_2-1) + \cdots + n_1 \cdots n_{l-1} \frac{n}{n_1 \cdots n_{l-1}}(n_l - 1) = n \sum_{i=1}^{l} (n_i - 1).$$

For instance, if $n = 2^l$, the required number of multiplications will be $nl = n \operatorname{Log}_2 n$. It thus is clear that an order-of-magnitude improvement has been achieved over the naive method of evaluating (13.1-11).

There is a variant of (13.1-19) which sometimes leads to even greater economy. Let $m = k + lq$, where $k = 0, 1, \ldots, q-1$; $l = 0, 1, \ldots, p-1$. We then have

$$w_n^{-mj} = w_n^{-(k-lq)j} = w_n^{-kj} w_p^{-lj},$$

and because each sequence $y^{(j)}$ has period q, there follows

$$y_{k+q} = \frac{1}{\sqrt{p}} \sum_{j=0}^{p-1} w_p^{-lj} w_n^{-kj} y_k^{(j)}, \qquad k = 0, 1, \ldots, q-1, \quad l = 0, 1, \ldots, p-1. \tag{13.1-21}$$

Equations (13.1-21) are evaluated by:

(a) Forming the $(p-1)q$ products

$$z_k^{(j)} = w_n^{-kj} y_k^{(j)}, \qquad j = 1, 2, \ldots, p-1, \quad k = 0, 1, \ldots, q-1.$$

(b) Forming the $(p-1)^2 q$ products

$$w_p^{-lj} z_k^{(j)}, \qquad l, j = 1, 2, \ldots, p-1, \quad k = 0, 1, \ldots, q-1.$$

We again have ignored the triviality of the multiplications for $k=0$. The steps from the $\mathbf{y}^{(j)}$ to $\mathbf{y}$ now requires a total of

$$(p-1)q + (p-1)^2 q = pq(p-1)$$

multiplications. Thus, in general, there is no saving in comparison to the use of (13.1-19). If, however, $p = 2^l$, then $w_p = -1$, and the multiplications of type (b) need not be counted. The step from the $\mathbf{y}^{(j)}$ to $\mathbf{y}$ then requires only $q(p-1)\mu$, and if $n = 2^l$ and (13.1-21) is used recursively, then the total number of μ to evaluate $\mathscr{F}_n\mathbf{x}$ is no more than

$$l\frac{n}{2}=\frac{1}{2}n\ \mathrm{Log}_2\, n.$$

THEOREM 13.1b. *If n is factored in the form* (13.1-20), *not more than*

$$n\sum_{i=1}^{l}(n_i-1)$$

complex multiplications are required to evaluate $\mathscr{F}_n\mathbf{x}$ for a given $\mathbf{x}\in\Pi_n$. In the special case $n=2^l$ the required number of multiplications does not exceed $\frac{1}{2}n\ \mathrm{Log}_2\, n$.

Any algorithm using the formulas (13.1-19) or (13.1-21) for the recursive evaluation of $\mathbf{y}=\mathscr{F}_n\mathbf{x}$ by definition is a **fast Fourier transform** (FFT). We do not discuss in detail the (nontrivial) problem of constructing such implementations, which may be arranged such that at no stage of the computation more than n numbers must be stored. If $n=2^l$, such implementations require the **bit inversion function** p_j, which is defined as follows. If m is an integer, $0\leqq m<2^j$, whose binary representation is

$$m=m_0+2m_1+\cdots+2^{j-1}m_{j-1},$$

then

$$p_j(m)=m_{j-1}+2m_{j-2}+\cdots+2^{j-1}m_0.$$

It will be shown in § 13.7 how to evaluate $\mathscr{F}_n\mathbf{x}$ for $\mathbf{x}\in\Pi_n$, where n is an arbitrary, not necessarily composite integer, in such a manner that the number of arithmetic operations is likewise bounded by $\gamma n\ \mathrm{Log}\ n$, where γ is a constant independent of n.

III. The Stability of the Discrete Fourier Transform

As was seen in our discussion of the quotient-difference algorithm (§ 7.6), the performance of an algorithm that looks nice on paper can be seriously impaired if the algorithm turns out to be *unstable.* We therefore insert here a brief discussion of the stability of the discrete Fourier transform.

Two kinds of stability have to be considered: (a) **mathematical stability** (or **condition**), which is the sensitivity of the result of the algorithm to changes in the data; (b) **numerical stability**, which is the sensitivity of the result to rounding errrors during the performance of the algorithm. While the mathematical stability is a property only of the function that maps the data on the result, the numerical stability may depend on the algorithm used to evaluate that function, and on the computing equipment on which the algorithm is carried out.

The *mathematical stability* of the discrete Fourier transform is easily discussed. The data here are the elements x_k of a sequence $\mathbf{x} \in \Pi_n$; the result of the discrete Fourier transform consists of the elements y_k of the sequence $\mathbf{y} := \mathcal{F}_n \mathbf{x}$. If the data are changed by $\delta\mathbf{x} = \{\delta x_k\}$, the result in view of the linearity of $\mathcal{F}_n$ changes by precisely

$$\delta\mathbf{y} := \mathcal{F}_n \, \delta\mathbf{x}.$$

If the size of these changes is to be expressed numerically, a choice has to be made with regard to the norms in which the changes are being measured. A formally nice appraisal is obtained if the Euclidean norm is used, for then by property (F_1) we have

$$\|\delta\mathbf{y}\| = \|\delta\mathbf{x}\|.$$

Thus in the Euclidean norm the discrete Fourier operator does not amplify changes in the data at all. However, in some applications bounds for the individual elements of $\delta\mathbf{y}$ in terms of bounds on the individual elements of $\delta\mathbf{x}$ are required. If all $|\delta x_k| \leqq \varepsilon$, it immediately follows from (13.1-11) that the maximum change in any δy_k satisfies

$$|\delta y_k| \leqq \sqrt{n}\,\varepsilon. \tag{13.1-22}$$

The constant $\sqrt{n}$, which is easily seen to be best possible, is called the **condition number** (in the supremum norm) of the discrete Fourier transform.

The *numerical stability* of the discrete Fourier transform will be studied for two different algorithms for carrying out the transform:

(a) The *conventional algorithm*, by which we mean using Horner's algorithm (see § 6.1) to evaluate the polynomial

$$p(w) := x_0 + x_1 w + x_2 w^2 + \cdots + x_{n-1} w^{n-1}$$

at the points $w_k := \exp(2\pi i k/n)$, so that

$$y_k = \frac{1}{\sqrt{n}} p(w_k).$$

(b) A variant of the *FFT* for $n = 2^l$, where l arrays $\mathbf{x}^{(m)}$ of length n are successively generated by formulas of the type

$$x_k^{(m+1)} = x_{k'}^{(m)} + w x_{k''}^{(m)},$$

where k', k'' are appropriate indices and $|w| = 1$, and by finally computing

$$y_k = \frac{1}{\sqrt{n}} x_k^{(l)}.$$

Using $x^{(m)}$ as a generic symbol for an element of the mth level of the computation, and denoting by w any complex number of modulus 1, both algorithms are described by similar sets of formulas, namely,

$$\begin{aligned} &x^{(0)} := x; \\ &x^{(m)} := wx^{(m-1)} + x^{(0)}, \qquad m = 1, 2, \ldots, n-1; \\ &y := \frac{1}{\sqrt{n}} x^{(n-1)} \end{aligned} \tag{13.1-23a}$$

for the Horner algorithm, and

$$\begin{aligned} &x^{(0)} := x; \\ &x^{(m)} := wx^{(m-1)} + x^{(m-1)}, \qquad m = 1, 2, \ldots, l; \\ &y := \frac{1}{\sqrt{n}} x^{(l)} \end{aligned} \tag{13.1-23b}$$

for the FFT.

In digital computation, formulas such as (13.1-23) are rarely executed exactly. As is well known, any computing machine M can exactly represent only a certain discrete set $\mathbb{C}_M$ of complex numbers. Thus in machine computation a given $x \in \mathbb{C}$ will generally be replaced by a number $x^\Delta \in \mathbb{C}_M$. Under certain simplifying but reasonable assumptions (unlimited exponent range) the precision of this replacement can be characterized by a single real number $\eta > 0$, called the **precision constant** of M, which has the property that for all $x \in \mathbb{C}$,

$$x^\Delta = x(1+u), \qquad u \in \mathbb{C}, \quad |u| \leqq \eta.$$

Ideally, if the real and imaginary parts of x are represented as N-digit floating numbers in the base β, η could be as small as $2^{-1/2}\beta^{-N+1}$.

Even if $x, y \in \mathbb{C}_M$, the numbers $x+y$ and xy in general are not in $\mathbb{C}_M$. In their place, the machine returns numbers $(x+y)^\Delta$ and $(xy)^\Delta$ in $\mathbb{C}_M$, which satisfy

$$(x+y)^\Delta = (x+y)(1+u), \qquad (xy)^\Delta = xy(1+u),$$

where again $u \in \mathbb{C}$, $|u| \leqq \eta$. Equivalently one may write

$$(x+y)^\Delta = x+y+|x+y|u, \qquad (xy)^\Delta = xy+|xy|u. \tag{13.1-24}$$

Frequently in the analysis of such expressions, it is necessary to replace $|x+y|$ and $|xy|$ by known a priori bounds.

Similar remarks hold for the evaluation of elementary functions f. In place of $f(x)$, which even for $x \in \mathbb{C}_M$ in general is not in $\mathbb{C}_M$, the machine returns a value $f^{\Delta}(x) \in \mathbb{C}_M$ satisfying

$$f^{\Delta}(x) = f(x)(1+u), \qquad u \in \mathbb{C}, \quad |u| \leqq \eta.$$

Thus for instance in the discrete Fourier transform, the root of unity w is replaced by a number $w^{\Delta} \in \mathbb{C}_M$ such that

$$w^{\Delta} = w + u, \qquad u \in \mathbb{C}, \quad |u| \leqq \eta.$$

If several arithmetic operations and function evaluations are performed sequentially, the rounding errors committed at each step propagate through the computation, and in place of the mathematical quantities x defined by the algorithm, the machine generates certain numbers $\tilde{x} \in \mathbb{C}_M$. In general, $\tilde{x} \neq x^{\Delta}$; however, there usually exist $\zeta > 0$ independent of η such that $|\tilde{x} - x| \leqq \zeta\eta$ for all sufficiently small η. In the **linear model of error propagation** one computes, under appropriate assumptions on the data, bounds for ζ working with the foregoing assumption modulo η^2, that is, by neglecting powers of u higher than the first. Numbers ζ obtained in this manner are called **coefficients of error propagation**. In this model, there result bounds of the form

$$|\tilde{x} - x| \leqq (\zeta + O(\eta))\eta = \zeta\eta \mod \eta^2.$$

In the present context, our goal is to find coefficients of error propagation under the assumption that the initial x are numbers in $\mathbb{C}$ satisfying $|x| \leqq \gamma$. It is clear that the following a priori bounds then hold for the exact numbers $x^{(m)}$:

$$|x^{(m)}| \leqq (m+1)\gamma, \qquad m = 0, 1, \ldots, n-1,$$

for the Horner algorithm, and

$$|x^{(m)}| \leqq 2^m \gamma, \qquad m = 0, 1, \ldots, l,$$

for the FFT. Actually in the analysis that follows, bounds for $u|\tilde{x}^{(m)}|$ are required. Replacing $|\tilde{x}^{(m)}|$ by bounds on $|x^{(m)}|$ introduces errors that are $O(\eta^2)$ only, and therefore is permissible in the linear model.

We represent the numbers $\tilde{x}^{(m)}$ generated by the two algorithms in the form

$$\tilde{x}^{(m)} = x^{(m)} + \xi_m \gamma u, \qquad m = 0, 1, \ldots. \tag{13.1-25}$$

We have $\xi_0 = 1$, and we seek recurrence relations for the ξ_m.

(a) *Horner Algorithm.* Here the basic relation (13.1-23a) is executed as

$$\tilde{x}^{(m)} = [(w^{\Delta}\tilde{x}^{(m-1)\Delta} + \tilde{x}^{(0)}]^{\Delta},$$

which by the rules of the linear model, using the notation (13.1-25), may be expressed as

$$x^{(m)}+\xi_m\gamma u=(w+u)(x^{(m-1)}+\xi_{m-1}\gamma u)+m\gamma u+x^{(0)}+(m+1)\gamma u$$
$$=wx^{(m-1)}+x^{(0)}+(\xi_{m-1}+3m+1)\gamma u.$$

There follows

$$\xi_m=\xi_{m-1}+3m+1,$$

which in view of $\xi_0=1$ yields

$$\xi_m=1+\sum_{k=1}^{m}(3m+1)=1+\tfrac{3}{2}m(m+1)+m<\tfrac{3}{2}(m+1)^2, \qquad m\geqq 0,$$

In particular, $\xi_{n-1}\leqq\frac{3}{2}n^2$, thus

$$|\tilde{x}^{(n-1)}-x^{(n-1)}|\leqq\tfrac{3}{2}n^2\gamma\eta \mod \eta^2.$$

The final division by $\sqrt{n}$ is numerically executed as

$$\tilde{y}=\left(\frac{1}{\sqrt{n}^{\Delta}}\tilde{x}^{(n-1)}\right)^{\Delta}=\frac{1}{\sqrt{n}}(1+u)(x^{(n-1)}+\xi_{n-1}\gamma u)+\sqrt{n}\,\gamma u$$
$$=y+\frac{1}{\sqrt{n}}(\xi_{n-1}+2n)\gamma u=y+\sqrt{n}(\tfrac{3}{2}n+2)\gamma u,$$

and we find

$$|\tilde{y}-y|\leqq\sqrt{n}(\tfrac{3}{2}n+2)\gamma\eta \mod \eta^2. \qquad (13.1\text{-}26a)$$

(b) *Fast Fourier Transform.* Here the basic recurrence relation

$$\tilde{x}^{(m)}=[(w^{\Delta}\tilde{x}^{(m-1)})^{\Delta}+\tilde{x}^{(m-1)}]^{\Delta}$$

translates into

$$x^{(m)}+\xi_m\gamma u=(w+u)(x^{(m-1)}+\xi_{m-1}\gamma u)+2^{m-1}\gamma u+x^{(m-1)}+\xi_{m-1}\gamma u+2^m\gamma u$$
$$=wx^{(m-1)}+x^{(m-1)}+(2\xi_{m-1}+2^{m+1})\gamma u.$$

There follows

$$\xi_m=2\xi_{m-1}+2^{m+1},$$

which for $\xi_0=1$ yields $\xi_m=(2m+1)2^m$. For $m=n=2^l$ we get

$$|\tilde{x}^{(l)}-x^{(l)}|\leqq\xi_l\gamma\eta=n(2\,\mathrm{Log}_2\,n+1)\gamma\eta \mod \eta^2.$$

The final division by $\sqrt{n}$ is dealt with as before, and we find

$$|\tilde{y}-y|\leqq\sqrt{n}(2\,\mathrm{Log}_2\,n+3)\gamma\eta \mod \eta^2. \qquad (13.1\text{-}26b)$$

In summary, we have proved:

THEOREM 13.1c. *In the linear model of error propagation, if all elements of* $\mathbf{x}=\{x_k\}\in\Pi_n$ *satisfy* $|x_k|\leqq\gamma$, *the coefficients of error propagation for the elements* y_k *of* $\mathbf{y}:=\mathscr{F}_n\mathbf{x}$ *are at most*

$$\sqrt{n}(\tfrac{3}{2}n+2)\gamma$$

for the Horner algorithm, and

$$\sqrt{n}(2\,\mathrm{Log}_2\,n+3)\gamma$$

for the FFT if $n=2^l$.

Since the two bounds are $O(n^{3/2})$ and $O(n^{1/2}\,\mathrm{Log}_2\,n)$, respectively, it is seen that the order-of-magnitude advantage of the FFT over the conventional algorithm is one of numerical stability as well as speed.

PROBLEMS

1. Show that the discrete Fourier transform of the sequence $x_k=k$, $k=0,1,\ldots,n-1$, periodically repeated, is $\{\sqrt{n}\,y_m\}$, where

$$y_m=\begin{cases}\dfrac{n-1}{2}, & m\equiv 0 \mod n\\[2mm] -\dfrac{1}{2}+\dfrac{i}{2}\cot\dfrac{\pi m}{n}, & m\not\equiv 0 \mod n\end{cases}$$

2. Show that for $n=1,2,\ldots$ and $m=0,1,2,\ldots,n-1$,

$$\binom{n-1}{m}=\frac{2^n}{n}\left\{\frac{1}{2}+\sum_{k=1}^{(n-1)/2}(-1)^k\left(\cos\frac{k\pi}{n}\right)^{n-1}\cos\left[k(2m+1)\frac{\pi}{n}\right]\right\}.$$

Verify analytically or by means of numerical tests that the "series" on the right converges very rapidly. (Apply the discrete Fourier transform to the function $f(\tau)=(1+e^{2\pi i\tau})^{n-1}$; see Good [1969].)

3. Let $n=2^l$, $w:=\exp(2\pi i/n)$, and

$$\mathbf{x}:=\{w^{k^2}\}.$$

If $\mathbf{y}=\{y_m\}:=\mathscr{F}\mathbf{x}$, show that

$$y_m=\begin{cases}\sqrt{2}\,w^{2-(m/2)^2}, & m \text{ even}\\ 0, & m \text{ odd.}\end{cases}$$

(The y_m are special **Gaussian sums**; see Hardy and Wright [1954], p. 54. A careful analysis of small-order cases may be helpful for solving this problem.)

4. *Simultaneous transformation of two real sequences.* Let $\mathbf{x}^{(1)}, \mathbf{x}^{(2)}$ be two real sequences in Π_n, $\mathbf{x} := \mathbf{x}^{(1)} + i\mathbf{x}^{(2)}$. If $\mathbf{a} := \mathscr{F}\mathbf{x}$, prove that

$$\mathscr{F}\mathbf{x}^{(1)} = \tfrac{1}{2}(\mathbf{a} + \overline{R\mathbf{a}}), \qquad \mathscr{F}\mathbf{x}^{(2)} = \frac{1}{2i}(\mathbf{a} - \overline{R\mathbf{a}}).$$

5. Let $\mathbf{x} \in \Pi_n$, $\mathbf{a} = \{a_m\} := \mathscr{F}\mathbf{x}$. Show that if $\mathbf{x}$ is real,

$$a_{-m} = \overline{a_m},$$

and if $\mathbf{x}$ is pure imaginary,

$$a_{-m} = -\overline{a_m}.$$

6. *Subtracting a constant.* Let $\mathbf{1} = \{1\}$ denote a sequence of which all elements are 1. Show that

$$\mathscr{F}_n \mathbf{1} = \sqrt{n}\,\mathbf{e}$$

and consequently, if $\mathbf{x} \in \Pi_n$ and $c \in \mathbb{C}$,

$$(\mathscr{F}(\mathbf{x} - c\mathbf{1}))_m = \begin{cases} (\mathscr{F}\mathbf{x})_m, & m \not\equiv 0 \mod n \\ (\mathscr{F}\mathbf{x})_m - \sqrt{n}\,c, & m \equiv 0 \mod n. \end{cases}$$

7. Let $\mathbf{x} \in \Pi_n$, $\mathbf{y} := \mathscr{F}\mathbf{x}$. Show that if the sequence $\mathbf{x}$ is *even*, that is, if $R\mathbf{x} = \mathbf{x}$, then $\mathbf{y}$ is even, and if the sequence $\mathbf{x}$ is *odd*, that is, if $R\mathbf{x} = -\mathbf{x}$, then $\mathbf{y}$ is odd.

Problems 8–10 deal with a matrix interpretation of the FFT; see Theilheimer [1969].

8. For numerical purposes we may identify a sequence $\mathbf{x} \in \Pi_n$ with one of its period segments written as a column vector:

$$\mathbf{x} = \begin{pmatrix} x_0 \\ x_1 \\ \vdots \\ x_{n-1} \end{pmatrix}.$$

The relation $\mathbf{y} = \mathscr{F}_n \mathbf{x}$ then is expressed in matrix form as

$$\mathbf{y} = \frac{1}{\sqrt{n}} \mathbf{W}_n \mathbf{x},$$

where $\mathbf{W}_n$ is the **Fourier matrix** of order n,

$$\mathbf{W}_n := (w_n^{-ij})_{i,j=0}^{n-1},$$

and $w_n := \exp(2\pi i/n)$.

9. If n is even, $n = 2m$, let $\mathbf{P}_{2m}$ denote the permutation matrix that moves the $2k$th row of a matrix into kth place, and the $(2k+1)$th row into $(k+m)$th place. Show that

$$\mathbf{W}_{2m}\mathbf{P}_{2m} = \begin{pmatrix} \mathbf{I} & \mathbf{D}_m \\ \mathbf{I} & -\mathbf{D}_m \end{pmatrix} \begin{pmatrix} \mathbf{W}_m & \mathbf{O} \\ \mathbf{O} & \mathbf{W}_m \end{pmatrix}$$

where $\mathbf{I}$ is the unit matrix of order m, and

$$\mathbf{D}_m = \operatorname{diag}(w_n^{-j})_{j=0}^{m-1} = \begin{pmatrix} 1 & & & 0 \\ & w_n^{-1} & & \\ & & \ddots & \\ 0 & & & w_n^{-m+1} \end{pmatrix}.$$

10. Let $n=2^l$. By iterating the factorization given in the preceding problem, show how to obtain a new explanation (or representation) of the FFT. Show the factorization explicitly for $n=8$. In the general case, show that the operations count agrees with Theorem 13.1b.

NOTES

The discrete Fourier transform is a classical topic of numerical analysis, where it is usually dealt with in connection with trigonometric interpolation; see Runge and König [1924]. The original publication of Cooley and Tukey [1965] simultaneously proved the existence of an FFT and provided an implementation for it. The treatment based on reduction formulas given here follows (in the case $n=2^l$) Cooley, Lewis, and Welch [1967b]; see also H. R. Schwarz [1978]. For the origin of the formulas see Cooley, Lewis and Welch [1967c]; in the meantime, Goldstine [1977] has traced the formulas to Gauss [1866]. Other approaches to FFTs are based on a factoring of the matrix representing the Fourier operator (Good [1958, pre-FFT!], Theilheimer [1969], Glassman [1970], McClellan and Parks [1972], and H. R. Schwarz [1977]), on determining remainders in the division of polynomials (Fiduccia [1972], Aho, Hopcroft, and Ullmann [1974], and Kahaner [1978]), or on interpolation at roots of unity (Meinardus [1978]). The literature on implementations of the FFT is abundant; in addition to some of the above see Singleton [1968], Uhrich [1969], Cooley, Lewis, and Welch [1970a], and de Boor [1980]. Some of the more sophisticated implementations avoid the bit reversion function (or "presorting") by means of a "twist," that is, by interpreting one-dimensional arrays as suitable multidimensional arrays; see de Boor [1980] and, for a survey, Temperton [1983]. The elementary implementation of Henrici [1982a] follows Gander and Mazzario [1972]. For a vector implementation see Fornberg [1981b]. Merz [1983] has another survey of several possible implementations.

It is an outstanding problem in the theory of computation whether the $O(n \operatorname{Log} n)$ operations count achieved by the FFT is best possible. This stands to be conjectured, but to date has not been proved.

The linear model of error propagation used in Section III is an extension to $\mathbb{C}$ of the model given in Henrici [1980]. A different roundoff analysis for the FFT is given by Ramos [1971]. For a model of roundoff propagation in complex arithmetic that does not neglect $O(u^2)$ terms, see Olver [1983].

§ 13.2. NUMERICAL HARMONIC ANALYSIS: FOURIER SERIES

Here we discuss some applications of discrete Fourier analysis to continuous Fourier analysis which, although classical, have become feasible only because of the availability of FFT algorithms.

I. The Trapezoidal Values of the Fourier Coefficients

Let Π denote the class of periodic, complex-valued functions of $\mathbb{R}$ with period 1. If $x \in \Pi$ is integrable, its (complex) Fourier coefficients are defined by

$$a_m := = \int_0^1 x(\tau)\, e^{-2\pi i m\tau}\, d\tau, \qquad m = 0, \pm 1, \pm 2, \ldots. \qquad (13.2\text{-}1)$$

With these coefficients the Fourier series

$$\sum_{m=-\infty}^{\infty} a_m\, e^{2\pi i m\tau}$$

may be formed. Under certain conditions—see, for example, § 10.7—this series converges to $x(\tau)$ for some or all values of τ.

In many applications the integrals (13.2-1) cannot be evaluated in closed form. This is true, in particular, if the function x is known only empirically, or if it can be calculated only on a discrete set of values of τ. The points τ at which x can be evaluated are known as the **sampling points**. If the point $\tau = 0$ plays no distinguished role, it is reasonable to assume that the sampling points are equidistant, and that their distance is commensurate with a period. We thus assume the sampling points to be

$$\tau_k = kh, \qquad h := \frac{1}{n},$$

where $k \in \mathbb{Z}$ and $n \geqq 1$ is a fixed integer. The values

$$x_k := x(\tau_k), \qquad k \in \mathbb{Z},$$

are called the **sampling values** of x. They form a sequence

$$\mathbf{x} = \{x_k\}_{k=-\infty}^{\infty}$$

which evidently belongs to Π_n.

What kind of integration rule should be used to evaluate the integrals (13.2-1) under these circumstances? The fact that the τ_k are equidistant rules out all Gaussian rules. As to the high-order Newton–Cotes formulas, it is obvious that they bring no advantage for functions x that are not smooth. But even if x is smooth, it now will be argued that the trapezoid rule is as good as any Newton–Cotes formula. For let g be any function of period 1 that has a continuous $(n+1)$th derivative. By constructing the polynomial p of degree $\leqq n$ such that $p(\tau_k) = g(\tau_k)$, $k = 0, 1, \ldots, n$, and by integrating p in place of g we obtain

$$\int_0^1 g(\tau)\, d\tau = \int_0^1 p(\tau)\, d\tau + r_n[g],$$

where for some constant γ_n depending only on n and for

$$\mu_{n+1} := \sup_{0 \leq \tau \leq 1} |g^{(n+1)}(\tau)|$$

there holds

$$|r_n[g]| \leq \gamma_n \mu_{n+1}.$$

The integral of the interpolating polynomial can be evaluated and expressed in terms of the sampling values $g_k := g(\tau_k)$:

$$\int_0^1 p(\tau)\, d\tau = \sum_{k=0}^{n} \omega_k g_k.$$

Here the coefficients ω_k are the Newton-Cotes coefficients of order n. Because the integration formula would yield the exact value of the integral if g happened to be a polynomial of degree $\leq n$, we have

$$\omega_0 + \omega_1 + \cdots + \omega_n = 1. \tag{13.2-2}$$

However, since g is periodic with period 1,

$$\int_0^1 g(\tau)\, d\tau = \int_0^1 g(\tau - \tau_j)\, d\tau, \qquad j = 0, 1, \ldots, n-1.$$

Evaluating the integral on the right as above, we obtain

$$\int_0^1 g(\tau)\, d\tau = \sum_{k=0}^{n} \omega_k g_{k-j} + \theta_j \gamma_n \mu_{n+1}, \qquad j = 0, 1, \ldots, n-1,$$

where $|\theta_j| \leq 1$. Taking the average of all these formulas and making use of the periodicity of the sequence $\{g_k\}$, we find

$$\int_0^1 g(\tau)\, d\tau = \frac{1}{n} \sum_{k=0}^{n} \omega_k \sum_{m=0}^{n-1} g_m + \theta \gamma_n \mu_{n+1},$$

where θ is the average of the θ_j and therefore $|\theta| \leq 1$. Using (13.2-2) this becomes

$$\int_0^1 g(\tau)\, d\tau = \frac{1}{n} \sum_{j=0}^{n-1} g_j + \theta \gamma_n \mu_{n+1}.$$

In view of $g_0 = g_n$, the sum on the right is just the trapezoidal value of the integral (see § 11.11). It thus turns out that *even for periodic functions that are sufficiently differentiable, the trapezoid formula is as accurate as any Newton-Cotes formula using the same number of sampling points.*

Applying this to the integral (13.2-1) and letting $w := \exp(2\pi i/n)$ as in § 13.1, we obtain

$$a_m = \frac{1}{n} \sum_{k=0}^{n-1} x_k w^{-km} + \theta \gamma_n \mu_{m,n+1} \tag{13.2-3}$$

where $|\theta| \leqq 1$ and

$$\mu_{m,n+1} := \sup_{\tau} |(x(\tau)\, e^{-2\pi i m \tau})^{(n+1)}|.$$

Neglecting the error term, we regard

$$\hat{a}_m := \frac{1}{n} \sum_{k=0}^{n-1} x_k w^{-km} \tag{13.2-4}$$

as an approximation to the mth Fourier coefficient a_m. We note that the sequence

$$\hat{\mathbf{a}} := \{\hat{a}_m\}$$

is given by

$$\hat{\mathbf{a}} = \frac{1}{\sqrt{n}} \mathscr{F}_n \mathbf{x}, \tag{13.2-5}$$

and thus may be evaluated rapidly by an FFT.

It is clear that the error estimate implied by (13.2-3),

$$|\hat{a}_m - a_m| \leqq \gamma_n \mu_{m,n+1},$$

is practical only for very small values of n and $|m|$. Fortunately a more explicit estimate is available under weaker assumptions. Let $x \in \Pi$ be the sum of an absolutely (and therefore uniformly) convergent Fourier series,

$$x(\tau) = \sum_{m=-\infty}^{\infty} a_m\, e^{2\pi i m \tau}, \qquad \sum_{m=-\infty}^{\infty} |a_m| < \infty.$$

Substituting the series into (13.2-4) and interchanging summations we find

$$\hat{a}_m = \sum_{k=-\infty}^{\infty} a_{m+kn} \tag{13.2-6}$$

and therefore:

THEOREM 13.2a. *If $x \in \Pi$ is the sum of an absolutely convergent Fourier series, then*

$$\hat{a}_m - a_m = \sum_{\substack{k=-\infty \\ k \neq 0}}^{\infty} a_{m+kn} = \sum_{k=1}^{\infty} (a_{m+kn} + a_{m-kn}). \tag{13.2-7}$$

II. Convergence Acceleration

In § 11.11 we have seen how to use the Romberg algorithm to speed up the convergence as $n \to \infty$ of the trapezoidal values of definite integrals if the integrand is sufficiently smooth. We mention two consequences of Theorem

13.2a which are relevant to this technique although the conclusions at first sight appear to be at variance with the results of § 11.11.

(a) Let x be an *analytic* function of the real variable τ. Then x can be continued into the complex t plane as a function $x(t)$ that is analytic (and still periodic with period 1) in a strip $|\text{Im } t| \leqq \eta$ where $\eta > 0$. By setting $z = e^{2\pi i t}$ and defining

$$f(z) := x\left(\frac{1}{2\pi i} \log z\right),$$

we obtain a single-valued function f that is analytic in the annulus $e^{-2\pi\eta} < |z| < e^{2\pi\eta}$. We know from § 4.5 that the Fourier coefficients of x are just the Laurent coefficients of f. For the latter we have, using Cauchy's estimate,

$$|a_m| \leqq \mu\, e^{-2\pi\eta|m|}, \qquad m \in \mathbb{Z},$$

where μ is the maximum of $|f(z)|$ in the annulus. Using this estimate in Theorem 13.2a, we obtain for $|m| \leqq n$

$$|\hat{a}_m - a_m| \leqq 2\mu \cosh(2\pi m\eta) \frac{e^{-2\pi n\eta}}{1 - e^{-2\pi n\eta}}. \tag{13.2-8}$$

The essential conclusion is:

COROLLARY 13.2b. *If m is fixed and $n \to \infty$, the error of the mth discrete Fourier coefficient of an analytic periodic function tends to zero at a geometric rate.*

Thus if n is doubled, the error is (approximately) squared. It follows that, despite the smoothness of the integrand, the convergence of the trapezoid values *cannot* be speeded up by the Romberg algorithm. Already the trapezoidal values themselves are as accurate as any speeded up values will ever be.

(b) Let now the function $x \in \Pi$ be *piecewise analytic*. This means that the interval $[0, 1]$ can be broken up into a finite number of subintervals such that, on each subinterval, x may be extended to a function that is analytic on the closed subinterval. By breaking up the integral (13.2-1) accordingly and applying integration by parts to each subinterval, one finds that a_m admits an asymptotic expansion,

$$a_m \approx \frac{c_p}{m^{p+1}} + \frac{c_{p+1}}{m^{p+2}} + \cdots, \qquad m \to \pm\infty,$$

where p is the smallest integer such that $x^{(p)}$ has a discontinuity. Using this expansion (with the appropriate remainder term) in (13.2-7), there follows for every $r > p$ the existence of a constant c_r' such that, if m is fixed and

$|m| < n$,

$$\hat{a}_m - a_m = \sum_{q=p}^{r-1} c_q \sum_{k=1}^{\infty} \left\{ \frac{1}{(m+kn)^{q+1}} + \frac{1}{(m-kn)^{q+1}} \right\}$$

$$+ \theta c_r' \sum_{k=1}^{\infty} \left\{ \frac{1}{(m+kn)^{r+1}} + \frac{1}{(m-kn)^{r+1}} \right\},$$

$|\theta| \leqq 1$. Defining ad hoc for $q = 0, 1, 2, \ldots$ and $-1 < \xi < 1$,

$$\phi_q(\xi) := \sum_{k=1}^{\infty} \left\{ \frac{1}{(k+\xi)^{q+1}} + \frac{1}{(k-\xi)^{q+1}} \right\},$$

this may be written

$$\hat{a}_m - a_m = \sum_{q=p}^{r-1} \frac{c_q}{n^{q+1}} \phi_q\left(\frac{m}{n}\right) + \theta \frac{c_r'}{n^{r+1}} \phi_r\left(\frac{m}{n}\right),$$

where always $|\theta| \leqq 1$. The functions ϕ_q being analytic at $\xi = 0$, the expressions $\phi_q(m/n)$ may be expanded in convergent power series in the variable $1/n$. It follows that for each fixed value of m the error $\hat{a}_m - a_m$ admits an asymptotic series of the form

$$\hat{a}_m - a_m \approx \sum_{q=p}^{\infty} \frac{b_{m,q}}{n^{q+1}}, \qquad n \to \infty. \tag{13.2-9}$$

COROLLARY 13.2c. *If $x \in \Pi$ is piecewise analytic, and if p is the lowest order of a discontinuous derivative of x, then the error of the mth discrete Fourier coefficient of x satisfies an asymptotic expansion of the form* (13.2-9) *where $b_{m,p} \neq 0$.*

From the theory of Romberg integration it now follows that the convergence of the coefficients $\hat{a}_m$ to the a_m may be accelerated in the following way. Let $\hat{a}_{m,k,0}$, $k = 0, 1, 2, \ldots$, denote the trapezoidal value computed with $n = 2^k$ sampling points. If accelerated values $\hat{a}_{m,k,j}$ are computed according to

$$\hat{a}_{m,k,j} := \hat{a}_{m,k,j-1} + \frac{1}{2^{p+j} - 1} (\hat{a}_{m,k,j-1} - \hat{a}_{m,k-1,j-1}),$$

$k = 0, 1, 2, \ldots; j = 1, 2, \ldots, k$, then for each fixed m and for $j \geqq 1$,

$$\hat{a}_{m,k,j} - a_m = O(2^{-(p+j+1)k})$$

as $k \to \infty$. Thus in the calculation of approximate Fourier coefficients, the Romberg principle is effective just in the case where x is *not smooth*.

III. Attenuation Factors

Satisfactory as the foregoing results may be to explain the convergence behavior of the approximate Fourier coefficients $\hat{a}_m$ if m is fixed and n, the number of sampling points per period, tends to ∞, they cannot vitiate a basic flaw of the coefficients $\hat{a}_m$ as functions of m when n is fixed. Considered as functions of m, these coefficients are *periodic* with period n. This already follows from the fact that the sequence $\hat{\mathbf{a}}$ is proportional to $\mathscr{F}_n\mathbf{x}$, and hence belongs to Π_n; the periodicity is confirmed by relation (13.2-6), where the expression on the right does not change if m is replaced by $m+n$. This behavior, which in signal processing is known as **aliasing**, is completely at variance with that of the exact Fourier coefficients, which by virtue of the Riemann-Lebesgue lemma (Theorem 10.6a) satisfy

$$a_m \to 0$$

as $m \to \pm\infty$.

Fortunately the deficiency may be corrected by a simple device known as *attenuation factors.* We continue to assume that the only values of the function $x \in \Pi$ that we know are the sampling values x_m. The idea now is to interpolate or otherwise approximate the sequence $\mathbf{x} = \{x_m\}$ by a function $P\mathbf{x} \in \Pi$, and to compute the Fourier coefficients of $P\mathbf{x}$ exactly. This computation turns out to be simple if the approximation operator P has the following properties, which seem very natural:

(i) P is *linear.*

(ii) P is *translation invariant,* that is, if E denotes the shift operator defined in Π_n by $(E\mathbf{x})_k = x_{k+1}$ and in Π by $(E\mathbf{x})(\tau) := x(\tau + 1/n)$, then

$$PE\mathbf{x} = EP\mathbf{x}$$

for all $\mathbf{x} \in \Pi_n$.

To compute, for a given sequence $\mathbf{x} \in \Pi_n$, the exact Fourier coefficients of $P\mathbf{x}$,

$$b_m := \int_0^1 (P\mathbf{x})(\tau)\, e^{-2\pi i m \tau}\, d\tau,$$

we introduce the **delta sequence** $\boldsymbol{\delta} := n\mathbf{e} = \|\!: n, 0, \ldots, 0 :\!\|$. Then

$$\mathbf{x} = \frac{1}{n} \sum_{k=0}^{n-1} x_k E^{-k} \boldsymbol{\delta}$$

and by (i) and (ii) we have

$$P\mathbf{x} = \frac{1}{n} \sum_{k=0}^{n-1} x_k P E^{-k} \boldsymbol{\delta} = \frac{1}{n} \sum_{k=0}^{n-1} x_k E^{-k} P \boldsymbol{\delta}.$$

Letting $p := P\boldsymbol{\delta}$, there follows

$$b_m = \frac{1}{n}\sum_{k=0}^{n-1} x_k \int_0^1 (E^{-k}p)(\tau)\, e^{-2\pi i m\tau}\, d\tau$$
$$= \frac{1}{n}\sum_{k=0}^{n-1} x_k \int_0^1 p\left(\tau - \frac{k}{n}\right) e^{-2\pi i m\tau}\, d\tau.$$

Because $p \in \Pi$,

$$\int_0^1 p\left(\tau - \frac{k}{n}\right) e^{-2\pi i m\tau}\, d\tau = \int_{-k/n}^{1-k/n} p(\tau)\, e^{-2\pi i m(\tau+k/n)}\, d\tau = w^{-km} p_m,$$

where $w := \exp(2\pi i/n)$ as always and where

$$p_m := \int_0^1 p(\tau)\, e^{-2\pi i m\tau}\, d\tau$$

is the mth Fourier coefficient of $p = P\boldsymbol{\delta}$. Hence there results

$$b_m = p_m \cdot \frac{1}{n}\sum_{k=0}^{n-1} x_k w^{-km} = p_m \hat{a}_m,$$

where $\hat{a}_m$ is the discrete Fourier coefficient considered earlier.

THEOREM 13.2d. *If the operator $P: \Pi_n \to \Pi$ is linear and translation invariant, the Fourier coefficients of $P\mathbf{x}$ are $p_m\hat{a}_m$, where $\hat{\mathbf{a}} := n^{-1/2}\mathcal{F}_n\mathbf{x}$ and p_m is the mth Fourier coefficient of $P\boldsymbol{\delta}$, $\boldsymbol{\delta} := n\mathbf{e}$.*

For any reasonable operator P, the function $P\boldsymbol{\delta}$ will be integrable, and the coefficients p_m thus satisfy $p_m \to 0$ for $m \to \pm\infty$. The p_m are known as the **attenuation factors** defined by the process P.

Theorem 13.2d delegates the responsibility for the accuracy of the Fourier coefficients entirely to the choice of the approximation operator P. This choice will be influenced by what the user subjectively knows about the function x. We consider two examples.

EXAMPLE **1. Linear interpolation**

Here the data $\{x_k\}$ are interpolated by a piecewise linear function. It is clear that this process of approximation is linear and translation invariant. In particular, the delta sequence $\boldsymbol{\delta}$ is approximated by the function p defined in $(-\frac{1}{2}, \frac{1}{2})$ by

$$p(\tau) = \begin{cases} n - n^2|\tau|, & |\tau| \leqq \dfrac{1}{n} \\ 0, & \text{otherwise.} \end{cases}$$

For its Fourier coefficients we easily find $p_0 = 1$,

$$p_m = \left(\frac{n}{\pi m}\right)^2 \left(\sin\frac{\pi m}{n}\right)^2, \qquad m \neq 0, \qquad (13.2\text{-}10)$$

confirming that $p_m \to 0$ for $m \to \pm\infty$.

EXAMPLE **2. Interpolation by periodic cubic splines**

Here we interpolate the data $\mathbf{x}$ by a function $y = P\mathbf{x} \in \Pi$ satisfying the following conditions:

(i) $y(\tau_k) = x_k$ for all k.

(ii) y' is continuous, and in each interval (τ_k, τ_{k+1}), y'' is continuous with finite one-sided limits at the endpoints.

(iii) The integral

$$\int_0^1 |y''(\tau)|^2 \, d\tau$$

is as small as possible.

It can be shown that y is uniquely determined by these postulates. The resulting approximation operator P obviously is linear and translation invariant. Without going into the details of the computation of y, we mention that the attenuation factors are given by $p_0 = 1$,

$$p_m = \left(\frac{n}{\pi m}\right)^4 \left(\sin\frac{\pi m}{n}\right)^4 \frac{3}{2 + \cos(2\pi m/n)}, \qquad m \neq 0. \qquad (13.2\text{-}11)$$

IV. Fourier Synthesis

Here we take a brief look at the problem of evaluating a terminating Fourier series (or a partial sum of an infinite Fourier series) for one or several values of the argument. Let $c_{-m}, c_{-m+1}, \ldots, c_m$ be given complex numbers, and let

$$t(\tau) := \sum_{j=-m}^{m} c_j \, e^{2\pi i j \tau}.$$

To evaluate t at a single point τ, we note that $t(\tau) = p_1(z) + p_2(\bar{z})$, where $z := e^{2\pi i \tau}$, and where p_1 and p_2 are ordinary polynomials,

$$p_1(z) = c_0 + c_1 z + \cdots + c_m z^m,$$
$$p_2(z) = c_{-1} z + c_{-1} z^2 + \cdots + c_{-m} z^m,$$

which may be evaluated efficiently and (since $|z| = 1$) stably by using the

Horner algorithm (see § 6.1) suggested by writing

$$p_1(z)=c_0+z(c_1+z(c_2+\cdots+z(c_{n-1}+zc_n)\cdots)).$$

If $t(\tau)$ is to be evaluated at $n>2m$ equidistant points $\tau=\tau_k:=k/n$, $k=0,1,\ldots,n-1$, we note that

$$t(\tau_k)=\sum_{j=-m}^{m} c_j w_n^{jk}$$

where $w_n:=\exp(2\pi i/n)$. Continuing the coefficient sequence as a sequence $\mathbf{c}=\{c_j\}$ in Π_n by setting $c_j:=0$ for j not congruent to any of $-m, -m+1,\ldots, m \bmod n$, in view of $w_n^n=1$ we also have

$$t(\tau_k)=\sum_{j=0}^{n-1} c_j w_n^{jk}.$$

We thus see that the sequence $\mathbf{t}:=\{t(\tau_k)\}$ is related to $\mathbf{c}$ by

$$\mathbf{t}=\sqrt{n}\,\bar{\mathcal{F}}_n\mathbf{c} \tag{13.2-12}$$

and thus may be evaluated in $O(n \operatorname{Log} n)$ operations by an FFT algorithm.

PROBLEMS

1. For the functions

(a) $$x(\tau)=\frac{1}{a-e^{2\pi i\tau}},\qquad |a|\neq 1$$

(b) $$x(\tau)=|1-2\tau|,\qquad 0\leqq\tau\leqq 1$$

compute analytically the exact Fourier coefficients a_m as well as their discrete counterparts $\hat{a}_m$, and study in detail how the latter converge to the former as $n\to\infty$.

2. For $z\in\mathbb{C}$ and $m\in\mathbb{Z}$, the Bessel functions $J_m(z)$ may be defined by

$$e^{iz\cos 2\pi\tau}=\sum_{m=-\infty}^{\infty} i^m J_m(z)\,e^{2\pi i m\tau}$$

(see § 4.5).

(a) Use an FFT program to generate numerical values of $J_m(z)$ by computing approximate Fourier coefficients of the function on the left, and study the numerical limitations of this method.

(b) If m is fixed and $n\to\infty$, the numerical effectiveness of the method described under (a) is even better than predicted by Corollary 13.2b. Explain.

3. Let P be the process of interpolation which computes $x(\tau)$ for $\tau_k\leqq\tau\leqq\tau_{k+1}$ by the cubic polynomial interpolating x at the points $\tau_{k-1}, \tau_k, \tau_{k+1}, \tau_{k+2}$. Show that

the attenuation factors of P are

$$p_m = \left(\frac{n}{\pi m}\right)^4 \left(\sin \frac{\pi m}{n}\right)^4 \left\{1 + \frac{8}{3}\left(\frac{m}{n}\right)^2\right\}.$$

4. Use Theorem 13.2d to compute the attenuation factors for the following interpolation process. P associates with the sequence $\{x_k\}$ the balanced trigonometric polynomial of degree $[(n+1)/2]$ interpolating the values x_k at the points τ_k (see § 13.6). Explain the result.

5. Establish (13.2-11) by using the following facts. The cubic spline function is the piecewise cubic polynomial represented in $[\tau_k, \tau_{k+1}]$ by

$$p(\tau) = x_k + (3t^2 - 2t^3)(x_{k+1} - x_k) + t(1-t)^2 s_k - t^2(1-t)s_{k+1},$$

$k = 0, 1, \ldots, n-1$, where $t := n(\tau - \tau_k)$, and where the sequence $\mathbf{s} = \{s_k\} \in \Pi_n$ is the solution of

$$s_{k-1} + 4s_k + s_{k+1} = 3(x_{k+1} - x_{k-1}), \qquad k \in \mathbb{Z}. \qquad (*)$$

(The system (*) may be solved by convolution techniques; see § 13.7.)

6. A method due to Filon [1929] for computing integrals of the form

$$\int_\alpha^\beta f(\tau)\, e^{-2\pi i \omega \tau}\, d\tau,$$

if applied to the calculation of Fourier coefficients, is closely related to the use of attenuation factors. Let $x \in \Pi$, and let x be known on the set of $2n$ equidistant points $\tau_k := k/2n$, $k = 0, 1, \ldots, 2n-1$, $x_k := x(\tau_k)$. Filon's method consists of interpolating x on each interval $[\tau_{2k}, \tau_{2k+2}]$ by a quadratic polynomial and evaluating the resulting integrals exactly.

(a) Show that the approximation operator P implied by Filon's method is linear.

(b) Show that P, although not translation invariant, satisfies

$$PE^{2k} = E^{2k}P, \qquad PE^{2k+1} = E^{2k}PE, \qquad k = 0, 1, \ldots.$$

(c) As a consequence, show that the exact Fourier coefficients b_m of $P\mathbf{x}$ may be computed as follows. Let

$$b'_m := \frac{1}{n}\sum_{k=0}^{n-1} x_{2k} w_{2n}^{-2km}, \qquad b''_m := \frac{1}{n}\sum_{k=0}^{n-1} x_{2k+1} w_{2n}^{-(2k+1)m}.$$

Then

$$b_m = \alpha_m b'_n + \beta_m b''_m$$

where $\alpha_0 = \frac{1}{2}$, $\beta_0 = \frac{2}{3}$, and

$$\alpha_m = \frac{n^3}{(\pi m)^3}\left\{\frac{\pi m}{n}\left[1 + \left(\cos\frac{\pi m}{n}\right)^2\right] - \sin\frac{2\pi m}{n}\right\},$$

$$\beta_m = \frac{2n^3}{(\pi m)^3}\left\{\sin\frac{\pi m}{n} - \frac{\pi m}{n}\cos\frac{\pi m}{n}\right\}, \qquad m \neq 0.$$

(d) Experiment with $x(\tau)=\tau(1-\tau)$, periodically repeated, where the b_m should be the *exact* Fourier coefficients.

NOTES

Much of Section I is folklore. Corollary 13.2b is due to Davis [1959]. Henrici [1982a] has numerical demonstrations for the effectiveness of convergence acceleration in case (b). The theory of attenuation factors is due to Gautschi [1972], who proves a converse of Theorem 13.2d and provides many additional examples. Example 1 is due to Dällenbach [1921]. Marti [1978] has an algorithm for computing Fourier coefficients which is based on B splines and which is particularly effective for nonuniform grids and piecewise analytic functions.

§ 13.3. NUMERICAL HARMONIC ANALYSIS: FOURIER INTEGRALS

I. The Spectrum

Adjusting the notation to be in tune with that of the present chapter, the *Fourier integral theorem* (Theorem 10.6d) reads as follows. Let F be a complex-valued Riemann integrable function on $\mathbb{R}$ such that

$$\int_{-\infty}^{\infty} |F(\tau)|\, d\tau < \infty,$$

and let, for $\omega \in \mathbb{R}$,

$$G(\omega) := \int_{-\infty}^{\infty} F(\tau)\, e^{-2\pi i \omega \tau}\, d\tau. \tag{13.3-1}$$

Then at every τ where F satisfies condition C as defined in § 10.6, there holds

$$F(\tau) = \mathrm{PV} \int_{-\infty}^{\infty} G(\omega)\, e^{2\pi i \tau \omega}\, d\omega.$$

In applications where τ denotes time, functions F such as those considered above are referred to as **signals**, and the function G defined by (13.3-1) is called the **spectrum** of F. Many physical applications call for the determination of the spectrum of a given signal. Because two limits are involved in the mathematical definition of a spectrum—first integration on a finite interval, and then letting the limits of integration tend to ∞—the problem is one step more difficult than the mere approximation of Fourier coefficients. However, the following naive approach is at least computationally feasible. Truncate the Fourier integral (13.3-1) to the finite interval $[-\lambda, \lambda]$, where λ is "sufficiently large," choose an integration step h such that $m := \lambda h^{-1}$ is an integer, and approximate the integral by its trapezoidal value. This yields the approximation

$$\tilde{G}(\omega) := h \sum_{k=-m}^{m}{}' F(kh)\, e^{-2\pi i \omega k h}, \tag{13.3-2}$$

where the prime indicates that the terms corresponding to $k = \pm m$ are to be multiplied by $\frac{1}{2}$.

The connection with discrete Fourier analysis becomes apparent if $\tilde{G}$ is evaluated at the points

$$\omega = \omega_j := \frac{j}{2\lambda}, \qquad j = -m, -m+1, \ldots, m.$$

We then have

$$e^{-2\pi i \omega_j k h} = e^{-2\pi i j k/2m} = w_{2m}^{-jk},$$

where $w_{2m} := \exp(2\pi i/2m)$. Consequently,

$$\tilde{G}(\omega_j) = \frac{\lambda}{m} \sum_{k=-m}^{m}{}' F\left(\frac{k\lambda}{m}\right) w_{2m}^{-jk},$$

and we see that

$$\tilde{G}(\omega_j) = 2\lambda \hat{g}_j,$$

where the $\hat{g}_j$ are the discrete Fourier coefficients of the function $f \in \Pi$ defined for $\tau \in [-\frac{1}{2}, \frac{1}{2}]$ by

$$f(\tau) = F(2\lambda\tau).$$

(Here it is necessary to replace the values of $F(\pm\lambda)$ by their arithmetic mean.) Thus to compute the values of the approximate spectrum $\tilde{G}(\omega)$ on the set of points ω_j, all one needs to do is to compute the discrete Fourier transform of the sequence $f = \{f_k\}$ defined by

$$f_k = f\left(\frac{k}{2m}\right) = F\left(\frac{k}{m}\lambda\right)$$

for $-m \leqq k \leqq m$ and continued with period $2m$. Using the FFT, this can be done in $O(m \operatorname{Log} m)$ operations. Because the $\hat{g}_j$ approximate Fourier coefficients, their accuracy may be enhanced by any of the tricks mentioned in § 13.2.

II. The Error of the Approximate Spectrum $\tilde{G}$

The vehicle for the discussion of the error $\tilde{G}(\omega) - G(\omega)$ of the approximate spectrum discussed above is the *Poisson sum formula* (Theorem 10.6e) which, if adapted to the present notation, states that if F has the spectrum

G, then for any $h>0$ and for all real τ

$$h \sum_{k=-\infty}^{\infty} F(\tau+kh) = \mathrm{PV} \sum_{n=-\infty}^{\infty} G\left(\frac{n}{h}\right) e^{2\pi i n\tau/h},$$

provided F is such that the series of the left converges uniformly with respect to τ, and that its sum (which represents a function of period h) equals its own Fourier series. Since for any $\alpha \in \mathbb{R}$ the function $F(\tau)\, e^{-2\pi i\alpha\tau}$ has the spectrum $G(\omega+\alpha)$, the formula also states that

$$h \sum_{k=-\infty}^{\infty} F(\tau+kh)\, e^{-2\pi i\alpha(\tau+kh)} = \mathrm{PV} \sum_{n=-\infty}^{\infty} G\left(\frac{n}{h}+\alpha\right) e^{2\pi i n\tau/h}.$$

In particular for $\tau=0$ this yields

$$h \sum_{k=-\infty}^{\infty} F(kh)\, e^{-2\pi i\alpha kh} = \mathrm{PV} \sum_{n=-\infty}^{\infty} G\left(\frac{n}{h}+\alpha\right). \qquad (13.3\text{-}3)$$

Letting $\alpha=\omega$ and limiting the sum on the left to values $|k| \leqq m$, we obtain the approximate spectrum $\tilde{G}(\omega)$. On the right, the term $n=0$ yields $G(\omega)$. We thus obtain the following general representation of the error:

THEOREM 13.3a. *If the signal F is such that Poisson's formula holds, then*

$$\tilde{G}(\omega)-G(\omega) = \sum_{n=1}^{\infty} \left\{ G\left(\omega+\frac{n}{h}\right) + G\left(\omega-\frac{n}{h}\right) \right\} - \sum_{k \geqq m}{}' \{F(kh)\, e^{-2\pi i\omega kh} + F(-kh)\, e^{2\pi i\omega kh}\} \qquad (13.3\text{-}4)$$

holds for all real ω.

Equation (13.3-4) neatly puts in evidence the error due to discretizing the Fourier integral (expressed by the first sum), and the error due to truncating the Fourier integral (expressed by the second sum).

Of particular interest are the cases where one of these errors is zero. If the signal F is **time-limited**, that is, if there exists λ such that $F(\tau)=0$ if $|\tau| \geqq \lambda$, then the second sum is vacuous, and if $mh \geqq \lambda$, the error is simply

$$\tilde{G}(\omega)-G(\omega) = \sum_{n=1}^{\infty} \left\{ G\left(\omega+\frac{n}{h}\right) + G\left(\omega-\frac{n}{h}\right) \right\}. \qquad (13.3\text{-}5)$$

The more rapidly $G(\omega) \to 0$ as $\omega \to \pm\infty$, the more rapidly the error will tend to zero with h. If the signal F is **band-limited**, that is, if there exists $\beta>0$ such that $G(\omega)=0$ if $|\omega| \geqq \beta$, then if $|\omega|<\beta$ and if $h^{-1}>2\beta$ or $h<1/2\beta$, the first sum on the right of (13.3-4) remains vacuous, and we have

$$\tilde{G}(\omega)-G(\omega) = -\sum_{k \geqq m}{}' \{F(kh)\, e^{-2\pi i\omega kh} + F(-kh)\, e^{2\pi i\omega kh}\}. \qquad (13.3\text{-}6)$$

The sum on the right is comparable to

$$\int_{\lambda}^{\infty} [|F(\tau)|+|F(-\tau)|]\, d\tau,$$

which permits appraising the accuracy of $\tilde{G}$ in terms of the speed with which F tends to zero.

Are there signals which are both time-limited and band-limited? If F is band-limited with bandwidth 2β, and if G is the spectrum of F, then

$$F(\tau)=\int_{-\beta}^{\beta} G(\omega)\, e^{2\pi i\tau\omega}\, d\omega.$$

Thus $F(\tau)=f(2\pi i\tau)$, where, for $z\in\mathbb{C}$,

$$f(z):=\int_{-\beta}^{\beta} G(\omega)\, e^{\omega z}\, d\omega.$$

Since G is continuous, it follows by Theorem 4.1a that f is an entire analytic function. Such a function can vanish on a nonempty interval only if it is identically zero.

THEOREM 13.3b. *The only signal that is both time-limited and band-limited is the zero signal.*

PROBLEMS

1. For the function $F(\tau):=e^{-|\tau|}$ compute analytically both the exact spectrum and the approximate spectrum and show how they are related. What is the relation between λ and h such that the errors due to truncating the Fourier integral and to discretizing its finite part are approximately equal?
2. Although there are no signals that are both time-limited and band-limited, some come pretty close. Elucidate this statement by determining the spectrum of the Gaussian signal $F(\tau):=e^{-\pi\tau^2}$, which is numerically 0 (mathematically, $<10^{-12}$) for $|\tau|\geqq 3$.

NOTES

The notion of a band-limited signal (more so than that of a time-limited signal) is of fundamental importance in digital signal processing. For illuminative remarks in connection with numerical computation see Hamming [1962]. Mathematically, band-limited signals are restrictions to $\mathbb{R}$ of functions of exponential type (see § 10.9). More generally, the spectrum of signal F decreases exponentially as $\omega\to\pm\infty$ if and only if F is the restriction to $\mathbb{R}$ of a function that is analytic in a strip about $\mathbb{R}$. This is a consequence of the Paley–Wiener theorems; see Katznelson [1968], ch. 6, § 7. Boas [1972] establishes a connection between the Poisson summation formula and the sampling theorem (Problem 14 of § 10.6).

§ 13.4. LAURENT SERIES; RESIDUES; ZEROS

The basic results of complex integration theory have been known for at least 150 years. Here we show how the discrete Fourier transform, combined with FFT algorithms, infuses new algorithmic life even into such seemingly old-fashioned topics.

I. The Laurent Series

Let f be analytic in the annulus A: $\rho_1 < |z| < \rho_2$, where $0 \leqq \rho_1 < 1 < \rho_2$. (Any annulus may be reduced to an annulus of this special type by a linear change of variables.) Then f is represented in A by its Laurent series (see § 4.4),

$$f(z) = \sum_{m=-\infty}^{\infty} a_m z^m, \qquad z \in A,$$

where the a_m are the Laurent coefficients of f for A,

$$a_m = \frac{1}{2\pi i} \int_{|z|=1} z^{-m-1} f(z)\, dz.$$

On setting $z = e^{2\pi i \tau}$ this becomes

$$a_m = \int_0^1 e^{-2\pi i m \tau} f(e^{2\pi i \tau})\, d\tau,$$

and we see that a_m also is the mth *Fourier* coefficient of the function $\tau \to f(e^{2\pi i \tau}) \in \Pi$. Thus for instance if $n = 2^l$, $w := \exp(2\pi i/n)$, $f_m := f(w^m)$, and $\mathbf{f} := \{f_m\}$, then $\mathbf{f} \in \Pi_n$, and the coefficients a_m with $|m|$ sufficiently small will be approximated by those of the sequence

$$\hat{\mathbf{a}} := \frac{1}{\sqrt{n}} \mathcal{F}_n \mathbf{f}, \tag{13.4-1}$$

which by an FFT can be calculated in $\frac{1}{2} n \operatorname{Log}_2 n$ multiplications. The error in this approximation by Theorem 13.2a is given by

$$\hat{a}_m - a_m = \cdots + a_{m-2n} + a_{m-n} + a_{m+n} + a_{m+2n} + \cdots. \tag{13.4-2}$$

If $\rho_1 < \rho < \rho_2$ and

$$\mu(\rho) := \max_{|z|=\rho} |f(z)|,$$

then by Cauchy's estimate,

$$|a_{m+kn}| \leqq \frac{\mu(\rho)}{\rho^{m+kn}}.$$

Thus if $\mu^*(\rho) := \max(\mu(\rho), \mu(\rho^{-1}))$ and $\rho_1 < \rho^{-1} < 1 < \rho < \rho_2$, we obtain the following estimate for the error committed in approximating Laurent coefficients by discrete Fourier transforms:

$$|\hat{a}_m - a_m| \leqq \mu^*(\rho)(\rho^m + \rho^{-m}) \frac{1}{\rho^n - 1}, \qquad m \in \mathbb{Z}. \tag{13.4-3}$$

Again we see that for m fixed and $n \to \infty$, $\hat{a}_m$ tends to a_m with geometric convergence.

We point out some applications of the implied algorithm for numerically calculating Laurent coefficients.

EXAMPLE 1. **Numerical differentiation of analytic functions**

Let g be analytic in the disk $|z - z_0| < \sigma$. Then the function

$$f(z) := g(z_0 + \rho z)$$

satisfies the above hypotheses for every ρ such that $0 < \rho < \sigma$, and

$$a_m = \begin{cases} 0, & m < 0 \\ \dfrac{\rho^m}{m!} g^{(m)}(z_0), & m \geqq 0. \end{cases} \tag{13.4-4}$$

Thus the coefficients $\hat{a}_m$ may be used to approximate the derivatives of g at z_0. If g is a polynomial of degree $< n$, (13.4-2) shows that $\hat{a}_m = a_m$ for $0 \leqq m < n$, and the exact values of the derivatives may be obtained from (13.4-4). For general analytic functions, an adaptation of (13.4-3) shows that every derivative of fixed order is obtained with geometric convergence as $n \to \infty$.

The algorithm in all its simplicity highlights some of the typical differences between analytic and nonanalytic functions.

(a) The points where g is evaluated do not lie on a straight line, as in real numerical differentiation, but on a circle. There is no preferred direction in the complex plane. The derivative, if defined as limit of a difference quotient, is independent of the manner in which the increment tends to zero.

(b) Even if the derivatives at z_0 are to be calculated with full precision, the points where g is evaluated do not tend to z_0. The values of g on the circle $|z - z_0| = \rho$ completely determine g.

(c) A large number of derivatives (theoretically all) are evaluated simultaneously. There is no question about the existence of derivatives. If the first derivative exists in a neighborhood of z_0, all derivatives exist.

Readers versed in numerical analysis will also note that due to the stability of the FFT (see § 13.1) the foregoing algorithm for numerical differentiation is stable no matter how high the accuracy of the formulas. This contrasts favorably with high-order formulas for real numerical differentiation, which tend to become unstable due to cancellation of large terms with opposite signs.

EXAMPLE **2. Generating functions**

If $\mathbf{a}=\{a_m\}_{m=-\infty}^{\infty}$ is a sequence (not assumed to be periodic) of complex numbers, and if the series

$$f(t):=\sum_{m=-\infty}^{\infty} a_m t^m$$

converges in a suitable annulus, then f is called the **generating function** of the sequence $\mathbf{a}$. (Frequently in applications the a_k, and hence f, depend on additional parameters.) The discrete Fourier transform may be used to generate the elements of $\mathbf{a}$ by evaluations of the generating function of $\mathbf{a}$. Experiments show that in the well-known example

$$f(t,x)=e^{(x/2)(t-t^{-1})}=\sum_{m=-\infty}^{\infty} J_m(x)t^m \tag{13.4-5}$$

(where J_m is the Bessel function of order m) the method is competitive with the familiar device of using the recurrence relation backward. In an example like

$$f(t,x)=\frac{t\,e^{xt}}{e^t-1}=\sum_{m=0}^{\infty}\frac{B_m(x)}{m!}t^m$$

(where $B_m(x)$ is the mth Bernoulli polynomial, see (11.11-9)) the usual recurrence relation requires $O(n^2)$ operations to generate $B_0(x), B_1(x),\ldots, B_n(x)$, whereas an FFT requires only $O(n \operatorname{Log} n)$ operations to generate approximations to the same numbers.

Some fine-tuning will be required in all these applications to determine the optimal choices of n and ρ. In addition, difficulties will appear if the exact Laurent coefficients tend to zero rapidly as $m\to\infty$. Although the absolute errors of the approximations $\hat{a}_m$ produced by the FFT method still will be small, the relative errors then may increase rapidly with $|m|$.

II. Numerical Inversion of Laplace Transforms

The theory of the Laplace transform was discussed extensively in Chapter 10. There we also mentioned briefly the problem of *inverting* the Laplace transform, that is, of computing values of an original function $F(\tau)$ from its image function

$$f(s) := \int_0^\infty e^{-s\tau} F(\tau)\, d\tau.$$

FFTs are essential for the numerical implementation of a technique (see § 10.5) which makes use of the following analytical properties of any Laplace transform f:

(i) f is analytic (at least) in a half-plane $\operatorname{Re} s > \gamma_F$, where γ_F is the growth indicator of F.
(ii) $f(s) \to 0$ as $s \to \infty$ in the half-plane just described.

By considering $e^{-\gamma_F \tau} F(\tau)$ in place of F, we may assume that $\gamma_F = 0$. Every function analytic in $\operatorname{Re} s > 0$ can be expanded as a series of powers of

$$z := \frac{s-\alpha}{s+\alpha},$$

where $\alpha > 0$, because for any such choice of α, $|z| < 1$ corresponds precisely to $\operatorname{Re} s > 0$. It thus seems natural to expand[1] $f \bullet\!\!-\!\!\!-\!\!\circ\, F$ in powers of z and thus to obtain an expansion of F in terms of the original functions of z^m. However, no such original functions exist, because z^m as a function of s does not satisfy (ii). On the other hand, the functions

$$l_m(s) := \frac{z^m}{s+\alpha} = \frac{(s-\alpha)^m}{(s+\alpha)^{m+1}}$$

are image functions, and their original functions are readily expressed in terms of the *Laguerre polynomials*

$$L_m(\tau) := \frac{1}{m!} e^{\tau} \frac{d^m}{d\tau^m} (e^{-\tau} \tau^m).$$

The precise correspondence is

$$l_m(s) \bullet\!\!-\!\!\!-\!\!\circ\, e^{-\alpha\tau} L_m(2\alpha\tau).$$

Thus the inversion problem is reduced to the problem of determining the coefficients a_m in the expansion

$$f(s) = \sum_{m=0}^{\infty} a_m \frac{(s+\alpha)^m}{(s+\alpha)^{m+1}}.$$

[1] The symbol $\bullet\!\!-\!\!\!-\!\!\circ$ denotes a Laplacian correspondence; see § 10.2.

These coefficients are also given by

$$g(z)=\sum_{m=0}^{\infty} a_m z^m,$$

where

$$g(z):=(s+\alpha)f(s)=\frac{2\alpha}{1-z}f\left(\alpha\frac{1+z}{1-z}\right).$$

If the a_m have been found, then at least formally

$$F(\tau)=e^{-\alpha\tau}\sum_{m=0}^{\infty} a_m L_m(2\alpha\tau) \tag{13.4-6}$$

For conditions of validity of this expansion see Theorem 10.5d. The values of $L_m(2\alpha\tau)$ required for the numerical evaluation of the series (13.4-6) may be generated rapidly by the forward recurrence

$$L_0(2\alpha\tau)=1$$

$$L_1(2\alpha\tau)=1-2\alpha\tau$$

$$L_m(2\alpha\tau)=\frac{1}{m}\{(2m-1-2\alpha\tau)L_{m-1}(2\alpha\tau)-(m-1)L_{m-2}(2\alpha\tau)\},$$

$$m=2,3,\ldots,$$

which turns out to be stable. Alternatively, the series may be evaluated by the *Clenshaw algorithm*, to be discussed in another context.

What makes the method feasible is the fact that the Taylor coefficients a_m may be evaluated efficiently by the technique that was discussed at the outset of this section. Let $n=2^l$ be a (large) power of 2, $w:=\exp(2\pi i/n)$, select $\rho\in(0,1)$, and define $\mathbf{g}=\{g_m\}$, where

$$g_m:=\frac{2\alpha}{1-\rho w^m}f\left(\alpha\frac{1+\rho w^m}{1-\rho w^m}\right).$$

The sequence

$$\mathbf{b}=\{b_m\}:=\frac{1}{\sqrt{n}}\mathcal{F}_n\mathbf{g}$$

can be evaluated in $O(n \operatorname{Log} n)$ operations, and the desired coefficients a_m are approximated by $\rho^{-m}b_m$ with an error that tends to zero geometrically as $n\to\infty$.

The algorithm again requires some fine-tuning to determine optimal values of the parameters α and ρ. Also it should be mentioned that the *mathematical* convergence of the series (13.4-6) (which has nothing to do with the numerical performance of the algorithm) is seriously impaired if the function F or one of its low-order derivatives is discontinuous.

III. The Zeros of a Function in a Disk

Here we describe an algorithm due to Geiger [1981] for solving the following problem. We are given a function g which is analytic on the closed disk $|z-z_0| \leqq \rho$ and different from zero on the boundary $|z-z_0| = \rho$. We wish to construct the polynomial p whose zeros are precisely the zeros of g inside $|z-z_0| = \rho$.

The efficient solution of this problem makes feasible a new approach to the task of determining the zeros of high-degree polynomials. Instead of determining the zeros one after the other and deflating, as in the conventional approach, or of isolating the zeros by a Weyl-type exclusion algorithm (see § 6.11), the task could be broken down into smaller subtasks by covering the plane by disks and determining the polynomials corresponding to the zeros in each disk. If necessary, the process could even be iterated, much like in the well-known Lehmer method (§ 6.10). Even if not carried to its conclusion where each disk contains at most one zero, the algorithm would yield information about the distribution or "density" of zeros.

By a shift of variable, it suffices to consider the special case $z_0 = 0$, $\rho = 1$. Moreover we may assume that there exists $\gamma \in (0, 1)$ such that g is analytic and free of zeros in the annulus A: $\gamma < |z| < \gamma^{-1}$. By the algorithm described earlier, an FFT may be used to generate approximate values $\hat{a}_m$ of the coefficients a_m of the Laurent series of the function

$$f(z) := \frac{g'(z)}{g(z)}$$

in A,

$$f(z) = \sum_{m=-\infty}^{\infty} a_m z^m, \qquad z \in A.$$

On the other hand, some of these coefficients may be calculated explicitly. If the zeros of g inside $|z| = 1$ are $z_1, z_2, \ldots, z_d$ (zeros of multiplicity m occurring m times in this enumeration), we have

$$f(z) = \sum_{i=1}^{d} \frac{1}{z - z_i} + f_0(z),$$

where f_0 has no poles in $|z| \leqq 1$. Hence f_0 does not contribute to the Laurent coefficients a_m where $m < 0$, and from

$$\frac{1}{z - z_i} = \frac{1}{z} \frac{1}{1 - (z_i/z)} = \sum_{m=0}^{\infty} \frac{z_i^m}{z^{m+1}}$$

we see that the a_m for $m<0$ are related to the *power sums* of the zeros z_i,

$$a_{-m-1} = s_m := \sum_{i=1}^{d} z_i^m, \qquad m = 0, 1, 2, \ldots. \tag{13.4-7}$$

In particular,

$$a_{-1} = s_0 = d,$$

the number of zeros inside $|z| = 1$; furthermore, if there is only one such zero,

$$a_{-2} = s_1 = z_1.$$

In general if the power sums $s_0, s_1, \ldots$ are known to sufficient accuracy, the polynomial

$$p(z) = \prod_{i=1}^{d} (z - z_i) = z^d + b_1 z^{d-1} + \cdots + b_d$$

having precisely the zeros $z_1, z_2, \ldots, z_d$ may be constructed as follows. We consider the reciprocal polynomial

$$q(z) := z^d p(z^{-1}) = \prod_{i=1}^{d} (1 - z z_i) = 1 + b_1 z + \cdots + b_d z^d$$

which satisfies

$$\frac{q'(z)}{q(z)} = -\sum_{i=1}^{d} \frac{z_i}{1 - z z_i} = -\sum_{m=0}^{\infty} s_{m+1} z^m.$$

If the s_m are known, the b_j may be determined by comparing coefficients in

$$\sum_{j=1}^{d} j b_j z^{j-1} = -\sum_{m=1}^{\infty} s_m z^{m-1} \sum_{j=0}^{d} b_j z^j$$

$(b_0 := 1)$, which yields the recurrence relation

$$b_m = -\frac{1}{m}(s_m b_0 + s_{m-1} b_1 + \cdots + s_1 b_{m-1}), \qquad m = 1, 2, \ldots. \tag{13.4-8}$$

The accuracy of the approximation $\hat{a}_{-m-1}$ to s_m may be assessed by Theorem 13.2a. In addition, the correct choice of n is facilitated by the following considerations:

(a) Because $s_0 = d$, the approximate value $\hat{a}_{-1}$ should be, within the permitted tolerance, an integer.
(b) For the integer d approximately determined by $\hat{a}_{-1}$, the coefficients b_i resulting from (13.4-8) should be zero for $i > d$.

The computation of the Laurent coefficients by a discrete Fourier transform requires the evaluation of the function $f = g'/g$ at the points w_n^k where

$w_n := \exp(2\pi i/n)$, $k=0, 1, \ldots, n-1$, for some large value of n. If g is a polynomial,

$$g(z) = c_0 + c_1 z + \cdots + c_l z^l,$$

the sequence $\mathbf{g} = \{g(w_n^k)\} \in \Pi_n$, if $n > l$, evidently is

$$\mathbf{g} = \sqrt{n}\, \bar{\mathscr{F}}_n \mathbf{c},$$

where $\mathbf{c} := \|: c_0, c_1, \ldots, c_l, 0, \ldots, 0 :\| \in \Pi_n$. A similar algorithm may be applied to g'. Thus the required values of f may be generated in $O(n \operatorname{Log} n)$ operations.

PROBLEMS

1. Find numerical values of the coefficients c_k in

$$\frac{1}{\Gamma(z)} = \sum_{k=1}^{\infty} c_k z^k,$$

where Γ denotes the gamma function. (Accurate values of $\Gamma(z)$ may be generated by Stirling's formula; see § 8.5. For numerical values of the c_k, see Abramowitz and Stegun [1965], p. 256.)

2. If the algorithm of Section III is used iteratively, subsequent steps will require values of $r(z) := p'(z)/p(z)$. Show how to obtain a continued fraction representation for r by applying the *qd* algorithm (§ 7.6) to the power series

$$r(z) = \sum_{m=0}^{\infty} \frac{s_m}{z^{m+1}}.$$

3. Show by means of numerical tests that the construction of the polynomial $p(z)$ by means of the recurrence relation (13.4-8) can be unstable.
4. Apply Geiger's algorithm to the problem of determining the solutions of the equation

$$\operatorname{erf}(z) = \varepsilon,$$

where

$$\operatorname{erf}(z) := \frac{2}{\sqrt{\pi}} \int_0^z e^{-t^2}\, dt,$$

and where $\varepsilon \in [0, 1]$ is a real parameter. (For the cases $\varepsilon = 0, 1$ see Geiger [1981].)

NOTES

This section closely follows Henrici [1979a]. Using the FFT for the differentiation of analytic functions was proposed by Lyness and Moler [1967]; see also Lyness and Sandee [1971]. Fornberg [1981c, d] combines it with the Richardson extrapolation. The technique for inverting Laplace transforms is discussed from a practical point of view by Wing [1967]; see also Weeks [1966]. Talbot [1979] has an effective method for inverting Laplace transforms which uses the analytic continuation into Re $s < \gamma_F$. For Geiger's method see Geiger [1981], who reports extensive numerical experiments.

The problem considered by Geiger contains as a special case the problem of *spectral factorization* of a *self-reciprocal* polynomial. Given a real polynomial p, of degree $2n$, such that $z^n p(z^{-1}) = z^{-n} p(z)$ (and therefore the zeros of p lie symmetric about the unit circle) and such that $p(z) \neq 0$ for $|z| = 1$, find a polynomial q such that $q(z) \neq 0$ for $|z| \geqq 1$ and $z^{-n} p(z) = q(z) q(z^{-1})$. This problem occurs in filter design and in the statistical analysis of moving-average processes; various algorithms are given by Levinson [1949], Bauer [1955], Wilson [1969], Vostry [1975], Clements and Anderson [1976], and Laurie [1980], and for rational functions by Smith [1983] and Gutknecht [1983b]. For generalizations to matrix-valued functions, important in systems theory, see Tuel [1968], Wilson [1972], Rissanen [1973], and Anderson, Hitz, and Diem [1974].

§ 13.5. THE CONJUGATE PERIODIC FUNCTION

Here we digress to anticipate some results which, although elementary and basic to complex analysis, will be presented systematically only in Chapter 14. Let D: $|z| < 1$ denote the unit disk, and let u_0 be a real continuous function defined on the boundary of D. We may then consider the following Dirichlet-type boundary value problem. To find a real function u, defined on the closure D' of D, such that:

(i) u is continuous on D'.

(ii) u is harmonic in D, that is, u has continuous partial derivatives up to order 2 in D and there holds

$$\Delta u := \frac{\partial^2 u}{\partial x^2} + \frac{\partial^2 u}{\partial y^2} = 0$$

at every point of D.

(iii) For z on the boundary of D,

$$u(z) = u_0(z).$$

It is well known (and will be proved in § 14.2) that the foregoing problem has a solution u, which is unique. Given any harmonic function u in a simply connected region it is possible, by using the Cauchy–Riemann equations, to construct another harmonic function v, called the conjugate harmonic function of u, such that the complex-valued function $f(z) := u(z) + iv(z)$ is analytic in the same region. The conjugate harmonic function is determined by u up to an additive constant. Thus since D is simply connected, the conjugate harmonic function for the solution of the above Dirichlet problem exists and is made unique by the condition

$$v(0) = 0. \tag{13.5-1}$$

Frequently (not always) it is possible to extend v to a function which, like u, is continuous on the closure D' of D. If so, and if $u_0(e^{2\pi i \tau}) =: \phi(\tau)$, $0 \leqq$

$\tau \leqq 1$, the function

$$\psi(\tau) := v(e^{2\pi i \tau})$$

is called the **conjugate periodic function** of ϕ, and we write

$$\psi = \mathcal{K}\phi.$$

We now consider one case where the conjugate periodic function of a given function $\phi \in \Pi$ is easily determined. Let ϕ be the sum of an absolutely convergent Fourier series,

$$\phi(\tau) = \sum_{m=-\infty}^{\infty} a_m \, e^{2\pi i m \tau}, \tag{13.5-2}$$

so that

$$\sum_{m=-\infty}^{\infty} |a_m| < \infty. \tag{13.5-3}$$

Since ϕ is real, there holds

$$a_{-m} = \overline{a_m}, \qquad m \in \mathbb{Z}; \tag{13.5-4}$$

in particular, a_0 is real. The solution of the Dirichlet problem for D with boundary data $u_0(e^{2\pi i \tau}) = \phi(\tau)$ is easily constructed. By the methods of advanced calculus, if $z = \rho \, e^{2\pi i \tau}$, the solution

$$u(z) = \sum_{m=-\infty}^{\infty} a_m \rho^{|m|} \, e^{2\pi i m \tau} \tag{13.5-5}$$

is found. (Verification: each term of the series obviously is harmonic. Since the differentiated series converges uniformly on each compact subset of D, the sum of the series is likewise harmonic. By (13.5-3), the series converges uniformly on the *closed* disk $\rho \leqq 1$ and therefore represents a continuous function, which for $\rho = 1$ evidently agrees with (13.5-2).)

On account of (13.5-4), the solution u may also be written

$$u(z) = a_0 + \sum_{m=1}^{\infty} (a_m z^m + \overline{a_m z^m})$$

$$= a_0 + 2 \operatorname{Re} \sum_{m=1}^{\infty} a_m z^m.$$

We therefore have $u(z) = \operatorname{Re} f(z)$, where

$$f(z) := a_0 + 2 \sum_{m=1}^{\infty} a_m z^m,$$

and since a_0 is real, it follows that the conjugate harmonic function of u satisfying (13.5-1) is

$$v(z) = \operatorname{Im} f(z) = 2 \operatorname{Im} \sum_{m=1}^{\infty} a_m z^m.$$

Obviously on account of (13.5-3), v is continuous on D'. For the conjugate *periodic* function of ϕ we therefore find

$$\psi(\tau) = 2 \operatorname{Im} \sum_{m=1}^{\infty} a_m e^{2\pi i m \tau} = \sum_{m=1}^{\infty} (-ia_m e^{2\pi i m \tau} + ia_{-m} e^{-2\pi i m \tau}).$$

Defining the **signum function** σ_m by

$$\sigma_m := \begin{cases} +1, & m > 0 \\ 0, & m = 0 \\ -1, & m < 0 \end{cases} \tag{13.5-6}$$

we therefore have:

THEOREM 13.5a. *If $\phi \in \Pi$ is the sum of an absolutely convergent Fourier series,*

$$\phi(\tau) = \sum_{m=-\infty}^{\infty} a_m e^{2\pi i m \tau},$$

the conjugate periodic function $\psi = \mathcal{K}\phi$ is given by

$$\psi(\tau) = -i \sum_{m=-\infty}^{\infty} \sigma_m a_m e^{2\pi i m \tau}. \tag{13.5-7}$$

From Theorem 13.5a, or directly from the definition of $\mathcal{K}$ there follow the relations

$$\mathcal{K} \cos 2\pi m \tau = \sin 2\pi m \tau, \qquad \mathcal{K} \sin 2\pi m \tau = -\cos 2\pi m \tau.$$

For Fourier series written in real form, we therefore have:

COROLLARY 13.5b. *If $\phi \in \Pi$ is the sum of a real Fourier series,*

$$\phi(\tau) = \tfrac{1}{2}a_0 + \sum_{m=1}^{\infty} [a_m \cos(2\pi m \tau) + b_m \sin(2\pi m \tau)],$$

where $\sum (|a_m| + |b_m|) < \infty$, the conjugate periodic function of ϕ is

$$\psi(\tau) = \sum_{m=1}^{\infty} [-b_m \cos(2\pi m \tau) + a_m \sin(2\pi m \tau)].$$

Many problems in applied analysis may be reduced to the computation of the conjugate periodic function of a given function $\phi \in \Pi$. Theorem 13.5a points the way to a viable algorithm for performing this computation. Determine (approximately) the Fourier series of ϕ by one of the methods of § 13.2. Then apply (13.5-7) to obtain the Fourier series of $\psi = \mathcal{K}\phi$. In § 13.6 we discuss some refinements of this technique. Moreover in § 14.2 we show how to compute, by means of singular integrals, conjugate periodic functions also for functions ϕ that are not represented by absolutely convergent Fourier series.

PROBLEMS

1. The purely mathematical treatment of conjugate periodic functions proceeds best in the space of square integrable functions, that is, functions ϕ for which the norm

$$\|\phi\| := \left(\int_0^1 |\phi(\tau)|^2 \, d\tau \right)^{1/2}$$

is finite. (This space includes functions that are not bounded.)

(a) Show that for functions ϕ satisfying (13.5-3),

$$\|\phi\|^2 = \sum_{m=-\infty}^{\infty} |a_m|^2.$$

(b) Show that the conjugation operator does not increase the norm, that is, that

$$\|\mathcal{K}\phi\| \leqq \|\phi\| \qquad (*)$$

for all ϕ satisfying (13.5-3).

(c) For which ϕ does equality hold in (*)?

2. Does every continuous $\phi \in \Pi$ have a continuous conjugate $\psi \in \Pi$? Show that the answer is negative by establishing the following:

(a) The function

$$\phi(\tau) := \sum_{n=2}^{\infty} \frac{\sin(2\pi n \tau)}{n \operatorname{Log} n}$$

is continuous. (Use Theorem 1.3 of Chapter V of Zygmund [1959].)

(b) Show that the conjugate of ϕ,

$$\psi(\tau) = -\sum_{n=2}^{\infty} \frac{\cos(2\pi n \tau)}{n \operatorname{Log} n},$$

is unbounded near $\tau = 0$, hence discontinuous.

3. Does a bounded function $\phi \in \Pi$ with a finite number of discontinuities have a bounded conjugate? Show that the answer is negative by computing the conjugate function of

$$\phi(\tau) := \pi(1 - 2\tau), \qquad 0 < \tau < 1.$$

NOTES

Conjugate periodic functions are a central topic in the theory of Fourier series; see Zygmund [1959], ch. II, IV, VII, or Katznelson [1968], ch. III. Although they may be defined directly by means of the singular integral (14.2-15), we find the present more roundabout approach didactically preferable.

§ 13.6. TRIGONOMETRIC INTERPOLATING POLYNOMIALS AND THEIR CONJUGATES

A **trigonometric polynomial** of degree m (and period 1) is a function of the form

$$t(\tau) := \sum_{k=-m}^{m} c_k e^{2\pi i k\tau}, \tag{13.6-1}$$

where $c_k \in \mathbb{C}$. The trigonometric polynomial (13.6-1) is called *balanced* if $c_{-m} = c_m$. If a balanced trigonometric polynomial is expressed in terms of sines and cosines, the term involving $\sin(2\pi m\tau)$ is missing. In this section we consider the problem of interpolating a given function $f \in \Pi$ on a set of equidistant sampling points by a trigonometric polynomial, possibly balanced, of lowest possible degree.

I. Polynomial Representation of the Interpolating Polynomial

Proceeding heuristically, let $f \in \Pi$ be represented by its Fourier series,

$$f(\tau) = \sum_{k=-\infty}^{\infty} a_k e^{2\pi i k\tau}. \tag{13.6-2}$$

For some integer n, let $\mathbf{f} = \{f_k\} = \{f(\tau_k)\}$ where $\tau_k = k/n$, and let

$$\mathbf{c} = \{c_k\} = \frac{1}{\sqrt{n}} \mathscr{F}_n \mathbf{f} \tag{13.6-3}$$

be the sequence of discrete Fourier coefficients of f. If it is our aim to approximate f by a trigonometric polynomial, using only the sampling values f_k, we might consider a partial sum of the Fourier series (13.6-2) where the a_k are replaced by the c_k. We know from § 13.2 that the c_k are not related to the a_k if $|k| > m$ where $m := n/2$. If n is even, $n = 2m$, then $c_{-m} = c_m$ by the periodicity of the sequence $\mathbf{c}$, and the terms where $k = \pm m$ both approximate the same term of the Fourier series. A reasonable approximation to $f(\tau)$ by a trigonometric polynomial thus might be

$$t(\tau) := \sum_{k=-m}^{m}{}' c_k e^{2\pi i k\tau}. \tag{13.6-4}$$

Here and in the following the prime indicates that if n is even, $n=2m$, the terms corresponding to $k=\pm m$ are to be multiplied by $\frac{1}{2}$. For n even, the polynomial (13.6-4) is balanced.

Two errors are being committed in approximating $f(\tau)$ by $t(\tau)$: (a) the error committed in replacing the exact Fourier coefficients a_k by their approximations c_k; (b) the error committed in truncating the Fourier series. Rather surprisingly it turns out that, at least at the sampling points τ_k, these two errors just cancel each other:

$$t(\tau_k)=f_k, \qquad k\in\mathbb{Z}. \tag{13.6-5}$$

There even holds

THEOREM 13.6a. *Let $n \geqq 1$ be an integer, $m:=[n/2]$. The function* (13.6-4) *(with the coefficients given by* (13.6-3)) *is the unique (if n is even: the unique balanced) trigonometric polynomial of degree m that interpolates the function f at all sampling points $\tau_k := k/n$.*

Proof. The interpolating property (13.6-5), far from being a deep result, is merely a consequence of the fact that the discrete Fourier operator is unitary. In fact, let

$$\mathbf{t}:=t(\tau_k)\in\Pi_n.$$

Then, as follows from (13.2-12).

$$\mathbf{t}=\sqrt{n}\,\overline{\mathscr{F}_n}\mathbf{c}. \tag{13.6-6}$$

Using (13.6-3) there follows

$$\mathbf{t}=\overline{\mathscr{F}_n}\mathscr{F}_n\mathbf{f}$$

or, since $\overline{\mathscr{F}_n}=\mathscr{F}_n^{-1}$, $\mathbf{t}=\mathbf{f}$, proving the interpolating property of t. To show that t is the only (balanced, if n is even) trigonometric polynomial with the interpolating property, assume that there are two such polynomials. Then their difference likewise is a trigonometric polynomial (balanced, if n is even) of degree m. Let $\mathbf{d}=\{d_k\}$ be the sequence of its coefficients, periodically continued into Π_n. Because the values of the difference at the sampling points are 0, we again have by (13.2-12)

$$\mathbf{0}=\sqrt{n}\,\mathscr{F}_n\mathbf{d}.$$

The operator $\mathscr{F}_n$, being unitary, is invertible, and there follows $\mathbf{d}=\mathbf{0}$. Thus any two interpolating polynomials with the indicated properties are identical. —

We note that by using an FFT the coefficients c_k in the representation of the interpolating trigonometric polynomial can (at any rate if $n=2^l$) be computed in $O(n \operatorname{Log} n)$ operations. Using Horner's rule, the evaluation of $t(\tau)$ at any point τ requires another $O(n)$ operations. Thus given the sampling values f_k, the interpolating polynomial can, at any τ, be evaluated in $O(n \operatorname{Log} n)$ operations. This excellent result will yet be improved upon below.

II. The Error of the Interpolating Polynomial

We know that $t(\tau_k) = f(\tau_k)$. But how big is the error $f(\tau) - t(\tau)$ at points τ that are not sampling points? Error formulas involving high-order derivatives of f exist but are seldom useful. Instead we assume that f is the sum of an absolutely convergent Fourier series,

$$f(\tau) = \sum_{k=-\infty}^{\infty} a_k e^{2\pi i k \tau}. \tag{13.6-7}$$

Letting $z := e^{2\pi i \tau}$, this becomes

$$f(\tau) = \sum_{k=-\infty}^{\infty} a_k z^k.$$

Writing (13.6-4) similarly and using Theorem 13.2a to represent the c_k in terms of the a_k, there follows

$$\begin{aligned} f(\tau) - t(\tau) &= \sum_{k=-m}^{m}{}' z^k \sum_{l=-\infty}^{\infty} a_{ln+k}(z^{ln} - 1) \\ &= \sum_{l=-\infty}^{\infty} (z^{ln} - 1) \sum_{k=-m}^{m}{}' a_{ln+k} z^k. \end{aligned}$$

Letting

$$\alpha_l := \sum_{k=-m}^{m}{}' |a_{ln+k}| \tag{13.6-8}$$

and observing that

$$|z^{ln} - 1| = 2|\sin(ln\pi\tau)|$$

which is zero for $l = 0$, we obtain:

THEOREM 13.6b. *If $f \in \Pi$ is represented by the absolutely convergent Fourier series (13.6-7) and if t is the unique (and if n is even, balanced) trigonometric polynomial interpolating f at the sampling points $\tau_k = k/n$, $k \in \mathbb{Z}$), then for all real τ*

$$|f(\tau) - t(\tau)| \leq 2 \sum_{l=1}^{\infty} (\alpha_l + \alpha_{-l}) |\sin(ln\pi\tau)|.$$

Replacing the sines by 1 and using (13.6-8), we also have:

COROLLARY 13.6c. *Under the above hypotheses,*

$$|f(\tau)-t(\tau)| \leqq 2 \sum_{|k| \geqq m}' |a_k|. \tag{13.6-9}$$

Again the prime is taken to mean that for n even, the factor $\frac{1}{2}$ is to be inserted in the terms $k = \pm m$.

For many classes of functions $f \in \Pi$, the rate at which the coefficients a_k tend to zero as $k = \pm\infty$ is well established. Theorem 13.6b, and in particular its corollary, can then be used to obtain information on how fast $t(\tau)$ approaches $f(\tau)$ as $n \to \infty$.

III. Barycentric Representation of $t(\tau)$

Let $z = e^{2\pi i \tau}$, $z_k := w^k$, where $w := \exp(2\pi i/n)$. By using the formulas (13.2-4) for c_k and reversing the order of summation, we can express $t(\tau)$ directly in terms of the sampling values f_k,

$$\begin{aligned} t(\tau) &= \sum_{j=-m}^{m}{}' z^j \frac{1}{n} \sum_{k=0}^{n-1} f_k (z_k^{-1})^j \\ &= \sum_{k=0}^{n-1} l_k(\tau) f_k, \end{aligned}$$

where

$$l_k(\tau) := \frac{1}{n} \sum_{j=-m}^{m}{}' (z_k^{-1} z)^j.$$

Using the well-known formula for terminating geometric sums, we have for n even, $n = 2m$,

$$l_k(\tau) = \frac{1}{2n} \frac{z^n - 1}{z_k^m z^m} \frac{z + z_k}{z - z_k}.$$

Using $z_k^m = (-1)^k$ and resubstituting $z = e^{2\pi i \tau}$, there follows

$$l_k(\tau) = \frac{\sin(n\pi\tau)}{n} (-1)^k \cot[\pi(\tau - \tau_k)].$$

If n is odd, $n = 2m+1$, a similar computation yields

$$\begin{aligned} l_k(\tau) &= \frac{1}{n} \frac{z^n - 1}{z_k^m z^m} \frac{z}{z - z_k} \\ &= \frac{\sin(n\pi\tau)}{n} (-1)^k \csc[\pi(\tau - \tau_k)] \end{aligned}$$

($\csc := \sin^{-1}$). We thus find the following classical formulas expressing the trigonometric interpolating polynomials in terms of the f_k:

$$t(\tau) = \frac{\sin(n\pi\tau)}{n} \sum_{k=0}^{n-1} (-1)^k f_k \cot[\pi(\tau - \tau_k)], \qquad n \text{ even}, \qquad (13.6\text{-}10a)$$

$$t(\tau) = \frac{\sin(n\pi\tau)}{n} \sum_{k=0}^{n-1} (-1)^k f_k \csc[\pi(\tau - \tau_k)], \qquad n \text{ odd}. \qquad (13.6\text{-}10b)$$

Simple as these formulas look, they are not well suited to numerical computation because the numerical values both of $\sin(n\pi\tau)$ and of the cot and csc factors may have large relative errors near the sampling points τ_k. Since the two errors are independent, the resulting values of $t(\tau)$ may be grossly in error. Stable formulas are obtained, however, by observing the identities

$$1 = \frac{\sin(n\pi\tau)}{n} \sum_{k=0}^{n-1} (-1)^k \cot[\pi(\tau - \tau_k)], \qquad n \text{ even}, \qquad (13.6\text{-}11a)$$

$$1 = \frac{\sin(n\pi\tau)}{n} \sum_{k=0}^{n-1} (-1)^k \csc[\pi(\tau - \tau_k)], \qquad n \text{ odd}, \qquad (13.6\text{-}11b)$$

which hold by virtue of the fact that for any $n \geqq 1$ the trigonometric polynomial of degree $[n/2]$ reproduces the constant 1 exactly. Dividing (13.6-10) by (13.6-11), the troublesome factor $\sin(n\pi\tau)$ cancels, and we obtain:

THEOREM 13.6d. *The interpolating trigonometric polynomial described in Theorem* 13.6a *is also given by*

$$t(\tau) = \frac{\sum_{k=0}^{n-1} (-1)^k f_k \cot[\pi(\tau - \tau_k)]}{\sum_{k=0}^{n-1} (-1)^k \cot[\pi(\tau - \tau_k)]} \qquad (13.6\text{-}12a)$$

if n is even, and by

$$t(\tau) = \frac{\sum_{k=0}^{n-1} (-1)^k f_k \csc[\pi(\tau - \tau_k)]}{\sum_{k=0}^{n-1} (-1)^k \csc[\pi(\tau - \tau_k)]} \qquad (13.6\text{-}12b)$$

if n is odd.

The formulas (13.6-12) are called *barycentric* because they formally resemble the well-known expressions for the center of mass (barycenter) of a system of points of mass. If the same numerical values of the cot and csc functions are used in the numerator and in the denominator, the formulas are numerically stable even very close to the sampling points τ_k.

The expression (13.6-12a) lends itself nicely to a recursive algorithm for successively computing the polynomials $t_l(\tau)$, $l = 0, 1, 2, \ldots$, interpolating at $n = 2^l$ sampling points. For each such n we split the numerator in (13.6-12a) into the sums of the terms where k is even and those where k is odd:

$$a_l(\tau) := \sum_{\substack{k=0 \\ k \text{ even}}}^{n-2} f\left(\frac{k}{n}\right) \cot\left[\pi\left(\tau - \frac{k}{n}\right)\right],$$

$$r_l(\tau) := \sum_{\substack{k=1 \\ k \text{ odd}}}^{n-1} f\left(\frac{k}{n}\right) \cot\left[\pi\left(\tau - \frac{k}{n}\right)\right], \qquad n = 2^l,$$

and similarly for the denominator:

$$b_l(\tau) := \sum_{\substack{k=0 \\ k \text{ even}}}^{n-2} \cot\left[\pi\left(\tau - \frac{k}{n}\right)\right],$$

$$s_l(\tau) := \sum_{\substack{k=1 \\ k \text{ odd}}}^{n-1} \cot\left[\pi\left(\tau - \frac{k}{n}\right)\right], \qquad n = 2^l.$$

Then clearly,

$$t_l(\tau) = \frac{a_l(\tau) - r_l(\tau)}{b_l(\tau) - s_l(\tau)}. \tag{13.6-13}$$

If we now proceed to the polynomial $t_{l+1}(\tau)$, which uses twice as many points, then evidently

$$\begin{aligned} a_{l+1}(\tau) &= a_l(\tau) + r_l(\tau), \\ b_{l+1}(\tau) &= b_l(\tau) + s_l(\tau), \end{aligned} \tag{13.6-14}$$

and the only sums that need to be computed are those with the odd numbered terms,

$$\begin{aligned} r_{l+1}(\tau) &= \sum_{\substack{0<k\delta<1 \\ k \text{ odd}}} f(k\delta) \cot[\pi(\tau - k\delta)], \\ s_{l+1}(\tau) &= \sum_{\substack{0<k\delta<1 \\ k \text{ odd}}} \cot[\pi(\tau - k\delta)], \end{aligned} \tag{13.6-15}$$

where $\delta := 2^{-l-1}$. The equations (13.6-13) and (13.6-14) are correct for $l = 0$ if we put

$$\begin{aligned} &a_0(\tau) := f(0) \cot \pi\tau, \qquad b_0(\tau) := \cot \pi\tau, \\ &r_0(\tau) := 0, \qquad s_0(\tau) := 0. \end{aligned} \tag{13.6-16}$$

THEOREM 13.6e. *The algorithm defined by equations* (13.6-13) *through* (13.6-16) *computes, for* $l=1,2,\ldots$, *the balanced trigonometric polynomials* $t_l(\tau)$ *of degree* $m=2^{l-1}$ *interpolationg a given function* $f\in\Pi$ *at* $n=2^l$ *equidistant points* $\tau_k=k2^{-l}$, $k=0,1,\ldots,2^l-1$.

Numerical tests show this algorithm to be stable even for values of n as large as $2^{14}=16{,}384$.

IV. The Conjugate of the Interpolating Polynomial

Let $t^*(\tau)$ denote the periodic conjugate of the trigonometric interpolating polynomial $t(\tau)$ defined by (13.6-4). Because a trigonometric polynomial is its own Fourier series, we have by Theorem 13.5a:

THEOREM 13.6f. *The conjugate of* $t(\tau)$ *is*

$$t^*(\tau)=-i\sum_{k=-m}^{m}{}' \sigma_k c_k\, e^{2\pi i k\tau}, \tag{13.6-17}$$

where $\sigma_k=\operatorname{sign}(k)$ *is the sign function defined by* (13.5-6).

Often one approximates the conjugate function $f^*:=\mathcal{H}f$ by t^*. For the error thus committed one finds by the method used in proving Theorem 13.6b, using the abbreviations

$$\alpha_{l0}:=|a_{-ln}|+|a_{ln}|,$$

$$\alpha_{l1}:=\sum_{k=1}^{m}{}' (|a_{ln+k}|+|a_{-ln-k}|),$$

$$\alpha_{l2}:=\sum_{k=1}^{m}{}' (|a_{ln-k}|+|a_{-ln+k}|),$$

the following result:

THEOREM 13.6g. *Under the hypotheses of Theorem* 13.6b, *there holds for all real* τ

$$|f^*(\tau)-t^*(\tau)|\leqq\sum_{l=1}^{\infty}(\alpha_{l0}+2\alpha_{l1}|\sin(ln\pi\tau)|+2\alpha_{l2}|\cos(ln\pi\tau)|). \tag{13.6-18}$$

Replacing all sines and cosines by 1, there follows:

COROLLARY 13.6h. *Under the hypotheses of Theorem* 13.6b,

$$|f^*(\tau)-t^*(\tau)|\leqq 2\sum_{|k|\geqq m}{}' |a_k|. \tag{13.6-19}$$

We finally seek to express $t^*(\tau)$ directly in terms of the sampling values f_k. Because the conjugation operator is linear, there evidently holds

$$t^*(\tau) = \sum_{k=0}^{n-1} l_k^*(\tau) f_k,$$

where $l_k^*(\tau) := \mathcal{K} l_k$, and hence by Theorem 13.5a,

$$l_k^*(\tau) = -\frac{i}{n} \sum_{j=-m}^{m}{}' \sigma_j (z_k^{-1} z)^j,$$

$z := e^{2\pi i \tau}$, $z_k := e^{2\pi i \tau_k}$. If n is even, $n = 2m$, we find on summing two geometric sums

$$l_k^*(\tau) = \frac{i}{n} \frac{z + z_k}{z - z_k} \left\{ 1 - \frac{1}{2} z_k^m (z^m + z^{-m}) \right\}.$$

By virtue of $z_k^m = (-1)^k$ there results

$$t^*(\tau) = \frac{i}{n} \sum_{k=0}^{n-1} \frac{z + z_k}{z - z_k} f_k - \frac{i}{2n} (z^m + z^{-m}) \sum_{j=0}^{n-1} (-1)^j \frac{z + z_j}{z - z_j} f_j.$$

This expression may be further simplified by noting that the conjugate of a constant is zero. Thus if all $f_k = 1$, there results

$$0 = \frac{i}{n} \sum_{k=0}^{n-1} \frac{z + z_k}{z - z_k} - \frac{i}{2n} (z^m + z^{-m}) \sum_{j=0}^{n-1} (-1)^j \frac{z + z_j}{z - z_j}.$$

We use this to eliminate

$$\frac{1}{2}(z^m + z^{-m}) = \left[\sum_{k=0}^{n-1} \frac{z + z_k}{z - z_k} \right] \Big/ \left[\sum_{j=0}^{n-1} (-1)^j \frac{z + z_j}{z - z_j} \right]$$

and obtain

$$t^*(\tau) = \frac{i}{n} \sum_{k=0}^{n-1} \frac{z + z_k}{z - z_k} \left\{ f_k - \left[\sum_{j=0}^{n-1} (-1)^j \frac{z + z_j}{z - z_j} f_j \right] \Big/ \left[\sum_{j=0}^{n-1} (-1)^j \frac{z + z_j}{z - z_j} \right] \right\}.$$

By Theorem 13.6d, the second term in braces equals $t(\tau)$. Reverting to the variable τ, we thus have

$$t^*(\tau) = \frac{1}{n} \sum_{k=0}^{n-1} \cot[\pi(\tau - \tau_k)] \{ f_k - t(\tau) \}. \tag{13.6-20}$$

The same formula is found to hold if n is odd. We thus have proved:

THEOREM 13.6i. *The periodic conjugate t^* of the trigonometric interpolating polynomial t described in Theorem* 13.6a *is given by* (13.6-20).

If we suppose that for $n \to \infty$, $t(\tau) \to f(\tau)$ and $t^*(\tau) \to f^*(\tau)$, then making this passage to the limit in (13.6-20) yields

$$f^*(\tau) = \int_0^1 \cot[\pi(\tau - \sigma)]\{f(\sigma) - f(\tau)\}\, d\sigma. \tag{13.6-21}$$

It is difficult to justify the passage to the limit by virtue of the singularity of the integrand, but the formula will be shown to be correct for certain classes of functions $f \in \Pi$ in § 14.2.

Theorem 13.6i can be made the basis of an algorithm for computing the conjugates t_l^* of the polynomials t_l interpolating at $n = 2^l$ points. In the notation of Section III we have, for $n = 2^l$,

$$\sum_{k=0}^{n-1} f_k \cot[\pi(\tau - \tau_k)] = a_l(\tau) + r_l(\tau),$$

$$\sum_{k=0}^{n-1} \cot[\pi(\tau - \tau_k)] = b_l(\tau) + s_l(\tau).$$

Hence from (13.6-20), using (13.6-13) and omitting arguments,

$$\begin{aligned} t^* &= \frac{1}{n}\left\{ a_l + r_l - (b_l + s_l)\frac{a_l - r_l}{b_l - s_l} \right\} \\ &= \frac{1}{n}\,\frac{2r_l b_l - 2a_l s_l}{b_l - s_l}. \end{aligned}$$

We thus find:

THEOREM 13.6j. *With the quantities defined in Theorem* 13.6e, *one also has*

$$t_l^*(\tau) = 2^{1-l}\frac{r_l(\tau)b_l(\tau) - a_l(\tau)s_l(\tau)}{b_l(\tau) - s_l(\tau)}. \tag{13.6-22}$$

Again this method for computing t^* is found to be stable even for very large values of n.

PROBLEMS

1. Use a programmable calculator to find numerical values of $t_l(\tau)$, $t_l^*(\tau)$, $l = 0, 1, 2, \ldots$, where

$$f(\rho) := \begin{cases} 0, & \tau = 0 \\ 1 - 2\tau, & 0 < \tau < 1. \end{cases}$$

2. Let n be even, $n = 2m$. Show that the most general trigonometric polynomial of degree m which interpolates given values f_k at the equidistant points $\tau_k = k/n$,

$k=0, 1, \ldots, n-1$, is

$$t(\tau) := t_0(\tau) + c\sin(n\pi\tau),$$

where t_0 is the balanced polynomial defined by (13.6-4), and where $c \in \mathbb{C}$. Show, however, that among all these polynomials $t_0(\tau)$ has the smallest L_2 norm.

3. *Differentiation of the interpolating polynomial.* If t is given by (13.6-12), show that, if n is even, $n=2m$,

$$t'(0) = \pi \sum_{k=1}^{m-1} (-1)^{k+1}(f_k - f_{-k}) \cot\frac{k\pi}{n},$$

and if n is odd, $n=2m+1$,

$$t'(0) = \pi \sum_{k=1}^{m} (-1)^{k+1}(f_k - f_{-k}) \csc\frac{k\pi}{n}.$$

Also show that, if $n=2m$,

$$t''(0) = 2\pi^2 \sum_{k=1}^{m} (-1)^{k+1}(f_k - 2f_0 + f_{-k})\left(\csc\frac{k\pi}{n}\right)^2,$$

and if $n=2m+1$,

$$t''(0) = 2\pi^2 \sum_{k=1}^{m} (-1)^{k+1}(f_k - 2f_0 + f_{-k}) \cos\frac{k\pi}{n}\left(\csc\frac{k\pi}{n}\right)^2.$$

4. *Osculatory trigonometric interpolation.* Let $n=1$ be an integer. Show that the most general trigonometric polynomial t of degree $2n$ satisfying the $2n$ conditions

$$t(\tau_k) = f_k, \qquad t'(\tau_k) = f'_k, \qquad k=0, 1, \ldots, n-1$$

($\tau_k := k/n$) for given f_k and f'_k has the form

$$t(\tau) = t_0(\tau) + c[1-\cos(n\pi\tau)],$$

where $c \in \mathbb{C}$ is arbitrary and

$$t_0(\tau) := \sum_{k=-n}^{n}{}' \; a_k \, e^{2\pi i k\tau},$$

where the terms $k = \pm n$ are taken with the factor $\frac{1}{2}$, and where, letting $w := \exp(2\pi i/n)$,

$$a_0 := \frac{1}{n}\sum_{k=0}^{n-1} f_k,$$

$$a_p := \frac{1}{n^2}\sum_{k=0}^{n-1} [(n-p)f_k + (2\pi)^{-1}f'_k]w^{-pk},$$

$$a_{-p} := \frac{1}{n^2}\sum_{k=0}^{n-1} [(n-p)f_k - (2\pi)^{-1}f'_k]w^{pk},$$

$p = 1, 2, \ldots, n.$

5. (Continuation) Letting $\mathbf{f} := \{f_k\}$, $\mathbf{f}' := \{f'_k\}$, show how to express the coefficients a_p in terms of the Fourier transforms

$$\mathbf{c} := \mathscr{F}_n \mathbf{f}, \qquad \mathbf{c}' := \mathscr{F}_n \mathbf{f}',$$

which may be calculated in $O(n \operatorname{Log} n)$ operations.

6. *Trigonometric interpolation on nonuniform grids.* Let $\tau_1, \tau_2, \ldots, \tau_n$ be distinct real numbers modulo 1, and let $f_1, f_2, \ldots, f_n$ be given.

(a) If n is odd, $n = 2m+1$, show that the function

$$t(\tau) = \sum_{k=1}^{n} \left\{ \prod_{\substack{j=1 \\ j \neq k}}^{n} \frac{\sin[\pi(\tau - \tau_j)]}{\sin[\pi(\tau_k - \tau_j)]} \right\} f_k$$

is a trigonometric polynomial of degree m such that

$$t(\tau_k) = f_k, \qquad k = 1, 2, \ldots, n. \tag{$*$}$$

Also show that t is uniquely determined by these properties.

(b) If n is even, $n = 2m$, show that

$$t(\tau) := \sum_{k=1}^{n} \left\{ \prod_{\substack{j=1 \\ j \neq k}}^{n} \frac{\sin[\pi(\tau - \tau_j)]}{\sin[\pi(\tau_k - \tau_j)]} \right\} \cos[\pi(\tau - \tau_k)] f_k + c t_0(\tau),$$

where

$$t_0(\tau) := \prod_{j=1}^{n} \sin[\pi(\tau - \tau_j)]$$

for any $c \in \mathbb{C}$ is a trigonometric polynomial of degree m (not necessarily balanced) which again satisfies the interpolating conditions $(*)$ (Gauss [1866]; see Berrut [1984] for computational aspects).

NOTES

The basic facts on trigonometric interpolation, even in the case of nonequidistant sampling points, were known to Gauss [1866] and Hermite [1885]; see Goldstine [1977]. Salzer [1948] provides numerical values for the interpolating coefficients and gives a barycentric formula for nonequidistance points. Theorems 13.6d and 13.6i, and the algorithms for t and t^*, are due to Henrici [1979b]. The error estimates of Theorems 13.6b and 13.6g are given in Henrici [1982b], but the corrolaries are due to Gaier [1974b]. Price [1984] gives generalizations. Brass [1982] gives best possible error bounds for the determination of conjugates of functions with a bounded derivative. The use of the FFT to compute conjugate trigonometric polynomials was proposed by Henrici [1976]. Gutknecht [1979] has an algorithm for computing either the even or the odd numbered components of t^* in less than $\frac{1}{2}n \operatorname{Log}_2 n$ multiplications. In Gutknecht [1983b] he solves the problem of obtaining the conjugate of a *rational* trigonometric function, that is, of a function represented as a quotient of two trigonometric polynomials.

§ 13.7. CONVOLUTION

Here we introduce two kinds of multiplications in the space Π_n, discuss their behavior under discrete Fourier transformation, and describe some of their applications.

I. Multiplication in Π_n

Let $\mathbf{x}=\{x_k\}$, $\mathbf{y}=\{y_k\}$ be sequences in Π_n. We already have agreed to define scalar multiplication in Π_n by

$$c\mathbf{x} := \{cx_k\}$$

for any $c \in \mathbb{C}$. We next define a product of two sequences in Π_n, called the **Hadamard product**, by

$$\mathbf{x} \cdot \mathbf{y} := \{x_k y_k\}.$$

(We always write the dot for clarity.) To form a Hadamard product, all that is required is to form the n products $x_k y_k$, $k=0, 1, \ldots, n-1$, at a total cost of $n\mu$. It is clear that the Hadamard product is commutative, associative, and (with respect to the addition already defined in Π_n) distributive. Under addition and Hadamard multiplication the space Π_n thus becomes a commutative ring. For $n>1$, this ring has divisors of zero.

EXAMPLE 1

Let $w := \exp(2\pi i/n)$, and let the sequence $\mathbf{w} \in \Pi_n$ be defined by

$$\mathbf{w} := \{w^k\}. \tag{13.7-1}$$

Powers of $\mathbf{w}$ are always to be understood in the sense of Hadamard multiplication. It thus is clear what is to be meant by $\mathbf{w}^m$ for any integer m. The basic property (F_2) of the discrete Fourier operator (see § 13.1) may then simply be expressed thus. Let

$$\mathscr{F}_n \mathbf{x} =: \hat{\mathbf{x}}.$$

If E denotes the shift operator, then

$$\mathscr{F}_n E\mathbf{x} = \mathbf{w} \cdot \hat{\mathbf{x}}.$$

More generally for any integer k,

$$\mathscr{F}_n E^k \mathbf{x} = \mathbf{w}^k \cdot \hat{\mathbf{x}}. \tag{13.7-2}$$

EXAMPLE 2

Let $\mathbf{x}=\{x_k\}$ be represented in the form

$$\mathbf{x} = x_0\mathbf{e} + x_1 E^{-1}\mathbf{e} + \cdots + x_{n-1}E^{-n+1}\mathbf{e},$$

where **e** denotes the unit sequence defined by (13.1-3). Then since

$$\mathscr{F}_n\mathbf{e} = \frac{1}{\sqrt{n}}\|{:}1, 1, \ldots, 1{:}\| = \frac{1}{\sqrt{n}}\mathbf{w}^0,$$

we have, in view of (13.7-2),

$$\hat{\mathbf{x}} := \mathscr{F}_n\mathbf{x} = \frac{1}{\sqrt{n}}\sum_{k=0}^{n-1} x_k\mathbf{w}^{-k}. \tag{13.7-3}$$

We next introduce another kind of multiplication in Π_n which occurs frequently in applications, although sometimes in disguised form. If again $\mathbf{x} = \{x_k\}$ and $\mathbf{y} = \{y_k\}$ are in Π_n, their **convolution product** or simply **convolution** is the sequence $\mathbf{z} = \{z_k\}$, where

$$z_k = \frac{1}{\sqrt{n}}\sum_{m=0}^{n-1} x_m y_{k-m}. \tag{13.7-4}$$

(The factor $1/\sqrt{n}$, unimportant from the point of view of applications, is inserted to simplify formal relationships.) The terms **cyclic convolution** or **convolution on the circle** are also used. The notation

$$\mathbf{z} = \mathbf{x} * \mathbf{y}$$

will be used to denote the convolution product. In the convolution product, each element is the sum of n products. Thus if the formulas (13.7-4) are taken literally, $n^2\mu$ and almost as many α are required to form the convolution of two sequences in Π_n. (We neglect the divisions by $\sqrt{n}$ because most applications do not require it anyway.) Convolution, if performed directly, thus is an expenseive operation from the numerical point of view.

EXAMPLE 3. **Smoothing**

Often in applications one is required to *smooth* a given sequence $\mathbf{x} = x_k \in \Pi_n$. By this is meant the forming of a new sequence $\mathbf{y} = \{y_k\}$, where

$$y_k := \sum_{j=-d}^{d} s_j x_{k+j}, \qquad k \in \mathbb{Z}.$$

Here d is an integer, $0 < d < \frac{1}{2}n$, and the s_j are fixed nonnegative numbers satisfying

$$\sum_{j=-d}^{d} s_j = 1.$$

Defining the **smoothing sequence** $\mathbf{s} = \{s_k\}$ where $\mathbf{s} \in \Pi_n$, $s_k = 0$ if k is not congruent modulo n to an integer $j \in [-d, d]$, we obviously have

$$\mathbf{y} = \sqrt{n}R\mathbf{s} * \mathbf{x}.$$

If, as is usually the case, the smoothing sequence is *even*, $R\mathbf{s}=\mathbf{s}$, then

$$\mathbf{y}=\sqrt{n}\mathbf{s}*\mathbf{x}. \tag{13.7-5}$$

EXAMPLE **4. Numerical differentiation**

Let $f\in\Pi$, and let $\mathbf{f}=\{f_k\}$ denote a sequence of sampling values of f, $f_k:=f(k/n)$. It is shown in numerical analysis that if f is sufficiently smooth, its derivatives may be approximated by linear combinations of the sampling values. For instance, the second derivative at the sampling point τ_k is approximated by

$$h^{-2}\delta^2 f_k := h^{-2}(f_{k+1}-2f_k+f_{k-1}).$$

Defining $\boldsymbol{\delta}=\{\delta_k\}\in\Pi_n$ by

$$\begin{aligned} \delta_k &:= -2, && k\equiv 0 \mod n,\\ \delta_k &:= 1, && k\equiv \pm 1 \mod n,\\ \delta_k &:= 0, && \text{otherwise,} \end{aligned}$$

the sequence $\{\delta^2 f_k\}$ obviously equals $\sqrt{n}\boldsymbol{\delta}*\mathbf{f}$. Similar expressions for approximate derivatives of any order and for more accurate differentiation formulas can obviously be constructed.

II. The Convolution Theorem

Let $\mathbf{x},\mathbf{y}\in\Pi_n$, $\hat{\mathbf{x}}:=\mathscr{F}_n\mathbf{x}$, $\hat{\mathbf{y}}:=\mathscr{F}_n\mathbf{y}$. We try to express the Fourier transform of $\mathbf{x}*\mathbf{y}$ in terms of $\hat{\mathbf{x}}$ and $\hat{\mathbf{y}}$. From the fact that

$$\mathbf{x}*\mathbf{y}=y_0\mathbf{x}+y_{-1}E\mathbf{x}+\cdots+y_{-n+1}E^{n-1}\mathbf{x}$$

we have, using (13.7-2),

$$\begin{aligned} \mathscr{F}_n(\mathbf{x}*\mathbf{y}) &= \frac{1}{\sqrt{n}}(y_0\hat{\mathbf{x}}+y_{-1}\mathbf{w}^1\cdot\hat{\mathbf{x}}+\cdots+y_{-n+1}\mathbf{w}^{n-1}\cdot\hat{\mathbf{x}})\\ &= \frac{1}{\sqrt{n}}\hat{\mathbf{x}}\cdot(y_0\mathbf{w}^0+y_{-1}\mathbf{w}^1+\cdots+y_{-n+1}\mathbf{w}^{n-1}). \end{aligned}$$

Because $y_{-k}\mathbf{w}^k=y_{n-k}\mathbf{w}^{k-n}$ by the periodicity of the sequences $\mathbf{y}$ and $\mathbf{w}$,

$$\frac{1}{\sqrt{n}}\sum_{k=0}^{n-1} y_{-k}\mathbf{w}^k=\frac{1}{\sqrt{n}}\sum_{k=0}^{n-1} y_k\mathbf{w}^k=\hat{\mathbf{y}},$$

and there follows

$$\mathscr{F}_n(\mathbf{x}*\mathbf{y})=\mathscr{F}_n\mathbf{x}\cdot\mathscr{F}_n\mathbf{y}.$$

There is also a reciprocal relation. Given $\mathbf{x}, \mathbf{y} \in \Pi_n$, let $\mathbf{u} := R\mathscr{F}_n\mathbf{x}$, $\mathbf{v} := R\mathscr{F}_n\mathbf{y}$. Then by the above,

$$\mathbf{x} \cdot \mathbf{y} = \mathscr{F}_n\mathbf{u} \cdot \mathscr{F}_n\mathbf{v} = \mathscr{F}_n(\mathbf{u} * \mathbf{v}) = \mathscr{F}_n(R\mathscr{F}_n\mathbf{x} * R\mathscr{F}_n\mathbf{y}).$$

Using $R\mathbf{r} * R\mathbf{s} = R(\mathbf{r} * \mathbf{s})$, this yields

$$\mathbf{x} \cdot \mathbf{y} = \mathscr{F}_n R(\mathscr{F}_n\mathbf{x} * \mathscr{F}_n\mathbf{y}).$$

Operating on this with $\mathscr{F}_n$, using $\mathscr{F}_n^2 R = I$, we get

$$\mathscr{F}_n(\mathbf{x} \cdot \mathbf{y}) = \mathscr{F}_n\mathbf{x} * \mathscr{F}_n\mathbf{y}.$$

We thus have established:

THEOREM 13.7a (Convolution theorem). *For arbitrary sequences* $\mathbf{x}, \mathbf{y} \in \Pi_n$ *there hold the identities*

$$\mathscr{F}_n(\mathbf{x} * \mathbf{y}) = \mathscr{F}_n\mathbf{x} \cdot \mathscr{F}_n\mathbf{y}, \tag{13.7-6}$$

$$\mathscr{F}_n(\mathbf{x} \cdot \mathbf{y}) = \mathscr{F}_n\mathbf{x} * \mathscr{F}_n\mathbf{y}. \tag{13.7-7}$$

By operating on (13.7-6) with $\mathscr{F}_n^{-1} = \overline{\mathscr{F}_n}$, we obtain:

COROLLARY 13.7b

$$\mathbf{x} * \mathbf{y} = \overline{\mathscr{F}_n}(\mathscr{F}_n\mathbf{x} \cdot \mathscr{F}_n\mathbf{y}). \tag{13.7-8}$$

The last identity is of enormous practical importance, because it shows that the convolution of two sequences, which if carried out naively is an expensive operation costing $n^2\mu$, can be performed by taking three discrete Fourier transforms and carrying out one Hadamard multiplication. Thus by Theorem 13.1b there follows:

COROLLARY 13.7c. *If* $n = 2^l$, *the convolution of two sequences in* Π_n *can be carried out in* $n(\frac{3}{2}\operatorname{Log}_2 n + 1)$ *multiplications.*

Surprisingly, a similar result also holds for arbitrary n. Let $\mathbf{x} = \{x_k\}$, $\mathbf{y} = \{y_k\}$ be two sequences in Π_n where n is arbitrary, and let p be the smallest power of 2 which is $\geqq 2n-1$. Defining two auxiliary sequences $\mathbf{u}, \mathbf{v}$ in Π_p by

$$\mathbf{u} := \|: x_0, x_1, \ldots, x_{n-1}, 0, \ldots, 0 :\|,$$

$$\mathbf{v} := \|: y_0, y_1, \ldots, y_{n-1}, 0, \ldots, 0, y_1, y_2, \ldots, y_{n-1} :\| \tag{13.7-9}$$

and letting $\mathbf{z} = \{z_k\} := \mathbf{u} * \mathbf{v}$, we have, for $k = 0, 1, \ldots, n-1$,

$$z_k = \frac{1}{\sqrt{p}} \sum_{j=0}^{n-1} x_j y_{k-j} = \sqrt{\frac{n}{p}}(\mathbf{x} * \mathbf{y})_k.$$

(The remaining elements of $\mathbf{z}$ are not related to $\mathbf{x} * \mathbf{y}$.) Thus we can form the convolution of two sequences of arbitrary period by suitably enlarging them and picking out certain elements of the convolution of the enlarged sequences. Since $p \leqq 4n-3$, a careful application of Corollary 13.7c yields:

COROLLARY 13.7d. *The convolution of two sequences of arbitrary period n can be carried out in less than $6n(\mathrm{Log}_2\, n+2)$ multiplications.*

Numerous applications will be shown later on. Here we mention just one consequence of some practical importance. We are always speaking about FFTs, but the existence of a really fast Fourier transform has been demonstrated only in the case where $n = 2^l$, or at any rate where n is a highly composite number. We now show that there exists a constant γ such that, for any n, the discrete Fourier transform of a sequence $\mathbf{x} \in \Pi_n$ can be calculated in less than $\gamma n \,\mathrm{Log}_2\, n$ multiplications.

Let $\mathbf{y} = \{y_k\} = \mathscr{F}_n \mathbf{x}$. Then by definition of $\mathscr{F}_n$,

$$y_m = \frac{1}{\sqrt{n}} \sum_{k=0}^{n-1} x_k w_n^{-mk}.$$

We now use the trivial fact that

$$w_n^{-mk} = w_{2n}^{-2mk} = w_{2n}^{(m-k)^2 - m^2 - k^2}.$$

We thus have

$$w_{2n}^{m^2} y_m = \frac{1}{\sqrt{n}} \sum_{k=0}^{n-1} (w_{2n}^{-k^2} x_k) w_{2n}^{(m-k)^2}, \qquad m = 0, 1, \ldots, n-1. \quad (13.7\text{-}10)$$

Again let p be the smallest power of 2 satisfying $p \geqq 2n-1$, and define sequences $\mathbf{u}, \mathbf{v}$ in Π_p by

$$u_k := \begin{cases} w_{2n}^{-k^2} x_k, & k \equiv 0, 1, \ldots, n-1 \mod p \\ 0, & \text{otherwise}, \end{cases}$$

$$v_k := \begin{cases} w_{2n}^{k^2}, & k \equiv 0, \pm 1, \pm 2, \ldots, \pm(n-1) \mod p \\ 0, & \text{otherwise}. \end{cases}$$

Then (13.7-10) shows that

$$y_m = w_{2n}^{-m^2} \left(\frac{p}{n}\right)^{1/2} (\mathbf{u} * \mathbf{v})_m, \qquad m = 0, 1, \ldots, n-1. \quad (13.7\text{-}11)$$

We thus may calculate the y_m by (a) computing $w_{2n}^{-k^2} x_k$, $k = 0, 0, 1, \ldots, n-1$, at a cost of $n\mu$; (b) forming the convolution $\mathbf{u} * \mathbf{v}$ at a cost of at most $6n(\mathrm{Log}_2\, n+2)\,\mu$; (c) computing the y_m by (13.7-11), again at a cost of $n\mu$. All in all we thus find:

THEOREM 13.7e. *For any n, the discrete Fourier transform of a sequence in Π_n can be computed in at most $n(6 \operatorname{Log}_2 n+14)$ multiplications.*

From a pragmatic point of view, the above results concerning $O(n \operatorname{Log} n)$ operations counts could well be meaningless if it would turn out that the resulting algorithms are numerically unstable. We therefore emphasize that as a result of Theorem 13.1c the FFT is extremely stable—more stable in fact than carrying out the discrete Fourier transform by means of the Horner algorithm. As a result, the convolution algorithm for carrying out the transform for arbitrary n likewise is numerically stable.

III. Time Series Analysis

Historically one of the moving forces behind the development of FFT methods was the need to economize numerical operations in time series analysis. Although they are not directly connected with complex analysis, time series are of such paramount importance in all of applied mathematics that it would be a serious omission not to discuss them at least briefly in our survey.

Let $x_0, x_1, \ldots, x_{n-1}$ be a finite sequence of real numbers. In applications, x_k may be the result of sampling a physical quantity x at time $\tau_k=\tau_0+k\Delta\tau$. The quantity x may be a brain current as recorded in an electroencephalogram, the displacement of a seismograph during an earthquake, or the elevation of an oceanic tide. In electroencephalograms $\Delta\tau$ is on the order of 1/100 of a second. Any such sequence $\{x_k\}$ obtained by sampling at equidistant time intervals is called a **time series**. The length n of a time series in electroencephalographic applications can be on the order of 2^{11} or 2^{12}.

By analyzing a time series, important conclusions on the nature of the underlying physical process can often be drawn. For instance in electroencephalography, epilepsy can be detected. The analysis of a time series usually requires the calculation of the following new sequences from a given time series $\{x_k\}$:

(a) The **covariance function**, that is, the sequence **r** with tth element

$$r_t := \sum_{k=0}^{n-t-1} x_k x_{k+t}, \qquad t=0,1,\ldots. \tag{13.7-12}$$

(b) The **power spectrum**, that is, the sequence **f** with qth element

$$f_q := \frac{1}{2n}\sum_{t=0}^{n-1} r_t\, e^{-(2\pi i/2n)tq}, \qquad q=0,1,\ldots. \tag{13.7-13}$$

(c) The **smoothed power spectrum g** defined by

$$g_q := \sum_{j=-h}^{h} c_j f_{q-j}, \qquad q=0,1,\ldots, \tag{13.7-14}$$

where h is a (small) positive integer and the c_j are constants satisfying

$$c_j \geqq 0, \qquad |j| \leqq h; \qquad \sum_{j=-h}^{h} c_j = 1. \tag{13.7-15}$$

It was recognized independently by Tukey and Bartlett in 1944/1945 that smoothing the power spectrum is necessary in order to obtain meaningful physical interpretations. It soon become clear that discrete Fourier analysis provides a common basis for the numerical analysis of all kinds of noisy data.

First of all, discrete Fourier analysis permits one to express the above operations concisely. Let the given time series be followed by n zeros, and let $\mathbf{x}$ denote the sequence in Π_{2n} obtained by repeating these $2n$ elements periodically. The other sequences $\mathbf{r}, \mathbf{f}, \mathbf{g}$ considered above will likewise be imbedded in Π_{2n}.

(a) For $t<0$, $t=-s$, we extend the definition (13.7-12) by setting

$$r_{-s} := \sum_{k=0}^{n-1+s} x_k x_{k-s}.$$

Omitting the terms that are zero, we obtain

$$r_{-s} = \sum_{k=s}^{n-1} x_k x_{k-s} = \sum_{m=0}^{n-1-s} x_{m+s} x_m = r_s,$$

and the sequence $\mathbf{r}$ becomes symmetric, $\mathbf{r} = R\mathbf{r}$. Setting $\mathbf{y} := R\mathbf{x}$, we have

$$r_t = r_{-t} = \sum_{k=s}^{n-1} x_k x_{k-t} = \sum_{k=0}^{n-1} x_k y_{t-k} = \sqrt{2n}(\mathbf{x} * \mathbf{y})_t,$$

and we see that

$$\mathbf{r} = \sqrt{2n}\mathbf{x} * R\mathbf{x}. \tag{13.7-16}$$

(b) Here and below let $\mathscr{F} := \mathscr{F}_{2n}$. The power spectrum evidently is

$$\mathbf{f} = \frac{1}{\sqrt{n}}\mathscr{F}\mathbf{r} = \mathscr{F}(\mathbf{x} * R\mathbf{x}). \tag{13.7-17}$$

(c) We have already seen in Example 3 how to express the smoothing operation as a convolution. Defining $\mathbf{c}$ to be the sequence with elements c_j in positions $\equiv j \bmod 2n$ and with zeros elsewhere, we have

$$\mathbf{g} = \sqrt{2n}\mathbf{c} * \mathbf{f}. \tag{13.7-18}$$

It was observed already before the advent of FFTs that the calculation of $\mathbf{g}$ could be simplified by performing the smoothing in the time domain. Let

$$\mathbf{s} := \mathscr{F}^{-1}\mathbf{c} = \bar{\mathscr{F}}\mathbf{c}.$$

Because $\mathbf{f} = (2n)^{-1/2}\mathscr{F}\mathbf{r}$, we have in view of the convolution theorem (Theorem 13.7a)

$$\mathbf{c} * \mathbf{f} = \mathscr{F}(\bar{\mathscr{F}}^{1}\mathbf{c} \cdot \mathscr{F}^{-1}\mathbf{f}) = \frac{1}{\sqrt{2n}}\mathscr{F}(\mathbf{s} \cdot \mathbf{r}).$$

Thus (13.7-18) may be replaced by

$$\mathbf{g} = \mathscr{F}(\mathbf{s} \cdot \mathbf{r}). \tag{13.7-19}$$

The sequence $\mathbf{s}$ is called the **time window** of the smoothing process defined by the constants c_j. If the smoothing process is simple enough, the time window may be calculated analytically. If the smoothing operation is performed in the time domain, it requires n multiplications only, compared with the $(2h+1)n$ multiplications that are required by smoothing in the frequency domain.

However, there still remain the roughly $\frac{1}{2}n^2$ multiplications required for forming $\mathbf{r}$. Here much greater savings are achieved by the FFT. Let $\mathbf{a} := \mathscr{F}^{-1}\mathbf{x}$. Then $\mathbf{y} = R\mathbf{x} = \mathscr{F}R\mathbf{a}$, and the determination of $\mathbf{r}$ requires forming

$$\mathbf{x} * \mathbf{y} = \mathscr{F}\mathbf{a} * \mathscr{F}R\mathbf{a}.$$

By the convolution theorem,

$$\mathbf{x} * \mathbf{y} = \mathscr{F}(\mathscr{F}^2\mathbf{a} \cdot \mathscr{F}^2 R\mathbf{a}).$$

In view of $\mathscr{F}^2 = R$, $R^2 = I$, we thus have

$$\mathbf{r} = \sqrt{2n}\mathbf{x} * \mathbf{y} = \sqrt{2n}\,\mathscr{F}(R\mathbf{a} \cdot \mathbf{a}). \tag{13.7-20}$$

In view of $\mathbf{f} = (2n)^{-1/2}\mathscr{F}\mathbf{r}$, it follows further that

$$\mathbf{f} = \mathbf{a} \cdot R\mathbf{a}. \tag{13.7-21}$$

Thus in order to find $\mathbf{r}$ and $\mathbf{f}$, it is best to compute $\mathbf{a} = \mathscr{F}^{-1}\mathbf{x}$ by one application of the FFT; $\mathbf{f}$ is then obtained trivially from (13.7-21), and $\mathbf{r}$ is obtained from (13.7-20) by one further application of the FFT. The smoothed power spectrum can now be obtained from (13.7-19) or, if h is small enough, directly from (13.7-18).

PROBLEMS

1. If the norm $\|\mathbf{x}\|$ of a sequence $\mathbf{x} \in \Pi_n$ is defined as in § 13.1, show that for any $\mathbf{x}, \mathbf{y} \in \Pi_n$,

   (a) $\|\mathbf{x} \cdot \mathbf{y}\|^2 \leqq \|\mathbf{x}\| \ \|\mathbf{y}\|$,

   (b) $\|\mathbf{x} * \mathbf{y}\|^2 \leqq \|\mathbf{x}\| \ \|\mathbf{y}\|$.

2. If R denotes the reversion operator, show that

   (a) $R\mathbf{x} \cdot R\mathbf{y} = R(\mathbf{x} \cdot \mathbf{y})$,

   (b) $R\mathbf{x} * R\mathbf{y} = R(\mathbf{x} * \mathbf{y})$.

3. A sequence $\mathbf{x} \in \Pi_n$ is called *even* if $R\mathbf{x} = \mathbf{x}$. It is called *odd* if $R\mathbf{x} = -\mathbf{x}$. Prove that the following statements hold for both the Hadamard product and the convolution product:
   (a) The product of two even sequences is even.
   (b) The product of an even and an odd sequence is odd.
   (c) The product of two odd sequences is even.

4. Show that the divisors of zero with respect to convolution are precisely those elements $\mathbf{x} \in \Pi_n$ for which $(\mathscr{F}_n\mathbf{x})^{-1}$ does not exist in the sense of Hadamard multiplication.

5. What are the necessary and sufficient conditions on a smoothing sequence $\mathbf{s}$ such that, for any $\mathbf{x} \in \Pi_n$, $\mathbf{x} \neq \mathbf{0}$, repeated application of the smoothing operation yields a nonzero constant sequence, that is, that

$$\lim_{k \to \infty} n^{k/2} \mathbf{s}^{*k} * \mathbf{x} = \text{const} \cdot \mathbf{w}^0?$$

   What is the value of the constant?

6. Show that if E denotes the shift operator,

   (a) $E\mathbf{x} \cdot E\mathbf{y} = E(\mathbf{x} \cdot \mathbf{y})$,

   (b) $E\mathbf{x} * \mathbf{y} = \mathbf{x} * E\mathbf{y} = E(\mathbf{x} * \mathbf{y})$.

7. *Systems of linear equations with a circulant matrix.* Let $\mathbf{A}$ be a circulant matrix of order n (see § 1.3), written in the form

$$\mathbf{A} = \begin{pmatrix} a_0 & a_{-1} & \cdots & a_{-n+1} \\ a_1 & a_0 & \cdots & a_{-n+2} \\ & & \vdots & \\ a_{n-1} & a_{n-2} & \cdots & a_0 \end{pmatrix}$$

   where $a_{n+k} = a_k$, and let

$$\mathbf{x} = \begin{pmatrix} x_0 \\ \vdots \\ x_{n-1} \end{pmatrix}, \qquad \mathbf{b} = \begin{pmatrix} b_0 \\ \vdots \\ b_{n-1} \end{pmatrix}$$

be vectors with n components. We consider the system of linear equations

$$\mathbf{A}\mathbf{x} = \mathbf{b}. \qquad (*)$$

(a) By defining sequences $\mathbf{a}, \mathbf{b}, \mathbf{x} \in \Pi_n$ in the obvious manner, show that (*) is equivalent to

$$\sqrt{n}\mathbf{a} * \mathbf{x} = \mathbf{b}.$$

(b) Letting $\hat{\mathbf{a}} := \mathscr{F}_n\mathbf{a}$, $\hat{\mathbf{b}} := \mathscr{F}_n\mathbf{b}$, $\hat{\mathbf{x}} := \mathscr{F}_n\mathbf{x}$ and applying the convolution theorem, show that the system is also equivalent to

$$\sqrt{n}\hat{\mathbf{a}} \cdot \hat{\mathbf{x}} = \hat{\mathbf{b}}.$$

(c) Thus formulate a necessary and sufficient condition for (*) to have a unique solution for every choice of $\mathbf{b}$.

(d) Show that if (*) is solvable, the solution is given by the formula (in a sense to be made precise)

$$\mathbf{x} = \frac{1}{\sqrt{n}}\mathscr{F}\left(\frac{\hat{\mathbf{b}}}{\hat{\mathbf{a}}}\right).$$

(e) Apply the above to the solution of Problem 5, § 13.2.

8. *Round-off analysis for convolution.* Let $\mathbf{x}, \mathbf{y} \in \Pi_n$, $|x_i| \leqq \gamma$, $|y_i| \leqq \gamma$. Show that in the linear model of error propagation the coefficient of error propagation for the elements z_k of $\mathbf{z} := \mathbf{x} * \mathbf{y}$ is $\leqq [\sqrt{n}(n+11)/2]\gamma^2$ if the convolution is computed directly, and $\leqq 24n^{3/2}(2\,\mathrm{Log}_2\, n + 7)\gamma^2$ if the FFT is used. Conclude that although the two bounds have almost the same order of magnitude, direct computation of the convolution is considerably more stable than using the FFT.
9. Show that in the linear model of error propagation the coefficient of error propagation of the algorithm suggested in Theorem 13.7e (called *Bluestein algorithm*), assuming all $|x_k| \leqq \gamma$, does not exceed

$$\sqrt{n}[4 + 96n(\mathrm{Log}_2\, n + 4)]\gamma.$$

NOTES

The name Hadamard product was coined in Henrici [1979a] in analogy to a similar product in the theory of power series. A name seemed to be required for didactic purposes. On the convolution of sequences see Cooley, Lewis, and Welch [1967c] and Aho, Hopcroft, and Ullmann [1974]. For the algorithm implied in Theorem 13.7e, see Bluestein [1970], although he seems to have considered only the case where n is a perfect square. For the general case the author is indebted to a personal communication by A. Schönhage. Winograd [1976, 1978] proposes algorithms to perform the discrete Fourier transform in $\gamma n\,\mathrm{Log}_2\, n$ operations for a smaller value of γ, but these suffer from the disadvantage that a new program is required for each value of n. Daeppen [1982] compares the effectiveness of Bluestein's algorithm with the Horner scheme and finds that the crossover occurs approximately at $n = 63$.

The literature on time series enormous. For comprehensive introductions see Blackman and Tukey [1959] (pre-FFT), Tukey [1967], Cooley, Lewis, and Welch [1970b], and Bloomfield [1976]. For a random sample of applications in geophysics, all using the FFT, see Plutchok and Broome [1969], Fuchs and Müller [1971], Ku, Telford, and Lim [1971], Joyner and Chen [1975], Joyner, Warrick, and Oliver III [1976], Chapman [1978], and Boatwright [1978].

In much of the literature (including some of the author's own papers) the main formulas of this section have a superficially different appearance due to our choice of the numerical factors in the definitions of convolution and of the discrete Fourier operator $\mathscr{F}_n$.

§ 13.8. ONE-SIDED CONVOLUTION

Here we discuss a variant of cyclic convolution which as a special case contains the multiplication of two polynomials.

I. Definition. Reduction to Cyclic Convolution

Let $\{a_0, a_1, \ldots, a_{m-1}\}$, $\{b_0, b_1, \ldots, b_{m-1}\}$ be two finite sequences of length m. The **one-sided convolution** of these sequences is the sequence $\{c_0, c_1, \ldots, c_{2m-2}\}$ of length $2m-1$ defined by

$$c_0 := a_0 b_0,$$

$$c_1 := a_0 b_1 + a_1 b_0,$$

$$c_2 := a_0 b_2 + a_1 b_1 + a_2 b_0,$$

and generally

$$c_k := a_0 b_k + a_1 b_{k-1} + \cdots + a_k b_0, \tag{13.8-1}$$

where $a_k := 0$ and $b_k := 0$ if $k \geqq m$. An obvious application occurs in the multiplication of two polynomials. Let

$$p(x) := \sum_{k=0}^{m-1} a_k x^k, \qquad q(x) := \sum_{k=0}^{m-1} b_k x^k$$

be two polynomials of degree $< m$. Then the coefficients c_k of the product polynomial

$$r(x) := p(x) q(x) = \sum_{k=0}^{2m-2} c_k x^k$$

evidently arise by one-sided convolution of the coefficient sequences $\{a_k\}$ and $\{b_k\}$.

It is easy to express the one-sided convolution as an ordinary convolution. If the sequences $\{a_k\}$, $\{b_k\}$, $\{c_k\}$ are embedded in a space Π_n where $n \geqq 2m$ by filling up undefined positions with zeros, then clearly

$$\mathbf{c} = \sqrt{n}\, \mathbf{a} * \mathbf{b}, \tag{13.8-2}$$

where the star denotes ordinary convolution. If n is a power of 2, then by Corollary 13.7c less than $\frac{3}{2} n \operatorname{Log}_2 2n\mu$ are required to carry out the convolution. We thus have:

THEOREM 13.8a. *Let m be a power of 2. The one-sided convolution of two sequences of length* $\leqq m$, *and hence the multiplication of two polynomials of degree* $< m$, *can be carried out in less than*

$$\phi(m) := 3m \operatorname{Log}_2 4m \tag{13.8-3}$$

complex multiplications.

The function ϕ defined in (13.8-3) will be used in § 13.9.

II. Multiplication of Large Integers

The foregoing result has an immediate application in the multiplication of large integers. Let $b > 0$ be a (small) integer, and let p and q be (large) integers, represented in the number system with base b as

$$p = \sum_{j=0}^{m-1} p_j b^j, \qquad q = \sum_{j=0}^{m-1} q_j b^j,$$

where the p_j and q_j are integers, $0 \leqq p_j, q_j < b$. Then clearly

$$r := pq = \sum_{j=0}^{2m-2} r_j b^j, \tag{13.8-4}$$

where the sequence $\{r_j\}$ is the one-sided convolution of the sequences $\{p_j\}$ and $\{q_j\}$. Although the definition implies that the r_j are integers, they are not necessarily the correct digits in the representation of r in the base b, because they need not satisfy the inequality $0 \leqq r_j < b$. However, they do satisfy

$$0 \leqq r_j < mb^2$$

and thus possess representations in base b,

$$r_j = \sum_{l=0}^{h} r_{j,l} b^l, \tag{13.8-5}$$

where $h := [\operatorname{Log}_b m] + 2$ is, in general, much smaller than m. The correct representation of r in base b is then easily obtained from (13.8-5).

It is evident that by calculating the convoluted sequence $\mathbf{r}$ by the FFT, the product pq can be computed much faster than by the conventional method, which requires m^2 multiplications of integers $< b$. However, it would not be correct to conclude from Theorem 13.8a that pq can be computed in $\phi(m)$ multiplications, because the multiplications required to form the convolution by FFT are not multiplications of integers, but multiplications of complex numbers whose real and imaginary parts are arbitrary real numbers. In that sense, forming the product pq requires only 1 multiplication.

For a correct appraisal of the FFT method we note that the r_j are known to be integers. Thus there is no need to compute them with high precision. Rather, it is sufficient to compute them with errors $\leqq \frac{1}{4}$, say. We now may use Theorem 13.1c to establish the relative precision η with which the individual operations in the FFT are to be performed. We have

$$\mathbf{r} = \sqrt{n}\,\overline{\mathcal{F}_n}(\mathcal{F}_n\mathbf{p} \cdot \mathcal{F}_n\mathbf{q}), \tag{13.8-6}$$

where the elements of the sequences $\mathbf{p}$ and $\mathbf{q}$ satisfy $|p_i| < b$, $|q_i| < b$. Working with a precision constant η, the elements of $\mathcal{F}_n\mathbf{p}$ and $\mathcal{F}_n\mathbf{q}$ by (13.1-26) will be in error by at most

$$\sqrt{n}(2\,\mathrm{Log}_2\, n+3)b\eta \mod \eta^2,$$

and those of $\mathcal{F}_n\mathbf{p} \cdot \mathcal{F}_n\mathbf{q}$ will have an error not exceeding

$$2n(2\,\mathrm{Log}_2\, n+3)b^2\eta \mod \eta^2.$$

In carrying out $\overline{\mathcal{F}_n}$, there thus is an inherited error of at most

$$2n^{3/2}(2\,\mathrm{Log}_2\, n+3)b^2\eta \mod \eta^2.$$

In addition, there will be rounding error in carrying out $\overline{\mathcal{F}_n}$. Because the initial elements now can be as large as nb^2, Theorem 13.1c for the accumulated error yields the bound

$$n^{3/2}(2\,\mathrm{Log}_2\, n+3)b^2\eta \mod \eta^2.$$

The total error after carrying out $\overline{\mathcal{F}_n}$ thus is bounded by

$$3n^{3/2}(2\,\mathrm{Log}_2\, n+3)b^2\eta \mod \eta^2,$$

and in view of the factor $\sqrt{n}$ in (13.8-6) the error in each component of $\mathbf{r}$ thus is bounded by

$$3n^2(2\,\mathrm{Log}_2\, n+3)b^2\eta \mod \eta^2.$$

Here n may be as large as $4m$. Requiring the above bound to be $\leqq \frac{1}{4}$ thus yields:

THEOREM 13.8b. *Neglecting terms of order η^2, a precision constant η satisfying*

$$\eta \leqq \frac{1}{192m^2(2\,\mathrm{Log}_2\, m+7)b^2} \tag{13.8-7}$$

is sufficient to obtain, by means of the discrete Fourier transform and the convolution theorem, the representation in base b of the product of two integers with a given m-digit representation in base b.

The machine on which the discrete Fourier transform is carried out enters the picture only through its precision constant. For $m=1000, b=10$ Theorem 13.8b yields the manageable requirement $\eta \leqq 1.93\times 10^{-12}$. Although certain second-order effects have been neglected, experience shows that the errors in the r_i are always much smaller than predicted by our worst case analysis.

III. Graeffe's Method

Theorem 13.8a concerning the rapid multiplication of polynomials provides us with the opportunity to take a modern look at a classical method, ascribed to Graeffe, for the simultaneous determination of all zeros of a polynomial provided the moduli of the zeros are *distinct.*

We assume the polynomial given in the form

$$p(z)=1+a_1z+a_2z^2+\cdots+a_dz^d, \tag{13.8-8}$$

where the degree $d\geqq 2$, $a_d\neq 0$, and let its zeros z_i be numbered such that

$$|z_1|\leqq|z_2|\leqq\cdots\leqq|z_d|.$$

We define the **separation ratio** of p by

$$\gamma_p:=\max_{1\leqq i<d}\frac{|z_i|}{|z_{i+1}|},$$

and we assume that $\gamma_p<1$.

Graeffe's method is based on two observations. The first concerns the fact that if the separation ratio of a polynomial is very small, the values of the zeros can be approximately read off from the coefficients. More precisely, setting $a_0:=1$ we have:

LEMMA 13.8c. *If the separation ratio* γ *of the polynomial* (13.8-8) *satisfies*

$$\gamma\left\{\binom{d}{k-1}-1\right\}<1, \tag{13.8-9}$$

then for $k=1,2,\ldots,d,$

$$z_k=-\frac{a_{k-1}}{a_k}\{1+\theta\lambda_k\gamma\}, \tag{13.8-10}$$

where

$$|\theta|\leqq 1 \quad \textit{and} \quad \lambda_k:=2\left[\binom{d+1}{k}-2\right].$$

Proof. By the product representation of p,

$$p(z)=\prod_{i=1}^{d}\left(1-\frac{z}{z_i}\right), \tag{13.8-11}$$

there follows for $k=1,2,\ldots,d$,

$$a_k=(-1)^k\sum_{\sigma_k}\prod_{i\in\sigma_k} z_i^{-1},$$

where σ_k runs through all subsets of k distinct elements of the set $1,2,\ldots,d$. There are $\binom{d}{k}$ such subsets. We factor out the product pertaining to the subset $\{1,2,\ldots,k\}$. Each of the remaining $\binom{d}{k}-1$ products has at least one term that is smaller by a factor $\leqq\gamma$ than one of $z_1,\ldots,z_k$. Thus we have

$$a_k=(-1)^k(z_1\cdots z_k)^{-1}\left\{1+\left[\binom{d}{k}-1\right]\theta\gamma\right\},$$

where $|\theta|\leqq 1$. If (13.8-9) holds, there follows

$$z_k=-\frac{a_{k-1}}{a_k}\left\{1+\left[\binom{d}{k}-1\right]\theta\gamma\right\}\Big/\left\{1+\left[\binom{d}{k-1}-1\right]\theta'\gamma\right\},$$

$|\theta'|\leqq 1$. The last ratio varies between the numbers

$$\left\{1-\left[\binom{d}{k}-1\right]\gamma\right\}\Big/\left\{1+\left[\binom{d}{k-1}-1\right]\gamma\right\}$$

and

$$\left\{1+\left[\binom{d}{k}-1\right]\gamma\right\}\Big/\left\{1-\left[\binom{d}{k-1}-1\right]\gamma\right\},$$

which, using (13.8-9), can be expressed in the form

$$1+2\left[\binom{d+1}{k}-2\right]\theta\gamma,$$

where $|\theta|\leqq 1$. —

The second observation underlying Graeffe's method is this. Given a polynomial p with separation ratio $\gamma<1$, it is easy to construct, for any integer h, a polynomial G_hp with separation ratio γ^h the zeros of which are simply related to the zeros of p. This construction is implied in:

LEMMA 13.8d. *If the polynomial* (13.8-8) *has the zeros* $z_1,\ldots,z_d$, *then for any integer* $h>1$, *if* $w_h:=\exp(2\pi i/h)$, *the function*

$$G_hp(z):=p(z^{1/h})p(w_h^{-1}z^{1/h})\cdots p(w_h^{-h+1}z^{1/h}) \tag{13.8-12}$$

is a polynomial of the same form with the zeros $z_1^h, z_2^h,\ldots,z_d^h$.

Proof. We consider $G_h p(z^h)$. Using the representation (13.8-11), we have

$$G_h p(z^h) = \prod_{j=0}^{h-1} \prod_{i=1}^{d} \left(1 - \frac{z}{w_h^j z_i}\right) = \prod_{i=1}^{d} \prod_{j=0}^{h-1} \left(1 - \frac{z}{w_h^j z_i}\right).$$

Here the inner product is a polynomial of degree h with constant term 1 and with the zeros $z_i, w_h z_i, \ldots, w_h^{h-1} z_i$, and therefore equals $1 - z_i^{-h} z^h$. There follows

$$G_h p(z) = \prod_{i=1}^{d} \left(1 - \frac{z}{z_i^h}\right),$$

tantamount to the assertion. —

It is now clear how **Graeffe's method** operates. Starting with a polynomial p in the form (13.8-8), whose separation ratio γ is <1, we select an integer $h>1$ (in classical versions of the method, $h=2$ is used) and construct, for $j=1,2,\ldots,$ the polynomials

$$(G_h)^j p(z) = 1 + a_1^{(j)} z + \cdots + a_d^{(j)} z^d, \qquad (13.8\text{-}13)$$

which have the zeros $z_k^{h^j}$, $k=1,2,\ldots, d$. By Lemma 13.8c, if

$$2\gamma^{h^j}\left[\binom{d}{k-1} - 1\right] < 1,$$

there holds

$$z_k^{h^j} = -\frac{a_{k-1}^{(j)}}{a_k^{(j)}}(1 + \theta\lambda_k \gamma^{h^j}), \qquad (13.8\text{-}14)$$

and thus we can at least determine the moduli of the zeros very rapidly by the relation

$$|z_k| = \lim_{j\to\infty} \left|\frac{a_{k-1}^{(j)}}{a_k^{(j)}}\right|^{h^{-j}}, \qquad k = 1, 2, \ldots, d. \qquad (13.8\text{-}15)$$

If the arguments of the zeros are known a priori (for instance, if the zeros are known to be positive), then of course the zeros themselves are determined by the above relation.

Graeffe's method is often said to suffer from the handicap that the coefficients $a_k^{(j)}$ will become very large even for moderate j. This can be avoided by scaling the polynomial so that all zeros satisfy $|z_i| \geqq 1$. The coefficients $a_k^{(j)}$ then will become *small* as j increases, and in numerical computation some $a_k^{(j)}$ where k is near d will eventually be replaced by 0. This does not disturb the validity of the above result for those k where $a_k^{(j)}$ is not yet zero. Thus some of the smaller zeros will still be found, and groups of larger zeros will be found in a similar manner on deflating the polynomial.

There still remains the problem of efficiently evaluating

$$G_h p(z) = 1 + a_1' z + a_2' z^2 + \cdots + a_d' z^d.$$

By definition of G_h, the one-sided convolution of the h coefficient sequences

$$\{1, w_h^{-j} a_1, w_h^{-2j} a_2, \ldots, w_h^{-dj} a_d\}, \qquad j = 0, 1, \ldots, h-1, \tag{13.8-16}$$

yields a coefficient sequence $\{b_i\}$, where $b_{kh} = a_k'$, $k = 0, 1, \ldots, d$, and $b_i = 0$ otherwise. These one-sided convolutions can be written as two-sided convolutions by embedding the sequences (13.8-16) and $\{b_i\}$ into a space Π_n, as follows. Let $n \geqq h(d+1)$, and define sequences of period n,

$$\mathbf{a}^{(j)} := \|: 1, w_h^{-j} a_1, w_h^{-2j} a_2, \ldots, 0, 0, \ldots, 0 :\|,$$

$$\mathbf{b} := \|: b_0, b_1, \ldots, 0, 0, \ldots, 0 :\|,$$

where undefined elements are zero. Then clearly by the definition of convolution

$$\mathbf{b} = n^{h/2} \mathbf{a}^{(0)} * \mathbf{a}^{(1)} * \cdots * \mathbf{a}^{(h-1)}. \tag{13.8-17}$$

By the convolution theorem, using the symbol $\hat{}$ to denote Fourier transforms,

$$\hat{\mathbf{b}} = n^{h/2} \hat{\mathbf{a}}^{(0)} \cdot \hat{\mathbf{a}}^{(1)} \cdot \cdots \cdot \hat{\mathbf{a}}^{(h-1)},$$

and by transforming back,

$$\mathbf{b} = n^{h/2} \overline{\mathcal{F}_n} (\hat{\mathbf{a}}^{(0)} \cdot \hat{\mathbf{a}}^{(1)} \cdot \cdots \cdot \hat{\mathbf{a}}^{(h-1)}).$$

Considerable simplifications are yet possible by the two following observations. Let $n = hm$, where $m \geqq d+1$. Then

$$w_h = w_n^m,$$

and it is clear that

$$\mathbf{a}^{(j)} = \mathbf{w}^{-jm} \cdot \mathbf{a},$$

where $\mathbf{w} := \{w_n^k\}$ as usual. By axiom (F_2) of § 13.1 we have

$$\hat{\mathbf{a}}^{(j)} = \mathcal{F}\mathbf{a}^{(j)} = \mathcal{F}(\mathbf{w}^{-jm} \cdot \mathbf{a}) = E^{-jm} \mathcal{F}\mathbf{a}.$$

Thus if $\hat{\mathbf{a}}^{(j)} = \{\hat{a}_k^{(j)}\}$, $\hat{\mathbf{b}} = \{\hat{b}_k\}$, there follows

$$\hat{b}_k = n^{h/2} \prod_{j=0}^{h-1} \hat{a}_{k+jm}.$$

Thus only $\hat{\mathbf{a}}$ needs to be computed. The sequence $\hat{\mathbf{b}}$ has period m. The elements b_{kh} that interest us are given by

$$b_{kh} = \frac{1}{\sqrt{n}} \sum_{j=0}^{n-1} \hat{b}_j w_n^{khj} = \left(\frac{h}{m}\right)^{1/2} \sum_{j=0}^{m-1} \hat{b}_j w_m^{kj}.$$

That is, the contracted sequence $\{b_{kh}\}$ (zeros omitted), which equals the sequence $\mathbf{a}' \in \Pi_m$ which we seek, is given by

$$\mathbf{a}' = \sqrt{h}\,\mathcal{F}_m \hat{\mathbf{b}}.$$

Thus only a short Fourier transform (of length m in place of $n = hm$) is required to obtain the coefficients of $G_h p$. We summarize:

THEOREM 13.8e. *One Graeffe step G_h for a polynomial of degree $< m$ can be carried out at the expense of one $\mathcal{F}_{mh}$, one $\mathcal{F}_m$, and $(m-1)h$ multiplications.*

In practice, powers of 2 will be selected for h as well as for m. Then the operations count of Theorem 13.1b is applicable and yields a total of

$$\tfrac{1}{2}m[(h+1)\log m + h(\log h + 2)]$$

multiplications, where m is the smallest power of 2 such that $m > d$.

PROBLEM

1. Let p be a polynomial of degree $d < m$ with separation ratio $\gamma < 1$. We wish to achieve, by means of repeated Graeffe transformations G_h, a polynomial with separation ratio γ^N, where N is large. Choosing h large will achieve this in fewer iterations; on the other hand every individual Graeffe transformation will be more expensive. What is the optimal choice of h? Check your conclusion by experiments.

NOTES

For a pre-FFT treatment of the multiplication of large numbers see Karatsuba and Ofman [1962]. Schönhage and Strassen [1971] give a rigorous treatment in the framework of complexity theory. Graeffe's method is ordinarily presented only for $h = 2$, and the one-sided convolutions of the coefficient sequences are done without FFT. See Runge and König [1924] and Hildebrand [1956], who also consider the case where several zeros have the same modulus.

§ 13.9. FAST ALGORITHMS FOR POWER SERIES

The integral domain $\mathcal{P}$ of formal power series (fps) was introduced in Chapter 1. The reader will recall the various algebraic operations that can be carried out with fps without regard to convergence. Here we review some of these operations from the point of view of economizing arithmetic operations, paying particular attention to the possibility of applying the FFT.

The coefficient field is always the field of complex numbers. No mentioning will be made of the obvious simplifications that arise if the coefficients are real. If $P = a_0 + a_1 x + a_2 x^2 + \cdots$ is a fps, we denote for $n = 1, 2, \ldots$ by

$$P_n(x) = a_0 + a_1 x + \cdots + a_{n-1} x^{n-1}$$

its partial sum of degree $n-1$, consisting of the first n terms. (This notation is at variance with the usage of Chapter 11.) We call P_n a **polynomial of length** n.

If P is any fps such that $P_n = O$, we write

$$P = O(x^n).$$

I. Multiplication

While nothing needs to be said about addition and subtraction, the FFT immediately furnishes a result on the multiplication of two fps. For any fps P and Q,

$$(PQ)_n = (P_n Q_n)_n.$$

As a corollary of Theorem 13.8a we thus have:

THEOREM 13.9a. *If $n = 2^l$ and P, Q are fps, the computation of $(PQ)_n$ by the FFT requires at most*

$$\phi(n) := 3n \operatorname{Log}_2 4n \tag{13.9-1}$$

complex multiplications.

If performed in the conventional manner, the computation of $(PQ)_n$ would require $\frac{1}{2}n^2$ complex multiplications. It may be verified that the smallest power of 2 for which $\phi(n) < \frac{1}{2}n^2$ is $n = 64$.

II. Newton's Method for Formal Power Series

Many fast algorithms for power series can be based on an extension of Newton's method to nonlinear equations in formal power series. We shall see that Newton's method in this context enjoys the nice property that it always behaves as it should.

Let $P = a_0 + a_1x + a_2x^2 \cdots \in \mathscr{P}$, and let $Q = b_1x + b_2x^2 + \cdots$ be a nonunit in $\mathscr{P}$. We recall that the composition of P with Q is defined by substituting Q for x in P,

$$P \circ Q = a_0 + a_1Q + a_2Q^2 + \ldots,$$

and collecting coefficients of equal powers. Because Q has constant coefficient zero, only finitely many terms can arise for each power, and the process of composition thus is algebraically well defined. We further recall that the *almost units* in $\mathscr{P}$ (that is, the series $P = a_1x + \ldots$ where $a_1 \neq 0$) form a group under composition with unit element $X = 1 \cdot x + 0 \cdot x^2 + \cdots$. The inverse $P^{[-1]}$ of a given almost unit P is called the *reversion* of P.

Here we consider the equation

$$Q \circ W - R = O, \tag{13.9-2}$$

where Q and R are given almost units, and where W is sought. The solution clearly is given by the formula

$$W = Q^{[-1]} \circ R,$$

but there remains the problem of constructing W.

Suppose W_k is an approximate solution to (13.9-2) in the sense that

$$W_k = W + x^k W^* \tag{13.9-3}$$

for a suitable $W^* \in \mathcal{P}$, where $k > 0$. This implies that W_k is a nonunit. Applying, in a purely formal sense, Newton's method to (13.9-2) we would expect to improve the approximation by forming

$$W^+ := W_k - \frac{Q \circ W_k - R}{Q' \circ W_k}.$$

Here the quotient is well defined, for because Q is an almost unit, Q' is a unit, and so is $Q' \circ W_k$. That W^+ is a better approximation to W than W_k is easily confirmed as follows. By the formal analog of Taylor's formula,

$$\begin{aligned} Q \circ W_k &= Q \circ (W + x^k W^*) \\ &= Q \circ W + (Q' \circ W) x^k W^* + O(x^{2k}) \\ &= R + (Q' \circ W) x^k W^* + O(x^{2k}), \end{aligned}$$

and again by Taylor's formula,

$$Q' \circ W_k = Q' \circ W + O(x^k).$$

Because $Q' \circ W$ is a unit, we get

$$\begin{aligned} W^+ &= W + x^k W^* - \frac{(Q' \circ W) x^k W^* + O(x^{2k})}{Q' \circ W + O(x^k)} \\ &= W + x^k W^* - x^k W^* (1 + O(x^k)) \\ &= W + O(x^{2k}). \end{aligned}$$

Thus in passing from W_k to W^+, the number of correct coefficients has been precisely doubled. We thus may construct a sequence of approximations $W^{(m)}$ to W by the following algorithm. Let

$$W^{(0)} := O, \tag{13.9-4a}$$

and for $m=0, 1, 2, \ldots$ form

$$W^{(m+1)} := W^{(m)} - \left(\frac{Q \circ W^{(m)} - R}{Q' \circ W^{(m)}}\right)_{2n}, \tag{13.9-4b}$$

where $n := 2^m$. Because $W^{(0)}$ satisfies (13.9-3) for $k=1$, there follows:

THEOREM 13.9b. *For $m=0, 1, 2, \ldots$,*

$$W^{(m)} = (Q^{[-1]} \circ R)_{2^m}.$$

Let $\omega(n)$ denote the number of multiplications required to compute W_n, and $\nu(n)$ the number of multiplications required to compute the Newton correction

$$\left(\frac{Q \circ W^{(m)} - R}{Q' \circ W^{(m)}}\right)_{2n}.$$

Then clearly if $n=2^m$, $\omega(2n) = \omega(n) + \nu(n)$, hence we have:

COROLLARY 13.9c. For $l=1, 2, \ldots$,

$$\omega(2^l) = \nu(2^{l-1}) + \nu(2^{l-2}) + \cdots + \nu(2) + \nu(1). \tag{13.9-5}$$

III. Division

Here we discuss the determination of

$$P^{-1} = c_0 + c_1 x + c_2 x^2 + \cdots,$$

where

$$P = a_0 + a_1 x + a_2 x^2 + \cdots$$

is a *unit* in $\mathcal{P}$, $a_0 \neq 0$. The conventional algorithm for determining the c_i was discussed in § 1.2. It is based on comparing coefficients in the identity $P^{-1}P = I$, that is,

$$(c_0 + c_1 x + c_2 x^2 + \cdots)(a_0 + a_1 x + a_2 x^2 + \cdots) = 1,$$

which yields the recurrence relation

$$c_0 = \frac{1}{a_0}, \qquad c_n = -\frac{1}{a_0}(a_1 c_{n-1} + a_2 c_{n-2} + \cdots + a_n c_0).$$

To compute the first n coefficients c_i by this method, $n\delta$ and $1+2+\cdots+(n-1) = \frac{1}{2}(n-1)n\mu$ are required.

To compute P^{-1} by Newton's method, we recall (in analogy to elementary numerical analysis, where c^{-1} may be calculated as the solution of $1/y - c =$

0) that $Y := P^{-1}$ is the solution of

$$Y^{-1} - P = O. \tag{13.9-6}$$

This is of the form (13.9-2) if we set

$$P = a_0 + R, \qquad Y = \frac{1}{a_0} + W,$$

and

$$Q = \frac{1}{1/a_0 + W} - a_0 = -\frac{a_0^2 W}{1 + a_0 W}.$$

We now apply Newton's algorithm to the resulting equation

$$\frac{a_0^2 W}{1 + a_0 W} + R = O.$$

Especially simple formulas are obtained by writing $Y^{(m)} := 1/a_0 + W^{(m)}$. The iteration (13.9-4) then reads

$$Y^{(0)} := \frac{1}{a_0}, \tag{13.9-7a}$$

$$Y^{(m+1)} := [Y^{(m)}(2 - PY^{(m)})]_{2n}, \qquad m = 0, 1, 2, \ldots . \tag{13.9-7b}$$

In view of Theorem 13.9b we have:

THEOREM 13.9d. *For* $m = 0, 1, 2, \ldots,$

$$Y^{(m)} = (P^{-1})_{2^m}. \tag{13.9-8}$$

Thus $Y^{(m)}$ is a polynomial of length 2^m whose coefficients agree with the first 2^m coefficients of P^{-1}. Each iteration step doubles the number of correct coefficients.

To appraise the cost of division by Corollary 13.9c, we count the number of multiplications required to carry out one step of the recurrence (13.9-7b). Because the final result is truncated to length $2n$, all intermediate results may be truncated likewise. Thus in actual calculation (13.9-7b) is replaced by

$$Y^{(m+1)} := \{Y^{(m)}[2 - (P_{2n} Y^{(m)})_{2n}]\}_{2n}$$

$(n := 2^m)$. This can be evaluated by forming two products of polynomials of length $2n$, which requires $2\phi(2n)\mu$. (This crude operations count could be somewhat refined by taking advantage of coefficients that are a priori known to be 0.) In the notation of Corollary 13.9c we thus have $\nu(n) = 2\phi(2n) = 12n \operatorname{Log}_2 8n$. Using the formula

$$\sum_{k=1}^{l-1} 2^k k = 2^l(l-2)+2, \tag{13.9-9}$$

the sum (13.9-5) is easily evaluated and yields

$$\omega(n) \leqq 12n \operatorname{Log}_2 2n \leqq 4\phi(n).$$

If Q is any fps and $(Q/P)_n$ is required, this may be computed as $(Q_n P_n^{-1})_n$, which requires another $\phi(n)$ multiplications. Thus in toto we have:

THEOREM 13.9e. *To compute $(Q/P)_n$, where P is a unit and Q is arbitrary, requires for $n = 2^l$ no more than $5\phi(n)$ multiplications.*

IV. Composition: Some Special Cases

Here we treat some special cases of the composition problem that can be dealt with by taking advantage of functional relationships.

Consider, for example, the logarithmic series

$$L := \operatorname{Log}(1+x) = x - \tfrac{1}{2}x^2 + \tfrac{1}{3}x^3 - \cdots.$$

If Q is any nonunit, then $Y := L \circ Q$ satisfies

$$Y' = \frac{Q'}{1+Q}$$

and therefore, if n is any integer > 1,

$$(Y')_{n-1} = \left(\frac{(Q')_{n-1}}{1+Q_{n-1}}\right)_{n-1}. \tag{13.9-10}$$

To compute $(Q')_{n-1}$ for a given Q requires $n-2$ multiplications. By Theorem 13.9e the computation of the quotient (13.9-10) requires no more than $5\phi(n)$ multiplications if $n = 2^l$. Y_n can be recovered unambiguously from $(Y')_{n-1}$ because the constant term of Y_n is known to be zero. This requires another $n-1$ multiplications. In view of $n < \phi(n)$ we thus find:

THEOREM 13.9f. *If $n = 2^l$, to compute $(L \circ Q)_n$ for any nonunit Q requires no more than $6\phi(n)$ multiplications.*

We next turn to the problem of computing the exponential of a given nonunit Q,

$$E \circ Q = 1 + \frac{1}{1!}Q + \frac{1}{2!}Q^2 + \cdots.$$

If

$$W := E \circ Q - 1,$$

then

$$L \circ W - Q = O, \tag{13.9-11}$$

which is an equation of the form (13.9-2) that is solvable by Newton's method. The algorithm (13.9-4) in this case yields

$$\begin{aligned} W^{(0)} &:= O \\ W^{(m+1)} &:= W^{(m)} - \{(1 + W^{(m)})(L \circ W^{(m)} - Q)\}_{2n}, \end{aligned} \tag{13.9-12}$$

$n := 2^m$, $m = 0, 1, 2, \ldots$. One step of the algorithm requires computing a logarithm to precision $O(x^{2n})$, and a multiplication to the same precision. By the Theorems 13.9f and 13.9a we thus have

$$\nu(n) \leqq 7\phi(n).$$

Evaluating the sum (13.9-5) we get:

THEOREM 13.9g. *If $n = 2^l$, if Q is any nonunit and if E is the exponential series, the evaluation of $(E \circ Q)_n$ by the algorithm (13.9-12) requires no more than $14\phi(n)$ multiplications.*

As an immediate application we consider the computation of $(1+Q)^\alpha$ for a given nonunit Q, where α is an arbitrary complex number. In view of

$$(1+Q)^\alpha = E \circ (\alpha L \circ Q),$$

this is reduced to forming $L \circ Q$, scalar multiplication by α, and exponentiation. By the foregoing results we have:

THEOREM 13.9h. *To compute $[(1+Q)^\alpha]_n$, where $n = 2^l$, requires no more than $21\phi(n)$ multiplications.*

This result is asymptotically much better than the $\frac{1}{2}n^2\mu$ that are required by Euler's already ingenious algorithm described in Theorem 1.6c. Even if α is a large integer, the result is more favourable than what would be obtained, say, by successive squaring and using the binary representation of α. However, for special values of α such as $\alpha = \pm\frac{1}{2}$ direct application of Newton's method produces a yet more favorable $O(n \operatorname{Log} n)$ result.

PROBLEMS

1. *A fallacious algorithm.* Why does the following simple-minded approach to the problem of constructing the reciprocal $Q = P^{-1}$ of a given unit $P \in \mathcal{P}$ not

work? We seek a series Q such that $PQ=I$ or, on truncating, $(PQ)_n=I_n$. Introducing the coefficient sequences $\mathbf{p}$ and $\mathbf{q}$ in Π_{2n} in the usual manner, we want $\mathbf{p}*\mathbf{q}=(2n)^{-1/2}\mathbf{e}$. On Fourier transforming, this yields $\hat{\mathbf{p}}\cdot\hat{\mathbf{q}}=(2n)^{-1/2}\hat{\mathbf{e}}$, thus

$$\hat{\mathbf{q}}=(2n)^{-1/2}\frac{\hat{\mathbf{e}}}{\hat{\mathbf{p}}}$$

(we know that a solution exists) and therefore

$$\mathbf{q}=(2n)^{-1/2}\mathscr{F}_{2n}^{-1}\frac{\hat{\mathbf{e}}}{\hat{\mathbf{p}}}.$$

2. Carry out the algorithm (13.9-7) for the series $P=1-x$ and show that

$$Y^{(m)}=(1+x)(1+x^2)(1+x^4)\cdots(1+x^{2^{m-1}}),\qquad m=0,1,2,\ldots.$$

3. Compute $(P^{-1})_1$, $(P^{-1})_2$, $(P^{-1})_4$ by the algorithm implied in Theorem 13.9e, where

$$P:=-\frac{\operatorname{Log}(1-x)}{x}=1+\tfrac{1}{2}x+\tfrac{1}{3}x^2+\cdots.$$

(The coefficients of P^{-1} are required for certain processes of numerical integration.)

4. Let P be a given almost unit, $Q:=P^{[-1]}=b_1x+b_2x^2+\cdots$. Show that any single coefficient b_k can be computed in $O(k\operatorname{Log}k)$ multiplications (Lagrange-Bürmann formula).

5. In the Geiger algorithm (§ 13.4) it was necessary to determine the coefficients of the polynomial

$$p(z)=z^n+a_1z^{n-1}+\cdots+a_n=\prod_{i=1}^{n}(z-z_i)$$

from the power sums of its zeros,

$$s_k:=\sum_{i=1}^{n}z_i^k,\qquad k=1,2,\ldots,n.$$

Show that this can be done in $O(n\operatorname{Log}n)$ operations. (Let $q(z):=z^np(z^{-1})=1+a_1z+\cdots+a_nz^n$, use the fact that

$$-\operatorname{Log}q(z)=\sum_{k=1}^{\infty}\frac{s_k}{k}z^k$$

and hence

$$q(z)=\exp\left\{-\sum_{k=1}^{\infty}\frac{s_k}{k}z^k\right\}.)$$

6. *Volterra integral equations of the convolution type.* Let f, g be given continuous functions on $[0,1]$, $f(0)=0$. We consider the Volterra integral equation of the

convolution type,

$$\int_0^x u(t)g(x-t)\,dt = f(x), \qquad 0 \leqq x \leqq 1,$$

for an unknown function u.

(a) Discretize the integral equation by selecting an integration step $h = n^{-1}$, introducing $x_k := kh$, $u_k := u(x_k)$, etc., and approximating the integral by the trapezoidal rule. Show that there results a system of linear equations for the u_k that can be expressed as a one-sided convolution.

(b) Conclude that the system can be solved in $O(n \text{ Log } n)$ operations, and give a more precise operations count.

7. Suppose $(PQ)_n$ has been computed by the FFT. What advantage can be taken of the available results to compute $(PQ)_{2n}$?

8. To compute the series $Y := (1+Q)^{1/2}$, where Q is a given nonunit, show that Newton's method may be used in the form

$$Y^{(m)} := 1,$$

$$Y^{(m+1)} := \frac{1}{2}\left[Y^{(m)} + \frac{1+Q}{Y^{(m)}}\right]_{2n}, \qquad m = 0, 1, 2, \ldots, \quad n := 2^m.$$

Conclude that for $n = 2^l$, Y_n can be computed in no more than $24n \text{ Log}_2 n$ multiplications.

9. *Polynomial division.* Let

$$f(x) := f_0 + f_1 x + \cdots + f_m x^m,$$

$$g(x) := g_0 + g_1 x + \cdots + g_{n-1} x^{n-1}$$

be polynomials, and consider the problem of determining the unique polynomials q and r such that

$$\frac{g}{f} = q + \frac{r}{f},$$

where the degree of r is $< m$. Show that the computation of q and r can be performed in $O(n \text{ Log } n)$ operations, and estimate the precise number of multiplications. (If $F := f_m + f_{m-1}x + \cdots$, $G := g_{n-1} + g_{n-2}x + \cdots$, $Q := q_{n-m-1} + q_{n-m-2}x + \cdots$, $R := r_{m-1} + r_{m-2}x + \cdots$, then

$$(F^{-1}Q)_n = Q + x^{n-m}Q^*, \qquad R = (FQ^*)_m.)$$

10. Let

$$p(x) := \sum_{i=0}^{m-1} b_i x^i$$

be a polynomial, and let z be a complex number. Show that the set of values $p(z^j)$, $j = 0, 1, \ldots, m-1$, can be evaluated in $O(n \text{ Log } n)$ operations. (Note that

$$\sum b_i z^{ij} = \sum b_i z^{-(j-i)^2} z^{i^2/2} z^{j^2/2}$$

can be evaluated by convolution; see Aho, Steiglitz, and Ullmann [1975].)

NOTES

The use of Newton's method for formal power series is due to Newton; see Whiteside [1968], pp. 206–247. This was used for a variety of purposes by Brent [1976]. Algorithm (13.9-7) (in a different notation) is due to Sieveking [1972]. Kung [1974] showed that this is just Newton's method, and studied other root-finding methods to construct reciprocals. For the algorithms to compute $L \circ Q$ and $E \circ Q$ see Brent and Kung [1976, 1978]. These papers also present a $O((n \operatorname{Log} n)^{3/2})$ algorithm to construct $P \circ Q$ for arbitrary P and Q. Ritzmann [1984], working in a slightly different model which takes the word length into account, improves on this result. For the complexity of generalized composition see Brent and Traub [1980], and for the solution of Toeplitz systems, Brent, Gustavson, and Yun [1980]. The problem of evaluating $E \circ Q$ is of importance in modern actuarial mathematics; see Bühlmann [1984].

§ 13.10. MULTIVARIATE DISCRETE FOURIER ANALYSIS

Certain applications to be discussed in later chapters (see § 15.13 and § 16.6) require a multidimensional (or multivariate) generalization of the one-dimensional discrete Fourier analysis that has been discussed so far. This generalization is, in principle, easy to establish; the difficulty is mainly one of notation. Our presentation therefore aims in the main at establishing a serviceable notation, providing the required definitions, and stating multivariate analogs to some theorems that were given earlier; proofs will be omitted. Some of the notation introduced here will be used again in the context of multivariate formal power series.

I. The Multivariate Discrete Fourier Operator

We already have used the symbol $\mathbb{Z}$ for the set of all integers, and the symbol $\mathbb{Z}_+$ for the set $\{1, 2, \ldots\}$ of all positive integers. For any $d \in \mathbb{Z}_+$, we now denote by $\mathbb{Z}^d$ the set of all d-tuples $(n_1, n_2, \ldots, n_d)$ where each $n_i \in \mathbb{Z}$. The elements of $\mathbb{Z}^d$ will be called d-dimensional **index vectors**, and will be denoted by boldface lowercase letters such as **i, j, k, n**, ... An index vector $\mathbf{n} = (n_1, n_2, \ldots, n_d)$ will be called **positive** if all $n_i \in \mathbb{Z}_+$. The set of all positive d-dimensional index vectors is denoted by $\mathbb{Z}_+^d$.

We consider d-fold indexed sequences

$$\mathbf{x} = \{x_{k_1, k_2, \ldots, k_d}\}$$

with complex elements where each index k_i ranges from $-\infty$ to ∞. (Speaking abstractly, such a sequence is merely a function from $\mathbb{Z}^d$ to $\mathbb{C}$.) The elements of such a sequence are more briefly denoted by $x_{\mathbf{k}}$.

Let $\mathbf{n} = (n_1, n_2, \ldots, n_d \in \mathbb{Z}_+^d$. The d-fold indexed sequence $\mathbf{x} = \mathbf{x_k}$ is called **periodic with period vector n** if for all $\mathbf{k} = (k_1, k_2, \ldots, k_d) \in \mathbb{Z}^d$ and for all $\mathbf{l} = (l_1, l_2, \ldots, l_d) \in \mathbb{Z}^d$ there holds

$$x_{k_1 + l_1 n_1, \ldots, k_d + l_d n_d} = x_{k_1, \ldots, k_d}.$$

The condition of periodicity can be expressed more concisely if we define the Hadamard product of the vectors $\mathbf{l}$ and $\mathbf{n}$ by

$$\mathbf{l}\cdot\mathbf{n} := (l_1 n_1, l_2 n_2, \ldots, l_d n_d), \tag{13.10-1}$$

which is again in $\mathbb{Z}^d$. The dot is always written for clarity. The condition of periodicity then simply is

$$x_{\mathbf{k}+\mathbf{l}\cdot\mathbf{n}} = x_{\mathbf{k}}, \qquad \forall\mathbf{k}, \quad \forall\mathbf{l}.$$

The set of all d-fold indexed sequences with complex elements that are periodic with period vector $\mathbf{n}$ is denoted by $\Pi_{\mathbf{n}}$. It is clear that under the obvious definitions of addition and scalar multiplication, $\Pi_{\mathbf{n}}$ is a linear space of dimension

$$[\mathbf{n}] := n_1 n_2 \ldots n_d. \tag{13.10-2}$$

If $\mathbf{n}$ is any period vector, we denote by $Q_{\mathbf{n}}$ the **period cube** in $\mathbb{Z}^d$ defined by $\mathbf{n}$, that is, the set of all $\mathbf{k} = (k_1, \ldots, k_d) \in \mathbb{Z}^d$ such that

$$0 \leqq k_i < n_i, \qquad i = 1, 2, \ldots, d.$$

The **norm** of an element $\mathbf{x} \in \Pi_{\mathbf{n}}$ is then defined by

$$\|\mathbf{x}\| := \left[\sum_{\mathbf{k}\in Q_{\mathbf{n}}} |x_{\mathbf{k}}|^2\right]^{1/2}.$$

Let now

$$\mathbf{w}_{\mathbf{n}} := (w_{n_1}, w_{n_2}, \ldots, w_{n_d}),$$

where

$$w_{n_j} := \exp\left(\frac{2\pi i}{n_j}\right), \qquad j = 1, 2, \ldots, d,$$

as always, and put, for any $\mathbf{k} \in \mathbb{Z}^d$,

$$\mathbf{w}_{\mathbf{n}}^{\mathbf{k}} := w_{n_1}^{k_1} w_{n_2}^{k_2} \cdots w_{n_d}^{k_d}. \tag{13.10-3}$$

We define the discrete Fourier operator on $\Pi_{\mathbf{n}}$ by

$$\mathbf{y} := \mathscr{F}_{\mathbf{n}}\mathbf{x}, \tag{13.10-4a}$$

where $\mathbf{y} = \{y_{\mathbf{m}}\} \in \Pi_{\mathbf{n}}$,

$$y_{\mathbf{m}} := \frac{1}{[\mathbf{n}]^{1/2}} \sum_{\mathbf{k}\in Q_{\mathbf{n}}} x_{\mathbf{k}} \mathbf{w}_{\mathbf{n}}^{-\mathbf{k}\cdot\mathbf{m}}, \qquad \mathbf{m} \in \mathbb{Z}^d. \tag{13.10-4b}$$

THEOREM 13.10a. *Let* $\mathbf{n}$ *be any period vector. The discrete Fourier operator* $\mathscr{F}_{\mathbf{n}}$ *defined by* (13.10-4) *is then an operator from* $\Pi_{\mathbf{n}}$ *onto* $\Pi_{\mathbf{n}}$ *with the following* *properties. If* $\mathbf{x} \in \Pi_{\mathbf{n}}$ *and* $\mathbf{y} := \mathscr{F}_{\mathbf{n}} x$, *then*

(i) $\|\mathbf{y}\| = \|\mathbf{x}\|$.
(ii) $\mathbf{x} = \overline{\mathscr{F}_{\mathbf{n}}}\mathbf{y}$, *where*

$$(\overline{\mathscr{F}_{\mathbf{n}}}\mathbf{y})_{\mathbf{m}} = [\mathbf{n}]^{-1/2} \sum_{\mathbf{k} \in Q_{\mathbf{n}}} y_{\mathbf{k}} \mathbf{w}^{+\mathbf{k} \cdot \mathbf{n}}.$$

Concerning the numerical evaluation of $\mathscr{F}_{\mathbf{n}}$, there are two possibilities:

(a) By writing (13.10-4b) in the form

$$y_{m_1 m_2 \cdots m_d} = \frac{1}{\sqrt{n_1}} \sum_{k_1=0}^{n_1-1} w_{n_1}^{-k_1 m_1} \cdot \frac{1}{\sqrt{n_2}} \sum_{k_2=0}^{n_2-1} w_{n_2}^{-k_2 m_2} \cdot \cdots \cdot$$
$$\times \frac{1}{\sqrt{n_d}} \sum_{k_d=0}^{n_d-1} w_{n_d}^{-k_d m_d} x_{k_1 k_2 \cdots k_d},$$

we see that $\mathscr{F}_{\mathbf{n}}$ can be evaluated by iterating one-dimensional transforms. To this end one would have to compute

$$\frac{[\mathbf{n}]}{n_1} \text{ transforms ot type } \mathscr{F}_{n_1},$$

$$\frac{[n]}{n_2} \text{ transforms of type } \mathscr{F}_{n_2},$$

and so on. If all n_i are powers of 2, $n_i = 2^{l_i}$, this can be done, by Theorem 13.1b, in

$$\frac{[\mathbf{n}]}{n_1} \frac{1}{2} n_1 \operatorname{Log}_2 n_1 + \cdots + \frac{[\mathbf{n}]}{n_d} \frac{1}{2} n_d \operatorname{Log}_2 n_d = \frac{1}{2} [\mathbf{n}] \operatorname{Log}_2 [\mathbf{n}] \quad (13.10\text{-}5)$$

complex multiplications.

(b) An alternate approach to evaluating $\mathscr{F}_{\mathbf{n}}$ consists in mimicking the reduction formulas (13.1-19). Let

$$\mathbf{n} = \mathbf{p} \cdot \mathbf{q},$$

where $\mathbf{p} \in \mathbb{Z}_+^d$, $\mathbf{q} \in \mathbb{Z}_+^d$. The $[\mathbf{p}]$ sequences

$$\mathbf{x}^{\mathbf{j}} := \{x_{\mathbf{j}+\mathbf{p} \cdot \mathbf{h}}\}_{\mathbf{h} \in \mathbb{Z}^d}, \qquad \mathbf{j} \in Q_{\mathbf{p}}$$

are in $\Pi_{\mathbf{q}}$. We assume that their Fourier transforms

$$\mathbf{y}^{\mathbf{j}} := \mathscr{F}_{\mathbf{q}} \mathbf{x}^{\mathbf{j}}, \qquad \mathbf{j} \in Q_{\mathbf{p}},$$

are known. Because every vector $\mathbf{k} \in Q_{\mathbf{n}}$ can be written in the form

$$\mathbf{k} = \mathbf{j} + \mathbf{p} \cdot \mathbf{h}, \qquad \mathbf{j} \in Q_{\mathbf{p}}, \qquad \mathbf{h} \in Q_{\mathbf{q}},$$

in exactly one way, we have

$$y_{\mathbf{m}} = \frac{1}{[\mathbf{n}]^{1/2}} \sum_{\mathbf{j}\in Q_{\mathbf{p}}} \sum_{\mathbf{h}\in Q_{\mathbf{q}}} \mathbf{w}_{\mathbf{n}}^{-(\mathbf{j}+\mathbf{p}\cdot\mathbf{h})\cdot\mathbf{m}} x_{\mathbf{j}+\mathbf{p}\cdot\mathbf{h}}$$

$$= \frac{1}{[\mathbf{p}]^{1/2}} \sum_{\mathbf{j}\in Q_{\mathbf{p}}} \mathbf{w}_{\mathbf{n}}^{-\mathbf{j}\cdot\mathbf{m}} \frac{1}{[\mathbf{q}]^{1/2}} \sum_{\mathbf{h}\in Q_{\mathbf{q}}} \mathbf{w}_{\mathbf{q}}^{-\mathbf{h}\cdot\mathbf{m}} x_{\mathbf{j}+\mathbf{p}\cdot\mathbf{h}},$$

where we have used $\mathbf{w}_{\mathbf{n}}^{\mathbf{p}} = \mathbf{w}_{\mathbf{q}}$. Because

$$\frac{1}{[\mathbf{q}]^{1/2}} \sum_{\mathbf{h}\in Q_{\mathbf{q}}} \mathbf{w}_{\mathbf{q}}^{-\mathbf{h}\cdot\mathbf{m}} x_{\mathbf{j}+\mathbf{p}\cdot\mathbf{h}} = y_{\mathbf{m}}^{\mathbf{j}},$$

by definition, we have

$$y_{\mathbf{m}} = \frac{1}{[\mathbf{p}]^{1/2}} \sum_{\mathbf{j}\in Q_{\mathbf{p}}} \mathbf{w}_{\mathbf{n}}^{-\mathbf{j}\cdot\mathbf{m}} y_{\mathbf{m}}^{\mathbf{j}}, \qquad (13.10\text{-}6)$$

which is the multivariate analog of (13.1-19).

To obtain the anlog of (13.1-21), we note that every $\mathbf{m}\in Q_{\mathbf{n}}$ may be represented in exactly one way also in the form

$$\mathbf{m} = \mathbf{k} + \mathbf{q}\cdot\mathbf{l}, \qquad \mathbf{k}\in Q_{\mathbf{q}}, \quad \mathbf{l}\in Q_{\mathbf{p}}.$$

Because the sequences $\mathbf{y}^{\mathbf{j}}$ admit the period vector $\mathbf{q}$,

$$y_{\mathbf{m}}^{\mathbf{j}} = y_{\mathbf{k}+\mathbf{q}\cdot\mathbf{l}}^{\mathbf{j}} = y_{\mathbf{k}}^{\mathbf{j}},$$

and observing that

$$\mathbf{w}_{\mathbf{n}}^{-\mathbf{j}\cdot\mathbf{m}} = \mathbf{w}_{\mathbf{n}}^{-\mathbf{j}\cdot(\mathbf{k}+\mathbf{q}\cdot\mathbf{l})} = \mathbf{w}_{\mathbf{n}}^{-\mathbf{j}\cdot\mathbf{l}} \mathbf{w}_{\mathbf{n}}^{-\mathbf{k}\cdot\mathbf{j}},$$

we find the representation

$$y_{\mathbf{k}+\mathbf{q}\cdot\mathbf{l}} = \frac{1}{[\mathbf{p}]^{1/2}} \sum_{\mathbf{j}\in Q_{\mathbf{p}}} \mathbf{w}_{\mathbf{p}}^{-\mathbf{l}\cdot\mathbf{j}} \mathbf{w}_{\mathbf{n}}^{-\mathbf{k}\cdot\mathbf{j}} y_{\mathbf{k}}^{\mathbf{j}}. \qquad (13.10\text{-}7)$$

If all n_i are powers of 2, one may select for $\mathbf{p}$ an index vector the components of which are either 1 or 2. Then all factors

$$\mathbf{w}_{\mathbf{p}}^{-\mathbf{l}\cdot\mathbf{j}} = \pm 1,$$

and the only products that need to be formed when stepping from $\Pi_{\mathbf{q}}$ to $\Pi_{\mathbf{n}}$ are

$$\mathbf{w}_{\mathbf{n}}^{-\mathbf{k}\cdot\mathbf{j}} y_{\mathbf{k}}^{\mathbf{j}}, \qquad \mathbf{j}\in Q_{\mathbf{p}}, \quad \mathbf{j}\neq\mathbf{0}, \quad \mathbf{k}\in Q_{\mathbf{q}}.$$

There are

$$([\mathbf{p}]-1)[\mathbf{q}] = (1-[\mathbf{p}]^{-1})[\mathbf{n}]$$

such products. In the special case where all $n_i = 2^l$ independently of i, we have $\mathbf{p} = (2, 2, \ldots, 2)$ l times, and there follows:

THEOREM 13.10b. *If $n_i = 2^l$, $i = 1, 2, \ldots, d$, the number of complex multiplications required to evaluate $\mathscr{F}_{\mathbf{n}}\mathbf{x}$ does not exceed*

$$\frac{1}{d}(1-2^{-d})[\mathbf{n}]\,\mathrm{Log}[\mathbf{n}]. \tag{13.10-8}$$

Already for $d = 2$ the operation count (13.10-8) compares favorably with the count (13.10-5) for iterated one-dimensional transforms, and the ratio gets more favorable as d increases.

II. Multiplication in $\Pi_{\mathbf{n}}$.

For sequences $\mathbf{x} = \{x_{\mathbf{k}}\}$ and $\mathbf{y} = \{y_{\mathbf{k}}\}$ in $\Pi_{\mathbf{n}}$ we define the **Hadamard product** by

$$\mathbf{x} \cdot \mathbf{y} := \{x_{\mathbf{k}} y_{\mathbf{k}}\} \tag{13.10-9}$$

and the **convolution** by

$$\mathbf{x} * \mathbf{y} = \mathbf{z},$$

where $\mathbf{z} = \{z_{\mathbf{k}}\}$ and

$$z_{\mathbf{k}} = \frac{1}{[\mathbf{p}]^{1/2}} \sum_{\mathbf{m} \in Q_{\mathbf{n}}} x_{\mathbf{m}} y_{\mathbf{k}-\mathbf{m}}. \tag{13.10-10}$$

With these definitions there holds:

THEOREM 13.10c (Convolution theorem in $\Pi_{\mathbf{n}}$). *If $\mathbf{n}$ is any period vector, then there holds for arbitrary sequences $\mathbf{x}$, $\mathbf{y}$ in $\Pi_{\mathbf{n}}$,*

$$\begin{aligned} \mathscr{F}_{\mathbf{n}}(\mathbf{x} * \mathbf{y}) &= \mathscr{F}_{\mathbf{n}}\mathbf{x} \cdot \mathscr{F}_{\mathbf{n}}\mathbf{y}, \\ \mathscr{F}_{\mathbf{n}}(\mathbf{x} \cdot \mathbf{y}) &= \mathscr{F}_{\mathbf{n}}\mathbf{x} * \mathscr{F}_{\mathbf{n}}\mathbf{y}. \end{aligned} \tag{13.10-11}$$

As a consequence of (13.10-11),

$$\mathbf{x} * \mathbf{y} = \overline{\mathscr{F}_{\mathbf{n}}}(\mathscr{F}_{\mathbf{n}}\mathbf{x} \cdot \mathscr{F}_{\mathbf{n}}\mathbf{y}). \tag{13.10-12}$$

Thus convolution can be reduced to Hadamard multiplication also in the multivariate case at the cost of three Fourier transforms. In view of Theorem 13.10b we thus have:

THEOREM 13.10d. *If $\mathbf{n} = (2^l, 2^l, \ldots, 2^l)$, the convolution of two sequences in $\Pi_{\mathbf{n}}$ can be carried out in no more than*

$$[\boldsymbol{n}]\{3d^{-1}(1-2^{-d})\,\mathrm{Log}[\mathbf{n}] + 1\}$$

complex multiplications.

III. Digital Image Processing

Two-dimensional discrete Fourier analysis, and fast algorithms for computing two-dimensional discrete Fourier transforms, are fundamental tools in modern image processing techniques. An **image** in a technical sense is a positive real function f, called the **light intensity function**, which is defined in a planar region Q, usually assumed to be rectangular.

In order to be in a form that is suitable for processing by computer, an image f must be discretized both in space and in intensity. Thus if Q is the square $0 \leqq x, y \leqq 1$, the image function f is replaced by the $n \times n$ array

$$\mathbf{f} = \{f_{i,j}\}, \qquad f_{i,j} := f\left(\frac{2i+1}{2n}, \frac{2j+1}{2n}\right),$$

where $i, j = 0, 1, \ldots, n-1$. Almost always in digital image processing, n is selected to be a power of 2, $n = 2^l$; for instance, $n = 512$ is frequently used. The elements f_{ij} of the array $\mathbf{f}$ are called **pixels**. The number of discretization levels of the values of each pixel f_{ij} between $f_{\min}$ (= black) and $f_{\max}$ (= white) likewise is selected as a power of 2. For instance, $2^8 = 256$ levels give satisfactory black-and-white pictures in most cases.

By periodic continuation in each index, the array $\mathbf{f}$ becomes a member of $\Pi_{n,n}$, and its discrete Fourier transform can be defined. The fact that the qualitative properties of an image reside in its spectrum is decisive in digital image processing. For instance, sharp edges in a picture show up in strong high-frequency components of its spectrum. Thus in order to emphasize edges, the high-frequency components must be amplified; in order to smoothe or blur an image, these components should be weakened. In both cases, appropriate filters must be applied to the spectrum. The design of appropriate filters, which to some extent is an empirical matter, is the subject of a large literature.

Mathematically this type of processing of a given digital image $\mathbf{f}$ proceeds as follows:

(a) The discrete Fourier transform $\hat{\mathbf{f}}$ is computed by an FFT algorithm.
(b) The Hadamard product $\mathbf{h} \cdot \hat{\mathbf{f}}$ is formed, where $\mathbf{h}$ is an appropriate filtering array.
(c) By an inverse Fourier transform, we obtain the restored picture

$$\mathbf{f}^+ := F_{n,n}^{-1}(\mathbf{h} \cdot \hat{\mathbf{f}}).$$

If n is a power of 2, then the entire process by Theorem 13.10b requires no more than $(\frac{3}{2} \operatorname{Log}_2 n + 1)n^2$ complex multiplications. For instance, if $n = 2^9 = 512$, the total number of multiplications does not exceed $15n^2$, where n^2 is the number of pixels in the image.

By similar techniques it is also possible to restore images that have been **degraded** in some qualitatively known way, either in transmission or in the process of taking the picture, for instance by a motion of the camera relative to the subject of the image.

PROBLEMS

1. Let E_i, $i = 1, 2, \ldots, d$, denote the unit shift operator in the direction of the ith component of the index vector, and for any $\mathbf{k} = (k_1, k_2, \ldots, k_d) \in \mathbb{Z}^d$ write

   $$\mathbf{E}^{\mathbf{k}} := E_1^{i_1} E_2^{k_2} \cdots E_d^{k_d}.$$

   Prove the following analog of axiom (F_2) of § 13.1. If $\mathbf{x} \in \Pi_{\mathbf{n}}$ and $\mathbf{y} = \{y_{\mathbf{m}}\} = \mathcal{F}_{\mathbf{n}}\mathbf{x}$, then for any $\mathbf{k} \in \mathbb{Z}^d$,

   $$\mathcal{F}_{\mathbf{n}}(\mathbf{E}^{\mathbf{k}}\mathbf{x}) = \{\mathbf{w}_{\mathbf{n}}^{\mathbf{k}\cdot\mathbf{m}} y_{\mathbf{m}}\}.$$

2. Develop a theory of attenuation factors for two-dimensional Fourier series, and compute these factors for piecewise bilinear interpolation between the data points.

NOTES

Formulas for multivariate discrete Fourier analysis were given by Henrici [1979a], who deals only with the case where all $n_i = n$. For reduction formulas similar to (13.10-6) and (13.10-7) see Harris et al. [1977] and Temperton [1983].

For the material on picture processing see Rosenfeld and Kak [1976], Carasso et al. [1978], and especially Gonzales and Wintz [1977].

14

CAUCHY INTEGRALS

Let the curve Γ be simple, closed, regular (§ 3.5), and positively oriented, and let the function f be analytic in the interior D of Γ and continuous in $D \cup \Gamma$. *Cauchy's integral formula* (Theorem 4.7b)

$$f(z) = \frac{1}{2\pi i} \int_\Gamma \frac{f(t)}{t-z}\, dt, \qquad z \in D,$$

then expresses the values of f in D in terms of the values of f on the boundary Γ of D.

At first thought, Cauchy's integral formula seems to afford a simple solution of the following *boundary value problem.* Find a function f which is analytic in D and which for $z \to t \in D$ tends to prescribed values $h(t)$, where h is a given function on the boundary. Is the solution of this problem not given by the function

$$f(z) := \frac{1}{2\pi i} \int_\Gamma \frac{h(t)}{t-z}\, dt?$$

On reflection, however, it is easy to see that the problem as stated can have a solution only in exceptional cases. By the Cauchy–Riemann equations, the function f is determined, up to a purely imaginary constant, by its real part alone. The real part of f is a harmonic function, and as such it is determined by the real parts of the boundary values. Thus the real part of h already determines f up to an imaginary constant, and the imaginary part of h thus cannot be prescribed arbitrarily.

The very fact that in the case of an arbitrarily given h the integral above is not merely a form of Cauchy's formula makes the study of such integrals interesting. As it turns out, such integrals are powerful tools for the solutions of various boundary value problems for analytic functions, for harmonic functions, and for the construction of conformal maps. In addition, Cauchy integrals have important applications in the theory of airfoils, in elasticity, and in digital signal processing.

NOTES

The basic reference for this chapter is Muschelishvili [1965]. A more recent account, containing many applications that are not presented here, is due to Meister [1983].

§ 14.1. THE FORMULAS OF SOKHOTSKYI

In the following, Γ will always denote a simple regular curve (see § 3.5), not necessarily closed. If closed, Γ will be assumed to be positively oriented. Let h be a complex-valued function defined on Γ, which for the moment we assume piecewise continuous; later h will be subjected to somewhat stronger restrictions. Let $z \notin \Gamma$. The function

$$f(z) := \frac{1}{2\pi i} \int_\Gamma \frac{h(t)}{t-z}\, dt \tag{14.1-1}$$

is known as a **Cauchy integral**. (It is clear that in the definition of the Cauchy integral the factor $(2\pi i)^{-1}$ could have been absorbed into h. However, the theory becomes neater if the factor is retained.) The present section is devoted to studying the properties of functions f defined by Cauchy integrals.

Let D denote the complement of Γ. For each $t \in \Gamma$, the integrand in (14.1-1) as a function of z is analytic in D. The integrand moreover depends (piecewise) continuously on t. By Theorem 4.1a, integration with respect to the parameter t preserves analyticity. It follows that f is analytic at each point of D. If Γ is not closed, then D is connected, and (14.1-1) defines a single analytic function f. If Γ is closed, then D has two components, the interior and the exterior of Γ. Then in each component of D (14.1-1) defines an analytic function. These functions have, in general, nothing to do with each other.

EXAMPLE 1

Let Γ be closed, $h = 1$. The Cauchy integral

$$f(z) := \frac{1}{2\pi i} \int_\Gamma \frac{1}{t-z}\, dt \tag{14.1-2}$$

has the values 1 if z is in the interior of Γ, and 0 if z is in the exterior of Γ. —

If $z \in \Gamma$, the integral (14.1-1), if interpreted in the usual sense, becomes an improper integral which in general does not exist. (Consider, for example,

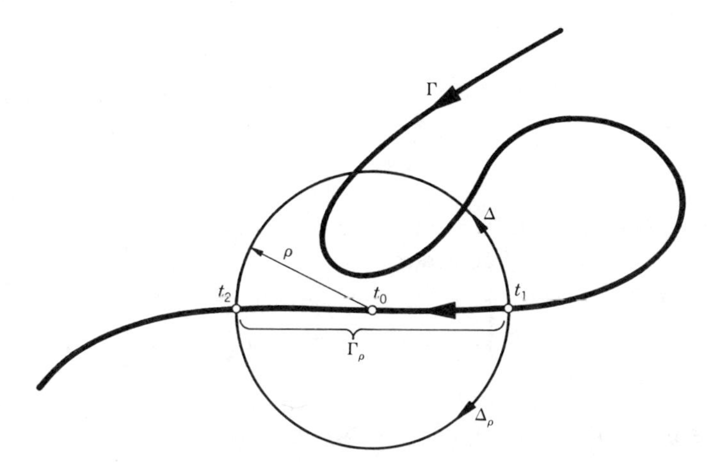

Fig. 14.1a. Definition of principal value.

the case where Γ is a straight line segment, and $h=1$.) However, we now shall ascribe to it a value by a special construction, as follows.

Let t_0 be a point on Γ which, if Γ is not closed, is not an endpoint. Such a point t_0 will be called an **interior point** of Γ; this phraseology is of course totally different from "a point in the interior of Γ." For $\rho>0$ sufficiently small, Γ will then intersect the circle Δ: $|z-t_0|=\rho$ in at least two points. We denote by t_1 the last point of intersection before t_0 and by t_2 the first point of intersection after t_0 (see Fig. 14.1a), and by Γ_ρ the portion of Γ lying between t_1 and t_2. The integral

$$\frac{1}{2\pi i}\int_{\Gamma-\Gamma_\rho}\frac{h(t)}{t-z}\,dt$$

then exists for every $\rho>0$. If its limit for $\rho\to 0$ also exists, the limit is called the **principal value** of the integral (14.1-1) for $z=t_0$. (This definition is consistent with that given in § 4.8 for the special case where Γ is a segment of the real line.) If the integral happens to exist as an ordinary improper integral (with t_1 and t_2 both approaching t_0, but not necessarily symmetrically), then it clearly also exists as a principal value integral, and the principal value agrees with the value of the ordinary improper integral. Thus no ambiguity can arise if we agree that for $z\in\Gamma$ the symbol (14.1-1) should denote the principal value of the integral, if the principal value exists, on in other words that for $z\in\Gamma$,

$$\frac{1}{2\pi i}\int_{\Gamma}\frac{h(t)}{t-z}\,dt := \lim_{\rho\to 0}\frac{1}{2\pi i}\int_{\Gamma-\Gamma_\rho}\frac{h(t)}{t-z}\,dt.$$

EXAMPLE 2

Let $\Gamma := [-1, 1]$, and let $-1 < \xi < 1$. As an ordinary improper integral,

$$\frac{1}{2\pi i}\int_\Gamma \frac{1}{t-\xi}\,dt$$

evidently does not exist. To see whether the principal value exists we assume $\rho < \min(\xi+1, 1-\xi)$ and consider

$$\int_{\Gamma-\Gamma_\rho} \frac{1}{t-\xi}\,dt = \operatorname{Log}\frac{1-\xi}{1+\xi}.$$

Clearly the limit of the result of the integration as $\rho \to 0$ exists (in fact, the result is independent of ρ), and thus we may write

$$\frac{1}{2\pi i}\int_{-1}^{1} \frac{1}{x-\xi}\,dx = \frac{1}{2\pi i}\operatorname{Log}\frac{1-\xi}{1+\xi}.$$

EXAMPLE 3

Here we calculate the principal value of (14.1-2) when $z = t_0 \in \Gamma$. Let $\rho > 0$ be sufficiently small; we first have to determine the value of the ordinary integral

$$I_\rho := \frac{1}{2\pi i}\int_{\Gamma-\Gamma_\rho} \frac{1}{t-t_0}\,dt.$$

Let Δ_ρ denote the circular arc from t_1 to t_2 centered at t_0 and running through the interior of Γ (see Fig. 14.1a). Then

$$I_\rho = \frac{1}{2\pi i}\int_{\Gamma-\Gamma_\rho+\Delta_\rho} \frac{1}{t-t_0}\,dt - \frac{1}{2\pi i}\int_{\Delta_\rho} \frac{1}{t-t_0}\,dt.$$

The first integral vanishes, since t_0 lies in the exterior of the closed curve $\Gamma-\Gamma_\rho+\Delta_\rho$. To evaluate the second integral, we set $t - t_0 = \rho\, e^{i\theta}$. Here θ varies from a value α of $\arg(t_1 - t_0)$ to a value β of $\arg(t_2 - t_0)$, which must be selected in the interval $\alpha - 2\pi < \beta < \alpha$. Then

$$\frac{1}{2\pi i}\int_{\Delta_\rho} \frac{1}{t-t_0}\,dt = \frac{1}{2\pi}\int_\alpha^\beta d\theta = \frac{1}{2\pi}(\beta-\alpha).$$

Now as $\rho \to 0$, both ratios

$$\frac{t_0 - t_1}{\rho} \quad \text{and} \quad \frac{t_2 - t_0}{\rho}$$

tend to the unit tangent vector of Γ at the point t_0, hence

$$\lim_{\rho\to 0} e^{i(\beta-\alpha)} = \lim_{\rho\to 0} \frac{t_2 - t_0}{t_1 - t_0} = -1$$

and $\lim_{\rho\to 0}(\beta - \alpha) = -\pi$. We thus find that the limit of I_ρ as $\rho \to 0$ equals $\frac{1}{2}$. In connection with Example 1 we see that for a closed curve Γ,

$$\frac{1}{2\pi i}\int_\Gamma \frac{1}{t-z}\,dt = \begin{cases} 1 & \text{if } z \text{ is in the interior of } \Gamma \\ \frac{1}{2} & \text{if } z \text{ is on } \Gamma \\ 0 & \text{if } z \text{ is in the exterior of } \Gamma. \end{cases} \tag{14.1-3}$$

Our examples show that the principal value integral may exist in cases where the ordinary integral does not. We now give a simple sufficient condition for the existence of a principal value Cauchy integral.

Let h be a complex-valued function defined on a closed set $S \subset \mathbb{C}$. If $t_0 \in S$, we say that h satisfies a **Hölder condition** at t_0 if there exist positive constants μ and γ such that

$$|h(t) - h(t_0)| \leqq \mu |t - t_0|^\gamma \tag{14.1-4}$$

for all $t \in S$ sufficiently close to t_0. The constant γ is called the **exponent** of the Hölder condition. If h satisfies (14.1-4) at every $t_0 \in S$ with the same constants μ and γ, then h is said to satisfy a **uniform Hölder condition** on S. If $0 < \gamma \leqq 1$, the class of all functions h defined on S that satisfy a uniform Hölder condition with exponent γ is called the **Lipschitz class of order** γ (for the set S) and is denoted by Lip γ. Clearly, any function $h \in \text{Lip } \gamma$ is uniformly continuous on S, but the converse is not true; see Problem 1.

THEOREM 14.1a. *Let Γ be a regular simple arc or curve, and let z_0 be an interior point of Γ. Let h be a piecewise continuous complex-valued function on Γ, and let it satisfy a Hölder condition at z_0. Then the Cauchy integral (14.1-1) exists for $z = z_0$ as a principal value integral. If Γ is closed and positively oriented, then*

$$f(z_0) = \frac{1}{2\pi i}\int_\Gamma \frac{h(t) - h(z_0)}{t - z_0}\,dt + \frac{1}{2}h(z_0), \tag{14.1-5}$$

where the integral now exists as an ordinary improper integral.

Proof. We first assume that Γ is closed. In the notation used earlier, we have

$$\int_{\Gamma-\Gamma_\rho} \frac{h(t)}{t-z_0}\,dt = \int_{\Gamma-\Gamma_\rho} \frac{h(t)-h(z_0)}{t-z_0}\,dt + h(z_0)\int_{\Gamma-\Gamma_\rho} \frac{1}{t-z_0}\,dt.$$

The limit of the last integral as $\rho \to 0$ exists (see Example 3) and equals $i\pi$.

The limit of the first integral exists as an ordinary improper integral since

$$\left|\frac{h(t)-h(z_0)}{t-z_0}\right| \leqq \mu|t-z_0|^{\gamma-1}$$

by virtue of the Hölder condition. Letting $\rho \to 0$ we thus obtain (14.1-5).

If Γ is not closed, we may join its endpoints by a curve Γ_1 such that $\Gamma+\Gamma_1$ is a simple closed curve. By defining h to be zero on Γ_1, we can appeal to the result just established, and the Theorem is proved. —

Examples 1 and 3 show that the limits of a Cauchy integral f as z approaches a point on Γ from the two sides of Γ need not be equal, and that neither limit necessarily equals the principal value taken by the Cauchy integral on Γ itself. Our next task is to study these limits in the general case. Some terminology concerning limits at points of Γ is required.

Let Γ: $z=z(\tau)$, $\alpha \leqq \tau \leqq \beta$, be a regular arc, and let $z_0 := z(\tau_0)$, $\alpha < \tau_0 < \beta$, be an interior point of Γ. The derivative $z'(\tau_0)$ then exists and is $\neq 0$ by the definition of regularity, and the complex number $z'(\tau_0)$ represents a tangent vector of Γ at z_0 that points in the direction of increasing parameters. The complex number $iz'(\tau_0)$ represents a vector perpendicular to the tangent vector which points to what we shall call the **positive** (or left) **side** of Γ; the complex number $-iz'(\tau_0)$ points to the **negative** (or right) **side** of Γ. If D is a sufficiently small disk centered at z_0, the set $D \cap \Gamma$ will have precisely two components. One component, which we call D^+, contains the points $z_0 + iz'(\tau_0)\sigma$, where $\sigma > 0$ is sufficiently small; the other component, D^-, contains the points $z_0 - iz'(\tau_0)\sigma$ for sufficiently small $\sigma > 0$.

Let f be a function defined on a set having z_0 as a point of accumulation. Provided that the limits involved exist, we shall write

$$\begin{aligned} f^+(z_0) &:= \lim_{\substack{z \to z_0 \\ z \in D^+}} f(z), \\ f^-(z_0) &:= \lim_{\substack{z \to z_0 \\ z \in D^-}} f(z). \end{aligned} \tag{14.1-6}$$

These will be called the **unrestricted one-sided limits** of f at $z_0 \in \Gamma$ from the left and from the right, respectively.

On occasion it is necessary to restrict the approach of z to z_0 further. Let $0 < \alpha < \pi/2$, and denote by V^+ the set of all $z \neq z_0$ such that

$$\left|\arg \frac{z-z_0}{iz'(\tau_0)}\right| \leqq \alpha$$

(see Fig. 14.1b). If

$$f^{V^+}(z_0) := \lim_{\substack{z \to z_0 \\ z \in V^+}} f(z) \tag{14.1-7}$$

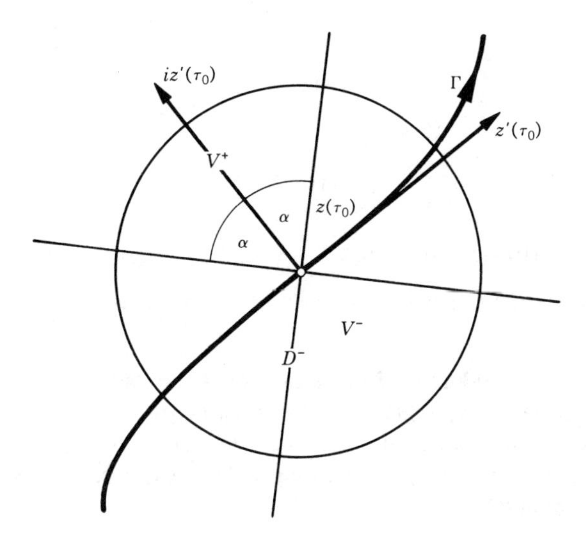

Fig. 14.1b. One-sided limits.

exists for every $\alpha < \pi/2$ (although not necessarily for $\alpha = \pi/2$), we shall say that the **nontangential one-sided limit** of f at z_0 exists; similarly for $f^{V^-}(z_0)$. The manner of approach implied in (14.1-7) will referred to as a **nontangential approach**.

EXAMPLE **4**

For the function

$$f(z) = \frac{1}{2\pi i} \int_\Gamma \frac{1}{t-z}\, dt, \qquad \Gamma \text{ closed}$$

considered in Example 1 we have for $z_0 \in \Gamma$

$$f^{V^+}(z_0) = f^+(z_0) = 1,$$
$$f^{V^-}(z_0) = f^-(z_0) = 0.$$

EXAMPLE **5**

Let Γ be the straight line segment $[-1, 1]$, and let

$$f(z) := \frac{|z|^2}{\operatorname{Im} z}.$$

Then $f^{V^+}(0)$ exists and equals zero. However, if the origin is approached on the curve

$$z = z(\tau) := \tau + i\tau^3,$$

then

$$f(z(\tau)) = \frac{1+\tau^4}{\tau},$$

and the limit $f^+(0)$ does not exist. —

We can now state the two main results of this section:

THEOREM 14.1b (Sokhotskyi formulas, local version). *Let h be piecewise continuous on the simple regular arc Γ, let z_0 be an interior point of Γ, and let h satisfy a Hölder condition at z_0. Then the one-sided limits of the Cauchy integral* (14.1-1) *exist at z_0 for nontangential approach, and there hold the* **formulas of Sokhotskyi,**

$$\begin{aligned} f^{V^+}(z_0) &= f(z_0) + \tfrac{1}{2}h(z_0), \\ f^{V^-}(z_0) &= f(z_0) - \tfrac{1}{2}h(z_0). \end{aligned} \tag{14.1-8}$$

THEOREM 14.1c (Sokhotskyi formulas, global version). *Let Γ be a simple regular curve, and let h be a complex-valued function defined on Γ which belongs to* Lip γ *for some $\gamma \in (0, 1)$. If Γ is closed, let $\Gamma' := \Gamma$; if Γ is not closed, let Γ' be any subarc of Γ whose endpoints are interior points of Γ. Then the restriction of the principal value of the Cauchy integral* (14.1-1) *to Γ' likewise belongs to* Lip γ. *Moreover, the one-sided limits f^+ and f^- exist uniformly for $z_0 \in \Gamma'$ without restriction on the approach. The Sokhotskyi formulas hold in the form*

$$\begin{aligned} f^+(z_0) &= f(z_0) + \tfrac{1}{2}h(z_0), \\ f^-(z_0) &= f(z_0) - \tfrac{1}{2}h(z_0), \end{aligned} \tag{14.1-9}$$

and the functions f^+ and f^- belong to Lip γ *on Γ'.*

For the *proof of Theorem* 14.1b we may assume, by defining $h(t) := 0$ where necessary, that Γ is closed and positively oriented. We write f in the form

$$f(z) = g(z) + h(z_0)\frac{1}{2\pi i}\int_\Gamma \frac{1}{t-z}\,dt,$$

where

$$g(z) := \frac{1}{2\pi i}\int_\Gamma \frac{h(t)-h(z_0)}{t-z}\,dt. \tag{14.1-10}$$

By Example 3, the limit of the second term equals $h(z_0)$ if $z \to z_0$ from the left of Γ, and it equals 0 if $z \to z_0$ from the right. We also know from Theorem 14.1a that

$$g(z_0) = \frac{1}{2\pi i} \frac{h(t) - h(z_0)}{t - z_0} dt$$

exists as as ordinary improper integral, and that

$$g(z_0) + \tfrac{1}{2} h(z_0) = f(z_0).$$

Thus the Sokhotskyi formulas (14.1-8) will be established if it is shown that

$$g^{V^+}(z_0) = g^{V^-}(z_0) = g(z_0). \qquad (14.1\text{-}11)$$

To prove (14.1-11), let let $d(z) := g(z) - g(z_0)$. We have

$$d(z) = \frac{1}{2\pi i} \int_\Gamma [h(t) - h(z_0)] \left(\frac{1}{t - z} - \frac{1}{t - z_0} \right) dt$$

$$= \frac{1}{2\pi i} \int_\Gamma \frac{z - z_0}{t - z} \frac{h(t) - h(z_0)}{t - z_0} dt.$$

For $\rho > 0$, let Γ_ρ denote the arc described in Fig. 14.1a. In the above integral for $d(z)$ we estimate separately the contributions d_1 and d_2 arising from Γ_ρ and from $\Gamma - \Gamma_\rho$. Using the Hölder condition,

$$|d_1(z)| \leqq \frac{1}{2\pi} \int_{\Gamma_\rho} \left| \frac{z - z_0}{t - z} \right| \left| \frac{h(t) - h(z_0)}{t - z_0} \right| |dt|$$

$$\leqq \frac{\mu}{2\pi} \int_{\Gamma_\rho} \frac{|z - z_0|}{\eta(z)} |t - z_0|^{\gamma - 1} |dt|,$$

where $\eta(z)$ denotes the distance from z to Γ. If z is restricted to V^+ or to V^-, then there exists $\kappa > 0$ (depending on the opening $2\alpha < \pi$ of V but not on z or z_0) such that

$$\frac{|z - z_0|}{\eta(z)} < \kappa.$$

Thus

$$|d_1(z)| \leqq \frac{\kappa\mu}{2\pi} \int_{\Gamma_\rho} |t - z_0|^{\gamma - 1} |dt|.$$

Since $|t - z_0|^{\gamma - 1}$ is improperly integrable, we may choose ρ so small that, given $\varepsilon > 0$, the above expression is $< \tfrac{1}{2}\varepsilon$ (under the hypotheses of Theorem 14.c, uniformly for all $z_0 \in \Gamma'$). With the ρ thus chosen, we consider

$$d_2(z) = \frac{1}{2\pi i} \int_{\Gamma - \Gamma_\rho} \frac{z - z_0}{t - z} \frac{h(t) - h(z_0)}{t - z_0} dt.$$

This function is analytic in the complement of $\Gamma - \Gamma_\rho$, thus in particular at $z = z_0$, and it moreover satisfies $d_2(z_0) = 0$. Thus there exists $\delta > 0$ such that $|d_2(z)| < \frac{1}{2}\varepsilon$ for all z satisfying $|z - z_0| < \delta$ (and under the hypotheses of Theorem 14.1c, this estimate holds uniformly in z_0 for all $z_0 \in \Gamma'$). It follows that $|d(z)| < \varepsilon$ for all $z \in V^+ \cup V^-$ such that $|z - z_0| < \delta$ (and at all $z_0 \in \Gamma'$ under the hypothesis of Theorem 14.1c). We thus have established the relations (14.1-11), and thus completed the proof of Theorem 14.1b.

We have even shown that under the hypotheses of Theorem 14.1c the limits (14.1-11), and hence the limits (14.1-8), exist uniformly for $z_0 \in \Gamma'$. To show the uniform existence of the limits f^+ and f^-, we note that these limits, if they exist at all, must equal the limits f^{V^+} and f^{V^-} whose existence just has been established. We denote, for any z sufficiently close to Γ, by z' the point on Γ closest to z (the existence of a unique such point follows from the regularity of Γ). If $z_0 \in \Gamma'$ and $z \notin \Gamma$ we have, for instance,

$$f(z) - f^{V^+}(z_0) = f(z) - f^{V^+}(z') + [f^{V^+}(z') - f^{V^+}(z_0)],$$

or, by virtue of (14.1-8) and (14.1-5),

$$f(z) - f^{V^+}(z_0) = f(z) - f^{V^+}(z') + [g(z') - g(z_0)] + [h(z') - h(z_0)],$$

or finally,

$$|f(z) - f^{V^+}(z_0)| \leqq |f(z) - f^{V^+}(z')| + |g(z') - g(z_0)| + |h(z') - h(z_0)|.$$

Let now $\varepsilon > 0$ be given. By the definition of Γ' and by the uniform continuity of h there exists $\delta_3 > 0$ such that (a) for all $z_0 \in \Gamma'$, all points z' such that $|z' - z_0| < \delta_3$ still form a proper subarc Γ'' of Γ; (b) for all such z',

$$|h(z') - h(z_0)| < \tfrac{1}{3}\varepsilon.$$

By the uniform existence of the limits (14.1-8) there also exists $\delta_1 > 0$ such that for those z on the left side of Γ with the property that $z' \in \Gamma''$ and $|z - z'| < \delta_1$,

$$|f(z) - f^{V^+}(z')| < \tfrac{1}{3}\varepsilon.$$

It remains to prove the existence of $\delta_2 > 0$ such that $|z' - z_0| < \delta_2$ implies

$$|g(z') - g(z_0)| < \tfrac{1}{3}\varepsilon.$$

This is a consequence of:

LEMMA 14.1d. *Under the hypotheses of Theorem* 14.1c, *the restriction of* g *to* Γ' *belongs to* Lip γ.

Proof. Without loss of generality, we may assume that Γ is closed. For if necessary, we may close Γ by an arc Γ_1 and define h on Γ_1 such that it is in Lip γ on $\Gamma+\Gamma_1$. The contribution of Γ_1 to the restriction of g to Γ' then is analytic on Γ'. It thus certainly is in Lip γ, and thus may be ignored.

It is to be shown that there exists $\nu>0$ such that for any two points $t_0, t_1 \in \Gamma$ such that $|t_1-t_0|$ is sufficiently small,

$$|g(t_1)-g(t_0)| \leqq \nu |t_1-t_0|^\gamma. \tag{14.1-12}$$

Let $\rho := |t_1-t_0|$. Due to the smoothness of Γ we may assume ρ so small that the circle of radius 2ρ about t_0 intersects Γ in exactly two points. We denote by $\Gamma_{2\rho}$ the subarc joining these points and containing t_0. In the integral

$$g(t_1)-g(t_0)=\frac{1}{2\pi i}\int_\Gamma \left[\frac{h(t)-h(t_1)}{t-t_1}-\frac{h(t)-h(t_0)}{t-t_0}\right] dt$$

we let

$$\int_\Gamma = \int_{\Gamma_{2\rho}} + \int_{\Gamma-\Gamma_{2\rho}} = I_0 + I.$$

As to I_0, using the Hölder condition on h in each term separately,

$$|I_0| \leqq \frac{\mu}{2\pi}\left\{\int_{\Gamma_{2\rho}} |t-t_1|^{\gamma-1}|dt| + \int_{\Gamma_{2\rho}} |t-t_0|^{\gamma-1}|dt|\right\}.$$

Because Γ has a continuously turning tangent we may assume that ρ is so small that the length of the arc between any two points does not exceed twice the length of the chord. We then have, for instance,

$$\int_{\Gamma_{2\rho}} |t-t_0|^{\gamma-1}|dt| \leqq \int_{-4\rho}^{4\rho} |\sigma|^{\gamma-1}\, d\sigma = \frac{2\cdot 4^\gamma}{\gamma}\rho^\gamma$$

and thus see that I_0 is bounded by a quantity of the form (14.1-12). As to the remaining integral

$$I=\frac{1}{2\pi i}\int_{\Gamma-\Gamma_{2\rho}} \left\{\frac{h(t)-h(t_1)}{t-t_1}-\frac{h(t)-h(t_0)}{t-t_0}\right\} dt,$$

we write this as $I = I_1 + I_2$, where

$$I_1 = \frac{1}{2\pi i}\int_{\Gamma-\Gamma_{2\rho}} \frac{h(t_0)-h(t_1)}{t-t_0}\, dt,$$

$$I_2 := \frac{1}{2\pi i}\int_{\Gamma-\Gamma_{2\rho}} \{h(t)-h(t_1)\}\frac{t_1-t_0}{(t-t_1)(t-t_0)}\, dt.$$

The integral I_1 equals $h(t_0)-h(t_1)$ times

$$\frac{1}{2\pi i}\int_{\Gamma-\Gamma_{2\rho}} \frac{1}{t-t_0}\, dt,$$

which expression approaches $\frac{1}{2}$ for $\rho \to 0$ and thus is bounded by 1 if ρ is sufficiently small, uniformly for $t_0 \in \Gamma$. Thus, using the Hölder condition,

$$|I_1| \leqq \frac{\mu}{2\pi}\rho^\gamma,$$

which again is of the form (14.1-12). To estimate I_2, we once more use the Hölder condition and get

$$|I_2| \leqq \frac{\mu\rho}{2\pi}\int_{\Gamma-\Gamma_{2\rho}} |t-t_1|^{\gamma-1}|t-t_0|^{-1}|dt|.$$

For $t \in \Gamma - \Gamma_{2\rho}$, $|t-t_1| \geqq \frac{1}{2}|t-t_0|$. There follows

$$|I_2| \leqq \frac{2^{1-\gamma}\mu\rho}{2\pi}\int_{\Gamma-\Gamma_{2\rho}} |t-t_0|^{\gamma-2}|dt|.$$

Let the length of Γ be denoted by 2λ, and let σ denote the arclength on Γ, measured from t_0. Due to the smoothness of Γ there exists a constant $\kappa > 0$ such that for $t = t(\sigma)$, if $-\lambda \leqq \sigma \leqq \lambda$, the length of the chord $|t-t_0|$ is not less than κ times the length of the arc,

$$|t-t_0| \geqq \kappa|\sigma|.$$

This constant may be chosen independent of ρ and t_0. There follows

$$|I_2| \leqq \frac{2^{1-\gamma}\mu\rho}{2\pi}\kappa^{\gamma-2}2\int_{2\rho}^{\lambda} \sigma^{\gamma-2}\,d\sigma$$

$$\leqq \frac{\mu\kappa^{\gamma-2}}{\pi(1-\gamma)}\rho^\gamma,$$

which once more has the form (14.1-12), thus proving Lemma 14.1d. —

Having proved Lemma 14.1d, the uniform existence of the one-sided limit $f^+(z_0)$ for $z_0 \in \Gamma'$ has now been established without restriction on the manner of approach. By virtue of the representation

$$f^+(z_0) = f^{V^+}(z_0) = g(z_0) + h(z_0)$$

the lemma also shows that $f^+ \in \text{Lip}\ \gamma$. Analogous considerations apply to f^-, and Theorem 14.1c is thus proved. —

The preceding theorems make no statements concerning the behavior of f near the endpoints of Γ if there are such. Indeed, as is shown by Example 2, even if h is constant, the Cauchy integral formed with h need not be bounded, let alone continuous, at the endpoints of Γ. This will be discussed more thoroughly in § 14.7.

PROBLEMS

1. Show that the function

$$f(x) := \begin{cases} 0, & x = 0 \\ \dfrac{1}{\operatorname{Log}|x|^{-1}}, & x \neq 0 \end{cases}$$

is continuous, but not Hölder continuous, at $x = 0$.

2. Let f and g be complex-valued functions.

(a) If g is in Lip β, and if f is in Lip α on the range of g, show that $f \circ g$ is in Lip $\alpha\beta$.

(b) If g is in Lip β, and if f is analytic on a closed set containing the range of g, show that $f \circ g$ is in Lip β.

3. Let Γ denote the unit circle, Γ: $t = e^{i\tau}$, $0 \leqq \tau \leqq 2\pi$. Show that if $|z_0| = 1$ and n is an integer,

$$\frac{1}{2\pi i} \text{PV} \int_\Gamma \frac{t^n}{t - z_0}\, dt = \begin{cases} \frac{1}{2} z_0^n, & n \geqq 0 \\ -\frac{1}{2} z_0^n, & n < 0. \end{cases}$$

(Evaluate

$$f(z) := \frac{1}{2\pi i} \int_\Gamma \frac{t^n}{t - z}\, dt$$

by Cauchy's formula and use the Sokhotskyi formulas.)

4. For $\alpha > -1$, $\beta > -1$, and for $z \notin [0, 1]$, let

$$f(z) := \frac{1}{2\pi i} \int_0^1 \frac{\tau^\alpha (1 - \tau)^\beta}{\tau - z}\, d\tau.$$

(a) Assuming $|z| > 1$, express f in terms of a hypergeometric series by expanding the integrand in powers of z^{-1} and integrating term by term.

(b) Find a representation of f for arbitrary $z \notin [0, 1]$ by applying appropriate linear transforms of the hypergeometric function (see Table 9.9a).

(c) Verify the Sokhotskyi formulas.

(d) Study the behavior of f near the points $z = 0, 1$. How is this behavior related to the behavior of $h(\tau) := \tau^\alpha (1 - \tau)^\beta$?

5. Show by means of an example that

$$\text{PV} \int_\Gamma \frac{h(t)}{t - t_0}\, dt, \qquad t_0 \in \Gamma$$

does not exist for every continuous function h.

6. Formulate conditions under which

$$\text{PV} \int_\Gamma \frac{\sum h_n(t)}{t - t_0}\, dt = \sum \text{PV} \int_\Gamma \frac{h_n(t)}{t - t_0}\, dt.$$

7. This exercise is devoted to the proof of the following proposition. *Let* $-\infty<\alpha<\beta<\infty$ *and put*

$$\phi(x):=\int_\alpha^\beta \operatorname{Log}|x-t|h(t)\,dt, \tag{1}$$

where h is in Lip. *Then* $\phi'(x)$ *exists at every interior point of* $[\alpha,\beta]$, *and*

$$\phi'(x)=\mathrm{PV}\int_\alpha^\beta \frac{h(t)}{x-t}\,dt. \tag{2}$$

(a) For $\eta>0$, define

$$\phi_\eta(x):=\int_{|t-x|\geqq\eta,\,t\in[\alpha,\beta]} \operatorname{Log}|x-t|h(t)\,dt,$$

and show that $\phi_\eta(x)\to\phi(x)$ uniformly in x.

(b) Show that ϕ'_η exists for every $\eta>0$, that

$$\phi'_\eta(x)=\int_{|t-x|=\eta,\,t\in[\alpha,\beta]} \frac{h(t)}{x-t}\,dt+[h(x+\eta)-h(x-\eta)]\operatorname{Log}\eta$$

and that

$$\phi'_\eta(x)\to\psi(x):=\mathrm{PV}\int_\alpha^\beta \frac{h(t)}{x-t}\,dt \tag{3}$$

uniformly in x in every closed subinterval of (α,β).

(c) Show that (3) implies (2) either directly or by appealing to a general theorem on interchanging differentiation and passing to the limit.

NOTES

Our presentation follows Muschelishvili [1965], §§ 16–18, who attributes Theorem 14.1c to Plemelj and Privalov. For a less formal presentation see Lavrentiev and Shabat [1967] or Carrier, Krook, and Pierson [1966]. What we call Sokhotskyi's formulas is called the **Plemelj formulas** in much of the western literature. Our terminology follows Lavrentiev and Shabat [1967], and indeed Muschelishvili [1965] provides a reference to Sokhotskyi which precedes that to Plemelj by 35 years.

The theory of Cauchy integrals has taken great strides in recent years. For instance, Calderon [1977] has proved the almost everywhere existence of nontangential boundary values for the function (14.1-1) if Γ is only rectifiable, and if h is integrable with respect to the arc length of Γ. In the same paper, Calderon has proved that for certain Lipschitzian curves Γ, the principal value of (14.1-1) defines a bounded operator from $L_2(\mathbb{R})$ to $L_2(\mathbb{R})$. This result has been extended by Coifman, McIntosh, and Meyer [1982] to *all* Lipschitzian curves. For further information see Havin et al. [1984]. Compared to these technically very advanced matters, the class Lip rather serves as a panacea which allows for a rigorous yet elementary presentation.

§ 14.2. THE HILBERT TRANSFORM ON THE CIRCLE

In this section we study Cauchy integrals in the special case where Γ is the unit circle, Γ: $t=e^{2\pi i\tau}$, $0\leqq\tau\leqq 1$. We shall use the Sokhotskyi formulas to solve the following boundary value problem. Given a real, Hölder continuous function u on Γ, find a function f which is analytic inside Γ, real at 0, and which on Γ satisfies $\operatorname{Re} f^+ = u$. Further consideration of the same problem yields a concise representation, called the Hilbert transform, of the conjugation operator $\mathscr{K}$ introduced in § 13.5. We then consider some alternative versions of the Hilbert transform and conclude by discussing the numerical evaluation of Cauchy integrals on the circle.

I. Cauchy Integrals on the Unit Circle

Let h be a *real* function defined on Γ, $h \in \text{Lip}\ \alpha$ for some $\alpha \in (0, 1)$, and let

$$f(z) := \frac{1}{2\pi i}\int_\Gamma \frac{h(t)}{t-z}\,dt \tag{14.2-1}$$

be the Cauchy integral formed with h. We recall that

$$f(0) = \frac{1}{2\pi i}\int_\Gamma \frac{h(t)}{t}\,dt = \int_0^1 h(e^{2\pi i\tau})\,d\tau := \hat{h} \tag{14.2-2}$$

equals the (linear) mean value of h.

Given any function f, we call the function f^* defined by

$$f^*(z) := \overline{f(1/\bar{z})} \tag{14.2-3}$$

the **associate** of f. If f is analytic at a point z_0, then clearly f^* is analytic at $\bar{z}_0^{-1}$. We now establish a connection between the Cauchy integral (14.2-1) and its associate. Since h is real,

$$\begin{aligned}
f^*(z) &= \left[\frac{1}{2\pi i}\int_\Gamma \frac{h(t)}{t-\bar{z}^{-1}}\,dt\right]^- \\
&= \left[\frac{1}{2\pi i}\int_\Gamma h(t)\frac{\bar{z}}{\bar{z}-\bar{t}}\frac{dt}{t}\right]^- \\
&= \left[\frac{1}{2\pi i}\int_\Gamma h(t)\left\{1+\frac{\bar{t}}{\bar{z}-\bar{t}}\right\}\frac{dt}{t}\right]^- \\
&= \left[\frac{1}{2\pi i}\int_\Gamma \frac{h(t)}{t}\,dt\right]^- + \left[\frac{1}{2\pi i}\int_\Gamma h(t)\frac{\bar{t}}{\bar{z}-\bar{t}}\frac{dt}{t}\right]^- .
\end{aligned}$$

In view of $\bar{t}\,dt + t\,\overline{dt} = 0$ the second term equals

$$\left[\frac{1}{2\pi i}\int_\Gamma h(t)\frac{1}{\bar{t}-\bar{z}}\overline{dt}\right]^- = -\frac{1}{2\pi i}\int_\Gamma \frac{h(t)}{t-z}\,dt = -f(z),$$

and we therefore find:

LEMMA 14.2a. *Let h be a real function on the unit circle Γ with mean value $\hat{h}$. Then between the Cauchy integral* (14.2-1) *and its associate f^* there holds the relation*

$$f^*(z) = \hat{h} - f(z). \tag{14.2-4}$$

If z approaches a point $t \in \Gamma$ from the outside, $\bar{z}^{-1}$ approaches the same point from the inside. Hence it follows from (14.2-3) that for all $t \in \Gamma$,

$$f^{*-}(t) = \overline{f^+(t)}.$$

By (14.2-4) this implies

$$\hat{h} - f^-(t) = \overline{f^+(t)}.$$

On the other hand, by the Sokhotskyi formulas,

$$f^+(t) - f^-(t) = h(t),$$

$$f^+(t) + f^-(t) = 2f(t),$$

where $f(t)$ denotes the principal value of (14.2-1) for $z = t$. Using (14.2-4) to express $f^-(t)$ in terms of $\overline{f^+(t)}$, we obtain relations for $f^+(t) \pm \overline{f^+(t)}$, and hence for $\operatorname{Re} f^+(t)$ and $\operatorname{Im} f^+(t)$. These relations are

$$\operatorname{Re} f^+(t) = \tfrac{1}{2}[h(t) + \hat{h}], \tag{14.2-5}$$

$$\operatorname{Im} f^+(t) = \frac{i}{2}\hat{h} - if(t). \tag{14.2-6}$$

II. Schwarz Formula

Let now u be another given real function on Γ, $u \in \operatorname{Lip} \alpha$, let

$$\hat{u} := \frac{1}{2\pi i}\int_\Gamma \frac{u(t)}{t}\,dt = \int_0^1 u(e^{2\pi i\tau})\,d\tau \tag{14.2-7}$$

denote the mean value of u, and form the Cauchy integral (14.2-1) with

$$h(t) := 2u(t) - \hat{u}. \tag{14.2-8}$$

Evidently, $\hat{h} = \hat{u}$. We thus see from (14.2-5) that

$$\operatorname{Re} f^+(t) = u(t), \qquad t \in \Gamma. \tag{14.2-9}$$

Therefore the Cauchy integral (14.2-1), with the choice (14.2-8) for h, solves the following *boundary value problem* (B). Given a real function $u \in \operatorname{Lip} \alpha$ on Γ, find a function f such that

(i) f is analytic in $|z| < 1$.
(ii) The boundary values $f^+(t)$ exist for $t \in \Gamma$ and satisfy $\operatorname{Re} f^+(t) = u(t)$.
(iii) $f(0)$ is real.

It is easy to see that problem (B) cannot have more than one solution. For let d denote the difference of any two solutions, and consider $g(z) := \exp\{d(z)\}$. Then g is analytic and $\neq 0$ in $|z|<1$, and for $|z|=1$ there holds $|g^+(z)|=1$. Thus by the principle of the maximum, $|g(z)|=1$ for all $|z|<1$, which implies that g is a constant of absolute value 1, which in turn implies that $d(z)$ is a constant which is pure imaginary. But $d(0)$ is real, and we conclude that $d(z)=0$.

On expressing f in terms of u, we find:

THEOREM 14.2b (Schwarz formula). *Given any real function $u \in \text{Lip}\,\alpha$, $0<\alpha \leqq 1$, on the unit circle Γ, the boundary value problem (B) has a unique solution f. This solution is given by*

$$f(z) = \frac{1}{i\pi}\int_\Gamma \frac{u(t)}{t-z}\,dt - \hat{u} \tag{14.2-10}$$

or also by

$$f(z) = \frac{1}{2\pi i}\int_\Gamma u(t)\frac{t+z}{t-z}\frac{1}{t}\,dt. \tag{14.2-11}$$

III. The Hilbert Transform

We next turn to (14.2-6). For a given $u \in \text{Lip}\,\alpha$ where $0<\alpha<1$, let f denote the unique solution of problem (B), and let $v(t) := \operatorname{Im} f^+(t)$, $t \in \Gamma$. By virtue of Theorem 14.1c, $v \in \text{Lip}\,\alpha$. From (14.2-6) we get for $t_0 \in \Gamma$,

$$v(t_0) = \frac{i}{2}\hat{u} - i\,\text{PV}\frac{1}{2\pi i}\int_\Gamma \frac{2u(t)-\hat{u}}{t-t_0}\,dt.$$

On using

$$\text{PV}\frac{1}{2\pi i}\int_\Gamma \frac{1}{t-t_0}\,dt = \frac{1}{2}$$

and on expressing $\hat{u}$ in terms of u, we find:

THEOREM 14.2c. *For a given real function $u \in \text{Lip}\,\alpha$, $0<\alpha<1$, on the unit circle Γ, let f denote the unique solution of the boundary value problem (B). Then the function $v(t) := \operatorname{Im} f^+(t)$, $t \in \Gamma$, is likewise in Lip α, and is given by*

$$v(t_0) = i\hat{u} - \frac{1}{\pi}\,\text{PV}\int_\Gamma \frac{u(t)}{t-t_0}\,dt \tag{14.2-12}$$

or also by

$$v(t_0) = -\frac{1}{2\pi}\,\text{PV}\int_\Gamma u(t)\frac{t+t_0}{t-t_0}\frac{1}{t}\,dt. \tag{14.2-13}$$

In the foregoing representation formulas for f and v, let $t = e^{2\pi i\tau}$ and put, for all real τ,

$$\phi(\tau) := u(e^{2\pi i\tau}), \qquad \psi(\tau) := v(e^{2\pi i\tau}).$$

The functions ϕ and ψ have domain $\mathbb{R}$ and period 1, and are both in Lip α. The Schwarz formula (14.2-11) now reads

$$f(z) = \int_0^1 \phi(\tau) \frac{e^{2\pi i\tau} + z}{e^{2\pi i\tau} - z} \, d\tau, \qquad |z| < 1. \tag{14.2-14}$$

As to (14.2-13), we let $t_0 = e^{2\pi i\sigma}$ and use

$$\frac{t + t_0}{t - t_0} = \frac{e^{2\pi i\tau} + e^{2\pi i\sigma}}{e^{2\pi i\tau} - e^{2\pi i\sigma}} = \frac{1}{i} \cot[\pi(\tau - \sigma)].$$

This yields

$$\psi(\sigma) = \mathrm{PV} \int_0^1 \phi(\tau) \cot[\pi(\sigma - \tau)] \, d\tau, \tag{14.2-15}$$

where for $\sigma \equiv 0 \bmod 1$ the definition of the principal value requires an obvious modification.

We now recall some concepts from § 13.5. The above construction, which led from ϕ to ψ, is identical with the construction which in § 13.5 led from a real function of period 1 to its conjugate periodic function. The above formula thus provides an explicit representation of the conjugation operator $\mathcal{K}$ defined in § 13.5, and it extends the definition of this operator to functions that are in Lip α for some $\alpha \in (0, 1)$ but are not necessarily represented by an absolutely convergent Fourier series.

THEOREM 14.2d. *Let ϕ be a real function on $\mathbb{R}$ with period* 1, $\varphi \in \mathrm{Lip}\,\alpha$ *for some $\alpha \in (0, 1)$. Then the conjugate periodic function $\psi = \mathcal{K}\phi$ of ϕ exists, is likewise in* Lip α, *and is represented by* (14.2-15).

The functional transformation defined by (14.2-15) is called the **Hilbert transform on the circle.**

Let f be the unique solution of the boundary value problem (B) where $u(e^{2\pi i\tau}) = \phi(\tau)$. Then

$$f^+(e^{2\pi i\tau}) = \phi(\tau) + i\psi(\tau).$$

Consequently $g(z) := -if(z)$ satisfies

$$g^+(e^{2\pi i\tau}) = \psi(\tau) - i\phi(\tau)$$

and since

$$g(0) = -if(0) = -i\hat{\phi},$$

where $\hat{\phi}$ is the mean value of ϕ, the function $g(z)+i\hat{\phi}$ is the solution of problem (B) for the boundary data $u(e^{2\pi i\tau})=\psi(\tau)$. There follow the **Hilbert reciprocity relations**, here stated as:

COROLLARY 14.2e. *If $\phi \in \text{Lip}\,\alpha$ and $\psi := \mathcal{H}\phi$, then $\phi = -\mathcal{H}\psi + \hat{\phi}$.*

We also note the following formula, which is a restatement of (14.2-12). If h is real, $h \in \text{Lip}\,\alpha$ on Γ, $\phi(\tau)=h(e^{2\pi i\tau})$, then the principal value of the Cauchy integral (14.2-1) at a point $t_0=e^{2\pi i\sigma}$ may be expressed as

$$\text{PV}\frac{1}{2\pi i}\int_\Gamma \frac{h(t)}{t-t_0}\,dt = \frac{i}{2}\mathcal{H}\phi(\sigma)+\tfrac{1}{2}\hat{\phi}.$$

By the Sokhotskyi formulas the limits of f at t_0 are therefore given by

$$\begin{aligned} f^+(t_0) &= \frac{i}{2}\mathcal{H}\phi(\sigma)+\tfrac{1}{2}\hat{\phi}+\tfrac{1}{2}\phi(\sigma), \\ f^-(t_0) &= \frac{i}{2}\mathcal{H}\phi(\sigma)+\tfrac{1}{2}\hat{\phi}-\tfrac{1}{2}\phi(\sigma). \end{aligned} \tag{14.2-16}$$

To state other properties of $\mathcal{H}$, we define the inner product of two real functions ϕ_1 and ϕ_2 in Π (the space of functions on $\mathbb{R}$ with period 1) by

$$(\phi_1, \phi_2) := \int_0^1 \phi_1(\tau)\phi_2(\tau)\,d\tau,$$

whenever the integral exists, and the norm of a $\phi \in \pi$ by

$$\|\phi\| := (\phi, \phi)^{1/2},$$

as usual. We then have:

THEOREM 14.2f. *$\mathcal{H}$ has norm* 1, *that is, for any real $\phi \in \text{Lip}\,\alpha$, $0<\alpha\leqq 1$, with period* 1 *there holds*

$$\|\mathcal{H}\phi\| \leqq \|\phi\|, \tag{14.2-17}$$

and equality holds for every ϕ such that its mean value $\hat{\phi}=0$. Moreover, the functions ϕ and $\mathcal{H}\phi$ are orthogonal, that is,

$$(\mathcal{H}\phi, \phi) = 0. \tag{14.2-18}$$

Proof. Let f be the unique solution of the boundary value problem (B) where $u(e^{2\pi i\tau})=\phi(\tau)$. Then $\text{Im}\, f^+(e^{2\pi i\tau})=\psi(\tau)$, where $\psi=\mathcal{H}\phi$. We evaluate the integral

$$I := \frac{1}{2\pi i}\int_\Gamma \frac{[f^+(t)]^2}{t}\,dt$$

in two ways. On the one hand, we get by direct evaluation

$$I=\int_0^1 [\phi(\tau)+i\psi(\tau)]^2\,d\tau=\|\phi\|^2-\|\psi\|^2+2i(\phi,\psi).$$

On the other hand, because f^+ is the continuous extension to Γ of f, we find by the residue theorem (in the form of Theorem 4.7d)

$$I=[f(0)]^2\geqq 0.$$

The two relations of Theorem 14.2f now follow by comparing the real and the imaginary parts of the two expressions for I. —

Some further facts concerning the operator $\mathcal{K}$ that will occasionally be required are now stated without proofs.

(a) If taken in Lebesgue's sense, the integral (14.2-15) exists for every function $\phi\in L_2(0,1)$ for almost all $\sigma\in[0,1]$ and defines a function $\psi\in L_2(0,1)$.
(b) Theorem 14.2f holds if Lip α is replaced by $L_2(0,1)$.
(c) If ϕ is absolutely continuous, and if ϕ' (which then exists almost everywhere) is in $L_2(0,1)$, then $(\mathcal{K}\phi)'=\mathcal{K}\phi'$ almost everywhere.

IV. Notational Conventions and Extensions

Functions of Arbitrary Period

The fact that we have defined the operator $\mathcal{K}$ for functions of period 1 has no particular significance. If, for instance, $\phi\in\text{Lip }\alpha$ has period 2π, one defines $\psi=\mathcal{K}\phi:=\text{Im}\,f^+(e^{i\tau})$, where f is the solution of boundary value problem (B) where $u(e^{i\tau})=\phi(\tau)$. The basic equation (14.2-15) then holds in the form

$$\psi(\sigma)=\frac{1}{2\pi}\,\text{PV}\int_0^{2\pi}\phi(\tau)\cot\frac{\sigma-\tau}{2}\,d\tau. \qquad (14.2\text{-}19)$$

No confusion should arise from using the same symbol $\mathcal{K}$ for the conjugation operators acting in spaces of functions of different periods.

Functions Defined on the Circle

Let u be a real function $\in$Lip on the unit circle Γ. For $s\in\Gamma$ we write

$$v(s)=\mathcal{K}u(s) \qquad (14.2\text{-}20)$$

as a shorthand for the statement "$v(e^{2\pi i\sigma})=\psi(\sigma)$, where $\psi=\mathcal{K}\phi$ and $\phi(\tau)=u(e^{2\pi i\tau})$." Again no confusion should arise from using the symbol $\mathcal{K}$ in these technically different meanings. In this new notation, relation

(14.2-12) simply reads

$$\mathscr{K}u(s) = i\hat{u} - \frac{1}{\pi} \text{PV} \int_\Gamma \frac{u(t)}{t-s}\, dt. \tag{14.2-21}$$

Conversely, principal values of Cauchy integrals on the unit circle Γ may be expressed by means of the formula

$$\text{PV} \frac{1}{2\pi i} \int_\Gamma \frac{u(t)}{t-s}\, dt = \frac{1}{2}\hat{u} + \frac{i}{2}\mathscr{K}u(s). \tag{14.2-22}$$

For the Cauchy integral

$$k(z) := \frac{1}{2\pi i} \int_\Gamma \frac{u(t)}{t-z}\, dt, \qquad z \notin \Gamma,$$

the Sokhotskyi formulas thus imply for $s \in \Gamma$,

$$k^+(s) = \frac{i}{2}\mathscr{K}u(s) + \tfrac{1}{2}\hat{u} + \tfrac{1}{2}u(s),$$

$$k^-(s) = \frac{i}{2}\mathscr{K}u(s) + \tfrac{1}{2}\hat{u} - \tfrac{1}{2}u(s).$$

Complex Functions

If u, $v \in \text{Lip}\ \alpha$, $\alpha > 0$, are real and $h := u + iv$, one adopts the natural definition

$$\mathscr{K}h := \mathscr{K}u + i\mathscr{K}v.$$

If h is defined on the unit circle Γ, and if

$$k(z) := \frac{1}{2\pi i} \int_\Gamma \frac{h(t)}{t-z}\, dt \tag{14.2-23}$$

is the resulting Cauchy integral, one easily finds for $s \in \Gamma$,

$$k(s) = \frac{i}{2}\mathscr{K}h(s) + \tfrac{1}{2}\hat{h},$$

$$k^+(s) = \frac{i}{2}\mathscr{K}h(s) + \tfrac{1}{2}\hat{h} + \tfrac{1}{2}h(s), \tag{14.2-24}$$

$$k^-(s) = \frac{i}{2}\mathscr{K}h(s) + \tfrac{1}{2}\hat{h} - \tfrac{1}{2}h(s).$$

Here $\hat{h}$ as always denotes the (linear) mean of h. For the associate of k,

$$k^*(z) := \overline{k(\bar{z}^{-1})},$$

we obtain from Lemma 14.2a the representation

$$k^*(z) = \hat{\bar{h}} - \frac{1}{2\pi i}\int_\Gamma \frac{\bar{h}(t)}{t-z}\,dt. \qquad (14.2\text{-}25)$$

From this there follow for $s \in \Gamma$,

$$\begin{aligned} k^*(s) &= -\frac{i}{2}\mathcal{K}\bar{h}(s) + \tfrac{1}{2}\hat{\bar{h}} \\ k^{*+}(s) &= -\frac{i}{2}\mathcal{K}\bar{h}(s) + \tfrac{1}{2}\hat{\bar{h}} - \tfrac{1}{2}\bar{h}(s), \\ k^{*-}(s) &= -\frac{i}{2}\mathcal{K}\bar{h}(s) + \tfrac{1}{2}\hat{\bar{h}} + \tfrac{1}{2}\bar{h}(s). \end{aligned} \qquad (14.2\text{-}26)$$

Some of these formulas will be useful in the explicit solution of Riemann problems (see § 14.5) and in numerical conformal mapping (see § 16.8).

For complex-valued functions g, h on Γ the scalar product is defined by

$$(g, h) := \int_0^1 g(e^{2\pi i\tau})\overline{h(e^{2\pi i\tau})}\,d\tau,$$

so that $(h, g) = \overline{(g, h)}$. With the usual definition of the norm we then still have

$$\|\mathcal{K}h\| \leqq \|h\|. \qquad (14.2\text{-}27)$$

However, in place of the relation $(\mathcal{K}h, h) = 0$, which is no longer true in general, there now holds

$$(\mathcal{K}h, \bar{h}) = 0. \qquad (14.2\text{-}28)$$

V. Cauchy Integrals on the Circle: Numerical Evaluation

Let h be a complex, Hölder continuous function on the unit circle Γ, and let

$$h(e^{2\pi i\tau}) = \sum_{k=-\infty}^{\infty} a_k\, e^{2\pi ik\tau} \qquad (14.2\text{-}29)$$

be its Fourier series. To evaluate the resulting Cauchy integral

$$f(z) = \frac{1}{2\pi i}\int_\Gamma \frac{h(t)}{t-z}\,dt = \int_0^1 h(e^{2\pi i\tau})\frac{1}{1-e^{-2\pi i\tau}z}\,d\tau \qquad (14.2\text{-}30)$$

we distinguish the cases $|z| < 1$, $|z| = 1$, $|z| > 1$. If $|z| < 1$, substituting (14.2-29) and using

$$\frac{1}{1-e^{-2\pi i\tau}z} = \sum_{l=0}^{\infty} e^{-2\pi il\tau}z^l$$

yields

$$f(z) = \sum_{k=0}^{\infty} a_k z^k, \qquad |z| < 1. \tag{14.2-31}$$

If $|z| > 1$, expanding the kernel

$$\frac{1}{1-e^{-2\pi i\tau}z} = -\frac{e^{2\pi i\tau}}{z(1-e^{2\pi i\tau}z^{-1})} = -\sum_{l=1}^{\infty} e^{2\pi i l\tau} z^{-l},$$

we obtain

$$f(z) = -\sum_{l=1}^{\infty} a_{-l} z^{-l}, \qquad |z| > 1. \tag{14.2-32}$$

If finally $|z| = 1$, we use (14.2-24) combined with $a_0 = \hat{h}$ and the fact that

$$\mathscr{K}h(z) = -i \sum_{k=-\infty}^{\infty} \operatorname{sgn}(k) a_k z^k$$

to obtain

$$f(z) = \tfrac{1}{2}a_0 + \tfrac{1}{2} \sum_{k=1}^{\infty} (a_k z^k - a_{-k} z^{-k}), \qquad |z| = 1. \tag{14.2-33}$$

It will be seen that the three representations (14.2-31), (14.2-32), and (14.2-33) formally satisfy the Sokhotskyi relations.

To implement these relations numerically, we choose a (large) integer n and let $h_k := h(e^{2\pi ik/n})$, $k \in \mathbb{Z}$. The sequence $\mathbf{h} = \{h_k\}$ then is in Π_n. The discrete Fourier coefficients based on these sampling values then are the elements of the sequence $\mathbf{c} = \{c_k\}$, where

$$\mathbf{c} = \frac{1}{\sqrt{n}} \mathscr{F}_n \mathbf{h},$$

which by means of an FFT can be computed in $O(n \operatorname{Log} n)$ operations. We know that the c_l are reasonable approximations to the a_k only for $|k| \leqq m := [\frac{n}{2}]$. Thus the following function $\tilde{f}$, called the **discrete Cauchy integral**, may be considered a reasonable approximation to the actual Cauchy integral f:

$$\tilde{f}(z) := \begin{cases} \sum_{k=0}^{m}{}' c_k z^k, & |z| < 1 \\ \tfrac{1}{2}c_0 + \tfrac{1}{2} \sum_{k=1}^{m}{}' (c_k z^k - c_{-k} z^{-k}), & |z| = 1 \\ -\sum_{k=1}^{m}{}' c_{-k} z^{-k}, & |z| > 1. \end{cases} \tag{14.2-34}$$

Here the prime indicates that if n is even, $n=2m$, the terms where $k=m$ are to be taken with an (additional) factor $\frac{1}{2}$. On any circle $|z|=\rho$, these expressions may be evaluated simultaneously at the n points $z_k=\rho\, e^{2\pi i k/n}$ by an FFT in another $O(n \operatorname{Log} n)$ operations.

The error $\tilde{f}(z)-f(z)$ of the discrete Cauchy integral may be estimated by the method of Theorem 13.6b. Some insight as to how $\tilde{f}$ approximates f is also obtained by expressing $\tilde{f}$ directly in terms of the sampling values h_k. It is clear that for each fixed z, $\tilde{f}(z)$ is a linear function of the h_k. Thus let

$$f(z)=\sum_{k=0}^{n-1} l_k(z) h_k.$$

We first determine $l_0(z)$. The Fourier image of the sequence $\|: 1, 0, \ldots, 0 :\|$ is the sequence $n^{-1/2}\|: 1, 1, \ldots, 1 :\|$. If n is even, $n=2m$, there follow

$$l_0(z)=-\frac{1}{n}(1+z+\cdots+z^{m-1}+\tfrac{1}{2}z^m)$$

$$=\frac{1}{n}\frac{1}{1-z}\left(1-\frac{1+z}{2}z^m\right), \qquad |z|<1,$$

$$l_0(z)=-\frac{1}{n}(z^{-1}+z^{-2}+\cdots+z^{-m+1}+\tfrac{1}{2}z^{-m})$$

$$=\frac{1}{n}\frac{1}{1-z}\left(1-\frac{1+z}{2}z^{-m}\right), \qquad |z|>1,$$

$$l_0(z)=\frac{1}{n}\frac{1}{1-z}\left(1-\frac{1+z}{2}\frac{z^m+z^{-m}}{2}\right), \qquad |z|=1.$$

If n is odd, $n=2m+1$, we have

$$l_0(z)=\frac{1}{n}(1+z+\cdots+z^m)$$

$$=\frac{1}{n}\frac{1-z^{m+1}}{1-z}, \qquad |z|<1,$$

$$l_0(z)=-\frac{1}{n}(z^{-1}+z^{-2}+\cdots+z^{-m})$$

$$=\frac{1}{n}\frac{1-z^{-m}}{1-z}, \qquad |z|>1,$$

$$l_0(z)=\frac{1}{n}\frac{1}{1-z}\left[1-\frac{1}{2}(z^{m+1}+z^{-m})\right], \qquad |z|=1.$$

In view of $l_k(z) = l_0(z_k^{-1}z)$, the resulting formulas for $\tilde{f}(z)$ are now easily written down:

THEOREM 14.2g. *Let* $z_k := e^{2\pi ik/n}$, $h_k := h(z_k)$. *Then the approximate Cauchy integral* (14.2-34) *is represented as follows:*

(a) *If n is even,* $n = 2m$

$$\tilde{f}(z) = \frac{1}{n}\sum_{k=0}^{n-1}\frac{z_k}{z_k - z}h_k - g_1(z)\frac{1}{2n}\sum_{k=0}^{n-1}(-1)^k\frac{z_k + z}{z_k - z}h_k, \qquad (14.2\text{-}35a)$$

where

$$g_1(z) := \begin{cases} z^m, & |z| < 1 \\ \frac{1}{2}(z^m + z^{-m}), & |z| = 1 \\ z^{-m}, & |z| > 1. \end{cases}$$

(b) *If n is odd,* $n = 2m+1$,

$$\tilde{f}(z) = \frac{1}{n}\sum_{k=0}^{n-1}\frac{z_k}{z_k - z}h_k - g_2(z)\frac{1}{2n}\sum_{k=0}^{n-1}(-1)^k\frac{z_k^{1/2}}{z_k - z}h_k, \qquad (14.2\text{-}35b)$$

where

$$g_2(z) := \begin{cases} z^{m+1}, & |z| < 1 \\ \frac{1}{2}(z^{m+1} + z^{-m}), & |z| = 1 \\ z^{-m}, & |z| > 1. \end{cases}$$

Evidently in the representations (14.2-35), the first term is a discretized form of

$$\frac{1}{2\pi i}\int_\Gamma \frac{h(t)}{t - z}\,dt,$$

while the second term is a correction, noticeable only in the immediate vicinity of $|z| = 1$, which keeps $\tilde{f}$ from becoming infinite as z approaches one of the points z_k.

PROBLEMS

1. (a) If $\phi = \text{constant}$, show that $\mathcal{H}\phi = 0$.
   (b) As a consequence, show that the Hilbert transform may be written as an ordinary improper integral,

$$\psi(\sigma) = \int_0^1 \{\phi(\tau) - \phi(\sigma)\}\cot[\pi(\sigma - \tau)]\,d\tau.$$

   (Compare Theorem 13.6i.)

2. By taking real parts in the Schwarz formula, establish *Poisson's formula*,

$$u(z)=\int_0^1 \phi(\tau)\frac{1-z\bar{z}}{1-2\,\mathrm{Re}\,z\,e^{-2\pi i\tau}+z\bar{z}}\,d\tau,$$

solving Dirichlet's problem for the unit disk.

3. Using the fact that the Hilbert transform of a constant is zero, show that

$$\mathrm{PV}\int_{-1}^{1}\frac{1}{\sqrt{1-t^2}(t-x)}\,dt=0$$

for all $x\in(-1,1)$.

4. Suppose f is analytic in $|z|<1$ and continuous in $|z|\leqq 1$, and let

$$\phi(\tau):=\mathrm{Re}\,f(e^{2\pi i\tau}),\qquad \psi(\tau):=\mathrm{Im}\,f(e^{2\pi i\tau}).$$

Show that although ϕ is not necessarily Hölder continuous, the principal value integral

$$\mathrm{PV}\int_0^1 \phi(\tau)\cot[\pi(\sigma-\tau)]\,d\tau$$

still exists for every $\sigma\in\mathbb{R}$ and equals $\psi(\sigma)$. (Let $t_0:=e^{2\pi i\sigma}$ and integrate

$$f(z)\frac{z+t_0}{z-t_0}\frac{1}{z}$$

around the unit circle indented at t_0; see Gaier [1964], p. 62.)

5. Show that if condition (iii) is omitted from the boundary value problem (B), its general solution may be written

$$f(z)=\frac{1}{i\pi}\int_\Gamma \frac{u(z)}{t-z}\,dt-\overline{f(0)}.$$

6. Devise an algorithm, similar to that described in Theorem 13.6j, for evaluating the approximate Cauchy integrals (14.2-34) when $n=2, 4, 8, 16, \ldots$ data points are used.

7. Let $\alpha>0, \beta>0$. Show that the periodic conjugate of

$$\phi(\tau):=\mathrm{Log}\sqrt{\alpha^2(\cos\tau)^2+\beta^2(\sin\tau)^2}$$

is

$$\psi(\tau)=\arctan\left(\frac{\beta}{\alpha}\tan\tau\right)-\tau,$$

where the arctan is to be defined as a continuous function on $\mathbb{R}$, $\psi(0)=0$.

NOTES

The Schwarz formula holds under the assumption that the boundary function $u(z)$ is merely continuous; see H. A. Schwarz [1890], p. 186. For the origin of the Hilbert transform on the circle see Hilbert [1912], ch. IX, eqs. (28) and (29). For historical perspective see Hellinger

[1935]. The L_2 theory of the Hilbert transform is an important topic in the theory of trigonometric series; in addition to Zygmund [1959], ch. II, IV, VII, X, see also Katznelson [1968], ch. III, and Hardy and Rogosinski [1944], Theorem 89.

The fact that Hilbert transforms can be evaluated efficiently by means of the FFT was realized only some time after 1965; see Bauer et al. [1975], Ives [1976], and Henrici [1976, 1979a].

§ 14.3. CAUCHY INTEGRALS ON CLOSED CURVES

In this section Γ denotes a regular, simple, closed, positively oriented curve. The set D^+ is the interior of Γ, and D^- is the exterior of Γ. We continue to write Lip for the union of all classes Lip α (of functions defined on Γ), where $\alpha > 0$.

I. The Analytic Continuation of Data on the Curve

We return to a question suggested at the outset of § 14.1: Which complex functions $h \in$ Lip can be continued analytically into D^+? Or, to put it more explicitly, for which $h \in$ Lip does there exist a function f that is analytic in D^+, continuous in $D^+ \cup \Gamma$, and for which $f^+ = h$ on Γ?

Let $h \in$ Lip be the boundary function of a function f which is analytic in D^+. Since $f^+(t) = h(t)$ for $t \in \Gamma$, we have, by Cauchy's integral formula, if $z \in D^+$,

$$f(z) = \frac{1}{2\pi i} \int_\Gamma \frac{f^+(t)}{t-z}\, dt = \frac{1}{2\pi i} \int_\Gamma \frac{h(t)}{t-z}\, dt.$$

On the other hand, if $z \in D^-$, then because $t \to (t-z)^{-1} f(t)$ is analytic in D^+ and continuous on $D^+ \cup \Gamma$,

$$\frac{1}{2\pi i} \int_\Gamma \frac{f^+(t)}{t-z}\, dt = \frac{1}{2\pi i} \int_\Gamma \frac{h(t)}{t-z}\, dt = 0.$$

Thus

$$\frac{1}{2\pi i} \int_\Gamma \frac{h(t)}{t-z}\, dt = 0, \qquad z \in D^-, \tag{14.3-1}$$

is a *necessary condition* for h to be analytically continuable into D^+.

We now show that this condition is also *sufficient*. Let (14.3-1) hold, and define

$$f(z) := \frac{1}{2\pi i} \int_\Gamma \frac{h(t)}{t-z}\, dt \tag{14.3-2}$$

for all z not on Γ. Then $f^-(t) = 0$ for all $t \in \Gamma$. Hence by the Sokhotskyi formulas,

$$h(t) = f^+(t) - f^-(t) = f^+(t),$$

showing that h is the boundary function of f.

Condition (14.3-1) may be cast in a form which is more amenable to computation. Clearly $f(z)=0$ for all $z \in D^-$ if and only if its Laurent series at ∞,

$$f(z)=\sum_{n=0}^{\infty} a_n z^{-n-1}$$

($|z|$ sufficiently large) is the zero series. From

$$f(z)=-\frac{1}{2\pi i}\frac{1}{z}\int_{\Gamma}\frac{h(t)}{1-t/z}\,dt=-\sum_{n=0}^{\infty} z^{-n-1}\frac{1}{2\pi i}\int_{\Gamma} t^n h(t)\,dt$$

we see that

$$a_n=-\frac{1}{2\pi i}\int_{\Gamma} t^n h(t)\,dt.$$

Thus a condition equivalent to (14.3-1) is

$$\frac{1}{2\pi i}\int_{\Gamma} t^n h(t)\,dt=0, \qquad n=0,1,2,\ldots. \tag{14.3-3}$$

THEOREM 14.3a. *A function $h \in$ Lip can be continued analytically into D^+ if and only if*

$$\frac{1}{2\pi i}\int_{\Gamma}\frac{h(t)}{t-z}\,dt=0 \quad \text{for all} \quad z \in D^-$$

or, equivalently, if (14.3-3) *holds. If these conditions are met, the analytic continuation is given by* (14.3-2).

Given $h \in$ Lip, we may also ask for conditions in order that h possesses an analytic continuation f into D^-, the exterior of Γ, in the sense that $f^- = h$. Such conditions will naturally depend on the required behavior of f at ∞. Let us stipulate that f at ∞ has at most a pole. It then can be represented in the form

$$f(z)=f_0(z)+p(z),$$

where f_0 is analytic at ∞, $f_0(\infty)=0$, and where p is a polynomial.

A necessary condition for h follows as before from Cauchy's formula. Let Γ_ρ denote the circle of radius ρ about 0. If $z \in D^-$, and if ρ is large enough so that Γ_ρ encloses both z and Γ, then by a form of Cauchy's formula, if f is the required continuation,

$$f(z)=\frac{1}{2\pi i}\int_{\Gamma_\rho}\frac{f(t)}{t-z}\,dt-\frac{1}{2\pi i}\int_{\Gamma}\frac{f(t)}{t-z}\,dt.$$

On the other hand, if $z \in D^+$ and Γ_ρ is as before, then

$$0 = \frac{1}{2\pi i}\int_{\Gamma_\rho} \frac{f(t)}{t-z}\,dt - \frac{1}{2\pi i}\int_{\Gamma} \frac{f(t)}{t-z}\,dt.$$

By letting $\rho \to \infty$, we see that the first term just equals $p(z)$, while in the second $f(t)$ may be replaced by $h(t)$. Thus if the required analytic continuation exists, with a pole of order $\leqq n$ at ∞, then necessarily

$$\frac{1}{2\pi i}\int_{\Gamma} \frac{h(t)}{t-z}\,dt = p(z), \qquad z \in D^+, \tag{14.3-4}$$

where p is a polynomial of degree not exceeding n.

To see that this condition is also sufficient, let (14.3-4) be satisfied, and let

$$f_0(z) := -\frac{1}{2\pi i}\int_{\Gamma} \frac{h(t)-p(t)}{t-z}\,dt$$

for all z not on Γ. Then $f_0^+(t) = 0$, $t \in \Gamma$, and by Sokhotzkyi's formulas

$$p(t) - h(t) = f_0^+(t) - f_0^-(t)$$

or

$$f_0^-(t) = h(t) - p(t), \qquad t \in \Gamma.$$

The function f_0 is analytic at ∞, $f_0(\infty) = 0$. It follows that

$$f(z) := f_0(z) + p(z) \tag{14.3-5}$$

has a pole of order $\leqq n$ at ∞ and satisfies

$$f^-(t) = h(t), \qquad t \in \Gamma,$$

thus solving the continuation problem.

Again the condition (14.3-4) may be cast in a computationally more useful form. Let c be a point in D^+. If the expression on the left of (14.3-4) is a polynomial of degree $\leqq n$, then in its Taylor expansion

$$\frac{1}{2\pi i}\int_{\Gamma} \frac{h(t)}{t-z}\,dt = \sum_{k=0}^{\infty} a_k (z-c)^k$$

all a_k, where $k > n$, must vanish. Now from

$$\frac{1}{2\pi i}\int_{\Gamma} \frac{h(t)}{t-z}\,dt = \frac{1}{2\pi i}\int_{\Gamma} \frac{h(t)}{(t-c)[1-(z-c)/(t-c)]}\,dt$$

$$= \sum_{k=0}^{\infty} (z-c)^k \frac{1}{2\pi i}\int_{\Gamma} \frac{h(t)}{(t-c)^{k+1}}\,dt$$

we see that

$$a_k = \frac{1}{2\pi i}\int_\Gamma \frac{h(t)}{(t-c)^{k+1}}\,dt.$$

Thus (14.3-4) is equivalent to the condition that

$$\frac{1}{2\pi i}\int_\Gamma \frac{h(t)}{(t-c)^{k+1}}\,dt = 0, \qquad k > n,$$

for some $c \in D^+$.

THEOREM 14.3b. *A function $h \in \mathrm{Lip}$ can be continued to a function f analytic in D^-, having a pole of order $\leqq n$ at ∞, and satisfying $f^- = h$ on Γ if and only if for $z \in D^+$,*

$$p(z) := \frac{1}{2\pi i}\int_\Gamma \frac{h(t)}{t-z}\,dt \tag{14.3-6}$$

is a polynomial of degree not exceeding n or, equivalently, if for some $c \in D^+$,

$$\frac{1}{2\pi i}\int_\Gamma \frac{h(t)}{(t-c)^{k+1}}\,dt = 0, \qquad k > n.$$

If these conditions are met, the continuation is given by

$$f(z) = p(z) - \frac{1}{2\pi i}\int_\Gamma \frac{h(t)}{t-z}\,dt, \qquad z \in D^-, \tag{14.3-7}$$

where p denotes the polynomial in (14.3-6).

The representation (14.3-7) is not identical with (14.3-5), but it follows from it by virtue of the fact that for $z \in D^-$, because p is a polynomial,

$$f_0(z) = \frac{1}{2\pi i}\int_\Gamma \frac{p(t) - h(t)}{t-z}\,dz = -\frac{1}{2\pi i}\int_\Gamma \frac{h(t)}{t-z}\,dt.$$

II. The Inversion of the Cauchy Principal Value Integral

Let $g \in \mathrm{Lip}$ be given. We wish to find $h \in \mathrm{Lip}$ such that for all $t_0 \in \Gamma$,

$$\frac{1}{i\pi}\,\mathrm{PV}\int_\Gamma \frac{h(t)}{t-t_0}\,dt = g(t_0). \tag{14.3-8}$$

To see what a solution h would have to look like, let us assume that one exists, and let, for all z not on Γ,

$$f(z) := \frac{1}{2\pi i}\int_\Gamma \frac{h(t)}{t-z}\,dt. \tag{14.3-9}$$

Then at any $t_0 \in \Gamma$ the limits $f^+(t_0)$ and $f^-(t_0)$ exist and define functions in Lip. By the Sokhotskyi formulas,

$$f^+(t_0) - f^-(t_0) = h(t_0), \tag{14.3-10a}$$

$$f^+(t_0) + f^-(t_0) = g(t_0), \tag{14.3-10b}$$

for all $t_0 \in \Gamma$. By Cauchy's formula,

$$\frac{1}{2\pi i}\int_\Gamma \frac{f^+(t)}{t-z}\,dt = \begin{cases} f(z), & z \in D^+ \\ 0, & z \in D^-, \end{cases}$$

and because $f(\infty) = 0$, taking account of Theorem 14.3b,

$$\frac{1}{2\pi i}\int_\Gamma \frac{f^-(t)}{t-z}\,dt = \begin{cases} 0, & z \in D^+ \\ -f(z), & z \in D^-. \end{cases}$$

Adding these relations and using (14.3-10b), we get

$$\frac{1}{2\pi i}\int_\Gamma \frac{g(t)}{t-z}\,dt = \begin{cases} f(z), & z \in D^+ \\ -f(z), & z \in D^-. \end{cases}$$

If we here let $z \to t_0 \in \Gamma$ and interpret the integral as a principal value integral, then again by Sokhotskyi's formulas,

$$\begin{aligned} \frac{1}{2\pi i}\,\mathrm{PV}\int_\Gamma \frac{g(t)}{t-t_0}\,dt &= \tfrac{1}{2}\{f^+(t_0) + [-f^-(t_0)]\} \\ &= \tfrac{1}{2}\{f^+(t_0) - f^-(t_0)\} = \tfrac{1}{2}h(t_0), \end{aligned}$$

where in the last step we have used (14.3-10a).

We thus see that *if* (14.3-8) has a solution h for a given $g \in \mathrm{Lip}$, then that solution is unique and is given by

$$h(t) = \frac{1}{i\pi}\,\mathrm{PV}\int_\Gamma \frac{g(s)}{s-t}\,ds. \tag{14.3-11}$$

It remains to be shown that if h is defined by (14.3-11), it actually solves (14.3-8). For z not on Γ, let

$$k(z) := \frac{1}{i\pi}\int_\Gamma \frac{g(s)}{s-z}\,ds.$$

By Sokhotskyi, if $t \in \Gamma$,

$$h(t) = k^+(t) + k^-(t),$$

and it is enough to show that, for all $t_0 \in \Gamma$,

$$\frac{1}{i\pi}\,\mathrm{PV}\int_\Gamma \frac{k^+(t) + k^-(t)}{t-t_0}\,dt = g(t_0).$$

The expression on the left may be written

$$\frac{1}{i\pi}\,\mathrm{PV}\int_\Gamma \frac{k^+(t)}{t-t_0}\,dt+\frac{1}{i\pi}\,\mathrm{PV}\int_\Gamma \frac{k^-(t)}{t-t_0}\,dt,$$

which is a sum of Cauchy integrals formed with boundary values of analytic functions. Thus by Sokhotskyi's formula this equals

$$k^+(t_0)+k^-(t_0)=g(t_0),$$

as desired. We thus have proved:

THEOREM 14.3c. *For any given $g\in \mathrm{Lip}$ the singular integral equation*

$$\frac{1}{i\pi}\,\mathrm{PV}\int_\Gamma \frac{h(t)}{t-t_0}\,dt=g(t_0)$$

has a unique solution $h\in \mathrm{Lip}$. This solution is given by

$$h(t_0)=\frac{1}{i\pi}\,\mathrm{PV}\int_\Gamma \frac{g(t)}{t-t_0}\,dt, \qquad t_0\in\Gamma. \tag{14.3-12}$$

PROBLEMS

1. As a corollary of Theorem 14.3c, obtain the *Poincaré–Bertrand formula*, valid for any $g\in \mathrm{Lip}$ and any $t_0\in\Gamma$,

$$\frac{1}{i\pi}\,\mathrm{PV}\int_\Gamma \frac{1}{t-t_0}\left[\frac{1}{i\pi}\,\mathrm{PV}\int_\Gamma \frac{g(s)}{s-t}\,ds\right]dt=g(t_0).$$

2. Obtain Hilbert's reciprocity relations (Corollary 14.2e) as a special case of Theorem 14.3c.
3. Let $\rho>1$, and let Γ denote the ellipse

$$t=t(\tau):=\tfrac{1}{2}(\rho\, e^{i\tau}+\rho^{-1}\,e^{-i\tau}), \qquad 0\leqq\tau\leqq 2\pi.$$

   To evaluate

$$g(t):=\frac{1}{\pi i}\,\mathrm{PV}\int_\Gamma \frac{h(s)}{s-t}\,ds, \qquad t_0\in\Gamma, \tag{$*$}$$

   we expand

$$h(s)=\hat{h}(\sigma)=\sum_{n=-\infty}^{\infty} h_n\, e^{in\sigma}.$$

   (a) If $t=t(\tau)$, show that

$$g(t)=\sum_{n=0}^{\infty} h_n\, e^{in\tau}+\sum_{n=1}^{\infty}(2h_n\rho^{-2n}-h_{-n})\,e^{-in\tau}.$$

(b) If

$$g(s)=\hat{g}(\sigma)=\sum_{n=-\infty}^{\infty} g_n e^{in\sigma},$$

is given, show that

$$h(s)=\hat{h}(\sigma)=\sum_{n=0}^{\infty} g_n e^{in\sigma}+\sum_{n=1}^{\infty}(2g_n\rho^{-2n}-g_{-n})\, e^{-in\sigma}$$

solves $(*)$.

NOTES

See Muschelishvili [1965], § 28–29. It is true, but not trivial, that a function f which is analytic in D^+ and continuous in $D^+\cup\Gamma$ with values on Γ which are in Lip α, $0<\alpha<1$, is in Lip α throughout $D^+\cup\Gamma$; see Gehring, Hayman, and Hinkkanen [1982].

§ 14.4. THE PRIVALOV PROBLEM FOR A CLOSED CURVE: THEORY

One of the principal applications of Cauchy integrals occurs in the solution of the so-called *problem of Privalov.* Let Γ be either a simple regular arc or a simple regular closed curve which is positively oriented. In either case we denote by D the complement of Γ. Let a and b be two complex-valued functions on Γ, both belonging to Lip α for some $\alpha\in(0,1)$. It will also be assumed throughout that $a(t)\neq 0$ for $t\in\Gamma$. The **problem of Privalov** consists in finding a function f which is analytic in D, has one-sided limits f^+ and f^- at every interior point of Γ, and satisfies the relation

$$f^+(t)=a(t)f^-(t)+b(t) \tag{P}$$

at every interior point $t\in\Gamma$. The problem is called **homogeneous** if $b=0$, that is, if (P) has the form

$$f^+(t)=a(t)f^-(t). \tag{P$_0$}$$

The homogeneous Privalov problem is, in the context of boundary problems for analytic functions, also called the **problem of Hilbert**.

The difference of two solutions of a given Privalov problem clearly is a solution of the corresponding Hilbert problem. The solutions of a Hilbert problem form a linear space in the sense that they may be multiplied not only by arbitrary constants, but also by arbitrary entire functions without damage to the condition (P_0). Unless the corresponding homogeneous problem has only the zero solution, the solution of a Privalov problem thus cannot be unique. In order to restrict the variety of solutions, only those functions will henceforth be called a **solution of a problem of Privalov** which, in addition to satisfying the conditions already stated, are *analytic* at $z=\infty$.

If Γ is an arc, we shall also impose conditions at the endpoints of the arc, where we want the solution to be as weakly singular as possible (see § 14.8).

In the present section we study the general problem of Privalov for a closed curve Γ. By the Jordan curve theorem the complement of Γ then has two components. We denote by D^+ the *interior* and by D^- the *exterior* of Γ.

If Γ is closed, existence and uniqueness of the solution of the Privalov problem hinge on a topological quantity, namely, on the winding number (see § 4.6) with respect to O of the curve $a(\Gamma)$, which by hypothesis does not pass through O. This number in the present context is called the **index** of a. We recall that the index is always an integer. It is given by the formula

$$n = \frac{1}{2\pi}[\arg a(t)]_\Gamma = \frac{1}{2\pi i}\int_\Gamma d[\log a(t)]$$

or, if a is differentiable, by

$$n = \frac{1}{2\pi i}\int_\Gamma \frac{a'(t)}{a(t)}\,dt.$$

We begin by studying the homogeneous Privalov problem and distinguish the cases $n=0$, $n>0$, and $n<0$.

If $n=0$, the argument of $a(t)$, if continued continuously along Γ, returns to its initial value as t moves along the entire curve Γ. Therefore $\log a(t)$ can be defined (in infinitely many ways) as a continuous function on Γ. Moreover, this function is again in Lip α. With one such definition, let

$$l(z) := \frac{1}{2\pi i}\int_\Gamma \frac{\log a(t)}{t-z}\,dt \tag{14.4-1}$$

$$f_0(z) := e^{l(z)}. \tag{14.4-2}$$

LEMMA 14.4a. *If the index of a is zero, then the general solution of the Hilbert problem*

$$f^+(t) = a(t)f^-(t), \qquad t\in\Gamma, \tag{P$_0$}$$

is $f(z) = cf_0(z)$, where c is an arbitrary complex constant.

Proof. (i) Since l clearly is analytic in $D\cup\{\infty\}$, so is f. The one-sided limits l^+ and l^- exist on Γ, and by the Sokhotskyi formulas they satisfy

$$l^+(t) - l^-(t) = \log a(t).$$

Consequently the limits f^+ and f^- exist, and

$$\begin{aligned} f^+(t) &= c\,e^{l^+}(t) = c\,e^{l^-(t)} + \log a(t) \\ &= a(t)c\,e^{l^-(t)} = a(t)f^-(t), \end{aligned}$$

that is, f solves the Hilbert problem.

(ii) Now let f denote any solution of (P_0). Evidently $f_0(z)\neq 0$ for $z\in D$, $f_0(\infty)=1$, and also the one-sided limits $f_0^+(t)$ and $f_0^-(t)$ are $\neq 0$ for $t\in\Gamma$. We thus may consider

$$g(z):=\frac{f(z)}{f_0(z)}.$$

The function g is analytic in $D\cup\{\infty\}$, and since for $t\in\Gamma$ we have

$$g^+(t)=\frac{f^+(t)}{f_0^+(t)}=\frac{a(t)f^-(t)}{a(t)f_0^-(t)}=g^-(t),$$

g has a continuous extension into Γ. By the principle of continuous continuation (Theorem 5.11a) it follows that g is analytic also on Γ, and hence entire. Because g is analytic also at ∞, g is a constant, $g(z)=c$ for all z. We conclude that

$$f(z)=cf_0(z)=c\,e^{l(z)},$$

as was to be shown. —

We next consider the case where the index n of a is positive. Without loss of generality we may assume that the origin lies in the interior of Γ. The function t then has the index 1, and consequently t^n has the index n. It follows that the index of

$$a_1(t):=t^{-n}a(t)$$

is zero. By Lemma 14.4a the Hilbert problem

$$f_1^+(t)=a_1(t)f_1^-(t),\qquad t\in\Gamma, \tag{14.4-3}$$

has the particular solution

$$f_1(z):=e^{l_1(z)}, \tag{14.4-4}$$

where

$$l_1(z):=\frac{1}{2\pi i}\int_\Gamma \frac{\log a_1(t)}{t-z}\,dt. \tag{14.4-5}$$

The function f_1 is never zero, nor are its boundary values f_1^+ and f_1^-. Let now f be any solution of (P_0). The function

$$k(z):=\frac{f(z)}{f_1(z)}$$

then is analytic in $D\cup\{\infty\}$. Moreover, on using (P_0) and (14.4-3) we see that

$$k^+(t)=t^nk^-(t),\qquad t\in\Gamma. \tag{14.4-6}$$

It follows that

$$p(z):=\begin{cases} k(z), & z\in D^+\\ z^nk(z), & z\in D^-\end{cases}$$

is analytic in D and on Γ satisfies

$$p^+(t) = k^+(t) = t^n k^-(t) = p^-(t).$$

Thus p is continuous on Γ and hence, by the principle of continuous continuation, entire. Since $p(z)$ grows at most like z^n as $z \to \infty$, p is a polynomial of degree $\leqq n$. Conversely, if p is any such polynomial, the function

$$k(z) := \begin{cases} p(z), & z \in D^+ \\ z^{-n} p(z), & z \in D^- \end{cases}$$

satisfies (14.4-6), and consequently $f := kf_1$ solves (P_0). Thus if the index $n > 0$, the Hilbert problem has the general solution

$$f(z) := \begin{cases} p(z) f_1(z), & z \in D^+ \\ z^{-n} p(z) f_1(z), & z \in D^-, \end{cases} \tag{14.4-7}$$

where p is any polynomial of degree $\leqq n$. Since such a polynomial contains $n+1$ arbitrary constants, the solution space has the dimension $n+1$.

We finally show that if $n < 0$, the only solution of the Hilbert problem is the trivial solution. Let f denote any solution of the problem, and let f_1 denote the solution of (P_0) for the function

$$a_1(t) := t^{-n} a(t)$$

defined by (14.4-4). Again f_1 is never zero. Letting as before

$$k(z) := \frac{f(z)}{f_1(z)},$$

we see again that k is analytic in $D \cup \{\infty\}$ and on Γ satisfies

$$k^+(t) = t^n k^-(t).$$

Considering now

$$q(z) := \begin{cases} z^{-n} k(z), & z \in D^+ \\ k(z), & z \in D^-, \end{cases}$$

we see in view of

$$q^+(t) = t^{-n} k^+(t) = t^{-n}[t^n k^-(t)] = k^-(t) = q^-(t)$$

that q is continuous on Γ, hence entire. Because q is analytic also at ∞, q is constant, and because $q(0) = 0$ by virtue of $n < 0$, q is the zero function. There follows $k = 0$ and $f = kf_1 = 0$.

We summarize the foregoing results on the Hilbert problem:

THEOREM 14.4b. *Let Γ be a simple, closed, regular, positively oriented curve enclosing O, and let a be a nonvanishing function on Γ which belongs to* Lip α *for some $\alpha \in (0, 1)$. If the index n of a satisfies $n \geqq 0$, then the general solution of the Hilbert problem $f^+(t) = a(t)f^-(t)$, $t \in \Gamma$, is given by*

$$f(z) := \begin{cases} p(z)f_1(z), & z \in D^+ \\ z^{-n}p(z)f_1(z), & z \in D^-, \end{cases}$$

where p is any polynomial of degree $\leqq n$, and where

$$f_1(z) := e^{l_1(z)}, \qquad l_1(z) := \frac{1}{2\pi i}\int_\Gamma \frac{\log[t^{-n}a(t)]}{t-z}\,dt.$$

If $n < 0$, the only solution of the Hilbert problem is the trivial solution.

Turning to the nonhomogeneous problem, we note that since the difference of any two solutions of (P) is a solution of (P_0), we obtain the general solution of (P) by merely adding to the general solution of (P_0) a particular solution of (P).

If the index $n \geqq 0$, we obtain such a solution from a solution of (P_0) by the method of variation of constants. Let f_2 denote the function (14.4-7), where $p(z) = 1$. The function

$$f(z) := k(z)f_2(z)$$

then solves the nonhomogeneous problem if and only if $k(z)$ is analytic in D, $k(z) = O(z^n)$, $z \to \infty$, and

$$k^+(t) = k^-(t) + \frac{b(t)}{f_2^+(t)}, \qquad t \in \Gamma. \tag{14.4-8}$$

Using the Sokhotskyi formulas, the problem (14.4-8) is easily solved by the Cauchy integral

$$k(z) := \frac{1}{2\pi i}\int_\Gamma \frac{b(t)\, e^{-l_1(t)}}{t-z}\,dt, \tag{14.4-9}$$

which has the required behavior at ∞. Thus if $n \geqq 0$, the totality of solutions of the nonhomogeneous problem is represented by

$$f(z) := \begin{cases} [p(z) + k(z)]\, e^{l_1(z)}, & z \in D^+ \\ z^{-n}[p(z) + k(z)]\, e^{l_1(z)}, & z \in D^-, \end{cases} \tag{14.4-10}$$

where l_1 is given by (14.4-5) and p is an arbitrary polynomial of degree $\leqq n$.

If, finally, $n < 0$, let once again

$$f_1(z) := e^{l_1(z)},$$

where

$$l_1(z) := \frac{1}{2\pi i} \int_\Gamma \frac{\log[t^{-n} a(t)]}{t - z}\, dt \tag{14.4-11}$$

for some continuous choice of the logarithm. There holds

$$f_1^+(t) = t^{-n} a(t) f_1^-(t), \qquad t \in \Gamma. \tag{14.4-12}$$

Let f be any solution of the nonhomogeneous problem. Because f_1 is never zero, f can be represented as

$$f(z) = k(z) f_1(z).$$

The analytic function k then satisfies

$$k^+(t) = t^n k^-(t) + \frac{b(t)}{f_1^+(t)}, \qquad t \in \Gamma. \tag{14.4-13}$$

If

$$k_1(z) := \frac{1}{2\pi i} \int_\Gamma \frac{t^{-n} b(t)}{f_1^+(t)(t - z)}\, dt, \tag{14.4-14}$$

then by the Sokhotskyi formulas,

$$k_1^+(t) = k_1^-(t) + \frac{b(t)}{t^n f_1^+(t)}.$$

Hence

$$g(z) := \begin{cases} z^{-n} k(z) - k_1(z), & z \in D^+ \\ k(z) - k_1(z), & z \in D^- \end{cases} \tag{14.4-15}$$

satisfies $g^+(t) = g^-(t)$ for $t \in \Gamma$. Since g is analytic in $D \cup \{\infty\}$ by virtue of $n < 0$, it follows from the principle of continuous continuation combined with Liouville's theorem that g is constant. The value of the constant is

$$g(0) = -k_1(0).$$

We now can use (14.4-15) to express k in terms of k_1:

$$k(z) = z^n [k_1(z) - k_1(0)], \qquad z \in D^+. \tag{14.4-16}$$

If $n < -1$, then from the fact that both k and k_1 are analytic at 0 it follows that in the Taylor series of k_1 at 0,

$$k_1(z) = \sum_{m=0}^{\infty} b_m z^m$$

say, we must have $b_1 = b_2 = \cdots = b_{-n-1} = 0$. From (14.4-14) we see by expanding the Cauchy kernel in a geometric series that

$$b_m = \frac{1}{2\pi i}\int_\Gamma t^{-n-1-m} b(t)\, e^{-l_1^+(t)}\, dt.$$

hence if $n < -1$,

$$\int_\Gamma t^h b(t)\, e^{-l_1^+(t)}\, dt = 0, \qquad h = 0, 1, \ldots, -n-2, \tag{14.4-17}$$

is a necessary condition for the existence of a solution of the nonhomogeneous Privalov problem.

Conversely, if these conditions are met and if k is defined in terms of k_1 by (14.4-16), then k satisfies (14.4-13), and the function $f = kf_1$, that is,

$$f(z) := \begin{cases} z^n[k_1(z) - k_1(0)]\, e^{l_1(z)}, & z \in D^+ \\ [k_1(z) - k_1(0)]\, e^{l_1(z)}, & z \in D^- \end{cases} \tag{14.4-18}$$

is a solution of the nonhomogeneous problem. No nonzero solution of the homogeneous problem is to be added, because none exists. In summary, we have obtained:

THEOREM 14.4c. *Let $a(t)$ and $b(t)$ be Hölder continuous functions on the regular, positively oriented Jordan curve Γ enclosing the origin, $a(t) \neq 0$, $t \in \Gamma$, and let n denote the index of a. If $n \geq 0$, the Privalov problem $f^+(t) = a(t)f^-(t) + b(t)$, $t \in \Gamma$, has the general solution*

$$f(z) := \begin{cases} [p(z) + k(z)]\, e^{l_1(z)}, & z \in D^+ \\ z^{-n}[p(z) + k(z)]\, e^{l_1(z)}, & z \in D^-, \end{cases}$$

where p is any polynomial of degree $\leq n$, and where $l_1(z)$ is defined as in Theorem 14.4b. *If $n = -1$, the problem has precisely one solution. If $n < -1$, the problem has a solution only if the function b satisfies the condition* (14.4-17). *In either case, the solution (if it exists) is given by* (14.4-18).

PROBLEMS

1. Let Γ be the unit circle, and let a have the index $n \geq 0$. The function $\log t^{-n}a(t)$ then is a periodic function of τ, where $t = e^{2\pi i\tau}$. Assuming that it is represented by an absolutely convergent Fourier series,

$$\log t^{-n}a(t) = \sum_{n=-\infty}^{\infty} b_n t^n,$$

represent the solution of the Hilbert problem (P_0) in terms of the Fourier coefficients b_n.

2. Let $a(t)$ have an index $-n < 0$. Show that if f is allowed to have a pole of order $\leq n$ at infinity, the Hilbert problem (P_0) has exactly one solution.

3. In (P) let the index of a be $-n$, where $n>0$. Find the totality of functions f satisfying the relation (P) and having poles of order $\leqq n$ at $z=0$ and at $z=\infty$. (These f are not solutions of (P) in the strict sense.)

NOTES

See Lavrentiev and Shabat [1967], § 53; Muschelishvili [1965], ch. II. In the first reference, the meanings of f^+ and f^- are interchanged.

§ 14.5. THE PRIVALOV PROBLEM FOR A CLOSED CURVE: APPLICATIONS

I. A Problem of Riemann

Let S be a simply connected region in the complex plane whose boundary Γ is a smooth Jordan curve, and let α, β, and γ be three *real* functions defined on Γ, all belonging to some Lipschitz class. We assume that α and β never vanish simultaneously. Riemann in his dissertation considered the problem of finding all functions $f=u+iv$ that are analytic in S and have boundary values $f^+=u^++iv^+$ on Γ which satisfy

$$\alpha(t)u^+(t)-\beta(t)v^+(t)=\gamma(t), \qquad t\in\Gamma.$$

The problem of constructing a function from the real part of its boundary values is contained in Riemann's problem as the special case $\alpha=1, \beta=0$.

Letting

$$c(t):=\alpha(t)+i\beta(t), \tag{14.5-1}$$

Riemann's condition may be written as

$$\operatorname{Re}[c(t)f^+(t)]=\gamma(t), \qquad t\in\Gamma,$$

or also as

$$c(t)f^+(t)+\overline{c(t)f^+(t)}=2\gamma(t), \qquad t\in\Gamma. \tag{R}$$

We note that the difference of any two solutions of (R) is a function analytic in S which satisfies the homogeneous boundary condition

$$c(t)f^+(t)+\overline{c(t)f^+(t)}=0, \qquad t\in\Gamma. \tag{R_0}$$

Furthermore, if f is a solution of (R_0), then so is κf for any *real* constant κ.

We also note that solutions of Riemann's problem are invariant under conformal transplantation, in the following sense. By the Osgood–Carathéodory theorem (Theorem 5.10e) the Jordan region S is the conformal image of $|z|<1$ under a map g which can be extended to a homeomorphism of $|z|\leqq 1$ to $S\cup\Gamma$. We assume that S is such that the extended map g

preserves Hölder continuity on the boundary. The function $f_1 := f \circ g$ solves the Riemann problem for the unit disk with the boundary condition

$$c_1(t)f_1^+(t) + \overline{c_1(t)f_1^+(t)} = 2\gamma_1(t),$$

where $c_1 := c \circ g$ and $\gamma_1 := \gamma \circ g$.

It thus suffices to solve the problem (R) for the unit disk. We begin by considering the existence of a solution. Let f be a solution of (R) for the unit disk. Denoting the unit disk by D^+ and the exterior of $|z|=1$ by D^-, we extend f from D^+ to $D := D^+ \cup D^-$ by setting

$$f(z) := \overline{f\left(\frac{1}{\bar{z}}\right)}, \qquad |z| > 1. \tag{14.5-2}$$

Clearly the extended function is analytic in $D \cup \{\infty\}$. We now show that the extended function solves a certain Privalov problem. If z approaches a certain point t_0 on $|t|=1$ from D^+, then $1/\bar{z}$ approaches t_0 from D^-. Therefore (14.5-2) implies

$$f^+(t) = \overline{f^-(t)}, \qquad |t| = 1, \tag{14.5-3}$$

and (R) may be written

$$c(t)f^+(t) + \overline{c(t)}f^-(t) = 2\gamma(t), \qquad |t| = 1.$$

Because $c(t) \neq 0$ for all t, we conclude that *f solves the special Privalov problem*

$$f^+(t) = a(t)f^-(t) + b(t), \qquad |t| = 1, \tag{P}$$

where

$$a(t) := -\frac{\overline{c(t)}}{c(t)}, \qquad b(t) := \frac{2\gamma(t)}{c(t)}. \tag{14.5-4}$$

Not every solution of (P) also solves (R), because not every such solution satisfies the relation (14.5-3) which is required to deduce (R) from (P). However, from any solution f_1 of (P) we may construct a solution satisfying (14.5-3), as follows. Let f_1 be a solution of (P) and let, for $|z| \neq 1$, f_1^* be the associate of f_1,

$$f_1^*(z) := \overline{f_1\left(\frac{1}{\bar{z}}\right)}. \tag{14.5-5}$$

We assert that f_1^* likewise is a solution of (P). Evidently, f_1^* is analytic in $D \cup \{\infty\}$, and its boundary values are related to those of f_1 by

$$f_1^{*+}(t) = \overline{f_1^-(t)}, \qquad f_1^{*-}(t) = \overline{f_1^+(t)}, \qquad |t| = 1. \tag{14.5-6}$$

Consequently, if f_1 satisfies (P), then

$$f_1^{*+}(t) = \frac{1}{\overline{a(t)}} \overline{f^+(t)} - \frac{\overline{b(t)}}{\overline{a(t)}}$$

$$= a_1(t) f_1^{*-}(t) + b_1(t), \tag{14.5-7}$$

where

$$a_1(t) := \frac{1}{\overline{a(t)}}, \qquad b_1(t) = -\frac{\overline{b(t)}}{\overline{a(t)}}.$$

Because $a_1(t) = a(t)$, $b_1(t) = b(t)$, as follows easily from (14.5-4), (14.5-7) is identical with (P).

Because f_1 and f_1^* both satisfy the nonhomogeneous relation (P), the same relation is also satisfied by

$$f(z) := \tfrac{1}{2}[f_1(z) + f_1^*(z)]. \tag{14.5-8}$$

This function also satisfies (14.5-3) because in view of (14.5-6), we have

$$f^+(t) = \tfrac{1}{2}[f_1^+(t) + f_1^{*+}(t)] = \tfrac{1}{2}[\overline{f_1^{*-}(t)} + \overline{f_1^-(t)}] = \overline{f^-(t)}.$$

It follows that the restriction of f to D^+ is a solution of (R) for the unit disk.

In essence we have proved:

THEOREM 14.5a. *Let the Jordan region S be the conformal image of $|z| < 1$ under a map g whose extension to $|z| \leqq 1$ preserves Hölder continuity. Then the Riemann problem* (R) *for S has a solution if and only if the Privalov problem* (P) *has a solution for the unit disk, where*

$$a := -\frac{\overline{c \circ g}}{c \circ g}, \qquad b := \frac{2\gamma \circ g}{c \circ g}. \tag{14.5-9}$$

By the theory developed in §14.4, the solvability of the problem (P) depends on the index of the function a, defined with the aid of a continuous argument $\phi(t) := \arg a(e^{i\tau})$ by means of

$$n := \frac{1}{2\pi}[\phi(2\pi) - \phi(0)].$$

As τ increases from 0 to 2π, the point $g(e^{i\tau})$ describes Γ, and the point $c \circ g(e^{i\tau})$ therefore describes the image of Γ under the map c. Since $\arg a(e^{i\tau}) = -2 \arg c \circ g(e^{i\tau}) + \pi$, it follows that $n = -2m$, where m is the index of $c(t)$. We conclude that the index n is always even.

Theorem 14.4c now implies that if $m > 0$ (that is, if the index of c is positive), problem (P) does not have a solution for every function $b(t)$. It follows a fortiori that problem (R) does not have a solution for every function $\gamma(t)$. If $m \leqq 0$, the solution of (P) is determined only up to a solution of the homogeneous problem. The general solution of the homogeneous problem was found in §14.4. This will now be used to find the structure of the general solution of the Riemann problem if $n = -2m \geqq 0$.

We begin by constructing a special solution of (P_0) in terms of which all others will be expressed. Let

$$\psi(t) := \arg[t^{-n}a(t)], \qquad t \in \Gamma,$$

the argument being defined continuously. Since $|t^n a(t)| = 1$, the function $l_1(z)$ occurring in (14.4-4) is

$$l(z) = \frac{1}{2\pi}\int_\Gamma \frac{\psi(t)}{t-z}\,dt. \tag{14.5-10}$$

By (14.4-7), the function

$$f_0(z) := \begin{cases} c_0\, e^{l(z)}, & z \in D^+ \\ c_0 z^{-n}\, e^{l(z)}, & z \in D^- \end{cases}$$

is a solution of (P_0) for every value of the constant c_0. We choose c_0 in such a way that there is a simple connection between $f(z)$ and its associate

$$f_0^*(z) := \overline{f_0\left(\frac{1}{\bar{z}}\right)},$$

which, as we already know, likewise is a solution of (P_0). Evidently if $z \in D^+$, then $\bar{z}^{-1} \in D^-$, and

$$f_0^*(z) = \overline{c_0} z^n\, e^{l^*(z)}, \tag{14.5-11}$$

where

$$l^*(z) := \overline{l(\bar{z}^{-1})}.$$

If $z \in D^-$, then $\bar{z}^{-1} \in D^+$, and the relation (14.5-11) is again seen to hold with the same definition of $l^*(z)$. The function

$$h(z) := -il(z) = \frac{1}{2\pi i}\int_\Gamma \frac{\psi(t)}{t-z}\,dt \tag{14.5-12}$$

is a Cauchy integral on the unit circle formed with a real function ψ. Hence by Lemma 14.2a,

$$h^*(z) = \hat{\psi} - h(z),$$

where $\hat{\psi}$ is the mean value of ψ, and consequently

$$l^*(z) = l(z) - i\hat{\psi}. \tag{14.5-13}$$

Relation (14.5-11) thus turns into

$$f_0^*(z) = \overline{c_0} z^n e^{l(z) - i\hat{\psi}} = \frac{\overline{c_0}}{c_0} e^{-i\hat{\psi}} z^n f_0(z),$$

and on choosing $c_0 := e^{-i\hat{\psi}/2}$ we see that the function

$$f_0(z) := \begin{cases} e^{-i\hat{\psi}/2} e^{l(z)}, & z \in D^+ \\ e^{-i\hat{\psi}/2} z^{-n} e^{l(z)}, & z \in D^- \end{cases} \tag{14.5-14}$$

is a solution of (P_0) satisfying

$$f_0^*(z) = z^n f_0(z). \tag{14.5-15}$$

This will be the **canonical solution** of (P_0) in the case under consideration.

Let now f be any solution of (P). We know from Theorem 14.4b that f can be represented in the form

$$f(z) = p(z) f_0(z), \tag{14.5-16}$$

where p is a polynomial of degree not exceeding n. The solution (14.5-16) also solves (R) if and only if it satisfies

$$f^*(z) = f(z)$$

or

$$p^*(z) f_0^*(z) = p(z) f_0(z),$$

where p^* is the associate of p. Using (14.5-15) this amounts to

$$z^n p^*(z) = z^n \overline{p(\bar{z}^{-1})} = p(z), \tag{14.5-17}$$

that is, to the condition that the polynomial p is **self-reciprocal**. If p is expressed in terms of its coefficients,

$$p(z) = \sum_{k=0}^{n} p_k z^k,$$

condition (14.5-17) means that

$$p_k = \overline{p_{n-k}}, \qquad k = 0, 1, \ldots, n. \tag{14.5-18}$$

We thus have obtained:

THEOREM 14.5b. *Let Γ denote the unit circle, let the functions α, β, and γ be in* Lip *on Γ, let $c := \alpha + i\beta$, $c(t) \neq 0$ for $t \in \Gamma$, and let m denote the index of c. If $m > 0$, the Riemann problem* (R) *does not have a solution for every function γ. If $m \leqq 0$, every solution of* (R) *has the form*

$$f(z) = p(z) f_0(z) + \tfrac{1}{2}[f_1(z) + f_1^*(z)] \tag{14.5-19}$$

where f_0 *is the canonical solution of the associated Privalov problem* (P_0), $p(z)$ *is a self-reciprocal polynomial of degree not exceeding* $n := -2m$, *and where* f_1 *is any particular solution of the associated Privalov problem* (P).

The actual construction of f is facilitated by some formulas from § 14.2.

The function (14.5-10) has the form (14.2-23), where $h(t) = i\psi(t)$. It therefore follows from (14.2-24) that

$$l^+(t) = -\frac{1}{2}\,\mathcal{H}\psi(t) + \frac{i}{2}\hat{\psi} + \frac{1}{2}\,\psi(t),$$

and therefore

$$f_0^+(t) = e^{-i\hat{\psi}/2}\, e^{l^+(t)} = e^{-\mathcal{H}\psi(t)/2 + i\psi(t)/2}. \tag{14.5-20}$$

Letting

$$h(t) := \frac{b(t)}{f_0^+(t)} = b(t)\, e^{\mathcal{H}\psi(t)/2 - i\psi(t)/2}, \tag{14.5-21}$$

the special solution f_1 of the nonhomogeneous problem that is supplied by Theorem 14.4c where $p = 0$ then is

$$f_1(z) = f_0(z) k(z),$$

where k is given by

$$k(z) := \frac{1}{2\pi i} \int_\Gamma \frac{h(t)}{t - z}\, dt. \tag{14.5-22}$$

In some applications, the values $f^+(t)$ of the solution of the Riemann problem are of interest. From (14.5-19) we have

$$f^+(t) = p(t) f_0^+(t) + \tfrac{1}{2}[f_0^+(t) k^+(t) + f_0^{*+}(t) k^{*+}(t)].$$

Using (14.2-24) this becomes

$$f^+(t) = f_0^+(t)\left\{ p(t) + \frac{1}{2}\, k^+(t) + \frac{t^n}{2}\, k^{*+}(t) \right\}. \tag{14.5-23}$$

Here $f_0^+(t)$ is obtained via (14.5-20) by evaluating $\mathcal{H}\psi$, and k^+ and k^{*+} both are obtained from (14.2-24) and (14.2-26) by evaluating $\mathcal{H}h$. Thus if $\psi(t)$ is given, f^+ can be obtained by two applications of the Hilbert operator $\mathcal{H}$.

II. Representations of Analytic Functions by Cauchy Integrals

A representation of a set A of mathematical objects by the elements of another set B is a (constructively realizable) one-to-one map from B onto A. In many representations essential structures (such as linearity) are preserved when passing from A to B.

Here we apply the theory of the Riemann problem to obtain representations of families of analytic functions by families of real functions. Let Γ be a regular Jordan curve with interior D^+ and exterior D^-, let Lip be the class of functions that belong to some class Lip α on Γ, $0<\alpha<1$, and let H be the class of functions f which are analytic in D^+ and continuous in $D\cup\Gamma$, with boundary values f^+ that are in Lip. We seek to represent the elements $f\in H$ by the real functions $\mu\in$ Lip by means of the Cauchy integral

$$f(z)=\frac{1}{2\pi i}\int_\Gamma \frac{h(t)\mu(t)}{t-z}\,dt+c, \tag{14.5-24}$$

where h is a fixed function in Lip which is never zero. The constant c is added because without it, as we shall see, the desired representation would, in general, not exist.

Our first step is to establish a connection between the existence of a representation of the form (14.5-24) and the existence of a solution of a certain Riemann problem. Suppose (14.5-24) holds. Then we also have

$$f(z)=\frac{1}{2\pi i}\int_\Gamma \frac{h(t)\mu(t)+c}{t-z}\,dt, \qquad z\in D^+. \tag{14.5-25}$$

The formula (14.5-25) can be used to define $f(z)$ for $z\in D^-$. By the Sokhotskyi formulas the boundary values of the extended function f satisfy

$$f^+(t)-f^-(t)=h(t)\mu(t)+c, \qquad t\in\Gamma.$$

If

$$g(z):=f(z)+c, \qquad z\in D^-,$$

then g is analytic in $D^-\cup\{\infty\}$, and the last relation becomes

$$h(t)\mu(t)=f^+(t)-g^-(t), \qquad t\in\Gamma. \tag{14.5-26}$$

Because μ is real and h is never zero, there follows

$$\operatorname{Re}\left\{\frac{i}{h(t)}[f^+(t)-g^-(t)]\right\}=0$$

or

$$\operatorname{Re}\left\{\frac{i}{h(t)}g^-(t)\right\}=\operatorname{Re}\left\{\frac{i}{h(t)}f^+(t)\right\}, \qquad t\in\Gamma. \tag{14.5-27}$$

Because h and f^+ are given, this relation defines a Riemann problem for a function g analytic in $D^-\cup\{\infty\}$. We conclude: If the function $f\in H$ can be represented in the form (14.5-24), then the Riemann problem (14.5-27) has a solution g.

Conversely, let the Riemann problem (14.5-27) have the solution g. We define

$$c := g(\infty) \tag{14.5-28}$$

and

$$\mu(t) := \frac{1}{h(t)}[f^+(t) - g^-(t)], \qquad t \in \Gamma. \tag{14.5-29}$$

By (14.5-27), μ is real. We assert that with these definitions of c and μ, the expression

$$\frac{1}{2\pi i}\int_\Gamma \frac{h(t)\mu(t) + c}{t - z}\,dt$$

equals $f(z)$ for $z \in D^+$. Since by Cauchy's formula

$$f(z) = \frac{1}{2\pi i}\int_\Gamma \frac{f^+(t)}{t - z}\,dt,$$

our assertion amounts to showing that

$$\frac{1}{2\pi i}\int_\Gamma \frac{h(t)\mu(t) + c - f^+(t)}{t - z}\,dt = 0, \qquad z \in D^+.$$

By (14.5-28) and (14.5-29), the numerator equals $g(\infty) - g^-(t)$. Hence we are to show that

$$\frac{1}{2\pi i}\int_\Gamma \frac{g^-(t) - g(\infty)}{t - z}\,dt = 0, \qquad z \in D^+. \tag{14.5-30}$$

For $z \in D^+$ the function

$$t \mapsto \frac{g(t) - g(\infty)}{t - z}$$

is analytic for $t \in D^-$. Hence the path Γ in (14.5-30) may be replaced by any other path surrounding D^+. In particular, it may be replaced by an arbitrarily large circle. Because

$$\frac{g(t) - g(\infty)}{t - z} = O(t^{-2}) \qquad \text{as } t \to \infty,$$

the integral along a circle of radius ρ is $O(\rho^{-1})$ and thus, being independent of ρ, equals zero.

We thus have shown:

LEMMA 14.5c. *The function $f \in H$ can be represented in the form* (14.5-24) (*where $\mu \in$* Lip) *if and only if the Riemann problem* (14.5-27) *has a solution*

g which is analytic in $D^- \cup \{\infty\}$. If g is a solution of the Riemann problem, the representation (14.5-24) *holds, where c is defined by* (14.5-28) *and μ is defined by* (14.5-29).

We examine in more detail the conditons for the solvability of the Riemann problem (14.5-27). It follows from Theorem 14.5a that this problem has a solution if and only if the Privalov problem (14.5-7) has a solution, where

$$a(t) = -\frac{\overline{c(t)}}{c(t)} = -\frac{-ih(t)}{\overline{ih(t)}} = \frac{h(t)}{\overline{h(t)}}.$$

As was shown in § 14.4, the solvability of the Privalov problem depends on the index of the function a. When calculating the index, the orientation of Γ must be reversed, because the domain of definition of g is the exterior of Γ. Thus

$$2\pi n = [\arg a(t)]_{-\Gamma} = [2 \arg h(t)]_{-\Gamma} = -2[\arg h(t)]_{\Gamma}.$$

Thus the index of a equals -2 times the index of h, and it follows from Theorem 14.4c that the Riemann problem is solvable if the index of h is negative or zero.

We pursue the case where the index of h is zero. By Theorem 14.5b the Riemann problem then has the general solution $g = g_1 + \gamma g_0$, where g_1 is a particular solution, g_0 is a nonzero solution of the homogeneous problem, and where γ is a real constant. By imposing some additional condition on g, it may be possible to standardize h and c in (14.5-24) such that they are uniquely determined by f. We consider three special cases.

(a) $h(t) = 1$. Clearly this function has index 0. The homogeneous problem corresponding to (14.5-27) reads

$$\operatorname{Re}[ig^-(t)] = 0, \qquad t \in \Gamma.$$

It obviously has the particular solution $g(z) = 1$, $z \in D^-$. Therefore the general solution of the nonhomogeneous problem is $g(z) = g_1(z) + \gamma$, where γ is any real number. By choosing $\gamma = -\operatorname{Re} g_1(\infty)$ we can achieve that $g(\infty)$ is purely imaginary. We thus obtain:

THEOREM 14.5d. *Every function $f \in H$ can be represented in the form*

$$f(z) = \frac{1}{2\pi i} \int_\Gamma \frac{\mu(t)}{t - z}\, dt + i\delta, \tag{14.5-31}$$

where $\mu \in$ Lip and δ is real. Both μ and δ are uniquely determined by f.

(b) $h(t) = i$ again has index 0. The homogeneous problem (14.5-27) has the particular solution $g(z) = i$. It follows as above that the nonhomogeneous problem has exactly one solution that is real at ∞. We obtain:

THEOREM 14.5e. *Every function $f \in H$ can also be represented in the form*

$$f(z) = \frac{1}{2\pi i} \int_\Gamma \frac{\mu(t)}{t - z} \, dt + \gamma, \tag{14.5-32}$$

where $\mu \in \text{Lip}$ and the real constant γ are uniquely determined by f.

(c) A third choice of h is prompted by the desire to represent f in the form

$$f(z) = \int_\Gamma \frac{\mu(t)}{t - z} |dt| + c. \tag{14.5-33}$$

This can easily be written in the form (14.5-24). If $u(t)$ denotes the unit tangent vector of Γ at the point t (pointing in the positive direction of Γ), then

$$|dt| = \overline{u(t)} \, dt.$$

Hence (14.5-33) has the form (14.5-24), where $h(t) = 2\pi i \overline{u(t)}$.

This function h, however, has index -1. Consequently the solution of the Riemann problem contains several arbitrary constants which remain undetermined. However, uniqueness can be restored as follows. Assuming $0 \in D^+$, we choose

$$h(t) := 2\pi i t \overline{u(t)}.$$

This function h has index 0, and the general solution of the Riemann problem has the form $g = g_1 + \gamma g_0$ described above. Integrating the relation

$$\text{Re}\left[\frac{1}{t\overline{u(t)}} g_0^-(t)\right] = 0$$

satisfied by g_0 along Γ, we find, using $\bar{u}^{-1} = u$,

$$\begin{aligned} 0 &= \int_\Gamma \text{Re}\left\{\frac{u(t)}{t} g_0^-(t)\right\} |dt| \\ &= \text{Re}\left\{\int_\Gamma \frac{u(t)}{t} g_0^-(t) |dt|\right\} \\ &= \text{Re} \int_\Gamma \frac{g_0^-(t)}{t} \, dt \\ &= \text{Re}\{-2\pi i g_0(\infty)\}. \end{aligned}$$

Hence $g_0(\infty)$ is real, and since $g_0(\infty) \neq 0$, γ may be chosen so that $g(\infty)$ is pure imaginary. Thus in this case there likewise exists a distinguished solution of (14.5-27) which renders μ and c unique. We thus have obtained the following representation theorem (due to I.N. Vekua):

THEOREM 14.5f. *Let $f \in H$. Then there exist a real function $\mu \in$ Lip and a real constant γ, both uniquely determined by f, such that*

$$f(z) = \int_\Gamma \frac{\mu(t)}{1 - z/t} |dt| + i\gamma, \qquad z \in D^+. \tag{14.5-34}$$

PROBLEM

1. *The skew derivative problem.* Let Γ be a regular closed curve with interior D^+, and let $\alpha(t)$, $\beta(t)$, $\gamma(t)$ be given real functions on Γ, $\alpha, \beta, \gamma \in \text{Lip}$, $\alpha^2 + \beta^2 = 1$. We seek a real function u which is harmonic in D^+ and continuously differentiable up to the boundary, and whose derivative on Γ in the direction of the vector (α, β) equals $\gamma(t)$, that is, which satisfies

$$\alpha(t)\frac{\partial u}{\partial x} + \beta(t)\frac{\partial u}{\partial y} = \gamma(t), \qquad t \in \Gamma.$$

Show that the problem of finding u can be reduced to a Riemann problem by setting $u = \text{Re} f$.

NOTES

For the Riemann problems, see Muschelishvili [1965], § 41–43, or Lavrentiev and Shabat [1967], § 55. For integral representations of analytic functions see Muschelishvili [1965], § 66–69, which is based on several papers by I.N. Vekua. For an application of a Riemann problem in hydrodynamics see Gerhold [1979].

The name Riemann problem is to be traced to a posthumous paper by Riemann [1953], p. 379, where a related problem (involving a system of unknown functions) is treated in the theory of differential equations.

§ 14.6. CAUCHY INTEGRALS ON STRAIGHT LINE SEGMENTS

If the support Γ of the function h in the basic Cauchy integral (14.1-1) is a straight line segment, the integral after a linear change of variable may be assumed in the form

$$f(z) = \frac{1}{2\pi i} \int_0^\beta \frac{h(t)}{t - z} dt, \tag{14.6-1}$$

where $0 < \beta < \infty$. Such integrals occur frequently in mathematical models of elasticity and aerodynamics. Here we relate the integral (14.6-1) to the theory of continued fractions and to orthogonal polynomials, and thereby discover a class of algorithms for their efficient numerical evaluation.

I. The Connection with the Stieltjes–Perron Formula

Let ψ be a nondecreasing real function on $[0, \beta]$, continuous from the left, and let

$$\mu_n := \int_0^\beta t^n \, d\psi(t), \qquad n = 0, 1, 2, \ldots,$$

denote the moments of ψ. In Chapter 12 we were concerned with the Stieltjes transform of ψ, that is, with the function

$$z \to \int_0^\beta \frac{1}{t+z} \, d\psi(t).$$

If ψ can be represented in the form

$$\psi(t) = \int_0^t \omega(\sigma) \, d\sigma,$$

where ω is nonnegative, continuous for $\sigma > 0$, and integrable at 0, then on replacing z by $-z$ the Stieltjes transform of ψ becomes

$$f(z) = \int_0^\beta \frac{\omega(t)}{t-z} \, dt, \tag{14.6-2}$$

which is a mere Cauchy integral, formed with the function $h(t) := 2\pi i \omega(t)$. In this normalization, the Stieltjes transform is analytic in the complement of $[0, \beta]$. We infer from the Plemelj-Privalov theorem (Theorem 14.1c) that, if ω on some interval $[\alpha, \beta]$ is in Lip, then $f^+(x)$, $f^-(x)$, and the Cauchy principal value $f(x)$ all exist for $x \in (\alpha, \beta)$. Moreover, if ω is in Lip γ, $0 < \gamma < 1$, then f^+, f^-, and f are in Lip γ on every interval $[\alpha', \beta']$, where $\alpha < \alpha' < \beta' < \beta$. Finally, since f is real for real z not on $[0, \beta]$, the reflection principle shows that at every $x \in (0, \beta)$,

$$f^-(x) = \overline{f^+(x)}. \tag{14.6-3}$$

In § 12.10 we learned how to determine ψ if f is given. Under the present assumptions the Stieltjes-Perron inversion formula (Theorem 12.10d) asserts that for real $\sigma, \tau \in (0, \beta)$,

$$\int_\sigma^\tau \omega(x) \, dx = \frac{1}{\pi} \lim_{y \to 0+} \operatorname{Im} \int_\sigma^\tau f(x+iy) \, dx.$$

In view of the uniform existence of f^+, limit and integration may be interchanged, and the above becomes

$$\int_\sigma^\tau \omega(x)\,dx = \frac{1}{\pi}\,\mathrm{Im}\int_\sigma^\tau f^+(x)\,dx$$

or, on differentiating with respect to the upper limit of integration,

$$\omega(x) = \frac{1}{\pi}\,\mathrm{Im}\,f^+(x), \qquad 0 < x < \beta.$$

By (14.6-3),

$$\mathrm{Im}\,f^+(x) = \frac{1}{2i}[f^+(x) - f^-(x)],$$

and we have

$$2\pi i\omega(x) = f^+(x) - f^-(x),$$

which is one of Sokhotskyi's formulas. We thus see that under the present assumptions the Stieltjes–Perron formula is merely an integrated version of Sokhotskyi's formula.

II. Continued Fraction Expansion

In § 12.11 it was shown that if the support of $d\psi$ is a bounded interval $[0, \beta]$, the Stieltjes transform of ψ can be represented as a continued fraction that converges, with a geometric rate of convergence, for all z off $[0, \beta]$. In the present normalization the even part of this continued fraction is

$$f(z) = -\frac{\alpha_1 |}{| z+\beta_1} - \frac{\alpha_2 |}{| z+\beta_2} - \frac{\alpha_3 |}{| z+\beta_3} - \cdots, \tag{14.6-4}$$

where the α_k and β_k are defined in terms of the elements $q_k^{(m)}$, $e_k^{(m)}$ of the quotient-difference scheme associated with the formal power series

$$\frac{\mu_0}{z} - \frac{\mu_1}{z^2} + \frac{\mu_2}{z^3} - \cdots$$

by means of the relations

$$\alpha_1 = \mu_0, \qquad \beta_1 = -q_1^{(0)} \tag{14.6-5}$$

and for $k = 2, 3, 4, \ldots$,

$$\alpha_k = q_{k-1}^{(0)} e_{k-1}^{(0)}, \qquad \beta_k = e_{k-1}^{(0)} + q_k^{(0)}. \tag{14.6-6}$$

The numerators p_n and the denominators q_n of the continued fraction

(14.6-4) satisfy the recurrence relations

$$\begin{pmatrix} p_n \\ q_n \end{pmatrix} = (z+\alpha_n)\begin{pmatrix} p_{n-1} \\ q_{n-1} \end{pmatrix} - \beta_n \begin{pmatrix} p_{n-2} \\ q_{n-2} \end{pmatrix}, \qquad n = 1, 2, 3, \ldots, \tag{14.6-7a}$$

subject to the initial conditions

$$\begin{aligned} p_{-1} &= 1, & p_0 &= 0, \\ q_{-1} &= 0, & q_0 &= 1, \end{aligned} \tag{14.6-7b}$$

(see (12.1-8)). The denominator q_n as a function of z evidently is a polynomial of degree n with leading coefficient 1; p_n is a polynomial of degree $n-1$ with leading coefficient $-\alpha_1$. If these polynomials are *defined* by means of (14.6-7), then $f(z)$ for z not on $[0, \beta]$ may be computed by the formulas

$$f(z) = \lim_{n\to\infty} \frac{p_n(z)}{q_n(z)} \tag{14.6-8}$$

$$= -\sum_{m=1}^{\infty} \frac{\alpha_1\alpha_2 \cdots \alpha_m}{q_{m-1}(z)q_m(z)}, \tag{14.6-9}$$

the second of which follows from (12.1-23) on account of $a_k = -\alpha_k$, $k = 1, 2, \ldots$.

III. A Series Expansion for the Principal Value Integral

In § 12.10 (where the notation was slightly different) the polynomials q_n were shown to be orthogonal with respect to the weight function $\omega(t)$, that is, there holds

$$\int_0^\beta q_m(t)q_n(t)\omega(t)\, dt = 0, \qquad n \neq m, \tag{14.6-10a}$$

whereas

$$\int_0^\beta [q_n(t)]^2\omega(t)\, dt = \nu_n,$$

where

$$\nu_n := \mu_0 \prod_{k=1}^{n} q_k^{(0)} e_k^{(0)}. \tag{14.6-10b}$$

The orthogonality of the sequence of polynomials $\{q_n\}$ implies that *if* a given function ϕ can be expanded in a series of the q_n, that is, *if* numbers λ_n exist such that the representation

$$\phi(t) = \sum_{n=0}^{\infty} \lambda_n q_n(t) \tag{14.6-11}$$

converges uniformly on the support of ω, then the coefficients λ_n are given by

$$\lambda_n = \frac{1}{\nu_n} \int_0^\beta q_n(t)\phi(t)\omega(t)\,dt, \qquad n = 0, 1, 2, \ldots. \tag{14.6-12}$$

Conditions under which the expansion (14.6-11) actually holds are the subject of extensive study in the theory of orthogonal polynomials. Generally the convergence results are of the "equiconvergence type," that is, if ω is sufficiently smooth in a neighborhood of the point $t_0 \in (0, \beta)$, then the representation (14.6-11) holds at t_0 if and only if the Fourier series for the function

$$\theta \to \phi\left(\frac{\beta}{2}(1+\cos\theta)\right)$$

converges at θ_0, where

$$t_0 = \frac{\beta}{2}(1+\cos\theta_0).$$

Assuming the uniform validity of (14.6-11) on $[0, \beta]$, the principal value of the Cauchy integral

$$g(z) := \int_0^\beta \frac{\phi(t)\omega(t)}{t-z}\,dt \tag{14.6-13}$$

at a point $z = x \in (0, \beta)$ may now be calculated as follows. Substituting (14.6-11) and proceeding formally yields

$$g(x) = \sum_{n=0}^{\infty} \lambda_n g_n(x), \tag{14.6-14}$$

where

$$g_n(x) := \mathrm{PV} \int_0^\beta \frac{q_n(t)\omega(t)}{t-x}\,dt, \qquad n = 0, 1, 2, \ldots. \tag{14.6-15}$$

The point now is that the evaluation of the functions $g_n(x)$ is no more difficult than the evaluation of the $q_n(x)$. For z off $[0, \beta]$, let

$$g_n(z) := \int_0^\beta \frac{q_n(t)\omega(t)}{t-z}\,dt, \tag{14.6-16}$$

so that, using Sokhotskyi's formulas,

$$g_n(x) = \tfrac{1}{2}[g_n^+(x) + g_n^-(x)]. \tag{14.6-17}$$

Adapting Theorem 12.11b to the present notation, we have for all complex z

$$\int_0^\beta \frac{q_n(t)-q_n(z)}{t-z}\,\omega(t)\,dt=-p(z), \qquad n=0,1,2,\ldots.$$

Therefore if $z\notin[0,\beta]$,

$$g_n(z)=q_n(z)\int_0^\beta \frac{\omega(t)}{t-z}\,dt-p_n(z)=q_n(z)f(z)-p_n(z),$$

where f is defined by (14.6-2). Thus the sequence $\{g_n(z)\}$ is a linear combination of two solutions of the difference equation (14.6-7a), and it therefore itself satisfies that difference equation,

$$g_n(z)=(z+\beta_n)g_{n-1}(z)-\alpha_n g_{n-2}(z), \qquad n=1,2,\ldots. \tag{14.6-18a}$$

The initial conditions in view of (14.6-7b) are

$$g_{-1}(z)=-1, \qquad g_0(z)=f(z). \tag{14.6-18b}$$

Let now $z=x\in(0,\beta)$. Since the limits $g_n^+(x)$ and $g_n^-(x)$ exist for $n=-1$ and $n=0$, the same is true for all n. Thus the principal values $g_n(x)$ satisfy

$$g_n(x)=(x+\beta_n)g_{n-1}(x)-\alpha_n g_{n-2}(x), \qquad n=1,2,\ldots, \tag{14.6-19a}$$

and in view of

$$\tfrac{1}{2}[f^+(x)+f^-(x)]=\operatorname{Re} f^+(x),$$

the initial values are

$$g_{-1}(x)=-1, \qquad g_0(x)=\operatorname{Re} f^+(x). \tag{14.6-19b}$$

THEOREM 14.6a. *Let the function $\omega\in$ Lip be nonnegative on $[0,\beta]$. If q_n denotes the nth denominator of the continued fraction representation (14.6-4) of the function f defined by (14.6-2), then the principal value integrals g_n at every $x\in(0,\beta)$ satisfy the recurrence relation (14.6-19a) and the initial conditions (14.6-19b).*

Theorem 14.6a does not express what from a numerical point of view may be its most significant aspect. It is well known that recurrence relations such as (14.6-19a) can be numerically unstable, in which case they are useless for the determination of a solution for given starting values. It may be shown that this instability takes place, in particular, if the solution sought is a *recessive* solution of the difference equation, that is, if $g_n(z)/h_n(z)\to 0$ for any solution $h_n(z)$ that is linearly independent from $g_n(z)$. It follows from (14.6-8) that this condition is satisfied whenever the continued fraction (14.6-4) converges. We know from Corollary 12.11h that convergence takes place for all $z\notin[0,\beta]$. Thus for such z, the difference equation (14.6-18) cannot be used to generate the $g_n(z)$. On the other hand, because every point $z = x \in (0, \beta)$ where $\omega(x) > 0$, is a point of accumulation of zeros of the denominators $q_n(z)$ of the fraction, the continued fraction cannot converge at any such point. Thus it is precisely for such x where (14.6-19) may be used to generate the singular integrals $g_n(x)$, and hence to expand the function g.

The procedure is summarized in:

ALGORITHM 14.6b. *To evaluate*

$$g(x) = \mathrm{PV} \int_0^\beta \frac{\phi(t)\omega(t)}{t - x} \, dt,$$

(i) *Find the coefficients* λ_n *in the expansion*

$$\phi(t) = \sum_{n=0}^{\infty} \lambda_n q_n(t).$$

(*In principle, these coefficients are given by* (14.6-12); *quadrature rules may be used.*)

(ii) *Generate the values* $g_n(x)$ *by means of* (14.6-19).

(iii) *Form the sum*

$$g(x) = \sum_{n=0}^{\infty} \lambda_n g_n(x).$$

Steps (ii) and (iii) may be performed simultaneously and economically by using the *Clenshaw algorithm*, to be discussed elsewhere.

IV. Arbitrary Interval

After a change of variables $z^* = l(z)$, $t^* = l(t)$, where l is linear, the foregoing results can be applied to the evaluation of principal value integrals of the general form

$$g(z) = \int_\alpha^\delta \frac{\phi(t)\omega(t)}{t - z} \, dt,$$

where $\alpha = l(0)$, $\delta = l(\beta)$. It is found that the polynomials $q_n(l^{[-1]}(z))$ now are orthogonal with respect to the transformed weight function $\omega(t)$ on $[\alpha, \delta]$. For several widely used weight functions a set of corresponding orthogonal polynomials $\hat{q}_n$ is well known. Because there exists, up to scalar factors, only one set of orthogonal polynomials for a given ω, we have $\hat{q}_n = \kappa_n q_n(l^{[-1]}(z))$ with certain nonzero factors κ_n. Depending on the normalization used, the $\hat{q}_n$ in place of (14.6-7) will in general satisfy a slightly more general relation

$$\hat{q}_n(z) = (\gamma_n z + \beta_n)\hat{q}_{n-1}(z) - \alpha_n \hat{q}_{n-2}(z), \qquad n = 1, 2, \ldots. \quad (14.6\text{-}20)$$

The principal value integrals

$$\hat{g}_n(x) := \mathrm{PV} \int_\alpha^\delta \frac{q_n(t)\omega(t)}{t - x}\, dt$$

then satisfy the same recurrence relation,

$$\hat{g}_n(x) = (\gamma_n x + \beta_n)\hat{g}_{n-1}(x) - \alpha_n \hat{g}_{n-2}(x), \qquad n = 2, 3, \ldots. \quad (14.6\text{-}21)$$

For $n = 1$, we have

$$\begin{aligned} \hat{g}_1(x) &= \mathrm{PV} \int_\alpha^\delta \frac{\hat{q}_1(t)\omega(t)}{t - x}\, dt \\ &= \mathrm{PV}\, \hat{q}_0 \int_\alpha^\delta \frac{(\gamma_1 t + \beta_1)\omega(t)}{t - x}\, dt \\ &= \hat{q}_0\, \mathrm{PV} \int_\alpha^\delta \frac{\gamma_1(t - x) + \gamma_1 x + \beta_1}{t - x}\, \omega(t)\, dt \\ &= \hat{q}_0 \left\{ \gamma_1 \int_\alpha^\delta \omega(t)\, dt + (\gamma_1 x + \beta_1) f(x) \right\}, \end{aligned}$$

which in view of

$$\int_\alpha^\delta \omega(t)\, dt = \mu_0$$

simplifies to

$$\hat{g}_1(x) = \hat{q}_0 \gamma_1 \mu_0 + \hat{q}_0(\gamma_1 x + \beta_1) f(x).$$

Thus (14.6-21) requires the initial conditions

$$\hat{g}_0(x) = \hat{q}_0 f(x), \quad \hat{g}_1(x) = \hat{q}_0 \gamma_1 \mu_0 + \hat{q}_0(\gamma_1 x + \beta_1) f(x), \quad (14.6\text{-}22)$$

where

$$f(x) = \mathrm{PV} \int_\alpha^\delta \frac{\omega(t)}{t - x}\, dt.$$

EXAMPLE 1

Let $[\alpha, \delta] = [-1, 1]$, $\omega(t) = 1$. A set of orthogonal polynomials is then formed by the *Legendre polynomials*

$$P_n(z) := F\left(-n, n+1; 1; \frac{1 - z}{2}\right),$$

which satisfy the recurrence relation $P_0(z)=1$,

$$P_n(z)=\left(2-\frac{1}{n}\right)zP_{n-1}(z)-\left(1-\frac{1}{n}\right)P_{n-2}(z), \qquad n=1,2,\ldots.$$

It follows that the functions

$$Q_n(z):=\int_{-1}^{1}\frac{P_n(t)}{t-z}\,dt, \qquad n=0,1,2,\ldots$$

(called *Legendre functions of the second kind*) likewise satisfy

$$Q_n(z)=\left(2-\frac{1}{n}\right)zQ_{n-1}(z)-\left(1-\frac{1}{n}\right)Q_{n-2}(z), \tag{14.6-23a}$$

subject to the initial values

$$\begin{aligned} Q_0(z)&=f(z)=\int_{-1}^{1}\frac{1}{t-z}\,dt=\operatorname{Log}\frac{z-1}{z+1},\\ Q_1(z)&=2+zf(z). \end{aligned} \tag{14.6-23b}$$

Thus if a function $\phi\in\text{Lip}$ is expanded in a series of Legendre polynomials,

$$\phi(t)=\sum_{n=0}^{\infty}\lambda_n P_n(t),$$

and if, for $-1<x<1$,

$$g(x):=\text{PV}\int_{-1}^{1}\frac{\phi(t)}{t-x}\,dt,$$

then (assuming that termwise singular integration is permissible)

$$g(x)=\sum_{n=0}^{\infty}\lambda_n Q_n(x),$$

where the Q_n may be calculated in a stable manner from (14.6-23a), $z:=x$, subject to the initial conditions

$$\begin{aligned} Q_0(x)&:=\text{PV}\int_{-1}^{1}\frac{1}{t-x}\,dt=\operatorname{Log}\frac{1-x}{1+x},\\ Q_1(x)&:=2+x\operatorname{Log}\frac{1-x}{1+x}. \end{aligned}$$

V. Reduction to Hilbert Transform

The principal value integral

$$g(x):=\frac{1}{\pi}\text{PV}\int_{-1}^{1}\frac{h(t)}{t-x}\,dt, \tag{14.6-24}$$

where $h \in \text{Lip}\,\alpha$, $0<\alpha<1$, may be reduced to a Hilbert transform by a clever transformation of variables. This reduction is possible even in two ways. In either case, we start with

$$t=\cos\tau, \qquad x=\cos\xi,$$

where τ and ξ are selected in $[0, \pi]$. In view of $dt=-\sin\tau\, d\tau$, we then get

$$g(\cos\xi)=\frac{1}{\pi}\,\text{PV}\int_0^{\pi}\frac{h(\cos\tau)\sin\tau}{\cos\tau-\cos\xi}\,d\tau. \tag{14.6-25}$$

We next recall the trigonometric identities

$$\begin{aligned}\frac{2\sin\tau}{\cos\tau-\cos\xi}&=\cot\frac{\xi-\tau}{2}-\cot\frac{\xi+\tau}{2},\\ \frac{2\sin\xi}{\cos\tau-\cos\xi}&=\cot\frac{\xi-\tau}{2}+\cot\frac{\xi+\tau}{2}.\end{aligned} \tag{14.6-26}$$

Using the first identity, the integral (14.6-25) becomes

$$g(\cos\xi)=\frac{1}{2\pi}\,\text{PV}\int_0^{\pi}\phi_0(\tau)\left[\cot\frac{\xi-\tau}{2}-\cot\frac{\xi+\tau}{2}\right]d\tau, \tag{14.6-27}$$

where

$$\phi_0(\tau):=h(\cos\tau), \qquad 0<\tau<\pi. \tag{14.6-28}$$

We continue ϕ_0 as an *odd* function of period 2π:

$$\phi_0(-\tau):=-\phi_0(\tau), \qquad \phi_0(\tau+2\pi):=\phi_0(\tau).$$

The function ϕ_0 thus continued is in Lip α, provided

$$h(1)=h(-1)=0. \tag{14.6-29}$$

The integral (14.6-27) then becomes

$$g(\cos\xi)=\frac{1}{2\pi}\,\text{PV}\int_0^{2\pi}\phi_0(\tau)\cot\frac{\xi-\tau}{2}\,d\tau, \tag{14.6-30}$$

which is to say that $\psi_0=\mathscr{H}\phi_0$, where

$$\psi_0(\xi):=g(\cos\xi). \tag{14.6-31}$$

This ψ_0 thus defined is *even*, and $\psi_0\in\text{Lip}\,\alpha$ by Theorem 14.2d.

To use (14.6-30) numerically, we expand ϕ_0 in a *sine* series,

$$\phi_0(\tau)=\text{sgn}\,\tau\cdot h(\cos\tau)=\sum_{k=1}^{\infty}\beta_k\sin k\tau, \tag{14.6-32}$$

and in view of $\mathscr{H} \sin k\tau = -\cos k\tau$, we obtain

$$\psi_0(\xi) = g(\cos \xi) = -\sum_{k=1}^{\infty} \beta_k \cos k\xi. \qquad (14.6\text{-}33)$$

By means of an FFT, approximate values of $\beta_1, \beta_2, \ldots, \beta_n$ using $2n$ data points are computed in $O(n \operatorname{Log} n)$ operations, and the same device permits the evaluation of ψ_0 at $2n$ equidistant points in a similar number of operations.

If condition (14.6-29) is violated, the convergence of (14.6-32), and hence of (14.6-33), will be very slow. It is then better to use the second identity (14.6-26). This yields

$$g(\cos \xi) = \frac{1}{2\pi \sin \xi} \text{PV} \int_0^{\pi} \phi_1(\tau) \left\{ \cot \frac{\xi - \tau}{2} + \cot \frac{\xi + \tau}{2} \right\} d\tau, \qquad (14.6\text{-}34)$$

where

$$\phi_1(\tau) := h(\cos \tau) \cdot \sin \tau, \qquad 0 \leqq \tau \leqq \pi.$$

Continuing ϕ_1 as an *even* function of period 2π,

$$\phi_1(-\tau) := \phi_1(\tau), \qquad \phi_1(\tau + 2\pi) := \phi_1(\tau),$$

which is in Lip α regardless of whether (14.6-29) holds, this becomes

$$g(\cos \xi) = \frac{1}{2\pi \sin \xi} \text{PV} \int_0^{2\pi} \phi_1(\tau) \cot \frac{\xi - \tau}{2} \, d\tau. \qquad (14.6\text{-}35)$$

This is to say that $\psi_1 = \mathscr{H}\phi_1$, where

$$\psi_1(\xi) := \sin \xi \cdot g(\cos \xi).$$

The function ψ_1 thus defined is *odd*, $\psi_1 \in \text{Lip } \alpha$.

To use (14.6-35) numerically, we expand

$$\phi_1(\tau) = |\sin \tau| h(\cos \tau) = \sum_{k=0}^{\infty} \alpha_k \cos k\tau$$

and obtain

$$g(\cos \xi) = \frac{\psi_1(\xi)}{\sin \xi} = \frac{1}{\sin \xi} \sum_{k=1}^{\infty} \alpha_k \cos k\xi.$$

Using $2n$ sampling values of ϕ_1, the corresponding values of g may again be computed in $O(n \operatorname{Log} n)$ operations.

In summary, we have obtained:

THEOREM 14.6c. *Let h be in* Lip α *on* $[-1, 1]$, *and let g be defined by* (14.6-24). *Then* $\psi_i = \mathcal{H}\phi_i$, $i = 0, 1$, *where*

$$\phi_0(\tau) := \operatorname{sgn} \tau \cdot h(\cos \tau), \qquad \psi_0(\xi) = g(\cos \xi)$$

and

$$\phi_1(\tau) := h(\cos \tau)|\sin \tau|, \qquad \psi_1(\xi) = g(\cos \xi) \cdot \sin \xi.$$

These relations may be reversed. Let ϕ be a 2π-periodic function which is either even or odd, $\phi \in \text{Lip } \alpha$, and let $\psi := \mathcal{H}\phi$. Let

$$\begin{aligned} h(t) &:= \phi(\arccos t), \qquad -1 \leqq t \leqq 1, \quad 0 \leqq \arccos t \leqq \pi, \\ g(x) &:= \psi(\arccos x), \qquad -1 \leqq X \leqq \perp, \quad 0 \leqq \arccos x \leqq \pi. \end{aligned} \tag{14.6-36}$$

Since $\arccos \in \text{Lip } \frac{1}{2}$, the functions h and g thus defined are at least in Lip $(\alpha/2)$. If ϕ is *odd*, the above transformation yields

$$g(x) = \frac{1}{\pi} \text{ PV} \int_{-1}^{1} \frac{h(t)}{t - x} \, dt. \tag{14.6-37}$$

If ϕ is *even*, we obtain in a similar manner

$$g(x) = \frac{1}{\pi} \sqrt{1 - x^2} \text{ PV} \int_{-1}^{1} \frac{h(t)}{\sqrt{1 - t^2}(t - x)} \, dt. \tag{14.6-38}$$

PROBLEMS

1. In the framework of Example 1, let

$$\phi(\tau) := \frac{1}{\sqrt{1 - 2t\tau + t^2}},$$

where t is a parameter, $-1 < t < 1$.

(a) Show that

$$\phi(\tau) = \sum_{n=0}^{\infty} t^n P_n(\tau)$$

(b) If

$$g(\xi) := \text{PV} \int_{-1}^{1} \frac{1}{\sqrt{1 - 2t\tau + t^2}} \frac{1}{\tau - \xi} \, d\tau,$$

conclude that

$$g(\xi) = \sum_{n=0}^{\infty} t^n Q_n(\xi). \tag{*}$$

(c) Demonstrate the numerical stability of Algorithm 14.6b by comparing the values obtained by summing (*) with the exact values

$$g(\xi) = \frac{1}{\xi_0} \log \frac{1-t+\xi_0}{1+t+\xi_0} \frac{1+t-\xi_0}{\xi_0+t-1},$$

where $\xi_0 := (1-2t\xi+t^2)^{1/2}$.

2. Here we consider the interval $[-1, 1]$ with the weight function

$$\omega(\tau) := \pi^{-1}(1-\tau^2)^{-1/2}.$$

(a) Show that a set of orthogonal polynomials is provided by the Chebyshev polynomials

$$\begin{aligned} T_n(z) &:= \cos[n \arccos z] \\ &= \tfrac{1}{2}\{(z+\sqrt{z^2-1})^n + (z-\sqrt{z^2-1})^n\} \\ &= F\left(-n, n; \frac{1}{2}; \frac{1-z}{2}\right). \end{aligned}$$

(b) From the fact that

$$T_n(z) = 2zT_{n-1}(z) - T_{n-2}(z), \qquad n \geqq 2,$$

show that the same recurrence relation is satisfied by the functions

$$U_n(z) := \frac{1}{\pi} \int_{-1}^{1} \frac{T_n(\tau)}{\sqrt{1-\tau^2}(\tau - z)} \, d\tau.$$

(c) Show that, for $z \notin [-1, 1]$,

$$U_0(z) = -\frac{1}{\sqrt{z^2-1}}, \qquad U_1(z) = 1 - \frac{z}{\sqrt{z^2-1}},$$

and conclude that

$$U_n(z) = -\frac{(z-\sqrt{z^2-1})^n}{\sqrt{z^2-1}}, \qquad n = 0, 1, \ldots.$$

(d) Show that the principal value integrals

$$U_n(\xi) := \frac{1}{\pi} \mathrm{PV} \int_{-1}^{1} \frac{T_n(\tau)}{\sqrt{1-\tau^2}(\tau-\xi)} \, d\tau,$$

where $\xi \in (-1, 1)$, satisfy $U_0(\xi) = 0$, $U_1(\xi) = 1$,

$$U_n(\xi) = 2\xi U_{n-1}(\xi) - U_{n-2}(\xi), \qquad n \geqq 2.$$

(e) Conclude that if

$$\phi(\tau) = \sum_{n=0}^{\infty} \lambda_n T_n(\tau), \qquad -1 \leqq \tau \leqq 1,$$

then

$$g(\xi) := \frac{1}{\pi} \text{PV} \int_{-1}^{1} \frac{\phi(\tau)}{\sqrt{1-\tau^2}(\tau-\xi)} d\tau = \sum_{n=1}^{\infty} \lambda_n U_n(\xi).$$

(f) What is the connection with the method of inverting the Hilbert transform by means of Fourier series discussed in Section V?

3. The Legendre polynomials

$$P_n(x) = F\left(-n, n+1; 1; \frac{1-x}{2}\right)$$

satisfy the differential equation

$$[(1-x^2)y']' + n(n+1)y = 0.$$

Show that the same differential equation is satisfied by the functions

$$Q_n(x) = \text{PV} \int_{-1}^{1} \frac{P_n(t)}{t-x} dt.$$

By discussing the behavior of $Q_n(\xi)$ at ∞, express $Q_n(z)$ as a Legendre function of the second kind.

4. For $\alpha > -1, \beta > -1$ let

$$\omega(\tau) := \tau^{\alpha}(1-\tau)^{\beta}, \qquad 0 < \tau < 1,$$

$\omega(\tau) = 0$ elsewhere. Show that the moments of $\omega(\tau)\, d\tau$ are

$$\mu_n = \frac{\Gamma(\alpha+\beta+2)\Gamma(\beta+1)}{\Gamma(\alpha+1)} \frac{(\alpha+1)_n}{(\alpha+\beta+2)_n},$$

and conclude that for the resulting orthogonal polynomials $q_n(z)$ the relations (14.6-7) hold where $\alpha_1 = \mu_0$,

$$\alpha_{k+1} = \frac{k(\alpha+k)(\beta+k)(\alpha+\beta+k)}{(\alpha+\beta+2k-1)(\alpha+\beta+2k)(\alpha+\beta+2k+1)}, \qquad k = 1, 2, \ldots,$$

$$\beta_{k+1} = \frac{1}{2}\left\{1 + \frac{\alpha^2-\beta^2}{(\alpha+\beta+2k)(\alpha+\beta+2k+2)}\right\}, \qquad k = 0, 1, 2, \ldots.$$

Also show that

$$q_n(z) = (-1)^n \frac{n!}{(\alpha+\beta+n+1)_n} P_n^{(\alpha,\beta)}(1-2z),$$

where $P_n^{(\alpha,\beta)}$ is the Jacobi polynomial (see (12.12-3)), or equivalently,

$$q_n(z) = (-1)^n \frac{(\alpha+1)_n}{(\alpha+\beta+n+1)_n} F(-n, \alpha+\beta+n+1; \alpha+1; z).$$

5. Prove Theorem 14.6a without continued fraction theory by writing

$$g_n(\xi) = \text{PV}\int_0^\infty \frac{q_n(\tau)}{\tau-\xi}\,\omega(\tau)\,d\tau$$
$$= \text{PV}\int_0^\infty \frac{(\tau-\xi+\xi-\beta_n)q_{n-1}(\tau)-\alpha_n q_{n-2}(\tau)}{\tau-\xi}\,\omega(\tau)\,d\tau$$

and using the fact that, by virtue of orthogonality,

$$\int_0^\infty q_{n-1}(\tau)\omega(\tau)\,d\tau = 0, \qquad n>1.$$

6. Extend the method described in Algorithm 14.6b to Cauchy integrals of the form

$$g(\xi) = \int_0^\infty \frac{\phi(\tau)\omega(\tau)}{\tau-\xi}\,d\tau,$$

where ω is such that

$$\mu_n := \int_0^\infty \tau^n\omega(\tau)\,d\tau < \infty$$

for all n. (Example: $\omega(\tau)=e^{-\tau}$.)

7. For $\alpha>-1$, let

$$\omega(\tau) := \tau^\alpha e^{-\tau}, \qquad \tau>0.$$

Show that for z not real and positive,

$$f(z) := \int_0^\infty \frac{\omega(\tau)}{\tau-z}\,d\tau = \Gamma(\alpha)\,{}_1F_1(1;1-\alpha;-z) - \frac{\pi}{\sin\pi\alpha}(-z)^\alpha e^{-z}.$$

(For integers α this and subsequent expressions must be replaced by their limits.) Show that in the recurrence relations (14.6-7) for the orthogonal polynomials q_n,

$$\alpha_{k+1} = k(\alpha+k), \qquad \beta_{k+1} = \alpha+2k+1, \qquad k=0,1,2,\ldots.$$

Also show that

$$q_n(z) = (-1)^n n!\,L_n^{(\alpha)}(z),$$

where $L_n^{(\alpha)}$ is a Laguerre polynomial,

$$L_n^{(\alpha)}(z) = \frac{(\alpha+1)_n}{n!}\,{}_1F_1(-n;\alpha+1;z).$$

From the fact that the Laguerre polynomials satisfy the recurrence relation

$$L_n^{(\alpha)}(z) = \frac{1}{n}\{(2n+\alpha-1-z)L_{n-1}^{(\alpha)}(z) - (\alpha+1-n)L_{n-2}^{(\alpha)}(z)\},$$

conclude that the same recurrence relation is satisfied by the functions

$$G_n^{(\alpha)}(\xi) := \mathrm{PV} \int_0^\infty \frac{\tau^\alpha e^{-\tau} L_n^{(\alpha)}(\tau)}{\tau - \xi} \, d\tau, \qquad n = 0, 1, 2, \ldots.$$

NOTES

On orthogonal polynomials, see Szegö [1959], where in ch. XIII a number of "equiconvergence" results are given. The literature on the numerical evaluation of principal value integrals is large. For surveys, see Rabinowitz [1978] or Gautschi [1981], § 3.2.4. The method described here is similar to that proposed by Paget and Elliott [1972] and Elliot and Paget [1975]. Longman [1958] proposes to convert a principal value integral into an ordinary improper integral by the device discussed in § 14.1.

§ 14.7. CAUCHY INTEGRALS ON ARCS: BEHAVIOR NEAR ENDPOINTS

Our next major task is to discuss the existence and uniqueness of the solution of the problem of Privalov (see § 14.4) if Γ is an arc. While solutions will be shown to exist always, these solutions are never unique unless their behavior near the endpoints of Γ is specified. For this reason we now turn to a discussion of the behavior of a Cauchy integral $f(z)$ near a terminal point of Γ. This discussion is more difficult than might be expected, for two reasons:

(a) Because not only the behavior of f but also that of f^{-1} is of interest, the behavior of f must be described by means of asymptotic formulas. Mere bounds on $|f(z)|$ are not sufficient.
(b) The Privalov problem requires us to deal with functions $h(t)$ which themselves are singular at the endpoints of Γ.

Our discussion is carried on by means of examples which from simple special situations lead up to the general case.

EXAMPLE 1

Let Γ: $t = \tau$, $0 \leqq \tau \leqq 1$ (see Fig. 14.7a), $h(t) := 1$. If $|\arg(-z)| < \pi$, the Cauchy integral

$$f(z) = \frac{1}{2\pi i} \int_\Gamma \frac{h(t)}{t - z} \, dt = \frac{1}{2\pi i} \int_0^1 \frac{1}{\tau - z} \, d\tau$$

is immediately evaluated as

$$f(z) = \frac{1}{2\pi i} \{\mathrm{Log}(1 - z) - \mathrm{Log}(-z)\},$$

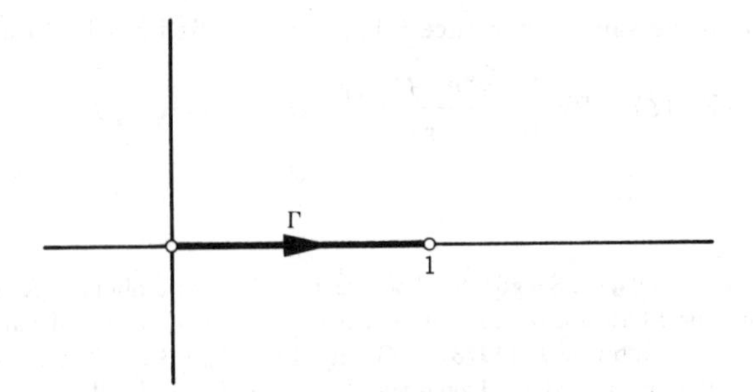

Fig. 14.7a. Simple arc.

and we evidently have

$$f(z) = -\frac{1}{2\pi i}\,\mathrm{Log}(-z) + g(z), \tag{14.7-1}$$

where g is analytic at 0.

EXAMPLE **2**

Let Γ: $t = t(\tau)$, $\alpha \leqq \tau \leqq \beta$, now be any regular simple arc that starts at $t = 0$ and initially proceeds in the direction of the positive real axis ($t(\alpha) = 0$, $t'(\alpha) > 0$, see Fig. 14.7b). We again let $h(t) := 1$.

We decompose Γ into two arcs, $\Gamma = \Gamma_1 + \Gamma_2$, such that Γ_1 starts out at $t = 0$ and lies entirely in $\mathrm{Re}\, z \geqq 0$. Let Γ_3 be the straight line segment which joins the terminal point of Γ_1 to $t = 1$, and let Γ_0 be the straight line arc

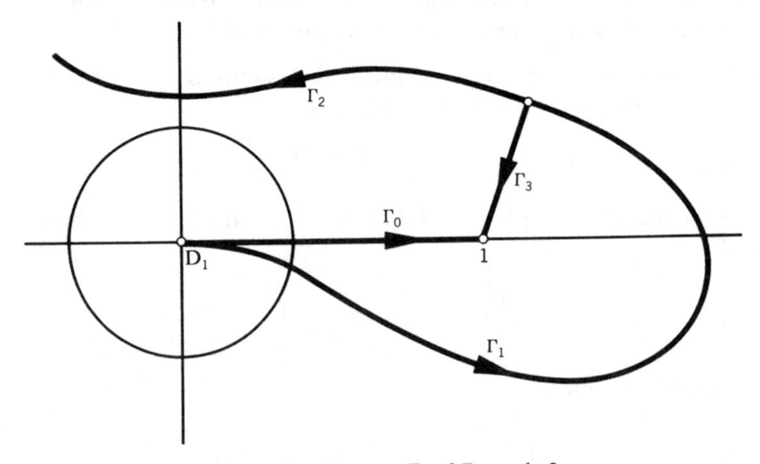

Fig. 14.7b. The arcs Γ_i of Example 2.

from $t=0$ to $t=1$. For z not on Γ_k let

$$f_k(z) := \frac{1}{2\pi i}\int_{\Gamma_k} \frac{1}{t-z}\,dt. \tag{14.7-2}$$

For fixed z such that $\operatorname{Re} z<0$, the function

$$t \to \frac{1}{t-z}$$

is analytic in $\operatorname{Re} t \geqq 0$. Because $\Gamma_1+\Gamma_3$ is homotopic to Γ_0 in $\operatorname{Re} t \geqq 0$, one form of Cauchy's theorem yields

$$f_0(z)=f_1(z)+f_3(z), \qquad \operatorname{Re} z<0.$$

There follows for the same z

$$f(z)=f_1(z)+f_2(z)=f_0(z)-f_3(z)+f_2(z)$$

or by Example 1, because f_2 and f_3 are analytic at 0,

$$f(z)=-\frac{1}{2\pi i}\operatorname{Log}(-z)+g(z), \qquad \operatorname{Re} z<0, \tag{14.7-3}$$

where g is analytic at 0.

Let now D: $|z|<\delta$ be a disk so small that it does not intersect Γ_2 and Γ_3, and that its boundary has precisely one point in common with Γ_1. The region

$$D_1 := D\backslash\Gamma_1$$

then is simply connected, and because $-z\neq 0$ in D_1, a single-valued branch of

$$s_0(z) := -\frac{1}{2\pi i}\log(-z) \tag{14.7-4}$$

can be defined in D_1 (Theorem 4.3f), which agrees with $-(1/2\pi i)\operatorname{Log}(-z)$ for $\operatorname{Re} z<0$. Thus all functions in (14.7-3) can be extended to functions analytic in D_1, and the principle of analytic continuation implies that

$$f(z)=s_0(z)+g(z), \qquad z\in D_1. \tag{14.7-5}$$

Since g is analytic, the singular behavior of f near the initial point $z=0$ of Γ thus is described by $s_0(z)$. We shall see that the behavior in the general case, where h is any function in Lip γ, is essentially the same.

Let now $t\in D\cap\Gamma_1$, $t\neq 0$. We know from the Sokhotskyi formulas that $f(t)$, defined as a principal value, satisfies

$$f(t)=\tfrac{1}{2}\{f^+(t)+f^-(t)\}.$$

Thus from (14.7-5),

$$f(t)=\tfrac{1}{2}\{s_0^+(t)+s_0^-(t)\}+g(t).$$

From the definition of s_0,

$$s_0^\pm(t)=-\frac{1}{2\pi i}[\operatorname{Log} t \mp i\pi],$$

thus for $t\in\Gamma$, t sufficiently close to 0, the asymptotic behavior of f is described by the formula

$$f(t)=-\frac{1}{2\pi i}\operatorname{Log} t+g(t), \tag{14.7-6}$$

with the same g as in (14.7-5), which is an even more explicit description than (14.7-5).

EXAMPLE 3

Here we let Γ as in Example 2, but h now is any function in Lip γ, $0<\gamma<1$,

$$|h(t_1)-h(t_2)|\leqq \mu|t_1-t_2|^\gamma \tag{14.7-7}$$

for any t_1, $t_2\in\Gamma$. In order to study

$$f(z)=\frac{1}{2\pi i}\int_\Gamma \frac{h(t)}{t-z}\,dt,$$

we naturally write

$$h(t)=h(0)+h(t)-h(0)$$

and accordingly define

$$f_1(z):=\frac{1}{2\pi i}\int_\Gamma \frac{h(0)}{t-z}\,dt,$$

$$f_2(z):=\frac{1}{2\pi i}\int_\Gamma \frac{h_0(t)}{t-z}\,dt,$$

where

$$h_0(t):=h(t)-h(0)$$

again satisfies (14.7-7). By Example 2,

$$f_1(z)=h(0)s_0(z)+g(z),$$

where g is analytic at 0. We shall show that, apart from a constant, this represents the main contribution to f.

In order to study f_2, we extend Γ backward along the axis of negative reals so that $z=0$ becomes an interior point of Γ. If $h_0(t):=0$ on the extended part of the arc, then h_0 is still in Lip γ, and there still holds

$$f_2(z)=\frac{1}{2\pi i}\int_\Gamma \frac{h_0(t)}{t-z}\,dt,$$

where Γ now denotes the extended arc.

For any z, let now z_0 denote a point on Γ closest to z. Because Γ is regular, z_0 is unique if z is sufficiently close to Γ and thus, in particular, if $|z|$ is sufficiently small (see Fig. 14.7c).

Writing $h_0(t)=h_0(z_0)+h_0(t)-h_0(z_0)$, we now have

$$f_2(z)=f_{21}(z)+f_{22}(z),$$

where

$$f_{21}(z):=\frac{1}{2\pi i}h_0(z_0)\int_\Gamma \frac{1}{t-z}\,dt,$$

$$f_{22}(z):=\frac{1}{2\pi i}\int_\Gamma \frac{h_0(t)-h_0(z_0)}{t-z}\,dt.$$

From $|z-z_0|\leqq|z-0|=|z|$ there follows $|z_0|\leqq 2|z|$. Hence, using Hölder continuity and the fact that $h_0(0)=0$,

$$|h_0(z_0)|\leqq \mu|2z|^\gamma,$$

and there follows

$$|f_{21}(z)|\leqq \text{const}|z|^\gamma|s_0(z)|=O(|z|^{\gamma-\varepsilon}), \qquad z\to 0,$$

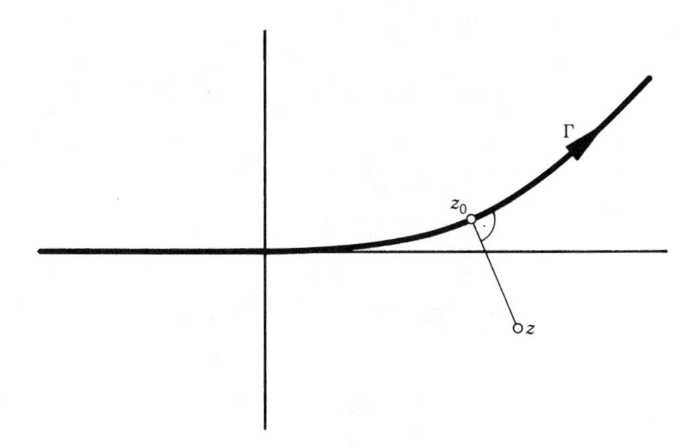

Fig. 14.7c. The point z_0.

for every $\varepsilon > 0$. As for $f_{22}(z)$, we write

$$f_{22}(z) = f_{22}(z_0) + d(z),$$

where

$$\begin{aligned} d(z) &= f_{22}(z) - f_{22}(z_0) \\ &= \frac{1}{2\pi i}\int_\Gamma [h_0(t) - h_0(z_0)]\left\{\frac{1}{t-z} - \frac{1}{t-z_0}\right\} dt. \end{aligned}$$

Concerning $f_{22}(z_0)$ we know from the proof of Theorem 14.1b (where the same function was denoted by $g(z_0)$) that it is in Lip γ on every interior subarc of Γ. Hence because 0 now is an interior point,

$$f_{22}(z_0) = f_{22}(0) + O(|z|^\gamma), \qquad z \to 0.$$

To estimate $d(z)$, let $|z|$ be so small that the circle of radius $3|z|$ around 0 intersects Γ in precisely two points. Let Γ_1 and Γ_2 denote those parts of Γ which lie, respectively, outside and inside the circle, and let

$$d(z) = d_1(z) + d_2(z),$$

where

$$d_k(z) := \frac{1}{2\pi i}\int_{\Gamma_k} \frac{h_0(t) - h_0(z_0)}{t - z_0}\,\frac{z - z_0}{t - z}\,dt.$$

If $t \in \Gamma_1$, we have $|t| \geqq 3|z|$. Consequently,

$$|t - z| \geqq |t| - |z| \geqq \tfrac{2}{3}|t|,$$

and on the other hand, because $|z_0| \leqq 2|z|$,

$$|t - z_0| \geqq |t| - 2|z| \geqq \tfrac{1}{3}|t|.$$

Thus using the Hölder condition (where $\gamma < 1$ without loss of generality),

$$\left|\frac{h_0(t) - h_0(z_0)}{t - z_0}\right| \leqq \mu|t - z_0|^{\gamma-1} \leqq \mu\left(\frac{1}{3}|t|\right)^{\gamma-1},$$

hence

$$\begin{aligned} |d_1(z)| &\leqq \frac{1}{2\pi}\int_{\Gamma_1} \mu\left(\frac{1}{3}|t|\right)^{\gamma-1} \cdot \frac{3}{2}\,\frac{|z - z_0|}{|t|}\,|dt| \\ &= \text{const}\,|z - z_0| \int_{\Gamma_1} |t|^{\gamma-2}\,|dt| \\ &\leqq \text{const}\,|z - z_0| \int_{3|z|}^{\infty} \sigma^{\gamma-2}\,d\sigma \\ &= \text{const}\,|z - z_0|\,|z|^{\gamma-1}, \end{aligned}$$

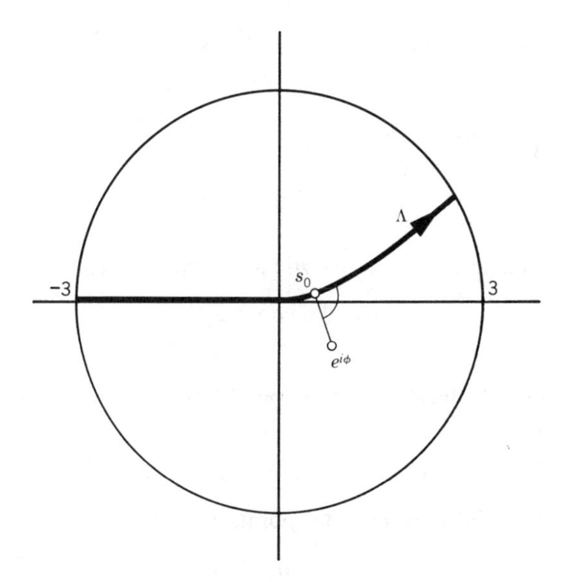

Fig. 14.7d. Notations to estimate d_2.

implying

$$d_1(z) = O(|z|^\gamma), \qquad z \to 0.$$

To estimate d_2, we let $t = s|z|$, $z_0 = s_0|z|$, $z = e^{i\phi}|z|$, $\Gamma_2 = |z|\Lambda$. The arc Λ then begins and ends on the circle $|s| = 3$; as $z \to 0$, it approaches the segment $[-3, 3]$ (see Fig. 14.7d). We then have

$$d_2(z) = \frac{1}{2\pi i} \int_\Lambda \frac{h_0(s|z|) - h_0(s_0|z|)}{s - s_0} \frac{e^{i\phi} - s_0}{s - e^{i\phi}} \, ds.$$

Thus using the Hölder condition,

$$|d_2(z)| \leqq \frac{1}{2\pi} \mu |z|^\gamma \int_\Lambda |s - s_0|^{\gamma - 1} \left| \frac{e^{i\phi} - s_0}{e^{i\phi} - s} \right| |ds|.$$

The ratio under the integral is $\leqq 1$, because s_0 is the point on Λ closest to $e^{i\phi}$, and there follows

$$|d_2(z)| \leqq \text{const} |z|^\gamma \int_\Lambda |s - s_0|^{\gamma - 1} |ds|$$

or, since the integral is bounded by a constant depending only on γ,

$$d_2(z) = O(|z|^\gamma), \qquad z \to 0.$$

Collecting our results, there follows for Example 3

$$f(z)=h(0)s_0(z)+g(z)+r(z), \tag{14.7-8}$$

where $s_0(z)$ is the branch of

$$-\frac{1}{2\pi i}\log(-z)$$

described in Example 2, g is analytic at 0, and r satisfies

$$|r(z)|\leqq \text{const}\,|z|^{\gamma-\varepsilon}$$

for every $\varepsilon>0$.

After Example 3, the general case is dealt with easily by a transformation of variables. Let

$$\Gamma:\quad t=t(\tau),\qquad \tau_0\leqq\tau\leqq\tau_1,$$

be an arc with initial and terminal points $t_k := t(\tau_k)$, $k=0, 1$, and let

$$\psi_k := \arg t'(\tau_k)$$

be the angle between the tangent vectors at the endpoints and the axis of positive reals (see Fig. 14.7e).

We wish to study the endpoint behavior of

$$f(z) := \frac{1}{2\pi i}\int_\Gamma \frac{h(t)}{t-z}\,dt, \tag{14.7-9}$$

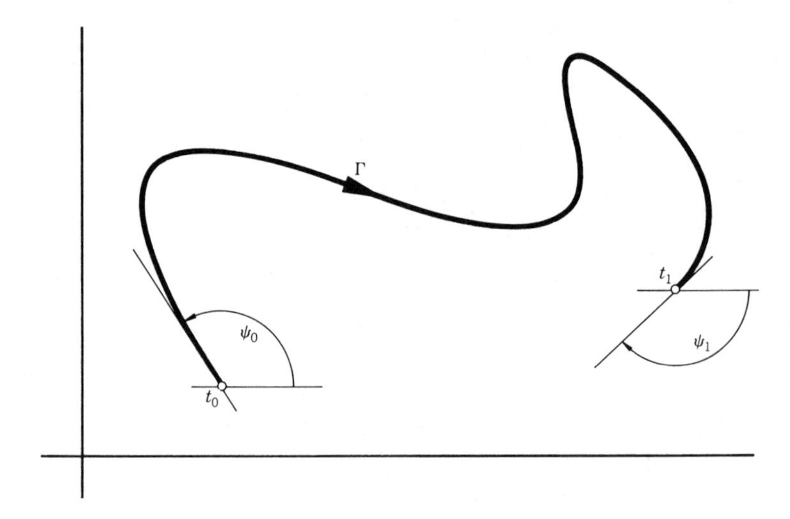

Fig. 14.7e. Notations for Theorem 14.7a.

assuming that h is in Lip γ, $0<\gamma<1$. For the initial point this is reduced to Example 3 by the transformation of variables

$$z = t_0 + e^{i\psi_0}\hat{z}, \qquad t = t_0 + e^{i\psi_0}\hat{t}.$$

The behavior at the terminal point is reduced to that at the initial point by replacing Γ by $-\Gamma$, which implies a change of sign in the tangent vector.

THEOREM 14.7a. *Let* Γ: $t=t(\tau)$, $\tau_0 \leqq \tau \leqq \tau_1$, *be a simple regular arc,* $t_k := t(\tau_k)$, $\psi_k := \arg t'(\tau_k)$, $k=0,1$. *Let the function h on Γ be in* Lip γ, *where* $0<\gamma<1$. *Then there exist functions g_k and r_k such that g_k is analytic at t_k and*

$$r_k(z) = O(|z-t_k|^{\gamma-\varepsilon}), \qquad z \to t_k,$$

for every $\varepsilon>0$, $k=0,1$, *such that the Cauchy integral* (14.7-9) *satisfies:*

(a) *If z is near* t_0,

$$f(z) = g_0(z) + r_0(z) + \begin{cases} h(t_0)s_0(e^{-i\psi_0}(t_0-z)), & z \notin \Gamma \\ -\dfrac{1}{2\pi i}h(t_0)\,\mathrm{Log}(e^{-i\psi_0}(z-t_0)), & z \in \Gamma. \end{cases} \tag{14.7-10a}$$

(b) *If z is near* t_1,

$$f(z) = g_1(z) + r_1(z) + \begin{cases} -h(t_1)s_0(-e^{-i\psi_1}(t_1-z)), & z \notin \Gamma \\ h(t_1)\dfrac{1}{2\pi i}\,\mathrm{Log}(-e^{-i\psi_1}(z-t_1)), & z \in \Gamma. \end{cases} \tag{14.7-10b}$$

It remains to discuss Cauchy integrals where the function h itself is singular at the endpoints of Γ. The singularities to be considered are of the type

$$h(t) = (t-t_k)^{-a}h^*(t),$$

where $a=\alpha+i\beta \neq 0$, $0 \leqq \alpha < 1$, and where h^* is in Lip γ, $0<\gamma<1$. We again proceed by examples.

EXAMPLE **4**

Γ is as in Example 1, $h^*(t)=1$. Thus

$$f(z) = \frac{1}{2\pi i}\int_0^1 \frac{\tau^{-a}}{\tau - z}\,d\tau.$$

For convenience we consider

$$f(-z) = \frac{1}{2\pi i}\int_0^1 \frac{\tau^{-a}}{\tau + z}\,d\tau, \tag{14.7-11}$$

where $|\arg z| < \pi$. The integral on the right could be studied by means of hypergeometric functions. However, the following elementary approach is equally instructive. Expanding in powers of z^{-1} and integrating term by term, we have for $|z| > 1$,

$$\int_0^1 \frac{\tau^{-a}}{\tau+z}\,d\tau = \frac{1}{z}\int_0^1 \tau^{-a}\sum_{n=0}^{\infty}(-1)^n\left(\frac{\tau}{z}\right)^n d\tau = \sum_{n=0}^{\infty}\frac{(-1)^n}{n+1-a}z^{-n-1}.$$

To discuss the behavior near $z=0$, we use the fact that

$$\frac{1}{n+1-a} = z^{n+1-a}\int_z^{\infty} t^{a-2-n}\,dt, \qquad n=0,1,2,\ldots,$$

where all powers have their principal values, and where the integration is to be performed along the ray $\arg t = \arg z$. This yields

$$\int_0^1 \frac{\tau^{-a}}{\tau+z}\,d\tau = z^{-a}\sum_{n=0}^{\infty}(-1)^n\int_z^{\infty} t^{a-2-n}\,dt,$$

and on interchanging summation and integration once more,

$$\begin{aligned}\int_0^1 \frac{\tau^{-a}}{\tau+z}\,d\tau &= z^{-a}\int_z^{\infty} t^{a-2}\frac{1}{1+t^{-1}}\,dt\\ &= z^{-a}\int_z^{\infty} t^{a-1}\frac{1}{1+t}\,dt.\end{aligned}$$

We let

$$\int_z^{\infty} = \int_0^{\infty} - \int_0^{z}$$

and note that by Cauchy's theorem the first integration on the right may be performed along the real axis. By using the well-known result (see § 4.8, Example 6)

$$\int_0^{\infty}\frac{\tau^{a-1}}{1+\tau}\,d\tau = \frac{\pi}{\sin \pi a},$$

we thus obtain

$$\int_0^1 \frac{\tau^{-a}}{\tau+z}\,d\tau = z^{-a}\left\{\frac{\pi}{\sin \pi a} - \int_0^z \frac{t^{a-1}}{1+t}\,dt\right\}.$$

This result, derived under the assumption that $|z|>1$, by analytic continuation holds for all z such that $|\arg z| < \pi$. Using the geometric series once more, there follows for $|z|<1$, $|\arg z| < \pi$,

$$z^{-a}\int_0^z \frac{t^{a-1}}{1+t}\,dt = \sum_{n=0}^{\infty}(-1)^n\frac{z^n}{n+a},$$

which is analytic at 0. We thus find altogether

$$f(z) = s_a(z) + g(z), \tag{14.7-12}$$

where g is analytic at 0, and where

$$s_a(z) := \frac{1}{2\pi i} \frac{\pi}{\sin \pi a} (-z)^{-a}, \tag{14.7-13}$$

the power being defined by the principal value of $\arg(-z)$.

EXAMPLE **5**

Γ is as in Example 2, $h^*(t) = 1$. The technique of Example 2 is applicable and shows that (14.7-12) still holds, except that $s_a(z)$ is now defined by analytic continuation of the function (14.7-13) from $\operatorname{Re} z < 0$ into $D \backslash \Gamma$, where D denotes a sufficiently small disk about $z = 0$. For $z \in \Gamma$, $|z|$ sufficiently small, $z \neq 0$, we have by the Sokhotskyi formula

$$f(z) = \frac{1}{2\pi i} \frac{\pi}{\sin \pi a} z^{-a} + g(z), \tag{14.7-14}$$

where z^{-a} has its principal value.

EXAMPLE **6**

Γ is as in Example 5,

$$h(t) = t^{-a} h^*(t),$$

where $a = \alpha + i\beta \neq 0$, $0 \leqq \alpha < 1$, and where h^* is in Lip γ, $0 < \gamma < 1$. We assume that t^{-a} is initially defined by its principal value, and thereafter by continuous continuation.

The technique of example 3 literally applies if $\alpha = 0$, because then

$$t^{-a} = t^{-i\beta} = e^{-i\beta \log t},$$

and consequently

$$|t^{-a}| = e^{\beta \arg t}$$

is bounded on Γ. For z not on Γ there results

$$f(z) = h^*(0) s_{i\beta}(z) + g(z) + r(z), \tag{14.7-15}$$

where g is analytic at 0, and where

$$r(z) = O(|z|^{\gamma}), \qquad z \to 0. \tag{14.7-16}$$

The technique of Example 3 essentially applies also in the case $0 < \alpha < 1$. The sole gap occurs in the estimation of $f_{22}(z_0)$, where the method used in proving Theorem 14.1b may be applied to show that

$$f_{22}(z_0) = O(|z|^{-\alpha+\gamma}), \qquad z \to 0.$$

The remaining estimates can be carried through as before and yield the asymptotic formula

$$f(z) = g(z) + h^*(0) s_a(z)\{1 + r(z)\}, \tag{14.7-17}$$

where g is analytic at 0, and where r satisfies (14.7-16).

The case of a general arc is reduced to Example 6 by a trivial transformation of variables. The results are summarized in the following theorem:

THEOREM 14.7b. *Let* Γ *be as in Theorem* 14.7a, *and let*

$$h(t) = (t - t_0)^{-a} h^*(t), \tag{14.7-18}$$

where $a = \alpha + i\beta \neq 0$, $0 \leqq \alpha < 1$, *where, for* t *close to* t_0, $(t - t_0)^{-a}$ *is defined by means of* $\arg(t - t_0) \sim \psi_0$, *and where* h^* *is in* Lip γ *on* Γ, *with* $0 < \gamma < 1$. *Then there exist functions* g_0 *and* r_0 (depending on a) *such that* g_0 *is analytic at* t_0 *and* r_0 *satisfies*

$$r_0(z) = O(|z - t_0|^{\gamma}), \qquad z \to t_0, \tag{14.7-19}$$

with the property that the Cauchy integral

$$f(z) = \frac{1}{2\pi i} \int_\Gamma \frac{h(t)}{t - z}\, dt$$

can, for z *near* t_0, *be represented as follows:*

(a) *If* Re $a = 0$,

$$f(z) = g_0(z) + r_0(z) + e^{-ia\psi_0} h^*(t_0) \begin{cases} s_a(e^{-i\psi_0}(z - t_0)), & z \notin \Gamma \\ s_a(e^{-i\psi_0}(t_0 - z)), & z \in \Gamma. \end{cases} \tag{14.7-20a}$$

(b) *If* $0 <$ Re $a < 1$,

$$f(z) = g_0(z) + e^{-ia\psi_0} h^*(t_0)[1 + r_0(z)] \begin{cases} s_a(e^{-i\psi_0}(z - t_0)), & z \notin \Gamma \\ s_a(e^{-i\psi_0}(t_0 - z)), & z \in \Gamma. \end{cases} \tag{14.7-20b}$$

The behavior at the terminal point of Γ where h has a singularity of the form (14.7-18) may be obtained from Theorem 14.7b by replacing Γ by $-\Gamma$.

For convenience we use the following manner of speaking. Let f be defined (but not necessarily analytic) on a simply connected set S having t_0 as a boundary point. The function f will be called **weakly singular at** t_0 if there exist a constant a, $0 \leqq$ Re $a < 1$, and two Hölder continuous functions

c_1 and c_2 such that for $z \in S$, $|z-t_0|$ sufficiently small,

$$f(z)=c_1(z)s_a(z-t_0)+c_2(z), \tag{14.7-21}$$

where s_a denotes the singular function used in the foregoing, that is,

$$s_0(z)=\text{const}\log(-z),$$

$$s_a(z)=\text{const}(-z)^{-a},\ a\neq 0,$$

the branches chosen in (14.7-21) being continuous for $z\neq t_0$. The number a will be called the **exponent** of the singularity, $\alpha := \operatorname{Re} a$ its **order**. A weak singularity of order 0 will also be called a **logarithmic singularity**.

In this terminology, the contents of the two foregoing theorems may somewhat loosely be stated thus. *If h has a weak singularity, but not a logarithmic singularity with exponent* 0, *at an endpoint of* Γ, *then f has a weak singularity with the same exponent. If h is continuous at an endpont of* Γ, *then f there has a weak singularity with exponent zero.*

PROBLEM

1. Obtain the result of Example 4 by means of the theory of hypergeometric functions.

NOTES

For the material of this section see Muschelishvili [1965], § 22–25. Although cumbersome and therefore omitted in many treatments of Cauchy integral theory, these results are indispensible for a rigorous treatment of the Privalov problem.

§ 14.8. THE PRIVALOV PROBLEM FOR AN ARC

Let Γ be a simple regular arc with initial point t_0 and terminal point t_1, and let a and b be two complex-valued functions on Γ which are in Lip γ for some $\gamma>0$, $a(t)\neq 0$ for $t\in\Gamma$. We recall from § 14.4 the *problem of Privalov*, which consists in finding a function f which is analytic in $\mathbb{C}\backslash\Gamma$, and which at every interior point $t\in\Gamma$ has one-sided limits $f^+(t)$ and $f^-(t)$ satisfying

$$f^+(t)=a(t)f^-(t)+b(t). \tag{P}$$

A **solution of the problem of Privalov** here is any function f satisfying the foregoing conditions and having at most a pole at ∞. (This usage deviates slightly from that of § 14.4, where analyticity at ∞ was required.) It will be seen that with this definition the solutions are never unique. However, it is possible to define a distinguished, "canonical" solution by specifying the singular behavior of f at the end points of Γ and at ∞.

In order to construct the totality of solutions, we begin by considering the homogeneous problem

$$f^+(t) = a(t) f^-(t). \tag{P$_0$}$$

We proceed as in § 14.4. There now are no difficulties in defining a Hölder continuous logarithm of $a(t)$. Because Γ is not closed, the problem of not returning to the initial value on continuous continuation does not pose itself. Let $\log a(t)$ denote any continuous logarithm, and let

$$l(z) := \frac{1}{2\pi i} \int_\Gamma \frac{\log a(t)}{t-z} \, dt \tag{14.8-1}$$

be the corresponding Cauchy integral. By the Sokhotskyi formulas,

$$l^+(t) = l^-(t) + \log a(t)$$

at every interior point of Γ. Consequently the function

$$f_0(z) := e^{l(z)} \tag{14.8-2}$$

satisfies

$$\begin{aligned} f_0^+(t) &= e^{l^+(t)} = e^{l^-(t) + \log a(t)} \\ &= a(t)\, e^{l^-(t)} = a(t) f_0^-(t) \end{aligned}$$

at every interior point $t \in \Gamma$. Evidently f_0 is analytic at ∞. The function (14.8-2) thus is a solution of the homogeneous Privalov problem.

How about uniqueness? Let f_1 be any solution of the homogeneous Privalov problem. We consider

$$s(z) := \frac{f_1(z)}{f_0(z)}.$$

As is evident from (14.8-2), $f_0(z) \neq 0$ for $z \in \mathbb{C} \backslash \Gamma$ and also for $z = \infty$. The quotient s thus is analytic in $\mathbb{C} \backslash \Gamma$ and has at most a pole at ∞. At interior points t of Γ, the limits $f_0^+, f_1^+, f_0^{-0}, f_1^-$ all exist, and the limits f_0^+ and f_0^- are $\neq 0$. It thus follows from (P$_0$) that at interior points $t \in \Gamma$,

$$s^+(t) = s^-(t).$$

The principle of continuous continuation (Theorem 5.11a) now implies that s is analytic at every interior point of Γ. The only finite points where s can cease to be analytic thus are t_0 and t_1, the endpoints of Γ, and we conclude that

$$f_1(z) = s(z) f_0(z), \tag{14.8-3}$$

where s is an analytic function with isolated singularities at t_0, t_1, and ∞, the singularity at ∞ being at most a pole. Conversely, if s is any such function, and if f_1 is defined by (14.8-3) where f_0 is given by (14.8-2), then f_1 is a solution of the homogeneous Privalov problem.

THEOREM 14.8a. *The totality of solutions of the homogeneous Privalov problem* (P_0) *for the arc* Γ *are given by* (14.8-3), *where* f_0 *is the special solution given by* (14.8-2), *and where* s *is any analytic function in* $\mathbb{C}$ *with isolated singularities at the endpoints of* Γ *and having at most a pole at* ∞.

To solve the nonhomogeneous problem, we use the method of variation of constants as in § 14.4. That is, we try to satisfy (P) by a function of the form

$$f(z)=k(z)f_1(z), \tag{14.8-4}$$

where f_1 is a solution of the homogeneous Privalov problem, and where k is to be determined from the condition

$$k^+(t)f_1^+(t)=a(t)k^-(t)f_1^-(t)+b(t)$$

to be satisfied at every interior point t of Γ, which on dividing by f_1^+ and using (P_0) yields

$$k^+(t)=k^-(t)+\frac{b(t)}{f_1^+(t)}. \tag{14.8-5}$$

It is easy enough to satisfy (14.8-5) by means of the Cauchy integral

$$k(z):=\frac{1}{2\pi i}\int_\Gamma \frac{b(t)}{f_1^+(t)(t-z)}\,dt \tag{14.8-6}$$

provided—and this is a crucial restriction—that the function $b(t)[f_1^+(t)]^{-1}$ is integrable on Γ for every Hölder continuous function b. This will be the case if $f_1^+(t)\neq 0$ at interior points $t\in\Gamma$, and if $[f_1^+(t)]^{-1}$ has (at most) weak singularities at the endpoints of Γ.

We shall be able to define the required solution f_1 in an essentially unique manner by requiring that at the endpoints it remains bounded or has at most a weak singularity with exponent 0, and that it has a pole of smallest possible order at ∞. A solution with these properties will be called a **canonical solution**. To single out the canonical solutions among the many solutions provided by Theorem 14.8a, the detailed discussions of § 14.7 are required.

Let f_0 be defined by (14.8-2), and let the values of $-\log a(t)$ at the endpoints of Γ be denoted by

$$c_j:=-\log a(t_j), \qquad j=0,1. \tag{14.8-7}$$

These numbers depend on the chosen branch of the logarithm. The values

c_j, c_j' obtained for different choices of the logarithm differ by $2k\pi i$, where k is the same for both values of j. By Theorem 14.7b,

$$
\begin{aligned}
l(z) &= \frac{1}{2\pi i} c_0 \log(z - t_0) + r_0(z), \qquad z \to t_0, \\
l(z) &= -\frac{1}{2\pi i} c_1 \log(z - t_1) + r_1(z), \qquad z \to t_1,
\end{aligned}
\tag{14.8-8}
$$

with the usual gloss concerning the branch of the logarithms, where the functions r_j are Hölder continuous on the sets $D_j \backslash \Gamma$ and $D_j \cap \Gamma$, D_j denoting as usual a sufficiently small disk about t_j. Letting

$$
d_j := \frac{1}{2\pi i} c_j, \qquad j = 0, 1,
\tag{14.8-9}
$$

we thus see that

$$
\begin{aligned}
f_0(z) &= (z - t_0)^{d_0} e^{r_0(z)}, \qquad z \in D_0, \\
f_0(z) &= (z - t_1)^{-d_1} e^{r_1(z)}, \qquad z \in D_1.
\end{aligned}
\tag{14.8-10}
$$

In these relations d_0 and d_1 can be arbitrary complex numbers. For different choices of $\log a(t)$, d_0 and d_1 differ by one and the same integer.

Our task now is to find a function q with isolated singularities at t_0 and t_1 and (at most) a pole at ∞ such that

$$
f_1(z) := q(z) f_0(z)
$$

has at most logarithmic singularities, and f_1^{-1} has at most weak singularities, at the endpoints of Γ. Furthermore we want $f_1(t) \neq 0$ at every interior point of Γ.

It is clear that the singularities of q at the points t_j can be at most poles, for if q had an essential singularity at, say, $z = t_0$, an estimate of the form

$$
|q(z)| \leqq \mu |z - t_0|^{\nu}
$$

by Theorem 4.4e could hold for no value of the exponent ν if $z \in D_0$. By Theorem 4.4h, q thus must be rational.

Let the integers n_0 and n_1 be uniquely determined by the conditions

$$
\begin{aligned}
0 &\leqq \operatorname{Re} d_0 + n_0 < 1, \\
0 &\leqq -\operatorname{Re} d_1 + n_1 < 1.
\end{aligned}
\tag{14.8-11}
$$

We let

$$
s(z) := (z - t_0)^{n_0} (z - t_1)^{n_1}
\tag{14.8-12}
$$

and define

$$
f_1(z) := s(z) f_0(z).
\tag{14.8-13}
$$

THEOREM 14.8b. *The function f_1 defined by* (14.8-13) *is a canonical solution of the homogeneous Privalov problem* (P_0), *and any other canonical solution is of the form cf_1, where c is a constant, $c \neq 0$.*

Proof. It is clear from Theorem 14.8a that f_1 is a solution of the homogeneous problem. By (14.8-11) and the definition of the n_j,

$$\begin{aligned} |f_1(z)| &= |z-t_0|^{\lambda_0} e^{\operatorname{Re} r_0(z)}, \qquad z \in D_0, \\ |f_1(z)| &= |z-t_1|^{\lambda_1} e^{\operatorname{Re} r_1(z)}, \qquad z \in D_1, \end{aligned} \tag{14.8-14}$$

where $0 \leqq \lambda_j < 1$. Thus f_1 is bounded near the endpoints of Γ, and f_1^{-1} has at most weak singularities. It is clear that $f_1(z) \neq 0$ for $z \in \mathbb{C} \backslash \Gamma$, and that $f_1^{+}(t) \neq 0$ at interior points t of Γ. Since $f_0(\infty) = 1$, the order of the pole of f_1 at ∞ is $n := n_0 + n_1$. Let now f_2 be another canonical solution, and let

$$g := \frac{f_2}{f_1}.$$

By the theorem on continuous continuation (Theorem 5.11a), g is analytic everywhere in the complex plane except possibly at t_0, t_1. At these exceptional points g in view of (14.8-14) grows at most like

$$|z-t_j|^{-\lambda},$$

where $0 \leqq \lambda < 1$. Since the singularities at t_j are isolated, this by Laurent's theorem (Theorem 4.4a) implies that the singularities are removable. Thus g is entire. Because f_2 is canonical, the order of the pole of f_2 at ∞ is $\leqq n$. It follows that g is analytic also at ∞ and hence, by Liouville's theorem, constant. If the order of the pole of f_2 were $< n$, that constant would be 0, and it would follow that $f_2 = 0$, a contradiction. Thus the order of the pole of f_2 at ∞ is precisely n, and it follows that $f_2 = cf_1$ with $c \neq 0$, as was to be shown. —

Having constructed a canonical solution, the process initiated earlier for solving the nonhomogeneous problem can now be carried through. If

$$k(z) := \frac{1}{2\pi i} \int_\Gamma \frac{b(t)}{f_1^{+}(t)(t-z)}\, dt, \tag{14.8-15}$$

then the function

$$f(z) := k(z) f_1(z), \tag{14.8-16}$$

as was seen earlier, satisfies (P) and hence is a special solution of the nonhomogeneous Privalov problem. The general solution of the nonhomogeneous problem is obtained by adding the general solution of the homogeneous problem. By virtue of Theorem 14.8a we obtain:

THEOREM 14.8c. *The general solution of the nonhomogeneous Privalov problem* (P) *is*

$$f(z)=[k(z)+s(z)]f_1(z), \tag{14.8-17}$$

where f_1 is the canonical solution of the homogeneous problem described in Theorem 14.8b, *k is given by* (14.8-15), *and s has isolated singularities at t_0, t_1, has at most a pole at ∞, and is analytic elsewhere.*

The choice of the function s in applications is often dictated by a prescribed singular behavior of the solution at the endpoints of Γ and at ∞. By definition of the canonical solution, f_1 at the endpoint t_0 behaves like

$$f_1(z)=(z-t_0)^{\hat{d}_0}\, e^{g_0(z)+r_0(z)},$$

where g_0 is analytic and r_0 is Hölder continuous, $r_0(t_0)=0$, and where $\hat{d}_0 := d_0+n_0$ satisfies $0 \leqq \operatorname{Re} \hat{d}_0 < 1$. By the theorems of § 14.7, a similar representation holds for $f_1^+(t)$, $t \in \Gamma$. Thus the Cauchy integral defining k is formed with a function

$$h(t)=\frac{b(t)}{f_1^+(t)},$$

which near t_0 behaves like $(t-t_0)^{-\hat{d}_0}h^*(t)$, where h^* is Hölder continuous and (if $b(t_0)\neq 0$) satisfies $h^*(t_0)\neq 0$. By § 14.7 we conclude that the singular behavior of k near t_0 may be described as follows:

(a) If $\hat{d}_0=0$,

$$k(z)\sim \text{const} \log(z-t_0).$$

(b) If $0\leqq \operatorname{Re}\hat{d}_0<1$, $\hat{d}_0\neq 0$,

$$k(z)\sim \text{const}(z-t_0)^{-\hat{d}_0}.$$

We conclude that the special solution kf_1 of the nonhomogeneous problem has a logarithmic singularity if $\hat{d}_0=0$, that is, if

$$a(t_0)=1,$$

and that it is bounded if $a(t_0)\neq 1$. The same holds, of course, for the behavior near the terminal point t_1.

Concerning the behavior at ∞, we see from (14.8-15) that $k(z)$ is analytic at ∞, $k(\infty)=0$. It thus follows that kf_1 has a pole at ∞ whose order (possibly <0) is at most one less than the order of the pole of the canonical solution f_1.

EXAMPLE 1

Let Γ: $t=\tau$, $-1\leqq\tau\leqq 1$, and let a be a complex constant, $a\neq 0$. We consider the special Privalov problem

$$f^{+}(t)=af^{-}(t)+b(t), \qquad t\in\Gamma, \tag{14.8-18}$$

where b is any Hölder continuous function. Let c be any value of $\log a$. The function l defined by (14.8-1) then is

$$l(z)=\frac{c}{2\pi i}\int_{-1}^{1}\frac{1}{\tau-z}\,d\tau=\frac{c}{2\pi i}\operatorname{Log}\frac{z-1}{z+1}.$$

Consequently if $d:=c/2\pi i$,

$$\begin{aligned} f_0(z)&=e^{d\operatorname{Log}[(z-1)/(z+1)]}\\ &=(z-1)^{d}(z+1)^{-d}\end{aligned}$$

is a solution of the homogeneous problem. To determine the canonical solution, let n be the unique integer such that

$$0\leqq\operatorname{Re} d+n<1.$$

If $\operatorname{Re} d$ is not an integer, then

$$0\leqq-\operatorname{Re} d-n+1<1,$$

and on writing $\hat{d}:=d+n$ the canonical solution is found to be

$$f_1(z):=(z-1)^{\hat{d}}(z+1)^{1-\hat{d}}. \tag{14.8-19}$$

If $\operatorname{Re} d$ is an integer, the canonical solution is

$$f_1(z):=(z-1)^{\hat{d}}(z+1)^{-\hat{d}}$$

with the same definition of $\hat{d}$.

For $t=\tau$, $-1<\tau<1$, $f_1^{+}(t)$ is the limit of $f_1(z)$ as $z\to t$ from above. We thus are to evaluate the above expressions with $\arg(z-1)=\pi$, which, if $\operatorname{Re} d$ is not an integer, yields

$$f_1^{+}(t)=(1-t)^{\hat{d}}(1+t)^{1-\hat{d}}\,e^{i\pi\hat{d}}.$$

Hence

$$k(z)=e^{-i\pi\hat{d}}\frac{1}{2\pi i}\int_{-1}^{1}\frac{(1-\tau)^{-\hat{d}}(1+\tau)^{\hat{d}-1}b(\tau)}{\tau-z}\,d\tau,$$

and with this definition of k the function

$$f(z):=k(z)(z-1)^{\hat{d}}(z+1)^{1-\hat{d}}$$

is a special solution of (P).

To see what happens if Re d is an integer, let $a=1$ in (14.8-18). Then $d=c/2\pi i$ may be taken as 0, $\hat{d}=0$, and the canonical solution is

$$f_1(z)=1,$$

as indeed it must be. Thus

$$k(z)=\frac{1}{2\pi i}\int_{-1}^{1}\frac{b(\tau)}{\tau-z}\,d\tau$$

is a particular solution of (14.8-18), as is clear from the Sokhotskyi formulas.

PROBLEM

1. Generalize the results of this section to a situation where Γ conists of several nonintersecting arcs $\Gamma_1,\ldots,\Gamma_k$, none of which is closed.

NOTES

See Muschelishvili [1965], § 77–80.

§ 14.9. SOME SINGULAR INTEGRAL EQUATIONS

I. The Integral Equation of Thin Airfoil Theory

The singular integral equation

$$\frac{1}{\pi}\mathrm{PV}\int_{-1}^{1}\frac{h(t)}{t-x}\,dt=g(x),\qquad -1<x<1,$$

where g is given and h is sought, occurs in the theory of thin airfoils. In this theory, the airfoil intersects the (x, y) plane near the segment $[-1, 1]$ of the x axis. The components of the velocity vector on the top side of the airfoil are $U+u(x)$ and $v(x)$, where U is the velocity at ∞ and where u, v are small corrections. If quadratic terms in u, v are neglected, the relation between u and v is given by (T), where $h=u$ and $g=v$. On the bottom side, the sign of u is reversed. If $y=y(x)$ describes the skeleton line of the airfoil, the function v is connected to the geometry of the foil, as follows. Because at the surface the velocity vector has the direction of the airfoil,

$$y'(x)=\frac{v(x)}{U+u(x)}$$

or, neglecting second-order terms,

$$y'(x)=\frac{v(x)}{U}. \qquad (14.9\text{-}1)$$

The function u, on the other hand, is connected with the forces exerted on the airfoil. By Bernoulli's law, the pressure equals $\rho/2$ times the square of the velocity ($\rho :=$ density of air). Thus at the top side of the airfoil the pressure equals

$$p(x)=\frac{\rho}{2}[(U+u(x))^2+(v(x))^2]$$

or, reduced to first-order terms,

$$p(x)=\frac{\rho}{2}U^2+Uu(x).$$

At the bottom side, the same pressure acts with the sign of u reversed. Thus the pressure differential at x is

$$\delta p(x)=2\rho Uu(x).$$

We thus see that (T) relates the pressure exerted on the airfoil to its geometry. If the pressure $(\sim h)$ is given, the geometry $(\sim g)$ may be evaluated simply by evaluating the integral on the left. If the geometry is given, the pressure is found by solving (T) for h.

Here we study the solution of (T), called **integral equation of thin airfoil theory**, as a mathematical problem. We assume that $g \in \operatorname{Lip} \alpha$ on $[-1, 1]$, and we call **solution** of the equation any function h which is in $\operatorname{Lip} \alpha$ on every closed subinterval of $(-1, 1)$ and which is at most weakly singular at the endpoints. Our aim is to determine the totality of all solutions of (T).

Let h be any solution, and let

$$f(z):=\frac{1}{\pi}\int_{-1}^{1}\frac{h(t)}{t-z}\,dt. \tag{14.9-2}$$

Then by the Sokhotskyi formulas, if $-1<x<1$,

$$h(x)=\frac{1}{2i}[f^+(x)-f^-(x)], \tag{14.9-3}$$

whereas

$$f(x)=\tfrac{1}{2}[f^+(x)+f^-(x)]=g(x).$$

Thus f solves the Privalov problem

$$f^+(x)=-f^-(x)+2g(x). \tag{14.9-4}$$

This is the special case $a=-1$, $b(t)=2g(t)$ of Example 1, § 14.8. We select $c=\log a=i\pi$ and find $\hat{d}=c/2\pi i=\frac{1}{2}$. This yields the canonical solution of the homogeneous problem

$$f_1(z)=\sqrt{z^2-1}, \tag{14.9-5}$$

where the square root is analytic for $z \notin [-1, 1]$, $f_1(z) \sim z$ for $z \to \infty$. For this choice of the root, if $-1 < x < 1$,

$$f_1^+(x) = -f_1^-(x) = i\sqrt{1-x^2}. \tag{14.9-6}$$

A particular solution of the nonhomogeneous equation now is given by $f_2(z) = k(z) f_1(z)$, where

$$\begin{aligned} k(z) &:= \frac{1}{2\pi i} \int_{-1}^{1} \frac{2g(t)}{f_1^+(t)(t-z)}\, dt \\ &= -\frac{1}{\pi} \int_{-1}^{1} \frac{g(t)}{\sqrt{1-t^2}(t-z)}\, dt. \end{aligned} \tag{14.9-7}$$

By Theorem 14.8c, the general solution of (14.9-4) now is

$$f(z) = \{s(z) + k(z)\}\sqrt{z^2-1}, \tag{14.9-8}$$

where $s(z)$ is an analytic function with isolated singularities at ± 1 and at most a pole at ∞. Our next task is to identify s.

For $-1 < x < 1$,

$$f^\pm(x) = [s(x) + k^\pm(x)] f_1^\pm(x).$$

In view of (14.9-6) there follows

$$\begin{aligned} f^+(x) - f^-(x) &= i\sqrt{1-x^2}\{2s(x) + k^+(x) + k^-(x)\} \\ &= 2i\sqrt{1-x^2}\{s(x) + k(x)\}, \end{aligned} \tag{14.9-9}$$

where $k(x)$ is the principal value of the integral (14.9-7) for $z = x$. By (14.9-3) the above equals $2ih(x)$, which is at most weakly singular at $x = \pm 1$. By Theorem 14.7b, $k(x)$ is at most weakly singular at ± 1. It follows that $s(z)$ has at most poles at $z = \pm 1$, and that the order of these poles is at most 1. Thus s has the form

$$s(z) = \frac{a}{z-1} + \frac{b}{z+1} + p(z), \tag{14.9-10}$$

where p is a polynomial.

As $z \to \infty$, it follows from (14.9-2) that $f(z) = O(z^{-1})$. Thus surely $p(z) = 0$. Moreover from (14.9-8) and (14.9-10), since

$$k(z) = \frac{c}{z} + O(z^{-2}),$$

where

$$c := \frac{1}{\pi} \int_{-1}^{1} \frac{g(t)}{\sqrt{1-t^2}}\, dt, \tag{14.9-11}$$

we have

$$f(z)=[z+O(z^{-1})]\left\{\frac{a+b}{z}+\frac{c}{z}+O(z^{-2})\right\}$$
$$=a+b+c+O(z^{-1}).$$

It follows that $a+b+c=0$, or that

$$a=-\frac{c}{2}+d, \qquad b=-\frac{c}{2}-d,$$

where d is a new free parameter. This yields

$$f(z)=\frac{1}{\sqrt{z^2-1}}(2d-cz)+\sqrt{z^2-1}\,k(z),$$

where c is given by (14.9-11). For h we now find from (14.9-9)

$$h(x)=\frac{1}{\sqrt{1-x^2}}(cx-2d)+\sqrt{1-x^2}\,k(x). \tag{14.9-12}$$

We finally show that for every choice of d, (14.9-12) indeed is a solution of (T). It is well known (see Problem 3, § 14.2) that

$$h_0(x):=\frac{1}{\sqrt{1-x^2}}$$

solves the homogeneous equation (T). It thus suffices to consider $d=0$. By the definition of c,

$$h(x)=\frac{1}{\pi\sqrt{1-x^2}}\,\mathrm{PV}\int_{-1}^{1}\frac{g(t)}{\sqrt{1-t^2}}\left(x-\frac{1-x^2}{t-x}\right)dt.$$

In view of

$$x-\frac{1-x^2}{t-x}=\frac{tx-1}{t-x}=-t-\frac{1-t^2}{t-x},$$

this becomes

$$h(x)=-\frac{1}{\pi\sqrt{1-x^2}}\int_{-1}^{1}\frac{tg(t)}{\sqrt{1-t^2}}\,dt-\frac{1}{\pi\sqrt{1-x^2}}\,\mathrm{PV}\int_{-1}^{1}\frac{\sqrt{1-t^2}\,g(t)}{t-x}\,dt.$$

Here the first term is of the form const $h_0(x)$ and may therefore be omitted. In the second term we set $t=\cos\tau$, $x=\cos\xi$,

$$\phi_0(\tau):=\sin\tau\, g(\cos\tau).$$

By virtue of Theorem 14.6c we then have

$$h(x) = -\frac{1}{|\sin \xi|}\psi_0(\xi),$$

where $\psi_0 := \mathcal{K}\phi_0$. By virtue of Corollary 14.2e, since the mean value of ϕ_0 is zero, there follows $\phi_0 = -\mathcal{K}\psi_0$. Since ψ_0 is even, this by (14.6-38) implies

$$\sqrt{1-x^2}g(x) = \frac{1}{\pi}\sqrt{1-x^2}\,\mathrm{PV}\int_{-1}^{1}\frac{h(t)}{t-x}\,dt,$$

which in turn implies (14.9-1).

THEOREM 14.9a. *Let g be in* Lip α, $0<\alpha<1$, *on* $[-1, 1]$. *Then the general solution h of the equation of thin airfoil theory* (T) *is of the form*

$$h(x) = ah_0(x) + h_1(x),$$

where a is an arbitrary constant,

$$h_0(x) := \frac{1}{\sqrt{1-x^2}} \tag{14.9-13}$$

$$h_1(x) := -\frac{1}{\pi\sqrt{1-x^2}}\,\mathrm{PV}\int_{-1}^{1}\frac{\sqrt{1-t^2}\,g(t)}{t-x}\,dt. \tag{14.9-14}$$

Depending on the parametrization of $s(z)$, various other forms of the solution may be obtained, but the above lends itself most easily to verification.

II. Carleman's Integral Equation

Let $a \in \mathbb{C}$, $0 < \mathrm{Re}\, a < 1$, and let g be continuous for $x \geqq 0$, $g(0) = 0$. In § 10.4 we briefly considered **Abel's integral equation**, which is the special case $a = \frac{1}{2}$ of the equation for an unknown function h,

$$\int_0^x \frac{h(t)}{(x-t)^a}\,dt = g(x), \qquad x > 0, \tag{A}$$

which we call the **generalized Abel equation. Carleman's integral equation** is the analog of (A) where the limits of integration are fixed,

$$\int_0^1 \frac{h(t)}{|x-t|^a}\,dt = g(x), \qquad 0 \leqq x \leqq 1. \tag{C}$$

To prepare for the solution of Carleman's equation, we have to solve the generalized Abel equation. If g is assumed to possess a generalized derivative in the sense of § 10.2, this is easily accomplished by Laplace transforms. Suppose (A) has a solution h which is an original function in the sense of the theory of Laplace transforms, and let $g \circ\!\!-\!\!\bullet\, G$, $h \circ\!\!-\!\!\bullet\, H$. Because the integral in (A) is the convolution of h with t^{-a}, the Laplacian image of (A) in view of

$$t^{-a} \circ\!\!-\!\!\bullet\, \frac{\Gamma(1-a)}{s^{1-a}}$$

is

$$\Gamma(1-a)H(s)s^{a-1} = G(s), \tag{14.9-15}$$

which yields

$$H(s) = \frac{1}{\Gamma(1-a)} s^{1-a} G(s). \tag{14.9-16}$$

The function s^{1-a} is not an image function, hence the expression on the right is not the image of any convolution with $g(t)$. However,

$$s^{-a} \circ\!\!-\!\!\bullet\, \frac{1}{\Gamma(a)} t^{a-1},$$

and since $sG(s)$ is the image of the generalized derivative g', we find on account of

$$\frac{1}{\Gamma(a)\Gamma(1-a)} = \frac{\sin \pi a}{\pi}$$

that if the generalized Abel equation has a solution that has a Laplace transform, this solution is given by

$$h(x) = \frac{\sin \pi a}{\pi} \int_0^x \frac{g'(t)}{(x-t)^{1-a}}\, dt. \tag{14.9-17}$$

It remains to be shown that the function thus defined indeed solves (A). By Theorem 10.4a, h is an original function that is continuous at all $x > 0$. By the convolution theorem (Theorem 10.4b) its Laplace transform is given by (14.9-16), which implies (14.9-15). Since the functions $H(s)$, s^{1-a}, and $G(s)$ all are Laplace transforms, going to the original space implies (A). We have proved:

THEOREM 14.9b. *If g is continuous for $x \geqq 0$, $g(0) = 0$, and if g possesses a generalized derivative which is an original function in the sense of* § 10.2, *then the generalized Abel integral equation* (A) *possesses a unique solution h which is continuous for $x > 0$. This solution is given by* (14.9-17).

We proceed to our main task, which is the solution of Carleman's equation (C). Let g be uniformly Hölder continuous on $[0, 1]$, and lets its derivative g' exist and be uniformly Hölder continuous on every closed subinterval of $(0, 1)$. We call **solution of Carleman's equation** any Hölder continuous function h, possibly weakly singular of order $<1-\operatorname{Re} a$ at $x=0$ and $x=1$, such that (C) is satisfied.

To solve the equation, we follow our usual procedure of assuming that a solution exists. This will lead to a representation of the alleged solution. It then has to be verified that the function h thus determined actually satisfies the equation.

Thus suppose h satisfies (C). For $|\arg(z-1)|<\pi$ we define

$$f(z) := \int_0^1 \frac{h(t)}{(z-t)^a}\, dt, \tag{14.9-18}$$

where the power has its principal value. As z approaches a point $x<1$ on the real axis, we have

$$[(z-t)^a]^+ = \begin{cases} e^{i\pi a}(t-x)^a, & x<t \\ (x-t)^a, & x>t. \end{cases} \tag{14.9-19}$$

For the approach from below $e^{i\pi a}$ is replaced by $e^{-i\pi a}$. For $x<0$ we thus have

$$f^{\pm}(x) = e^{\mp i\pi a} \int_0^1 \frac{h(t)}{(t-x)^a}\, dt,$$

where the upper and lower signs go together, and thus

$$f^+(x) = e^{-2i\pi a} f^-(x), \qquad x<0. \tag{14.9-20}$$

For $0<x<1$ we find, using (14.9-19),

$$f^+(x) = \int_0^x \frac{h(t)}{(x-t)^a}\, dt + e^{-i\pi a} \int_x^1 \frac{h(t)}{(t-x)^a}\, dt,$$

$$f^-(x) = \int_0^x \frac{h(t)}{(x-t)^a}\, dt + e^{i\pi a} \int_x^1 \frac{h(t)}{(t-x)^a}\, dt.$$

These equations may be solved for the two integrals to yield

$$\int_0^x \frac{h(t)}{(x-t)^a}\, dt = \frac{1}{2i \sin \pi a}\{e^{i\pi a} f^+(x) - e^{-i\pi a} f^-(x)\} \tag{14.9-21}$$

$$\int_x^1 \frac{h(t)}{(t-x)^a}\, dt = -\frac{1}{2i \sin \pi a}\{f^+(x) - f^-(x)\}. \tag{14.9-22}$$

The first equation is a generalized Abel equation and thus permits us to find h if f^+ and f^- are known and have the required properties.

To find f^+ and f^- we show that f, or a function closely related to f, solves a Privalov problem. By adding the two equations, using the fact that the sum of the two integrals equals the left member of (C), we get

$$g(x)=\frac{1}{2i\sin \pi a}(e^{i\pi a}-1)f^+(x)-(e^{-i\pi a}-1)f^-(x)$$

or, solving for f^+,

$$f^+(x)=-e^{-i\pi a}f^-(x)+(1+e^{-i\pi a})g(x), \qquad 0<x<1. \tag{14.9-23}$$

Together with (14.9-20) this defines a Privalov problem for f where Γ is the half-line $-\infty<x<1$. To get a Privalov problem for a finite interval, we introduce the function

$$s(z):=[z(z-1)]^{(a-1)/2}f(z). \tag{14.9-24}$$

For $x<0$ we have

$$s^{\pm}(x)=[(-x)(1-x)]^{(a-1)/2}\, e^{\pm i\pi(a-1)}f^{\pm}(x).$$

Hence from (14.9-20) there follows

$$s^+(x)=s^-(x), \qquad x<0. \tag{14.9-25}$$

For $0<x<1$, on the other hand, we have

$$s^{\pm}(x)=k(x)\, e^{\pm i\pi(a-1)/2}f^{\pm}(x),$$

where

$$k(x):=[x(1-x)]^{(a-1)/2}. \tag{14.9-26}$$

Hence if (14.9-23) is expressed in terms of s, we find

$$s^+(x)=s^-(x)+g_0(x), \tag{14.9-27}$$

where

$$g_0(x):=-2i\cos\frac{\pi a}{2}k(x)g(x). \tag{14.9-28}$$

We also note that on account of (14.9-25), using the principle of continuous continuation, s is analytic at all points $x<0$, and hence in $\mathbb{C}\backslash[0,1]$. As to the behavior for $z\to\infty$, it follows from (14.9-24) and $f(z)=O(z^{-a})$ that

$$s(z)=O(z^{-1}), \qquad z\to\infty,$$

without restriction on the manner of approach. If, at the points $z=0, 1$, f has at most weak singularities of order $<\frac{1}{2}$, (14.9-24) shows that s has at most weak singularities of order <1.

By the uniqueness theorem on the solution of Privalov's problem there exists but one function s with all the indicated properties. It is given by the Cauchy integral

$$s(z):=\frac{1}{2\pi i}\int_0^1 \frac{g_0(t)}{t-z}\,dt. \tag{14.9-29}$$

Having constructed s, we can find f from (14.9-24) and then may express the function on the right of (14.9-21) in terms of a known function which in turn permits us to find h via the solution of a generalized Abel's equation. This program is carried out in the following theorem:

THEOREM 14.9c. *Let* $a \in \mathbb{C}$, $0<\operatorname{Re} a<1$. *Let g be uniformly Hölder continuous on* $[0, 1]$, *and let g' exist and be uniformly Hölder continuous on every closed subinterval of* $(0, 1)$. *Let*

$$s(z):=\frac{1}{2\pi i}\int_0^1 \frac{g_0(t)}{t-z}\,dt, \tag{14.9-29}$$

where

$$g_0(t):=-2i\cos\frac{\pi a}{2}k(t)g(t), \tag{14.9-28}$$

$$k(t):=[t(1-t)]^{(a-1)/2}. \tag{14.9-26}$$

Let h denote the unique solution of the generalized Abel equation

$$\int_0^x \frac{h(t)}{(x-t)^a}\,dt=u(x), \qquad 0<x\leqq 1, \tag{14.9-30}$$

where

$$u(x):=\frac{1}{2\sin(\pi a/2)}k^{-1}(x)s(x)+\frac{1}{2}g(x), \tag{14.9-31}$$

$s(x)$ denoting the principal value of the integral (14.9-29). *Then h is a solution of Carleman's equation* (C), *and it is the only solution.*

Proof. We first check whether the functions defined exist, and how singularly they behave at $z=0, 1$. Since g is regular, g_0 is weakly singular of order $\leqq \frac{1}{2}(1-\operatorname{Re} a)$. By Theorem 14.7b, the same holds for s. In the definition of u, the singularities of s are again removed. Because g_0 has a derivative which is uniformly Hölder continuous on every closed subinterval of $(0, 1)$, the same may be shown to hold for $s(x)$, and therefore for u. By (14.9-17) it follows that h is weakly singular of order at most $1-\operatorname{Re} a$ at $x=0, 1$.

With the h defined by (14.9-30), let

$$f_1(z) := \int_0^1 \frac{h(t)}{(z-t)^a}\, dt,$$

where the power has its principal value. As in the derivation of (14.9-20), we have

$$f_1^+(x) = e^{-2\pi i a} f_1^-(x), \qquad x<0, \tag{14.9-32}$$

and

$$f_1^\pm(x) = u(x) + e^{\mp i\pi a}\, v(x), \qquad 0<x<1,$$

where u is defined by (14.9-31) and

$$v(x) := \int_x^1 \frac{h(t)}{(t-x)^a}\, dt. \tag{14.9-33}$$

There follows

$$e^{i\pi a} f_1^+(x) - e^{-i\pi a} f_1^-(x) = 2i \sin \pi a\, u(x), \tag{14.9-34}$$

$$f_1^+(x) - f_1^-(x) = 2i \sin \pi a\, v(x). \tag{14.9-35}$$

We now let

$$s_1(z) := [z(z-1)]^{(a-1)/2} f_1(z)$$

and assert that, for all $z \in \mathbb{C}\backslash[0, 1]$,

$$s_1(z) = s(z), \tag{14.9-36}$$

where s is defined by (14.9-29).

The proof of (14.9-36) is surprisingly involved. Expressed in terms of s_1, (14.9-32) becomes

$$s_1^+(x) = s_1^-(x), \qquad x<0,$$

while (14.9-34) is equivalent to

$$e^{i\pi a/2} s_1^+(x) + e^{-i\pi a/2} s_1^-(x) = 2k(x) \sin \pi a\, u(x)$$

or, on using the definition of u,

$$e^{i\pi a/2} s_1^+(x) + e^{-i\pi a/2} s_1^-(x) = 2 \cos \frac{\pi a}{2} s(x) + i \sin \frac{\pi a}{2} g_0(x), \qquad 0<x<1. \tag{14.9-37}$$

Here we take note of Sokhotskyi's formulas satisfied by $s(z)$,

$$s^+(x) + s^-(x) = 2s(x),$$

$$s^+(x) - s^-(x) = g_0(x),$$

where $0<x<1$. Multiplying the first relation by $\cos(\pi a/2)$ and the second by $i\sin(\pi a/2)$ and subtracting from (14.9-37), we get

$$e^{i\pi a/2}(s_1-s)^+(x)+e^{-i\pi a/2}(s_1-s)^-(x)=0, \qquad 0<x<1.$$

We conclude that s_1-s is the solution of a special homogeneous Privalov problem with the constant coefficient $-e^{-i\pi a}$, and that therefore by Theorem 14.8a,

$$s_1(z)-s(z)=q(z)z^{(1+a)/2}(z-1)^{-(1+a)/2},$$

where q has (at most) a pole at ∞ and isolated singularities at $z=0$ and $z=1$. Since both s_1 and s are at most weakly singular at $z=0$ and $z=1$ and since both s_1 and s are $O(z^{-1})$ as $z\to\infty$, q must be of the form cz^{-1}, where c is a complex constant.

We wish to show that $c=0$, and to this end shall prove that the leading coefficients in the Laurent series at ∞ of both s_1 and s are the same. As to the Laurent series for s, it directly follows from (14.9-29) that

$$s(z)=-\frac{1}{2\pi i}\int_0^1 g_0(t)\,dt\cdot z^{-1}+O(z^2). \tag{14.9-38}$$

As to the Laurent series of s_1, we find from the definition of s_1 and f_1 that

$$s_1(z)=\int_0^1 h(t)\,dt\cdot z^{-1}+O(z^{-2}). \tag{14.9-39}$$

To evaluate the integral we use Laplace transforms. Let $h \circ\!\!-\!\!\bullet\, H$, $u \circ\!\!-\!\!\bullet\, U$. We know from the proof of Theorem 14.9b that

$$H(s)=\frac{1}{\Gamma(1-a)}s^{1-a}U(s).$$

By an elementary rule on Laplace transforms,

$$\int_0^x h(t)\,dt \circ\!\!-\!\!\bullet\, \frac{1}{s}H(s)=\frac{1}{\Gamma(1-a)}s^{-a}U(s),$$

and therefore, using the convolution theorem (Theorem 10.4b),

$$\int_0^1 h(t)\,dt=\frac{\sin\pi a}{\pi}\int_0^1 (1-t)^{a-1}u(t)\,dt.$$

Using the definition of u, this yields

$$\int_0^1 h(t)\,dt=\frac{\cos(\pi a/2)}{\pi}\int_0^1\left(\frac{1-t}{t}\right)^{(a-1)/2}s(t)\,dt$$
$$-\frac{\sin(\pi a/2)}{2\pi i}\int_0^1\left(\frac{1-t}{t}\right)^{(a-1)/2}g_0(t)\,dt. \tag{14.9-40}$$

We evaluate the first integral on the right. The idea is to substitute for s the principal value integral (14.9-29) and to interchange the order of the integrations. However, to avoid justifying this interchange in a singular integral, we use the fact that on account of the Sokhotskyi formulas,

$$s(t)=\tfrac{1}{2}[s^+(t)+s^-(t)].$$

Because the limits on the right exist uniformly, we have

$$\int_0^1\left(\frac{1-t}{t}\right)^{(a-1)/s} s(t)\,dt=\lim_{\eta\to 0+}\frac{1}{2}\int_0^1\left(\frac{1-t}{t}\right)^{(a-1)/2}[s(t+i\eta)+s(t-i\eta)]\,dt.$$

For $\eta>0$ there holds

$$\begin{aligned}\int_0^1\left(\frac{1-t}{t}\right)^{(a-1)/2} &s(t-i\eta)\,dt\\ &=\frac{1}{2\pi i}\int_0^1\left(\frac{1-t}{t}\right)^{(a-1)/2}\left\{\int_0^1\frac{g_0(\tau)}{\tau-t-i\eta}\,d\tau\right\}dt\\ &=\frac{1}{2\pi i}\int_0^1 g_0(\tau)\int_0^1\left(\frac{1-t}{t}\right)^{(a-1)/2}\frac{1}{(\tau-i\eta)-t}\,dt\,d\tau.\end{aligned}$$

The interchange of integrations is now justifiable, for example, by Fubini's theorem. The inner integral is evaluated by letting $s:=\tau-i\eta$ and expanding around $s=\infty$, with the result

$$\int_0^1\left(\frac{1-t}{t}\right)^{(a-1)/2}\frac{1}{s-t}\,dt=\frac{\pi}{\cos(\pi a/2)}\left\{\left(\frac{s-1}{s}\right)^{(a-1)/2}-1\right\}.$$

For $s=t-i\eta$, where $0<t<1$ and $\eta>0$, the limit as $\eta\to 0$ is

$$\frac{\pi}{\cos(\pi a/2)}\left\{\left(\frac{1-t}{t}\right)^{(a-1)/2}e^{-(a-1)i\pi/2}-1\right\}.$$

If $s=t+i\eta$, where $\eta>0$, the limit as $\eta\to 0$ is

$$\frac{\pi}{\cos(\pi a/2)}\left\{\left(\frac{1-t}{t}\right)^{(a-1)/2}e^{(a-1)i\pi/2}-1\right\}.$$

There follows

$$\begin{aligned}\int_0^1\left(\frac{1-t}{t}\right)^{(a-1)/2}&s(t)\,dt\\ &=\frac{\pi}{\cos(\pi a/2)}\frac{1}{2\pi i}\int_0^1\left\{\left(\frac{1-t}{t}\right)^{(a-1)/2}\sin\frac{\pi a}{2}-1\right\}g_0(t)\,dt.\end{aligned}$$

Substituting this in (14.9-40), the desired equality of the leading terms in

(14.9-38) and (14.9-39) follows immediately. There follows $s_1(z)-s(z)=O(z^{-2})$, which is compatible with the formula

$$s_1(z)-s(z)=cz^{(a-1)/2}(z-1)^{-(a+1)/2}$$

only if $c=0$. This concludes the proof of (14.9-36), $s_1(z)=s(z)$.

The proof that h is a solution of Carleman's equation (C) will now be complete if we show that

$$u(x)+v(x)=g(x), \qquad 0<x<1.$$

To this end we use (14.9-35) to express v in terms of $s_1=s$. We know that

$$f_1(z)=[z(z-1)]^{-(a-1)/2}s(z)$$

and hence that for $0<x<1$,

$$f_1^{\pm}(x)=\pm i\, e^{\mp i\pi a/2}k^{-1}(x)s^{\pm}(x)$$

or using Sokhotskyi's formulas for s,

$$f_1^{\pm}(x)=\pm i\, e^{\mp i\pi a/2}\left\{k^{-1}(x)s(x)\mp i\cos\frac{\pi a}{2}g(x)\right\}.$$

There follows

$$\begin{aligned} v(x)&=-\frac{1}{2i\sin(\pi a/2)}\{f_1^{+}(x)-f_1^{-}(x)\}\\ &=-\frac{1}{2\sin(\pi a/2)}k^{-1}(x)s(x)+\frac{1}{2}g(x), \end{aligned}$$

and using (14.9-31) the desired relation $u(x)+v(x)=g(x)$ follows immediately.

It remains to establish the uniqueness of the solution h. This simply follows from the fact established earlier that any solution h necessarily satisfies the generalized Abel equation (A) and that the solution of this equation is unique.

This completes the proof of Theorem 14.9c. —

PROBLEMS

1. Show that in the general solution of (T) given in Theorem 14.9a, the function h_1 may be replaced by either

$$h_2(x):=-\frac{1}{\pi}\sqrt{\frac{1-x}{1-x}}\,\mathrm{PV}\int_{-1}^{1}\sqrt{\frac{1+t}{1-t}}\frac{g(t)}{t-x}\,dt$$

or

$$h_3(x):=-\frac{1}{\pi}\sqrt{\frac{1+x}{1-x}}\,\mathrm{PV}\int_{-1}^{1}\sqrt{\frac{1-t}{1+t}}\frac{g(t)}{t-x}\,dt.$$

2. Show that if $g(\pm 1) \neq 0$, there is precisely one solution of (T) such that $h(-1)$ is finite, and that there is precisely one solution such that $h(1)$ is finite. Also show that in general *no* solution of (T) is square integrable on $[-1, 1]$.
3. Show, by a method analogous to that employed to solve (C), that the equation

$$\int_0^1 h(t) \operatorname{Log}|x-t| \, dt = g(x)$$

has the solution

$$\begin{aligned} h(x) := & \frac{1}{\pi^2} \frac{1}{\sqrt{x(1-x)}} \operatorname{PV} \int_0^1 \frac{g'(t)\sqrt{t(1-t)}}{t-x} \, dt \\ & - \frac{1}{2\pi^2 \operatorname{Log} 2 \cdot \sqrt{x(1-x)}} \int_0^1 \frac{g(t)}{\sqrt{t(1-t)}} \, dt. \end{aligned}$$

NOTES

For background on the equation of thin airfoil theory, see Cheng and Rott [1954]. They include references to earlier work, among which Söhngen [1939] and Weissinger [1941] stand out. Nickel [1951, 1953, and other papers] has generalizations to "multifoils." For Carleman's equation see Carleman [1922]; that the function h defined by (14.9-30) actually solves (C) is verified by Carleman only for analytic g. Generalizations of equations (T) and (C) are considered by Peters [1969] and Heins and Camy [1957, 1958]. A related function-theoretic method for solving integral equations of the form

$$\lambda h(x) + \int_0^\infty h(t) k(x-t) \, dt = g(x)$$

is due to Wiener and Hopf [1931]; see also Heins [1956], Noble [1958], and Carrier, Krook, and Pearson [1966].

§ 14.10. A METHOD FOR SOLVING TRANSCENDENTAL EQUATIONS

Here we use the theory of the Privalov problem to discuss a method, designed by Burniston, Siewert, and others, for finding zeros of certain transcendental functions. The method consists in constructing, by means of quadratures, a polynomial whose zeros are identical with the zeros of the given function. In many cases the resulting polynomial is of low degree and thus can be solved explicitly. The method then furnishes not only numerical values, but also qualitative information about the zeros, say as functions of the parameters of the problem.

Let Γ: $t = t(\tau)$, $\alpha \leqq \tau \leqq \beta$, be a simple arc such that $t' \in \operatorname{Lip}$ and $t'(\tau) \neq 0$ for all $\tau \in [\alpha, \beta]$. The method to be discussed applies to functions f having the following properties:

(i) f is analytic in $\mathbb{C} \backslash \Gamma$.

(ii) f has at most a pole at ∞.

(iii) At the endpoints $t_0 := t(\alpha)$ and $t_1 := t(\beta)$ of Γ, f is not required to be meromorphic, let alone analytic. We do require, however, that the behavior of f be "polelike" in the sense that there exist real numbers δ, ε, and μ such that, if $|z-t_j|$ is sufficiently small, $z\in\mathbb{C}\backslash\Gamma$,

$$\mu|z-t_j|^{\delta} \leqq |f(z)| \leqq \mu|z-t_j|^{\varepsilon}, \qquad j=0,1. \tag{14.10-1}$$

(iv) At every interior point t of Γ, the one-sided limits $f^+(t), f^-(t)$ exist and are different from zero.

(v) The limits of the function

$$a(t) := \frac{f^+(t)}{f^-(t)} \tag{14.10-2}$$

exist for $t\to t_0$ and $t\to t_1$ and are different from zero, and $\log a(t)$ may be defined on the whole of Γ as a function that is in Lip γ for some $\gamma>0$.

We prove:

THEOREM 14.10a. (a) *The function f has only finitely many zeros.* (b) *A polynomial whose zeros are precisely the zeros of f can be constructed rationally in terms of finitely many of the Laurent coefficients of f at ∞, and of finitely many of the quantities*

$$m_k := \frac{1}{2\pi i}\int_\Gamma t^k \log a(t)\,dt, \qquad k=0,1,2,\ldots, \tag{14.10-3}$$

where $\log a(t)$ *denotes any continuous logarithm of* $a(t)$.

Proof and construction. By the hypotheses (i), (ii), (iv), and (v), f is a solution of the Privalov problem

$$f^+(t) = a(t)f^-(t), \qquad t\in\Gamma. \tag{14.10-4}$$

According to Theorem 14.8a, a special solution of (14.10-4) is given by

$$f_0(z) := e^{g(z)},$$

where g is a Cauchy integral,

$$g(z) := \frac{1}{2\pi i}\int_\Gamma \frac{\log a(t)}{t-z}\,dt. \tag{14.10-5}$$

The same theorem tells us that every solution of the Privalov problem, thus also our function f, is of the form

$$f(z) = s(z)f_0(z),$$

where s is a function that has isolated singularities (at most) at the points t_0, t_1, and ∞. By virtue of (ii) and (iii), the singularities of s at these points are at most poles, and s thus is rational.

Because $f_0(z) \neq 0$, $z \in \mathbb{C} \backslash \Gamma$, the zeros of f are those of s, and it remains to identify s.

Clearly,

$$s(z) = e^{-g(z)} f(z). \tag{14.10-6}$$

This relation enables us to compute the Laurent series of s at ∞. This is so because the Laurent series of f is assumed to be known,

$$f(z) = \sum_{n=-l}^{\infty} a_n z^{-n}$$

where l is the order of the pole of f at ∞, and the Laurent series

$$e^{-g(z)} =: \sum_{n=0}^{\infty} b_n z^{-n}$$

can be constructed from the fact that

$$\begin{aligned} -g(z) &= \frac{1}{2\pi i} \int_\Gamma \frac{\log a(t)}{z-t}\, dt \\ &= \frac{1}{z} \frac{1}{2\pi i} \int_\Gamma \frac{\log a(t)}{1-t/z}\, dt \\ &= \sum_{n=0}^{\infty} \frac{m_n}{z^{n+1}}, \end{aligned}$$

where the m_n are given by (14.10-3). The actual computation of the b_n would be carried out by comparing coefficients in the identity

$$(-e^{-g(z)})' = g'(z)\, e^{-g(z)},$$

that is,

$$\sum_{n=1}^{\infty} \frac{nb_n}{z^{n+1}} = \sum_{n=1}^{\infty} \frac{nm_{n-1}}{z^{n+1}} \sum_{k=0}^{\infty} \frac{b_k}{z^k},$$

which yields the recurrence relation

$$b_0 = 1,$$

$$nb_n = nm_{n-1}b_0 + (n-1)m_{n-2}b_1 + \cdots + 1 \cdot m_0 b_{n-1}, \qquad n = 1, 2, \ldots. \tag{14.10-7}$$

Knowing the Laurent series for f and for e^{-g} we can, by forming a Cauchy product, construct as many Laurent coefficients s_n in the expansion

$$s(z)=\sum_{n=-l}^{\infty} s_n z^{-n} \tag{14.10-8}$$

as we like. Knowing finitely many coefficients in this expansion is not sufficient, however, to identify s, even if s is known to be rational. If, on the other hand, a bound for the order of the poles of s were known, the identification could be made. Let the order of the poles of s at t_j be at most r_j, $j=0,1$. Then, because a rational function is the sum of its principal parts (Theorem 4.4h), s must be of the form

$$\begin{aligned}
s(z)=&\sum_{n=0}^{l} s_{-n}z^n \quad (s_0+\text{principal part at }\infty)\\
&+\sum_{n=1}^{r_0}\frac{a_{0,n}}{(z-t_0)^n} \quad (\text{principal part at } t_0)\\
&+\sum_{n=1}^{r_1}\frac{a_{1,n}}{(z-t_1)^n} \quad (\text{principal part at } t_1).
\end{aligned}\tag{14.10-9}$$

Here the coefficients $a_{j,n}$ are, as yet, unknown. However, we may expand these principal parts into their Laurent series at ∞ by using the formulas

$$\frac{1}{(z-t_j)^n}=z^{-n}\left(1-\frac{t_j}{z}\right)^{-n}=z^{-n}\sum_{k=0}^{\infty}\frac{(n)_k}{k!}\left(\frac{t_j}{z}\right)^k. \tag{14.10-10}$$

Combining these expansions, it emerges that the coefficient of z^{-n} in the Laurent series of s at ∞ can also be expressed as a linear combination of the $a_{j,k}$, where $k\leqq n$. Equating these expressions to s_n for $1\leqq n\leqq r_0+r_1$, a system of linear equations for the r_0+r_1 unknowns $a_{j,n}$ results, which suffices to determine these unknowns and hence, by virtue of (14.10-9), s.

To complete the construction of s, and hence the proof of Theorem 14.10a, it is thus necessary to find bounds r_j for the order of the poles of s at t_j, $j=0,1$. The behavior of f at these points is controlled by condition (iii). The behavior of

$$f_0^{-1}(z)=e^{-g(z)}$$

may be studied by means of Theorem 14.7a. The gist of that theorem is that for z near t_0, $z\notin\Gamma$, $g(z)$ behaves like

$$\frac{1}{2\pi i}\int_\Gamma \frac{\log a(t)}{t-z}\,dt,$$

that is, like

$$-\frac{1}{2\pi i}\log a(t_0)\log(t_0-z).$$

It follows that for $z\to t_0$, $z\notin\Gamma$,

$$e^{-g(z)}\sim \operatorname{const}(z-t_0)^{(1/2\pi i)a(t_0)}.$$

Similarly for $z\to t_1$, $z\notin\Gamma$,

$$e^{-g(z)}\sim \operatorname{const}(z-t_1)^{-(1/2\pi i)a(t_1)}.$$

It emerges that

$$|s(z)|\leqq \operatorname{const}|z-t_0|^{\varepsilon+\operatorname{Re}(1/2\pi i)\log a(t_0)},\qquad z\to t_0$$

and

$$|s(z)|\leqq \operatorname{const}|z-t_1|^{\varepsilon-\operatorname{Re}(1/2\pi i)\log a(t_1)},\qquad z\to t_1,$$

from which we see that bounds for the order of the poles at t_0 and t_1 are given by

$$r_0 := -\left[\varepsilon+\operatorname{Re}\frac{1}{2\pi i}\log a(t_0)\right],$$

$$r_1 := -\left[\varepsilon-\operatorname{Re}\frac{1}{2\pi i}\log a(t_1)\right]. \quad —$$

We note that the conclusions of Theorem 14.10a, as well as the construction implied in the foregoing proof, also hold if Γ is a finite collection of nonintersecting smooth arcs.

As an example for the Burniston–Siewert method we consider the equation

$$(z-\beta)e^z = z-\alpha, \tag{14.10-11}$$

where α and β are real, $\alpha<\beta$.

To apply Theorem 14.10a we write the equation in the form

$$e^z=\frac{z-\alpha}{z-\beta}$$

and take logarithms. For any choice of the logarithm, any solution of

$$z=\log\frac{z-\alpha}{z-\beta} \tag{14.10-12}$$

is a solution of (14.10-11); conversely, any solution of (14.10-11) is a solution of (14.10-12) for some choice of the logarithm.

We now take as Γ the straight line segment $[\alpha, \beta]$. Any branch of the logarithm on the right of (14.10-11) that is analytic outside $[\alpha, \beta]$ can be written in the form

$$\operatorname{Log}\frac{z-\alpha}{z-\beta}+2\pi ik,$$

where Log is the principal value of the logarithm, and k is an integer. Thus the totality of solutions of (14.10-12) is identical with the totality of zeros of the functions

$$f(z):=\operatorname{Log}\frac{z-\alpha}{z-\beta}-z+2k\pi i.$$

We verify that f satisfies the hypotheses of Theorem 14.10a. It is clear that f is analytic on $\mathbb{C}\backslash\Gamma$, that it has a pole (of order 1) at ∞, and that it is "polelike" at $z=\alpha$ and $z=\beta$, where it has logarithmic singularities. For $\alpha<t<\beta$,

$$f^{+}(t)=\operatorname{Log}\frac{t-\alpha}{\beta-t}-t+(2k-1)\pi i,$$

$$f^{-}(t)=\operatorname{Log}\frac{t-\alpha}{\beta-t}-t+(2k+1)\pi i.$$

Thus

$$a(t):=\frac{f^{+}(t)}{f^{-}(t)}=\frac{\phi(t)+(2k-1)\pi i}{\phi(t)+(2k+1)\pi i},$$

where

$$\phi(t):=\operatorname{Log}\frac{t-\alpha}{\beta-t}-t.$$

An elementary discussion shows that

$$\phi(t)\to-\infty \quad \text{as} \quad t\to\alpha^{+} \qquad \text{and} \qquad \phi(t)\to+\infty \quad \text{as} \quad t\to\beta^{-}.$$

Thus $\phi(t)$ moves from $-\infty$ to $+\infty$ as t moves from α to β; however, ϕ is not necessarily monotone.

To discuss the Hölder continuity of $\log a(t)$, we distinguish the cases $k=0$ and $k\neq 0$.

(a) If $k=0$, then

$$\operatorname{Log} a(t)=-2i\operatorname{Arg}[\phi(t)+i\pi]=2i\arctan\frac{\phi(t)}{\pi}-i\pi. \quad (14.10\text{-}13)$$

As t moves from α to β, Log $a(t)$ moves from $-2i\pi$ to 0. It is clear that Log $a(t)$ is in Lip (in fact, differentiable) on every closed subinterval of (α, β). For $t \to \beta^-$,

$$\arctan\frac{\phi(t)}{\pi} = \frac{\pi}{2} - \arctan\frac{\pi}{\phi(t)} \sim \frac{\pi}{2} - \frac{\pi}{\phi(t)} \sim \frac{\pi}{2} - \frac{\pi}{\text{Log}(\beta - t)}.$$

The latter function is *not* Hölder continuous at $t = \beta$. Similarly, Log $a(t)$ fails to be Hölder continuous at $t = \alpha$. We shall see, however, that the method of Theorem 14.10a still works.

(b) $k \neq 0$. For definiteness we assume $k > 0$. Then

$$\begin{aligned}\text{Log } a(t) &= \text{Log}\frac{\phi(t)+(2k-1)\pi i}{\phi(t)+(2k+1)\pi i} + i\,\text{Arg}\frac{\phi(t)+(2k-1)\pi i}{\phi(t)+(2k+1)\pi i}\\ &= \frac{1}{2}\text{Log}\frac{\phi(t)^2+(2k-1)^2\pi^2}{\phi(t)^2+(2k+1)^2\pi^2} - i\arctan\frac{2\pi\phi(t)}{\phi(t)^2+(4k-1)^2\pi^2}.\end{aligned} \tag{14.10-14}$$

As t moves from α to β, Log $a(t)$ now moves continuously from 0 to 0. Again there is the difficulty that the function fails to be Hölder continuous at the two endpoints of the interval.

We are now ready to tackle the construction of s. According to the general outline, a study of the singular parts of s at the three singularities $z = \alpha$, $z = \beta$, $z = \infty$ is required. Again we distinguish the cases $k = 0$ and $k \neq 0$.

(a) $k = 0$. At $z = \alpha$,

$$f(z) \sim \text{Log}(z - \alpha).$$

As to the behavior of $f_0^{-1}(z) = e^{-g(z)}$, Theorem 14.7a in view of Log $a(\alpha) = -2\pi i$ would yield

$$e^{-g(z)} \sim (z-\alpha)^{-1}h(z),$$

where h were Hölder continuous if Log $a(t)$ were Hölder continuous. Since this is not quite so, we can only assert that $h(z)$ is more weakly singular than $|z-\alpha|^{-\nu}$ for any $\nu > 0$. From $s(z) = e^{-g(z)}f(z)$ it then follows that

$$s(z) \sim \text{const}(z-\alpha)^{-1}$$

at $z = \alpha$. That is, s has a pole of order 1. At $z = \beta$ we find in view of Log $a(\beta) = 0$ in a similar way that s has a removable singularity.

At $z=\infty$, we conclude from

$$f(z)=\operatorname{Log}\frac{z-\alpha}{z-\beta}-z=-z+(\beta-\alpha)z^{-1}+\cdots,$$

$$f_0^{-1}(z)=1+m_0z^{-1}+\cdots$$

that s has a pole of order 1 with principal part $-z+m_0$. Altogether it thus follows that s has the form

$$s(z)=-z+m_0+\frac{a_{01}}{z-\alpha}.$$

To determine the remaining unknown a_{01}, we compare the Laurent series at ∞. Since there is only one unknown, only the term in z^{-1} needs to be compared. From

$$s(z)=-z+m_0+a_{01}z^{-1}+O(z^{-2}),$$

$$e^{-g(z)}f(z)=\left(1+m_0z^{-1}+\frac{m_0^2+2m_1}{2}z^{-2}+\cdots\right)$$
$$\times(-z+(\beta-\alpha)z^{-1}+\cdots)$$

there follows

$$a_{01}=\beta-\alpha-m_1-\tfrac{1}{2}m_0^2.$$

The zeros of f may now be found as solutions of the quadratic

$$(-z+m_0)(z-\alpha)+a_{01}=0,$$

where

$$m_k:=\frac{1}{2\pi i}\int_\alpha^\beta \operatorname{Log} a(t)\,dt, \qquad k=0,1,$$

$\operatorname{Log} a(t)$ being given by (14.10-13). For instance, for $\alpha=1$, $\beta=2$ we obtain in this manner the two real solutions

$$z_1=2.1342\ldots, \qquad z_2=-0.5085\ldots.$$

(b) $k\neq 0$. Without loss of generality we assume $k>0$. The solutions for k and $-k$ are complex conjugate. The function f now is

$$f(z)=\operatorname{Log}\frac{z-\alpha}{z-\beta}-z+2k\pi i,$$

$\operatorname{Log} a(t)$ is given by (14.10-14). At $z=\alpha$,

$$f(z)\sim\operatorname{Log}(z-\alpha).$$

Since $\text{Log}\, a(\alpha)=0$, $e^{-g(z)}$ is more weakly singular than $|z-\alpha|^{-\nu}$ for any $\nu>0$, and it follows that the singularity of s at $z=\alpha$ is removable. The same holds for $z=\beta$. Thus $s(z)$ equals its principal part at ∞ (including the constant term). From

$$f(z)=-z+2k\pi i+O(z^{-1}),$$
$$e^{-g(z)}=1+m_0 z^{-1}+O(z^{-2})$$

there follows

$$\begin{aligned} s(z)=e^{-g(z)}f(z)&=-z+(2k\pi i-m_0)+O(z^{-1})\\ &=-z+(2k\pi i-m_0), \end{aligned}$$

and we see immediately that the sole zero in this case is

$$z_0=2k\pi i-m_0,$$

where

$$m_0:=\frac{1}{2\pi i}\int_\alpha^\beta \text{Log}\, a(t)\,dt.$$

For instance, for $\alpha=1$, $\beta=2$ we obtain, for various values of k,

k	z_0
1	$-0.03829+6.130i$
5	$-0.0015+31.8341i$.

By discussing m_0 as a function of k, accurate asymptotic information about z_0 may be obtained.

PROBLEMS

1. Apply the Burniston–Siewert method to the equation

$$\tanh x=\alpha x, \tag{$*$}$$

where $0<\alpha<1$, by writing it in the form

$$2x=\text{Log}\,\frac{1+\alpha x}{1-\alpha x}$$

or, setting $z:=\alpha^{-2}x^{-2}$,

$$f(z)=0,$$

where

$$f(z):=1-\frac{1}{2}\alpha z^{1/2}\,\text{Log}\,\frac{1+z^{-1/2}}{1-z^{-1/2}}.$$

(a) Show that f satisfies the hypotheses of the method.

(b) Show that

$$f^{+}(t)=e^{2i\phi(t)}f^{-}(t), \qquad 0<t<1,$$

where

$$\begin{aligned}\phi(t)&=\arg\left\{1+\frac{\alpha}{2}\sqrt{t}\left(\operatorname{Log}\frac{1-\sqrt{t}}{1+\sqrt{t}}+i\pi\right)\right\}\\&=\frac{\pi}{2}-\arctan\frac{2+\alpha\sqrt{t}\operatorname{Log}[(1-\sqrt{t})/(1+\sqrt{t})]}{\sqrt{t}}.\end{aligned}$$

(c) Letting $f_0(z):=e^{l(z)}$,

$$l(z):=\frac{1}{\pi}\int_0^1\frac{\phi(t)}{t-z}\,dt,$$

show that

$$\begin{aligned}&f(z)\sim 1-\alpha, \qquad f_0(z)\sim 1, \qquad z\to\infty;\\&f(z)\sim 1, \qquad f_0(z)=O(1), \qquad z\to 0;\\&f(z)=O(\operatorname{Log}(z-1)), \qquad f_0(z)=O((z-1)^{1+\varepsilon}), \qquad z\to 1,\end{aligned}$$

and conclude that

$$s(z)=1-\alpha+\frac{c}{z-1}.$$

(d) By comparing Laurent series at ∞, show that

$$c=a_1+b_0a_0,$$

where

$$f(z)=\sum a_n z^{-n}, \qquad -g(z)=\sum b_n z^{-n-1}.$$

(e) Show that the two zeros $x\neq 0$ of $(*)$ are

$$x=\pm\frac{1}{\alpha\sqrt{z_0}},$$

where

$$z=1-b_0+\frac{\alpha}{3(1-\alpha)}$$

is the unique zero of f.

2. Apply the Burniston–Siewart method to the transcendental equation

$$\alpha x=e^{-x},$$

where α is a real parameter, $\alpha>0$, by writing the equation in the form

$$x+\operatorname{Log}x+\beta=0$$

($\beta := \operatorname{Log} \alpha$), which on introducing z by

$$x = \frac{z}{z-1}$$

becomes equivalent to $f(z) = 0$, where

$$f(z) := (1+\beta)z - \beta - (z-1) \operatorname{Log} \frac{z-1}{z}.$$

3. Find formulas for *all* solutions of the equation

$$z = \cosh(az + b)$$

by writing it in the form

$$az + b = \log(z + \sqrt{z^2 - 1})$$

and applying a Moebius transformation sending the three branch points $\infty, -1, +1$, respectively, to $-1, 0, 1$.

NOTES

See Burniston and Siewert [1972, 1973a, b, 1978], Siewert and Burniston [1972a, b, 1974], Siewert and Essig [1973], and Siewert and Phelps [1979]. Equations similar to (14.10-11) occur in connection with the difference-differential equations of control theory; see Bellmann and Cooke [1963]. For an asymptotic treatment of such equations see Wright [1961] and Schmidt and Meyer [1977].

§ 14.11. THE HILBERT TRANSFORM ON THE REAL LINE

I. Cauchy Integrals on the Real Line

Let the complex function h now be defined on the whole real line. Here we wish to study Cauchy integrals of the form

$$f(z) := \frac{1}{2\pi i} \int_{-\infty}^{\infty} \frac{h(t)}{t-z} \, dt. \tag{14.11-1}$$

For the general theory of Cauchy integrals to be applicable, we shall assume that h is uniformly Hölder continuous on every bounded interval. In addition, we have to take care that the improper integral (14.11-1) exists. By defining the integral as a "principal value integral at ∞,"

$$\int_{-\infty}^{\infty} := \lim_{\rho \to \infty} \int_{-\rho}^{\rho},$$

we could avoid requiring that h vanish at ∞; however, not much seems to be gained from this. Instead, we subject h to the condition

$$\lim_{t \to \pm\infty} h(t) = 0,$$

and in addition we require that h satisfy a **Hölder condition at** ∞ in the sense that the function

$$h^*(t) := \begin{cases} h\left(\dfrac{1}{t}\right), & t \neq 0 \\ 0, & t = 0 \end{cases}$$

satisfies a Hölder condition at 0. The existence of constants μ and $\gamma > 0$ such that

$$|h^*(t) - h^*(0)| = |h^*(t)| \leqq \mu |t|^\gamma$$

implies

$$|h(t)| \leqq \mu |t|^{-\gamma}$$

and thus guarantees the existence of the integral (14.11-1). The following is a restatement of the corresponding results of § 14.1:

THEOREM 14.11a. *Let h be uniformly Hölder continuous on every bounded interval, and let it satisfy a Hölder condition at ∞ with $h(\infty) = 0$. Then the integral*

$$f(z) := \frac{1}{2\pi i} \int_{-\infty}^{\infty} \frac{h(t)}{t - z}\, dt$$

exists as an ordinary improper integral for every nonreal z and as a principal value integral for real z. Moreover, the function f has the following properties:

(i) *f is analytic in the upper half-plane U and in the lower half-plane L.*
(ii) *The limits f^+ and f^- exist at every real t_0 without restriction on the manner of approach, and define functions of the real variable t_0 that are uniformly Hölder continuous on every bounded interval.*
(iii) *The Sokhotskyi formulas*

$$\begin{aligned} f^+(t) - f^-(t) &= h(t), \\ f^+(t) + f^-(t) &= 2f(t) \end{aligned} \tag{14.11-2}$$

hold at every real t.

We shall have to be able to describe the behavior of f as $|z| \to \infty$. This we can do by means of the results of § 14.7. We write

$$f = \int_{-\infty}^{\infty} = \int_{-\infty}^{-1} + \int_{-1}^{1} + \int_{1}^{\infty} = f_1 + f_2 + f_3.$$

The function

$$f_2(z) = \frac{1}{2\pi i}\int_{-1}^{1} \frac{h(t)}{t-z}\,dt$$

obviously is analytic at ∞, $f_2(\infty)=0$. As to f_3, we have

$$f_3\left(\frac{1}{z}\right) = \frac{1}{2\pi i}\int_{1}^{\infty} \frac{h(t)}{t-z^{-1}}\,dt$$

or, by using the h^* already defined,

$$f_3\left(\frac{1}{z}\right) = z\cdot\frac{1}{2\pi i}\int_{0}^{1} \frac{t^{-1}h^*(t)}{z-t}\,dt.$$

The function

$$g(z) := -\frac{1}{z}f_3\left(\frac{1}{z}\right)$$

consequently is given by

$$g(z) = \frac{1}{2\pi i}\int_{0}^{1} \frac{t^{-1}h^*(t)}{t-z}\,dt.$$

We have already assumed that $|h^*(t)| \leqq \mu t^{\gamma}$, where $\gamma > 0$. If we now make the somewhat stronger assumption that h^* can be represented in the form

$$h^*(t) = t^{\gamma}h_1(t), \qquad 0 \leqq t \leqq 1, \tag{14.11-3}$$

where h_1 is uniformly Hölder continuous, then g satisfies the hypotheses of Theorem 14.7b where $a = 1-\gamma$. This permits us to conclude that

$$g(z) = O(|z|^{\gamma-1}), \qquad z \to 0,$$

where the approach of z to 0 is not restricted. Going back to f_3, this implies the existence of a constant μ such that

$$|f_3(z)| \leqq \frac{\mu}{|z|^{\gamma}}$$

for all z such that $|z|$ is sufficiently large, and hence for all z. If h is such that also

$$h^*(t) = (-t)^{\gamma}h_2(t), \qquad -1 \leqq t \leqq 0, \tag{14.11-4}$$

for a certain uniformly Hölder continuous h_2, a similar result can be established for the function f_1.

If h is such that both conditions (14.11-3) and (14.11-4) hold, we shall say that h at ∞ **vanishes regularly of order** γ. With this terminology we have proved:

THEOREM 14.11b. *If, in addition to satisfying the hypotheses of Theorem* 14.11a, *h at ∞ vanishes regularly of order γ, where $0<\gamma<1$, then*

$$f(z)=O(|z|^{-\gamma}), \qquad z\to\infty \tag{14.11-5}$$

without restriction on the manner of approach.

II. The Schwarz Formula for the Upper Half Plane

We now consider for the upper half-plane U the boundary problems that were considered in § 14.2 for the unit disk. That is, given a real function u on the real line, we seek to find a function f that is analytic in U and continuous in the closure of U such that $\operatorname{Re} f(x)=u(x)$. Moreover we seek to express $v(x):=\operatorname{Im} f(x)$ directly in terms of $u(x)$. In order to apply the theory of Cauchy integrals, we assume that u vanishes regularly at infinity of some positive order and is uniformly Hölder continuous on every finite interval.

It is easy to see that the solution of the boundary value problem as posed is not unique, for the function $f(z):=iz$ satisfies $\operatorname{Re} f(x)=0$ without vanishing identically. Noticing that this particular function is unbounded, we seek the solution of the boundary value problem in the class of functions that, in addition to being analytic in U and continuous in the closure of U, are bounded.

To see what can be done with Cauchy integrals, let h, in addition to satisfying the conditions of Theorem 14.11b, be *real.* The Cauchy integral

$$f(z):=\frac{1}{2\pi i}\int_{-\infty}^{\infty}\frac{h(t)}{t-z}\,dt$$

for nonreal z then obviously satisfies

$$f(\bar{z})=\frac{1}{2\pi i}\int_{-\infty}^{\infty}\frac{h(t)}{t-\bar{z}}\,dt=-\overline{f(z)}.$$

If z approaches a real point x from above, $\bar{z}$ approaches the same point from below, and therefore

$$f^{-}(x)=-\overline{f^{+}(x)}.$$

The Sokhotskyi formulas (14.11-2) in the present context thus yield

$$h(x)=f^{+}(x)+\overline{f^{+}(x)},$$
$$2f(x)=f^{+}(x)-\overline{f^{+}(x)},$$

which may be written

$$h(x) = 2 \operatorname{Re} f^+(x), \tag{14.11-6}$$

$$f(x) = i \operatorname{Im} f^+(x). \tag{14.11-7}$$

From (14.11-6) we see that for $h(x) = 2u(x)$ the function f has the required properties. We thus have proved the existence part of the following theorem:

THEOREM 14.11c. *Let the real function u vanish regularly at ∞ and satisfy a uniform Hölder condition on every finite interval. Then*

$$f(z) := \frac{1}{i\pi} \int_{-\infty}^{\infty} \frac{u(t)}{t - z} \, dt \tag{14.11-8}$$

is the only function having the following properties:

(i) *f is analytic in the upper half-plane U and continuous on the closure of U.*

(ii) $$\lim_{z \to \infty} f(z) = 0.$$

(iii) $$\operatorname{Re} f(x) = u(x) \qquad \text{for all real } x.$$

Proof of uniqueness. If f_1 and f_2 are two functions as required,

$$g(z) := e^{f_1(z) - f_2(z)}$$

is analytic and without zeros in U, and it satisfies $|g(x)| = 1$ for real x. Because g is bounded, it follows from the principle of the maximum that $g(z) = e^{i\gamma}$, where γ is a real constant. We conclude that f_1 and f_2 can only differ by a constant that is purely imaginary. In view of (ii), that constant is zero. —

III. The Hilbert Transform on $\mathbb{R}$

Turning to (14.11-7), we see that $v(x) := \operatorname{Im} f^+(x)$ is given by

$$v(x) = -\frac{1}{\pi} \operatorname{PV} \int_{-\infty}^{\infty} \frac{u(t)}{t - x} \, dt.$$

The known facts on Cauchy integrals on $\mathbb{R}$ thus permit us to state:

THEOREM 14.11d. *Let u satisfy the conditions of the preceding theorem, and let f be the unique function characterized by properties* (i), (ii), *and* (iii) *of that theorem. Then the function $v(x) := \operatorname{Im} f^+(x)$ is uniformly Hölder continuous on every finite interval, satisfies a Hölder condition at ∞, and is* *represented by*

$$v(x)=\frac{1}{\pi}\,\mathrm{PV}\int_{-\infty}^{\infty}\frac{u(t)}{x-t}\,dt. \tag{14.11-9}$$

Whenever u is such that the integral (14.11-9) exists for all (or almost all) real x, the function v thus defined is called the **Hilbert transform** (on the real line) of u, and we write

$$v=\mathscr{H}u. \tag{14.11-10}$$

(In the definition of the Hilbert transform, the sign is sometimes reversed. However, the convention used here is consistent with the definition of the Hilbert transform on the circle.) Because v is the real part of the boundary values of $i^{-1}f$, it follows that under the conditions of Theorem 14.11d, (14.11-10) implies

$$u=-\mathscr{H}v.$$

We define the inner product of two real functions u and v on $\mathbb{R}$ by

$$(u,v):=\int_{-\infty}^{\infty}u(t)v(t)\,dt,$$

and the norm of a function u by

$$\|u\|:=(u,u)^{1/2}.$$

The analog of Theorem 14.2f then is as follows:

THEOREM 14.11e. *Let u vanish regularly of order $\gamma>\frac{1}{2}$ at ∞, and let it be uniformly Hölder continuous on every finite interval. Then*

$$\|\mathscr{H}u\|=\|u\|, \tag{14.11-11a}$$

$$(u,\mathscr{H}u)=0. \tag{14.11-11b}$$

Proof. Let f be the function associated with u in the manner of Theorem 14.11c. The integral

$$I:=\int_{-\infty}^{\infty}[f(t)]^2\,dt$$

exists by virtue of Theorem 14.11b. Closing the path of integration by means of a large semicircle and applying Cauchy's theorem, we see that $I=0$ since

f is analytic in U. On the other hand, letting $v = \mathcal{H}u$, we have

$$I = \int_{-\infty}^{\infty} [u^2(t) - v^2(t) + 2iu(t)v(t)]\, dt$$
$$= \|u\|^2 - \|v\|^2 + 2i(u, v),$$

which in view of $I = 0$ immediately yields the formulas (14.11-11). —

A deeper treatment of the Hilbert transform is possible in the framework of the class $L_2(-\infty, \infty)$ of functions that are square integrable on the real line. It then may be shown that if u is any such function, the integral (14.11-9) exists for almost all x and defines a function that is in the same class. In this manner the operator $\mathcal{H}$ is extended to $L_2(-\infty, \infty)$. The formulas (14.11-11) hold for the extended operator.

IV. The Hilbert Transform and the Fourier Transform

If ϕ is a function of period 1, we mean by the **spectrum** of ϕ the sequence $\{a_n\}$ of its complex Fourier coefficients

$$a_n = \int_0^1 \phi(\tau)\, e^{-2\pi i n \tau}\, d\tau,$$

where n ranges from $-\infty$ to ∞. In § 14.2 we have seen that the spectrum $\{b_n\}$ of the Hilbert transform $\mathcal{H}\phi$ of ϕ is related to the spectrum of ϕ by the formula

$$b_n = -i \operatorname{sgn} n \cdot a_n. \qquad (14.11\text{-}12)$$

If u is a function on $\mathbb{R}$ such that

$$\int_{-\infty}^{\infty} |u(t)|\, dt < \infty,$$

we mean by the spectrum of u its Fourier transform (in the normalization of § 13.3),

$$s(\omega) := (\mathcal{F}u)(\omega) = \int_{-\infty}^{\infty} u(t)\, e^{-2\pi i \omega t}\, dt. \qquad (14.11\text{-}13)$$

Under suitable conditions (see Theorem 10.6d) the function u can be reconstituted from its spectrum by means of the formula

$$u(t) = (\mathcal{F}^{-1}s)(t) = \int_{-\infty}^{\infty} s(\omega)\, e^{2\pi i \omega t}\, d\omega. \qquad (14.11\text{-}14)$$

We now ask the question that is answered in the case of the circle by (14.11-12): how is the spectrum of $\mathcal{H}u$ related to the spectrum of u?

It is easy to guess what the answer might be. Proceeding formally and ignoring the singularity, the Hilbert transform integral

$$v(x)=\frac{1}{\pi}\int_{-\infty}^{\infty}\frac{u(t)}{x-t}\,dt$$

may be considered as the *convolution* (as in § 10.4) of u with the singular function

$$\sigma(t):=\frac{1}{\pi t}.$$

Now for the Fourier transform, as for the Laplace transform, there holds a convolution theorem stating that the Fourier transform of the convolution $u_1 * u_2$ is the product of the transforms of the factors, that is,

$$\mathscr{F}(u_1 * u_2)=\mathscr{F}u_1\cdot\mathscr{F}u_2. \tag{14.11-15}$$

It thus remains to find the Fourier transform of the singular function $\sigma(t)$. In the strict sense, this transform does not exist because σ is not integrable. If, however, we interpret the Fourier integral as a principal value integral and apply the calculus of residues in the manner of § 4.8, we get

$$(\mathscr{F}\sigma)(\omega)=\frac{1}{\pi}\,\mathrm{PV}\int_{-\infty}^{\infty}\frac{e^{-i\omega t}}{t}\,dt=\begin{cases}-i, & \omega>0\\ +i, & \omega<0.\end{cases}$$

For $\omega=0$ we have, again in the sense of a principal value integral, $(\mathscr{F}\sigma)(0)=0$. We thus get

$$(\mathscr{F}\sigma)(\omega)=-i\operatorname{sgn}\omega \tag{14.11-16}$$

and thus, on the basis of the convolution formula (14.11-15), are led to conjecture that there holds the formula

$$(\mathscr{F}\mathscr{H}u)(\omega)=-i\operatorname{sgn}\omega\,(\mathscr{F}u)(\omega),$$

which is completely analogous to (14.11-12), or in an abbreviated notation,

$$\mathscr{F}\mathscr{H}=-i\operatorname{sgn}\cdot\mathscr{F}. \tag{14.11-17}$$

The analysis that we have performed is obviously invalid. Among other things, we have operated with Fourier transforms which do not in fact exist. However, the final result (14.11-17) is correct, even in the context of Fourier and Hilbert transforms of functions $u\in L_2(-\infty,\infty)$. For an exhaustive treatment we refer to chapter V of Titchmarsh [1937]. Here we content ourselves with presenting a version of (14.11-17) that makes strong hypotheses but can be established by elementary methods.

THEOREM 14.11f. *Let u be defined on $\mathbb{R}$, let this function have a continuous derivative, and let its support be compact (that is, let $u(x)=0$ outside some* *bounded interval). Then the spectrum of $\mathcal{H}u$ is related to the spectrum of u by the formula*

$$(\mathcal{F}\mathcal{H}u)(\omega) = -\operatorname{sgn}\omega \cdot (\mathcal{F}u)(\omega). \tag{14.11-18}$$

Proof. By the Fourier inversion theorem (Theorem 10.6d), the hypotheses of which are obviously satisfied, the relation to be proved is equivalent to the formula

$$(\mathcal{H}u)(x) = \mathcal{F}^{-1}(-i \operatorname{sgn}\omega \cdot \mathcal{F}u)(x), \tag{14.11-19}$$

which states that the Hilbert transform on the line can be calculated, as in the case of the circle, by taking two Fourier transforms. To establish (14.11-19) we transform the expressions on either side in such a way that their equality becomes obvious.

Letting $v := \mathcal{H}u$, we have

$$v(x) = \frac{1}{\pi} \operatorname{PV} \int_{-\infty}^{\infty} \frac{u(t)}{x-t}\, dt = \frac{1}{\pi} \lim_{\varepsilon \to 0} \left(\int_{-\infty}^{x-\varepsilon} + \int_{x+\varepsilon}^{\infty} \right).$$

For every $\varepsilon > 0$ we have, using integration by parts,

$$\begin{aligned} \int_{-\infty}^{x-\varepsilon} + \int_{x+\varepsilon}^{\infty} &= -u(x-\varepsilon) \operatorname{Log} \varepsilon + \int_{-\infty}^{x-\varepsilon} u'(t) \operatorname{Log}(x-t)\, dt \\ &\quad + u(x+\varepsilon) \operatorname{Log} \varepsilon + \int_{x+\varepsilon}^{\infty} u'(t) \operatorname{Log}(t-x)\, dt. \end{aligned}$$

Using $u(x \pm \varepsilon) = u(x) + O(\varepsilon)$, the limit as $\varepsilon \to 0$ yields

$$v(x) = \frac{1}{\pi} \int_{-\infty}^{\infty} u'(t) \operatorname{Log}|x-t|\, dt.$$

Using the fact that

$$\int_{-\infty}^{\infty} u'(t) \operatorname{Log}|x|\, dt = 0,$$

this may be written in the form

$$v(x) = \frac{1}{\pi} \int_{-\infty}^{\infty} u'(t) \operatorname{Log}\left|1 - \frac{t}{x}\right| dt. \tag{14.11-20}$$

Turning to the expression on the right of (14.11-19), we first consider

$$s(\omega) := (\mathcal{F}u)(\omega) = \int_{-\infty}^{\infty} u(t)\, e^{-2\pi i \omega t}\, dt.$$

We use integration by parts and choose as antiderivative of $e^{-2\pi i \omega t}$ the function

$$t \mapsto \begin{cases} \dfrac{e^{-2\pi i\omega t}-1}{-2\pi i\omega}, & \omega \neq 0 \\ t, & \omega = 0, \end{cases}$$

which for each t also is a continuous function of ω. This yields

$$s(\omega)=\frac{1}{2\pi i}\int_{-\infty}^{\infty} u'(t)\,\frac{e^{-2\pi i\omega t}-1}{\omega}\,dt.$$

Therefore,

$$-i \operatorname{sgn} \omega\, s(\omega)=\frac{1}{2\pi}\int_{-\infty}^{\infty} u'(t)\,\frac{1-e^{-2\pi i\omega t}}{|\omega|}\,dt.$$

Taking the inverse Fourier transform now yields

$$\begin{aligned}&\mathcal{F}^{-1}[-i \operatorname{sgn} \omega\, s(\omega)](x)\\ &\quad=\frac{1}{2\pi}\int_{-\infty}^{\infty}\left\{\int_{-\infty}^{\infty} u'(t)\,\frac{1-e^{-2\pi i\omega t}}{|\omega|}\,dt\right\} e^{2\pi i\omega x}\,d\omega.\end{aligned}$$

Reversing the order of the integrations (justifiable here because u has compact support), this becomes

$$\frac{1}{2\pi}\int_{-\infty}^{\infty} u'(t)\left\{\int_{-\infty}^{\infty}\frac{1-e^{-2\pi i\omega t}}{|\omega|}\,e^{2\pi i\omega x}\,d\omega\right\}dt.$$

The inner integral equals

$$I(x,t):=2\int_0^{\infty}\frac{\cos(2\pi\omega x)-\cos[2\pi\omega(x-t)]}{\omega}\,d\omega$$

and by Laplace transforms (see Example 5, § 10.2) may be evaluated to yield

$$I(x,t)=2\,\operatorname{Log}\left|\frac{x-t}{x}\right|, \qquad x\neq t.$$

We thus have

$$\mathcal{F}^{-1}[-i \operatorname{sgn} \omega\, s(\omega)](x)=\frac{1}{\pi}\int_{-\infty}^{\infty} u'(t)\,\operatorname{Log}\left|1-\frac{t}{x}\right|\,dt,$$

which is the same as (14.11-20) and thus proves Theorem 14.11f. —

The formulas (14.11-18) and (14.11-19) thus established are of considerable importance in digital signal processing (see the following section). Considering the fact that Fourier transforms can be evaluated rapidly by means of the FFT algorithm (see § 13.3), the identity (14.11-19) also provides a means for the efficient numerical evaluation of Hilbert transforms.

PROBLEMS

1. Let ϕ be a real function on the real line that vanishes regularly at ∞ and is uniformly Hölder continuous on every bounded interval. By taking real parts in (14.11-8), show that the unique function $u = u(x, y)$, which is

   (i) harmonic in U,
   (ii) bounded and continuous in the closure of U,
   (iii) equal to $\phi(x)$ for $y = 0$,

   is given by

$$u(x, y) = \frac{1}{\pi}\int_{-\infty}^{\infty} \frac{y}{(x-t)^2 + y^2}\phi(t)\,dt.$$

2. If $\hat{u} := \mathscr{H}u$, establish the following pairs of Hilbert transforms and their Fourier transforms:

   (a)
$$u(t) = \frac{\sin 2\pi\omega_0 t}{2\pi\omega_0 t}, \qquad \mathscr{F}u(\omega) = \begin{cases} (2\omega_0)^{-1}, & |\omega| < \omega_0 \\ 0, & \text{elsewhere}, \end{cases}$$

$$\hat{u}(t) = \frac{1 - \cos 2\pi\omega_0 t}{2\pi\omega_0 t}, \qquad \mathscr{F}\hat{u}(\omega) = \begin{cases} -i \operatorname{sgn} \omega \cdot (2\omega_0)^{-1}, & |\omega| < \omega_0 \\ 0, & \text{elsewhere}. \end{cases}$$

   (b)
$$u(t) = \begin{cases} 1, & |t| < T, \\ 0, & \text{elsewhere}, \end{cases} \qquad \mathscr{F}u(\omega) = \frac{\sin 2\pi\omega T}{\pi\omega}$$

$$\hat{u}(t) = \frac{1}{\pi}\operatorname{Log}\left|\frac{t+T}{t-T}\right|, \qquad \mathscr{F}\hat{u}(\omega) = -i\frac{\sin 2\pi\omega T}{\pi|\omega|}.$$

3. *Operational rules for the Hilbert transform.* Using the notation $u(t) \circ\!\!-\!\!\bullet\, v(x)$ to denote the relationship $v = \mathscr{H}u$, establish, under suitable hypotheses, the following relations:

   (a) $v(t) \circ\!\!-\!\!\bullet\, u(-x)$.

   (b) $u(t+a) \circ\!\!-\!\!\bullet\, v(x+a), \quad a \in \mathbb{R}$.

   (c) $u(at) \circ\!\!-\!\!\bullet\, v(ax), \quad a > 0$.

   (d) $u(-at) \circ\!\!-\!\!\bullet\, -v(-ax), \quad a > 0$.

   (e) $$tu(t) \circ\!\!-\!\!\bullet\, xv(x) + \frac{1}{\pi}\int_{-\infty}^{\infty} u(t)\,dt.$$

   (f) $$(t+a)u(t) \circ\!\!-\!\!\bullet\, (x+a)v(x) + \frac{1}{\pi}\int_{-\infty}^{\infty} u(t)\,dt.$$

   (g) $u'(t) \circ\!\!-\!\!\bullet\, v'(x)$.

4. *Numerical evaluation of Hilbert transforms.* For $h > 0$ and for functions u with compact support, define a discrete Fourier transform $\mathscr{F}_h u$ by

$$v(\omega) = (\mathscr{F}_h u)(\omega) := h \sum_{k=-\infty}^{\infty} u(kh) \exp\{-2\pi i k h \omega\}$$

and its approximate inverse by

$$(\mathscr{F}_h^{-1}v)(x) = h \sum_{k=-\infty}^{\infty} v(kh)\exp\{2\pi ikhx\}.$$

Approximate $\mathscr{H}$ by

$$\mathscr{H}_h := \mathscr{F}_h^{-1}(-i\,\mathrm{sgn}\cdot\mathscr{F}_h)$$

and show how $\mathscr{H}_h \to \mathscr{H}$ as $h \to 0$.

Problems 5–7 are concerned with the following problem posed by Keldysh and Sedov (see Lavrentiev and Shabat [1967]). Let $-\infty < \alpha_1 < \beta_1 < \alpha_2 < \beta_2 < \cdots < \alpha_n < \beta_n < \infty$, and let real Hölder continuous functions u_k and v_k be defined, respectively, on the intervals (α_k, β_k) and (β_k, α_{k+1}), $k = 1, 2, \ldots, n$. (Here (β_n, α_{n+1}) means the union of (β_n, ∞) and $(-\infty, \alpha_1)$.) We are to find a function f which is analytic in $\operatorname{Im} z > 0$, continuous on $\operatorname{Im} z \geqq 0$ (except possibly at the points α_k and β_k), and which satisfies the following requirements:

(i) For $k = 1, 2, \ldots, n$, if $\xi \in (\alpha_k, \beta_k)$,

$$\operatorname{Re} f(\xi) = u_k(\xi).$$

(ii) For $k = 1, 2, \ldots, n$, if $\xi \in (\beta_k, \alpha_{k+1})$,

$$\operatorname{Im} f(\xi) = v_k(\xi).$$

(iii) $\lim_{z\to\infty} f(z)$ exists and equals a given value γ.

(iv) Near the points α_k and β_k, the indefinite integral of f is bounded. —

5. Show that a *particular* solution of the problem of Keldysh and Sedov is

$$f(z) = \sum_{k=1}^{n} f_{1k}(z) + \sum_{k=1}^{n} f_{2k}(z) + \frac{\gamma}{h(z)}, \tag{14.11-21}$$

where

$$f_{1k}(z) := \frac{1}{i\pi h(z)} \int_{\alpha_k}^{\beta_k} \frac{u_k(\xi)h(\xi)}{\xi - z}\, d\xi,$$

$$f_{2k}(z) := \frac{1}{\pi h(z)} \int_{\beta_k}^{\alpha_{k+1}} \frac{v_k(\xi)h(\xi)}{\xi - z}\, d\xi,$$

and where h is any function of the form

$$h(z) := \prod_{k=1}^{n} (z-\alpha_k)^{\varepsilon_k}(z-\beta_k)^{\eta_k},$$

where the numbers ε_k and η_k are $\pm\frac{1}{2}$ and satisfy

$$\sum_{k=1}^{n} \varepsilon_k + \sum_{k=1}^{n} \eta_k = 0,$$

$h(\xi) > 0$ for $\xi > \beta_k$.

6. Show that the general solution of the problem of Keldysh and Sedov is $f+d$, where f is given by (14.11-21) and where

$$d(z) := \frac{p(z)}{\prod_{k=1}^{n} (z-\alpha_k)^{1/2}(z-\beta_k)^{1/2}},$$

p being a polynomial of degree $\leqq n$.

7. Show that the solution of the problem of Keldysh and Sedov can be made unique by requiring that it remain bounded at n points chosen from the set $\{\alpha_1, \beta_1, \alpha_2, \beta_2, \ldots, \alpha_n, \beta_n\}$.

NOTES

For a treatment of the Hilbert transform in the space $L_2(-\infty, \infty)$, see Titchmarsh [1937], ch. V; see also Achieser and Glasmann [1960], p. 120 et seq., and sec. 2.7 of Morrey [1966]. A rich collection of explicit pairs of Hilbert transforms is in Erdélyi [1954], vol. II, pp. 243–262. For the numerical evaluation of Hilbert transforms, see Kress and Martensen [1970]. In physics, Hilbert transforms first occurred in the form of the dispersion relations for the real and imaginary parts of magnetic susceptibility that were discovered independently by Kronig and Kramer around 1926. There they express the physical principle of causality; see Jackson [1975], p. 310; Pines [1963], app. B; and Källén [1964], p. 1966. Other physical applications occur in S-matrix theory; see Chew [1961] and Frautschi [1963].

§ 14.12. APPLICATIONS OF THE HILBERT TRANSFORM IN SIGNAL PROCESSING

In this section we revive the convention of Chapter 10 of denoting functions of time by capital letters $F, G, X, \ldots$, and their Fourier or Laplace transforms by the corresponding lowercase letters $f, g, x, \ldots$. The applications of the Hilbert transform to be discussed here concern functions both in the time and in the transform (or frequency) domains. The Hilbert transform $\mathscr{H}F$ of a time function F will also be denoted by $\hat{F}$.

I. Single-Sideband Modulation

By a signal S we mean a function of time such that its *spectrum* $s = \mathscr{F}S$ (with $\mathscr{F}$ as defined in § 13.3) is zero outside an interval $[-\omega_0, \omega_0]$, where $\omega_0 > 0$ is a constant that is fixed during the discussion. (For radio applications, $\omega_0 = 4.5$ kHz is typical.) By the Fourier inversion theorem, if the signal S is sufficiently smooth,

$$S(\tau) = \int_{-\omega_0}^{\omega_0} s(\omega)\, e^{2\pi i \omega \tau}\, d\omega$$

for all real τ.

Signals cannot be broadcast directly, for two reasons: (a) frequencies in the range $[-\omega_0, \omega_0]$ are not suited for transmission by electromagnetic waves; (b) to permit the simultaneous broadcasting by several stations, different emitters should use nonoverlapping frequency ranges. For these reasons, signals must be *modulated* before they are broadcast. Two different systems of modulation are in use: (a) amplitude modulation (AM), and (b) frequency modulation (FM). Here we are concerned with two different kinds of amplitude modulation.

Classically, amplitude modulation is performed by multiplying the given signal by $\cos 2\pi\omega_c\tau$, where ω_c is a constant, $\omega_c \gg \omega_0$, that is typical for a given emitter. The number ω_c is called the **carrier frequency**; this is the frequency that you dial on your receiver. (For instance, for the German-speaking Swiss radio, the carrier frequency is 529 kHz.) The modulated signal then is

$$F(\tau) = 2S(\tau)\cos 2\pi\omega_c\tau. \tag{14.12-1}$$

The factor 2 is inserted for later convenience. It is easy to calculate the spectrum of the modulated signal. It is

$$\begin{aligned} f(\omega) &= (\mathcal{F}F)(\omega) \\ &= \int_{-\infty}^{\infty} S(\tau)[e^{2\pi i\omega_c\tau} + e^{-2\pi i\omega_c\tau}]\, e^{-2\pi i\omega\tau}\, d\tau \\ &= s(\omega - \omega_c) + s(\omega + \omega_c) \end{aligned} \tag{14.12-2}$$

and occupies the two intervals $[-\omega_c - \omega_0, -\omega_c + \omega_0]$ and $[\omega_c - \omega_0, \omega_c + \omega_0]$. After the modulated signal has arrived at the receiver, it must be demodulated in order to become audible. This is achieved again by multiplying by $\cos 2\pi\omega_c\tau$. This yields the function

$$F_1(\tau) = F(\tau)\cos 2\pi\omega_c\tau = 2S(\tau)(\cos 2\pi\omega_c\tau)^2.$$

Its Fourier transform is

$$f_1(\omega) = \tfrac{1}{2}s(\omega + 2\omega_c) + s(\omega) + \tfrac{1}{2}s(\omega - 2\omega_c). \tag{14.12-3}$$

The high-frequency components $s(\omega \pm 2\omega_c)$ of the spectrum are easily filtered out, and the ear will automatically perform a Fourier synthesis with the remaining spectrum $s(\omega)$, that is, it will hear the original signal $S(\tau)$.

It should be noted that in this classical system of amplitude modulation the functions $\cos 2\pi\omega_c\tau$ used for modulation and demodulation must be strictly in phase, and must use the same carrier frequency ω_c, for if they are not, or if slightly different frequencies are used, the original spectrum $s(\omega)$ will be weakened or totally suppressed.

The total length of the spectrum of the modulated signal $F(\tau)$ is $4\omega_0$, as opposed to $2\omega_0$ for the original signal. For reasons of economy it would be desirable for the spectrum of the modulated signal to occupy less space; more radio stations could then be accommodated. Clearly, some of the information contained in $f(\omega)$ is redundant. Because the signal is real,

$$s(-\omega) = \overline{s(\omega)} \tag{14.12-4}$$

for all ω. It does not seem possible to take direct advantage of this relation, because we cannot physically isolate the positive and the negative parts of the spectrum. For the spectrum of the modulated signal, however, we have, if $|\omega| < \omega_0 \leqq \frac{1}{2}\omega_c$,

$$f(\omega_c + \omega) = s(\omega),$$
$$f(\omega_c - \omega) = s(-\omega),$$

and therefore by (14.12-4),

$$f(\omega_c - \omega) = \overline{f(\omega_c + \omega)}.$$

It thus is sufficient to know the function f in the interval $[\omega_c, \omega_c + \omega_0]$.

We now ask: Is it possible to associate with the given signal S a modulated signal F_0 whose spectrum f_0 coincides with f in the intervals $\pm[\omega_c, \omega_c + \omega_0]$ and is zero outside these intervals? In other words, we wish to find F_0 such that $f_0 := \mathscr{F}F_0$ is given by

$$f_0(\omega) = \tfrac{1}{2}[1 + \operatorname{sgn}(\omega - \omega_c)]s(\omega - \omega_c) + \tfrac{1}{2}[1 - \operatorname{sgn}(\omega + \omega_c)]s(\omega + \omega_c). \tag{14.12-5}$$

This type of modulation, called **single-sideband modulation**, can indeed be realized, and the appearance of the signum function indicates that the realization will involve the Hilbert transform.

Indeed, if $\mathscr{F}S =: s$, then the Hilbert transform $\hat{S} := \mathscr{H}S$ by Theorem 14.11f has the spectrum

$$(\mathscr{F}\hat{S})(\omega) = -i \operatorname{sgn}(\omega) s(\omega).$$

By the translation theorem of the Fourier transform, the time function

$$\tau \mapsto \hat{S}(\tau)\, e^{2\pi i \omega_c \tau}$$

therefore has the spectrum

$$\omega \mapsto -i \operatorname{sgn}(\omega - \omega_c) \cdot s(\omega - \omega_c).$$

It thus follows that

$$\tfrac{1}{2} \operatorname{sgn}(\omega - \omega_c) \cdot s(\omega - \omega_c)$$

is the spectrum of

$$\frac{i}{2} \hat{S}(\tau)\, e^{2\pi i \omega_c \tau}.$$

Similarly,

$$-\tfrac{1}{2} \operatorname{sgn}(\omega + \omega_c) \cdot s(\omega + \omega_c)$$

is the spectrum of

$$-\frac{i}{2}\hat{S}(\tau)\, e^{-2\pi i\omega_c\tau}.$$

We conclude that the function f_0 defined by (14.12-5) is the spectrum of

$$F_0(\tau) := \frac{1}{2}(e^{2\pi i\omega_c\tau} + e^{-2\pi i\omega_c\tau})S(\tau) + \frac{i}{2}(e^{2\pi i\omega_c\tau} - e^{-2\pi i\omega_c\tau})\hat{S}(\tau),$$

that is, of

$$F_0(\tau) = S(\tau)\cos 2\pi\omega_c\tau - \hat{S}(\tau)\sin 2\pi\omega_c\tau. \qquad (14.12\text{-}6)$$

The desired purpose is thus achieved by modulating the given signal S not according to (14.12-1) but according to (14.12-6). This requires forming the Hilbert transform $\hat{S}$ of the given signal S. An approximate Hilbert transform may be obtained simply by discretizing the integral defining it. That is, $\hat{S}(\tau)$ is approximated by the expression

$$\frac{1}{\pi}\sum_{k=-n}^{n}\frac{S(\tau + k\,\Delta\tau)}{k}, \qquad (14.12\text{-}7)$$

where $\Delta\tau$ is a small time delay, and where n is an integer such as 15 or 30. In terms of hardware, the expression (14.12-7) may be implemented by means of transversal filters. Evidently, the approximation (14.12-7) to $\hat{S}(\tau)$ becomes available only after a time delay $n\,\Delta\tau$. In order to evaluate the modulated signal (14.12-6), the original signal must be stored an equal amount of time.

Having arrived at the receiving end, the modulated signal must again be demodulated. As in classical amplitude modulation, this is done by forming

$$F_1(\tau) := 2F_0(\tau)\cos 2\pi\omega_c\tau. \qquad (14.12\text{-}8)$$

By the translation theorem of Fourier analysis, the demodulated signal is seen to have the spectrum

$$\begin{aligned} f_1(\omega) &= f_0(\omega - \omega_c) + f_0(\omega + \omega_c) \\ &= \tfrac{1}{2}[1 + \operatorname{sgn}(\omega - 2\omega_c)]s(\omega - 2\omega_c) \\ &\quad + \tfrac{1}{2}[1 + \operatorname{sgn}(\omega)]s(\omega) + \tfrac{1}{2}[1 - \operatorname{sgn}(\omega)]s(\omega) \\ &\quad + \tfrac{1}{2}[1 - \operatorname{sgn}(\omega + 2\omega_c)]s(\omega + 2\omega_c), \end{aligned}$$

which simplifies to

$$\begin{aligned} f_1(\omega) = {}& \tfrac{1}{2}[1 + \operatorname{sgn}(\omega - 2\omega_c)]s(\omega - 2\omega_c) + s(\omega) \\ & + \tfrac{1}{2}[1 - \operatorname{sgn}(\omega + 2\omega_c)]s(\omega + 2\omega_c). \end{aligned}$$

The first and third terms represent high-frequency components that are easily filtered out, and the ear then performs a spectral synthesis with the original spectrum $s(\omega)$.

For the technical realization it is a bonus that, unlike demodulation in the classical case, demodulation of single-sideband modulation is not sensitive to phase shifts. If in place of (14.12-8) demodulation takes place according to the formula

$$F_2(\omega) := 2F_1(\omega)(\alpha \cos 2\pi\omega_c\tau + \beta \sin 2\pi\omega_c\tau),$$

then a simple calculation shows that the central part of the spectrum is given by

$$s_2(\omega) = [\alpha + i\beta \operatorname{sgn}(\omega)]s(\omega).$$

Multiplication by $i \operatorname{sgn} \omega$ of $s(\omega)$ amounts to a phase shift in $e^{2\pi i\omega\tau}$ by 90°. Since the human ear can detect frequencies but not phases, $s_2(\omega)$ has the same effect on it as has $s(\omega)$.

II. Analytic Signals

Equation (14.12-6) defining the single-sideband modulated signal may be written

$$F_0(\tau) = \operatorname{Re}\{Z(\tau)\, e^{2\pi i\omega_c\tau}\}, \tag{14.12-9}$$

where

$$Z(\tau) := S(\tau) + i\hat{S}(\tau).$$

If X is any time function that satisfies the hypotheses of Theorem 14.11c (X vanishes regularly at ∞ and is uniformly Hölder continuous in every bounded interval), and if $\hat{X} := \mathscr{H}X$, then the function

$$Z(\tau) := X(\tau) + i\hat{X}(\tau) \tag{14.12-10}$$

is called the **analytic signal** associated with X. It follows from the theorems of § 14.11 that the analytic signal can be continued to a function $Z(t)$ that is analytic for $\operatorname{Im} t > 0$, Hölder continuous for $\operatorname{Im} t \geqq 0$, and that satisfies

$$\lim_{t\to\infty} Z(t) = 0. \tag{14.12-11}$$

There is an interesting characterization of analytic signals in terms of their Fourier transforms.

THEOREM 14.12a. *The complex-valued function Z of the real variable τ is an analytic signal if and only if its Fourier transform $z := \mathscr{F}Z$ satisfies $z(\omega) = 0$ for $\omega < 0$.*

Proof. (a) Let $Z = X + i\hat{X}$ be an analytic signal. If $x := X$, then by Theorem 14.11f the Fourier transform of Z is

$$\begin{aligned} z(\omega) &= \mathscr{F}X(\omega) + i\mathscr{F}\hat{X}(\omega) \\ &= x(\omega) + i[-i \operatorname{sgn} \omega\, x(\omega)] \\ &= \begin{cases} 2x(\omega), & \omega > 0 \\ 0, & \omega < 0. \end{cases} \end{aligned}$$

(b) Let the spectrum $z(\omega)$ be 0 for $\omega < 0$. By the Fourier inversion formula the corresponding signal is

$$Z(\tau) = \int_0^\infty z(\omega)\, e^{2\pi i\omega\tau}\, d\omega.$$

This is identical with the Laplace transform of z, evaluated at $s := -2\pi i\tau$. Because the Laplace transform may be continued analytically into $\operatorname{Re} s > 0$ and vanishes as $s \to \infty$, Z may be continued as an analytic function of $t = \tau + i\sigma$ that vanishes for $t \to \infty$. It follows that $\operatorname{Im} Z(\tau) = \mathscr{H} \operatorname{Re} Z(\tau)$, that is, Z is the analytic signal associated with $X = \operatorname{Re} Z$. —

THEOREM 14.12b. *The family of analytic signals forms an algebra.*

Proof. According to the definition given in § 1.1, it is to be shown that, given any two analytic signals Z_1, Z_2 and any two complex constants c_1, c_2, then the signals $c_1Z_1 + c_2Z_2$ as well as Z_1Z_2 are analytic. This is obvious for the linear combination. Two proofs will be given for the product, both equally simple.

(a) If the analytic continuations of Z_1 and Z_2 into $\operatorname{Im} t > 0$ are again denoted by Z_1 and Z_2, then Z_1Z_2 is an analytic function in $\operatorname{Im} t > 0$ that vanishes for $t \to \infty$ and that is Hölder continuous in $\operatorname{Im} t \geqq 0$. Thus the restriction of Z_1Z_2 to $\operatorname{Im} t = 0$ is an analytic signal.

(b) Let $z_k := \mathscr{F}Z_k$, $k = 1, 2$. Then by the convolution theorem for Fourier transforms,

$$(Z_1Z_2)(\omega) = \int_{-\infty}^\infty z_1(\sigma)z_2(\omega - \sigma)\, d\sigma.$$

Because $z_k(\omega) = 0$ for $\omega < 0$, $k = 1, 2$, this reduces to

$$\int_0^\omega z_1(\sigma)z_2(\omega - \sigma)\, d\sigma,$$

which is 0 for $\omega < 0$. Thus by Theorem 14.12a, the signal Z_1Z_2 is analytic. —

III. Frequency Admittance of Minimum Phase Networks

We recall the notation of the transfer function that was introduced in § 10.3. The transfer function $g(s)$ of a linear system such as a network describes the response of the system in the Laplace domain. If f_1 and f_2 denote, respectively, the transforms of the input and of the output, then

$$f_2(s) = g(s) f_1(s).$$

In particular, if the system is excited by a simple harmonic vibration of frequency ω, then

$$f_1(s) = \frac{1}{s - i\omega}$$

and consequently,

$$f_2(s) = \frac{1}{s - i\omega} g(s).$$

If the system is stable, that is, if $g(s)$ is analytic for $\operatorname{Re} s \geqq 0$, the steady state component of the response is easily determined as

$$F_{ss}(\tau) = g(i\omega)\, e^{i\omega\tau}.$$

The function $\omega \mapsto g(i\omega)$ is called the **frequency admittance** of the system. Assuming $g(i\omega) \neq 0$ for all real ω, we may write

$$g(i\omega) = \gamma(\omega)\, e^{i\psi(\omega)}, \tag{14.12-12}$$

where both $\gamma > 0$ and ψ are real and continuous. The functions γ and ψ are called, respectively, the **amplitude characteristic** and the **phase characteristic** of the system. The amplitude characteristic describes the amplification of a pure cosine wave caused by the system, while the phase characteristic describes the phase shift. If the frequency admittance of the system is known, then the inversion formula for the Laplace transform permits, in principle, to determine the shock response $G = \mathscr{L}^{-1} g$ of the system.

Given a physical system, it is easy in practice to determine experimentally its amplitude characteristic $\gamma(\omega)$. To measure the phase characteristic $\psi(\omega)$ is difficult or impossible. Thus there arises the problem of finding ψ knowing γ. After some preliminary work this may again be accomplished by taking Hilbert transforms.

From the assumption that $g(s)$ is analytic in the right half-plane, it follows that $g(iz)$, considered as a function of the complex variable $z = \omega + i\eta$, is analytic in the *lower* half-plane $\operatorname{Im} z < 0$. In addition to $g(i\omega) \neq 0$ for real ω we now assume that $g(s) \neq 0$ for $\operatorname{Re} s \geqq 0$. Networks with this property are called **minimum phase networks**. Then

$$l(z) := \log g(iz)$$

may be defined as an analytic function in Im $z<0$, with continuous boundary values

$$l(\omega)=\operatorname{Log}\gamma(\omega)+i\psi(\omega).$$

It thus looks as if $\psi(\omega)$ could be determined by taking the Hilbert transform of Log $\gamma(\omega)$. Unfortunately this is not necessarily so because by virtue of $\gamma(\omega)\to 0$ as $\omega\to\pm\infty$, the function Log $\gamma(\omega)$ tends to $-\infty$ for $\omega\to\pm\infty$ instead of vanishing regularly as required by Theorem 14.11c. However, the situation is saved by differentiation. Under suitable hypotheses on g (for instance, if g is a rational function vanishing at ∞), the function

$$l'(z)=i\frac{g'(iz)}{g(iz)}$$

has the required behavior, and from

$$l'(\omega)=\frac{\gamma'(\omega)}{\gamma(\omega)}+i\psi'(\omega)$$

it follows that γ'/γ and ψ' are, respectively, the real and imaginary parts of the boundary values of a function that is analytic in a half-plane and vanishes at ∞. Because the half-plane of analyticity is the *lower* half-plane, the connection between ψ' and γ'/γ is given by

$$\psi'=-\mathcal{H}\frac{\gamma'}{\gamma}. \tag{14.12-13}$$

The function ψ' thus determined even has a direct physical interpretation; it is called the **group velocity**. The phase characteristic now may be found by an integration,

$$\psi(\omega)=\psi(0)-\int_0^\omega\left(\mathcal{H}\frac{\gamma'}{\gamma}\right)(\sigma)\,d\sigma,$$

using the fact that (usually) $\psi(0)$ will be zero.

PROBLEMS

1. *The product theorem* (Bedrosian [1963]). Let U and S be two time functions such that their spectra u and s satisfy, for some $\omega_0>0$,

$$s(\omega)=0,\qquad |\omega|>\omega_0;$$

$$u(\omega)=0,\qquad |\omega|<\omega_0.$$

Assuming the validity of (14.11-18) for functions in $L_2(-\infty,\infty)$, show that

$$\mathcal{H}(s\cdot u)=s\mathcal{H}u.$$

(Prove the corresponding relation for the Fourier transforms.)

2. In the context of Section III, show that the hypothesis $g(s) \neq 0$ (Re $s > 0$) is guaranteed if $g(i\omega) \neq 0$ for real ω.

NOTES

The material in this section is based on Leuthold [1974]; see also ch. 7 of Oppenheim and Schafer [1975]. For a comparison of several methods to construct a transfer function from the real part of the frequency response, see Unbehauen [1972].

15

POTENTIAL THEORY IN THE PLANE

Potential theory is concerned with solutions u of **Poisson's equation,**

$$-\Delta u = \rho, \tag{P}$$

where Δ is the Laplacian, and where ρ is a given function, defined on some region $R \subset \mathbb{R}^n$. In the important special case where $\rho \equiv 0$, the solutions of (P) are called **harmonic**. We have already studied in Chapter 5 the distinguished role that harmonic functions play in the modeling of time-independent phenomena in electricity, magnetism, heat conduction, and of flows of ideal fluids. The general equation (P) likewise has many such applications.

Here we are concerned with the special case $n = 2$ of (P). Harmonic functions in two dimensions enjoy a special property which has no counterpart in higher dimensions. They are, at least locally, representable as real parts of analytic functions, and many of their properties thus are corollaries of properties of analytic functions. Also if ρ is not identically zero, the study of the solutions of (P) is often simplified by complex variable methods, for instance, by Pompeiu's formula.

Our treatment of potential theory in the plane as usual stresses constructive and computational aspects, with particular emphasis on methods linked to complex analysis.

§ 15.1. BASIC PROPERTIES OF HARMONIC FUNCTIONS

Let u be a real-valued function defined in some region $R \subset \mathbb{R}^2$. The function u is called **harmonic** in R if it possesses continuous second partial derivatives in R, and if

$$\Delta u := \frac{\partial^2 u}{\partial x^2} + \frac{\partial^2 u}{\partial y^2} = 0, \qquad (x, y) \in R. \tag{15.1-1}$$

Here we establish the basic connections between harmonic functions and analytic functions.

I. Harmonic Functions and Analytic Functions

Let $f = u + iv$ be an analytic function of $z = x + iy$, where the point (x, y) varies in some region R of $\mathbb{R}^2$. It is well known that the real and imaginary parts u and v of f are harmonic in R. Indeed, this follows immediately from the Cauchy-Riemann equations,

$$\frac{\partial u}{\partial x} = \frac{\partial v}{\partial y}, \qquad \frac{\partial u}{\partial y} = -\frac{\partial v}{\partial x}. \tag{15.1-2}$$

For instance, by differentiating the first equation with respect to x, the second with respect to y, and adding, using the fact that the two mixed derivatives are equal, we obtain (15.1-1).

EXAMPLE 1

Let R: $z \neq 0$, $f(z) := z^{-1}$. The functions

$$u(x, y) = \operatorname{Re} z^{-1} = \frac{x}{x^2 + y^2}, \qquad v(x, y) = \operatorname{Im} z^{-1} = -\frac{y}{x^2 + y^2}$$

are harmonic.

EXAMPLE 2

Let R be any simply connected region that does not contain O, and let f be defined on R as a single-valued analytic branch of $\log z$. Then the functions

$$u(x, y) = \operatorname{Re} \log z = \operatorname{Log}|z|$$

$$v(x, y) = \operatorname{Im} \log z = \arg z$$

are both harmonic in R. In fact, $u(x, y) = \operatorname{Log}|z|$ is harmonic in *any* region R not containing O, simply connected or not. However, the same is not true for v. —

The question arises whether *every* harmonic function is the real part of an appropriate analytic function. The answer is simple if the domain of definition is simply connected.

THEOREM 15.1a. *Let R be a simply connected region. Then every u which is harmonic in R is the real part of some function f which is analytic in R. The difference of any two analytic f such that* $\operatorname{Re} f = u$ *is a purely imaginary constant.*

Proof. To show the existence of f, let

$$g(z) := u_x(z) - iu_y(z). \tag{15.1-3}$$

Using the fact that u is harmonic, it is readily shown that the real and imaginary parts of g satisfy the Cauchy–Riemann equations:

$$(\operatorname{Re} g)_x = (u_x)_x = u_{xx} = -u_{yy} = (-u_y)_y = (\operatorname{Im} g)_y,$$

$$(\operatorname{Re} g)_y = (u_x)_y = u_{xy} = u_{yx} = -(-u_y)_x = -(\operatorname{Im} g)_x.$$

It follows that g is analytic in R. Since R is simply connected, g by Theorem 4.3e possesses an indefinite integral $f = \int g$, which may be calculated by

$$f(z) = f(z_0) + \int_{z_0}^{z} g(t)\, dt. \tag{15.1-4}$$

Here z_0 is a fixed point in R, the path of integration from z_0 to z is irrelevant, and $f(z_0)$ may be chosen arbitrarily. We subject $f(z_0)$ to the condition $\operatorname{Re} f(z_0) = u(z_0)$. Then

$$\operatorname{Re} f(z) = u(z_0) + \int_{z_0}^{z} (u_x\, dx + u_y\, dy) = u(z_0) + u(z) - u(z_0)$$

$$= u(z).$$

Thus for every choice of $\operatorname{Im} f(z_0)$ the f thus constructed satisfies $\operatorname{Re} f = u$, as desired.

To see that all functions f such that $\operatorname{Re} f = u$ are obtained in this manner, we note that any such f has the derivative

$$f'(z) = u_x + iv_x = u_x - iu_y.$$

All such f thus have the same derivative, and thus can differ only by a complex constant. Since moreover the real parts of all these f are the same, the constant must be pure imaginary. —

If u is harmonic in a region R, any real function v such that $f = u + iv$ is analytic in R is called a **harmonic conjugate function** of u. (We already have made passing use of this concept in § 13.5 and § 14.2.) Another way to state Theorem 15.1a is thus:

COROLLARY 15.1b. *If R is simply connected, every harmonic function in R has a harmonic conjugate function in R. Any two such harmonic conjugate functions differ by a constant.*

The proof of Theorem 15.1a shows that a harmonic conjugate v of a given harmonic function u may be calculated by the line integral

$$v(z) = v(z_0) + \int_{z_0}^{z} (-u_y\,dx + u_x\,dy), \qquad (15.1\text{-}5)$$

where the point $z_0 \in R$, the number $v(z_0)$, and the path of integration from z_0 to z all may be chosen arbitrarily.

Part of the argument to prove Theorem 15.1a still goes through if R is no longer simply connected. Thus let u now be harmonic in a region R of arbitrary connectivity. The function

$$g(z) := u_x(z) - iu_y(z)$$

then still is analytic in R, and it is still true that, for any $z_0 \in R$,

$$u(z) = u(z_0) + \int_{z_0}^{z} (u_x\,dx + u_y\,dy).$$

Furthermore, the integral is still the real part of

$$\int_{z_0}^{z} g(t)\,dt,$$

although this latter integral does not necessarily define a single-valued function in R. We thus have:

THEOREM 15.1c. *Let u be harmonic in a region R of arbitrary connectivity. Then u is the real part of the indefinite integral of an analytic function in R.*

As is shown by Example 2, a function that is harmonic in a multiply connected region R does not necessarily possess a conjugate harmonic function. The problem thus arises to find necessary and sufficient conditions for the existence of a conjugate harmonic function.

We shall need to consider only the case where the connectivity of R is finite. Thus let the complement of R on the Riemann sphere have the components $K_0, K_1, K_2, \ldots, K_n$, where K_0 is the (possibly empty) unbounded component (see Fig. 15.1a).

Let u be harmonic in R, and let v be a conjugate harmonic function of u so that $f := u + iv$ is analytic in R. Then $g := f'$ is likewise analytic in R. Thus by Theorem 4.3e,

$$\int_\Gamma g(z)\,dz = 0$$

for every closed curve Γ in R.

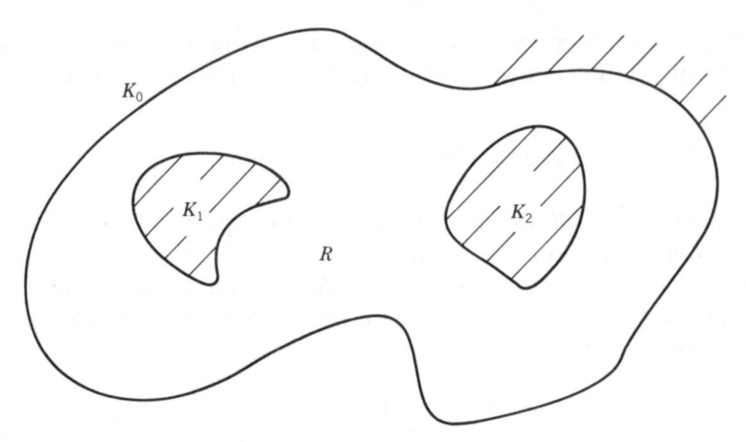

Fig. 15.1a. Multiply connected region.

We now take for granted the intuitively obvious fact that every closed curve Γ in R is homotopic to a system of curves

$$k_1\Gamma_1 + k_2\Gamma_2 + \cdots + k_n\Gamma_n,$$

where the k_i are integers, and where, for each i, Γ_i is a closed curve in R which has winding number $+1$ with respect to all points $z \in K_i$, and which has winding number 0 with respect to all $z \in K_j$, where $j \neq i$ (see Fig. 15.1b).

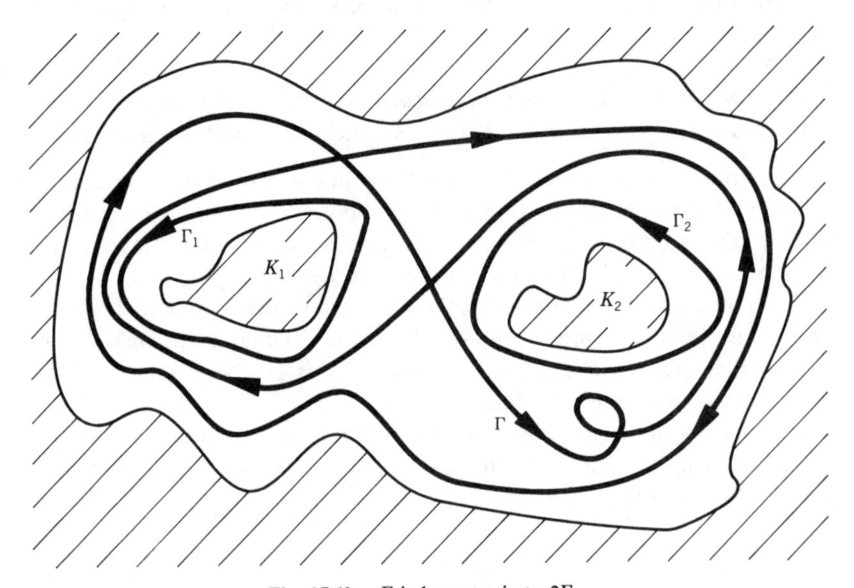

Fig. 15.1b. Γ is homotopic to $2\Gamma_1$.

By Theorem 4.3d,

$$\int_\Gamma g(z)\,dz = \sum_{i=1}^{n} k_i \int_{\Gamma_i} g(z)\,dz.$$

If this is to be zero for all closed Γ, there must necessarily hold

$$\int_{\Gamma_i} g(z)\,dz = 0, \qquad i = 1, 2, \ldots, n. \tag{15.1-6}$$

For any analytic function g the integrals on the left of (15.1-6) are called the **periods** of g with respect to K_i. In our situation, $g = f'$ is expressed in terms of u by (15.1-3). We thus have

$$\int_{\Gamma_i} g(z)\,dz = \int_{\Gamma_i} (u_x\,dx + u_y\,dy) + i \int_{\Gamma_i} (-u_y\,dx + u_x\,dy).$$

On the right, the first integral is automatically zero. *For the existence of a conjugate harmonic function of u it is thus necessary that the integrals*

$$\eta_i := \int_{\Gamma_i} (-u_y\,dx + u_x\,dy), \qquad i = 1, 2, \ldots, n, \tag{15.1-7}$$

are all zero. The numbers η_i are the periods of the differential of the conjugate function of u; more briefly, they will be called the **conjugate periods** of u.

The condition just mentioned is not only necessary, but also sufficient for the existence of a conjugate harmonic function. Let u be harmonic in R, and consider the analytic function $g := u_x - iu_y$. If the conjugate periods of u are all zero, the integral of g along any closed curve Γ in R is zero. By Theorem 4.3a, g possesses an indefinite integral f. We select f such that $\operatorname{Re} f(z_0) = u(z_0)$ for some selected point $z_0 \in R$. As in the proof of Theorem 15.1a, we can verify that $\operatorname{Re} f(z) = u(z)$ for all $z \in R$. The function $v(z) := \operatorname{Im} f(z)$ is a conjugate harmonic function of u. Again it is determined only up to an additive constant. We have proved:

THEOREM 15.1d. *A harmonic function u in a finitely connected region R has a conjugate harmonic function in R if and only if the conjugate periods with respect to all bounded components of the complement of R are zero.*

Given sufficient regularity, the expression (15.1-7) for the periods can be transformed so as to carry more intuitive appeal. Let the boundary of K_i be parametrizable as a curve with a piecewise continuously turning tangent, and let u have continuous first-order derivatives in the closure of R. We then may choose for Γ_i the boundary ∂K_i of K_i. If (dx, dy) is the direction of the tangent, $(-dy, dx)$ is the direction of the exterior normal. Thus the integrand in (15.1-7) is the scalar product of (u_x, u_y), the gradient of u, with the exterior normal unit vector times $|dz|$, that is, it equals $\partial u/\partial n\,|dz|$. Thus we find:

OBSERVATION 15.1e. *Given sufficient smoothness as indicated above, the conjugate period of u with respect to K_i may be calculated as*

$$\eta_i = \int_{\partial K_i} \frac{\partial u}{\partial n}\,|dz|, \tag{15.1-8}$$

where ∂K_i is the boundary of K_i, and where $\partial/\partial n$ denotes differentiation in the direction of the exterior normal.

II. The Mean Value Property

Let f be a real- or complex-valued function defined in a region $R \subset \mathbb{C}$. Then f is said to possess the **mean value property** in R if it is continuous in R and if for every $z \in R$ the identity

$$f(z) = \int_0^1 f(z + \rho\, e^{2\pi i \tau})\, d\tau \tag{15.1-9}$$

holds for all sufficiently small values of ρ. The formula (15.1-9) expresses the fact that the value of f at the center of a circle of radius ρ equals the arithmetic mean of the values on the circumference.

Let f be analytic in R. If $z \in R$, and if Γ denotes the positively oriented circle of radius ρ about z, then by Cauchy's formula, if ρ is sufficiently small,

$$f(z) = \frac{1}{2\pi i} \int_\Gamma \frac{f(t)}{t - z}\, dt,$$

which on introducing $t = z + \rho\, e^{2\pi i \tau}$ reduces to (15.1-9). Thus *analytic functions possess the mean value property.*

Let now u be harmonic in R. If R_1 is any simply connected subregion of R, then there exists an analytic function f in R_1 such that $u = \operatorname{Re} f$ in R_1. If $z \in R_1$ and ρ is sufficiently small, then by (15.1-9),

$$u(z) = \operatorname{Re} f(z) = \operatorname{Re} \int_0^1 f(z + \rho\, e^{2\pi i \tau})\, d\tau = \int_0^1 \operatorname{Re} f(z + \rho\, e^{2\pi i \tau})\, d\tau$$

$$= \int_0^1 u(z + \rho\, e^{2\pi i \tau})\, d\tau,$$

and we see that u possesses the mean value property. In brief, we have:

THEOREM 15.1f. *Let u be harmonic in the region R. Then u possesses the mean value property in R.*

In § 15.2 we shall establish the important converse that a real function which possesses the mean value property is harmonic. Here we state another consequence of the mean value property for real functions:

THEOREM 15.1g (Principle of the maximum). *Let the real function u possess the mean value property in the region R, and let*

$$\mu := \sup_{z \in D} u(z)$$

be finite. Then either

$$u(z) < \mu \quad \text{for all} \quad z \in R$$

or u is constant and

$$u(z) = \mu \quad \text{for all} \quad z \in R.$$

That is, a real function possessing the mean value property cannot take on its least upper bound on a connected open set without being identically equal to the least upper bound.

Proof. R is covered by the two sets

$$R_1 := \{z \in R: \quad u(z) < \mu\},$$

$$R_2 := \{z \in R: \quad u(z) = \mu\},$$

which both will be shown to be open. R_1 is open by the continuity of u. If R_2 were not open, there would exist a point $z_2 \in R_2$ such that in every neighborhood of z_2 there are points z_1 such that $u(z_1) < \mu$. Choose such a point z_1 with the additional property that $\rho = |z_1 - z_2|$ is sufficiently small for the relation (15.1-9) to hold for $z = z_2$ with $f = u$. The integrand then is $\leqq \mu$, and at the point z_1 it is $< \mu$. Since the integrand is continuous, we find

$$u(z_2) = \int_0^1 u(z_2 + |z_1 - z_2| \, e^{2\pi i \tau}) \, d\tau < \mu,$$

contradicting the fact that $z_2 \in R_2$. It follows that R_2 is open. Since $R = R_1 \cup R_2$ is connected and $R_1 \cap R_2 = \emptyset$, the definition of connectedness (see § 3.2) implies that either R_1 or R_2 is empty, which yields the alternative of the theorem. —

In an identical manner (or by applying the foregoing result to $-u$) one can prove that a real function possessing the mean value property cannot take its *infimum* in a region without being identically equal to it.

Frequently the maximum principle is used in the following form which is an immediate consequence:

COROLLARY 15.1h. *Let the real function u be continuous on the closure of the region R, and let it possess the mean value property in R. Then u takes its maximum and its minimum on the boundary of R.*

PROBLEMS

1. Derive the maximum principle for analytic functions from the maximum principle for harmonic functions. (Consider $\mathrm{Log}|f(z)|$.)
2. If the function $z \mapsto u(z)$ is harmonic in the region R, prove that $z \mapsto u(\bar{z})$ is harmonic in the complex conjugate region $\bar{R}$. Does the analogous statement hold for analytic functions?
3. If Γ is the circle with center a and radius $\rho > 0$, then the reflection of a point z at Γ is

$$z' := a + \frac{\rho^2}{\bar{z} - \bar{a}}.$$

   Show that harmonicity is preserved under reflection, that is, if u is harmonic in a region R not containing a, then the function $z \to u(z')$ is harmonic in the reflected region R'.
4. Let ϕ be a real function defined in a region R. If f is a conformal map of R, the function $\psi := \phi \circ f^{[-1]}$ is said to arise from ϕ by conformal transplantation. In § 5.6 we gave a computational proof of the fact that the conformal transplant of a harmonic function is harmonic. Use Theorem 15.1a to give a noncomputational proof.
5. Let u be harmonic in the disk D: $|z - a| < \rho$. Show that there exist real constants $a_0, a_1, b_1, a_2, b_2, \ldots$ such that for $z \in D$ the expansion

$$u(z) = a_0 + \sum_{n=1}^{\infty} (a_n r^n \cos n\phi + b_n r^n \sin n\phi)$$

   holds, where $z - a = re^{i\phi}$.
6. Let f be analytic in a multiply connected region R.

   (a) Show that f has an indefinite integral in R (that is, there exists an analytic function h in R such that $h' = f$) if and only if, in the notation of Theorem 15.1a, all periods

$$p_i := \int_{\Gamma_i} f(z)\, dz$$

   are zero.

   (b) In general, the indefinite integral of f is not single-valued. Show, however, that the indefinite integral of f can always be written as

$$h(z) = h_0(z) + \sum_{i=1}^{n} c_i \log(z - a_i),$$

   where h_0 is single-valued and analytic in R, where each a_i is an arbitrarily chosen point in K_i, and where the c_i are appropriate constants that do not depend on the a_i. (All periods of the function

$$f_0(z) := f(z) - \sum_{j=1}^{n} \frac{p_j}{2\pi i} \frac{1}{z - a_j}$$

are zero.)

7. Let u be harmonic in the annulus A: $\rho_1 < |z - a| < \rho_2$. Show that there exist real numbers a_n, b_n, c such that for $z - a = re^{i\phi}$, $\rho_1 < r < \rho_2$, u has the representation

$$u(z) = a_0 + c \operatorname{Log} r + \sum_{n=1}^{\infty} \{(a_n r^n + a_{-n} r^{-n}) \cos n\phi + (b_n r^n + b_{-n} r^{-n}) \sin n\phi\}.$$

When does the conjugate harmonic function exist, and what is its representation if it exists?

NOTES

For an excellent survey of potential theory see Gårding [1980]. This is physically motivated, as is Wermer [1974], who gives a full treatment of the case $n \geqq 3$. For a profound but difficult discussion of the two-dimensional case, see Tsuji [1959].

§ 15.2. THE DIRICHLET PROBLEM FOR A DISK AND FOR AN ANNULUS

I. The Dirichlet Problem

The problem of determining the "general" solution of Laplace's differential equation (15.1-1) has already been solved at the very beginning of the present chapter. The totality of all functions harmonic in a region R is equal to the totality of the real parts of the indefinite integrals of all functions that are analytic in R. As is generally the case with differential equation problems, practical applications require us to find not so much the general solution of a given equation, but rather a particular solution satisfying certain subsidiary conditions. In the case of harmonic functions (as for more general elliptic equations) these subsidiary conditions often take the form of *boundary conditions.* A major part of the present chapter is devoted to analytical methods for identifying solutions that satisfy given conditions on the boundary. We begin with Dirichlet's problem, which serves as a prototype for all such problems.

A simple version of the Dirichlet problem was defined in § 5.6. Here we consider the following more general form of the problem. Let R be a region with boundary ∂R, and let ϕ be a real continuous function defined on ∂R. The **problem of Dirichlet** consists in finding a function u satisfying the following conditions:

(i) u is continuous in $R \cup \partial R$.
(ii) u is harmonic in R.
(iii) $u = \phi$ on ∂R.

(In § 5.6 we considered the special case where R is a Jordan region.) We refer to Chapter 5 for a variety of mathematical models of physical situations that lead to Dirichlet problems.

THEOREM 15.2a. *The Dirichlet problem for a bounded region R has at most one solution.*

Proof. Let u_1 and u_2 be any two solutions. Then $u_1 - u_2$ is continuous on $R \cup \partial R$, harmonic in R, and zero on ∂R. By the maximum principle, $u_1 - u_2$ assumes its maximum as well as its minimum on ∂R, and hence is zero. —

The following example shows that the boundedness of R is essential for the uniqueness of the solution of the Dirichlet problem.

EXAMPLE 1

Let R: $\operatorname{Im} z > 0$, and let $\phi = 0$ on $\operatorname{Im} z = 0$. Then the functions $u = 0$ and $u = y$ both are solutions of the Dirichlet problem.

II. The Dirichlet Problem for a Disk

In the present section we prove the *existence* of a solution of the Dirichlet problem in the case where R is a disk. Actually, this existence was almost proved in § 14.2, for there it was shown that if R is the unit disk and ϕ is a given real function in Π (the functions of period 1), which is uniformly Hölder continuous, then there exists a unique function f such that f is analytic in $|z| < 1$, continuous in $|z| \leqq 1$, and satisfies

$$\operatorname{Re} f(e^{2\pi i\tau}) = \phi(\tau), \qquad 0 \leqq \tau \leqq 1,$$
$$\operatorname{Im} f(0) = 0.$$

This function is given by the *Schwarz formula* (14.2-14),

$$f(z) = \int_0^1 \phi(\tau) \frac{e^{2\pi i\tau} + z}{e^{2\pi i\tau} - z}\, d\tau.$$

Taking real parts and setting

$$\kappa(z, \tau) := \operatorname{Re} \frac{e^{2\pi i\tau} + z}{e^{2\pi i\tau} - z} = \frac{1 - |z|^2}{|e^{2\pi i\tau} - z|^2}, \tag{15.2-1}$$

we obtain the following explicit formula for the solution of Dirichlet's problem for the unit disk with boundary condition $u(e^{2\pi i\tau}) = \phi(\tau)$:

$$u(z) = \int_0^1 \phi(\tau)\kappa(z, \tau)\, d\tau. \tag{15.2-2}$$

The function $\kappa(z, \tau)$ is known as the **Poisson kernel**, and the formula (15.2-2) is called **Poisson's formula**. For a disk $|z| < \rho$ the solution of Dirichlet's problem with boundary data $u(\rho\, e^{2\pi i\tau}) = \phi(\tau)$ evidently is

$$u(z) = \int_0^1 \phi(\tau)\kappa\left(\frac{z}{\rho}, \tau\right) d\tau. \tag{15.2-3}$$

By the foregoing derivation, Poisson's formula solves the Dirichlet problem if the boundary data are Hölder continuous. In § 12.10 we derived the same formula under the assumption that the boundary data can be expanded in an absolutely convergent Fourier series. We now show that Poisson's formula solves the Dirichlet problem for arbitrary continuous boundary data ϕ. In fact, the following slightly more general result will be proved:

THEOREM 15.2b. *Let $\phi \in \Pi$ be real and Riemann integrable, and let u for $|z| < \rho$ be defined by Poisson's integral* (15.2-3). *Then:*

(i) *u is harmonic.*
(ii) *If ϕ is continuous at τ_0 and $z_0 := e^{2\pi i\tau_0}$, there holds*

$$\lim_{z \to z_0} u(z) = \phi(\tau_0). \tag{15.2-4}$$

Proof. It clearly suffices to prove the special case $\rho = 1$. Assertion (i) is intuitively clear from the fact that the Poisson kernel $\kappa(z, \tau)$ for each fixed τ is harmonic, and integration with respect to τ should not destroy the harmonicity. To prove this rigorously, we note that by Theorem 4.1a (integration with respect to a parameter) the Schwarz integral

$$f(z) = \int_0^1 \phi(\tau)\frac{e^{2\pi i\tau} + z}{e^{2\pi i\tau} - z}\, d\tau$$

is analytic in $|z| < 1$. Thus $u = \operatorname{Re} f$ indeed is harmonic.

Turning to (ii) we first note that Poisson's formula certainly yields the right answer for the boundary data $\phi(\tau) = 1$, for which the unique solution of the Dirichlet problem is $u(z) = 1$. Therefore

$$\int_0^1 \kappa(z, \tau)\, d\tau = 1 \tag{15.2-5}$$

for all z such that $|z| < 1$. By subtracting $\phi(\tau_0)$ times (15.2-5) from (15.2-4), it thus suffices to prove (15.2-4) for the special case $\phi(\tau_0) = 0$. By rotating the coordinate system we may further assume that $z_0 = 1$. It thus is to be proved that if ϕ is continuous at $\tau = 0$ and $\phi(0) = 0$, then

$$u(z) = \int_{-1/2}^{1/2} \phi(\tau)\kappa(z, \tau)\, d\tau \to 0 \tag{15.2-6}$$

as $z \to 1$. In order to appreciate the following proof, it helps to understand the mechanism by which the Poisson kernel works. If z is close to 1, then

$$\kappa(z, \tau) = \frac{1 - |z|^2}{|e^{2\pi i\tau} - z|^2}$$

is small (because of a small numerator) for all $\tau \in [-\frac{1}{2}, \frac{1}{2}]$ except those near 0. As $z \to 1$, the Poisson kernel thus acts as a filter which lets pass only values of ϕ taken near $\tau = 0$. To prove (15.2-6), then, we split the integral into two which are small for different reasons. Given $\varepsilon > 0$, we choose $\delta > 0$ so that

$$|\phi(\tau)| < \varepsilon, \qquad -\delta < \tau < \delta,$$

which is possible in view of the presumed continuity of ϕ. Writing

$$\int_{-1/2}^{1/2} = \int_{-\delta}^{\delta} + \int_{\delta \leqq |\tau| \leqq 1/2} = I_1 + I_2,$$

we have $|I_1| < \varepsilon$ in view of (15.2-5) and the fact that κ is $\geqq 0$. As to I_2, we note that κ can be extended to a continuous function on the compact set Σ described by the conditions

$$|z| \leqq 1, \qquad |z - 1| \leqq \frac{\delta}{2}, \qquad \delta \leqq |\tau| \leqq \frac{1}{2},$$

with value 0 for $|z| = 1$. Hence κ is uniformly continuous on Σ, and

$$\lim_{z \to 1} I_2 = 0.$$

It follows that for $|z - 1|$ sufficiently small, $|u(z)| < 2\varepsilon$. Since ε was arbitrary, this proves (15.2-6), and hence (ii). —

COROLLARY 15.2c. *Poisson's integral solves the Dirichlet problem for the disk $|z| < \rho$ for arbitrary continuous boundary data ϕ.*

For the numerical evaluation of Poisson's formula, it is probably best not to use the formula directly, which will involve an almost singular integrand if z is close to 1. Instead, we expand the kernel using

$$\operatorname{Re} \frac{e^{2\pi i\tau} + z}{e^{2\pi i\tau} - z} = \operatorname{Re} \left\{ 1 + 2 \sum_{m=1}^{\infty} z^m e^{-2\pi i m\tau} \right\},$$

which on letting $z = \rho\, e^{2\pi i\sigma}$ may be written

$$\operatorname{Re} \frac{e^{2\pi i\tau} + z}{e^{2\pi i\tau} - z} = \sum_{m=-\infty}^{\infty} \rho^{|m|}\, e^{2\pi i m(\sigma-\tau)}.$$

Using this in (15.2-2) we get

$$u(z) = \int_0^1 \phi(\tau) \sum_{m=-\infty}^{\infty} \rho^{|m|}\, e^{2\pi i m(\sigma-\tau)}\, d\tau,$$

and on interchanging integration and summation,

$$u(z) = \sum_{m=-\infty}^{\infty} a_m \rho^{|m|}\, e^{2\pi i m\tau}, \tag{15.2-7}$$

where the coefficients

$$a_m := \int_0^1 \phi(\tau)\, e^{-2\pi i m\tau}\, d\tau$$

are the complex Fourier coefficients of ϕ. The convergence of the series (15.2-7) for $|z| < 1$ takes place regardless of whether the Fourier series for ϕ converges, for the sequence $\{a_m\}$ certainly is bounded, by the Riemann-Lebesgue lemma (Theorem 10.6a), and thus (15.2-7) has the convergent majorant $\mu \sum \rho^{|m|}$.

To evaluate the series (15.2-7) one first computes the a_m by means of an FFT (see § 13.1), and then by another application of an FFT one can evaluate the expression (15.2-7) simultaneously at all points $z = \rho \exp(2\pi i k/n)$, $k = 0, 1, \ldots, n-1$, where n is the number of sampling points used to compute the Fourier coefficients.

III. The Dirichlet Problem for an Annulus

Let $0 < \rho < 1$, let A denote the annulus $\rho < |z| < 1$, and let ϕ_0 and ϕ_1 be two continuous functions in Π. To solve the Dirichlet problem for A with boundary conditions

$$\begin{aligned} u(e^{2\pi i\tau}) &= \phi_0(\tau), \\ u(\rho\, e^{2\pi i\tau}) &= \phi_1(\tau), \qquad 0 \leqq \tau \leqq 1, \end{aligned} \tag{15.2-8}$$

we seek the solution in the form

$$u = u_0 + u_1,$$

where u_0 and u_1 are the solutions of the Dirichlet problem for A with the special boundary conditions

$$u_0(e^{2\pi i\tau}) = \phi_0(\tau), \qquad u_0(\rho\, e^{2\pi i\tau}) = 0,$$

and

$$u_1(e^{2\pi i\tau})=0, \qquad u_1(\rho\, e^{2\pi i\tau})=\phi_1(\tau),$$

$0 \leqq \tau \leqq 1$, respectively.

To find u_0, we write

$$u_0 = u_{00} - u_{01},$$

where u_{00} is the solution of the Dirichlet problem for the full unit disk $|z|<1$ satisfying

$$u_{00}(e^{2\pi i\tau}) = \phi_0(\tau), \qquad 0 \leqq \tau \leqq 1,$$

and where u_{01} is the solution of the Dirichlet problem for A satisfying

$$u_{01}(e^{2\pi i\tau})=0, \qquad u_{01}(\rho\, e^{2\pi i\tau}) = u_{00}(\rho\, e^{2\pi i\tau}).$$

By Poisson's formula (15.2-2),

$$u_{00}(z) = \int_0^1 \phi_0(\tau)\kappa(z, \tau)\, d\tau,$$

where $\kappa(z, \tau)$ is the Poisson kernel. For $z = \rho\, e^{2\pi i\sigma}$ this by (15.2-7) yields

$$u_{00}(\rho\, e^{2\pi i\sigma}) = \sum_{m=-\infty}^{\infty} a_m \rho^{|m|}\, e^{2\pi i m\sigma}, \tag{15.2-9}$$

where

$$a_m := \int_0^1 \phi_0(\tau)\, e^{-2\pi i m\tau}\, d\tau \tag{15.2-10}$$

is the mth Fourier coefficient of ϕ_0, $a_{-m} = \overline{a_m}$. Because $\rho < 1$, the Fourier series (15.2-9) converges at least like a geometric series, and $u_{00}(\rho\, e^{2\pi i\sigma})$ is a very smooth function of σ.

We now seek to represent u_{01} in the form

$$u_{01}(z) = \sum_{m=-\infty}^{\infty} a_m \rho^{|m|} u_{01m}(z),$$

where u_{01m} is the solution of the special Dirichlet problem for A satisfying the boundary conditions

$$u_{01m}(e^{2\pi i\tau}) = 0, \qquad u_{01m}(\rho\, e^{2\pi i\tau}) = e^{2\pi i m\tau}, \qquad 0 \leqq \tau \leqq 1.$$

For $m = 0$, the explicit form of u_{10m} is easily found to be

$$u_{010}(z) = \frac{\operatorname{Log}|z|}{\operatorname{Log}\rho}.$$

For $m \neq 0$ we have, letting $z = r\,e^{2\pi i\sigma}$,

$$u_{01m}(z) = \frac{r^m - r^{-m}}{\rho^m - \rho^{-m}}\, e^{2\pi i m\sigma}.$$

Altogether this yields

$$u_{01}(z) = a_0 \frac{\operatorname{Log}|z|}{\operatorname{Log}\rho} + \sum_{m\neq 0} a_m \rho^{|m|} \frac{r^m - r^{-m}}{\rho^m - \rho^{-m}}\, e^{2\pi i m\sigma}, \qquad (15.2\text{-}11)$$

where the series converges, uniformly for $\rho \leqq |z| \leqq 1$, at least like a geometric series with ratio ρ. Thus u_{01} is clearly harmonic in A, continuous in the closure of A, and satisfies the required boundary conditions.

Using (15.2-10) and interchanging summation and integration, we have

$$u_{01}(z) = \int_0^1 \phi_0(\tau)\kappa_0(\rho, z, \tau)\, d\tau,$$

where, always letting $z = r\,e^{2\pi i\sigma}$,

$$\kappa_0(\rho, z, \tau) := \frac{\operatorname{Log} r}{\operatorname{Log}\rho} + \sum_{m\neq 0} \rho^{|m|} \frac{r^m - r^{-m}}{\rho^m - \rho^{-m}}\, e^{2\pi i m(\sigma-\tau)}. \qquad (15.2\text{-}12)$$

Using $u_0 = u_{00} - u_{01}$ and representing u_{00} by Poisson's integral, this yields the Poisson-like representation for u_0,

$$u_0(z) = \int_0^1 \phi_0(\tau)\kappa(\rho, z, \tau)\, d\tau, \qquad (15.2\text{-}13)$$

where

$$\kappa(\rho, z, \tau) := \kappa(z, \tau) - \kappa_0(\rho, z, \tau). \qquad (15.2\text{-}14)$$

Equation (15.2-13) represents the solution u_0 in a fashion resembling Poisson's integral, the only difference being that the kernel here cannot be represented in closed form.

It remains to determine u_1. Assuming the existence of u_1, we note that the function

$$u_1^*(z) := u_1\left(\frac{1}{\bar{z}}\right)$$

is a solution of Dirichlet's problem for the annulus $1 < |z| < \rho^{-1}$ satisfying the boundary conditions

$$u_1^*(e^{2\pi i\tau}) = 0, \qquad u_1^*(\rho^{-1}\, e^{2\pi i\tau}) = \phi_1(\tau).$$

It follows that the function $z \to u_1^*(\rho^{-1}z)$ is a solution for A having boundary values $\phi_1(\tau)$ on $|z|=1$ and 0 on $|z|=\rho$. Hence $u_1^*(\rho^{-1}z)$ is given by (15.2-13),

$$u_1^*(\rho^{-1}z)=\int_0^1 \phi_1(\tau)\kappa(\rho, z, \tau)\,d\tau$$

and $u_1(z)=u_1^*(1/\bar{z})$ consequently is given by

$$u_1(z)=\int_0^1 \phi_1(\tau)\kappa\left(\rho, \frac{\rho}{\bar{z}}, \tau\right)d\tau.$$

Altogether we have obtained:

THEOREM 15.2d. *The solution of Dirichlet's problem for the annulus A: $\rho<|z|<1$ with boundary data* (15.2-8) *is given by*

$$u(z)=\int_0^1 \left\{\phi_0(\tau)\kappa(\rho, z, \tau)+\phi_1(\tau)\kappa\left(\rho, \frac{\rho}{\bar{z}}, \tau\right)\right\}d\tau,$$

where $\kappa(\rho, z, \tau)$ is defined by (15.2-14), (15.2-12), *and* (15.2-1).

For the numerical evaluation of u it seems best to use the decomposition $u=u_0+u_1$ and to compute, say, u_0 by obtaining the Fourier coefficients a_m of ϕ_0 by an FFT (see § 13.1), to compute the Fourier coefficients of the series (15.2-11), and then to evaluate u_{01} on any circle $|z|=\text{const}$ by another FFT.

PROBLEMS

1. Let $\Gamma:=$ unit circle, f analytic on and in the interior of Γ.
   (a) Modify Cauchy's formula,

$$f(z)=\frac{1}{2\pi i}\int_\Gamma \left(\frac{1}{t-z}+\cdots\right)f(t)\,dt$$

   such that $(2\pi i)^{-1}(1/(t-z)+\cdots)\,dt$ becomes real.
   (b) Obtain Poisson's formula by taking real parts in the resulting expression,

$$f(z)=\frac{1}{2\pi i}\int_\Gamma \left(\frac{1}{t-z}+\frac{1}{t(1-t\bar{z})}-\frac{1}{t}\right)f(t)\,dt.$$

2. Let u be harmonic in D: $|z|<1$ and continuous in $|z|\leqq 1$. Let $|a|<1$. As is well known, the Moebius transformation

$$t\colon z\to w=t(z):=\frac{z+a}{1+\bar{a}z}$$

   maps D onto D and sends 0 into a.
   (a) Using the fact that $u(t(z))$ is harmonic, show that

$$u(t(0))=\int_0^1 u(t(e^{2\pi i\tau}))\,d\tau \qquad (*)$$

   (mean value property).

(b) Obtain Poisson's formula by letting $t(e^{2\pi i\tau}) = e^{2\pi i\sigma}$ in (*).

3. *Poisson's formula for the ellipse.* Let $\rho > 1$. To solve Dirichlet's problem for the interior E of the ellipse

$$\Gamma: \quad z = z(\tau) := \tfrac{1}{2}(\rho\, e^{2\pi i\tau} + \rho^{-1}\, e^{-2\pi i\tau}), \qquad 0 \leqq \tau \leqq 1,$$

with foci at ± 1, assume the boundary data

$$\psi(\tau) := \phi(z(\tau))$$

to be given as an absolutely convergent Fourier series,

$$\psi(\tau) = \sum_{n=-\infty}^{\infty} a_n\, e^{2\pi i n\tau}.$$

(a) If, for $n \in \mathbb{N}$, $u_n(z)$ is (complex-valued) harmonic in E, continuous in $E \cup \Gamma$, and such that $u_n(z(\tau)) = e^{2\pi i n\tau}$, then the solution of Dirichlet's problem is given by

$$u(z) = \sum_{n=-\infty}^{\infty} a_n u_n(z)$$

(principle of the maximum).

(b) If T_n denotes the nth Chebyshev polynomial (see § 14.6), prove that

$$T_n(z(\tau)) = \tfrac{1}{2}(\rho^n\, e^{2\pi i n\tau} + \rho^{-n}\, e^{-2\pi i n\tau}).$$

Conclude that $u_0(z) = 1$,

$$u_n(z) = \frac{\rho^n T_n(z) - \rho^{-n}\overline{T_n(z)}}{\frac{1}{2}(\rho^{2n} - \rho^{-2n})}, \qquad n \neq 0.$$

(c) Thus show that the solution of the Dirichlet problem is given by the Poisson-like formula

$$u(z) = \int_0^1 \psi(\tau)\varepsilon(\rho, z, \tau)\, d\tau,$$

where

$$\varepsilon(\rho, z, \tau) := 1 + \sum_{n \neq 0} \frac{\rho^n T_n(z) - \rho^{-n}\overline{T_n(z)}}{\frac{1}{2}(\rho^{2n} - \rho^{-2n})}\, e^{-2\pi i n\tau}.$$

NOTES

For the derivation of Poisson's formula given in Problem 2, see Ahlfors [1966], pp. 167–168. Theorem 15.2d is due to Villat [1912], where the function κ is expressed in terms of elliptic functions; see also Demchenko [1931].

§ 15.3. SOME CONSEQUENCES OF POISSON'S FORMULA

This section offers a quick perusal of some theoretical properties of harmonic functions. The analogies, as well as the differences, to the corresponding properties of analytic functions should be noted. The justification of all these properties requires Poisson's integral, or the powerful Theorem 15.3a following from it.

I. The Converse of the Mean Value Property

It was seen rather trivially in Theorem 15.1f that every harmonic function has the mean value property. Conversely, we have:

THEOREM 15.3a. *Let u be a real continuous function which has the mean value property in a region R. Then u is harmonic in R.*

The importance of this result lies in the fact that it permits us to verify the harmonicity of a function without computing any derivatives.

Proof. It evidently suffices to show that u is harmonic in any disk whose closure is contained in R. Let D be such a disk, and let u_0 be the harmonic function whose values on the boundary ∂D coincide with the values of u. (It is precisely the existence of u_0 where Poisson's integral is required for the proof.) Both functions u and u_0 possess the mean value property in D, the first by hypothesis, and the second because it is harmonic. It follows that $u - u_0$ possesses the mean value property in D. But now the maximum principle (in the form given as Corollary 15.1h) implies that $u - u_0$ takes both its maximum and its minimum on ∂D. Because the only value taken by $u - u_0$ on D is zero, it follows that $u = u_0$ in D. Hence u equals a harmonic function in D. —

II. Harmonic Continuation

Let the region R be a proper subregion of R_1, and let u be harmonic in R. A function u_1 harmonic in R_1 whose restriction to R equals u is called a **harmonic continuation** (or **harmonic extension**) of u. In the following we deal with some simple situations where harmonic continuations exist. Our first result is a rather precise analog of the reflection principle for analytic functions (Theorem 5.11b).

THEOREM 15.3b (Reflection principle for harmonic functions). *Let R be a region in the upper half-plane whose boundary contains a segment Γ of the real axis. Let u be harmonic in R, continuous in $R \cup \Gamma$, and equal to zero on Γ. Then the function u_1 defined by*

$$u_1(z) := \begin{cases} u(z), & z \in R \cup \Gamma \\ -u(\bar{z}), & z \in \bar{R} \end{cases}$$

is a harmonic continuation of u into the region $R_1 := R \cup \Gamma \cup \bar{R}$.

Proof. It is clear that u_1 is continuous in R_1. By construction it equals u in R. To prove that u_1 is harmonic in R_1, we show that u_1 possesses the mean value property in R_1. This is clear for points $z \in R$, and by the reversal of sign also for the points $z \in \bar{R}$. For the points $z \in \Gamma$ the mean value property follows from the fact that in the mean value integral (15.1-9) the contributions from points located symmetrically with respect to the real axis cancel each other because of opposite signs. Hence the value of the integral is zero, which equals the value of u_1 on Γ. —

For analytic functions we proved in Theorem 5.11a a far-reaching principle of continuous continuation which stated, in essence, that for analytic functions a continuous continuation equals an analytic continuation. The following example shows that an analogous theorem cannot hold for harmonic functions.

EXAMPLE 1

Let u be the bounded harmonic function in the upper half-plane, which on the real axis takes the continuous boundary values

$$u(x) = \min(1, |x|).$$

(Poisson's integral establishes the existence of u via conformal transplantation.) Setting $u(z) := u(\bar{z})$ ($\operatorname{Im} z < 0$), u is continued continuously into $\mathbb{C}$, and the continued function is harmonic in the lower half-plane. Yet u fails to be harmonic in $\mathbb{C}$, because it does not satisfy the mean value property at $z = 0$. —

In place of Theorem 5.11a we can offer only the following weaker result:

THEOREM 15.3c (Principle of harmonic continuation). *Let the boundary of the region R contain an analytic arc Γ. Let u be harmonic in R, continuous in $R \cup \Gamma$, and let the values taken by u on Γ depend analytically on the curve parameter on Γ. Then u can be extended harmonically into a region which contains Γ in its interior.*

Proof. To begin with we consider the case where Γ is a segment of the real axis. The function $u(x)$ being real-analytic for $x \in \Gamma$, it can be continued to a function f^* which is analytic in a region R^* containing Γ. We may assume that R^* is symmetric with respect to the real axis. The function $u^* := \operatorname{Re} f^*$ is harmonic in R^* and agrees with u on Γ. Thus $u_1 := u - u^*$ satisfies the hypotheses of the preceding theorem in the region $R \cap R^*$, and thus can be extended harmonically into $R_1 := (R \cap R^*) \cup \Gamma \cup (\overline{R \cap R^*})$. With the extended definition of u_1, we can define $u := u_1 + u^*$ as a harmonic function in R_1, and the desired continuation is accomplished (see Fig. 15.3a).

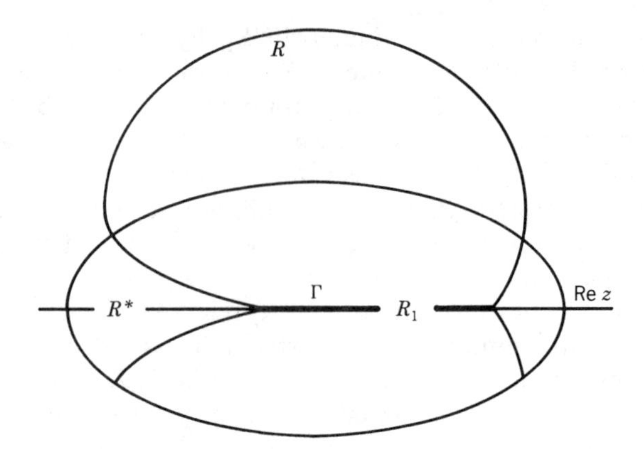

Fig. 15.3a. Proof of special case of Theorem 15.3c.

We now turn to the general case where Γ is an analytic arc. Let Γ be defined by $z = a(\tau)$, where τ runs through some real interval Γ_0, $a'(\tau) \neq 0$, $\tau \in \Gamma_0$. Since a is analytic, it can be continued to a function which is analytic in a region R_0^* containing Γ_0. The extended function maps R_0^* onto a region R^* containing Γ. Because $a'(t) \neq 0$ for $t \in \Gamma_0$, we may, by taking R_0^* narrow enough, assume that a is one-to-one and hence that $a^{[-1]}$ exists and is analytic. The function u being harmonic in $R \cap R^*$, its conformal transplant $u_0(t) := u(a(t))$ is harmonic in the region $a^{[-1]}(R \cap R^*)$ bordering on Γ_0. It is moreover continuous in $a^{[-1]}(R \cap R^*) \cup \Gamma_0$, and the values taken on Γ_0 depend analytically on t. Thus by the special case considered above, u_0 can

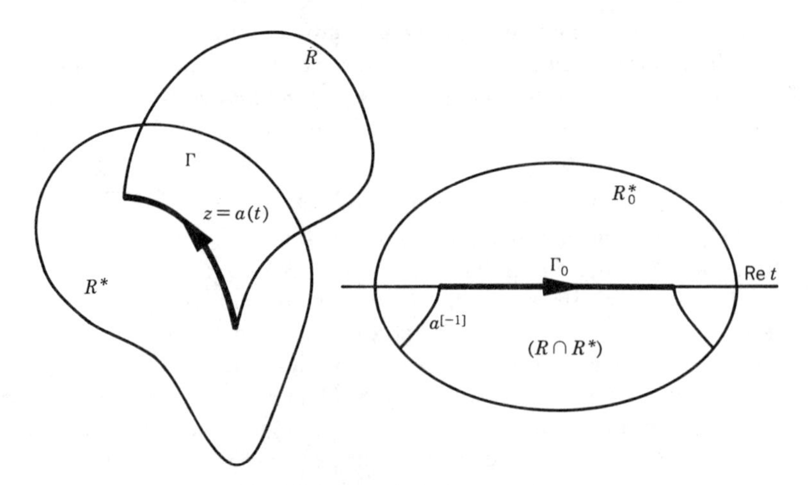

Fig. 15.3b. Proof of Theorem 15.3c.

be continued across Γ_0. Again by analytic transplantation, $u(z) = u_0(a^{[-1]}(z))$ is harmonic in R^* and thus represents a harmonic continuation of the original u (see Fig. 15.3b). —

In reflecting why continuous continuation should work for analytic functions and not for harmonic functions, one should keep in mind that the continuity of an analytic function implies the continuity not only of its real part, but also of the conjugate harmonic function, and thus is a far stronger requirement than continuity of a single real function.

III. Isolated Singularities of Harmonic Functions

Let u be harmonic in the punctured disk D_0: $0<|z-a|<\rho$, and let u be undefined for $z=a$. The harmonic function u is then said to have an **isolated singularity** at the point $z=a$. It is our object here to study the behavior of a harmonic function in the neighborhood of an isolated singularity. The reader will recall from § 4.4 that for *analytic* functions, isolated singularities are classified as removable, polelike, or essential, according to whether

$$\lim_{z\to a} f(z)$$

exists as a proper limit, as an improper limit, or does not exist at all. A similar classification is now given for harmonic functions, although the analogy is not complete.

For analytic functions, the Laurent series served as a general representation of the function in the punctured disk D_0. By Theorem 15.1c, u can in D_0 be represented as the real part of the indefinite integral of a function g which is analytic in D_0. Let r denote the residue of g at $z=a$. We assert that r is *real.* Using (15.1-3),

$$g(z) = \frac{\partial u}{\partial x} - i\frac{\partial u}{\partial y},$$

we have, if Γ is a circle in D_0 around a,

$$r = \frac{1}{2\pi i}\int_\Gamma g(z)\,dz = \frac{1}{2\pi i}\int_\Gamma \left(\frac{\partial u}{\partial x} - i\frac{\partial u}{\partial y}\right)(dx + i\,dy),$$

and hence

$$\operatorname{Im} r = -\frac{1}{2\pi}\int_\Gamma \left(\frac{\partial u}{\partial x}\,dx + \frac{\partial u}{\partial y}\,dy\right) = 0.$$

If $r \neq 0$, the indefinite integral of g is

$$f(z) := r\log(z-a) + f_1(z),$$

where f_1 is again analytic in D_0, with an isolated singularity at a. By virtue of the logarithm, f is multivalued, all possible values at a point $z \in D_0$ differing by integral multiples of the purely imaginary number $2\pi i r$. It follows that

$$h(z) := e^{r^{-1}f(z)} = (z-a)\, e^{r^{-1}h(z)}$$

is single-valued and analytic in D_0, still with an isolated singularity at $z = a$. Since $f = r \log h$, where r is real, the fact that $u = \operatorname{Re} f$ yields for u the representation

$$u(z) = \alpha \operatorname{Log}|h(z)| \tag{15.3-1}$$

in terms of a real number $\alpha = r^{-1}$ and a function h analytic in D_0.

If $r = 0$, we may put $h = e^{f_1}$ and obtain the similar representation

$$u(z) = \operatorname{Log}|h(z)|,$$

where h again has an isolated singularity at $z = a$. This evidently is contained in (15.3-1) as the special case $\alpha = 1$.

The known facts about isolated singularities of analytic functions are now readily reinterpreted in terms of u:

THEOREM 15.3d (On removable singularities of harmonic functions). *Let u be harmonic in the punctured disk D_0: $0 < |z-a| < \rho$, and let u be bounded. Then*

$$\beta := \lim_{z \to a} u(z)$$

exists, and the function u_1 defined by

$$u_1(z) := \begin{cases} u(z), & z \in D_0 \\ \beta, & z = a \end{cases}$$

is harmonic in the full disk D: $|z-a| < \rho$.

Proof. The boundedness of u implies that $|h|$ is bounded from above as well as bounded away from zero. The first fact by Riemann's theorem (Theorem 4.4d) implies that h has a removable singularity at $z = a$, and the second implies that $h(a) \neq 0$. Thus

$$u(z) = \alpha \operatorname{Log}|h(z)|$$

is continuous at $z = a$. Since $h(z) \neq 0$ in D by virtue of the boundedness of u, an analytic branch of $\log h$ can be defined (Theorem 4.3f). The fact that $u = \operatorname{Re} \log h$ now shows that the extended function u is harmonic in D. —

THEOREM 15.3e. *Let u be harmonic in D_0, and let*

$$\lim_{z \to a} u(z)$$

be either $+\infty$ *or* $-\infty$. *Then u can be represented in the form*

$$u(z) = \beta \operatorname{Log}|z-a| + u_1(z), \qquad z \in D_0, \tag{15.3-2}$$

where $\beta \neq 0$, *and where* u_1 *is harmonic in* D_0 *and can be continued harmonically into* $z = a$.

Proof. The representation (15.3-1) now shows that $|h|$ tends either to 0 or to infinity as $z \to a$. The analytic function h thus can be represented in the form

$$h(z) = (z-a)^n h_1(z),$$

where n is an integer (positive or negative, but not zero), and where h_1 is analytic and $\neq 0$ in D. The representation (15.3-1) now yields (15.3-2), where $\beta = n\alpha$ and where $u_1 := \alpha \operatorname{Log}|h_1|$. Since $h_1(z) \neq 0$, $z \in D$, an analytic branch of $\log h_1$ may be defined in D, and we can write $u_1 = \alpha \operatorname{Re} \log h_1$, showing that u_1 is harmonic in D. —

There remains the possibility that

$$\lim_{z \to a} u(z)$$

exists neither as a proper nor as an improper limit. It follows from the above that h then has an essential singularity at $z = a$. By the Casorati–Weierstrass theorem (Theorem 4.4g), given any complex number w, there exists a sequence of numbers $z_n \in D_0$ such that $z_n \to a$ and $h(z_n) \to w$. *A fortiori*, given any real number ω, there exists a similar sequence $\{z_n\}$ such that $\operatorname{Log}|h(z_n)| \to \omega$. We thus obtain the following analog of the Casorati–Weierstrass theorem for harmonic functions:

THEOREM 15.3f. *Let u be harmonic in the punctured disk* D_0: $0 < |z-a| < \rho$, *and let*

$$\lim_{z \to a} u(z)$$

exist neither as a proper nor as an improper limit. Then, given any real number ω, *there exists a sequence* $\{z_n\}$, $z_n \in D_0$, *such that*

$$\lim_{n \to \infty} u(z_n) = \omega. \tag{15.3-3}$$

In the realm of analytic functions, a function which has no limit (proper or improper) at an isolated singularity is necessarily transcendental. The following example shows that in the realm of harmonic functions very simple rational functions can have an essential singularity in the sense of Theorem 15.3f.

EXAMPLE 2

Let $z=\rho\, e^{i\phi}$. The function

$$u(z)=\operatorname{Re}\frac{1}{z}=\frac{1}{\rho}\cos\phi$$

is evidently harmonic for $z\neq 0$; moreover, u is a rational function of x and y. It is easily seen that $\lim_{z\to 0} u(z)$ exists neither as a proper nor as an improper limit. And indeed, if ω is any real number, in order to obtain a sequence $\{z_n\}$ satisfying (15.3-3), it suffices to select the points z_n on the curve $\rho=\omega^{-1}\cos\phi$ if $\omega\neq 0$ and on $x=0$ if $\omega=0$. —

IV. Positive Harmonic Functions

Let u be harmonic and positive in the disk D: $|z|<\rho$, and let u be continuous in the closure of D. Then by Poisson's integral there holds

$$u(z)=\int_0^1 u(\rho\, e^{2\pi i\tau})\frac{\rho^2-|z|^2}{|\rho\, e^{2\pi i\tau}-z|^2}\,d\tau$$

for any $z\in D$. Here we can estimate the Poisson kernel from above and from below by using

$$\rho-|z|\leqq|\rho\, e^{2\pi i\tau}-z|\leqq\rho+|z|,$$

which yields

$$\frac{\rho+|z|}{\rho-|z|}\geqq\frac{\rho^2-|z|^2}{|\rho\, e^{2\pi i\tau}-z|^2}\geqq\frac{\rho-|z|}{\rho+|z|}.$$

Since $u(\rho\, e^{2\pi i\tau})\geqq 0$, we thus have

$$u(z)\leqq\frac{\rho+|z|}{\rho-|z|}\int_0^1 u(\rho\, e^{2\pi i\tau})\,d\tau,$$

$$u(z)\geqq\frac{\rho-|z|}{\rho+|z|}\int_0^1 u(\rho\, e^{2\pi i\tau})\,d\tau.$$

By the mean value property,

$$\int_0^1 u(\rho\, e^{2\pi i\tau})\,d\tau=u(0),$$

and we obtain:

LEMMA 15.3g (Harnack's inequality). *If u is a positive harmonic function in the disk D: $|z|<\rho$, then for any $z\in D$,*

$$\frac{\rho-|z|}{\rho+|z|}u(0)\leqq u(z)\leqq\frac{\rho+|z|}{\rho-|z|}u(0). \tag{15.3-4}$$

Harnack's inequality implies that if E is any closed subset of D, then there exists a positive constant μ depending only on E such that for any positive harmonic function u in D,

$$\sup_{z,z' \in E} \frac{u(z)}{u(z')} \leqq \mu. \qquad (15.3\text{-}5)$$

In fact, let E be contained in the closed disk $|z| \leqq \rho_1$, where $\rho_1 < \rho$. Then by (15.3-4),

$$u(z) \leqq \frac{\rho + \rho_1}{\rho - \rho_1} u(0), \qquad z \in E,$$

as well as

$$u(z') \geqq \frac{\rho - \rho_1}{\rho + \rho_1} u(0), \qquad z' \in E,$$

which on division yields (15.3-5), where

$$\mu = \left(\frac{\rho + \rho_1}{\rho - \rho_1} \right)^2.$$

Rather surprisingly, a result analogous to (15.3-5) can be proved for functions that are harmonic and positive in *any* region R.

THEOREM 15.3h (Harnack's theorem). *Let R be a region, and let $P = P_R$ denote the family of functions that are harmonic and positive in R. Then for each compact subset $T \subset R$,*

$$\mu(T) := \sup_{u \in P} \sup_{z,z' \in T} \frac{u(z)}{u(z')} < \infty. \qquad (15.3\text{-}6)$$

Proof. Let $z_0 \in T$ be fixed. For each $z \in R$ we let

$$\mu(z) := \sup_{u \in P} \frac{u(z)}{u(z_0)}.$$

$\mu(z)$ is either a real number $\geqq 1$, or $\mu(z) = +\infty$. We assert that

$$\mu(z) < \infty \quad \text{for all} \quad z \in R. \qquad (15.3\text{-}7)$$

For the proof we define the two sets

$$S_1 := \{z \in R: \quad \mu(z) < \infty\},$$
$$S_2 := \{z \in R: \quad \mu(z) = \infty\},$$

which now are both shown to be open.

To show that S_1 is open, let $z_1 \in S_1$, and let the disk of radius ρ about z_1 belong to R. Then if $|z - z_1| \leqq \frac{1}{2}\rho$ and $u \in P$, the right half of Harnack's inequality shows that

$$u(z) \leqq \frac{2\rho + \rho}{2\rho - \rho} u(z_1) \leqq 3\mu(z_1)u(z_0),$$

$$\mu(z) \leqq 3\mu(z_1),$$

and thus that $z \in S_1$ for $|z - z_1| \leqq \frac{1}{2}\rho$. To show that S_2 is open, let $z_2 \in S_2$. Then there exists a sequence of functions $u_n \in P$ such that

$$\frac{u_n(z_2)}{u_n(z_0)} \to \infty.$$

Let again a disk of radius ρ about z_2 belong to R. Then by the left part of Harnack's inequality, if $|z - z_2| \leqq \frac{1}{2}\rho$,

$$u_n(z) \geqq \frac{2\rho - \rho}{2\rho + \rho} u_n(z_2) = \frac{1}{3} u_n(z_2),$$

and hence

$$\frac{u_n(z)}{u_n(z_0)} \to \infty.$$

The set S_1 is not empty, because it contains z_0, for which $\mu(z_0) = 1$. Because $S_1 \cap S_2 = \emptyset$, $S_1 \cup S_2 = R$, it follows that $S_1 = R$ by the definition of connectedness, establishing (15.3-7).

In exactly the same manner one can show that

$$\lambda(z) := \inf_{u \in P} \frac{u(z)}{u(z_0)} > 0 \quad \text{for all} \quad z \in R. \tag{15.3-8}$$

To prove the assertion of the theorem consider the covering of T by all disks with center $z \in R$ and radius $\rho = \rho(z)$ such that the disk of radius 2ρ about z still belongs to R. By Heine–Borel we may extract a finite subcovering consisting, say, of the disks $D_1, D_2, \ldots, D_k$ with centers $z_1, \ldots, z_k$. If $u \in P$ and $z \in D_j$, then by the foregoing

$$\frac{1}{3}\lambda(z_j) \leqq \frac{u(z)}{u(z_0)} \leqq 3\mu(z_j).$$

Thus if $z \in T$, $z' \in T$, then

$$u(z) \leqq 3u(z_0) \max_{1 \leqq j \leqq k} \mu(z_j),$$

$$u(z') \geqq \frac{1}{3} u(z_0) \min_{1 \leqq j \leqq k} \lambda(z_j),$$

and there follows

$$\frac{u(z)}{u(z')} \leq 9 \frac{\max \mu(z_j)}{\min \lambda(z_j)}.$$

The bound on the right being independent of u, this establishes (15.3-6). —

V. Sequences of Harmonic Functions

In Theorem 3.4b we proved the fundamental result that if the functions $f_0, f_1, \ldots$ all are analytic in a region R, and if the sequence $\{f_n\}$ converges locally uniformly in R (that is, converges uniformly on every compact subset of R), then the limit function $f := \lim f_n$ is necessarily analytic in R. The obvious analog for harmonic functions is as follows:

THEOREM 15.3i. *Let the functions $u_0, u_1, u_2, \ldots$ all be harmonic in the region R, and let the sequence $\{u_n\}$ converge locally uniformly in R. Then the limit function $u = \lim u_n$ is harmonic in R.*

Proof. By Theorem 15.3a it suffices to show that u possesses the mean value property in R. Thus let D: $|z-a| < \rho$ be any disk whose closure is contained in R. Because each u_n is harmonic, there holds

$$u_n(a) = \int_0^1 u_n(a + \rho\, e^{2\pi i \tau})\, d\tau.$$

Here we let $n \to \infty$. On the left we obtain $u(a)$. On the right, since $u_n(z)$ converges uniformly on the compact set $|z-a| = \rho$, we may interchange integration and passage to the limit and thus obtain

$$u(a) = \int_0^1 u(a + \rho\, e^{2\pi i \tau})\, d\tau,$$

establishing the mean value property. —

Theorem 15.3i will now be supplemented by a result which does not presuppose uniform convergence:

THEOREM 15.3j. *Let R be a region, and let $\{u_n\}$ be a sequence of harmonic functions in R that increases monotonically at each $z \in R$. If the sequence $\{u_n\}$ converges at one point of R, it converges in all of R. The convergence then is locally uniform, and the limit function is therefore harmonic. If the sequence converges at no point of R, the divergence to $+\infty$ is likewise locally uniform.*

Proof. Let z_0 be a point such that $\{u_n(z_0)\}$ is convergent, let T be a compact subset of R, $z_0 \in T$, and let $\varepsilon > 0$. By the Cauchy criterion there exists m_0

such that for all $m > m_0$ and all $n > m$,

$$|u_n(z_0) - u_m(z_0)| < \frac{\varepsilon}{\mu(T)},$$

where $\mu(T)$ denotes the number defined by (15.3-6). By virtue of the monotonicity, the last relation may be improved to

$$0 < u_n(z_0) - u_m(z_0) < \frac{\varepsilon}{\mu(T)}.$$

Now because $u_n - u_m$ is a positive harmonic function in R, there follows for $z \in T$ by the definition of $\mu(T)$,

$$0 < u_n(z) - u_m(z) \leqq \mu(T)[u_n(z_0) - u_m(z_0)] \leqq \mu(T)\frac{\varepsilon}{\mu(T)} = \varepsilon,$$

establishing the uniform convergence in T. That the limit function is harmonic follows from Theorem 15.3i. The last part of Theorem 15.3j is proved similarly. —

PROBLEMS

1. *Harnack's theorem for analytic functions.* Let R be a region, and let $F = F_R$ denote the class of analytic functions $f = u + iv$ in R such that $u \in P$.

   (a) Deduce from the Schwarz formula (14.2-11) that if $f \in F$ in a region containing $|z| = \rho$, then for $|z| < \rho$,

$$|f(z)| \leqq \frac{2\rho}{\rho - |z|}|f(0)|.$$

   (b) Prove that if $f \in F$, then $f^{-1} \in F$, and conclude that under the hypotheses of (a),

$$|f(z)| \geqq \frac{\rho - |z|}{2\rho}|f(0)|.$$

   (c) Proceeding as in the proof of Theorem 15.3h, prove:

THEOREM 15.3k. *Let R be a region, and let $F := F_R$. Then for each compact subset $T \subset R$,*

$$\mu(T) := \sup_{f \in F} \sup_{z,z' \in T} \left|\frac{f(z)}{f(z')}\right| < \infty.$$

2. *Liouville's theorem for harmonic functions* states: *If a function u is harmonic in $\mathbb{C}$ and bounded, it is necessarily constant.* Prove Liouville's theorem by representing $u(z) - u(0)$ by Poisson's formula for the disk $|z| = \rho$ and letting $\rho \to \infty$.
3. Prove Liouville's theorem by considering $f := e^{u+iv}$, where v is a harmonic conjugate of u.

NOTES

See Ahlfors [1966] and, for isolated singularities of harmonic functions, Lavrentiev and Shabat [1967].

§ 15.4. THE GENERAL DIRICHLET PROBLEM

I. Existence and Uniqueness of the Solution

The Dirichlet problem was formulated in § 15.2 for an arbitrary region R. However, the existence of a solution has been proved so far only in the very special cases where R is either a disk or a nondegenerate annulus. The uniqueness of a possible solution has been proved only if R is bounded.

It is easy to show that the general Dirichlet problem as formulated in § 15.2 need not have a solution at all.

EXAMPLE 1

Let R: $0<|z|<1$, and let the boundary function be $\phi(z)=0$, $|z|=1$; $\phi(0)=1$. Any solution u of this Dirichlet problem would have to be continuous on the closed disk $|z|\leqq 1$, and hence bounded. By Theorem 15.3d the isolated singularity at $z=0$ of the harmonic function u would thus be removable, that is, u would be harmonic in $|z|<1$. Since the values on $|z|=1$ are zero, u would have to be identically zero. It thus could not have the value 1 at 0. —

We conclude that the Dirichlet problem cannot, in general, be solved for regions whose complement contains isolated points. It turns out that this is the only obstacle to the existence of a solution of the Dirichlet problem for bounded regions, because the following result can be proved:

PROPOSITION 15.4a. *Let R be a bounded region such that no component of ∂R reduces to a single point. Let ϕ be a continuous real function on ∂R. Then there exists precisely one function u which is continuous in $R\cup\partial R$, harmonic in R, and equal to ϕ on ∂R.*

It is unlikely that a result of this generality can be proved by a constructive method. In a nonconstructive way, Proposition 15.4a can be proved, for instance, by Perron's device of maximizing a family of subharmonic functions.

Proposition 15.4a is not yet a statement on the solvability of Dirichlet's problem that takes into account all situations that may arise in practice. For one thing, the hypothesis that the boundary function ϕ is continuous is often too narrow. For instance, if electrostatic boundary value problems involving unbounded regions are solved by means of conformal transplantation, the transplanted boundary values may be discontinuous at the images of ∞. We are thus led to consider the following **generalized Dirichlet problem**:

Let R be a region in the complex plane with boundary ∂R, and let ϕ be a real-valued, bounded function on ∂R which is continuous except at the finitely many points $t_1, t_2, \ldots, t_n \in \partial R$. Let Γ_0 be the set ∂R with the points t_i removed. The generalized Dirichlet problem consists in finding a function u such that:

(i) u is continuous in $R \cup \Gamma_0$.
(ii) u is harmonic in R.
(iii) For all $t \in \Gamma_0$, $u(t) = \phi(t)$.
(iv) u is bounded.

Concerning (iv), the boundedness of u does, of course, not follow from the continuity, because the set $R \cup \Gamma_0$ is not closed. What, then, is the role of condition (iv)? The following example shows that without this condition the solution of the generalized Dirichlet problem need not be unique.

EXAMPLE 2

The function $\operatorname{Re} z$ is harmonic in the right half-plane and zero on the imaginary axis. Transplanting this function into $|z|<1$ by means of a Moebius transformation yields a function which is harmonic in $|z|<1$ and continuous, with boundary values 0, in $|z| \leqq 1$ except at the image of ∞, where it is unbounded. —

PROPOSITION 15.4b. *The generalized Dirichlet problem has a unique solution for every bounded region R such that no component of its complement reduces to a point.*

The *existence* part of Proposition 15.4b is again proved by Perron's method of subharmonic functions. The uniqueness statement is a consequence of the following result, which will be used again later on:

LEMMA 15.4c (Generalized maximum principle). *Let the function u be harmonic and bounded in a bounded region R, let it be continuous on the set $R \cup \Gamma_0$, where Γ_0 is the boundary of R with finitely many points $t_1, t_2, \ldots, t_n$ removed, and let $u(z) \geqq 0$ for $z \in \Gamma_0$. Then $u(z) \geqq 0$ throughout R.*

Proof. Let δ denote the diameter of R. We require the function

$$v(z) := \sum_{k=1}^{n} \operatorname{Log} \frac{2\delta}{|z - t_k|},$$

which is harmonic in R and positive and continuous in $R \cup \Gamma_0$. Furthermore if $z \to t_j$, then $v(z) \to +\infty$, $j = 1, 2, \ldots, n$.

For $\rho > 0$, denote by R_ρ the open set obtained from R by removing the disks $|z - t_j| \leqq \rho$, $j = 1, 2, \ldots, n$, and let Γ_ρ be the boundary of R_ρ. If ρ is sufficiently small, then R_ρ is a region. For every $\varepsilon > 0$, $u + \varepsilon v$ is harmonic in R_ρ and continuous in $R_\rho \cup \Gamma_\rho$. Furthermore, if ρ is chosen sufficiently small, then $u + \varepsilon v > 0$ on Γ_ρ. Thus by the ordinary maximum principle, $u + \varepsilon v > 0$ throughout R_ρ. Hence at any point $z \in R$, taking ρ so small that $z \in R_\rho$, the inequality

$$u(z) > -\varepsilon v(z)$$

holds. Because this is true for every $\varepsilon > 0$, it follows that $u(z) \geqq 0$, $z \in R$. —

To prove the uniqueness statement of Proposition 15.4b, let u_1 and u_2 be two solutions of the generalized Dirichlet problem. The hypotheses of Lemma 15.4c are satisfied for both $u := u_1 - u_2$ and $u := u_2 - u_1$, and it thus follows that $u_1 = u_2$ throughout R. —

Even Proposition 15.4b is not yet satisfactory from the point of view of applications, because we know from Chapter 5 that frequently in applications the region R is unbounded. The following example shows that for unbounded regions the solution of the Dirichlet problem cannot be expected to be unique, even if the boundary data are continuous.

EXAMPLE 3

Let R: $|z| > 1$, and let $\phi(z) = 0$, $|z| = 1$. For every real γ, the function

$$u(z) = \gamma \operatorname{Log}|z|$$

is a solution of this Dirichlet problem. —

All solutions of the Dirichlet problem of Example 3 are unbounded, with the exception of the solution $u = 0$. This again suggests that a boundedness condition may be sufficient to guarantee the uniqueness of the solution, even for discontinuous boundary data. The following theorem asserts this to be the case for an important class of regions.

THEOREM 15.4d. *Let R be a region* (bounded or unbounded) *such that the complement of R contains an inner point, and none of its components reduces to a point. Then the generalized Dirichlet problem for R always has a unique solution.*

Proof. If R is bounded, this is equivalent to Proposition 15.4b. (The unbounded component of the complement of a bounded region always contains a disk.) If R is unbounded, let z_0 be an interior point of the complement of R. By the inversion

$$z \to w := \frac{1}{z - z_0},$$

R is mapped onto a bounded region R_1, and the given boundary data ϕ are transformed into boundary data for a generalized Dirichlet problem for R_1. (If the point ∞ is a boundary point of R, its image 0 may have to be included in the points of discontinuity of the boundary data.) This problem has a unique solution by Proposition 15.4b. Transforming back we obtain a solution of the generalized Dirichlet problem for R. —

II. Solution Formulas

We now proceed to the construction of solution formulas for a variety of regions. The method, conformal transplantation, has already been exploited in Chapter 5 in many concrete situations; here we restate the method in general terms. To begin with, let R be a Jordan region, that is, a region that is bounded by a Jordan curve Γ. The essential tool is the Osgood-Carathéodory extension (Theorem 5.10e, for a proof see § 16.3) of the Riemann mapping theorem, which asserts not only the existence of an analytic function f mapping R bijectively onto the unit disk D: $|w| < 1$, but also the possibility of extending f to a homeomorphism of $R \cup \Gamma$ onto the closure D' of D. The extended map is again denoted by f.

Let u be the solution of the Dirichlet problem for R with boundary data ϕ. The function $u \circ f^{[-1]}$ then is continuous on D' and, by Problem 4, § 15.1, harmonic in D. On the boundary $w = e^{2\pi i\tau}$ of D it assumes the values

$$\psi(\tau) := u \circ f^{[-1]}(e^{2\pi i\tau}) = \phi \circ f^{[-1]}(e^{2\pi i\tau}), \qquad 0 \leqq \tau \leqq 1.$$

It thus solves the Dirichlet problem for D with boundary data ψ. Hence by Poisson's formula (Corollary 15.2c) we have for $|w| < 1$,

$$u \circ f^{[-1]}(w) = \int_0^1 \psi(\tau) \frac{1 - |w|^2}{|e^{2\pi i\tau} - w|^2} \, d\tau.$$

Letting $w = f(z)$, where $z \in R$, this becomes

$$u(z) = \int_0^1 \psi(\tau) \frac{1 - |f(z)|^2}{|e^{2\pi i\tau} - f(z)|^2} \, d\tau. \tag{15.4-1}$$

If we assume that the derivative f' of the mapping function can be extended continuously to the boundary Γ of R, the solution u can be represented as a line integral into which the boundary data enter more directly. Let, for $t \in \Gamma$,

$$e^{2\pi i\tau} = f(t).$$

Then

$$2\pi i\, e^{2\pi i\tau}\, d\tau = f'(t)\, dt,$$

hence

$$d\tau = \frac{1}{2\pi i}\frac{f'(t)}{f(t)}\, dt,$$

and by virtue of

$$\psi(\tau) = \phi \circ f^{[-1]}(e^{2\pi i\tau}) = \phi(t),$$

(15.4-1) appears in the form

$$u(z) = \frac{1}{2\pi i}\int_\Gamma \phi(t)\frac{1-|f(z)|^2}{|f(t)-f(z)|^2}\frac{f'(t)}{f(t)}\, dt. \tag{15.4-2}$$

EXAMPLE **4**

Let R be the upper half-plane $\operatorname{Im} z > 0$, and let the boundary data ϕ on the real line be bounded and piecewise continuous, with at most finitely many discontinuities. An appropriate map to D is given by

$$w = f(z) = \frac{i-z}{i+z}.$$

A straightforward application of (15.4-2) yields for the solution of the Dirichlet problem the formula

$$u(z) = \frac{1}{\pi}\int_{-\infty}^{\infty} \phi(t)\frac{y}{(t-x)^2+y^2}\, dt, \qquad z = x+iy, \tag{15.4-3}$$

which, by a different method, has already been obtained in § 14.11.

PROBLEMS

1. Show that the solution of the Dirichlet problem for the infinite strip

$$-\frac{\pi}{2} < \operatorname{Im} z < \frac{\pi}{2}$$

with boundary data

$$u\left(i\frac{\pi}{2}+\tau\right) = \phi_0(\tau), \qquad u\left(-i\frac{\pi}{2}+\tau\right) = \phi_1(\tau)$$

is given by

$$u(z) = \frac{1}{2\pi}\int_{-\infty}^{\infty} \phi_0(\tau)\frac{\cos y}{\cosh(\tau-x)-\sin y}\, d\tau + \frac{1}{2\pi}\int_{-\infty}^{\infty} \phi_1(\tau)\frac{\cos y}{\cosh(\tau-x)+\sin y}\, d\tau.$$

(Required map:

$$w = f(z) = \frac{e^z - 1}{e^z + 1}.)$$

2. The solution of the Dirichlet problem for the *exterior* of the unit circle with boundary data

$$u(e^{2\pi i\tau}) = \phi(\tau)$$

is given by

$$u(z) = \int_0^1 \phi(\tau) \frac{|z|^2 - 1}{|z - e^{2\pi i\tau}|^2} \, d\tau.$$

Prove this fact:

(a) From (15.4-2) by means of $f(z) = z^{-1}$.

(b) By noting that the Dirichlet problem for the *interior* of the circle with the same boundary data is solved by $u(\bar{z}^{-1})$.

3. Carry out explicitly the construction sketched in Example 2. How is the result related to the Poisson kernel?

4. Find, by the method of separating variables, unbounded solutions of Dirichlet's problem with zero boundary values for the following unbounded regions:

(a) $\operatorname{Re} z > 0, 0 < \operatorname{Im} z < \eta$.

(b) $0 < \arg z < \alpha$.

Show that, in these examples, the "narrower" the region, the "more unbounded" the solution becomes.

NOTES

For a proof of Proposition 15.4a by Perron's method see Ahlfors [1966], p. 240. The generalized Dirichlet problem is discussed by Lavrentiev and Shabat [1967], § 42.

§ 15.5. HARMONIC MEASURE

Here we use the existence theory for the generalized Dirichlet problem to introduce a special solution of Dirichlet's problem for a given region R called *harmonic measure.* This function (it has nothing to do with a measure in the sense of measure theory) will be useful in the estimation of certain domain functionals (see § 16.11) and in the construction of conformal maps of multiply connected regions (see § 17.1).

Let R be a region in the extended plane for which the generalized Dirichlet problem is uniquely solvable, and let the boundary ∂R be divided into two nonempty, nonintersecting subsets Γ_0 and Γ_1 such that each set Γ_i has only finitely many components. The solution of the Dirichlet problem for R with boundary values

$$\phi(t)=\begin{cases}1, & t\in\Gamma_1\\ 0, & t\in\Gamma_0\end{cases}$$

is called the **harmonic measure** of Γ_1 with respect to R, and its values will be denoted by $\omega(z, R, \Gamma_1)$. By the principle of the maximum the values of the harmonic measure in R lie strictly between 0 and 1.

EXAMPLE **1**

Let R be the upper half-plane. The function $\operatorname{Arg} z = \operatorname{Im} \operatorname{Log} z$ is clearly harmonic in R, and it equals 0 on the positive real line and π on the negative real line. Thus if Γ_1 is the negative real line,

$$\omega(z, R, \Gamma_1)=\frac{1}{\pi}\operatorname{Arg} z.$$

Similarly, if Γ_1 is the half-line, $x<x_0$, then

$$\omega(z, R, \Gamma_1)=\frac{1}{\pi}\operatorname{Arg}(z-x_0).$$

It follows that if Γ_1 is a finite interval, $\Gamma_1 := [x_0, x_1]$, then

$$\omega(z, R, \Gamma_1)=\frac{1}{\pi}[\operatorname{Arg}(z-x_1)-\operatorname{Arg}(z-x_0)],$$

that is, ω equals $1/\pi$ times the angle under which Γ_1 is seen from z. If Γ_1 is a finite collection of nonoverlapping intervals on the real line, then ω at z equals $1/\pi$ times the total angle under which these intervals are seen from z.

EXAMPLE **2**

Let $R=D$ be the unit disk, and let Γ_1 be an arc of the unit circle, Γ_1: $t=e^{i\tau}$, $\alpha_0\leqq\tau\leqq\alpha_1$. We seek to determine the harmonic measure $\omega(z, D, \Gamma_1)$. If $z=0$, then clearly from Poisson's formula,

$$\omega(0, D, \Gamma_1)=\frac{1}{2\pi}(\alpha_1-\alpha_0). \tag{15.5-1}$$

To determine the harmonic measure at an arbitrary point $a\in D$, we use conformal transplantation. The Moebius map

$$z\to w=\frac{z-a}{1-\bar{a}z}$$

maps a into 0 and Γ_1 into an arc Γ_1^* on the unit circle with endpoints

$$e^{i\beta_k} = \frac{e^{i\alpha_k} - a}{1 - \bar{a}\, e^{i\alpha_k}} = e^{-i\alpha_k} \frac{e^{i\alpha_k} - a}{e^{-i\alpha_k} - \bar{a}}, \qquad k = 0, 1.$$

We conclude that

$$\beta_k = 2 \arg(e^{i\alpha_k} - a) - \alpha_k.$$

The value of ω at a equals the value of the transplanted measure at 0. By (15.5-1) the latter equals

$$\frac{1}{2\pi}\left[2 \arg \frac{e^{i\alpha_1} - a}{e^{i\alpha_0} - a} - (\alpha_1 - \alpha_0)\right].$$

We thus have

$$\omega(a, D, \Gamma_1) = \frac{1}{2\pi}(2\theta - \alpha),$$

where α is the angle subtended by Γ_1 at 0, and where θ is the positive angle under which Γ_1 is seen at a.

EXAMPLE 3

If R is the annulus $\rho_0 < |z| < \rho_1 (0 < \rho_0 < \rho_1 < \infty)$, and if Γ_1 is the inner boundary circle $|z| = \rho_0$, then by advanced calculus

$$\omega(z, R, \Gamma_1) = \frac{\operatorname{Log}(\rho_1/|z|)}{\operatorname{Log}(\rho_1/\rho_0)}. \quad —$$

We shall want to compare the harmonic measures for several regions R and for different subsets Γ_1 of the boundary. To this end we introduce the following notation. Let R again be a region, and let T be a closed set in $\mathbb{C}$ which meets the closure of R. Provided that the expression on the right has a meaning, we define

$$\omega(z, R, T) := \omega(z, R \backslash T, T \cap \partial R).$$

In the electrostatic interpretation, T may be thought of as a conducting body which is kept at potential 1, and $\omega(z, R, T)$ is the potential in $R \backslash T$ that results from keeping that part of ∂R which does not intersect T at potential 0 (see Fig. 15.5a).

Consider now two pairs of sets (R, T) and (R^*, T^*) such that $R \subset R^*$ and $T \subset T^*$ (see Fig. 15.5b).

We assume that the functions $\omega(z) := \omega(z, R, T)$ and $\omega^*(z) := \omega(z, R^*, T^*)$ both are defined and have only finitely many discontinuities on the boundaries of their respective domains of definition. Now let z_0 be

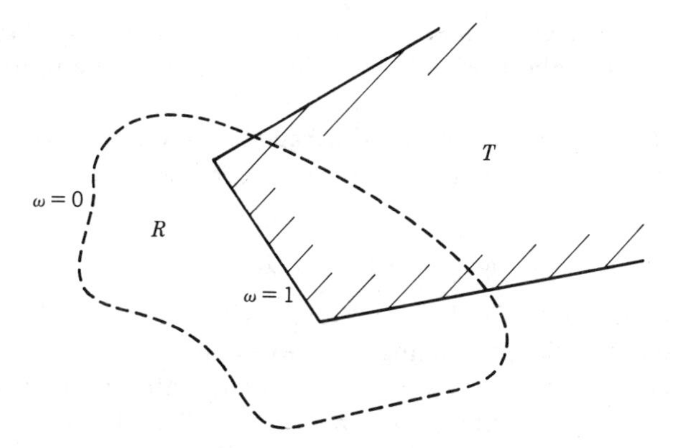

Fig. 15.5a. Harmonic measure.

a boundary point of the intersection $R \setminus T^*$ of the two domains of definition. Either z_0 is a boundary point of T^*, or z_0 is a boundary point of R which is not in T^*. We disregard the finitely many points where ω or ω^* is discontinuous. By the definition of harmonic measure we then have in the first case

$$\omega^*(z_0) = 1, \qquad \omega(z_0) \leqq 1$$

and in the second,

$$\omega^*(z_0) \geqq 0, \qquad \omega(z_0) = 0.$$

Thus in each component of $R \setminus T^*$, $\omega^* - \omega$ solves a generalized Dirichlet

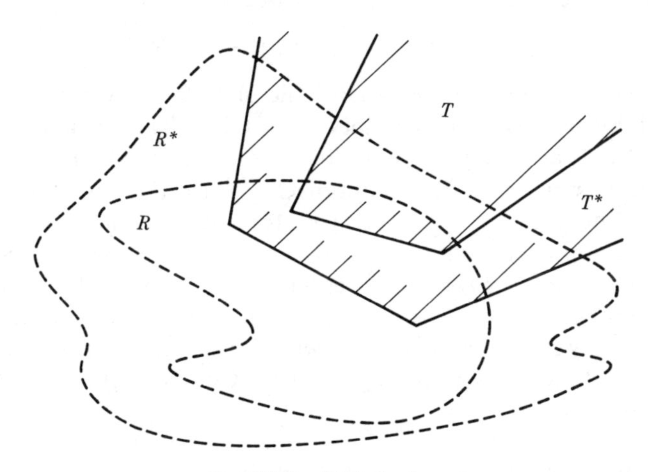

Fig. 15.5b. Majorization.

problem with nonnegative boundary values. By Lemma 15.4c it follows that $\omega^*(z) \geqq \omega(z)$ throughout $R \setminus T^*$. We thus have proved the following:

THEOREM 15.5a (Monotonicity of harmonic measure). *Let R, R^* be two regions and T, T^* two closed sets such that $R \subset R^*$, $T \subset T^*$. Then for all $z \in R \setminus T^*$,*

$$\omega(z, R, T) \leqq \omega(z, R^*, T^*).$$

That is, ω increases if either set R or T is enlarged. Simple illustrations are to be found in the preceeding three examples.

How does harmonic measure behave under conformal transplantation? Let f be analytic and one-to-one in a region R and continuous on the closure of R, and let T be a closed set contained in the closure of R. By conformal transplantation it is clear that

$$\omega(z, R, T) = \omega(f(z), f(R), f(T)) \tag{15.5-2}$$

for all $z \in R \setminus T$. But how about the case where f is not one-to-one?

EXAMPLE 4

Let R be the unit disk, and let T be that part of the boundary circle that lies in $\operatorname{Re} z \geqq 0$. Let $f(z) = z^2$. Then $f(R) = R$, but $f(T)$ now is the whole boundary circle. Therefore $\omega(f(z), f(R), f(T)) = 1$, while certainly $\omega(z, R, T) < 1$. —

We conclude that (15.5-2) cannot hold in general for functions f that are not one-to-one. However, the inequality that is suggested by Example 4 can be justified in general. Consider the function

$$\nu(z) := \omega(f(z), f(R), f(T)) - \omega(z, R, T).$$

Even if f is not one-to-one, ν is harmonic in $R \setminus T$ and continuous on the closure of this set except at the points of discontinuity of ν on the boundary of $R \setminus T$. (As always, we presuppose that there are only finitely many such points.) As z approaches T, $f(z)$ approaches $f(T)$; therefore both terms in the above difference are 1, and $\nu(z_0) = 0$ for boundary points of $R \setminus T$ that lie in T. As z approaches a point $z_0 \in \partial R$ that does not lie in T, $\omega(z, R, T) \to 0$, while $\omega(f(z), f(R), f(T))$ approaches 0 if $f(z_0) \notin f(T)$ and 1 if $f(z_0) \in f(T)$. (The second possibility exists because f is not one-to-one.) Thus $\nu(z) \geqq 0$ on the boundary of $R \setminus T$, and by the generalized maximum principle (Lemma 15.4c) there follows $\nu(z) \geqq 0$ throughout $R \setminus T$. Combining this with Theorem 15.5a, we get:

THEOREM 15.5b (Principle of majorization). *Let f be analytic in a region R and continuous on the closure of R, and let T be a closed set meeting the* *closure of R such that $\omega(z, R, T)$ is defined. If R^* is a region such that $f(R) \subset R^*$, and if T^* is a closed set such that $f(T) \subset T^*$ and $\omega(z, R^*, T^*)$ is defined, then*

$$\omega(z, R, T) \leqq \omega(f(z), R^*, T^*)$$

for all z such that $f(z) \in R^* \backslash T^*$.

As an application of the principle of majorization, we prove a result which replaces the Schwarz lemma if a function is known to be analytic merely in an annulus. To begin with, let f be analytic in the full disk $|z| < \rho_1$ and continuous on its closure. For $0 \leqq \rho \leqq \rho_1$ we define

$$\mu(\rho) := \max_{|z|=\rho} |f(z)|. \tag{15.5-3}$$

The principle of the maximum implies that $\mu(\rho)$ is an increasing function (strictly, unless f is constant). The Schwarz lemma in its unnormalized form asserts that if $\mu(0) = 0$, then

$$\mu(\rho) \leqq \mu_1 \frac{\rho}{\rho_1}, \tag{15.5-4}$$

where $\mu_1 := \mu(\rho_1)$, that is, the graph of $\mu(\rho)$ cannot lie above the straight line joining $(0, 0)$ and (ρ_1, μ_1) (see Fig. 15.5c). But what can be said about $\mu(\rho)$ if f is known to be analytic merely in the annulus A: $\rho_0 < |z| < \rho_1$, where $\rho_0 > 0$, and continuous on the closure of A?

Let $\mu_i := \mu(\rho_i)$, $i = 0, 1$. To fix ideas, we assume that $\mu_0 < \mu_1$. Let T: $|z| = \rho_0$. By Example 3,

$$\omega(z, A, T) = \frac{\operatorname{Log}(\mu_1/|z|)}{\operatorname{Log}(\mu_1/\mu_0)}.$$

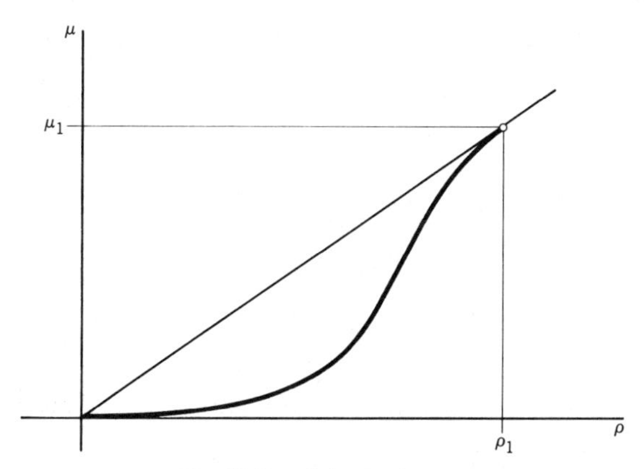

Fig. 15.5c. Schwarz lemma.

By the principle of the maximum, $|f(z)| \leqq \mu_1$; by definition of $\mu_0, f(T)$ is contained in the set $|w| \leqq \mu_0$. The hypotheses of the principle of majorization thus are satisfied for R^*: $|z| < \mu_1$, T^*: $|z| \leqq \mu_0$, and in view of

$$\omega(w, R^*, T^*) = \frac{\operatorname{Log}(\mu_1/|w|)}{\operatorname{Log}(\mu_1/\mu_0)}$$

we get

$$\frac{\operatorname{Log}(\rho_1/|z|)}{\operatorname{Log}(\rho_1/\rho_0)} \leqq \frac{\operatorname{Log}(\mu_1/|f(z)|)}{\operatorname{Log}(\mu_1/\mu_0)}.$$

If $|z| = \rho$, where $\rho_0 < \rho < \rho_1$, this holds for every $|f(z)|$ where $|z| = \rho$, thus also for the value where $|f(z)|$ is largest. There follows

$$\frac{\operatorname{Log}(\rho_1/\rho)}{\operatorname{Log}(\rho_1/\rho_0)} \leqq \frac{\operatorname{Log}(\mu_1/\mu(\rho))}{\operatorname{Log}(\mu_1/\mu_0)}. \tag{15.5-5}$$

The same result is obtained for $\mu_0 \geqq \mu_1$. The result may be cast in the form

$$\operatorname{Log} \mu(\rho) \leqq \frac{(\operatorname{Log} \rho_1 - \operatorname{Log} \rho) \operatorname{Log} \mu_0 - (\operatorname{Log} \rho - \operatorname{Log} \rho_0) \operatorname{Log} \mu_1}{\operatorname{Log} \rho_1 - \operatorname{Log} \rho_0},$$

showing that the graph of $\operatorname{Log} \mu(\theta)$ plotted against $\operatorname{Log} \rho$ runs below the graph of the linear function of $\operatorname{Log} \rho$ passing through $(\operatorname{Log} \rho_0, \operatorname{Log} \mu_0)$ and $(\operatorname{Log} \rho_1, \operatorname{Log} \mu_1)$. Because the same statement obviously holds if ρ_0 and ρ_1 are replaced by any two numbers ρ_2 and ρ_3 such that $\rho_0 \leqq \rho_2 < \rho_3 \leqq \rho_1$, it follows that $\operatorname{Log} \mu(\rho)$ is a convex function of $\operatorname{Log} \rho$. Yet another way to express the same result is as follows:

THEOREM 15.5c (Hadamard three-circle theorem). *Let* $0 < \rho_0 < \rho_1$, *and let f be analytic in the annulus A:* $\rho_0 < |z| < \rho_1$ *and continuous on the closure of A. Let* $\mu(\rho)$ *be defined by* (15.5-3). *If* α *is the unique real number such that*

$$\mu_1 = \mu_0 \left(\frac{\rho_1}{\rho_0}\right)^{\alpha},$$

then for $\rho_0 \leqq \rho \leqq \rho_1$,

$$\mu(\rho) \leqq \mu_0 \left(\frac{\rho}{\rho_0}\right)^{\alpha}. \tag{15.5-6}$$

PROBLEMS

1. Obtain the Schwarz lemma as a limiting form of the Hadamard three-circle theorem.

2. Let g be a nonconstant analytic function in the strip $\eta_0 \leqq \operatorname{Im} z \leqq \eta_1$, which is periodic with period $\lambda > 0$. If

$$\nu(\eta) := \max_{\operatorname{Im} z = \eta} |g(z)|,$$

$\nu_i := \nu(\eta_i)$, $i = 0, 1$, and if β is defined by

$$\nu_1 = \nu_0 \exp\{\beta(\eta_1 - \eta_0)\},$$

show that

$$\nu(\eta) \leqq \nu_0 \exp\{\beta(\eta - \eta_0)\}.$$

Conclude that it is impossible for $\nu(\eta)$ to be linear between any two points. Confirm this fact by considering the examples e^{iz}, $\sin z$, $\cos z$.

NOTES

Although considered earlier, the notion of harmonic measure was first put to systematic use by R. Nevanlinna [1936]; see also Ahlfors [1973].

§ 15.6. GREEN'S FUNCTION

I. Fundamental Solutions

Let $R \subset \mathbb{C}$ be a region. A function v of two complex variables $(z, t) \in (R, R)$ is called a **fundamental solution** of Laplace's equation for R if for each $t \in R$, v as a function of z is harmonic in $R \backslash \{t\}$ and at $z = t$ has an isolated singularity such that

$$\lim_{z \to t} v(z, t) = +\infty.$$

We recall from Theorem 15.3e (isolated singularities) that any such function necessarily has the form

$$v(z, t) = \kappa \operatorname{Log} \frac{1}{|z - t|} + w(z, t),$$

where $\kappa > 0$, and where w is a harmonic function of z for $z \neq t$ that can be continued harmonically into $z = t$. Usually we shall only consider **normalized** fundamental solutions where $\kappa = (2\pi)^{-1}$. Thus a normalized fundamental solution is a function of the form

$$v(z, t) = \frac{1}{2\pi} \operatorname{Log} \frac{1}{|z - t|} + w(z, t), \tag{15.6-1}$$

where w is as described above. The point t is called the **pole** of the fundamental solution. A simple fundamental solution is given by

$$v(z, t) = \frac{1}{2\pi} \operatorname{Log} \frac{1}{|z-t|}.$$

However, we shall take advantage of the greater generality allowed for by (15.6-1).

We note that if v is a fundamental solution for a region R, it can be used as a fundamental solution for any subregion of R.

Now let R be a bounded region whose boundary ∂R consists of a finite number of differentiable Jordan curves. We then can assert the validity of Green's formula (integration by parts in two dimensions), which states that for any two functions u and v having the required number of continuous derivatives in $R \cup \partial R$ there holds

$$\iint_R \operatorname{grad} u \operatorname{grad} v \boxed{dz} = \int_{\partial R} u \frac{\partial v}{\partial n} |dz| - \iint_R u \, \Delta v \boxed{dz}. \qquad (15.6\text{-}2)$$

Here $\partial/\partial n$ denotes differentiation in the direction of the *exterior* normal, and $\boxed{dz} := dx\,dy$. A second version of Green's formula is obtained by interchanging the roles of u and v and subtracting, which yields

$$\int_{\partial R} \left(u \frac{\partial v}{\partial n} - v \frac{\partial u}{\partial n} \right) |dz| = \iint_R (u \, \Delta v - v \, \Delta u) \boxed{dz}. \qquad (15.6\text{-}3)$$

Now in addition to the above hypotheses, assume u to be harmonic in R, and let $t \in R$. We let $v = v(z, t)$ be a normalized fundamental solution and apply Green's second formula (15.6-3) to the region R_ε obtained from R by removing the disk $|z-t| \leqq \varepsilon$, assumed to be contained in R. The expression on the right of (15.6-3) then vanishes, since both u and v are harmonic for $z \in R_\varepsilon$, and there remains

$$\int_{\partial R} \left(u \frac{\partial v}{\partial n} - v \frac{\partial u}{\partial n} \right) |dz| = - \int_{\Gamma_\varepsilon} \left(u \frac{\partial v}{\partial n} - v \frac{\partial u}{\partial n} \right) |dz|, \qquad (15.6\text{-}4)$$

where Γ_ε denotes the circle $|z-t| = \varepsilon$. We can easily evaluate the limit of the integral on the right as $\varepsilon \to 0$. The normal derivative of u being bounded, we have for a suitable $\mu > 0$

$$\left| \int_{\Gamma_\varepsilon} v \frac{\partial u}{\partial n} |dz| \right| \leqq \frac{\mu}{2\pi} \int_{\Gamma_\varepsilon} \left| \operatorname{Log} \frac{1}{\varepsilon} + O(1) \right| |dz| = \mu \varepsilon \operatorname{Log} \frac{1}{\varepsilon} + O(\varepsilon),$$

which tends to 0 as $\varepsilon \to 0$. On the other hand, since $\partial/\partial n = -\partial/\partial \varepsilon$ on Γ_ε,

$$\frac{\partial v}{\partial n} = \frac{1}{2\pi\varepsilon} + O(1),$$

hence

$$\int_{\Gamma_\varepsilon} u\frac{\partial v}{\partial n}|dz| = \int_{\Gamma_\varepsilon} (u(t)+O(\varepsilon))\left(\frac{1}{2\pi\varepsilon}+O(1)\right)|dz| \to u(t)$$

as $\varepsilon \to 0$. Thus the limit of (15.6-4) as $\varepsilon \to 0$ yields the **boundary representation formula,**

$$u(t) = \int_{\partial R}\left[v(z,t)\frac{\partial u}{\partial n}(z) - u(z)\frac{\partial v(z,t)}{\partial n}\right]|dz|. \qquad (15.6\text{-}5)$$

This formula represents the value of a harmonic function u at a point $t \in R$ in terms of the values of u and of $\partial u/\partial n$ on the boundary ∂R, provided that both u and the fundamental solution v are sufficiently smooth up to the boundary for Green's formula to be valid.

II. Green's Function

The idea seems natural to use the boundary representation formula for the solution of Dirichlet's problem. This is not directly possible, because the formula involves the values of both u and $\partial u/\partial n$ on the boundary, whereas only the values of u are prescribed for the solution of Dirichlet's problem.

However, if we succeeded in finding a fundamental solution v such that, for all $t \in R$, $v(z,t)=0$ for all $z \in \partial R$, the boundary representation formula would reduce to

$$u(t) = -\int_{\partial R} u(z)\frac{\partial v(z,t)}{\partial n}|dz|,$$

and thus would express $u(t)$ directly in terms of the boundary data $u(z) = \phi(z)$ of the Dirichlet problem. Now from (15.6-1) we see that the fundamental solution has the desired form if the function w satisfies

$$w(z,t) = -\frac{1}{2\pi}\operatorname{Log}\frac{1}{|z-t|}, \qquad z \in \partial R, \quad t \in R. \qquad (15.6\text{-}6)$$

Thus for each $t \in R$, w as a function of z itself is a solution of a Dirichlet problem. We conclude that w exists for all regions R for which Dirichlet's problem can be solved. The fundamental solution thus obtained is called **Green's function** for the region R. Green's function is unique for bounded regions R, because the solution of Dirichlet's problem for w is unique for such regions. If Green's function is denoted by γ, and if R is such that the derivatives of γ are continuous up to the boundary, we have

$$u(z) = -\int_{\partial R} u(t)\frac{\partial \gamma(t,z)}{\partial n}|dt|. \qquad (15.6\text{-}7)$$

The representation (15.6-7) of the solution of Dirichlet's problem, while elegant, is of immediate use only if γ is known explicitly. Since the construction of γ itself amounts to the solution of infinitely many Dirichlet problems (one for each location of the pole t), this seemingly is a very strong requirement; however, see Sections III and IV.

Far from being a mere device for the solution of Dirichlet's problem, Green's function is an important concept in its own right. In addition, it has a direct physical interpretation (see § 5.7). Let R be the cross section of a cylindrical region whose walls are kept at zero potential. Imagine a thin wire carrying charge 1 per unit length that runs in the direction perpendicular to R and which pierces R at the point $t \in R$. Then $\gamma(z, t)$ is the value at z of the potential generated by the charged wire. The following **symmetry property** of Green's function is especially interesting from the point of view of this application. Let $\gamma(z, t)$ denote Green's function for a bounded region R, assumed to satisfy the hypotheses for the validity of Green's formulas. Then if t_1, t_2 are any two points of R, there holds

$$\gamma(t_1, t_2) = \gamma(t_2, t_1). \tag{15.6-8}$$

That is, the potential generated at t_1 by a wire through t_2 equals the potential at t_2 generated by a wire through t_1.

To prove (15.6-8) we apply Green's second formula (15.6-3), where $u(z) = \gamma(z, t_1)$, $v(z) = \gamma(z, t_2)$, to the region R_ε obtained by deleting from R the two small disks $|z - t_1| \leqq \varepsilon$, $|z - t_2| \leqq \varepsilon$. Since u and v are both harmonic in R_ε and zero on ∂R, all that remains from (15.6-3) are the integrals along the two circles $|z - t_j| = \varepsilon, j = 1, 2$. The limit as $\varepsilon \to 0$ can be computed as in the proof of the boundary representation formula, and yields (15.6-8).

III. Green's Function and the Riemann Mapping Function for Simply Connected Regions

To begin with, let R denote the unit disk $|w| < 1$. If the pole of γ is to be at s, we try to represent γ in the form

$$\gamma(w, s) = -\frac{1}{2\pi} \operatorname{Log}|f_s(w)|,$$

where f is analytic for $|w| < 1$. The function γ then will automatically be harmonic at all points where $f_s(w) \neq 0$, which we must require for all $w \neq s$. The desired pole at s will result if f_s has a simple zero at $w = s$, and γ vanishes on $|w| = 1$ if $|f_s(w)| = 1$ for $|w| = 1$. These properties are possessed by any function f which maps the disk $|w| < 1$ conformally onto itself and sends the point s into the origin, for instance, by

$$f_s(w) = \frac{w - s}{1 - \bar{s}w}.$$

Thus Green's function for the disk is given by

$$\gamma(w, s) = -\frac{1}{2\pi} \operatorname{Log} \left| \frac{w-s}{1-\bar{s}w} \right|. \tag{15.6-9}$$

For arbitrary Jordan regions R Green's function can now be found by the method of conformal transplantation. Let $w = f(z)$ map R conformally onto $|w| < 1$ and $R \cup \partial R$ continuously onto $|w| \leqq 1$. [The existence of such a function is assured by the Osgood-Carathéodory theorem (Theorem 16.3a).] If γ is Green's function for R, the function

$$(w, s) \mapsto \gamma(f^{[-1]}(w), f^{[-1]}(s))$$

has all properties of Green's function for the unit disk, and hence must agree with (15.6-9). It follows that the desired Green's function for R is given by

$$\gamma(z, t) = -\frac{1}{2\pi} \operatorname{Log} \left| \frac{f(z) - f(s)}{1 - f(z)\overline{f(s)}} \right|. \tag{15.6-10}$$

The symmetry property (15.6-8) is made explicit in this representation.

IV. Green's Function for Doubly Connected Regions

Let the doubly connected region R be such that no component of its complement reduces to a single point. We furthermore assume that the conformal map f of R onto a suitable annulus A: $\mu < |z| < 1$ (which always exists by a theorem to be proved in § 17.1) can be extended to a continuous one-to-one map of the closure of R onto the closure of A. By the Osgood-Carathéodory theorem this will be the case, for instance, if the boundary of R consists of two Jordan curves.

The problem of determining Green's function γ_R for R is then reduced to the problem of finding Green's function γ_A for A by the identity

$$\gamma_R(z, t) = \gamma_A(f(z), f(t)). \tag{15.6-11}$$

The function γ_A will now be determined by three different methods, each of which leads to a different analytical representation for γ_A. In all three methods we make use of the identity

$$\gamma_A(e^{i\alpha}z, e^{i\alpha}t) = \gamma_A(z, t), \tag{15.6-12}$$

stating that the values of γ_A depend only on the relative positions of z and t. We thus may assume, in particular, that t is real and positive, $t = \tau$ where $\mu < \tau < 1$.

(a) *Fourier Series.* Here the starting point is Green's function for the full unit disk,

$$\gamma_0(z, \tau) = -\frac{1}{2\pi} \operatorname{Log} \left| \frac{z-\tau}{1-\tau z} \right|.$$

This has the correct singularity at $z = \tau$ and assumes the correct boundary values 0 on the outer boundary $|z| = 1$ of the annulus A. In order to obtain Green's function for A, γ_0 must be amended by a function γ_1 that is harmonic in A with the boundary values

$$\gamma_1(z, \tau) = 0, \qquad |z| = 1;$$
$$\gamma_1(z, \tau) = -\gamma_0(z, \tau), \qquad |z| = \mu.$$

By the method of § 15.2 the solution

$$\gamma_1(z, \tau) = \frac{1}{2\pi} \frac{\operatorname{Log} \tau \operatorname{Log} \rho}{\operatorname{Log} \mu} + \frac{1}{2\pi} \sum_{n=1}^{\infty} \frac{\mu^n}{n} \frac{\tau^n - \tau^{-n}}{\mu^n - \mu^{-n}} (\rho^n - \rho^{-n}) \cos n\sigma, \qquad z = \rho\, e^{i\sigma}$$

is readily found, and Green's function for A thus has the representation

$$2\pi\gamma_A(z, \tau) = \left(\frac{\operatorname{Log} \rho}{\operatorname{Log} \mu} - 1 \right) \operatorname{Log} \tau - \operatorname{Log} \left| \frac{1 - \tau^{-1}\rho\, e^{i\sigma}}{1 - \tau\rho\, e^{i\sigma}} \right| + \sum_{n=1}^{\infty} \frac{\mu^n}{n} \frac{\tau^n - \tau^{-n}}{\mu^n - \mu^{-n}} (\rho^n - \rho^{-n}) \cos n\sigma, \tag{15.6-13}$$

where the infinite series converges, uniformly for $\mu \leqq |z| \leqq 1$ and $\mu \leqq \tau \leqq 1$, at least like a geometric series with ratio μ.

(b) *Infinite Product.* Here we start again with Green's function for the full unit disk. This may be written

$$\gamma_0(z, \tau) = -\frac{1}{2\pi} \operatorname{Log} |f_0(z)|,$$

where $f_0(z) := f(z)$,

$$f(z) := \frac{z - \tau}{1 - \tau z}. \tag{15.6-14}$$

As noted above, γ_0 has the correct singularity and assumes the correct boundary values on $|z| = 1$. To obtain the correct boundary values on $|z| = \mu$, we attach a factor as follows:

$$\gamma_1(z,\tau):=-\frac{1}{2\pi}\operatorname{Log}\left|\frac{f_0(z)}{f_1(z)}\right|,$$

where f_1 is the conformal transplant of f_0 under inversion at $|z|=\mu$,

$$f_1(z)=f_0\left(\frac{\mu^2}{z}\right).$$

If $|z|=\mu$, then $\mu^2z^{-1}=\bar{z}$, hence $f_1(z)=f_0(\bar{z})=\overline{f_0(z)}$,

$$|f_1(z)|=|f_0(z)|,$$

and γ_1 indeed assumes the correct value 0 on $|z|=\mu$; however, the boundary values 0 on $|z|=1$ have now been destroyed. To correct again, we consider

$$\gamma_2(z,\tau)=-\frac{1}{2\pi}\operatorname{Log}\left|\frac{f_0(z)f_2(z)}{f_1(z)}\right|,$$

where

$$f_2(z)=f_1\left(\frac{1}{z}\right),$$

the conformal transplant of f_1 under inversion at $|z|=1$. Continuing in a like manner, we are led to consider the sequence of functions

$$\gamma_n(z):=-\frac{1}{2\pi}\operatorname{Log}|p_n(z)|,\qquad n=0,1,2,\ldots,$$

where

$$p_{2n}:=\frac{f_0f_2\cdots f_{2n}}{f_1f_3\cdots f_{2n-1}},\qquad p_{2n+1}:=\frac{f_0f_2\cdots f_{2n}}{f_1f_2\cdots f_{2n+1}},$$

the functions f_k being defined recursively by

$$f_{2k}(z):=f_{2k-1}\left(\frac{1}{z}\right),\qquad f_{2k+1}(z):=f_{2k}\left(\frac{\mu^2}{z}\right),\qquad k=0,1,\ldots.$$

An easy induction shows that

$$f_{2k-1}(z)=f\left(\frac{\mu^{2k}}{z}\right),\qquad f_{2k}(z)=f(\mu^{2k}z),$$

hence that

$$\begin{aligned}\frac{f_{2k}(z)}{f_{2k+1}(z)}&=\frac{\mu^{2k}z-\tau}{1-\mu^{2k}\tau z}\cdot\frac{1-\mu^{2k+2}\tau z^{-1}}{\mu^{2k+2}z^{-1}-\tau}\\&=\frac{1-\mu^{2k}\tau^{-1}z}{1-\mu^{2k}\tau z}\cdot\frac{1-\mu^{2k+2}\tau z^{-1}}{1-\mu^{2k+2}\tau^{-1}z^{-1}}.\end{aligned}$$

Hoping to obtain the correct result by letting $n \to \infty$, we are led to consider the function

$$\gamma^*(z, \tau) := -\frac{1}{2\pi} \operatorname{Log}|p(z)|, \tag{15.6-15}$$

where p is the limit of the sequence of partial products $p_1, p_3, p_5, \ldots$, that is,

$$p(z) := \prod_{k=1}^{\infty} \frac{1-\mu^{2k-2}\tau^{-1}z}{1-\mu^{2k-2}\tau z} \frac{1-\mu^{2k}\tau z^{-1}}{1-\mu^{2k}\tau^{-1}z^{-1}}. \tag{15.6-16}$$

By Theorem 8.1c the product p converges uniformly on every compact set of the z plane that does not contain $z=0$. Thus in particular, p converges uniformly on the closure of A and represents an analytic function there. The function γ^* thus is harmonic in A except where p vanishes, which happens only at $z=\tau$, where γ^* has the correct singularity. By construction, moreover, $\gamma^*(z, \tau)=0$ for $|z|=\mu$. For $|z|=1$, on the other hand, we have

$$\left|\frac{f_{2k}(z)}{f_{2k-1}(z)}\right| = 1,$$

hence

$$|p(z)| = \frac{1}{\tau}|f_0(z)| = \frac{1}{\tau}.$$

Thus in order to obtain the correct Green's function, γ^* must be amended by a harmonic function which on $|z|=\mu$ is zero and on $|z|=1$ has the value $-(1/2\pi)|\operatorname{Log} \tau$. Clearly,

$$\frac{1}{2\pi}\left(\frac{\operatorname{Log} \rho}{\operatorname{Log} \mu} - 1\right) \operatorname{Log} \tau$$

is the solution of this trivial Dirichlet problem. Thus the desired Green's function for A with pole at τ has the representation

$$2\pi\gamma_A(z, \tau) = \left(\frac{\operatorname{Log} \rho}{\operatorname{Log} \mu} - 1\right) \operatorname{Log} \tau - \operatorname{Log}|p(z)|, \tag{15.6-17}$$

where $p(z)$ is given by (15.6-16).

(c) *Theta Series.* A representation for γ_A that converges much more rapidly is obtained by applying to the product p Jacobi's triple product identity (Theorem 8.2b). By that identity we have

$$\prod_{k=1}^{\infty} (1-\mu^{2k-2}\tau^{-1}z)(1-\mu^{2k}\tau z^{-1}) = \prod_{k=1}^{\infty}\left(1-\mu^{2k-1}\frac{z}{\mu\tau}\right)\left(1-\mu^{2k-1}\frac{\mu\tau}{z}\right)$$

$$= \frac{1}{q(\mu^2)} \sum_{k=-\infty}^{\infty} \mu^{k^2}\left(-\frac{z}{\mu\tau}\right)^k,$$

where the definition of q is not required, and similarly

$$\prod_{k=1}^{\infty}(1-\mu^{2k-2}\tau z)(1-\mu^{2k}\tau^{-1}z^{-1})=\prod_{k=1}^{\infty}\left(1-\mu^{2k-1}\frac{\tau z}{\mu}\right)\left(1-\mu^{2k-1}\frac{\mu}{\tau z}\right)$$

$$=\frac{1}{q(\mu^2)}\sum_{k=-\infty}^{\infty}\mu^{k^2}\left(-\frac{\tau z}{\mu}\right)^k.$$

There follows the representation

$$p(z)=\frac{\sum_{k=-\infty}^{\infty}(-1)^k\mu^{k^2-k}\tau^{-k}z^k}{\sum_{k=-\infty}^{\infty}(-1)^k\mu^{k^2-k}\tau^k z^k}, \qquad (15.6\text{-}18)$$

which for purposes of numerical evaluation is best written

$$p(z)=\frac{1+\sum_{k=1}^{\infty}(-1)^k\mu^{k^2}[(z/\mu\tau)^k+(\mu\tau/z)^k]}{1+\sum_{k=1}^{\infty}(-1)^k\mu^{k^2}[(\tau z/\mu)^k+(\mu/\tau z)^k]}. \qquad (15.6\text{-}19)$$

This allows to compute p very rapidly if μ is not very close to 1. For values of μ close to 1, Jacobi's identity (10.6-23) will convert (15.6-18) into a quotient of two series that converge rapidly.

PROBLEMS

1. If Green's function $g(z, \tau)$ is evaluated numerically by (15.6-17) and (15.6-19) at a point z such that $|z-\tau|=10^{-6}$, a value such as 2 results. What is the explanation of this paradoxical fact?
2. Use the method of § 15.2 to construct Green's function for the interior of the ellipse

$$E_\rho:\quad z=\tfrac{1}{2}(\rho\, e^{2\pi i\tau}+\rho^{-1}\, e^{-2\pi i\tau}),$$

where $\rho>1$ is fixed and $0\leqq\tau\leqq 1$.

NOTES

For collections of Green's functions, see Courant and Hilbert [1951], chap. V, § 15, or Butkovskyi [1982]. The notation $\boxed{dz}$ for the area element is borrowed from Teichmüller [1939], p. 19. An excellent, physically motivated account of Green's function is in Bergman and Schiffer [1953].

§ 15.7. THE NEUMANN PROBLEM

I. Definition and Motivation

Let R be a bounded region whose boundary ∂R consists of a finite number of regular Jordan curves, so that at every point of the boundary a normal is defined. Let ψ be a real-valued continuous function defined on ∂R. In its simplest form, the **Neumann problem** consists in finding a function u satisfying the following conditions:

(i) u is continuous and differentiable in $R \cup \partial R$.
(ii) u is harmonic in R.
(iii) If $\partial/\partial n$ denotes differentiation in the direction of the exterior normal, then

$$\frac{\partial u}{\partial n}(t) = \psi(t), \qquad t \in \partial R.$$

Applications of the Neumann problem, as of the Dirichlet problem, abound in classical mathematical physics. If u describes the electrostatic potential in a cylinder with cross section R, then the electric charge at the surface is given by $\varepsilon\, \partial u/\partial n$, where ε is the dielectric constant. Hence $\partial u/\partial n$ is proportional to the specific density of the electric charge sitting on the surface. If the charge distribution on the surface is assumed to be known, the determination of the potential leads to a Neumann problem. If u describes a stationary distribution of temperature in a homogeneous medium, then the flow of heat is proportional to grad u. Neumann-like boundary conditions thus result if the amount of heat forced across the boundary is prescribed rather than the temperature at the boundary. The potential of the flow around an airfoil may be regarded as the solution of a modified Neumann problem. Here the region R is the exterior of the cross section of the airfoil. The function u here must have a prescribed behavior at infinity and satisfy $\partial u/\partial n = 0$ on R.

As in the case of the Dirichlet problem, one may consider generalized forms of the Neumann problem where the boundary is only piecewise regular and/or the boundary function ψ is only piecewise continuous. Also, the requirement that the solution be continuously differentiable up to the boundary is often inconvenient. Therefore in mathematical treatments of the problem, conditions (i) and (iii) are replaced by the following:

(iv) If $(\cos\alpha, \sin\alpha)$ denotes the exterior normal unit vector at a point $t \in \partial R$, then

$$\lim_{z \to t} \left[\cos\alpha \frac{\partial u}{\partial x}(z) + \sin\alpha \frac{\partial u}{\partial y}(z) \right] = \psi(t) \qquad \forall t \in \partial R.$$

It is clear that if u possesses continuous partial derivatives in $R \cup \partial R$, this condition means the same as (iii). The problem of satisfying (ii) and (iv) will be called the **generalized Neumann problem**.

II. The Solvability of the Neumann Problem

It is easy to see that the Neumann problem need not have a solution at all. We recall Green's first formula (15.6-2),

$$\int_{\partial R} v \frac{\partial u}{\partial n} |dz| = \iint_R (v\,\Delta u + \operatorname{grad} u \operatorname{grad} v)\ \boxed{dz}$$

($n :=$ exterior normal) which holds for regions R with the properties stipulated above, and for functions u and v having the degree of smoothness required by the formula. If u is harmonic in R, it follows on putting $v=1$ that

$$\int_{\partial R} \frac{\partial u}{\partial n} |dz| = 0. \tag{15.7-1}$$

If u is a solution of the Neumann problem, then $\partial u/\partial n = \psi$ on ∂R, hence

$$\int_{\partial R} \psi(t)\,|dt| = 0 \tag{15.7-2}$$

is a necessary condition for the existence of a solution. The condition is easy to understand from the thermodynamic interpretation of Neumann's problem. If the thermodynamic equilibrium is to be maintained, the net flow of heat through the surface is zero.

On the other hand, it is also clear that the solution of the Neumann problem, if it exists, is not unique, because by adding to any solution a nonzero constant we immediately obtain another function satisfying the conditions of the problem. This, however, is as far as nonuniqueness goes.

THEOREM 15.7a. *The difference of any two solutions of a Neumann problem is a constant.*

Proof. Let u_1 and u_2 be two solutions. We apply Green's formula with $u = v := u_1 - u_2$. Since $\Delta u = 0$ in R and $\partial u/\partial n = 0$ on ∂R, we obtain

$$\int_{\partial R} |\operatorname{grad} u|^2\ \boxed{dz} = 0,$$

which in view of the connectedness of R implies $u = \text{const.}$ —

It is always possible to single out one special solution u of a given Neumann problem by imposing the extra condition

$$\int_{\partial R} u|dz| = 0. \tag{15.7-3}$$

We call this u the **normalized solution** of Neumann's problem.

As in the case of the Dirichlet problem, we shall not prove the existence of a solution of the Neumann problem in the general case. We simply state:

PROPOSITION 15.7b. *For boundary values satisfying* (15.7-2), *the generalized Neumann problem has a unique normalized solution for every bounded region R, the boundary of which consists of a finite number of regular Jordan curves.*

III. Construction of the Solution for Simply Connected Regions

To begin with, let R be the unit disk. We follow our usual method of deriving a solution formula under strong hypotheses and then show that the formula also holds under weaker assumptions. The strong assumption made is that the boundary function $\psi(e^{2\pi i\tau})$ can be represented as an absolutely convergent Fourier series,

$$\psi(e^{2\pi i\tau}) = \operatorname{Re} \sum_{n=1}^{\infty} a_n e^{2\pi in\tau}, \tag{15.7-4}$$

$\sum |a_n| < \infty$. By setting $a_0 = 0$ we have already taken into account condition (15.7-2). Trying to represent the normalized solution $u(z)$, where $z = \rho e^{2\pi i\tau}$, as a series of the special harmonic functions $\rho^n e^{2\pi in\tau}$, $n = 1, 2, \ldots$, written in the form

$$u(z) = \operatorname{Re} \sum_{n=1}^{\infty} b_n \rho^n e^{2\pi in\tau}, \tag{15.7-5}$$

we find from $\partial u/\partial n = \partial u/\partial\rho$ that

$$\frac{\partial u}{\partial n}(e^{2\pi i\tau}) = \operatorname{Re} \sum_{n=1}^{\infty} nb_n e^{2\pi in\tau}.$$

The Neumann boundary condition is satisfied if

$$b_n = \frac{1}{n} a_n = \frac{2}{n} \int_0^1 \psi(e^{2\pi i\sigma}) e^{-2\pi in\sigma} \, d\sigma.$$

Substituting this into (15.7-5) and interchanging summation and integration, we find

$$u(z) = \operatorname{Re} \int_0^1 \psi(e^{2\pi i\sigma}) \sum_{n=1}^{\infty} \frac{2}{n} \rho^n e^{2\pi in(\tau-\sigma)} \, d\sigma.$$

On noting

$$2 \operatorname{Re} \sum_{n=1}^{\infty} \frac{1}{n} \rho^n e^{2\pi in(\tau-\sigma)} = \operatorname{Log}|e^{2\pi i\sigma} - z|^{-2},$$

the solution formula becomes

$$u(z)=\int_0^1 \psi(e^{2\pi i\sigma})\mathrm{Log}|e^{2\pi i\sigma}-z|^{-2}\,d\sigma. \tag{15.7-6}$$

The solution is normalized, because the Fourier series (15.7-5) has no constant term. Formula (15.7-6) is called **Dini's formula**.

We now show that Dini's formula solves Neumann's problem in the sense of condition (iv):

THEOREM 15.7c. *Let ψ be a real continuous function on the unit circle with mean value zero, and let u be defined by Dini's formula. Then for all real τ,*

$$\lim_{z\to e^{2\pi i\tau}} \frac{\partial u}{\partial\rho}(z)=\psi(e^{2\pi i\tau}). \tag{15.7-7}$$

Proof. Writing Dini's formula as

$$u(z)=-\int_0^1 \psi(e^{2\pi i\sigma})\,\mathrm{Log}(1-2\rho\cos[2\pi(\tau-\sigma)]+\rho^2)\,d\sigma$$

and differentiating with respect to ρ, we find for $\rho<1$,

$$\frac{\partial u}{\partial\rho}(z)=-\int_0^1 \psi(e^{2\pi i\sigma})\frac{1}{\rho}\left\{1-\frac{1-\rho^2}{1-2\rho\cos[2\pi(\tau-\sigma)]+\rho^2}\right\}d\sigma$$

or, using the fact that ψ has mean value zero,

$$\rho\frac{\partial u}{\partial\rho}(z)=\int_0^1 \psi(e^{2\pi i\sigma})\frac{1-\rho^2}{1-2\rho\cos[2\pi(\tau-\sigma)]+\rho^2}\,d\sigma.$$

Thus $\rho\,\partial u/\partial\rho$ is represented in terms of ψ by Poisson's integral, and the assertion (15.7-7) follows from Theorem 15.2b. —

To evaluate Dini's integral numerically, it is best to use its Fourier series representation (15.7-5) where $b_n=a_n/n$ and to compute the coefficients a_n by an FFT.

By the method of conformal transplantation, a formula can now be given for the solution of Neumann's problem in an arbitrary simply connected region with a sufficiently smooth boundary. Let the region be R, let ∂R be its boundary, and let $w=f(z)$ map R conformally onto D: $|w|<1$. We assume that not only f but also f' can be extended to a function that is continuous in $R\cup\partial R$.

If u is a solution of Neumann's problem for R with the boundary condition

$$\frac{\partial u}{\partial n}(t)=\psi(t), \qquad t\in\partial R,$$

then the function $u_1 := u \circ f^{[-1]}$ is harmonic in D. Because the mapping f is conformal also at the boundary, derivation in the direction perpendicular to the boundary curve of R corresponds to derivation perpendicular to the unit circle. Thus

$$\frac{\partial u_1}{\partial n}(e^{2\pi i\tau}) = \frac{\partial u}{\partial n}(f^{[-1]}(e^{2\pi i\tau}))\left|\frac{dz}{dw}\right|$$
$$= \psi(f^{[-1]}(e^{2\pi i\tau})|f^{[-1]\prime}(e^{2\pi i\tau})|$$

Hence u_1 solves a Neumann problem for D with known boundary data. Thus by Dini's formula,

$$u_1(w) = \int_0^1 \psi(f^{[-1]}(e^{2\pi i\tau}))|f^{[-1]\prime}(e^{2\pi i\tau})| \operatorname{Log}|e^{2\pi i\tau} - w|^{-2}\, d\tau + \gamma,$$

where γ is a real constant. Here we let $w = f(z)$, which on the left yields $u_1(f(z)) = u(z)$. In view of

$$2\pi i f^{[-1]\prime}(e^{2\pi i\tau})\, e^{2\pi i\tau}\, d\tau = dt,$$

we have

$$|f^{[-1]\prime}(e^{2\pi i\tau})|\, d\tau = \frac{1}{2\pi}|dt|,$$

and we simply get

$$u(z) = \frac{1}{2\pi}\int_{\partial R} \psi(t) \operatorname{Log}|f(t) - f(z)|^{-2}\,|dt| + \gamma. \tag{15.7-8}$$

We thus have proved:

THEOREM 15.7d. *If the boundary data ψ are continuous and satisfy* (15.7-2), *the solution of Neumann's problem for a smoothly*[1] *bounded Jordan region R is given by* (15.7-8), *where f denotes a conformal map of R onto* $|w| < 1$.

EXAMPLE 1

Although the hypotheses of Theorem 15.7d are not fully applicable, we consider the Neumann problem for the upper half-plane. The function

$$w = f(z) = \frac{z - i}{z + i}$$

[1] That is, such that f' can be extended continuously to $R \cup \partial R$.

provides an appropriate mapping. The boundary function of the transformed problem is

$$\psi_1(e^{2\pi i\tau}) = \psi\left(i\frac{1+e^{2\pi i\tau}}{1-e^{2\pi i\tau}}\right)\frac{2}{|1-e^{2\pi i\tau}|^2}.$$

Assuming ψ to be continuous, ψ_1 is continuous at all points $w = e^{2\pi i\tau} \neq 1$. Continuity of ψ_1 at $w=1$ is not required; however, for the existence of Poisson's integral used in the proof of Theorem 15.7c we want ψ_1 to be bounded, which will be the case if $\xi^2\psi(\xi)$ is bounded on the interval $(-\infty, \infty)$. Under these strong hypotheses formula (15.7-8) is applicable. We have

$$f(\xi)-f(z) = \frac{2i(\xi - z)}{(\xi+i)(z+i)},$$

hence the formula yields

$$u(z) = -\frac{1}{2\pi}\int_{-\infty}^{\infty} \psi(\xi)\left\{\operatorname{Log}|\xi - z|^2 + \operatorname{Log}\frac{2}{|\xi+i|^2} + \operatorname{Log}\frac{2}{|z+i|^2}\right\} d\xi + \gamma.$$

Here

$$\int_{-\infty}^{\infty} \psi(\xi)\operatorname{Log}\frac{2}{|\xi+i|^2}\, d\xi$$

is independent of z and hence may be added to γ, while

$$\int_{-\infty}^{\infty} \psi(\xi)\operatorname{Log}\frac{2}{|z+i|^2}\, d\xi = \operatorname{Log}\frac{2}{|z+i|^2}\int_{-\infty}^{\infty} \psi(\xi)\, d\xi$$

is zero on account of (15.7-2). Hence the solution of Neumann's problem for the upper half-plane is

$$u(z) = \frac{1}{2\pi}\int_{-\infty}^{\infty} \psi(\xi)\operatorname{Log}|\xi - z|^{-2}\, d\xi + \gamma. \qquad (15.7\text{-}9)$$

IV. Neumann's Function

A glance at the boundary representation formula (15.6-5) shows that Neumann's problem could be solved elegantly if $v(z, t)$ were a normalized fundamental solution satisfying the additional condition

$$\frac{\partial v}{\partial n} = 0, \qquad z \in \partial R, \quad t \in R. \qquad (15.7\text{-}10)$$

The solution of Neumann's problem could then be expressed in the form

$$u(z) = \int_{\partial R} \psi(t)v(t, z)|dt|,$$

much as the solution of Dirichlet's problem is expressed in terms of Green's function.

However, no fundamental solution satisfying (15.7-10) exists. This is seen by deleting from R the disk $|z-t|\leqq \varepsilon$, where ε is sufficiently small. Denoting by R_ε the region thus obtained, we have by (15.7-1), since $v(z, t)$ as a function of z is harmonic in R_ε,

$$\int_{\partial R_\varepsilon} \frac{\partial v}{\partial n}(z, t)|dz| = 0.$$

If Γ_ε denotes the circle $|z-t|=\varepsilon$, we have, on account of the logarithmic singularity at t,

$$\int_{\Gamma_\varepsilon} \frac{\partial v}{\partial n}(z, t)|dz| = +1.$$

Thus for every normalized fundamental singularity there holds

$$\int_{\partial R} \frac{\partial v}{\partial n}(z, t)|dz| = -1. \tag{15.7-11}$$

Let λ denote the length of ∂R, assumed to be finite. The simplest way to satisfy (15.7-11) is to impose on v the condition

$$\frac{\partial v}{\partial n}(z, t) = -\frac{1}{\lambda}, \qquad z \in \partial R, \quad t \in R. \tag{15.7-12}$$

For a v satisfying (15.7-12), the boundary representation formula yields for the solution of the Neumann problem with boundary data ψ

$$u(z) = \int_{\partial R} \psi(t) v(t, z)|dt| + \frac{1}{\lambda}\int_{\partial R} u(t)|dt|.$$

The second term is a constant. Omitting it, we find that the function

$$u(z) := \int_{\partial R} \psi(t) v(t, z)|dt| \tag{15.7-13}$$

likewise is a solution of Neumann's problem.

Condition (15.7-12) does not determine the fundamental solution v uniquely, since adding a constant to v does not invalidate (15.7-12). The desired fundamental solution is made unique by imposing the normalizing condition

$$\int_{\partial R} v(z, t)|dz| = 0, \qquad t \in R. \tag{15.7-14}$$

It is shown below that for bounded regions R whose boundary consists of a finite number of regular Jordan curves a fundamental solution with the properties (15.7-12) and (15.7-14) in fact exists. This is called the **Neumann function** of R and will be denoted by $\nu(z, t)$. Thus to summarize, the Neumann function for R is a normalized fundamental solution with pole at t which satisfies the conditions

$$\frac{\partial \nu}{\partial n}(z, t) = -\frac{1}{\lambda}, \qquad z \in \partial R, \quad t \in R, \tag{15.7-15a}$$

where λ is the length of ∂R, and

$$\int_{\partial R} \nu(z, t)\,|dt| = 0, \qquad t \in R. \tag{15.7-15b}$$

The existence of Neumann's function is easily seen by writing

$$\nu(z, t) = \frac{1}{2\pi} \operatorname{Log} \frac{1}{|z-t|} + w(z, t),$$

where w as a function of z is harmonic. Condition (15.7-15a) yields a Neumann problem for w, the solution of which exists by Proposition 15.4a. As in the proof of (15.6-8), one can show that the symmetry relation

$$\nu(t_1, t_2) = \nu(t_2, t_1)$$

holds for arbitrary $t_1, t_2 \in R \cup \partial R$ $(t_1 \neq t_2)$. As shown above, the formula

$$u(z) = \int_{\partial R} \psi(t)\nu(t, z)\,|dt| \tag{15.7-16}$$

yields a solution of the Neumann problem with boundary data ψ. This turns out to be the normalized solution, because u is represented as a linear superposition of solutions satisfying (15.7-15b).

One curious property of Neumann's function remains to be mentioned. Let γ and ν denote, respectively, Green's and Neumann's function for a smoothly bounded region R as above. Let

$$\kappa(z, t) := \nu(z, t) - \gamma(z, t). \tag{15.7-17}$$

This is called the (real) **Bergman kernel function** of R. Since the fundamental singularity has been removed, this function for each fixed $t \in R$ as a function of z is harmonic in the whole region R and, on account of the symmetry of both γ and ν, the same holds for each fixed $z \in R$ when κ is considered as a function of t. Moreover, the kernel function enjoys the following reproducing property:

THEOREM 15.7e. *Let ω be harmonic in R and continuously differentiable in $R \cup \partial R$, and let $\int_{\partial R} \omega(z)|dz| = 0$. Then for $t \in R$,*

$$\iint_R \operatorname{grad}_z \kappa(z, t) \operatorname{grad}_z \omega(z) \boxed{dz} = \omega(t). \tag{15.7-18}$$

Proof. Let R_ε and Γ_ε have the same meanings as before. By Green's first identity, the quantity on the left equals the limit as $\varepsilon \to 0$ of

$$-\int_{\Gamma_\varepsilon} \gamma(z, t)\frac{\partial\omega}{\partial n}(z)|dz| + \int_{\Gamma_\varepsilon} \frac{\partial\nu}{\partial n}(z, t)\omega(z)|dz|.$$

The first term is $O(\varepsilon \operatorname{Log}(1/\varepsilon))$ and therefore has limit 0. The second term is

$$\frac{1}{2\pi}\frac{1}{\varepsilon}\omega(t) \cdot 2\pi\varepsilon + O(\varepsilon)$$

and therefore tends to $\omega(t)$. —

PROBLEMS

1. Use the method of Fourier series to show that Neumann's function for the unit disk is

$$\nu(z, t) = \frac{1}{2\pi} \operatorname{Log} \frac{1}{|z-t||1-\bar{z}t|}.$$

2. Construct Neumann's function for the annulus A: $\mu < |z| < 1$. Find representations analogous to those given for Green's function in § 15.6.
3. (a) Show that Neumann's function *is not* a conformal invariant. That is, if ν is Neumann's function for a smoothly bounded region R and if f maps the region S conformally onto R, then it is not true (except in trivial cases) that Neumann's function for S is $\nu(f(z), f(t))$.
   (b) Show, however, that it is nevertheless true that a solution of Neumann's problem for the region S is given by

$$u(z) = \int_{\partial S} \psi(t)\nu(f(t), f(z))|dt|.$$

4. Show that the real Bergmann kernel function for the unit disk is

$$\kappa(z, t) = \frac{1}{\pi} \operatorname{Log} \frac{1}{|1-\bar{z}t|},$$

and verify the reproducing property (15.7-18) by a Fourier series computation.

NOTES

The subject matter of this section is classical; see again Bergman and Schiffer [1953]. For a treatment of a "mixed" boundary value problem by means of the formula of Keldysh and Sedov, see Lavrentiev and Shabat [1967], § 54. Our treatment of Neumann's problem requires ∂R to be smooth, so that $\partial u/\partial n$ can be defined. If the problem is considered from a more abstract point of view, this condition can be relaxed; see Kral [1980].

§ 15.8. THE LOGARITHMIC POTENTIAL OF A LINE CHARGE

I. Single-Layer Potential

Imagine a thin wire in the form of a straight line perpendicular to the z plane which carries a uniformly distributed electric charge. If no other conductors are present, and if the wire pierces the z plane at the point $z = t$, the charged wire will give rise to an electric potential which is constant along any straight line perpendicular to the z plane and which in the z plane is described by

$$u(z) = \mu \operatorname{Log} \frac{1}{|z-t|}.$$

Here μ is a real constant which is proportional to, and carries the sign of, the charge per unit length on the wire.

Imagine next a cylindrical surface whose generators are again perpendicular to the z plane. We assume that each generator carries a uniform electric charge; contrary to intuition, however, the charges are not permitted to move freely on the surface. If the surface intersects the z plane along a curve Γ: $z = z(\tau)$, $\alpha \leqq \tau \leqq \beta$, then the resulting potential in the z plane is obtained by summing the contributions of each generator, and is thus given by

$$u(z) = \int_{\Gamma} \mu(t) \operatorname{Log} \frac{1}{|t-z|} |dt|. \tag{15.8-1}$$

Here the function μ describes the density of the charges as a function of $t \in \Gamma$ (see Fig. 15.8a). The expression (15.8-1) is called the **logarithmic potential** due to a **single layer of charges** $\mu(t)$ placed on Γ, or briefly the **logarithmic potential of a single layer**.

For the mathematical discussion we shall assume that Γ is piecewise regular, in the sense that Γ consists of a finite number of regular arcs (see definition in § 3.5). Since $u(z)$ is the sum of the contributions of each regular subarc, it suffices to study the contribution of a regular subarc which we again call Γ.

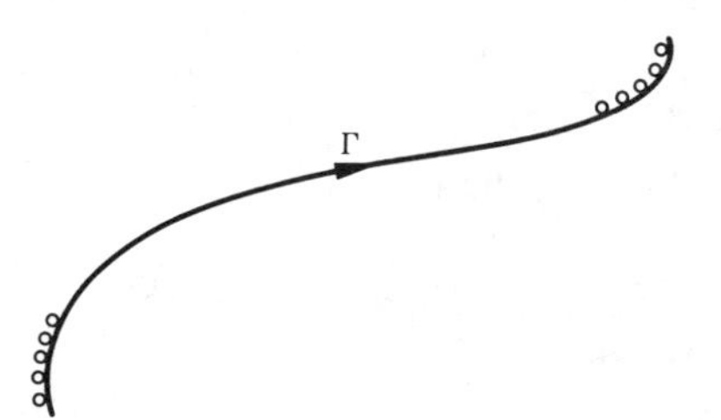

Fig. 15.8a. Single-layer potential.

For the existence of the integral (15.8-1) at a point z off Γ it is clearly sufficient that the integral

$$\int_\Gamma |\mu(t)||dt|$$

exists, that is, that μ belongs to the class $L_1(\Gamma)$. (This allows μ to be discontinuous, and even unbounded.) The value of the integral (15.8-1) then may be obtained constructively from a subdivision Δ of Γ into subarcs Γ_i, selecting a point t_i on each subarc, and putting

$$u_\Delta(z) := \sum_i \operatorname{Log} \frac{1}{|t_i - z|} \int_{\Gamma_i} \mu(t)|dt|.$$

If we denote by $|\Delta|$ the maximum of the lengths of the subarcs of the subdivision Δ, we have

$$u(z) = \lim_{|\Delta| \to 0} u_\Delta(z). \tag{15.8-2}$$

Let $R := \mathbb{C}\backslash\Gamma$, and let T be a compact subset of R. Since $\operatorname{Log}|z-t|^{-1}$ is bounded on T, uniformly with respect to $t \in \Gamma$, simple estimates show that the limit (15.8-2) exists uniformly with respect to $z \in T$. Since each function u_Δ is harmonic in R, it follows from Theorem 15.3i that the limit function $u(z)$ likewise is harmonic. We thus have obtained:

THEOREM 15.8a. *Let Γ be a piecewise regular curve, and let the real function $\mu \in L_1(\Gamma)$. Then the logarithmic potential of μ exists in the open set $R := \mathbb{C}\backslash\Gamma$ and is harmonic there.*

We now pass to the existence of the logarithmic potential at points $z \in \Gamma$. The integral (15.8-1) is then, in any case, improper. The following example shows that $\mu \in L_1(\Gamma)$ is no longer sufficient for the integral to exist.

EXAMPLE 1

Let $\Gamma := [0, \frac{1}{2}]$,

$$\mu(t) := \frac{1}{t(\operatorname{Log}(1/t))^2}, \qquad t \neq 0.$$

Then, as may be shown by simple substitutions, the integral

$$\int_0^{1/2} |\mu(t)|\, dt = \int_0^{1/2} \frac{1}{t(\operatorname{Log}(1/t))^2}\, dt$$

exists, whereas the integral defining the logarithmic potential at $z=0$,

$$\int_0^{1/2} \frac{1}{t \operatorname{Log}(1/t)}\, dt,$$

fails to exist. —

If we assume, however, that $\mu \in L_p(\Gamma)$ for some $p>1$, that is, that

$$\int_\Gamma |\mu(t)|^p\, dt < \infty, \tag{15.8-3}$$

then the integral (15.8-1) may be shown to exist even for points $z \in \Gamma$. This is a simple consequence of Hölder's inequality, which for nonnegative functions f and g asserts that

$$\int fg \leqq \left(\int f^p\right)^{1/p} \left(\int g^q\right)^{1/q}, \tag{15.8-4}$$

where p and q are related by $p^{-1}+q^{-1}=1$. In the present case, $f(t)=|\mu(t)|$, and the first factor on the right of (15.8-4) is finite by hypothesis. The existence of $\int g^q(t)\, dt$, where $g(t) := \operatorname{Log}|z-t|^{-1}$, follows from the existence of

$$\int_0^1 \left(\operatorname{Log}\frac{1}{t}\right)^q dt$$

for any $q<\infty$, using the regularity of Γ.

Thus if $\mu \in L_p(\Gamma)$ for some $p>1$, the logarithmic potential $u(z)$ exists for all $z \in \mathbb{C}$. We next wish to show that *under no additional hypotheses u is continuous at all points z*, including those on Γ.

Since continuity at points $z \notin \Gamma$ already follows from Theorem 15.8a ($\mu \in L_p$ with $p>1$ implies $\mu \in L_1$), it remains to establish the continuity at points $z_0 \in \Gamma$. Thus let $z_0 \in \Gamma$, and let $\varepsilon>0$. We are to show the existence of $\delta>0$ such that, for all z satisfying $|z-z_0|<\delta$ (whether or not $z \in \Gamma$), there holds

$$|u(z)-u(z_0)|<\varepsilon,$$

where u is given by (15.8-1).

If $\rho>0$ is sufficiently small, then since Γ is regular, the portion of Γ which lies in the disk $|z-z_0| \leqq \rho$ consists of a single arc. Let Γ_ρ be this arc. We write

$$\begin{aligned} d(z) := u(z)-u(z_0) &= \left(\int_{\Gamma_\rho} + \int_{\Gamma-\Gamma_\rho}\right) \mu(t) \operatorname{Log}\left|\frac{z_0-t}{z-t}\right| |dt| \\ &= d_1(z)+d_2(z). \end{aligned}$$

As to $d_1(z)$, we have by Hölder's inequality, if p is such that (15.8-3) holds and $p^{-1}+q^{-1}=1$,

$$|d_1(z)| \leqq \left\{\int_{\Gamma_\rho} |\mu(t)|^p\,|dt|\right\}^{1/p} \left\{\int_{\Gamma_\rho} \left|\operatorname{Log}\left|\frac{z_0-t}{z-t}\right|\right|^q |dt|\right\}^{1/q}.$$

The first factor on the right is bounded, independently of ρ, by (15.8-3). As to the second factor, we have

$$\int_{\Gamma_\rho} \left|\operatorname{Log}\left|\frac{z_0-t}{z-t}\right|\right|^q |dt| \leqq 2^q \int_{\Gamma_\rho} \left\{\left|\operatorname{Log}\frac{1}{|z-t|}\right|^q + \left|\operatorname{Log}\frac{1}{|z_0-t|}\right|^q\right\}|dt|.$$

The integral of the first term on the right is comparable (using the regularity of Γ_ρ) to the integral obtained by replacing z by t_z, the point on Γ_ρ closest to z. Thus the whole integral tends to 0 as $\rho \to 0$, uniformly in z. Thus there exists $\rho > 0$ such that

$$|d_1(z)| < \tfrac{1}{2}\varepsilon \quad \text{for all} \quad z.$$

Turning to $d_2(z)$, we note that this function is continuous at z_0, with value 0, by virtue of Theorem 15.8a. Thus there exists $\delta > 0$ such that

$$|d_2(z)| < \tfrac{1}{2}\varepsilon, \qquad |z-z_0| < \delta.$$

We thus have proved:

THEOREM 15.8b. *Let Γ be a piecewise regular curve, and let $\mu \in L_p(\Gamma)$ for some $p > 1$. Then the logarithmic potential of μ exists and is continuous on $\mathbb{C}$, and is harmonic in $\mathbb{C}\backslash\Gamma$.*

II. Double-Layer Potential

We next imagine two parallel straight lines carrying uniform charges of equal amount and opposite sign, piercing the z plane at the points $z=t$ and $z=t+\varepsilon e^{i\alpha}$, where α is real and where ε is a small positive number. By the foregoing the potential will be

$$\nu\left[\operatorname{Log}\frac{1}{|t-z+\varepsilon e^{i\alpha}|} - \operatorname{Log}\frac{1}{|t-z|}\right],$$

where ν is real. If we here let $\varepsilon \to 0$ and simultaneously increase the charges by a factor ε^{-1}, so that in the limit they will not cancel each other, there will result the potential

$$u(z) = \nu \lim_{\varepsilon\to 0} \frac{1}{\varepsilon}\left\{\operatorname{Log}\frac{1}{|t-z+\varepsilon e^{i\alpha}|} - \operatorname{Log}\frac{1}{|t-z|}\right\}. \qquad (15.8\text{-}5)$$

Apart from the factor ν, this is the derivative of the function

$$\lambda: \quad t \mapsto \operatorname{Log} \frac{1}{|t-z|} \tag{15.8-6}$$

at the point t in the direction $e^{i\alpha}$. By differential calculus, if $t = r + is$, the directional derivative is

$$\cos \alpha \frac{\partial \lambda}{\partial r} + \sin \alpha \frac{\partial \lambda}{\partial s}$$

or, in terms of the complex gradient grd $\lambda = \partial\lambda/\partial r + i\, \partial\lambda/\partial s$ introduced in § 5.6,

$$\operatorname{Re}\{e^{i\alpha}\, \overline{\operatorname{grd} \lambda(t)}\}.$$

For the function λ defined by (15.8-6),

$$\operatorname{grd} \lambda(t) = -\frac{1}{\bar{t} - \bar{z}},$$

thus the limit (15.8-5) is

$$u(z) = -\nu \operatorname{Re}\left\{\frac{e^{i\alpha}}{t-z}\right\}. \tag{15.8-7}$$

The idealized electrical device which we have described is called a **dipole**; the number $e^{i\alpha}$ indicates the orientation of the dipole.

Suppose now that the generators of the cylindrical surface described earlier carry dipoles oriented in the direction perpendicular to the surface. By the foregoing, the potential

$$u(z) = \int_\Gamma \nu(t) \frac{\partial}{\partial n_t} \operatorname{Log} \frac{1}{|t-z|} |dt| \tag{15.8-8}$$

will result. The orientation of the normal n_t at the point $t \in \Gamma$ is, in principle, arbitrary, but for definiteness we assume n_t to be the normal pointing to the left of Γ. The function (15.8-8) is called the **logarithmic potential** due to a **layer of dipoles** of density $\nu(t)$, or simply **double-layer potential** (see Fig. 15.8b).

We represent Γ by $t = t(\tau)$ and to make use of (15.8-7), let

$$\phi(t) := \arg t'(\tau), \qquad t = t(\tau).$$

On each regular subarc of Γ, $\phi(t)$ may be assumed to be continuous. If Γ is closed and regular throughout, then ϕ may be assumed to be continuous except at one point. The direction of the normal oriented to the left of Γ (that is, to the interior if Γ is closed) then is $ie^{i\phi(t)}$, and in view of

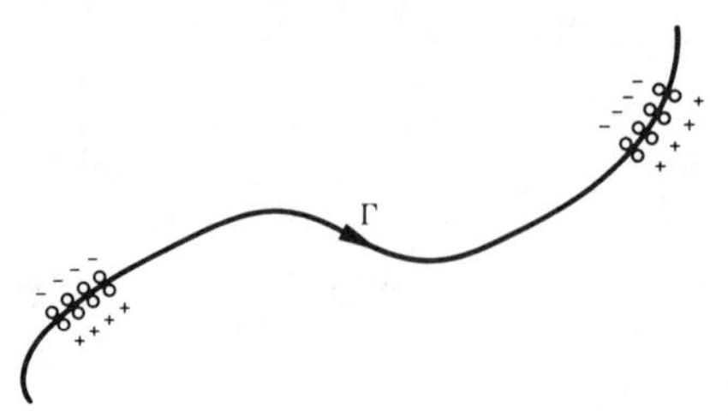

Fig. 15.8b. Double-layer potential.

$\operatorname{Re}(ia) = -\operatorname{Im} a$, we have

$$u(z) = \int_\Gamma \nu(t) \operatorname{Im} \frac{e^{i\phi(t)}}{t-z} |dt|,$$

which in view of $e^{i\phi(t)}|dt| = dt$ equals

$$u(z) = \operatorname{Im} \int_\Gamma \frac{\nu(t)}{t-z} dz$$

or

$$u(z) = \operatorname{Re} f(z), \tag{15.8-9}$$

where

$$f(z) := \frac{1}{2\pi i} \int_\Gamma \frac{2\pi\nu(t)}{t-z} dt. \tag{15.8-10}$$

The double-layer potential with density $\nu(t)$ thus equals the real part of the Cauchy integral formed with $h(t) = 2\pi\nu(t)$. To avail ourselves of the basic results on Cauchy integrals, we assume that ν and Γ satisfy the basic conditions of Chapter 14: ν is in Lip, and Γ is regular. Then by means of (15.8-9) u may also be defined at points $z_0 \in \Gamma$ by interpreting the integral (15.8-10) as a principal value. Theorems 14.1b and 14.1c (Sokhotskyi formulas) moreover imply:

THEOREM 15.8c. *Let Γ be regular, and let the real function ν be uniformly Hölder continuous on Γ. Then the double-layer potential* (15.8-8) *exists at all $z \in \mathbb{C}$ and is harmonic on $\mathbb{C}\backslash\Gamma$. At each interior point $t \in \Gamma$ the one-sided limits $u^+(t)$ and $u^-(t)$ exist, are uniformly Hölder continuous on each closed subarc of Γ, and satisfy*

$$\begin{aligned} u^+(t) - u^-(t) &= 2\pi\nu(t), \\ u^+(t) + u^-(t) &= 2u(t). \end{aligned} \tag{15.8-11}$$

PROBLEMS

1. Deduce the properties of the logarithmic single-layer potential

$$u(z)=\int_\Gamma \mu(t)\operatorname{Log}\frac{1}{|t-z|}|dt|$$

from properties of Cauchy integrals, assuming that the arc Γ is Hölder regular with initial point t_0 and terminal point t_1, and that

$$\mu^*(t):=\int_{t_0}^{t}\mu(t)|dt|, \qquad t\in\Gamma,$$

is uniformly Hölder continuous on Γ. Show in particular:

(a)
$$u(z)=\mu^*(t_1)\operatorname{Log}\frac{1}{|t_1-z|}+\operatorname{Re}f(z),$$

where

$$f(z):=\frac{1}{2\pi i}\int_\Gamma \frac{2\pi i\mu^*(t)}{t-z}\,dt.$$

(b) Conclude from Sokhotskyi's formulas that if $t\in\Gamma, t\neq t_0, t_1, u^+(t)=u^-(t)$, and hence that u is continuous at t.

(c) Apply Theorem 14.7b to conclude that u is continuous also at the points t_0 and t_1.

2. Let Γ be piecewise Hölder regular, and let μ be continuous on Γ except for a finite number of points $t_0, t_1, \ldots, t_k\in\Gamma$ in the neighborhood of which μ may be represented as

$$\mu(t)=|t-t_i|^{-\alpha}\mu_i(t)$$

with Hölder continuous functions μ_i, and with $\alpha\in(0,1)$ independent of i. Show that for every $\varepsilon>0$ the restriction of the logarithmic potential (15.8-1) to Γ satisfies a uniform Hölder condition with exponent $1-\alpha-\varepsilon$.

NOTES

For the continuity of the logarithmic single-layer potential see Gaier [1976]. He shows that if Γ is merely assumed to be rectifiable, a geometric condition (ε condition) must be imposed if the potential is to be continuous. The conditions imposed to establish the jump properties of the double-layer potential seem unnecessarily strong; a more abstract approach such as the one by Kral [1980] is probably required to achieve a satisfactory theory.

§ 15.9. THE INTEGRAL EQUATIONS OF POTENTIAL THEORY

Our constructive approaches to the Dirichlet problem for general regions R have so far required the availability of suitable conformal mapping functions for R. Here now we shall begin to discuss methods for the Dirichlet problem that do not require any mapping functions. As will be shown in later chapters, some of these approaches can even be used to *construct* mapping functions. In the present section we show how to reduce the problems of Dirichlet and of Neumann to a set of linear integral equations of a single real variable.

I. An Integral Equation for the Dirichlet Problem

Let R be a Jordan region with boundary Γ which we assume to be regular. The boundary data on Γ, here called $u_0(z)$, are assumed to be uniformly Hölder continuous. We try to represent the solution u as the potential of a double layer with density ν on Γ. By (15.8-9) we then have

$$u(z) = \operatorname{Re} f(z), \tag{15.9-1a}$$

where

$$f(z) = \frac{1}{2\pi i}\int_\Gamma \frac{2\pi\nu(t)}{t-z}\,dt. \tag{15.9-1b}$$

If z approaches a point z_0 on the boundary, then on the one hand

$$u(z) \to u_0(z_0)$$

and on the other, using the Sokhotskyi formula and supposing ν to be Hölder continuous,

$$f(z) \to f(z_0) + \pi\nu(z_0),$$

where $f(z_0)$ denotes the principal value of the integral (15.9-1b) for $z_0 \in \Gamma$. We thus get

$$u_0(z_0) = \operatorname{Re}\frac{1}{2\pi i}\,\mathrm{PV}\int_\Gamma \frac{2\pi\nu(t)}{t-z_0}\,dt + \pi\nu(z_0). \tag{15.9-2}$$

Let, as before, $\phi(t) := \arg t'(\tau)$, where $t = t(\tau)$, so that $dt = e^{i\phi(t)}|dt|$ and

$$\operatorname{Re}\frac{1}{i}\frac{\nu(t)}{t-z}\,dt = \nu(t)\operatorname{Im}\frac{e^{i\phi(t)}}{t-z}\,|dt|.$$

If for $t, z \in \Gamma$, $t \neq z$, we now define the real function

$$\kappa(t, z) := \frac{1}{\pi}\operatorname{Im}\frac{e^{i\phi(t)}}{t-z}, \tag{15.9-3}$$

called the **Neumann kernel** of Γ, then (15.9-2) becomes

$$\nu(z) = -\mathrm{PV}\int_\Gamma \nu(t)\kappa(t, z)\,|dt| + \frac{1}{\pi}u_0(z), \qquad z \in \Gamma. \tag{15.9-4}$$

We shall see in a moment that under certain conditions the symbol PV may be omitted from the integral (15.9-4). The equation then represents a **Fredholm integral equation of the second kind** for the unknown density ν; more about this later.

II. An Integral Equation for the Neumann Problem

To solve the Neumann problem

$$\Delta u = 0 \quad \text{in} \quad R,$$

$$\frac{\partial u}{\partial n} = v_0 \quad \text{on} \quad \Gamma,$$

($n :=$ exterior normal), where R and Γ are as above, and where v_0 is a Hölder continuous function satisfying the equation

$$\int_\Gamma v_0(t)\,|dt| = 0 \tag{15.9-5}$$

necessary for the existence of a solution, we seek to represent the solution u as the potential of a single layer,

$$u(z) = \int_\Gamma \mu(t) \operatorname{Log} \frac{1}{|t-z|}\,|dt|. \tag{15.9-6}$$

By differentiating under the integral sign if $z \in R$,

$$\operatorname{grd} u(z) = \int_\Gamma \mu(t) \frac{1}{\bar{t} - \bar{z}}\,|dt|.$$

If $z_0 \in \Gamma$, and if $e^{i\phi(z_0)}$ is the direction of the tangent of Γ at z_0, then the derivative of u at z in the direction of the exterior normal at z_0 is

$$\operatorname{Re}\{-ie^{i\phi(z_0)}\overline{\operatorname{grd} u(z)}\} = \operatorname{Re}\{e^{i\phi(z_0)} f(z)\},$$

where

$$f(z) := \frac{1}{2\pi i} \int_\Gamma \frac{2\pi\mu(t) e^{-i\phi(t)}}{t - z}\,dt$$

is a Cauchy integral. If u solves the Neumann problem, the limit of the directional derivative as $z \to z_0$ equals $v_0(z_0)$. Thus by the foregoing formula,

$$v_0(z_0) = \operatorname{Re}\{e^{i\phi(z_0)} f^+(z_0)\}. \tag{15.9-7}$$

By the Sokhotskyi formulas,

$$f^+(z_0) = f(z_0) + \pi\mu(z_0)\, e^{-i\phi(z_0)}.$$

Thus (15.9-7) becomes

$$v_0(z_0) = \operatorname{Re}\{e^{i\phi(z_0)} f(z_0)\} + \pi\mu(z_0).$$

Dropping the subscript in z, we get for $z \in \Gamma$,

$$\frac{1}{\pi} v_0(z) = \operatorname{Re}\left\{\frac{1}{i\pi} \operatorname{PV} \int_\Gamma \frac{\mu(t) e^{i\phi(t)}}{t-z} |dt| + \mu(z).\right.$$

Using

$$\operatorname{Re}\left\{\frac{1}{i\pi} \frac{e^{i\phi(z)}}{t-z}\right\} = -\frac{1}{\pi} \operatorname{Im} \frac{e^{i\phi(z)}}{z-t} = -\kappa(z, t),$$

where $\kappa(z, t)$ is the Neumann kernel defined by (15.9-3) (note the interchange of the arguments), we obtain

$$\mu(z) = \operatorname{PV} \int_\Gamma \mu(t) \kappa(z, t) |dt| + \frac{1}{\pi} v_0(z). \tag{15.9-8}$$

Again we shall see that the symbol PV may be dropped under certain conditions. Equation (15.9-8) then is an ordinary Fredholm integral equation of the second kind for the unknown charge density $\mu(t)$.

III. The Neumann Kernel

The Neumann kernel

$$\kappa(t, z) = \frac{1}{\pi} \operatorname{Im} \frac{e^{i\phi(t)}}{t-z},$$

occurring in both integral equations found above, is certainly continuous at all points $(t, z) \in \Gamma \times \Gamma$ except for $t = z$, where it is undefined. It will now be shown that if Γ is sufficiently smooth, κ can be defined for $t = z$ so as to become continuous.

THEOREM 15.9a. *Let the regular Jordan curve* Γ: $t = t(\tau)$, $\alpha \leqq \tau \leqq \beta$, *be such that* $t''(\tau)$ *exists and is continuous on* $[\alpha, \beta]$. *Then the limit of* $\kappa(t, z)$ *as* $z \to t$ *exists for every* $t \in \Gamma$, *uniformly in* t, *and*

$$\lim_{z \to t} \kappa(t, z) = \frac{1}{2\pi} \kappa(t), \tag{15.9-9a}$$

where

$$\kappa(t) := \frac{\operatorname{Im}(t''(\tau)\overline{t'(\tau)})}{|t'(\tau)|^3} \tag{15.9-9b}$$

is the **curvature** *of* Γ *at* t. *Defining* $\kappa(t, t)$ *by the limit* (15.9-9), $\kappa(t, z)$ *is continuous (jointly in both variables) on* $\Gamma \times \Gamma$.

Proof. Let $t = t(\tau)$, $z = t(\sigma)$. Then

$$\kappa(t, z) = -\frac{1}{\pi}\frac{1}{|t'(\tau)|}\operatorname{Im}\frac{t'(\tau)}{t(\sigma)-t(\tau)}.$$

Using

$$t(\sigma)-t(\tau) = t'(\tau)(\sigma-\tau)+\tfrac{1}{2}t''(\tau)(\sigma-\tau)^2+o((\sigma-\tau)^2),$$

we have

$$\frac{t'(\tau)}{t(\sigma)-t(\tau)} = \frac{1}{\sigma-\tau}\left\{1+\frac{1}{2}\frac{t''(\tau)}{t'(\tau)}(\sigma-\tau)+o(\sigma-\tau)\right\}^{-1}$$
$$= \frac{1}{\sigma-\tau}\left\{1-\frac{1}{2}\frac{t''(\tau)}{t'(\tau)}(\sigma-\tau)+o(\sigma-\tau)\right\},$$

hence

$$\lim_{\sigma\to\tau}\operatorname{Im}\frac{t'(\tau)}{t(\sigma)-t(\tau)} = -\frac{1}{2}\operatorname{Im}\frac{t''(\tau)}{t'(\tau)},$$

implying (15.9-9). It is obvious that the limit exists uniformly although we omit the technical details.

To prove the continuity of $\kappa(t, z)$ at a point (t_0, t_0), it must be shown that however $\varepsilon > 0$ is chosen,

$$|\kappa(t, z)-\kappa(t_0, t_0)| < \varepsilon \tag{15.9-10}$$

whenever (t, z) is sufficiently close to (t_0, t_0). By the triangle inequality (15.9-10) will hold if both

$$|\kappa(t, z)-\kappa(t, t)| < \frac{\varepsilon}{2}$$

and

$$|\kappa(t, t)-\kappa(t_0, t_0)| < \frac{\varepsilon}{2}.$$

The first inequality holds for all $t \in \Gamma$ whenever $|z-t| < \delta_1$ by virtue of the uniform existence of the limit (15.9-9). The second inequality holds whenever $|t-t_0| < \delta_2$, say, because the function $\kappa(t, t)$ as defined by (15.9-9a) is continuous on the compact set Γ, hence uniformly continuous. Both the above inequalities thus hold if $\max(|t-t_0|, |z-t_0|) < \frac{1}{2}\min(\delta_1, \delta_2)$. —

EXAMPLE **1**

If Γ is a circle, $t = \rho e^{i\tau}$, $0 \leqq \tau \leqq 2\pi$, then $t'(\tau) = i\rho e^{i\tau}$, $|t'(\tau)| = \rho$, and $e^{i\phi(t)} = it/\rho$. Hence

$$\kappa(t, z) = \frac{1}{\pi\rho}\operatorname{Im}\frac{it}{t-z} = \frac{1}{\pi\rho}\operatorname{Re}\frac{1}{1-e^{i(\sigma-\tau)}} = \frac{1}{2\pi\rho}.$$

Thus in the case of a circle, the Neumann kernel is constant, and (15.9-9) is trivial. —

More generally the hypotheses of Theorem 15.9a are satisfied if Γ is an analytic Jordan curve, or if Γ is piecewise analytic and t' and t'' are continuous at the corners.

IV. Fredholm Theory

If Γ satisfies the conditions of Theorem 15.9a, the integral equations for the densities ν and μ occurring in the representations (15.9-1) and (15.9-6) of the solutions of Dirichlet's and Neumann's problems may be written

$$\nu(z)+\int_\Gamma \nu(t)\kappa(t, z)|dt|=\frac{1}{\pi} u_0(z) \tag{15.9-11}$$

and

$$\mu(z)-\int_\Gamma \mu(t)\kappa(z, t)|dt|=\frac{1}{\pi} v_0(z), \qquad z\in\Gamma. \tag{15.9-12}$$

After parametrizing Γ, these equations will have the form

$$\phi(\sigma)-\lambda\int_\alpha^\beta \phi(\tau)\kappa(\sigma, \tau)\, d\tau=\gamma(\sigma) \tag{15.9-13}$$

or

$$\psi(\sigma)-\lambda\int_\alpha^\beta \psi(\tau)\kappa(\tau, \sigma)\, d\tau=\gamma(\sigma), \tag{15.9-14}$$

where α and β are finite, κ is a given continuous function of two real variables, γ denotes various given functions, λ is a parameter which here takes the values ± 1, and ϕ and ψ are to be determined.

We now recall the famous theorem of Fredholm regarding such integral equations (see Courant and Hilbert [1951], p. 99):

THEOREM 15.9b. *If λ is fixed, the equation* (15.9-13) *either possesses precisely one continuous solution for every continuous function γ (in particular, the solution $\phi=0$ for $\gamma=0$), or the corresponding homogeneous equation*

$$\phi(\sigma)-\lambda\int_\alpha^\beta \kappa(\sigma, \tau)\phi(\tau)\, d\tau=0 \tag{15.9-15}$$

possesses a positive, but finite, number r of linearly independent solutions $\phi_1, \phi_2, \ldots, \phi_r$ (**Fredholm alternative**). *In the first case, the* **transposed equation** (15.9-14) *likewise always has a unique solution; in the second case, the* *transposed homogeneous equation*

$$\psi(\sigma)-\lambda \int_{\alpha}^{\beta} \kappa(\tau, \sigma)\psi(\tau)\, d\tau = 0 \tag{15.9-16}$$

has the same number r of linearly independent solutions $\psi_1, \psi_2, \ldots, \psi_r$, *and the inhomogeneous equation* (15.9-13) *is solvable only if* γ *satisfies the r conditions*

$$\langle \gamma, \psi_i \rangle := \int_{\alpha}^{\beta} \gamma(\tau)\psi_i(\tau)\, d\tau = 0, \qquad i = 1, \ldots, r. \tag{15.9-17}$$

In this case, the solution of (15.9-13) *is determined only up to an arbitrary linear combination* $\alpha_1\phi_1 + \cdots + \alpha_r\phi_r$; *it can be made unique by requiring that*

$$\langle \phi, \phi_i \rangle = 0, \qquad i = 1, \ldots, r. \tag{15.9-18}$$

A number λ such that the second alternative holds is called an **eigenvalue** of (15.9-13). The number r of linearly independent solutions is the **multiplicity** of λ, and the functions ϕ_i and ψ_j, $i, j = 1, 2, \ldots, r$, are the **eigenfunctions** belonging to λ.

Fredholm's theorem holds for arbitrary continuous kernels. Much more is known if the kernel is symmetric, that is, if $\kappa(\sigma, \tau) = \kappa(\tau, \sigma)$. Then it can be shown, for instance, that all eigenvalues are real, and that there always exist eigenvalues.

The Neumann kernel defined by (15.9-3) is not symmetric. Its properties, however, have been studied extensively; see Blumenfeld and Mayer [1914]. A first important fact is that $\lambda = -1$ is not an eigenvalue of κ. The homogeneous equation corresponding to (15.9-11) thus has only the trivial solution, and by the Fredholm alternative the nonhomogeneous equation has exactly one continuous solution ν for any continuous boundary function u_0. If u_0 is Hölder continuous, then so is the corresponding solution ν. By virtue of the derivation of (15.9-11) the double-layer potential formed with the solution ν satisfies $u^+(z_0) = u_0(z_0)$ at every $z_0 \in \Gamma$ and hence is identical with the solution of the Dirichlet problem. We summarize:

THEOREM 15.9c. *If the Jordan curve* Γ *has a continuous curvature and if the boundary data* u_0 *are Hölder continuous, the solution of the Dirichlet problem for the interior of* Γ *is given by* (15.9-1), *where* ν *is the* (*unique*) *solution of the integral equation* (15.9-14).

As to the equation (15.9-12), the Fredholm theory requires us to consider the homogeneous equation with the transposed kernel,

$$\tilde{\mu}(z) - \int_{\Gamma} \tilde{\mu}(t)\kappa(t, z)|dt| = 0,$$

which by the definition of the Neumann kernel may be written

$$\tilde{\mu}(z) = \frac{1}{\pi} \operatorname{Im} \int_\Gamma \tilde{\mu}(t) \frac{1}{t-z}\, dt.$$

By Example 3, § 14.1, this equation has the solution $\tilde{\mu}(z) = 1$. It can be shown that there is no other linearly independent solution. Thus $\lambda = 1$ is an eigenvalue of κ with multiplicity 1. Fredholm's theory now states that the nonhomogeneous equation has a solution only for functions $v_0(z)$ satisfying

$$\int_\Gamma v_0(z)\tilde{\mu}(z)|dz| = \int_\Gamma v_0(z)|dz| = 0.$$

This is precisely the condition (15.9-5). Since the transposed homogeneous equation has a nontrivial solution, the same is true for the homogeneous equation corresponding to (15.9-12),

$$\mu(z) = \int_\Gamma \mu(t)\kappa(z, t)|dt|. \tag{15.9-19}$$

Denoting the solution by μ_0, the solution of (15.9-12) for functions v_0 satisfying (15.9-5) is determined only up to multiples of μ_0. It follows that the potential u is determined only up to constant multiples of the logarithmic potential

$$u_1(z) := \int_\Gamma \mu_0(t) \operatorname{Log} \frac{1}{|t-z|}|dt|. \tag{15.9-20}$$

We know, of course, that the solution of the Neumann problem is determined only up to an additive constant. It follows that u_1 is constant. In § 16.6 the function μ_0 will be expressed in terms of a certain conformal mapping function, and the fact that $u_1 = \text{const}$ will be proved independently of the theory of the Neumann problem.

In summary, we have:

THEOREM 15.9d. *If the Jordan curve* Γ *has a continuous curvature and if the boundary data* v_0 *are Hölder continuous and satisfy* (15.9-5), *then any solution of the Neumann problem for the interior of* Γ *is given by* (15.9-6), *where* μ *is any continuous solution of the integral equation* (15.9-12).

PROBLEMS

1. Represent the solution of the *exterior* Dirichlet problem as a double-layer potential, and obtain an integral equation for the density ν.
2. Solve the integral equations (15.9-11) and (15.9-12) for the case where R is the unit disk, using the fact (established in Example 1) that $\kappa(t, z) = 1/2\pi$. Substitute

the solutions constructed into (15.9-1) and (15.9-6) and thus obtain Poisson's and Dini's formulas.

NOTES

The integral equations of potential theory date back at least to C. Neumann [1877]. These and similar integral equations for the three-dimensional Dirichlet and Neumann problems supplied the impetus for the development of the theory of linear integral equations, which culminated in Hilbert's [1912] treatise. Under the name "boundary integral methods," integral equations more recently have met with renewed interest. See Jaswon and Symm [1977] or Blue [1978] for accounts stressing applications and computational aspects. Richter [1977] solves Dirichlet's problem by an integral equation of the first kind.

§ 15.10. WIRTINGER CALCULUS; POMPEIU'S FORMULA

I. Wirtinger Derivatives

Calculations in two-dimensional potential theory are often simplified by means of the symbolic differential operators, also called **Wirtinger derivatives,**

$$\frac{\partial}{\partial z}:=\frac{1}{2}\left(\frac{\partial}{\partial x}-i\frac{\partial}{\partial y}\right),\qquad \frac{\partial}{\partial \bar{z}}:=\frac{1}{2}\left(\frac{\partial}{\partial x}+i\frac{\partial}{\partial y}\right),$$

which were already briefly studied in § 5.5. These operators act on real or complex functions of the real variables x and y that possess continuous partial derivatives of the required order with respect to these variables. Thus for instance,

$$\frac{\partial f}{\partial z}=\frac{1}{2}\left(\frac{\partial f}{\partial x}-i\frac{\partial f}{\partial y}\right),\qquad \frac{\partial f}{\partial \bar{z}}=\frac{1}{2}\left(\frac{\partial f}{\partial x}+i\frac{\partial f}{\partial y}\right), \tag{15.10-1}$$

and if the second partial derivatives exist and are continuous so that $\partial^2 f/\partial x\,\partial y=\partial^2 f/\partial y\,\partial x$, then

$$\frac{\partial^2 f}{\partial z\,\partial \bar{z}}=\frac{1}{4}\left(\frac{\partial^2 f}{\partial x^2}+\frac{\partial^2 f}{\partial y^2}\right).$$

Let f be defined in a region $R\subset\mathbb{C}$. It was pointed out already in § 5.5 that the following two propositions are equivalent:

(i) f is *real-differentiable* (that is, differentiable as a function of the two real variables x and y) at the point $z_0\in R$, and

$$\frac{\partial f}{\partial z}(z_0)=a,\qquad \frac{\partial f}{\partial \bar{z}}(z_0)=b.$$

(ii) For all complex numbers h such that $|h|$ is sufficiently small,

$$f(z)-f(z_0)-ah-b\bar{h}=h\varepsilon(h), \tag{15.10-2a}$$

where $\varepsilon(h)\to 0$ as $h\to 0$.

If f is real, the condition (15.10-2a) may be written

$$f(z_0+h)-f(z_0)-2\,\mathrm{Re}\,ah=h\varepsilon(h). \tag{15.10-2b}$$

On the other hand, f is *complex-differentiable* at z_0, and hence satisfies the Cauchy–Riemann equations at that point if and only if $b=0$, that is, if

$$\frac{\partial f}{\partial \bar{z}}(z_0)=0.$$

In this case and in this case only, the complex derivative

$$f'(z_0)=\lim_{h\to 0}\frac{f(z_0+h)-f(z_0)}{h}$$

exists, and

$$f'(z_0)=\frac{\partial f}{\partial z}(z_0). \tag{15.10-3}$$

Although the symbolic derivatives $\partial f/\partial z$ and $\partial f/\partial \bar{z}$ have no direct interpretation as limits of difference quotients, the reader may easily convince himself that the ordinary rules of calculus, such as the rules for differentiating sums, products, and quotients, remain valid for the operators $\partial/\partial z$ and $\partial/\partial \bar{z}$. Also, since every function of x and y may formally be expressed as a function of z and $\bar{z}$ by virtue of the relations

$$x=\tfrac{1}{2}(z+\bar{z}),\qquad y=\frac{1}{2i}(z-\bar{z}),$$

one may symbolically differentiate with respect to z and $\bar{z}$ as if these quantities were in fact independent variables.

EXAMPLE **1**

For arbitrary integers m and n,

$$\frac{\partial}{\partial z}z^m\bar{z}^n=mz^{m-1}\bar{z}^n,$$

$$\frac{\partial}{\partial \bar{z}}z^m\bar{z}^n=nz^m\bar{z}^{n-1}.$$

EXAMPLE **2**

Let $a \in \mathbb{C}$ be fixed,

$$u(z) := \operatorname{Log} \frac{1}{|z-a|}.$$

By virtue of

$$\operatorname{Log} \frac{1}{|z-a|} = -\tfrac{1}{2} \operatorname{Log}[(z-a)(\bar{z}-\bar{a})] = -\tfrac{1}{2}\{\operatorname{Log}(z-a) + \operatorname{Log}(\bar{z}-\bar{a})\}$$

we have

$$\frac{\partial u}{\partial z}(z) = -\frac{1}{2}\frac{1}{z-a}, \qquad \frac{\partial u}{\partial \bar{z}}(z) = -\frac{1}{2}\frac{1}{\bar{z}-\bar{a}}. \quad —$$

Since the differentiation rules mentioned will be used in simple situations only, there is no need to establish them systematically.

II. Pompeiu's Generalization of Cauchy's Formula

Let R be a region, not necessarily simply connected, and let the boundary Γ of R consist of finitely many regular closed curves, oriented so that the winding number (see § 4.6) of a point $z \notin \Gamma$ is $+1$ if $z \in R$ and 0 if $z \notin R$. If u and v are two real, differentiable functions in a region containing the closure of R, then Stokes' theorem asserts that

$$\int_\Gamma u\,dx + v\,dy = \iint_R \left(\frac{\partial v}{\partial x} - \frac{\partial u}{\partial y}\right) \boxed{dz}$$

($\boxed{dz}$:= area element := $dx\,dy$). Thus if $f = u + iv$, then

$$\begin{aligned}\int_\Gamma f(z)\,dz &= \int_\Gamma (u\,dx - v\,dy) + i\int_\Gamma (v\,dx + u\,dy)\\ &= \iint_R \left\{-\left(\frac{\partial v}{\partial x} + \frac{\partial u}{\partial y}\right) + i\left(\frac{\partial u}{\partial x} - \frac{\partial v}{\partial y}\right)\right\} \boxed{dz}.\end{aligned}$$

By (15.10-1), the expression in braces is $2i\,\partial f/\partial\bar{z}$, and we thus have

$$\int_\Gamma f(z)\,dz = 2i \iint_R \frac{\partial f}{\partial \bar{z}}(z) \boxed{dz}. \tag{15.10-4}$$

This evidently is the form taken by Cauchy's theorem if f is not analytic but merely real-differentiable. In a similar manner we may establish the formula

$$\int_{\Gamma} f(z)\,\overline{dz} = -2i \iint_{R} \frac{\partial f}{\partial z}(z)\,\boxed{dz}. \qquad (15.10\text{-}5)$$

We next proceed as in the derivation of Cauchy's integral formula from Cauchy's theorem. Let R, f, and Γ as above, and let $a \in R$. We choose ρ so small that the disk D_ρ: $|z-a| \leqq \rho$ is contained in R, and we let Γ_ρ: $z = a + \rho\, e^{i\tau}$, $0 \leqq \tau \leqq 2\pi$. Applying formula (15.10-4) to the function

$$g(z) := \frac{f(z)}{z-a},$$

which in $R \backslash D_\rho$ evidently satisfies its hypotheses, and considering

$$\frac{\partial}{\partial \bar{z}} g(z) = \frac{1}{z-a} \frac{\partial f}{\partial \bar{z}}(z),$$

we obtain

$$\int_{\Gamma} \frac{f(z)}{z-a}\,dz - \int_{\Gamma_\rho} \frac{f(z)}{z-a}\,dz = 2i \iint_{R \backslash D_\rho} \frac{(\partial f/\partial \bar{z})(z)}{z-a}\,\boxed{dz}.$$

Here we let $\rho \to 0$. The first term on the left is independent of ρ, while the limit of the second term evidently exists and equals $-2\pi i f(a)$. Therefore the limit of the integral on the right exists. Writing t in place of z and z for a in order to stress that this point is arbitrary in R, we obtain:

THEOREM 15.10a (Pompeiu formula). *Let R be a region bounded by a system Γ of regular closed curves such that points in R have winding number 1 with respect to Γ. If f is a complex-valued function that is real-differentiable in a region containing $R \cup \Gamma$, then for any point $z \in R$ there holds*

$$f(z) = \frac{1}{2\pi i} \int_{\Gamma} \frac{f(z)}{t-z}\,dt - \frac{1}{\pi} \iint_{R} \frac{(\partial f/\partial \bar{z})(t)}{t-z}\,\boxed{dt}. \qquad (15.10\text{-}6)$$

If f is analytic, then $\partial f/\partial \bar{z} = 0$, and (15.10-6) reduces to Cauchy's formula.

EXAMPLE 3

Let m, n be integers $\geqq 0$, and let D_ρ: $|z| < \rho$. We wish to compute the integrals

$$I_{m,n} := \frac{1}{\pi} \iint_{D_\rho} \frac{t^m \bar{t}^n}{t-z}\,\boxed{dt} \qquad (15.10\text{-}7)$$

for $|z| < \rho$. Letting

$$f(z) := \frac{1}{n+1} z^m \bar{z}^{n+1}$$

in Pompeiu's formula so that $\partial f/\partial \bar{z} = z^m \bar{z}^n$, we get for $R = D_\rho$,

$$I_{m,n} = \frac{1}{2\pi i} \int_{\Gamma_\rho} \frac{f(t)}{t-z} \, dt - f(z).$$

The integral is evaluated by noting that on Γ_ρ, $t\bar{t} = \rho^2$. Hence

$$\frac{1}{2\pi i} \int_{\Gamma_\rho} \frac{t^m \bar{t}^{n+1}}{t-z} \, dt = \frac{\rho^{2n+2}}{2\pi i} \int_{\Gamma_\rho} \frac{t^{m-n-1}}{t-z} \, dt.$$

By residues, the last integral equals 0 if $m \leqq n$, and z^{m-n-1} if $m > n$. There follows

$$I_{m,n} = \begin{cases} -\dfrac{1}{n+1} z^m \bar{z}^{n+1}, & m \leqq n \\ \dfrac{1}{n+1} [\rho^{2n+2} z^{m-n-1} - z^m \bar{z}^{n+1}], & m > n. \end{cases}$$

III. A converse of Pompeiu's Theorem

Let the complex-valued function f be bounded and integrable in a region R. The integral

$$g(z) := -\frac{1}{\pi} \iint_R \frac{f(t)}{t-z} \boxed{dt} \tag{15.10-8}$$

then obviously exists for z in the complement of the closure R' of R and represents an analytic function there. The integral also exists for $z \in R'$, as is seen by passing to polar coordinates around the point $z \in \mathbb{C}$, because the functions

$$g_n(z) := -\frac{1}{\pi} \iint_{R_n(z)} \frac{f(t)}{t-z} \boxed{dt}$$

($R_n(z) := R \backslash \{t\colon |t-z| < \rho\}$) are obviously continuous, and $g_n(z) \to g(z)$ uniformly in z.

In the complement of R', $\partial g/\partial \bar{z} = 0$ because g is analytic there. We wish to show that, under certain conditions, the symbolic derivative $\partial g/\partial \bar{z}$ exists at points inside R. It is easy to see what the value of $\partial g/\partial \bar{z}$ at a point $z \in R$ must be if it exists. Without loss of generality we may assume $z = 0$. If $\partial g/\partial \bar{z}$ exists and is continuous in a neighborhood of $z = 0$, then it follows from (15.10-4) that

$$\frac{\partial g}{\partial \bar{z}}(0) = \lim_{\rho \to 0} \frac{1}{2\pi i \rho^2} \int_{\Gamma_\rho} g(z) \, dz. \tag{15.10-9}$$

Assuming ρ so small that $D_\rho \subset R$, we write

$$g(z) = g_1(z) + g_2(z),$$

where

$$g_1(z) = -\frac{1}{\pi}\iint_{D_\rho} \frac{f(t)}{t-z}\,\boxed{dt},$$

and g_2 is the corresponding integral over $R\backslash D_\rho$. Since g_2 is analytic in D,

$$\int_{\Gamma_\rho} g_2(z)\,dz = 0.$$

Since g_1 is analytic outside D_ρ,

$$\int_{\Gamma_\rho} g_1(z)\,dz = \int_{\Gamma_{2\rho}} g_1(z)\,dz = -\frac{1}{\pi}\int_{\Gamma_{2\rho}} \left(\iint_{D_\rho} \frac{f(t)}{t-z}\,\boxed{dt}\right) dz.$$

The order of the integrations may be interchanged due to the continuity of the integrand. Thus the last integral equals

$$\frac{1}{\pi}\iint_{D_\rho} f(t)\left(\int_{\Gamma_{2\rho}} \frac{1}{z-t}\,dz\right)\boxed{dt} = 2i\iint_{D_\rho} f(t)\,\boxed{dt}.$$

If f is continuous at 0, there follows

$$\lim_{\rho\to 0}\frac{1}{2\pi i\rho^2}\int_{\Gamma_\rho} g(z)\,dz = \lim_{\rho\to 0}\frac{1}{2\pi i\rho^2}\int_{\Gamma_\rho} g_1(z)\,dz = f(0).$$

Since the choice $z = 0$ was arbitrary, we obtain

$$\frac{\partial g}{\partial \bar{z}}(z) = f(z). \tag{15.10-10}$$

This result was obtained under the assumption that $\partial g/\partial \bar{z}$ exists and is continuous. Without assuming the existence of this derivative, the following may be proved:

THEOREM 15.10b. *Let f be a complex-valued, bounded, and integrable function in the region R, and let $g(z)$ be defined by* (15.10-8). *Then g is continuous on $\mathbb{C}$. Moreover, at any point z where f is real-differentiable, the symbolic derivative $\partial g/\partial \bar{z}$ exists and equals $f(z)$.*

Proof. Continuity on $\mathbb{C}$ has already been established. To prove (15.10-10), let $z = 0$ without loss of generality. Let $g = g_1 + g_2$ as above. Since g_2 is analytic in D_ρ, we have

$$\frac{\partial g_2}{\partial \bar{z}}(0)=0,$$

and it remains to show that

$$\frac{\partial g_1}{\partial \bar{z}}(0)=f(0)$$

or, using the definition of differentiability, that

$$g(z)-g(0)-\bar{z}f(0)=z(c+\varepsilon(z)), \qquad (15.10\text{-}11)$$

where $\varepsilon(z)\to 0$ as $z\to 0$, and where c is a constant. Using the integral

$$I_{0,0}=\bar{z}$$

established in Example 3, we have

$$\begin{aligned} g(z)-g(0)-\bar{z}f(0) &= -\frac{1}{\pi}\iint_{D_\rho}\left\{\frac{f(t)}{t-z}-\frac{f(t)}{t}-\frac{f(0)}{t-z}+\frac{f(0)}{t}\right\}\boxed{dt} \\ &= -\frac{1}{\pi}\iint_{D_\rho}\frac{h(t)}{t-z}\boxed{dt}. \end{aligned}$$

where

$$h(t):=\frac{f(t)-f(0)}{t}, \qquad t\neq 0.$$

If f is real-differentiable at 0, there exist constants a and b such that

$$f(t)-f(0)-at-b\bar{t}=t\eta(t),$$

where $\eta(t)\to 0$ as $t\to 0$. There follows

$$h(t)=a+b\frac{\bar{t}}{t}+\eta(t).$$

Although h is not necessarily continuous at 0, it is bounded in a neighborhood of 0. Thus by the proof given above, the function

$$z\to -\frac{1}{\pi}\iint_{D_\rho}\frac{h(t)}{t-z}\boxed{dt}$$

is continuous at 0, and hence may be represented as $c+\varepsilon(z)$ where $\varepsilon(z)\to 0$ as $z\to 0$. This establishes (15.10-11), and hence Theorem 15.10b. —

PROBLEMS

1. Compute the integrals $I_{m,n}$ considered in Example 3 for $|z|>\rho$, and verify that they represent continuous functions on $\mathbb{C}$.

2. Show that for the function

$$w(z) := \begin{cases} \dfrac{z}{|z| \operatorname{Log}|z|^{-1}}, & z \neq 0 \\ 0, & z = 0 \end{cases}$$

the limit (15.10-9) exists, although w is not real-differentiable at 0.

3. If g is defined by (15.10-8), show that

$$\frac{\partial g}{\partial z}(z) = -\frac{1}{\pi} \int\int_R \frac{f(t) - f(z)}{(t-z)^2} \boxed{dt}.$$

4. *Liouville's equation.* Use the Wirtinger calculus to show that a general solution of Liouville's equation,

$$\Delta u + \lambda\, e^u = 0, \qquad \lambda \in \mathbb{R},$$

in a region $R \subset \mathbb{C}$ is given by

$$u(z) = \operatorname{Log} \frac{|f'(z)|^2}{[1 + (\lambda/8)|f(z)|^2]^2},$$

where f is any meromorphic function in R such that $f'(z) \neq 0$, $z \in R$ with at most simple poles.

NOTES

Wirtinger derivatives were used by Wirtinger [1927] and Peschl [1932]. They may be interpreted in terms of differential forms; see, for example, Jänich [1977]. The relation (15.10-9) was used by Pompeiu [1912] to *define* $\partial w/\partial \bar{z}$. (Problem 2 shows that this limit may exist for functions that are not real-differentiable at z.) Using this definition, (15.10-6) is stated by Pompeiu [1913]. For a proof of Pompeiu's theorem and its converse under much weaker hypotheses see Vekua [1962], ch. 1.

§ 15.11. POISSON'S EQUATION: THE LOGARITHMIC POTENTIAL; SYMBOLIC INTEGRATION

Poisson's equation is the name usually given to the nonhomogeneous form of Laplace's equation, which we write in the form

$$-\Delta u = f, \tag{15.11-1}$$

where f is a given function.

Consider the problem to find a solution u of Poisson's equation in a region R which on the boundary ∂R satisfies prescribed Dirichlet, Neumann, or mixed boundary conditions. If u_0 is any solution of (15.11-1), we set $u = u_0 + w$ and find that the required correction w is a solution of Laplace's equation which satisfies boundary conditions of the same type as u, and thus can be found by methods discussed earlier. The new task on hand thus is the construction of particular solutions of Poisson's equation.

Here we describe two methods for constructing such particular solutions: (a) forming the logarithmic potential, and (b) symbolic integration. Some additional methods are discussed in the subsequent two sections.

I. The Logarithmic Potential

We proceed heuristically, trying to guess what a solution might look like. Then we shall verify that the function that we have guessed is, in fact, a solution of the equation.

It is necessary first to broaden the meaning of the term "solution of Poisson's equation." One form of the divergence theorem (obtained by setting $v=1$ in (15.6-3)) states that if u has continuous partial derivatives of order 2 up to the boundary (assumed to be smooth) of a bounded region R, then

$$\int_{\partial R} \frac{\partial u}{\partial n} |dz| = \iint_R \Delta u \, \boxed{dz},$$

where the differentiation is in the direction of the outward normal. Thus if u satisfies Poisson's equation (15.11-1), then

$$-\int_{\partial R} \frac{\partial u}{\partial n} |dz| = \iint_R f(z) \, \boxed{dz}. \tag{15.11-2}$$

Conversely, if a function u is twice continuously differentiable and satisfies (15.11-2) for every smoothly bounded region R, then from the fact that

$$\iint_R \Delta u \, \boxed{dz} = \iint_R f(z) \boxed{dz}$$

holds for arbitrarily small disks, it follows that

$$-\Delta u = f.$$

However, the condition (15.11-2) has a meaning also for functions u that are differentiable only once. For this reason, any function u which has a continuous gradient and which satisfies (15.11-2) for arbitrary smoothly bounded subregions R of its domain of definition will be called a **solution** of Poisson's equation **in the sense of the divergence theorem**. If such a solution has a continuous Laplacian, it automatically is a solution of Poisson's equation in the ordinary sense.

To construct a solution in the sense of the divergence theorem, we evaluate the integral

$$\int_{\partial R} \frac{\partial u}{\partial n} |dz|$$

for the case where u is the elementary potential, or fundamental singularity,

$$u(z)=u(z, z_0):=\frac{1}{2\pi}\operatorname{Log}\frac{1}{|z-z_0|} \tag{15.11-3}$$

already used in the definition of the potential of a single layer. We know that the complex gradient of u is

$$\operatorname{grd} u(z)=-\frac{1}{2\pi}\frac{1}{\bar{z}-\bar{z}_0},$$

thus the derivative of u in the direction $e^{i\phi}$ at the point z is

$$\operatorname{Re}\{e^{i\phi}\,\overline{\operatorname{grd} u(z)}\}=-\frac{1}{2\pi}\operatorname{Re}\left\{\frac{e^{i\phi}}{z-z_0}\right\}.$$

If R is a Jordan region whose boundary curve Γ (positively oriented) at the point z has the tangent vector $e^{i\phi(z)}$, then the direction of the outward normal is $-ie^{i\phi(z)}$, and thus if Γ does not pass through z_0,

$$-\int_{\partial R}\frac{\partial u}{\partial n}|dz|=-\frac{1}{2\pi}\int_\Gamma \operatorname{Re}\frac{ie^{i\phi(z)}}{z-z_0}|dz|=\operatorname{Re}\frac{1}{2\pi i}\int_\Gamma\frac{1}{z-z_0}\,dz.$$

For the fundamental singularity (15.11-3), the integral on the left of (15.11-2) thus has the value 1 or 0, depending on whether the pole z_0 is located in the interior or in the exterior of Γ. Similarly, if we consider several fundamental singularities with poles at the points $z_1, z_2, \ldots, z_n$, then for the function

$$u(z):=\sum_{k=1}^{n} u(z, z_k)=\frac{1}{2\pi}\sum_{k=1}^{n}\operatorname{Log}\frac{1}{|z-z_k|} \tag{15.11-4}$$

we will have

$$-\int_\Gamma\frac{\partial u}{\partial n}|dz|=\sum_{z_k\in\operatorname{int}\Gamma}1 \tag{15.11-5}$$

(the summation extended with respect to the z_k in the interior of the positively oriented curve Γ), provided that Γ does not pass through any of the points z_k.

We now let the number of poles increase in such a manner that the density of poles per unit area at the point z approaches $f(z)$, where the function f is continuous in R. In the limit (15.11-4) then becomes

$$u(z):=\frac{1}{2\pi}\iint_R f(t)\operatorname{Log}\frac{1}{|z-t|}\boxed{dt}. \tag{15.11-6}$$

If Γ is an arbitrary Jordan curve in R, the expression on the right of (15.11-5) approaches $\iint_{\text{int}\,\Gamma} f(z)\,\boxed{dz}$. If on the left the passage to the limit and the integration could be interchanged—a delicate operation since derivatives of u are involved—we would obtain

$$-\int_\Gamma \frac{\partial u}{\partial n}(z)\,|dz| = \iint_{\text{int}\,\Gamma} f(z)\,\boxed{dz},$$

and thus could conclude that the function u defined by (15.11-6) is a solution of Poisson's equation at least in the sense of the divergence theorem. Rather than trying to justify the interchange, we establish this fact directly:

THEOREM 15.11a. *Let the real function f be bounded and integrable in the bounded region R. Then*

$$u(z) := \frac{1}{2\pi}\iint_R f(t)\,\mathrm{Log}\frac{1}{|t-z|}\,\boxed{dt} \tag{15.11-7}$$

exists for all $z \in \mathbb{C}$. The function u as well as its derivatives of the first order are continuous, and u is a solution of Poisson's equation in the sense of the divergence theorem.

The function u defined by (15.11-7) is called the **logarithmic potential** of f.

Proof. We first establish the existence of the improper integral (15.11-7). Let ρ_1 denote the diameter of R, and let $|f(t)| \leqq \mu$, $t \in R$. Then by introducing polar coordinates around the point z we see that the integral is bounded by the convergent integral

$$\mu \int_0^{\rho_1} \rho \left|\mathrm{Log}\frac{1}{\rho}\right| d\rho$$

and hence is itself convergent.

We next show the continuity of u. For $z \in R$ and for $n = 1, 2, \ldots$, let $R_n(z) := R \backslash \{t: |t-z| \leqq 1/n\}$, and define

$$u_n(z) := \frac{1}{2\pi}\iint_{R_n(z)} f(t)\,\mathrm{Log}\frac{1}{|t-z|}\,\boxed{dt}.$$

The functions u_n are easily seen to be continuous, because both the integrand and the domain of integration depend continuously on z. By the definition of the improper integral, $u_n(z) \to u(z)$ for each fixed z. But in view of

$$|u_n(z) - u(z)| \leqq \mu \int_0^{1/n} \rho\,\mathrm{Log}\frac{1}{\rho}\,d\rho,$$

where the expression on the right is independent of z and tends to 0 as

$n \to \infty$, the convergence is uniform in z. It follows that u, being the uniform limit of a sequence of continuous functions, is itself continuous.

We use the Wirtinger calculus to establish the remaining assertions of Theorem 15.11a. Since we may write

$$u(z) = \frac{1}{4\pi} \iint_R f(t) \operatorname{Log}[(t-z)(\bar{t}-\bar{z})]^{-1} \boxed{dt}$$

and since, by Example 2 of § 15.10,

$$\frac{\partial}{\partial z} \operatorname{Log}[(t-z)(\bar{t}-\bar{z}]^{-1} = \frac{1}{t-z},$$

we may expect that

$$\frac{\partial u}{\partial z}(z) = \frac{1}{4} w(z), \tag{15.11-8}$$

where

$$w(z) := \frac{1}{\pi} \iint_R \frac{f(t)}{t-z} \boxed{dt}.$$

In any case, the integral defining w exists for all z and is continuous by Theorem 15.10b. We shall establish (15.11-8) by showing that at every point z_0 the number

$$a := = \tfrac{1}{4} w(z_0)$$

satisfies (15.10-2b). Without loss of generality we may assume $z_0 = 0$ because this can be achieved by a simple change of variables. We then have

$$\begin{aligned} \delta(z) &:= u(z) - u(0) - 2 \operatorname{Re} az \\ &= -\frac{1}{2\pi} \iint_R f(t) \operatorname{Re}\left\{ \operatorname{Log} \frac{t-z}{t} + \frac{z}{t} \right\} \boxed{dt} \end{aligned}$$

and thus

$$|\delta(z)| \leq \frac{\mu}{2\pi} \iint_R \left| \operatorname{Re}\left\{ \operatorname{Log} \frac{t-z}{t} + \frac{z}{t} \right\} \right| \boxed{dt}.$$

The integral may be estimated by passing to polar coordinates and splitting the range of integration into the two sets $0 \leq |t| \leq |z|$ and $|z| \leq |t| \leq \rho_1$ where ρ_1 is the diameter of R. There results

$$|\delta(z)| \leq \mu |z|^2 \left\{ 1 + \log \rho_1 + \operatorname{Log} \frac{1}{|z|} \right\}.$$

Thus clearly $z^{-1}\delta(z) \to 0$ as $z \to 0$, establishing (15.11-8).

We finally show that u solves Poisson's equation in the sense of the divergence theorem. From the fact that

$$\frac{\partial u}{\partial z}(z)=\frac{1}{4\pi}\iint_R \frac{f(t)}{t-z}\boxed{dt}$$

there follows by the definition of the complex gradient

$$\begin{aligned}\operatorname{grd} u(z)&=2\overline{\frac{\partial u}{\partial z}(z)}\\&=\frac{1}{2\pi}\iint_R f(t)\frac{1}{\bar t-\bar z}\boxed{dt}.\end{aligned}$$

Thus if Γ is a simple closed curve in R such that $\operatorname{int}\Gamma\subset R$, and if the tangent vector at $z\in\Gamma$ exists and has the direction $e^{i\phi(z)}$, then the derivative in the direction of the exterior normal is

$$\operatorname{Re}\{-i\,e^{i\phi(z)}\,\overline{\operatorname{grd} u(z)}\}=\operatorname{Re}\left\{\frac{1}{2\pi i}\,e^{i\phi(z)}\iint_R f(t)\frac{1}{t-z}\boxed{dt}\right\}.$$

Hence

$$-\int_\Gamma \frac{\partial u}{\partial n}|dz|=\int_\Gamma \operatorname{Re}\left\{\frac{1}{2\pi i}\,e^{i\phi(z)}\iint_R f(t)\frac{1}{z-t}\boxed{dt}\right\}|dz|$$

or on interchanging the order of the integrations,

$$-\int_\Gamma \frac{\partial u}{\partial n}|dz|=\iint_R f(t)\left[\operatorname{Re}\frac{1}{2\pi i}\int_\Gamma \frac{1}{z-t}\,dz\right]\boxed{dt}.$$

The expression in brackets has the value 1 or 0, depending on whether t is in the interior or in the exterior of Γ, and there follows

$$-\int_\Gamma \frac{\partial u}{\partial n}|dz|=\iint_{\operatorname{int}\Gamma} f(t)\boxed{dt},$$

establishing (15.11-2). —

It is of interest to know conditions under which u is an actual solution of $-\Delta u=f$ in the sense that the partial derivatives required for forming Δ exist.

THEOREM 15.11b. *In addition to the hypotheses of the preceding theorem, let the function f be real-differentiable at the point $z_0\in R$. Then the Laplacian of the logarithmic potential* (15.11-7) *exists at z_0, and there holds*

$$-\Delta u(z_0)=f(z_0). \tag{15.11-9}$$

Proof. It has been established in the proof of the preceding theorem that

$$4\frac{\partial u}{\partial z}(z)=w(z)=\frac{1}{\pi}\iint_R \frac{f(t)}{t-z}\boxed{dt},$$

and in view of $\Delta = 4\partial^2/\partial z\,\partial\bar{z}$ it remains only to show that at $z=z_0$,

$$\frac{\partial w}{\partial \bar{z}}(z_0)=-f(z_0).$$

This, however, was established in Theorem 15.10b. —

II. Symbolic Integration

In terms of Wirtinger derivatives, Poisson's equation takes the form

$$\frac{\partial^2 u}{\partial z\,\partial \bar{z}}=-\frac{1}{4}f(z). \tag{15.11-10}$$

If z and $\bar{z}$ were independent real variables, a solution would obviously be

$$u(z):=-\frac{1}{4}\int_0^z\left(\int_0^{\bar{z}} f(z)\,d\bar{z}\right)dz.$$

In the present context, this formula has no meaning because z is not real, and z and $\bar{z}$ are not independent, but complex conjugate. Thus the sense in which the above integrals are to be understood is not clear; in fact, these integrals have no constructive definition and thus cannot, not even in principle, be calculated by numerical integration.

However, it is still true that if g is a **symmetric antiderivative** of f in the sense that g is twice real-differentiable and satisfies

$$\frac{\partial^2 g}{\partial z\,\partial \bar{z}}(z)=f(z) \tag{15.11-11}$$

for all z in the domain of definition of f, then

$$u(z):=-\tfrac{1}{4}g(z)$$

is a solution of (15.11-10). Since the Wirtinger derivatives obey the formal rules of calculus, it may often be possible to find a symmetric antiderivative simply by formal, or symbolic, integration. The method does not, of course, require f to be real.

EXAMPLE 1

For $f(z):=z^m\bar{z}^n$, a symmetric antiderivative is

$$g(z):=\frac{1}{(m+1)(n+1)}z^{m+1}\bar{z}^{n+1},$$

thus

$$u(z) := -\frac{1}{4(m+1)(n+1)} z^{m+1}\bar{z}^{n+1}$$

is a corresponding solution of Poisson's equation. For $m=n=0$ we find, in particular, that $u(z) := -\frac{1}{4}z\bar{z} = -\frac{1}{4}(x^2+y^2)$ is a solution of $-\Delta u = 1$. —

Every polynomial in x and y may be written as a polynomial in z and $\bar{z}$. The example thus shows that Poisson's equation can be solved by symbolic integration whenever the inhomogeneous term is a polynomial.

EXAMPLE 2

For $f(z) := e^{az+b\bar{z}}$, $a, b \in \mathbb{C}$, symbolic integration yields the solution

$$u(z) := -\frac{1}{4ab} e^{az+b\bar{z}}. \quad —$$

There remains the theoretical question of how the result obtained by symbolic integration is related to the logarithmic potential of f. Thus let f be defined in a region R with boundary Γ, let g be a symmetric antiderivative of f, and let

$$u_0(z) := \frac{1}{2\pi}\iint_R f(t) \operatorname{Log}\frac{1}{|z-t|}\boxed{dt}$$

be the logarithmic potential of f. Since u_0 and $-\frac{1}{4}g$ both are solutions of (15.11-1), $u_0 + \frac{1}{4}g$ must be harmonic, and it is our purpose to identify this harmonic function. By (15.11-11) we have

$$u_0(z) = \frac{1}{2\pi}\iint_R \frac{\partial^2 g}{\partial z\,\partial\bar{z}}(t) \operatorname{Log}\frac{1}{|z-t|}\boxed{dt}.$$

The integrand equals

$$\frac{\partial}{\partial z}\left\{\frac{\partial g}{\partial \bar{z}} \operatorname{Log}\frac{1}{|z-t|}\right\} - \frac{1}{2}\frac{\partial g}{\partial\bar{z}}\frac{1}{t-z}.$$

Applying (15.10-4) to the first term and Pompeiu's formula to the second, we thus get

$$u_0(z) = \frac{i}{4\pi}\int_\Gamma \frac{\partial g}{\partial\bar{z}}(t) \operatorname{Log}\frac{1}{|z-t|}|dt| - \frac{1}{4}g(z) + \frac{1}{8\pi i}\int_\Gamma \frac{g(t)}{t-z}\,dt.$$

We thus may identify $u_0(z) + \frac{1}{4}g(z)$ as the sum of the potential of a single layer,

$$\frac{i}{4\pi}\int_\Gamma \frac{\partial g}{\partial\bar{z}}(t) \operatorname{Log}\frac{1}{|z-t|}|dt|,$$

and of the Cauchy integral

$$\frac{1}{8\pi i}\int_\Gamma \frac{g(t)}{t-z}\,dt,$$

both of which are harmonic in R.

PROBLEMS

1. Let $\gamma(z, t)$ be Green's function for the region R. Show that, under suitable conditions on f,

$$u(z) := \iint_R f(t)\gamma(t, z)\,\boxed{dt}$$

is a solution of Poisson's equation (possibly in the sense of the divergence theorem), which is zero on the boundary ∂R of R.

2. Assuming that the series

$$f(z) = \sum_{m,n=0}^{\infty} a_{m,n} z^m \bar{z}^n$$

converges absolutely for $|z| \leqq 1, |\bar{z}| \leqq 1$, construct the solution of

$$-\Delta u = f(z)$$

that vanishes for $|z| = 1$, and show how to calculate it rapidly by means of the FFT.

3. A function is called *biharmonic* if it satisfies the equation

$$\Delta\Delta u = 0.$$

(a) Let u be biharmonic in a bounded region R, and let $f := -\Delta u$. Show that for $z \in R$, u may be represented in the form

$$u(z) = \frac{1}{2\pi}\iint_R f(t)\,\text{Log}\,\frac{1}{|t-z|}\,\boxed{dt} + v(z), \tag{$*$}$$

where v is harmonic.

(b) Letting $\rho := |t - z|$ and using the identity

$$f(t)\,\text{Log}\,\rho^2 = \frac{\partial}{\partial t}\{f(t)[(t-z)\,\text{Log}\,\rho^2 - (t-z)]\} - \frac{\partial}{\partial \bar{t}}\left\{\frac{\partial f}{\partial t}\rho^2\,\text{Log}\,\rho^2\right\},$$

where f is harmonic, show that the integral in $(*)$ is given by

$$\frac{1}{8\pi i}\int_\Gamma f(t)\{(t-z)\,\text{Log}\,\rho^2 - (t-z)\}\,\overline{dt} + \frac{1}{8\pi i}\int_\Gamma \frac{\partial f}{\partial t}\rho^2\,\text{Log}\,\rho^2\,dt,$$

where Γ is the boundary curve of R.

4. Let u be biharmonic in a simply connected region R. Show that there exist *analytic* functions f and g in R such that

$$u(z) = \text{Re}\{f(z) + \bar{z}g(z)\}.$$

(Use Problem 3b.)

NOTES

Our ad hoc definition of a solution in the sense of the divergence theorem is related to the concept of a "weak" solution of a partial differential equation, and also to solutions in the sense of distribution theory; see John [1978]. The proof of Theorem 15.11b by real variable methods (see Petrovsky [1954]) is more involved than our complex variable proof. The theorem itself is due to Gauss. The solution of Poisson's equation by symbolic integration, although used by engineers, is not often dealt with in the mathematical literature. A satisfactory treatment is possible by I. N. Vekua's theory, to be dealt with in a later volume, of analytically extending f and u to *complex* x and y such that $x+iy \in R$, $x-iy \in \bar{R}$.

§ 15.12. OTHER METHODS FOR SOLVING POISSON'S EQUATION

Here we discuss, in an informal manner, some additional methods for solving Poisson's equation. These methods include: (a) finite differences, including the Mehrstellenverfahren; (b) variational methods, which in their modern form lead to the finite-element technique; (c) eigenfunction expansions. Tools of numerical analysis are required for the implementation of all of these methods.

I. Finite Differences

Because of its conceptual simplicity, the technique of finite differences frequently serves as a "method of last resort" in partial differential equations. Let Poisson's equation

$$-\Delta u = f, \qquad (15.12\text{-}1)$$

be solved for the bounded region R, with values (not necessarily zero) prescribed for u on the boundary ∂R. We choose a **discretization step** $h > 0$ and define, for integers i and j, $x_i := ih$, $y_j := jh$. A point $z_{ij} = (x_i, y_j)$ such that $z_{ij} \in R$ is called **active**; the set of all active points (which is finite by the boundedness of R) is denoted by R_h.

The method of finite differences aims at determining approximate values u_{ij} for the values $u(z_{ij})$ of the exact solution at the active points, and at these points only. To this end, a set of linear equations is set up which model the differential equation (15.12-1) at the active points. The modeling is easy at the points $z_{ij} \in R$ that are **regular**, i.e., which have the property that the four straight line segments of length h that emanate from z_{ij} in the directions parallel to the coordinate axes belong to R (see Fig. 15.12a). At the regular active points, the Laplacian Δu is usually modeled by the finite-difference expression

$$\frac{1}{h^2}(u_{i+1,j} + u_{i,j+1} + u_{i-1,j} + u_{i,j-1} - 4u_{ij})$$

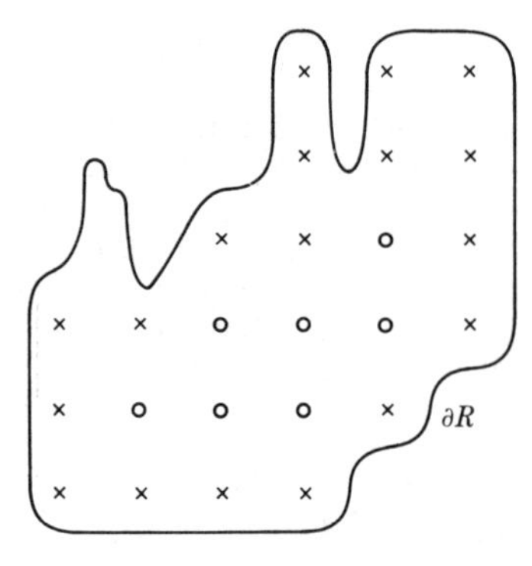

Fig. 15.12a. Regular (○) and irregular (×) active points.

well known from numerical analysis. In this way, an equation of the form

$$4u_{ij}-u_{i+1,j}-u_{i,j+1}-u_{i-1,j}-u_{i,j-1}=h^2 f_{ij}, \qquad (15.12\text{-}2)$$

where $f_{ij}:=f(z_{ij})$, is obtained for each regular point z_{ij}. We use the following "stencil" notation to write equation (15.12-2) in abbreviated form:

$$\begin{array}{|c|c|c|} \hline & -1 & \\ \hline -1 & 4 & -1 \\ \hline & -1 & \\ \hline \end{array} \; u_{ij}=h^2 f_{ij}. \qquad (15.12\text{-}3)$$

Similar equations, which take into account the given boundary values, may also be set up for the irregular points of R_h, where at least one of the four straight line segments mentioned above intersects ∂R. In this way, a system of as many linear equations is obtained as there are unknowns u_{ij}. Although the number of equations is large—it is asymptotic to αh^{-2} as $h\to 0$, where α denotes the area of R—the numerical solution of the system is practically feasible due to the sparsity of the coefficient matrix. Numerous iterative and direct methods have been devised for this purpose. For special geometries, for instance if R is a rectangle, the solution may be constructed by discrete Fourier methods, as shown in § 15.13.

The error of a finite-difference method can be estimated by substituting the exact solution $u(z)$, assumed to be sufficiently differentiable, into the finite-difference equation and expanding in powers of h. There will be a first term in the expansion that does not vanish identically; its order is called the **local order** of the finite-difference approximation. In this sense, the local order of the finite-difference approximation (15.12-2) is 4. It may be shown that if the exact solution u is in $C^4(R \cup \partial R)$ and if the approximation at irregular active points is done with sufficient care, the global error $u_{ij} - u(z_{ij})$ will be $O(h^2)$ uniformly on R.

There are, in principle, two ways to increase the accuracy of a finite-difference approximation. First, one may try to model the Laplacian $-\Delta$ by a finite-difference approximation that has a higher local order for any sufficiently differentiable function u. In this way one obtains, for instance, the stencil

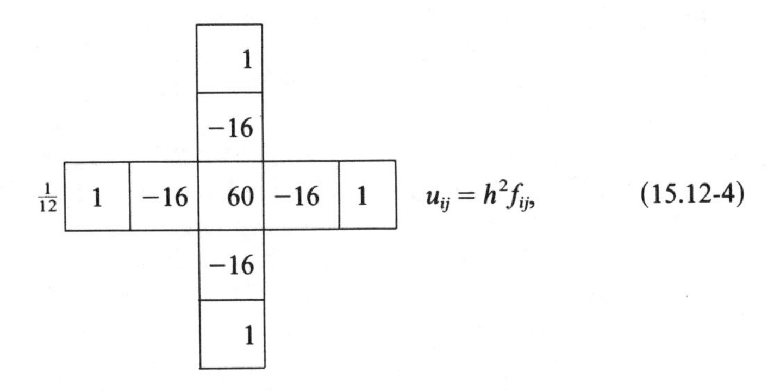

(15.12-4)

which has local order 6, and therefore global order 4. Second, one may try to find a linear combination of values u_{ij} and f_{ij} which for solutions of $-\Delta u = f$ (and only for such solutions, but these are the only functions which interest us) have a local error of high order. In this way we find, for instance, the stencil

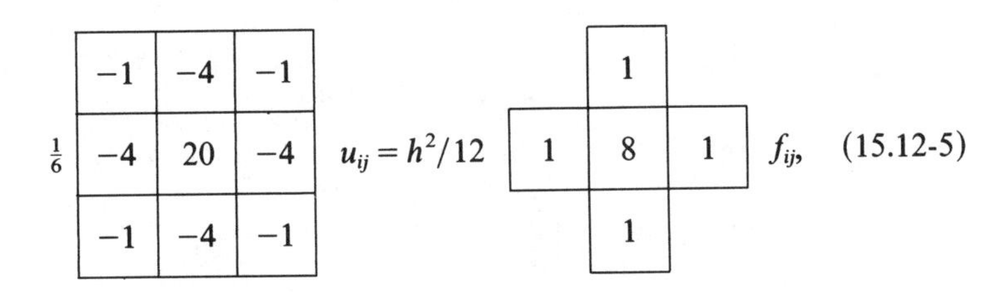

(15.12-5)

which is smaller than the stencil (15.12-4) but likewise has order 6. Methods based on stencils of this kind are called **Mehrstellenverfahren** (methods of several points) in the German literature; no equivalent term in English is generally accepted.

II. Variational Methods; Finite Elements

Let R be a bounded region with piecewise smooth boundary ∂R, and let X be the linear space of real continuous functions on $R \cup \partial R$. We define the **inner product** of two elements $u, v \in X$ by

$$(u, v) := \iint_R u(z)v(z) \boxed{dz},$$

and the **norm** of a single element $u \in X$ by

$$\|u\| := (u, u)^{1/2}.$$

With these definitions of inner product and norm, X becomes an **inner product space**; X is not a Hilbert space because it is not complete.

Now let $\mathring{D}$ be the subspace of X consisting of those functions $u \in X$ that vanish on ∂R and have piecewise continuous partial derivatives of the orders 1 and 2 in $R \cup \partial R$. (By this we mean that the closure of R can be represented as the union of a finite number of closed sets $R_1, \ldots, R_m$, no two of which have an interior point in common, such that on each set R_i, u can be extended to a function that is continuous and has continuous partial derivatives of the orders 1 and 2). The problem of finding a solution of Poisson's equation $-\Delta u = f$ assuming the values 0 on ∂R can then be formulated abstractly as follows. For a given $f \in X$, find $u \in \mathring{D}$ such that

$$Lu = f, \tag{15.12-6}$$

where $L := -\Delta$. The operator L: $\mathring{D} \to X$ thus defined enjoys the following simple properties, which are shared also by more general differential operators:

(i) L is linear.

(ii) L is symmetric in the sense that for any two elements $u, v \in \mathring{D}$ there holds

$$(Lu, v) = (Lv, u). \tag{15.12-7}$$

Proof. By Green's formula, if n denotes the exterior normal,

$$\begin{aligned}(Lu, v) &= -\iint_R v\, \Delta u \boxed{dz} \\ &= \iint_R \operatorname{grad} u \operatorname{grad} v \boxed{dz} - \int_{\partial R} v \frac{\partial u}{\partial n} |dz|.\end{aligned}$$

The second term on the right is 0 by virtue of $v = 0$ on ∂R, hence the above equals (Lv, u). —

(iii) L is positive definite in the sense that

$$(Lu, u) \geqq 0, \tag{15.12-8}$$

with equality only for $u = 0$.

Proof. The foregoing computation shows that

$$(Lu, u) = \iint_R |\text{grad } u|^2 \boxed{dz} \geqq 0.$$

The expression on the right can be zero only if grad u vanishes identically, which by virtue of $u = 0$ on ∂R can happen only if u vanishes identically. —

To solve (15.12-1) we consider the functional

$$\phi(v) := (Lv, v) - 2(f, v) \qquad (15.12\text{-}9)$$

defined for $v \in \mathring{D}$. If u is the solution of (15.12-1) vanishing on ∂R, then in view of $Lu = f$ this functional has the value

$$\phi(u) = -(f, u).$$

We assert that for any other $v \in \mathring{D}$, ϕ has a larger value. The following *proof* requires only properties (i), (ii), and (iii) of the operator L and thus holds for any nonhomogeneous problem that may be cast in the above form. Let $v = u + w$, where $w \in \mathring{D}$. Then

$$\begin{aligned}\phi(v) &= (L(u+w), u+w) - 2(f, u+w)\\ &= (Lu, u) + 2(Lu, f) + (Lw, w) - 2(f, u) - 2(f, w).\end{aligned}$$

Using $Lu = f$ and $-(f, u) = \phi(u)$, this becomes

$$\phi(v) = \phi(u) + (Lw, w),$$

which is $\geqq \phi(u)$ by virtue of (iii), with equality only for $w = 0$.

The problem of solving Poisson's equation thus is equivalent to the problem of minimizing ϕ. In order to minimize ϕ numerically, one selects a finite number of linearly independent basis functions $v_1, v_2, \ldots, v_n$ in $\mathring{D}$ and puts

$$v = \sum_{i=1}^{n} \xi_i v_i, \qquad (15.12\text{-}10)$$

where $\xi_1, \xi_2, \ldots, \xi_n$ are real parameters. Instead of minimizing ϕ in $\mathring{D}$, one now minimizes ϕ in the subspace $\hat{D}$ of $\mathring{D}$ spanned by $v_1, \ldots, v_n$. This is a finite-dimensional problem that can be solved by means of linear algebra. Indeed one has

$$\begin{aligned}\phi\left(\sum_i \xi_i v_i\right) &= \left(\sum_i \xi_i L v_i, \sum_j \xi_j v_j\right) - 2\left(f, \sum_j \xi_j v_j\right)\\ &= \sum_{i,j=1}^{n} a_{ij}\xi_i\xi_j - 2\sum_{j=1}^{n} \beta_j \xi_j,\end{aligned}$$

where

$$\begin{aligned} a_{ij} &:= (Lv_i, v_j) = \iint_R \text{grad } v_i \text{ grad } v_j \boxed{dz} \\ &= \alpha_{ji}, \\ \beta_j &:= (v_j, f) = \iint_R v_j f \boxed{dz}. \end{aligned}$$

Defining the positive definite symmetric matrix

$$\mathbf{A} := (\alpha_{ij}) \tag{15.12-11}$$

and the vectors $\boldsymbol{\beta} := (\beta_1, \ldots, \beta_n)^T$, $\boldsymbol{\xi} := (\xi_1, \ldots, \xi_n)^T$, this may be written

$$\phi\left(\sum_i \xi_i v_i\right) = \boldsymbol{\xi}^T \mathbf{A} \boldsymbol{\xi} - 2\boldsymbol{\beta}^T \boldsymbol{\xi},$$

and the problem of minimizing ϕ is the problem of minimizing a quadratic function of the n variables $\xi_1, \ldots, \xi_n$. By setting the partial derivatives with respect to the ξ_i equal to zero, one readily sees that the minimum occurs at the $\boldsymbol{\xi}$ for which

$$\mathbf{A}\boldsymbol{\xi} = \boldsymbol{\beta}, \tag{15.12-12}$$

and thus can be determined by solving a system of n equations in n unknowns.

It might be considered a drawback of the above method that it requires the computation of the $\frac{1}{2}n(n+1)$ integrals α_{ij}, where $1 \leqq i \leqq j \leqq n$. Especially if the region R is irregular, these integrations would have to be performed numerically. This drawback is avoided in the modern form of the method, called **method of finite elements**. In the simplest version of the method, the closure of R is represented as the union of a finite number of nonoverlapping polygonal **elements** R_j (usually, triangles or rectangles), and the space $\hat{D}$ is a space of **piecewise polynomial functions** in $\mathring{D}$ in the sense that the restriction of each $v \in \hat{D}$ to any R_i is represented by a polynomial in x and y. In its simplest manifestation, the space $\hat{D}$ has the further property that any function $v \in \hat{D}$ is fully determined by its values at the vertices $z_1, z_2, \ldots, z_n$ of all polygons bounding the elements $R_1, R_2, \ldots$. A convenient basis for $\hat{D}$ is then provided by the functions $v_i \in \hat{D}$, $i = 1, \ldots, n$, which have the property that

$$v_i(z_j) = \begin{cases} 1, & j = i \\ 0, & j \neq i. \end{cases} \tag{15.12-13}$$

Any $v \in \hat{D}$ then is represented in the form

$$v(z) = \sum_{i=1}^{n} v(z_i) v_i(z).$$

The parameters ξ_i appearing in (15.12-10) thus in this case equal the values of v at the vertices. The functions v_i characterized by (15.12-13) are $\neq 0$ only on those elements R_j which have z_i as one of their vertices, and the integrals α_{ij} are different from zero only for pairs (i, j) such that both z_i and z_j are vertices of a common element R_k. Since on each such R_k both v_i and v_j are represented by polynomials, the integrals that are not zero are easily evaluated. The integrals β_j may be similarly evaluated by interpolating f by

$$\tilde{f}(z) := \sum_{i=1}^{n} f(z_i) v_i(z).$$

The linear equations (15.12-12) that result from this method generally look similar to, but are not identical with, the equations that result from a Mehrstellenverfahren discussed in Section I. However, the method is more flexible than the finite-difference methods discussed earlier because the elements R_i may easily be chosen so as to accommodate irregular boundaries and irregular solution behavior.

EXAMPLE 1

Let R be the rectangle $0 \leqq x \leqq a$, $0 \leqq y \leqq b$, where $a = n_1 h$, $b = n_2 h$ for some (small) $h > 0$ and for suitable integers n_1, n_2. We consider the subdivision of R into square elements of side length h, and we take as our space $\hat{D}$ the functions v that on each element are represented by a polynomial of the form $\alpha + \beta x + \gamma y + \delta xy$. These functions are linear on each edge of each element, and they are determined by their values on the vertices which here are given by the points $z_{ij} = (ih, jh)$, where $0 < i < n_1$, $0 < j < n_2$. Application of the procedure outlined above yields a set of equations for the approximate values u_{ij} at the vertices which in stencil form may be written as follows:

$$\frac{1}{3}\begin{array}{|c|c|c|}\hline -1 & -1 & -1 \\ \hline -1 & 8 & -1 \\ \hline -1 & -1 & -1 \\ \hline \end{array}\ u_{ij} = h^2/36 \begin{array}{|c|c|c|}\hline 1 & 4 & 1 \\ \hline 4 & 16 & 4 \\ \hline 1 & 4 & 1 \\ \hline \end{array}\ f_{ij}. \qquad (15.12\text{-}14)$$

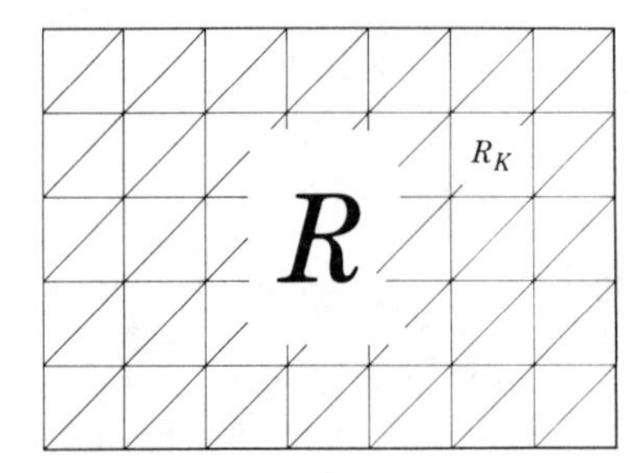

Fig. 15.12b. Triangular elements.

EXAMPLE **2**

Here we consider the same rectangle as above, but we now subdivide into triangular elements, as indicated in Fig. 15.12b. As our space $\hat{D}$ we now take the functions $v \in \mathring{D}$ that are *linear* in each element. Any function $v \in \hat{D}$ obviously is determined by its value at the vertices, which are the same as in Example 1. The general procedure outlined earlier now yields the equations

$$\begin{array}{|c|c|c|} \hline & -1 & \\ \hline -1 & 4 & -1 \\ \hline & -1 & \\ \hline \end{array} \; u_{ij} = h^2/12 \; \begin{array}{|c|c|c|} \hline 0 & 1 & 1 \\ \hline 1 & 6 & 1 \\ \hline 1 & 1 & 0 \\ \hline \end{array} \; f_{ij}. \tag{15.12-15}$$

III. Eigenfunction Expansions

A totally different, and theoretically very powerful, approach to the solution of Poisson's equation, which in principle works for any bounded region, is provided by the method of eigenfunction expansions. Let R be a bounded plane region of arbitrary shape and connectivity, and let ∂R be its boundary. As is well known, the **eigenvalue problem**

$$\begin{aligned} -\Delta u = \lambda u \quad &\text{in} \quad R, \\ u = 0 \quad &\text{on} \quad \partial R \end{aligned} \tag{15.12-16}$$

then has a denumerable set of eigenvalues $0 < \lambda_1 < \lambda_2 \leqq \lambda_3 \leqq \cdots$ and corresponding eigenfunctions $u_1, u_2, \ldots$ such that $u_n \neq 0$, and

$$\begin{aligned} -\Delta u_n = \lambda_n u_n \quad &\text{in} \quad R, \\ u_n = 0 \quad &\text{on} \quad \partial R \end{aligned} \tag{15.12-17}$$

holds for $n = 1, 2, \ldots$. Physically, the eigenfunctions describe the free vibrations of a membrane of shape R which is fixed at the boundary ∂R. The eigenfunctions may be taken to be **orthonormal**, that is, with the definition of the scalar product given earlier, the relations

$$(u_n, u_m) = \begin{cases} 1, & n = m \\ 0, & n \neq m \end{cases}$$

may be assumed to hold. For this orthonormal set of eigenfunctions there holds an expansion theorem. If g is an arbitrary square integrable function on R, and if

$$\gamma_n := (g, u_n), \tag{15.12-18}$$

then the series

$$\sum_{n=1}^{\infty} \gamma_n u_n$$

converges to g at least in norm, that is, its nth partial sums s_n satisfy $\|s_n - g\| \to 0$ as $n \to \infty$. If g is sufficiently smooth and zero on ∂R, the series converges pointwise, even uniformly so, and we then write

$$g = \sum_{n=1}^{\infty} \gamma_n u_n. \tag{15.12-19}$$

This expansion is unique in the sense that if a series of the form (15.12-19) represents the zero function, then necessarily all $\gamma_n = 0$.

If the eigenfunctions are orthogonal but not normalized, (15.12-19) is replaced by

$$g = \sum_{n=1}^{\infty} \frac{\gamma_n}{\omega_n} u_n,$$

where the γ_n are still given by (15.12-18), and where

$$\omega_n := (u_n, u_n). \tag{15.12-20}$$

In order to apply the above to the solution of Poisson's equation $-\Delta u = f$, we assume that the given function f is such that

$$f = \sum \alpha_n u_n,$$

where $\{u_n\}$ is an orthonormal set of eigenfunctions, and

$$\alpha_n := (f, u_n), \qquad n = 1, 2, \ldots. \tag{15.12-21}$$

Since the solution of Poisson's equation always is smoother than the non-homogeneous term, the solution u then likewise may be expanded,

$$u = \sum \beta_n u_n.$$

Applying $-\Delta$ termwise—we assume that this is permissible—and expressing the fact that u satisfies $-\Delta u = f$, we get, using (15.12-17),

$$-\Delta u = \sum \beta_n(-\Delta u_n) = \sum \beta_n \lambda_n u_n = f = \sum \alpha_n u_n$$

and thus, by virtue of the uniqueness of the eigenfunction expansion, $\beta_n \lambda_n = \alpha_n$ or

$$\beta_n = \frac{\alpha_n}{\lambda_n}, \qquad n = 1, 2, \ldots.$$

That is, the solution of Poisson's equation with zero boundary conditions is given in the explicit form

$$u = \sum_{n=1}^{\infty} \frac{\alpha_n}{\lambda_n} u_n, \tag{15.12-22}$$

where the α_n are given by (15.12-20). If the eigenfunctions are not normalized, (15.12-22) is replaced by

$$u = \sum_{n=1}^{\infty} \frac{\alpha_n}{\lambda_n \omega_n} u_n, \tag{15.12-23}$$

where ω_n is defined by (15.12-19).

The solution formula (15.20-23) holds for any bounded region R for which the eigenvalue problem (15.12-16) makes sense. However, if the formula is to be utilized algorithmically, the eigenfunctions and eigenvalues must be known. Practically this means that it must be possible to obtain the eigenfunctions by the method of separation of variables. We have alluded in Problem 9, § 5.6, to the fact that separation of variables is possible only for regions R that are the images of a rectangle under a conformal map g such that $|g'(w)|^2 = \phi(u) + \psi(v)$. Apart from the rectangle itself, this is the case only for regions that are bounded by concentric circles, or by systems of confocal ellipses and hyperbolas, or by parabolas.

EXAMPLE 3

For the rectangle R considered in the preceding two examples, it is clear that for each pair of positive integers $\mathbf{m} = (m_1, m_2)$ the function

$$u_{\mathbf{m}}(x, y) = -\frac{2}{(ab)^{1/2}} \sin \frac{\pi m_1 x}{a} \sin \frac{\pi m_2 y}{b} \tag{15.12-24}$$

is a normalized eigenfunction belonging to the eigenvalue

$$\lambda_{\mathbf{m}} = \pi^2 \left(\frac{m_1^2}{a^2} + \frac{m_2^2}{b^2} \right).$$

(The minus sign is inserted for later convenience.) It can be shown that there are no other eigenfunctions. Thus if

$$f(z)=\sum_{\mathbf{m}>\mathbf{0}} \alpha_{\mathbf{m}} u_{\mathbf{m}}(z),$$

where

$$\alpha_{\mathbf{m}} := -\frac{2}{(ab)^{1/2}} \int_0^a \int_0^b f(z) \sin\frac{\pi m_1 x}{a} \sin\frac{\pi m_2 y}{b} \boxed{dz}, \tag{15.12-25}$$

the solution of Poisson's equation $-\Delta u=f$ with values 0 on the boundary is

$$u(z)=\sum_{\mathbf{m}>\mathbf{0}} \frac{\alpha_{\mathbf{m}}}{\pi^2(a^{-2}m_1^2+b^{-2}m_2^2)} u_{\mathbf{m}}(z). \tag{15.12-26}$$

Here and in the preceding sum $\mathbf{m}>\mathbf{0}$ means that the summation is to be carried out with respect to all (m_1, m_2) such that $m_1>0$ and $m_2>0$.

PROBLEMS

1. Verify that the stencil given by (15.12-5) has local order 6.
2. Carry out the computations leading to the finite-element stencils (15.12-14) and (15.12-15).
3. *Clamped plate problem.* Show that the solution of the biharmonic problem

$$Lu := \Delta\Delta u = g \quad \text{in} \quad R,$$

$$u=\frac{\partial u}{\partial n}=0 \qquad \text{on } \partial R$$

   may be obtained by minimizing the functional $\phi(v)=(Lv, v)-2(g, v)$ in an appropriate function space.
4. Show that the solution of the boundary value problem

$$-\Delta u = g \quad \text{in} \quad R,$$

$$\alpha(z)u+\frac{\partial u}{\partial n}=0 \quad \text{on} \quad \partial R$$

   may be obtained by minimizing the functional

$$\phi(v)=(Lv, v)-2(g, v)+\int_{\partial R} \alpha v^2 \,|dz|,$$

   where $L:=-\Delta$, in the space of functions satisfying the required boundary conditions.
5. *Dirichlet integral.* Let u be continuous in the closure of a region R, and let it have piecewise continuous partial derivatives of order 1. The functional

$$D(u):=\iint_R |\text{grad } u|^2 \,\boxed{dz}$$

is called the **Dirichlet integral** of u. Show that the solution of the Dirichlet problem for the region R may be obtained by minimizing the Dirichlet integral in the class of functions that satisfy the required boundary conditions. How could this fact be used to construct approximate solutions by the method of finite elements?

6. In advanced calculus the eigenfunctions for a circular disk are shown to be expressible in terms of Bessel functions. Show what would be involved in using eigenfunction expansions for the solutions of Poisson's equation for the disk. To what extent could FFTs be used? Compare with the approach sketched in Problem 2, § 15.11.

NOTES

Finite-difference methods are described in the fine books by Milne [1953], Collatz [1960], and Forsythe and Wasow [1960]. Meanwhile, at least in problems with complicated geometries, finite-difference methods have frequently been displaced by the method of finite elements, on which the literature is enormous. The method is to be traced back to Courant [1943]; for early numerical uses of the method, see Hersch, Pfluger, and Schopf [1956] and Engeli, Ginsburg, Rutishauser, and Stiefel [1959]. For the current state of the art, see Ciarlet [1978] or, from a more practical point of view, Schwarz [1984]. All these sources deal with problems far more general than Poisson's equation.

The bottleneck in the application of both finite-difference and finite-element methods is the solution of the large linear systems that arise. For authoritative presentations of the classical iterative methods, see Varga [1962], Young [1971]. Engeli et al. [1959] have an especially fine section (due to Rutishauser) on the conjugate gradient method. Meanwhile direct methods again have gained prominence. The problem here is mainly one of handling the enormous sets of data; see Golub and van Loan [1983] and again Schwarz [1984] for recent developments. The *multigrid method* due to Brandt [1977] is an ingenious iterative method whose convergence is so fast that ordinarily only very few iteration steps are required; see also Hackbusch and Trottenberg [1982].

As to eigenfunction expansions, Courant and Hilbert [1951] still in the best presentation from the point of view of mathematical physics. On cases where the Helmholtz equation is separable, see Weber [1870].

§ 15.13. FAST POISSON SOLVERS

A fast Poisson solver is an algorithm for the efficient implementation of a method for solving Poisson's equation in a standardized region R. Here we use discrete Fourier analysis to obtain fast Poisson solvers for rectangular regions. Although formulated explicitly only for two dimensions, these algorithms can be used to solve Poisson's equation in a parallelepiped of arbitrary dimension.

I. Finite Differences

Let R denote the rectangle $0 \leqq x \leqq a = n_1 h$, $0 \leqq y \leqq b = n_2 h$. To solve a system of finite-difference equations such as (15.12-2) by discrete Fourier analysis, we write, for $\mathbf{k}=(i,j)$,

$$u_{\mathbf{k}}=u_{ij}, \qquad f_{\mathbf{k}}=f_{ij}.$$

Here the indices i and j originally range over the sets $0 \leqq i \leqq n_1$, $0 \leqq j \leqq n_2$. However, letting $\mathbf{n}=(n_1, n_2)$, we embed both arrays $u_{\mathbf{k}}$ and $f_{\mathbf{k}}$ in the space $\Pi_{2\mathbf{n}}$ of doubly periodic sequences (see § 13.10) by first continuing them as *odd* functions of i and j, that is, by defining

$$u_{-i,j}=u_{i,-j} := -u_{ij}, \qquad u_{-i,-j} := u_{ij},$$

and then continuing them as doubly periodic sequences with period vector $2\mathbf{n}$. The subspace of all sequences in $\Pi_{2\mathbf{n}}$ that are odd in this sense is denoted by $\Pi_{2\mathbf{n}}^{\circ}$. As a consequence of our embedding the arrays $\{u_{ij}\}$ and $\{f_{ij}\}$ in $\Pi_{2\mathbf{n}}^{\circ}$, the required boundary conditions

$$u_{0,j}=u_{i,0}=u_{n_1,j}=u_{i,n_2}=0 \qquad (15.13\text{-}1)$$

are satisfied automatically. Moreover, the difference equations (15.12-2) now hold for *all* index vectors $\mathbf{k}=(i,j)\in \mathbb{Z}^2$, and not only for those satisfying $0<i<n_1$, $0<j<n_2$.

To write the difference equations (15.12-2) in a compact manner, we define the array $\boldsymbol{\delta}=\{\delta_{\mathbf{k}}\}$ in $\Pi_{2\mathbf{n}}$ by setting

$$\delta_{\mathbf{k}} := 4$$

if $\mathbf{k} \equiv \mathbf{0} \bmod 2\mathbf{n}$, and

$$\delta_{\mathbf{k}} := -1$$

if $\mathbf{k}$ is congruent to one of the vectors $(1,0)$, $(0,1)$, $(-1,0)$, $(0,-1) \bmod 2\mathbf{n}$, and by setting all other $\delta_{\mathbf{k}}$ equal to zero. For symmetry we also introduce the array $\boldsymbol{\varepsilon}=\{\varepsilon_{\mathbf{k}}\}$ such that

$$\varepsilon_{\mathbf{k}} := h^2$$

if $\mathbf{k} \equiv \mathbf{0} \bmod 2\mathbf{n}$, and all other $\varepsilon_{\mathbf{k}} := 0$. Letting $\mathbf{u}=\{u_{\mathbf{k}}\}$, $\mathbf{f}=\{f_{\mathbf{k}}\}$, the difference equations (15.12-2) may then be written

$$\boldsymbol{\delta} * \mathbf{u} = \boldsymbol{\varepsilon} * \mathbf{f}. \qquad (15.13\text{-}2)$$

Far from being a mere notational device, the convolution notation points the way to the efficient solution of the difference equations. For let

$$\hat{\boldsymbol{\delta}} := \mathscr{F}_{2\mathbf{n}}\boldsymbol{\delta}, \qquad \hat{\boldsymbol{\varepsilon}} := \mathscr{F}_{2\mathbf{n}}\boldsymbol{\varepsilon}, \qquad \hat{\mathbf{u}} := \mathscr{F}_{2\mathbf{n}}\mathbf{u}, \qquad \hat{\mathbf{f}} := \mathscr{F}_{2\mathbf{n}}\mathbf{f}.$$

Operating on (15.13-2) with $\mathscr{F}_{2\mathbf{n}}$, we obtain by the multidimensional analog of the convolution theorem (Theorem 13.10c)

$$\hat{\boldsymbol{\delta}} \cdot \hat{\mathbf{u}} = \hat{\boldsymbol{\varepsilon}} \cdot \hat{\mathbf{f}}. \qquad (15.13\text{-}3)$$

This equation means the same as

$$\hat{\delta}_{\mathbf{m}}\hat{u}_{\mathbf{m}}=\hat{\varepsilon}_{\mathbf{m}}\hat{f}_{\mathbf{m}}, \qquad \mathbf{m}\in\mathbb{Z}^2,$$

and it thus may be used to calculate $\hat{u}_{\mathbf{m}}$ as long as $\hat{\delta}_{\mathbf{m}}\neq 0$. By the definition of the discrete Fourier operator,

$$\hat{\delta}_{\mathbf{m}}=(4n_1n_2)^{-1/2}\sum_{\mathbf{k}\in Q_{2\mathbf{n}}}\delta_{\mathbf{k}}\mathbf{w}_{2\mathbf{n}}^{-\mathbf{k}\cdot\mathbf{m}}.$$

Letting $\mathbf{m}=(m_1, m_2)$, using

$$\mathbf{w}_{2\mathbf{n}}^{-\mathbf{k}\cdot\mathbf{m}}=w_1^{-k_1m_1}w_2^{-k_2m_2},$$

where

$$w_j:=\exp\left(\frac{2\pi i}{2n_j}\right), \qquad j=1,2, \tag{15.13-4}$$

and introducing the explicit values of $\delta_{\mathbf{k}}$, we obtain

$$\begin{aligned}\hat{\delta}_{m_1m_2}&=\frac{1}{(4n_1n_2)^{1/2}}\{4-w_1^{-m_1}-w_1^{m_1}-w_2^{-m_2}-w_2^{m_2}\}\\&=\frac{2}{(4n_1n_2)^{1/2}}\left\{2-\cos\frac{m_1\pi}{n_1}-\cos\frac{m_2\pi}{n_2}\right\}.\end{aligned}$$

Letting

$$s_1:=\left(\sin\frac{m_1\pi}{2n_1}\right)^2, \qquad s_2:=\left(\sin\frac{m_2\pi}{2n_2}\right)^2, \tag{15.13-5}$$

this may be written

$$\hat{\delta}_{m_1m_2}=\frac{4}{(4n_1n_2)^{1/2}}(s_1+s_2).$$

We thus see that $\hat{\delta}_{\mathbf{m}}=0$ only if $\mathbf{m}\equiv\mathbf{0} \bmod 2\mathbf{n}$. Thus (15.13-3) may be solved for all $u_{\mathbf{m}}$ such that $\mathbf{m}\not\equiv\mathbf{0} \bmod 2\mathbf{n}$. Since both $\mathbf{u}$ and $\mathbf{f}$ are odd, we have $\hat{u}_{\mathbf{0}}=\hat{f}_{\mathbf{0}}=0$, and it is not necessary to use (15.13-3) for $\mathbf{m}=\mathbf{0}$.

We note that

$$\hat{\varepsilon}_{\mathbf{m}}=\frac{h^2}{2(n_1n_2)^{1/2}}, \qquad \mathbf{m}\in\mathbb{Z}^2.$$

Letting

$$\boldsymbol{\rho}:=\hat{\boldsymbol{\delta}}\cdot\hat{\boldsymbol{\varepsilon}}^{-1}, \tag{15.13-6}$$

that is

$$\rho_{m_1m_2}=\frac{4}{h^2}(s_1+s_2),$$

and agreeing to interpret $0^{-1}0=0$, we may write

$$\hat{\mathbf{u}}=\boldsymbol{\rho}^{-1}\cdot\hat{\mathbf{f}}$$

and thus obtain the solution formula

$$\mathbf{u}=\mathscr{F}_{2\mathbf{n}}^{-1}(\boldsymbol{\rho}^{-1}\cdot\hat{\mathbf{f}}). \tag{15.13-7}$$

As always, the exponent in $\boldsymbol{\rho}^{-1}$ is to be understood in the sense of the Hadamard product.

The solution of the system of difference equations (15.12-2) by discrete Fourier methods thus involves the following steps:

(i) Compute the discrete Fourier transform $\hat{\mathbf{f}}$ of the array $\mathbf{f}=\{f_{ij}\}=\{f(z_{ij})\}$, continued as an *odd* array in $\Pi_{2\mathbf{n}}$.
(ii) Compute the elements of the array $\boldsymbol{\rho}$ from (15.13-6).
(iii) Perform the Hadamard product $\boldsymbol{\rho}^{-1}\cdot\hat{\mathbf{f}}$.
(iv) Apply $\mathscr{F}_{2\mathbf{n}}^{-1}$; this will yield the array $\mathbf{u}=\{u_{ij}\}$ of the values of the desired finite difference solution at the points $z_{ij}=(x_i, y_j)$.

As always in numerical uses of the discrete Fourier transform, the various scalar factors that occur are initially ignored and only adjusted at the last step if necessary. The total cost of the computation, if both n_1 and n_2 are powers of 2, by Theorem 13.10b does not exceed

$$\tfrac{1}{4}n_1n_2(\mathrm{Log}(n_1n_2)+4)$$

complex multiplications. In addition, n_1+n_2 sine functions have to be evaluated for the computation of $\boldsymbol{\rho}$.

If the "large star" difference operator (15.12-4) is used, the solution formula (15.13-7) remains the same, except that here

$$\hat{\delta}_{m_1m_2}=\frac{1}{(4n_1n_2)^{1/2}}\frac{1}{12}\{60-16[w_1^{-m_1}+w_1^{m_1}+w_2^{-m_2}+w_2^{m_2}]$$
$$+[w_1^{-2m_1}+w_1^{2m_1}+w_2^{-2m_2}+w_2^{2m_2}]\}$$

and hence, using the abbreviations (15.13-5), $\boldsymbol{\rho}=\{\rho_{\mathbf{m}}\}$, where

$$\rho_{m_1m_2}=\frac{4}{h^2}\left\{s_1+s_2+\frac{1}{3}(s_1^2+s_2^2)\right\}. \tag{15.13-8}$$

If a Mehrstellenoperator is used, the only difference is that the array $\boldsymbol{\varepsilon}$ on the right of (15.13-2) must be replaced by an appropriate averaging sequence. For instance, in order to describe the operator (15.12-5), we let $\boldsymbol{\delta}=\{\delta_{\mathbf{k}}\}$ and $\boldsymbol{\mu}=\{\mu_{\mathbf{k}}\}$, where

$$\delta_{\mathbf{k}}:=\tfrac{20}{6},\qquad \mu_{\mathbf{k}}:=\tfrac{8}{12}h^2$$

if $\mathbf{k} \equiv \mathbf{0} \bmod 2\mathbf{n}$, where

$$\delta_{\mathbf{k}} := -\tfrac{4}{6}, \qquad \mu_{\mathbf{k}} := \tfrac{1}{12}h^2$$

if $\mathbf{k}$ is congruent to one of $(0,1)$, $(1,0)$, $(0,-1)$, $(-1,0) \bmod 2\mathbf{n}$, and where

$$\delta_{\mathbf{k}} := -\tfrac{1}{6}$$

if $\mathbf{k}$ is congruent to one of $(1,1), (-1,1), (1,-1), (-1,-1) \bmod 2\mathbf{n}$. All elements not yet defined are set equal to 0. The Mehrstellen equation (15.12-5) then is

$$\boldsymbol{\delta} * \mathbf{u} = \boldsymbol{\mu} * \mathbf{f}. \qquad (15.13\text{-}9)$$

Taking the discrete Fourier transform, this becomes

$$\hat{\boldsymbol{\delta}} \cdot \hat{\mathbf{u}} = \hat{\boldsymbol{\mu}} \cdot \hat{\mathbf{f}}$$

or

$$\hat{\mathbf{u}} = \boldsymbol{\rho}^{-1} \cdot \hat{\mathbf{f}},$$

where

$$\boldsymbol{\rho} := \hat{\boldsymbol{\delta}} \cdot \hat{\boldsymbol{\mu}}^{-1}.$$

Using these abbreviations, we obtain the solution formula

$$\mathbf{u} = \mathcal{F}_{2\mathbf{n}}^{-1}(\boldsymbol{\rho}^{-1} \cdot \hat{\mathbf{f}}) \qquad (15.13\text{-}10)$$

similar to (15.13-7). Computation yields $\boldsymbol{\rho} = \{\rho_{m_1 m_2}\}$, where

$$\rho_{m_1 m_2} = \frac{4}{h^2} \frac{s_1 + s_2 - \frac{2}{3}s_1 s_2}{1 - \frac{1}{3}(s_1 + s_2)} \qquad (15.13\text{-}11)$$

II. Finite Elements

The finite-element method, as described in § 15.12, determines approximate values u_{ij} for the solution $u(z_{ij})$ by means of systems of linear equations which have the same general structure as the equations obtained in the Mehrstellenverfahren. Therefore, with appropriate choices of the array $\boldsymbol{\rho}$, the solution formula

$$\mathbf{u} = \mathcal{F}_{2\mathbf{n}}^{-1}(\boldsymbol{\rho}^{-1} \cdot \hat{\mathbf{f}}) \qquad (15.13\text{-}12)$$

still holds. Computation yields for the method described in Example 1 of § 15.12 (square elements),

$$\rho_{m_1 m_2} = \frac{4}{h^2} \frac{s_1 + s_2 - \frac{4}{3}s_1 s_2}{1 - \frac{2}{3}(s_1 + s_2) + \frac{4}{9}s_1 s_2} \qquad (15.13\text{-}13)$$

and for the method described in Example 2 (triangular elements),

$$\rho_{m_1 m_2} = \frac{4}{h^2} \frac{s_1 + s_2}{1 - \frac{2}{3}(s_1 + s_2) + \frac{2}{3}s_1 s_2 - \frac{2}{3}[s_1(1-s_1)s_2(1-s_2)]^{1/2}}. \tag{15.13-14}$$

III. Eigenfunction Expansion

Discrete Fourier analysis plays a role also in the implementation of the method of eigenfunctions for the rectangle R. To begin with, we write the solution formula (15.12-26) as a series of complex exponentials. Since

$$u_{\mathbf{m}}(z) = \frac{1}{2(ab)^{1/2}} \left\{ \exp\left(\frac{\pi i m_1 x}{a}\right) - \exp\left(-\frac{\pi i m_1 x}{a}\right) \right\} \times \left\{ \exp\left(\frac{\pi i m_2 y}{b}\right) - \exp\left(-\frac{\pi i m_2 y}{b}\right) \right\},$$

we have

$$u(z) = \frac{1}{2(ab)^{1/2}} \sum_{\mathbf{m} \neq \mathbf{0}} \frac{\alpha_{\mathbf{m}}}{\lambda_{\mathbf{m}}} \exp\left(\frac{\pi i m_1 x}{a} + \frac{\pi i m_2 y}{b}\right), \tag{15.13-15}$$

where

$$\lambda_{\mathbf{m}} = \pi^2(a^{-2}m_1^2 + b^{-2}m_2^2),$$

and where the sum now comprises all $\mathbf{m} = (m_1, m_2)$ such that $\mathbf{m} \neq \mathbf{0}$, provided that the definition of $\alpha_{\mathbf{m}}$ is extended to all of $\mathbb{Z}^2$ as follows:

$$\alpha_{-m_1,m_2} = \alpha_{m_1,-m_2} = -\alpha_{m_1 m_2}, \qquad \alpha_{-m_1,-m_2} = \alpha_{m_1 m_2}, \qquad \alpha_{m_1,0} = \alpha_{0,m_2} = 0.$$

It is seen on inspection that (15.12-25) remains consistent with the extended definition for all integers m_1, m_2.

Again by replacing the sines by complex exponentials, and by continuing $f(z) = f(x, y)$ for x, y negative as a function that is odd in both x and y, we see that

$$\alpha_{\mathbf{m}} = \frac{1}{2(ab)^{1/2}} \int_{-a}^{a} \int_{-b}^{b} f(x, y) \exp\left(-\frac{\pi i m_1 x}{a} - \frac{\pi i m_2 y}{b}\right) dx\, dy. \tag{15.13-16}$$

To evaluate the series (15.13-15), we require numerical values of the $\alpha_{\mathbf{m}}$. We assume that values f_{ij} of the function f are available at the points $z_{ij} = (ih, jh)$, where $-n_1 \leqq i \leqq n_1$, $-n_2 \leqq j \leqq n_2$. By evaluating the integral (15.13-16) by the trapezoidal rule we obtain approximate values

$$\beta_{\mathbf{m}} = \frac{h^2}{2(ab)^{1/2}} \sum_{k_1=-n_1}^{n_1}{}' \sum_{k_2=-n_2}^{n_2}{}' f_{k_1 k_2} w_1^{-m_1 k_1} w_2^{-m_2 k_2},$$

where w_1 and w_2 are given by (15.13-4), and where the prime indicates that the terms corresponding to $k_1 = \pm n_1$, $k_2 = \pm n_2$ are to be taken with the weight $\frac{1}{2}$. The relation thus obtained may be expressed as a discrete Fourier transform. In fact, if $\mathbf{f} := \{f_{ij}\}$, $\boldsymbol{\beta} = \{\beta_{m_1 m_2}\}$, then

$$\beta = h\mathscr{F}_{2\mathbf{n}}\mathbf{f} = h\hat{\mathbf{f}}. \qquad (15.13\text{-}17)$$

We next have to truncate the series (15.13-15), where the $\alpha_{\mathbf{m}}$ are replaced by the $\beta_{\mathbf{m}}$. Since $\boldsymbol{\beta} \in \Pi_{2\mathbf{n}}$, it does not make sense to take into account any terms such that $|m_1| > n_1$ or $|m_2| > n_2$, and the terms where $m_1 = \pm n_1$ or $m_2 = \pm n_2$ should be taken with the weight $\frac{1}{2}$. We thus arrive at the approximate solution formula obtained by the eigenfunction method,

$$\tilde{u}(z) = \frac{1}{2(ab)^{1/2}} \sum_{m_1=-n_1}^{n_1}{}' \sum_{m_2=-n_2}^{n_2}{}' \frac{\beta_{\mathbf{m}}}{\lambda_{\mathbf{m}}} \exp\left(\frac{\pi i m_1 x}{a} + \frac{\pi i m_2 y}{b}\right). \qquad (15.13\text{-}18)$$

We finally evaluate the truncated series at the points $z_{ij} = (x_i, y_j)$. The sum then becomes $2(n_1 n_2)^{1/2}$ times the inverse Fourier operator $\mathscr{F}_{2\mathbf{n}}^{-1}$. All scalar factors cancel. Denoting by $\mathbf{u}^E$ the resulting array of values $u(z_{ij})$, and setting $\boldsymbol{\lambda} = \{\lambda_{\mathbf{m}}\}$, the result may be written

$$\mathbf{u}^E = \mathscr{F}_{2\mathbf{n}}^{-1}(\boldsymbol{\lambda}^{-1} \cdot \hat{\mathbf{f}}). \qquad (15.13\text{-}19)$$

The formula just obtained is very similar to the formulas for the approximations $\mathbf{u}^M$ due to finite difference, Mehrstellen, and finite-element methods,

$$\mathbf{u}^M = \mathscr{F}_{2\mathbf{n}}^{-1}(\boldsymbol{\rho}^{-1} \cdot \hat{\mathbf{f}}). \qquad (15.13\text{-}20)$$

In fact, if both $\mathbf{u}^E$ and $\mathbf{u}^M$ are to approximate the exact solution $u(z)$ as $h \to 0$, there must be a close relationship between the arrays $\boldsymbol{\rho}$ and $\boldsymbol{\lambda}$. What is it?

The answer is seen if we keep m_1, m_2 fixed and let $h \to 0$. For instance, if the ordinary five-point operator (15.12-2) is used, we have by (15.13-6), using $a = n_1 h$, $b = n_2 h$,

$$\begin{aligned}
\rho_{m_1 m_2} &= \frac{4}{h^2}\left\{\left(\sin\frac{m_1 \pi}{n_1}\right)^2 + \left(\sin\frac{m_2 \pi}{2n_2}\right)^2\right\} \\
&= \frac{4}{h^2}\left\{\left(\sin\frac{m_1 \pi h}{2a}\right)^2 + \left(\sin\frac{m_2 \pi h}{2b}\right)^2\right\} \\
&= \pi^2(a^{-2}m_1^2 + b^{-2}m_2^2) + O(h^2) \\
&= \lambda_{m_1 m_2} + O(h^2).
\end{aligned} \qquad (15.13\text{-}21)$$

In a similar manner one finds for the $\boldsymbol{\rho}$ arising from the "large star" operator (15.12-4) or from the Mehrstellen operator (15.12-5),

$$\rho_{m_1 m_2} = \lambda_{m_1 m_2} + O(h^4). \qquad (15.13\text{-}22)$$

The arrays $\boldsymbol{\rho}$ thus *approximate* the arrays $\boldsymbol{\lambda}$ as $h \to 0$, and the approximation becomes better if higher order difference or Mehrstellen operators are used. The arrays $\boldsymbol{\rho}$ arising from the finite-element method likewise approximate $\boldsymbol{\lambda}$. Here the error is $O(h^2)$ for both arrays (15.13-13) and (15.13-14).

Since the exact eigenvalues $\lambda_{m_1 m_2}$ are known, and are even more easily computed than the elements of the various arrays $\boldsymbol{\rho}$ because no trigonometric functions are involved, there does not seem to be any reason to approximate them. In that sense, a fast Poisson solver for the rectangle based on the exact eigenfunction expansion appears to be superior to the Poisson solvers based on the finite difference, Mehrstellen, or finite-element methods.

However, the accuracy of the solution $\mathbf{u}^E$ depends not only on the accuracy of the eigenvalues, but also on the accuracy of the Fourier coefficients. The approximations obtained via $\hat{\mathbf{f}}$ are accurate, *if the function obtained by continuing $f(x, y)$ as a function that is odd in both x and y and that has period $2a$ in x and period $2b$ in y is smooth.* If this condition is not met, even if f is smooth in the original rectangle, then the Fourier coefficients $\alpha_{\mathbf{m}}$ do not decrease rapidly, and the series (15.13-15) does not converge sufficiently rapidly for the truncation error to be negligible. In such situations, an approach based on a crude finite-difference approximation will work equally well.

PROBLEMS

1. Construct fast Poisson solvers for the rectangle R: $0 \leqq x \leqq a$, $0 \leqq y \leqq b$ under the boundary conditions:

$$u(0, y) = u(a, y) = 0, \qquad \frac{\partial u}{\partial n}(x, 0) = \frac{\partial u}{\partial n}(x, b) = 0.$$

(Continue u as a periodic function with period vector $(2a, 2b)$, which is odd in x and even in y.)

2. Using the methods discussed above, construct fast solvers for the **Helmholtz equation**

$$-\Delta u + \lambda u = g,$$

where $\lambda > 0$.

3. *Construction of Mehrstellen operators.* The connection between the arrays $\boldsymbol{\rho}$ and $\boldsymbol{\lambda}$ shown in (15.13-21) may be exploited systematically to construct Mehrstellen operators.

(a) Show that to every rational function $r(s_1, s_2)$ such that $r(0, 0) = 0$ there corresponds a Mehrstellen operator, expressible entirely in terms of central difference operators, with the following property. If the equations arising from the Mehrstellen operator are written in the form $\boldsymbol{\delta} * \mathbf{u} = \boldsymbol{\mu} * \mathbf{f}$ (see (15.13-9)), then

$$\boldsymbol{\rho} := \hat{\boldsymbol{\delta}} \cdot \hat{\boldsymbol{\mu}}^{-1} = \{\rho_{m_1 m_2}\},$$

where

$$\rho_{m_1 m_2} = \frac{4}{h^2} r(s_1, s_2).$$

(b) Show that the two following statements are equivalent:

(i)
$$\frac{4}{h^2} r(s_1, s_2) = \lambda_{m_1 m_2} + O(h^{2p}) \qquad \text{as } h \to 0. \qquad (*)$$

(ii) The associated Mehrstellen operator has order $2p$.

(c) Show that $(*)$ holds if and only if $r(s, t)$ approximates the function

$$f(s, t) := (\arcsin \sqrt{s})^2 + (\arcsin \sqrt{t})^2$$

to order $p+1$ in the sense that the Taylor series at O of r and of f agree through terms of order p.

(d) Use the known series

$$(\arcsin \sqrt{s})^2 = s\,{}_3F_2(1, 1, 1; \tfrac{3}{2}, 2; s) = s + \tfrac{1}{3}s^2 + \tfrac{8}{45}s^3 + \tfrac{4}{35}s^4 + \cdots$$

to construct (by means of the quotient-difference algorithm, or otherwise) rational functions $r(s)$ that approximate $(\arcsin \sqrt{s})^2$ to order $p+1$.

(e) Use the approximations obtained in (d) to approximate $f(s, t)$ to order $p+1$. For $p = 1, 2$ obtain the operators obtained in the text, and for $p = 3$ obtain the operator (showing the portion of the stencil lying in the first quadrant)

$\frac{1}{300}$

−11	−2	
−288	−4	−2
988	−288	−11

$u_{ij} = h^2/255$

22	4
121	22

f_{ij}.

NOTES

The literature on fast Poisson solvers is large. In addition to the methods based on Fourier analysis, there exist Poisson solvers based on tensor product methods (Lynch, Rice, and Thomas [1964]), cyclic reduction (a device related to the reduction formulas of FFT; see Buzbee, Golub, and Nielson [1970], Buzbee, Dorr, George, and Golub [1971], Swarztrauber [1977], and Sweet [1977]), and multigrid techniques (Schröder, Trottenberg, and Witsch [1978]). Discrete Fourier analysis is used by Hockney [1965, 1970, 1972b] and Bunemann [1969]. All these methods are based on simple finite-difference approximations. Higher order difference or Mehrstellen operators are considered by Houstis and Papatheodorou [1979] and Pickering [1977]. The method of eigenfunctions is discussed by Rosser [1974] and Sköllermo [1975]. For fast solvers for more general partial differential equations see Concus and Golub [1973], Buzbee and Dorr [1974], Fischer, Golub, Hald, Leiva, and Widlund [1974], and Proskurowski and Widlund [1976]. The last paper also deals with irregular geometries; Swarztrauber and Sweet [1973] have a fast Poisson solver for the disk. Surveys and comparisons between different methods are given by Dorr [1970], Houstis and Papatheodorou [1977], and Henrici [1984]. For some applications in physical problems see Hockney [1972a], Hockney, Warriner, and Reiser [1974], Hockney and Brownrigg [1974], Hockney and Brown [1975], and Hockney and Goel [1975]. The passage from eigenfunction expansions to difference methods via Padé approximations to the eigenvalues is discussed in Henrici [1985]. Seewald [1985] has accurate fast Poisson solvers for functions f that do not vanish on the boundary of R.

16

CONSTRUCTION OF CONFORMAL MAPS: SIMPLY CONNECTED REGIONS

Conformal mapping was already the subject of an earlier chapter (Chapter 5). There we stated, partly without proof, some basic facts on conformal maps, and we presented a number of applications in mechanics, aerodynamics, and electrostatics. The emphasis was on maps that could be calculated explicitly, that is, represented in terms of elementary analytic functions.

Here we begin a discussion of the general case, that is, we show how to construct mapping functions for arbitrary simply and multiply connected regions, usually subject only to mild conditions concerning the boundary. Maps of simply connected regions are discussed in this chapter; maps of regions of higher connectivity in Chapter 17. In both cases we start by proving the relevant mapping theorem along classical, nonconstructive lines. We then present constructive versions of the same theorems without making any assumptions regarding the boundary. In the simply connected case, there follows a discussion of the behavior of the mapping function on the boundary, including a study of the behavior at corners of piecewise analytic boundary curves. For sufficiently regular boundaries we then present the classical integral equations for the boundary correspondence function of maps of both simply and doubly connected regions, as well as certain integral equations of more modern origin due to Symm and Gaier. Then follows a discussion of methods based on conjugate functions, as well as (both in the simply and doubly connected cases) a discussion of the parameter problem in Schwarz-Christoffel maps.

In addition to the applications of conformal mapping to two-dimensional problems of classical mathematical physics that were described in Chapter 5, we should stress one application that has become feasible only through the enormous power of modern computers. The realistic modeling of certain problems in fluid dynamics requires the solution of systems of nonlinear partial differential equations. Naturally, the solution is possible only by purely numerical methods, using finite-difference techniques on a suitable grid. The formulation of these methods is much simplified if the meshes of the grid are approximately square, while at the same time the grid fits the geometry of the problem. In problems with an irregular geometry this property is not easily secured. However, if the region in question is mapped conformally onto a standard region, where an orthogonal grid fitting the boundary is easily defined, then the inverse image of the orthogonal grid will yield a curvilinear grid which fits the geometry and at the same time is locally square. In such cases it is especially important to be able to evaluate both the map and its inverse efficiently and with precision.

NOTES

To the extent to which it is used in the present chapter, the theory of conformal mapping is largely covered by Ahlfors [1966]. Some classical texts dealing primarily with conformal mapping are due to Julia [1931], Carathéodory [1932], Nehari [1950], Bieberbach [1953], and Goluzin [1969].

Some papers dealing with physical applications of conformal mapping that were not covered in chapter 5 are due to Segel [1961a, b], Goedbloed [1981], and Fornberg [1981a].

Gaier [1964] is the classical survey of numerical methods in conformal mapping. Koppenfels and Stallmann [1959] have a useful collection of explicitly computable maps. Additional surveys of numerical mapping techniques are given by Beckenbach [1952], Seidel [1952], Todd [1955], Opfer [1974a], Laura [1975], and Menikoff and Zemach [1980].

For numerical grid generation by conformal mapping and (mostly) other methods, see the surveys NASA [1980] and Thompson [1982]. Some individual papers using conformal maps are due to Davis [1979], and Ives [1982]. Thacker [1980] mentions 80 papers on grid generation, none of which uses conformal mapping. An additional such method is due to Steger and Chaussee [1980].

§ 16.1. THE RIEMANN MAPPING THEOREM

The **Riemann mapping theorem** was stated in § 5.10 as Corollary 5.10c. It forms the basis not only of the theory of conformal mapping but also for the numerical construction of mapping functions. We restate it here as follows:

THEOREM 16.1a. *Let D be a simply connected region which is not the whole plane, and let a be a point of D. Then there exists in D a unique analytic* *function f satisfying the conditions*

$$f(a)=0, \qquad f'(a)>0, \tag{16.1-1}$$

and assuming every value in the unit disk E: $|w|<1$ exactly once.

In other words, there exists a unique function f mapping D conformally onto E such that the image of a is the origin, and the directions of curves passing through a are unchanged. It was pointed out already in § 5.10 that the hypothesis that D is not the whole plane is necessary for the truth of the theorem. By Liouville's theorem, no such mapping can exist if $D=\mathbb{C}$.

As to the *proof* of the mapping theorem, we recall that the *uniqueness* of the mapping function was already proved in § 5.10, using the lemma of Schwarz (Theorem 5.10b). It thus remains to demonstrate the *existence* of a function f satisfying the required conditions. In addition to what is currently regarded the most elegant proof (which is also the least constructive), we present in § 16.2 a proof that can be used to actually construct the mapping function.

Proof of Theorem 16.1a (*Carathéodory-Ostrowski-Ahlfors*). We consider the family $\mathcal{F}$ of all functions g satisfying the following conditions:

(i) g is analytic and one-to-one in D.
(ii) $|g(z)|\leqq 1$ for $z\in D$.
(iii) $g(a)=0$, $g'(a)>0$.

We show:

(a) $\mathcal{F}$ is not empty.
(b) There exists an $f\in\mathcal{F}$ such that

$$g'(a)<f'(a) \quad \text{for all} \quad g\in\mathcal{F}, \quad g\neq f.$$

(c) The function f determined by (b) is the desired mapping function.

Within the family $\mathcal{F}$, the mapping function thus is characterized by the property that it maximizes the derivative at a.

To deal with (a), we recall from Theorem 4.3f that if h is analytic and $\neq 0$ in a simply connected region D, then a single-valued and analytic branch of $\log h$ can be defined in D. By letting $\sqrt{h}=e^{(\log h)/2}$, we can also define a single-valued and analytic branch of $\sqrt{h}$. (By this is meant nothing more than a function s that is analytic in D and satisfies $s^2=h$.) In the present context—and it is here that we use the hypothesis that $D\neq\mathbb{C}$—let $b\in\mathbb{C}$, $b\notin D$. Then $z-b\neq 0$ in D, and one can define an analytic branch $s(z)$ of $\sqrt{z-b}$. We assert that $s(z)$, in addition to being one-to-one in D, does not take any two values that lie symmetric to O. This is so because $s(z_1)=-s(z_2)$ upon squaring implies $z_1-b=z_2-b$ and hence $z_1=z_2$. The function s takes the value $w_0:=s(a)\neq 0$; hence it does not take the value $-w_0$. Let N be a neighborhood of a, $N \subset D$. By the local mapping theorem (Theorem 2.4f) there exists $\delta > 0$ such that s assumes in N all values w such that $|w - w_0| < \delta$. It therefore does not assume any value of the disk $|w - (-w_0)| = |w + w_0| < \delta$. There follows $|s(z) + w_0| \geqq \delta$ in D. Consequently,

$$h(z) := \frac{1}{s(z) + w_0}$$

is a one-to-one function in D that satisfies $|h(z)| \leqq \delta^{-1}$ there. Using h it is now easy to construct a function $g \in \mathcal{F}$. We map a onto 0 by subtracting $h(a)$, and we make the derivative positive by multiplying by $|h'(a)|/h'(a)$. It remains only to adjust the size. Evidently

$$g(z) := \frac{\delta}{2} \frac{|h'(a)|}{h'(a)} [h(z) - h(a)]$$

satisfies all conditions (i), (ii), and (iii).

To deal with (b), let

$$\beta := \sup_{g \in \mathcal{F}} g'(a).$$

Whether β is finite or not, there exists a sequence $\{g_n\}$ of functions of $\mathcal{F}$ such that $g_n'(a) \to \beta$. The family $\mathcal{F}$ consists of bounded functions and hence is *normal* by Montel's theorem (Theorem 12.8a). Hence by the definition of normality we can extract from the sequence $\{g_n\}$ a subsequence $\{g_{n_k}\}$ that converges to an analytic function f, uniformly on compact subsets of D. (This is a nonconstructive element in the proof.) From $|g_{n_k}(z)| \leqq 1$ there follows $|f(z)| \leqq 1$ for all $z \in D$, and in view of $g_{n_k}(a) = 0$, we have $f(a) = 0$. By the construction of the sequence $\{g_n\}$ we furthermore have $f'(a) = \beta$.

We now must show that f is one-to-one. It is clear that f is not constant, because $f'(0) = \beta > 0$. Let z_1 be a point in D, and consider the family $\mathcal{F}_1$ of functions $g_1(z) := g(z) - g(z_1)$, where $g \in \mathcal{F}$. The family $\mathcal{F}_1$ is bounded, hence again normal. In the region D_1 obtained from D by omitting the point z_1, $g_1(z) \neq 0$ for all $g_1 \in \mathcal{F}_1$. It follows from the theorem of Hurwitz (Corollary 4.10f) that any limit function of functions in $\mathcal{F}_1$ is either different from zero in D_1 or identically zero. But $f(z) - f(z_1)$ is a limit function, and it is not identically zero. Hence $f(z) \neq f(z_1)$ for all $z \in D_1$. Because $z_1 \in D_1$ was arbitrary, we have proved that f is one-to-one.

It remains to verify (c). The proof that f assumes every value in $|w| < 1$ is indirect. Let $E_0 := f(D)$. E_0 is simply connected by Theorem 3.5e. Assuming that $E_0 \neq E$, we shall arrive at a contradiction by constructing a function g which is in $\mathcal{F}$ and satisfies $g'(0) > \beta$, thus violating the defining property of β.

The definition of g is based on an auxiliary function, called the **Koebe function**, that is required also for a more constructive proof of the mapping theorem to be presented in § 16.2. If $E_0 \neq E$, there exists $b_0 \in E$ such that $b_0 \notin E_0$. Because $0 = f(a) \in f(D)$, we also have $b_0 \neq 0$. If $b_0 = -\rho\, e^{i\phi}$, then the rotation

$$w \mapsto e^{-i\phi} w$$

places b_0 at $-\rho$, a point on the negative real axis. Now the mapping

$$w \mapsto \frac{w+\rho}{1+\rho w}$$

is one-to-one in E, hence in $e^{-i\phi}E_0$, and it omits the value 0 in $e^{-i\phi}E_0$. Thus we can define in $e^{-i\phi}E_0$ an analytic branch of its square root,

$$s(w) = \sqrt{\frac{w+\rho}{1+\rho w}}.$$

This is made unique by requiring that

$$s(0) = \sqrt{\rho} > 0.$$

The function s is one-to-one in $e^{-i\phi}E_0$ and obviously satisfies $s(e^{-i\phi}E_0) \subset E$. The same holds for the function

$$k(w, \rho) := \frac{s(w) - \sqrt{\rho}}{1 - \sqrt{\rho}\, s(w)}.$$

Moreover, $k(0, \rho) = 0$. The function $w \mapsto k(w, \rho)$ is called the **Koebe function** formed with the parameter ρ. It should be noted, however, that the function also depends on the region $e^{-i\phi}E_0$, which entered into the definition of $s(w)$. For the following the value of $k'(0, \rho)$ (derivative with respect to w) will be essential. From

$$k'(w, \rho) = \frac{1-\rho}{[1 - \sqrt{\rho}\, s(w)]^2} s'(w),$$

$$2s(w)s'(w) = \frac{1-\rho^2}{(1+\rho w)^2},$$

we have in view of $s(0) = \sqrt{\rho}$,

$$k'(0, \rho) = \frac{1-\rho^2}{1-\rho} \cdot \frac{1}{2\sqrt{\rho}} = \frac{1+\rho}{2\sqrt{\rho}}. \tag{16.1-2}$$

Thus by the inequality of the arithmetic and the geometric means,

$$k'(0, \rho) > 1. \tag{16.1-3}$$

Now let

$$g(z) := e^{i\phi} k(e^{-i\phi} f(z), \rho).$$

Clearly, g is analytic and one-to-one in D, and its values lie in $|w|<1$. From $f(a)=0$ and $k(0,\rho)=0$ there follows $g(a)=0$. Furthermore,

$$g'(a)=e^{i\phi}k'(0,\rho)\,e^{-i\phi}f'(a)=k'(0,\rho)\beta,$$

which is positive. Thus g satisfies conditions (i), (ii), and (iii) and hence is in $\mathscr{F}$. However, in view of (16.1-3), $g'(a)>\beta$, contradicting the definition of β. Thus our assumption that f does not take every value in E is ill-founded, and f is indeed the desired mapping function. —

PROBLEM

1. Show that the inverse $k^{[-1]}$ of the Koebe function $k(z,\rho)$ can be extended to a function that is analytic (but not one-to-one) in E and satisfies $k^{[-1]}(E,\rho)\subset E$.

NOTES

The above proof of the Riemann mapping theorem has been current in textbooks for half a century; see Hurwitz [1929], p. 394; Bieberbach [1934], p. 6; Titchmarsh [1939], p. 207; Hille [1962], p. 320 (historical remarks); Ahlfors [1966], p. 221; Conway [1973], p. 156; and Jänich [1977], p. 152.

§ 16.2. OSCULATION METHODS

The proof of the Riemann mapping theorem given in § 16.1 does not permit the actual construction of the mapping function, for the following reasons:

(a) There is no prescription for constructing a sequence $\{f_n\}$ such that $f'_n(a)\to\beta$.

(b) The process of selecting a convergent subsequence from the sequence $\{f_n\}$ cannot actually be carried out.

In this section we give a prescription for constructing sequences $\{f_n\}$ of functions that are guaranteed to converge to the desired mapping theorem, and at the same time, a family of methods for obtaining the mapping function numerically. These methods are variants of the Schmiegungsverfahren (osculation method) invented by P. Koebe around 1910. They are more generally applicable than any of the methods to be discussed later inasmuch as they require no regularity whatsoever of the boundary of the region to be mapped. The price to be paid is slow convergence.

The osculation methods assume that the region D to be mapped is already contained in the unit disk E, and that the point a where the normalizing condition is given coincides with O. This is easily achieved by the preliminary maps that were discussed in connection with the proof given in § 16.1. The idea then is to push the boundary of D closer to the unit circle at each iteration step—hence the name osculation method.

I. Osculation Families

Let $\mathscr{D}$ be the set of all simply connected regions $D \subset E$ such that $0 \in D$, and for each $\rho \in (0, 1]$, let $\mathscr{D}(\rho)$ denote the subset of $\mathscr{D}$ consisting of all $D \in \mathscr{D}$ such that $\min\{|z|: \quad z \notin D\} = \rho$.

For each $D \in \mathscr{D}$, $D \neq E$, let there be given a nonempty family $\mathscr{F}(D)$ of conformal maps h of D with the following properties:

(A) For each $h \in \mathscr{F}(D)$ there holds $h(D) \in \mathscr{D}$, and

$$h(0) = 0, \qquad h'(0) > 0. \tag{16.2-1}$$

(B) The inverse map $h^{[-1]}$ can be extended analytically (but not necessarily as a one-to-one function) to E, and $h^{[-1]}(E) \subset E$.

(C) For any closed interval $I := [0, \beta]$ with $\beta < 1$ the function

$$\gamma(\rho) := \inf_{D \in \mathscr{D}(\rho)} \inf_{h \in \mathscr{F}(D)} h'(0) \tag{16.2-2}$$

satisfies

$$\inf_{\rho \in I} \gamma(\rho) > 1. \tag{16.2-3}$$

A family $\mathscr{F} = \bigcup_{D \in \mathscr{D}} \mathscr{F}(D)$ with the properties (A), (B), and (C) above is called an **osculation family**, and the function $\gamma(\rho)$ defined by (16.2-2) is the **dilatation measure** of $\mathscr{F}$.

EXAMPLE **1**

A simple osculation family is obtained by letting, for each $D \in \mathscr{D}$, $D \neq E$, $\mathscr{F}(D)$ consist of the Koebe function $k(z, \rho)$ already used in the proof of the Riemann mapping theorem. This is the special case $m = 2$ of the osculation family considered in Example 2.

EXAMPLE **2**

Let $D \in \mathscr{D}(\rho)$, $0 < \rho < 1$, and let z^* be a point of ∂D closest to the origin. By a preliminary rotation, we may assume that z^* lies on the negative real axis, $z^* = -\rho$. By

$$z \mapsto z_1 := \frac{z + \rho}{1 + \rho z}$$

D is mapped onto a simply connected region $D_1 \subset E$, which contains $z_1 = \rho$ and which has 0 as a boundary point. Thus for any integer $m \geqq 2$, the map

$$z_1 \mapsto z_2 := z_1^{1/m}$$

can be defined as a one-to-one analytic function in D_1 whose range D_2 is made unique by requiring that $\rho \mapsto \rho^{1/m} > 0$. Finally,

$$z_2 \mapsto w := \frac{z_2 - \rho^{1/m}}{1 - \rho^{1/m} z_2}$$

maps D_2 onto a region $D_3 \in \mathscr{D}$. The composition of the foregoing three maps,

$$z \mapsto w = h(z) := \frac{[(z+\rho)/(1+\rho z)]^{1/m} - \rho^{1/m}}{1 - \rho^{1/m}[(z+\rho)/(1+\rho z)]^{1/m}},$$

clearly satisfies (A). The inverse map is calculated to be

$$z = h^{[-1]}(w) = \frac{[(w+\rho^{1/m})/(1+\rho^{1/m} 2)]^m - \rho}{1 - \rho[(w+\rho^{1/m})/(1+\rho^{1/m} w)]^m},$$

and this evidently satisfies (B). Since

$$h'(0) = \frac{1}{m} \frac{\rho^{-1} - \rho}{\rho^{-1/m} - \rho^{1/m}}$$

independently of $D \in \mathscr{D}(\rho)$, it follows that

$$\gamma(\rho) = \frac{1}{m} \frac{\rho^{-1} - \rho}{\rho^{-1/m} - \rho^{1/m}}.$$

For $m = 2$,

$$\gamma(\rho) = \frac{1+\rho}{2\sqrt{\rho}} > 1,$$

and becasue $m(\rho^{-1/m} - \rho^{1/m})$ decreases to its limit $2 \operatorname{Log} 1/\rho$ as $m \to \infty$, it follows that (C) is satisfied for every $m \geqq 2$. Thus by letting $\mathscr{F}(D)$ consist of functions h where $m \geqq 2$, another osculation family is obtained.

EXAMPLE 3

The limit $m \to \infty$ of the functions h in Example 2 is

$$h(z) = \frac{\operatorname{Log} \rho - \log[(z+\rho)/(1+\rho z)]}{\operatorname{Log} \rho + \log[(z+\rho)/(1+\rho z)]},$$

where $\log[(z+\rho)/(1+\rho z)]$ is analytic in D and has the value $\operatorname{Log} \rho$ for $z = 0$. This again satisfies (A). The inverse function is

$$z = h^{[-1]}(w) = \frac{\rho^{(1-w)/(1+w)} - \rho}{1 - \rho \cdot \rho^{(1-w)/(1+w)}},$$

and this clearly satisfies (B). Computation shows

$$\gamma(\rho) = \inf_{D \in \mathscr{D}(\rho)} h'(0) = -\frac{\rho^{-1} - \rho}{2 \operatorname{Log} \rho^{-1}},$$

which is >1 for $0<\rho<1$.

II. Augmentation of an Osculation Family

Here we show that it is sometimes possible to add functions to a set $\mathscr{F}(D)$ without changing the dilatation measure $\gamma(\rho)$.

THEOREM 16.2a. *Let $D_1 \in \mathscr{D}(\rho)$, where $0<\rho<1$, and let f be the normalized function mapping D_1 onto E. Then f may be added to any set $\mathscr{F}(D)$ where $D \in \mathscr{D}(\rho)$, $D \subset D_1$, without changing $\gamma(\rho)$.*

Proof. Clearly, $f(D) \in \mathscr{D}$, and properties (A) and (B) are obvious for the restriction of f to any set $D \subset D_1$. To establish (C), let h_1 be any function in $\mathscr{F}(D_1)$. Then $h_1 \circ f^{[-1]}$ is analytic on E, zero at 0, and its values are in E. Thus by the Schwarz lemma,

$$|h_1 \circ f^{[-1]}(w)| \leqq |w|, \qquad w \in E,$$

$$h_1'(0) f^{[-1]\prime}(0) \leqq 1.$$

Hence,

$$h_1'(0) \leqq f'(0),$$

and since $h_1'(0) \geqq \gamma(\rho)$, the conclusion follows. —

EXAMPLE **4. The crescent map**

Let $D \in \mathscr{D}(\rho)$ be contained in the set $D_1 := E \backslash K \in \mathscr{D}(\rho)$, where K is a disk that intersects the boundary of E. Then $\mathscr{F}(D)$ may be augmented by the function f_1 (called the crescent map) mapping D_1 onto E. This map may be constructed elementarily as follows (see Fig. 16.2a). Let $c = e^{i\gamma}$ and $\bar{c} = e^{-i\gamma}$ be the points where the boundaries of E and of K intersect, and let $-\rho$ be the point in E where the boundary of K intersects the real axis, $0<\rho<1$. Then the Moebius transformation

$$z \mapsto z' = t(z) := \frac{z-c}{z-\bar{c}} \, \frac{\rho+\bar{c}}{\rho+c} \tag{16.2-4a}$$

maps D_1 onto the wedge $0 < \arg z' < \alpha$, where α and c are linked by

$$c = -e^{-i\alpha} \frac{1+e^{i\alpha}}{1+e^{-i\alpha}}.$$

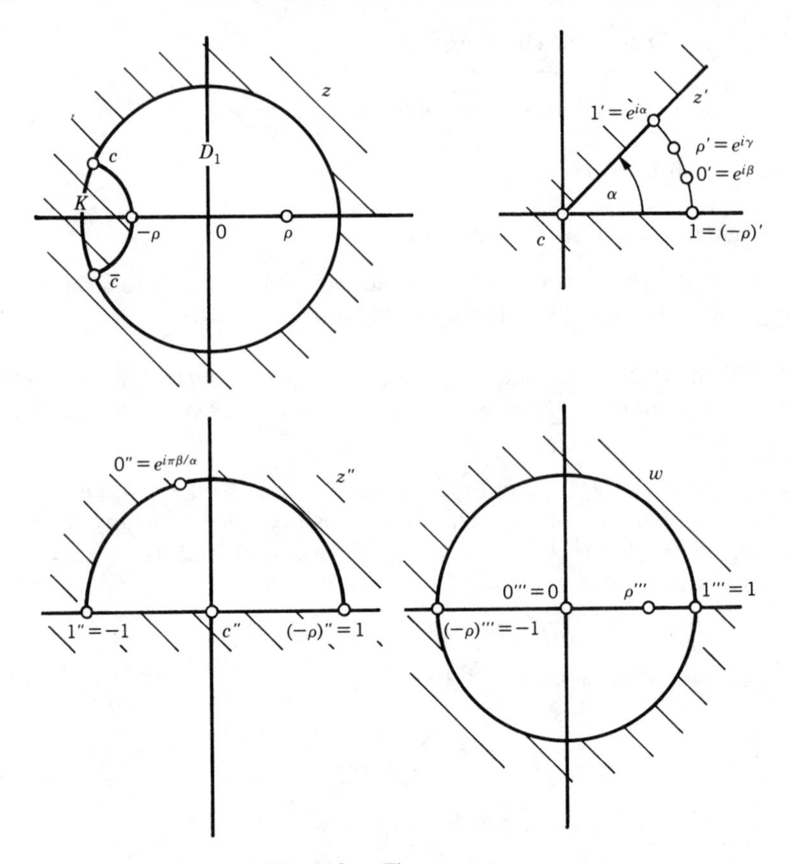

Fig. 16.2a. The crescent map.

The points $-\rho, 0, \rho, 1$ are mapped onto

$$(-\rho)' = 1,$$

$$0' = e^{i\beta} := \frac{1+e^{i\alpha}}{1+e^{-i\alpha}}$$

$$\rho' = e^{i\gamma} := \frac{1+\rho^2+2\rho\, e^{i\alpha}}{1+\rho^2+2\rho\, e^{-i\alpha}},$$

$$1' := e^{i\alpha},$$

respectively. The map

$$z' \mapsto z'' := (z')^{\pi/\alpha} \tag{16.2-4b}$$

opens the wedge into the upper half-plane, sending the four points to

$$1, \qquad e^{i\pi\beta/\alpha}, \qquad e^{i\pi\gamma/\alpha}, \qquad -1,$$

respectively. The final map

$$z'' \mapsto w := \frac{z'' - e^{i\pi\beta/\alpha}}{z'' - e^{-i\pi\beta/\alpha}} \cdot e^{-i\pi\beta/\alpha} \tag{16.2-4c}$$

maps the upper half-plane onto E while sending the four special points mentioned into

$$-1, \qquad 0, \qquad \rho''' := \frac{\sin[\pi(\gamma-\beta)/2\alpha]}{\sin[\pi(\gamma+\beta)/2\alpha]}, \qquad +1.$$

The complete crescent map is obtained by composing the three maps (16.2-4).

EXAMPLE **5. The slit map**

This map is applicable to any $D \in \mathcal{D}(\rho)$ that does not contain the straight line segment from -1 to $-\rho$. The map is again elementary (see § 5.11). It is given by

$$f_1(z) := \frac{s(z) - 1 + z}{s(z) + 1 - z}, \tag{16.2-5}$$

where

$$s(z) := \sqrt{(1+\rho z)(1+\rho^{-1} z)}, \qquad s(0) = 1.$$

It is of interest to note that

$$f_1'(0) = \frac{(1+\rho)^2}{4\rho} \tag{16.2-6}$$

is just the square of the derivative at 0 of the Koebe map.

III. An Upper Bound for the Dilatation Measure

For any $\rho \in (0, 1)$, let D_ρ denote the region considered in Example 5 (E with segment $[-1, -\rho]$ removed). Let $\mathcal{F}$ be an osculation family. Since, by Theorem 16.2a, the slit map may be added to $\mathcal{F}(D_\rho)$ without decreasing the dilation measure $\gamma(\rho)$ of $\mathcal{F}$, there follows in view of (16.2-6):

THEOREM 16.2b. *For any osculation family $\mathcal{F}$, the dilatation measure $\gamma(\rho)$ satisfies*

$$\gamma(\rho) \leqq \frac{(1+\rho)^2}{4\rho}. \tag{16.2-7}$$

If $\rho = 1-\varepsilon$, (16.2-7) means the same as

$$\gamma(1-\varepsilon) \leqq 1 + \tfrac{1}{4}\varepsilon^2(1-\varepsilon)^{-1},$$

from which we conclude:

COROLLARY 16.2c. *No dilatation measure can satisfy an inequality*

$$\gamma(1-\varepsilon) \geqq 1 + \alpha\varepsilon^2$$

for all $\varepsilon \in (0, 1)$, *where* $\alpha > \frac{1}{4}$.

The question arises whether the inequality (16.2-7) is best possible, or in other words, whether there exists an osculation family such that its dilatation measure satisfies

$$\gamma(\rho) = \frac{(1+\rho)^2}{4\rho}.$$

This is answered by the following example.

EXAMPLE **6. The Riemann family**

Here each set $\mathscr{F}(D)$ consists of a single element only, namely, of the normalized Riemann map of D onto E. It is clear that the Riemann family satisfies (A) and (B). To compute its dilation measure, let $D \in \mathscr{D}(\rho)$. Without loss of generality we assume that $z = -\rho$ is a point of ∂D closest to O. Let f be the normalized Riemann map of D, and let k denote the function

$$k(z) := \frac{z}{(1-z)^2}$$

familiar from the theory of univalent functions (see § 19.1). It is known that k defines a one-to-one map of E. Consider

$$g(w) := f'(0) \cdot k \circ f^{[-1]}(w).$$

This defines a one-to-one map of E satisfying $g'(0) = 1$. By the Koebe $\frac{1}{4}$ theorem (Theorem 19.1i) the image of E under g contains the disk $|z| < \frac{1}{4}$. Thus if $\{w_n\}$ is a sequence of points in E such that $z_n := f^{[-1]}(w_n) \to -\rho$, then

$$f'(0)|k(-\rho)| = \lim_{n\to\infty} f'(0)|k(z_n)| = \lim_{n\to\infty} |g(w_n)| = \tfrac{1}{4}.$$

In view of $k(-\rho) = -\rho(1+\rho)^{-2}$ this implies

$$f'(0) \geqq \frac{(1+\rho)^2}{4\rho}.$$

Equality is attained for the mapping function of Example 5. Thus for the Riemann family, equality holds in (16.2-7).

IV. The Osculation Algorithm

Let $\mathscr{F}$ be an osculation family, and let $h \in \mathscr{F}(D)$, where $D \in \mathscr{D}(\rho)$, $\rho < 1$. By (A) and (B), the function $h^{[-1]}$ satisfies the hypotheses of the Schwarz lemma. By (C), $h^{[-1]\prime}(0) \leqq \gamma(\rho)^{-1} < 1$; thus $h^{[-1]}$ is not a rotation. There follows

$$|h^{[-1]}(w)| < |w|, \qquad w \in E, \quad w \neq 0. \tag{16.2-8}$$

Letting $w = h(z)$, where $z \in D$, we obtain the first assertion of:

LEMMA 16.2d. *For $z \in D$, $z \neq 0$,*

$$|h(z)| > |z|; \tag{16.2-9}$$

furthermore

$$h(D) \in \mathscr{D}(\rho'), \tag{16.2-10}$$

where $\rho' > \rho$.

Proof of (16.2-10). This does not directly follow from (16.2-9), because h need not be defined on ∂D, and by taking limits, the inequality might degenerate into an equality. However, let w^* be a boundary point of $h(D)$ closest to O, so that $|w^*| = \rho'$. If $\rho' = 1$, then (16.2-10) is clear. If $w^* \in E$, let $\{w_k\}$ be a sequence of points in $h(D)$ that converges to w^*. Let $z_k := h^{[-1]}(w_k)$. Because $h^{[-1]}$ is analytic at w^*, $\{z_k\}$ converges, and the limit $z^* := h^{[-1]}(w^*)$ is necessarily a boundary point of D. By (16.2-8) there follows

$$\rho \leqq |z^*| < |w^*| \leqq \rho',$$

proving (16.2-10). —

Given an osculation family $\mathscr{F}$, the following **osculation algorithm** for mapping a given region $D \in \mathscr{D}$ onto E now makes sense. Let $D_1 := D \in \mathscr{D}(\rho_1)$. Select $h_1 \in \mathscr{F}(D_1)$ and put $f_1 := h_1$. By virtue of Lemma 16.2d,

$$D_2 := h_1(D_1) = f_1(D_1) \in \mathscr{D}(\rho_2),$$

where $\rho_2 > \rho_1$. Generally, having constructed f_{n-1} and having obtained the region $D_n := f_{n-1}(D_1)$, we select $h_n \in \mathscr{F}(D_n)$ and put $f_n := h_n \circ f_{n-1}$. Again by (16.2-10),

$$D_{n+1} = h_n(D_n) = f_n(D_1) \in \mathscr{D}(\rho_{n+1}),$$

where $\rho_{n+1} > \rho_n$. The functions f_n are also given by

$$f_n = h_n \circ h_{n-1} \circ \cdots \circ h_1, \qquad n = 1, 2, \ldots.$$

By virtue of (A) they satisfy

$$f_n(0) = 0. \tag{16.2-11}$$

Furthermore, using the chain rule,

$$f_n'(0) = h_n'(0)h_{n-1}'(0) \cdots h_1'(0) > 0. \tag{16.2-12}$$

All f_n thus are members of the family $\mathscr{F}$ considered in the proof of Theorem 16.1a. In fact, we have:

THEOREM 16.2e. *The whole sequence* $\{f_n\}$ *(and not merely a selected subsequence) converges. The convergence is locally uniform on* D_1, *and the limit function* f *maps* D_1 *conformally onto* E *such that* $f(0) = 0, f'(0) > 0$.

Proof. Each function $f_n(\rho_1 z)$ maps E into E and 0 onto 0. Therefore by the Schwarz lemma,

$$f_n'(0) < \frac{1}{\rho_1}, \qquad n = 1, 2, \ldots.$$

By (16.2-12), using (C), there follows

$$\gamma(\rho_1)\gamma(\rho_2) \cdots \gamma(\rho_n) < \frac{1}{\rho_1}, \qquad n = 1, 2, \ldots. \tag{16.2-13}$$

Since all $\gamma(\rho_k) \geqq 1$, there necessarily holds

$$\lim_{k \to \infty} \gamma(\rho_k) = 1,$$

which by (C) implies

$$\lim_{k \to \infty} \rho_k = 1.$$

The regions D_k thus tend to E. To prove the uniform convergence of the sequence $\{f_n\}$ on compact subsets $\hat{D}_1 \subset D_1$, we use a version of Harnack's theorem (compare Theorem 15.3h and Problem 1, § 15.3), which asserts the existence of a constant μ (depending only on D_1 and on $\hat{D}_1$) such that for any function p that is analytic in D_1, has a positive real part, and satisfies $\operatorname{Im} p(0) = 0$, there holds

$$|p(z)| \leqq \mu p(0) \quad \text{for all} \quad z \in \hat{D}_1.$$

To obtain a function with positive real part, we note that for $n > m$,

$$\frac{f_n(z)}{f_m(z)} = \frac{1}{f_m(z)} h_n \circ h_{n-1} \circ \cdots \circ h_{m+1}(f_m(z))$$

is analytic and $\neq 0$ in D_1, because both numerator and denominator vanish only at $z = 0$, where they have a simple zero. Moreover by Lemma 16.2d,

$$\left|\frac{f_n(z)}{f_m(z)}\right| > 1, \qquad z \in D_1. \tag{16.2-14}$$

Because D_1 is simply connected, it is possible to define an analytic logarithm

$$p(z) := \log \frac{f_n(z)}{f_m(z)},$$

which is made unique by requiring that

$$p(0) = \operatorname{Log} \frac{f'_n(0)}{f'_m(0)}$$

be real. In view of (16.2-13), $\operatorname{Re} p(z) > 0$. The function $f_n \circ f_m^{[-1]}$ maps a disk of radius ρ_{m+1} onto E and leaves the origin fixed, hence by the Schwarz lemma,

$$(f_n \circ f_m^{[-1]})'(0) = \frac{f'_n(0)}{f'_m(0)} < \frac{1}{\rho_{m+1}} < \frac{1}{\rho_m}.$$

By Harnack's theorem we thus conclude

$$\left| \log \frac{f_n(z)}{f_m(z)} \right| \leqq \mu \operatorname{Log} \frac{1}{\rho_m}, \qquad z \in \hat{D}_1.$$

To derive from this an estimate for $|f_n - f_m|$, we use the elementary fact that for arbitrary $\eta > 0$ and arbitrary complex u such that $|u| \leqq \eta$,

$$|e^u - 1| \leqq \eta\, e^{\eta}.$$

(This follows in view of

$$|e^u - 1| = \left| \int_0^u e^t\, dt \right| \leqq \int_0^u |e^t|\,|dt| \leqq e^{|u|}|u|.)$$

Using this with $u := p(z)$, $\eta = \mu \operatorname{Log} \rho_m^{-1}$, we get

$$\left| \frac{f_n(z)}{f_m(z)} - 1 \right| \leqq \frac{\mu}{\rho_m^{\mu}} \operatorname{Log} \rho_m^{-1}, \qquad z \in \hat{D}_1,$$

which on multiplying by $f_m(z)$, observing that $|f_m(z)| < 1$ and using the elementary inequality

$$\operatorname{Log} \frac{1}{\xi} \leqq \frac{1}{\xi} - 1, \qquad \xi > 0,$$

yields

$$|f_n(z) - f_m(z)| \leqq \frac{\mu}{\rho_m^{\mu+1}} (1 - \rho_m). \tag{16.2-15}$$

Since $\rho_m \to 1$, the expression on the right can be made as small as we please by choosing m large enough. The sequence $\{f_n\}$ for $z \in \hat{D}_1$ thus satisfies the Cauchy criterion, and the limit function f thus exists.

By the theorem of Hurwitz we see as in the proof of Theorem 16.1a that the limit function f is one-to-one. That f assumes every value in E follows from the fact that the domain of values of f contains every disk $|z| < \rho_m$, and thus by virtue of $\rho_m \to 1$ is the unit disk. This completes our constructive proof of the Riemann mapping theorem. —

V. An Estimate for the Speed of Convergence

A first indication of the speed of convergence of the osculation algorithm can be gleaned from the convergence of the product (16.2-12). Since the partial products are bounded by a quantity that does not depend on the osculation family chosen, it follows that convergence will be largest if the h_n are chosen such that $h_n'(0)$ is as large as possible. This explains the speedup of convergence, observed experimentally by Grassmann [1979], if maps such as the crescent and the slit map (see Examples 4 and 5) are used whenever applicable. Our observation would also predict that the logarithmic osculation family of Example 3 should yield faster convergence than the Koebe family. This as yet remains to be verified.

More precise statements concerning the speed of convergence are possible if the osculation family $\mathscr{F}$ is such that its dilatation measure $\gamma(\rho)$ decreases for increasing ρ. (All our examples are of this kind.) For convenience we set

$$\rho = 1 - \varepsilon,$$
$$\rho_n = 1 - \varepsilon_n, \qquad n = 1, 2, \ldots.$$

We then have:

THEOREM 16.2f. *Let the dilatation measure be decreasing, and let there exist $\alpha > 0$ such that*

$$\gamma(1-\varepsilon) \geqq 1 + \alpha\varepsilon^2, \qquad 0 < \varepsilon < 1. \tag{16.2-16}$$

(*We know from Corollary* 16.2c *that necessarily $\alpha \leqq \frac{1}{4}$.*) *Then*

$$\varepsilon_n \leqq \frac{8}{\alpha\rho_1^2}\frac{1}{n}, \qquad n = 1, 2, \ldots. \tag{16.2-17}$$

Proof. Because any D_n could serve as D_1, (16.2-13) implies

$$\gamma(\rho_n)\gamma(\rho_{n+1}) \cdots \gamma(\rho_{n+m}) < \rho_n^{-1}, \qquad n = 1, 2, \ldots, \quad m = 0, 1, \ldots.$$

Because $\{\rho_k\}$ is increasing, the sequence $\{\gamma(\rho_k)\}$ decreases, and there follows

$$[\gamma(\rho_{2n})]^{n+1} < \frac{1}{\rho_n}$$

or

$$\gamma(\rho_{2n}) < \rho_n^{-1/(n+1)} < \rho_n^{-1/n}.$$

By the hypothesis,

$$1 + \alpha\varepsilon_{2n}^2 \leqq \gamma(1-\varepsilon_{2n}) = \gamma(\rho_{2n}).$$

On the other hand, using Taylor's formula for $(1-x)^{-\beta}$,

$$\rho_n^{-1/n} = (1-\varepsilon_n)^{-1/n} \leqq 1 + \frac{1}{n}\rho_n^{-1-1/n}\varepsilon_n \leqq 1 + \frac{1}{n}\rho_1^{-2}\varepsilon_n.$$

There follows for $n = 1, 2, \ldots$,

$$\alpha\varepsilon_{2n}^2 \leqq \frac{1}{n}\rho_1^{-2}\varepsilon_n,$$

$$\alpha\rho_1^2 n\varepsilon_{2n}^2 \leqq \varepsilon_n,$$

$$\tfrac{1}{4}\alpha\rho_1^2(2n)^2\varepsilon_{2n}^2 \leqq n\varepsilon_n,$$

$$(2n)^2\left(\frac{\alpha\rho_1^2}{4}\right)^2\varepsilon_{2n}^2 \leqq n\frac{\alpha\rho_1^2}{4}\varepsilon_n.$$

Letting

$$\xi(n) := \tfrac{1}{4}\alpha\rho_1^2 n\varepsilon_n,$$

we thus see that $[\xi(2n)]^2 \leqq \xi(n)$ and hence that

$$[\xi(2^k)]^{2^k} \leqq \xi(1) = \tfrac{1}{4}\alpha\rho_1^2(1-\rho_1), \qquad k = 0, 1, 2, \ldots.$$

The last expression is $\leqq 1$ by hypothesis on α, and we therefore see that

$$\xi(2^k) \leqq 1, \qquad k = 0, 1, 2, \ldots,$$

or that

$$\varepsilon_n \leqq \frac{4}{\alpha\rho_1^2}\frac{1}{n}$$

whenever n is a power of 2. Since ε_n decreases, there holds, for all $n = 1, 2, \ldots$,

$$\varepsilon_n \leqq \frac{4}{\alpha\rho_1^2}\cdot\frac{1}{n'},$$

where n' is the greatest power of 2 that does not exceed n. In view of $n' > \frac{1}{2}n$, the result (16.2-17) follows. —

EXAMPLE 7

For the Koebe family (case $m = 2$ of Example 2),

$$\gamma(\rho) = \tfrac{1}{2}(\rho^{1/2} + \rho^{-1/2}).$$

Computation shows that

$$\gamma'(1)=0,$$

$$\gamma''(\rho)=\tfrac{1}{8}\rho^{-3/2}(3\rho^{-1}-1)\geqq\tfrac{1}{4}, \qquad 0<\rho\leqq 1.$$

Thus by Taylor's theorem

$$\gamma(1-\varepsilon)\geqq 1+\tfrac{1}{2}\gamma''(1-\theta\varepsilon)\varepsilon^2$$

for some $\theta\in(0,1)$. Hence,

$$\gamma(1-\varepsilon)\geqq 1+\tfrac{1}{8}\varepsilon^2,$$

and the hypothesis of Theorem 16.2f is seen to hold for $\alpha=\frac{1}{8}$. It thus follows that

$$\varepsilon_n\leqq\frac{64}{\rho_1^2}\frac{1}{n}, \qquad n=1,2,\ldots.$$

In a similar manner it can be shown that if the osculation family of Example 2 is used with arbitrary m,

$$\alpha=\tfrac{1}{6}(1-m^{-2}),$$

and if the osculation family of Example 3 is used,

$$\alpha=\tfrac{1}{6}.$$

Thus the error estimate of Theorem 16.2f becomes increasingly favorable with larger values of m and becomes best for $m=\infty$.

VI. Fast Initial Convergence

Theorem 16.2f provides an estimate for the speed of convergence of an osculation algorithm based on an osculation family with a monotonic dilatation function, but the predicted speed of convergence is excruciatingly slow. Numerical experiments do in fact show that the convergence for instance of the Koebe method is ultimately very slow. However, the same experiments also show that in the beginning (where the estimate (16.2-17) is useless because it yields bounds on ε_n that are >1) the convergence of the osculation method is much faster than the predicted $O(1/n)$ rate, especially if augmented osculation families are used. We now present some results aimed at understanding the phenomenon of fast initial convergence.

For any osculation family $\mathcal{F}$, and for $0<\rho<1$, let

$$\psi(\rho):=\inf_{D\in\mathcal{D}(\rho)}\ \inf_{h\in\mathcal{F}(D)}\ \inf_{|z|=\rho}|h(z)|. \qquad (16.2\text{-}18)$$

This is the radius of the largest disk centered at O that can be placed within the image of the circle of radius ρ about O under a map $h\in\mathcal{F}(D)$, where

$D \in \mathscr{D}(\rho)$. It follows from Lemma 16.2d that

$$\psi(\rho) > \rho, \qquad 0 < \rho < 1.$$

The function ψ can be computed, in principle, for any concretely given osculation family, and it usually turns out that ψ is monotonic.

EXAMPLE **8**

For the Koebe family (Example 1), $\psi(\rho)$ was computed by Julia [1931] with the result

$$\psi(\rho) = \sqrt{\rho}\frac{1+\rho}{\sqrt{2(1+\rho^2)}+1-\rho} = \sqrt{\rho}\frac{\sqrt{2(1+\rho^2)}-(1-\rho)}{1+\rho}. \tag{16.2-19}$$

The monotonicity of ψ is evident from the representation

$$[\psi(\rho)]^2 = \rho\left\{1 - \frac{2}{\sqrt{2(1+\rho^2)/(1-\rho)+1}}\right\}.$$

EXAMPLE **9**

For the family considered in Example 3 it can be verified that

$$\psi(\rho) = \frac{\operatorname{Log} 2 - \operatorname{Log}(1+\rho^2)}{\operatorname{Log}(1+\rho^{-2}) - \operatorname{Log} 2}. \tag{16.2-20}$$

Monotonicity follows by differentiation. —

If $D \in \mathscr{D}(\rho)$ and $h \in \mathscr{F}(D)$, then by the definition of ψ, $h(D) \in \mathscr{D}(\rho')$, where $\rho' \geqq \psi(\rho)$. Thus there follows:

THEOREM 16.2g. *If the function ψ defined by* (16.2-18) *is monotonic, then the radii ρ_n generated by an osculation algorithm based on $\mathscr{F}$ satisfy*

$$\rho_n \geqq \rho_n^*, \qquad n = 1, 2, \ldots,$$

where

$$\rho_1^* = \rho_1, \qquad \rho_{n+1}^* = \psi(\rho_n^*), \qquad n = 1, 2, \ldots. \tag{16.2-21}$$

In both examples considered, $\psi(0) = 0$ and $\psi'(0) = \infty$. Thus for small ρ, $\psi(\rho)/\rho$ is large, and the iteration sequence (16.2-21) increases rapidly in the beginning if ρ_1 is sufficiently small. For instance, if $\rho_1^* = 0.1$, only ten iterations are necessary to achieve $\rho_n^* > 0.4$ if the iteration function (16.2-19) is used, and only seven if the function (16.2-20) is used.

The computations required to obtain the explicit formulas for $\psi(\rho)$ given in the Examples 8 and 9 require a detailed knowledge of the functions h

of the osculation family and thus can be tedious. It would be desirable to compute, or at least to estimate, the function ψ solely on the basis of the dilatation measure of $\mathscr{F}$. In the following such an estimate is given for the more general function

$$\psi(\rho, \sigma) := \inf_{D \in \mathscr{D}(\rho)} \inf_{h \in \mathscr{F}(D)} \inf_{|z|=\sigma} |h(z)|, \tag{16.2-22}$$

where $0<\rho<1$, $0<\sigma \leqq \rho$. This function is required for the discussion of osculation algorithms for mapping doubly connected regions (see § 17.2).

It is easy to give an upper bound for ψ. Let $h \in \mathscr{F}(D)$, where $D \in \mathscr{D}(\rho)$. Because $z^{-1}h(z)$ is analytic and $\neq 0$ for $|z|<\rho$, the minimum of its modulus on every disk $|z| \leqq \sigma \leqq \rho$ is assumed on the boundary. The value at 0 being $h'(0)$, there follows

$$\inf_{|z|=\sigma} \frac{|h(z)|}{\sigma} \leqq h'(0).$$

Taking the infimum with respect to all $h \in \mathscr{F}(D)$ and all $D \in \mathscr{D}(\rho)$, there follows

$$\psi(\rho, \sigma) \leqq \sigma\gamma(\rho).$$

In particular,

$$\psi(\rho) = \psi(\rho, \rho) \leqq \rho\gamma(\rho),$$

and by virtue of Theorem 16.2b there follows

$$\psi(\rho) \leqq \tfrac{1}{4}(1+\rho)^2.$$

If $\rho = 1-\varepsilon$, this means the same as

$$\psi(1-\varepsilon) \leqq 1-\varepsilon+\tfrac{1}{4}\varepsilon^2. \tag{16.2-23}$$

It follows from the known behavior of an iteration sequence formed with an iteration function satisfying (16.2-23) that the sequence $\{\rho_n^*\}$ formed with ψ will at best satisfy $1-\rho_n^* = O(1/n)$. Thus for large n, no qualitative improvement over Theorem 16.2f results from our present approach.

As to lower bounds for ψ, Lemma 16.2d implies the trivial bound

$$\psi(\rho, \sigma) \geqq \sigma.$$

To state a less trivial bound, we consider, for given (ρ, σ) such that $0<\rho<1$, $0<\sigma \leqq \rho$, the equation for ξ,

$$\xi = \sigma\gamma^{(1-\gamma\xi)/(1+\gamma\xi)}, \tag{16.2-24}$$

where $\gamma := \gamma(\rho)$. As a function of ξ, the function on the right in the interval $[0, \gamma^{-1}]$ strictly decreases from $\sigma\gamma$ to σ. It thus is clear that (16.2-24) has a unique solution $\xi \in (0, \gamma^{-1})$. Let this solution be denoted by $\psi_0 = \psi_0(\rho, \sigma)$. By virtue of $\gamma^{-1} > \rho$, ψ_0 satisfies $\psi_0(\rho, \sigma) > \sigma$ for every $\sigma \in (0, \rho]$. Moreover for each fixed ρ, $\psi_0(\rho, \sigma)$ as a function of σ increases from $\psi_0(\rho, 0) = 0$ to

$$\psi_0(\rho) := \psi_0(\rho, \rho).$$

THEOREM 16.2h. *Let $\mathcal{F}$ be an osculation family with dilatation measure $\gamma(\rho)$. Then for $0 < \rho < 1$, $0 < \sigma \leqq \rho$,*

$$\psi(\rho, \sigma) \geqq \psi_0(\rho, \sigma). \tag{16.2-25}$$

Proof. Let $h \in \mathcal{F}(D)$ where $D \in \mathcal{D}(\rho)$. Then the function

$$w \mapsto \operatorname{Log}\left|\frac{w}{h^{[-1]}(w)}\right| \tag{16.2-26}$$

is harmonic and, by (16.2-9), positive at $w = 0$ and at every point $w \in E$ where $h^{[-1]}(w) \neq 0$. It is asserted that this set includes the disk

$$|w| < \frac{1}{h'(0)}.$$

Indeed, let $h^{[-1]}(w_1) = 0$, $w_1 \in E$, $w_1 \neq 0$. Then the modulus of the function

$$w \to \frac{h^{[-1]}(w)}{w} \frac{1 - \overline{w_1} w}{w - w_1}$$

is $\leqq 1$ on $|w| = 1$, and consequently also at 0. There follows

$$\frac{1}{h'(0)} \cdot \frac{1}{|w_1|} \leqq 1,$$

and consequently $|w_1| \geqq 1/h'(0)$. We may thus apply Harnack's inequality (Lemma 15.3g) to (16.2-26) in the disk $|w| \leqq \chi^{-1}$, where $\chi := h'(0)$. The result is

$$\operatorname{Log}\left|\frac{w}{h^{[-1]}(w)}\right| \geqq \frac{1 - \chi|w|}{1 + \chi|w|} \operatorname{Log} \chi.$$

For any w such that $|h^{[-1]}(w)| = \sigma$ there follows

$$\operatorname{Log}\frac{|w|}{\sigma} \geqq \frac{1 - \chi|w|}{1 + \chi|w|} \operatorname{Log} \chi. \tag{16.2-27}$$

In particular this holds if w, depending on h, is chosen such that

$$|w| = \inf_{|z|=\sigma} |h(z)|.$$

We now consider a sequence of regions $D_n \in \mathcal{D}(\rho)$ and functions $h_n \in \mathcal{F}(D_n)$ such that $h_n'(0) \to \gamma(\rho)$. Then the chosen values $w = w_n$ satisfy $|h_n(w_n)| \to \psi := \psi(\rho, \sigma)$, and since (16.2-27) holds for each h_n, there follows

$$\operatorname{Log} \frac{\psi}{\sigma} \geqq \frac{1-\gamma\psi}{1-\gamma\psi} \operatorname{Log} \gamma$$

or

$$\psi \geqq \sigma\gamma^{(1-\gamma\psi)/(1+\gamma\psi)}.$$

From the discussion of (16.2-24) given above, it is clear that $\psi \geqq \psi_0$, as asserted. —

With Theorem 16.2h we have achieved our goal of estimating the quantity $\psi(\rho, \sigma)$ in terms of $\gamma(\rho)$ alone. A result analogous to Theorem 16.2g is also available:

COROLLARY 16.2i. *If the function $\psi_0(\rho)$ is monotonic, the radii ρ_n generated by an osculation algorithm based on an osculation family $\mathcal{F}$ with dilatation measure $\gamma(\rho)$ satisfy*

$$\rho_n \geqq \rho_n', \qquad n = 1, 2, \ldots,$$

where

$$\rho_1' = \rho_1, \qquad \rho_{n+1}' := \psi_0(\rho_n'), \qquad n = 1, 2, \ldots.$$

Numerical tests indicate that for the special osculation families where the exact function $\psi(\rho)$ is available, the general estimate of Corollary 16.2i is only slightly inferior to that of Theorem 16.2g. For instance, if $\rho_1 = 0.1$, $\rho_n' > 0.4$ is now achieved in eleven iterations using the dilatation measure for the Koebe family, and in eight iterations using the dilatation measure for the family of Example 3.

PROBLEMS

1. *An upper bound for the speed of convergence of the Koebe algorithm.* Show that the radii ρ_n obtained by using the osculation algorithm with the Koebe family satisfy

$$1 - \rho_n \geqq 2^{-n}(1 - \rho_0).$$

(Thus at best, linear convergence is obtained.) What are the corresponding results for the families of Examples 2 and 3?

2. *Stability of the Koebe algorithm.* In the osculation algorithm based on the Koebe family, let the Koebe map be formed not with regard to a point of ∂D closest to O, but with regard to a point $z^* \in \partial D$ such that $\rho^* := |z^*|$, where

$D \in \mathscr{D}(\rho)$. Assuming $\gamma(\rho)$ to be decreasing, show:

(a) If $\rho^* \to 1$ only if $\rho \to 1$, the algorithm still converges.

(b) If the choice of ρ^* is such that

$$\varepsilon^* := 1 - \rho^* \geqq \theta\varepsilon, \qquad \varepsilon := 1 - \rho,$$

for some fixed $\theta \in (0, 1)$, then the estimate (16.2-17) holds in the form

$$\varepsilon_n \leqq \frac{8}{\alpha\theta^2\rho_1^2} \cdot \frac{1}{n}, \qquad n = 1, 2, \ldots.$$

NOTES

Koebe proposed his osculation method in [1912]. Ostrowski [1929] proved that $1 - \rho_n = O(1/n)$. Julia [1931] proved the result of Example 8. Although an asymptotically bad *estimate* of the speed of convergence does not of course imply that the actual speed of convergence is slow, these results seem to have been taken as an indication that the numerical performance of Koebe's method would be poor. Modifications of the method, aimed at improving the convergence in special cases, were suggested by Heinhold [1947, 1948] and Albrecht [1952]. Hübner [1964a, b] applied the idea of osculation to mappings onto the upper half-plane and implemented his method in simple cases. Little comprehensive experimentation with osculation methods was carried out until Grassmann [1979] by his extensive experiments obtained a more discriminating view of the method. Our present account, which follows Henrici [1983], tries to give theoretical support to Grassmann's experimental conclusions.

§ 16.3. BOUNDARY CORRESPONDENCE

The Riemann mapping theorem is a statement on mappings of simply connected *regions*, that is, simply connected open sets. Nothing at all is said about the behavior of the mapping function on the boundary of the region being mapped. Indeed, none of the processes discussed earlier even *defines* the mapping function on the boundary. This state of affairs cannot really surprise us, for we know from some of the examples given in § 5.10 that the boundary even of a simply connected region can be an extremely complicated point set.

On the other hand, we also know from Chapter 5 that in many applications of conformal mapping, such as the solution of boundary value problems, it is essential that one is able to extend the mapping function to the boundary. It will be seen in subsequent sections of the present chapter that the same holds for most of the more commonly used methods for the numerical *construction* of conformal maps of bounded regions D. For these reasons it is important to identify those bounded regions D for which the mapping function can be extended to a *homeomorphism* of the closure D' of D. Here a **homeomorphism** (or **topological map**) is a map that is one-to-one and continuous in both directions.

Let us see what at best can be expected. Let f map D conformally onto the unit disk E, and let f be extended to a homeomorphism of D'. If ∂D is the boundary of D, then $f(\partial D)$ must be the unit circle, and the restriction of f to ∂D defines a homeomorphism of ∂D. Since the inverse of a homeomorphism again is a homeomorphism, ∂D thus is the homeomorphic image of the unit circle, that is, a Jordan curve. It follows that a bounded region D for which f can be extended to a homeomorphism of D' necessarily is a **Jordan region**, that is, the interior int (Γ) of a Jordan curve Γ.

The following result (stated earlier as Theorem 5.10e, but not proved) therefore is the best one can hope for:

THEOREM 16.3a (Osgood–Carathéodory theorem). *Let D be a Jordan region, and let f be a conformal map of D onto the open unit disk E. Then f can be extended to a topological map of the closure of D onto the closure of E.*

It is clear that the extended map cannot, in general, be analytic. The inverse of a one-to-one analytic map being analytic, analyticity of the extended map would imply that Γ, the inverse image of the analytic curve $|w|=1$, were likewise analytic.

The concept of a topological map requires continuity in both directions. Actually, if the sets being mapped are compact, it suffices to establish continuity in one direction. For let f be a continuous one-to-one map from a compact set K to a compact set M. We assert that the inverse map $g := f^{[-1]}$ (which in any case exists) then automatically likewise is continuous. If this were not so, then there would exist a point $w_0 \in M$ and a sequence $\{w_n\}$ converging to w_0 such that the sequence of inverse images $\{z_n\}$, where $z_n := g(w_n)$, does not converge to $z_0 := g(w_0)$. The set K being compact, it then would have a limit point $z_0' \in K$, $z_0' \neq z_0$, and we could extract a subsequence $\{w_{n_k}\}$ such that $z_{n_k} \to z_0'$. Because f is continuous, the sequence of image points $f(z_{n_k}) = w_{n_k}$ would then converge to $w_0' := f(z_0')$, and because f is one-to-one, $w_0' \neq w_0$. This contradicts the fact that every subsequence of a convergent sequence $\{w_n\}$ converges to the limit of $\{w_n\}$.

Our proof of the Osgood–Carathéodory theorem requires several lemmas:

LEMMA 16.3b. *Let g be analytic and bounded on E, and let there exist α and $\beta > \alpha$ such that for all $\theta \in [\alpha, \beta]$,*

$$g(w) \to 0 \qquad as \quad w \to e^{i\phi}, \quad w \in E.$$

Then g is the zero function.

Proof. By rotating the plane we may assume that $\alpha = 0$. Choose n so that $2\pi/n < \beta$, and let $u := \exp(2\pi i/n)$. Consider the auxiliary function

$$h(w) := g(w)g(uw)\cdots g(u^{n-1}w).$$

This is still bounded on E but now tends to zero as $w \to e^{i\theta}$ for *all* θ (one of the factors tends to zero, and the others remain bounded). Thus by defining $h(w) := 0$ for $|w| = 1$, h becomes continuous on the closure on E. By the principle of the maximum, h is identically zero. It follows that g has infinitely many zeros in $|w| \leqq \frac{1}{2}$, say, and thus is likewise identically equal to zero. —

LEMMA 16.3c. *Let f be a topological map of a region D onto a region F, and let $\{z_n\}$, $z_n \in D$, be a sequence all of whose points of accumulation lie on the boundary of D. Then any point of accumulation of the sequence of images $\{f(z_n)\}$ lies on the boundary of F.*

Proof (*indirect*). Let the sequence $\{w_n\}$, $w_n := f(z_n)$, have a point of accumulation b in F. Then there exists a subsequence $\{w_{n_k}\}$ that converges to b. Because the inverse map of a topological map is continuous, the sequence of inverse images $z_{n_k} := f^{[-1]}(w_{n_k})$ then converges to $a := f^{[-1]}(b)$. Thus the sequence $\{z_n\}$ would have a point of accumulation $a \in D$, contrary to the hypothesis. —

Let D be a region. A **crosscut** of D is a simple arc Γ which lies in D, with the exception of its endpoints which are two *distinct* boundary points of D. Thus if the crosscut Γ is represented parametrically by $z = z(\tau)$, $\zeta \leqq \tau \leqq \beta$, then

$$z(\tau) \in D, \qquad \alpha < \tau < \beta,$$

and $z(\alpha) \in \partial D$, $z(\beta) \in \partial D$, $z(\alpha) \neq z(\beta)$.

LEMMA 16.3d. *Let $z = g(w)$ map the unit disk E conformally onto a Jordan region D, let Γ be a crosscut of E which is a piecewise regular arc, and suppose that the image of Γ under g has finite length,*

$$\int_\Gamma |g'(w)||dw| < \infty. \tag{16.3-1}$$

Then $\Gamma_1 := g(\Gamma)$ exists as an arc, and is a crosscut of D.

Proof. Let $w = w(\tau)$, $\alpha \leqq \tau \leqq \beta$, be a parametric representation of Γ. The function $z(\tau) := g(w(\tau))$ certainly is continuous at every $\tau \in (\alpha, \beta)$, and $z(\tau_1) \neq z(\tau_2)$ for $\alpha < \tau_1 < \tau_2 < \beta$ because g is one-to-one. We prove that $a := \lim_{\tau \to \alpha} z(\tau)$ exists. If $\{\tau_n\}$ is any sequence such that $\tau_n \in (\alpha, \beta)$ and $\tau_n \to \alpha$, then

$$|z(\tau_n) - z(\tau_m)| = |g(w(\tau_n)) - g(w(\tau_m))| = \left| \int_{\tau_m}^{\tau_n} g'(w(\tau))w'(\tau)\, d\tau \right|$$

$$\leqq \int_{\tau_m}^{\tau_n} |g'(w(\tau))||w'(\tau)|\, d\tau.$$

By (16.3-1) the last integral can be made arbitrarily small by choosing m and n large enough. Thus $a = \lim_{\tau\to\alpha} z(\tau)$ exists by Cauchy's criterion. The existence of $b := \lim_{\tau\to\beta} z(\tau)$ is proved similarly.

To show that Γ_1 is a crosscut, it remains to establish that $b \neq a$. Proceeding indirectly, suppose that $b = a$. Then Γ_1 is a Jordan curve. Let D_1 be its interior. D_1 is the image under g of one of the two subregions of E defined by the crosscut Γ. Let E_1 be that subregion; E_1 is a Jordan region whose boundary consists of Γ, and of a nondegenerate circular arc $w = e^{i\theta}$, $\alpha \leqq \theta \leqq \beta$, where $\beta > \alpha$. Let $\{w_n\}$ be any sequence of points in E_1 such that $|w_n| \to 1$. By Lemma 16.3c, the points of accumulation of the sequence of image points $z_n := g(w_n)$ lies on the boundary of D. But they must also lie in the closure of D_1 and because $b = a$ is the only point in the closure of D_1 that also lies on ∂D, we have $z_n \to a$. Thus the function $g(w) - a$ tends to zero as w tends to that part of the boundary of E_1 which coincides with $|w| = 1$. By Lemma 16.3b it follows that $g(w) - a$ vanishes identically. This is impossible because g is one-to-one in E. We conclude that $b \neq a$. —

LEMMA 16.3e (Lebesgue–Wolff lemma). *Suppose D is a region of finite area $\alpha := \alpha(D)$, and suppose g maps the unit disk E conformally onto D. Let w_0 denote a boundary point of E, and for $0 < \rho < 1$, let Γ_ρ denote the crosscut of E defined by the intersection of E with $|w - w_0| = \rho$. Then for every $\rho \in (0, 1)$ there exists ρ^*, $\rho^2 < \rho^* < \rho$, with the following property. If $\lambda(\rho)$ denotes the length of $\Lambda_\rho := g(\Gamma_\rho)$, then*

$$\lambda(\rho^*) \leqq \left[\frac{2\pi\alpha}{\operatorname{Log}(1/\rho^*)}\right]^{1/2}. \tag{16.3-2}$$

(*See Fig.* 16.3a.)

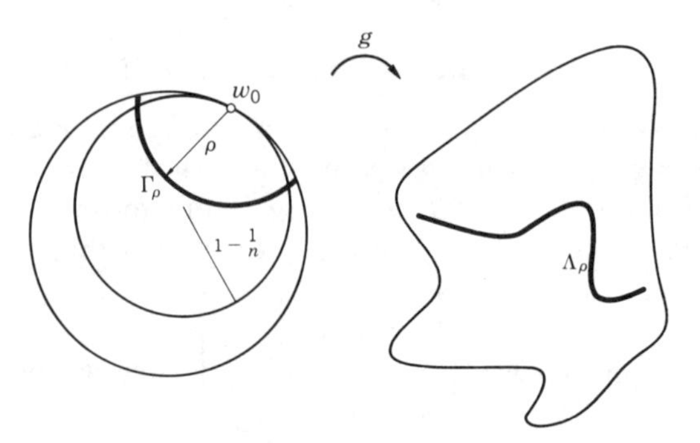

Fig. 16.3a. Notation of Lemma 16.3e.

Proof. The main difficulty is that one cannot a priori assert the existence of

$$\int_{\Gamma_\rho} |g'(w)|\,|dw|.$$

We thus consider a circle interior to $|w|=1$ of radius $1-1/n$ which touches $|w|=1$ at w_0, and we denote by $\Gamma_\rho^{(n)}$ the restriction of Γ_ρ to that interior circle. Then the length $\lambda_n(\rho)$ of the corresponding image arc $\Lambda_\rho^{(n)}$ satisfies

$$[\lambda_n(\rho)]^2 = \left[\int_{\Gamma_\rho^{(n)}} |g'(w)|\,|dw|\right]^2.$$

Applying the Schwarz inequality there follows

$$[\lambda_n(\rho)]^2 \leqq \int_{\Gamma_\rho^{(n)}} |g'(w)|^2 |dw| \cdot \int_{\Gamma_\rho^{(n)}} |dw|.$$

The last integral equals the length of $\Gamma_\rho^{(n)}$ and thus is bounded by $\pi\rho$. If $\Gamma_\rho^{(n)}$ is parametrized by $w = w_0 + \rho\, e^{i\theta}$, there follows

$$\frac{[\lambda_n(\rho)]^2}{\rho} \leqq \pi \int_{\Gamma_\rho^{(n)}} |g'(w)|^2\, \rho\, d\theta.$$

The function $\lambda_n(\rho)$ is continuous. We integrate the above between the limits ρ^2 and ρ,

$$\int_{\rho^2}^{\rho} \frac{[\lambda_n(\sigma)]^2}{\sigma}\, d\sigma \leqq \pi \int_{\rho^2}^{\rho} \left(\int_{\Gamma_\sigma^{(n)}} |g'(w)|^2\, \sigma\, d\theta\right) d\sigma.$$

The integral on the right equals the area of the image under g of a certain subset of E. It thus is bounded by α, and we have

$$\int_{\rho^2}^{\rho} \frac{[\lambda_n(\sigma)]^2}{\sigma}\, d\sigma \leqq \pi\alpha.$$

Letting $\phi_n(\xi) := [\lambda_n(e^{\xi})]^2$ this means that we have a sequence of functions $\phi_n(\xi)$, continuous in ξ for each n and monotonically increasing in n for each fixed ξ, such that

$$\int_{-2\nu}^{-\nu} \phi_n(\xi)\, d\xi \leqq \mu,$$

where $\mu := \pi\alpha$, $\nu := \mathrm{Log}(1/\rho)$. By the mean value theorem for continuous functions there exists, for every n, $\xi_n \in I := [-2\nu, -\nu]$ such that

$$\phi_n(\xi_n)\nu = \int_{-2\nu}^{-\nu} \phi_n(\xi)\, d\xi \leqq \mu.$$

From the sequence $\{\xi_n\}$ we extract a convergent subsequence $\{\xi_{n_k}\}$ which converges to some $\xi^* \in I$. For fixed m and for $n_k > m$ we then have

$$\phi_m(\xi_{n_k}) \leqslant \phi_{n_k}(\xi_{n_k}) \leqq \frac{\mu}{\nu}.$$

Letting $k \to \infty$, there follows in view of the continuity of ϕ_m

$$\phi_m(\xi^*) \leqq \frac{\mu}{\nu}.$$

The sequence $\phi_m(\xi^*)$ thus, in addition to being monotonic, is bounded, and hence has a limit $\leqq \mu/\nu$. Reverting to the function $\lambda_n(\sigma)$, this means that for some ρ^* such that $\rho^2 \leqq \rho^* \leqq \rho$,

$$\sup_n [\lambda_n(\rho^*)]^2 \leqq \frac{\mu}{\nu},$$

and thus that the arc $\Lambda_{\rho^*}^{(n)}$ has a finite length not exceeding

$$\left(\frac{\mu}{\nu}\right)^{1/2} = \left[\frac{\pi\alpha}{\operatorname{Log}(1/\rho)}\right]^{1/2},$$

as was to be shown. —

Taken together with Lemma 16.3d the Lebesgue–Wolff lemma yields:

COROLLARY 16.3f. $\Lambda_{\rho^*} := g(\Gamma_{\rho^*})$ *is a crosscut of D.*

LEMMA 16.3g. *Let Γ be a Jordan curve. Then for every $\varepsilon > 0$ there exists $\delta = \delta(\varepsilon)$ with the following property. If z_1 and z_2 are any two points on Γ such that $|z_1 - z_2| < \delta$, then one of the two arcs of Γ joining z_1 and z_2 has a diameter $< \varepsilon$.*

Here as usual the diameter of a set S, denoted by diam S, is defined as the supremum of $|z' - z''|$ as z' and z'' range over S independently of each other.

Proof of Lemma 16.3g. Let Γ have the parametric representation $z = z(\tau)$, $0 \leqq \tau \leqq \beta$. To avoid difficulties near $\tau = 0$, we extend z to a periodic function on $\mathbb{R}$ by setting $z(\tau + \beta) := z(\tau)$ for all τ. The function z then is continuous on the closed interval $[-\beta, \beta]$, hence uniformly continuous. Thus given any $\varepsilon > 0$, there exists $\eta = \eta(\varepsilon)$ such that

$$|z(\tau_1) - z(\tau_2)| < \varepsilon \qquad \text{whenever} \quad |\tau_1 - \tau_2| < \eta. \tag{16.3-3}$$

We require a form of uniform continuity of the inverse function of z, called property (A). Given any $\eta > 0$, there exists $\delta = \delta(\eta) > 0$ such that whenever two points z_1 and z_2 on Γ satisfy

$$|z_1 - z_2| < \delta, \tag{16.3-4a}$$

corresponding values of the parameters τ_1 and τ_2 can be found such that

$$|\tau_1 - \tau_2| < \eta. \tag{16.3-4b}$$

The proof of property (A) is indirect. Assuming the contrary, there would exist an exceptional $\eta_0 > 0$ such that for certain pairs of points (z_1, z_2) on Γ that are arbitrarily close, all possible pairs of corresponding parameter values would differ by more than η_0. In particular, for any $n > 0$ there would exist points $z_1^{(n)}$, $z_2^{(n)}$ on Γ such that

$$|z_1^{(n)} - z_2^{(n)}| < \frac{1}{n}$$

with corresponding parameter values $\tau_1^{(n)}$, $\tau_2^{(n)}$ in $[-\beta, \beta]$ satisfying

$$\eta_0 \leqq |\tau_1^{(n)} - \tau_2^{(n)}| \leqq \tfrac{1}{2}\beta.$$

By the Bolzano–Weierstrass theorem, we may assume that the sequences $\{\tau_1^{(n)}\}$, $\{\tau_2^{(n)}\}$ converge to limits τ_1 and τ_2, which again satisfy

$$\eta_0 \leqq |\tau_1 - \tau_2| \leqq \tfrac{1}{2}\beta. \tag{16.3-5}$$

But then the corresponding sequences of points on Γ also converge:

$$z_1^{(n)} \to z(\tau_1), \qquad z_2^{(n)} \to z(\tau_2).$$

In view of (16.3-5), $z(\tau_1) \neq z(\tau_2)$, yet on the other hand by (16.3-4), $z(\tau_1) = z(\tau_2)$. This contradiction proves property (A).

The assertion of Lemma 16.3g now follows by combining the two assertions just established. Given $\varepsilon > 0$, we first select $\eta > 0$ such that (16.3-3) holds, and then we select $\delta > 0$ such that (16.3-4) holds. Then if z_1 and z_2 are two points on Γ such that $|z_1 - z_2| < \varepsilon$, corresponding parameters τ_1 and τ_2 can be found such that, say, $0 < \tau_2 - \tau_1 < \eta$. Then, again by (16.3-3), $|z(\tau') - z(\tau'')| < \varepsilon$, corresponding parameters τ_1 and τ_2 can be found such that, say, $0 < \tau_2 - \tau_1 < \eta$. Then, again by (16.3-3), $|z(\tau') - z(\tau'')| < \varepsilon$ whenever $|\tau' - \tau''| < \eta$, thus in particular for all τ' and τ'' such that $\tau_1 \leqq \tau' < \tau'' \leqq \tau_2$. This indeed is to say that the arc $z = z(\tau)$, $\tau_1 \leqq \tau \leqq \tau_2$, has diameter $< \varepsilon$. —

We now begin with the proof of the Osgood-Carathéodory theorem proper. As observed earlier, it suffices to establish the existence of an extended one-to-one continuous mapping function in one direction. We choose to extend the map $g = f^{[-1]}$.

(i) Let w_0 be a point on the boundary of E, $|w_0| = 1$. Our first task is to define $g(w_0)$ such that the extended map is continuous.

We begin by establishing some notation. For $0<\rho<1$, let Γ_ρ denote the crosscut of E defined in Lemma 16.3e. Γ_ρ divides E into two subregions, E_1 and E_2. We denote by E_1 the subregion which has w_0 as a boundary point.

By virtue of the Lebesgue–Wolff lemma, there exist arbitrarily small ρ for which $\Lambda_\rho := g(\Gamma_\rho)$ has finite length $\lambda(\rho)$. The following discussion is restricted to those ρ for which $\lambda(\rho)<\infty$. For such ρ, Λ_ρ by Corollary 16.3f is a crosscut of D. Let a and b be its initial and terminal points. We denote the piece of Γ between b and a (in the sense of increasing parameter values) by $\Delta_{1,\rho}$, and the piece from a to b by $\Delta_{2,\rho}$ (see Fig. 16.3b). Because $a \neq b$, the curves $\Delta_{1,\rho}+\Gamma_\rho$ and $\Delta_{2,\rho}-\Gamma_\rho$ both are Jordan curves. Denoting their interiors by D_1 and D_2, respectively, it is clear that one of the sets $g(E_1)$ and $g(E_2)$ equals D_1, and the other equals D_2.

In the notation thus introduced, it is plain what must be done in order to prove that g has a continuous extension to w_0. It must be shown that if $\rho \to 0$, the region $g(E_1)$ shrinks to zero. Existence of $\lim g$ then follows by the Cauchy criterion. Our first task thus will be to estimate the diameter of either D_1 or D_2.

Let $\{\rho_n\}$ be a sequence that decreases monotonically to zero. By the Lebesgue–Wolff lemma (Lemma 16.3e) there exists a sequence $\{\rho_n^*\}$, $\rho_n^2 \leqq \rho_n^* \leqq \rho_n$, such that

$$\lambda(\rho_n^*) \to 0. \tag{16.3-6}$$

Let $\Lambda_n := \Lambda_{\rho_n^*}$, $\Delta_{j,n} := \Delta_{j,\rho_n^*}$, $j=1, 2$,

$$D_{1,n} := \operatorname{int}(\Delta_{1,n}+\Lambda_n),$$

$$D_{2n} := \operatorname{int}(\Delta_{2,n}-\Lambda_n).$$

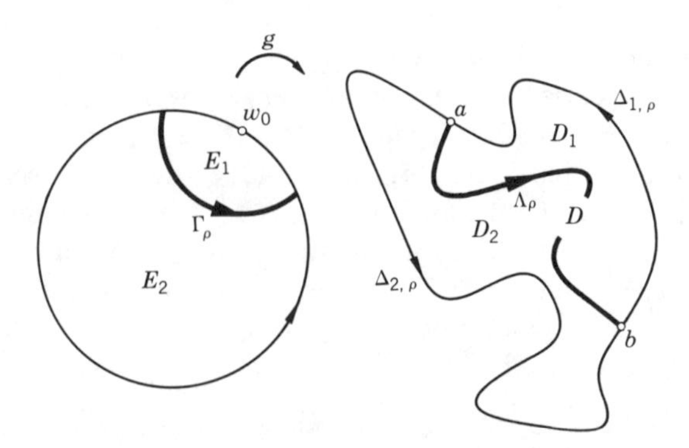

Fig. 16.3b. Notation for proof of Theorem 16.3a.

We estimate the diameter of $D_{1,n}$. This equals the diameter of $\Delta_{1,n}+\Lambda_n$, the boundary of $D_{1,n}$. Evidently,

$$\operatorname{diam}(\Delta_{1,n}+\Lambda_n) \leqq \operatorname{diam}\Delta_{1,n} + \operatorname{diam}\Lambda_n.$$

Here $\operatorname{diam}\Lambda_n$ is bounded by $\lambda(\rho_n^*)$, which just has been shown to tend to zero. By Lemma 16.3g, because the distance of the endpoints of Λ_n tends to zero, one of $\operatorname{diam}\Delta_{1,n}$ and $\operatorname{diam}\Delta_{2,n}$ must tend to zero. Because $\Delta_{1,n+1} \subset \Delta_{1,n}$,

$$\operatorname{diam} D_{1,n} \to 0, \qquad n \to \infty.$$

This means that given $\varepsilon > 0$, we can choose n such that $\operatorname{diam} D_{1,n} < \varepsilon$. Thus if w_1 and w_2 are in E and $|w_j - w_0| < \rho_n$, $j = 1, 2$, their images $z_j := g(w_j)$, which both lie in $D_{1,n}$, differ by less than ε. Therefore by Cauchy's criterion,

$$g(w_0) := \lim_{w \to w_0} g(w)$$

exists.

(ii) Having extended the definition of g to the boundary of E, we show that the boundary values are continuous also if approached from the boundary. Again let $|w_0| = 1$, and let $\varepsilon > 0$ be given. By the foregoing, $\delta > 0$ exists such that $w \in E$, $|w - w_0| < \delta$ implies $|g(w) - g(w_0)| < \varepsilon$. While maintaining the condition $|w - w_0| < \delta$, we now let $w \to w_1$, where $|w_1| = 1$, $|w_1 - w_0| < \delta$. Because we already know that $\lim g(w)$ exists, there follows $|g(w_1) - g(w_0)| \leqq \varepsilon$, establishing the continuity of the boundary values of g.

(iii) We next show that the extended function g is still one-to-one. In other words, if $|w_j| = 1, j = 1, 2$, but $w_1 \neq w_2$, we show that $g(w_1) \neq g(w_2)$.

Assuming the contrary, let $g(w_1) = g(w_2) =: z_0$. Consider the radii P_1: $w = \tau w_1$, P_2: $w = \tau w_2$, $0 \leqq \tau \leqq 1$. Their images $g(P_1)$ and $g(P_2)$ both begin at $g(0)$ and end at z_0; otherwise, they have no point in common. Thus $g(P_1) - g(P_2)$ is a Jordan curve. We let $D_1 := \operatorname{int}(g(P_1) - g(P_2))$, and $E_1 := g^{[-1]}(D_1)$. E_1 is one of the two subregions of E defined by the crosscut $P_1 - P_2$ (see Fig. 16.3c). By means of Lemma 16.3b it follows as in the proof of Lemma 16.3d that $g(w) - z_0$ vanishes identically. This contradiction shows that $g(w_1) \neq g(w_2)$.

(iv) We finally have to show that the extended function g assumes every value on Γ. Let $a \in \Gamma$, and choose a sequence $\{z_n\}$, $z_n \in D$, such that $z_n \to a$. We select a subsequence, which we denote again by $\{z_n\}$, such that $w_n := g^{[-1]}(z_n) \to b$ where $|b| \leqq 1$. By Lemma 16.3c we know that $|b| = 1$. Because g is defined and continuous at b,

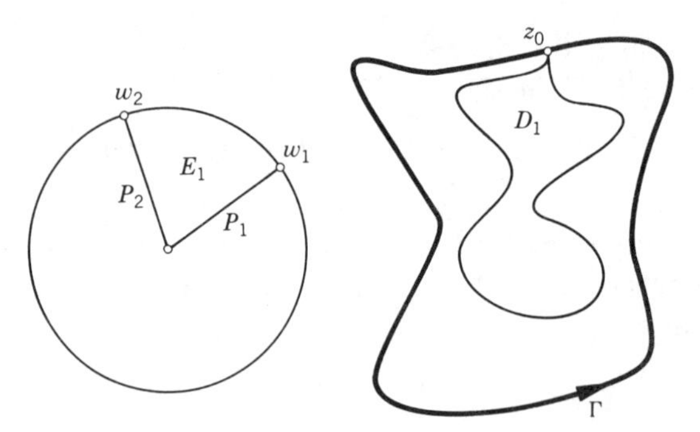

Fig. 16.3c. The extended map is one-to-one.

$z_n = g(w_n) \to g(b)$. In view of $z_n \to a$ there follows $g(b) = a$. Thus the value a is assumed by g. —

Having established the continuity of the mapping function on the boundary assuming nothing more than mere continuity of the boundary curve, one can reasonably ask whether more than mere continuity can be established under stronger hypotheses on the boundary. For instance, if the boundary curve Γ has a tangent at the point $z = a$, it is tempting to conjecture than the derivative of the mapping function can be extended continuously into the point $z = a$.

Disappointingly, examples exist showing that this need not be so. However, under the stated hypotheses the map f will at least be **angle preserving** in the sense that

$$\lim_{z \to a} \arg \frac{f(z) - f(a)}{z - a} \tag{16.3-7}$$

exists. This means, for instance, that if $z = z(\tau)$ is a differentiable arc in D emanating from $a = z(0)$, then the image arc $w(\tau) = f(z(\tau))$, although not necessarily differentiable, is such that

$$\lim_{\tau \to 0} \arg[f(z(\tau)) - f(a)]$$

exists and differs from

$$\arg z'(0) = \lim_{\tau \to 0} \arg[z(\tau) - a]$$

by a constant that is the same for all values of $z'(0)$. It also means that if t is a tangent vector of Γ at a, oriented such that it is the interior normal, then any set S: $|\arg z - \arg(it) \leq \beta$ where $0 \leq \beta < \pi/2$ is mapped by f onto a set $f(S)$ which is bounded by two curves that asymptotically make angles $\pm\beta$ with the interior normal at the image pont $f(a)$ (see Fig. 16.3d).

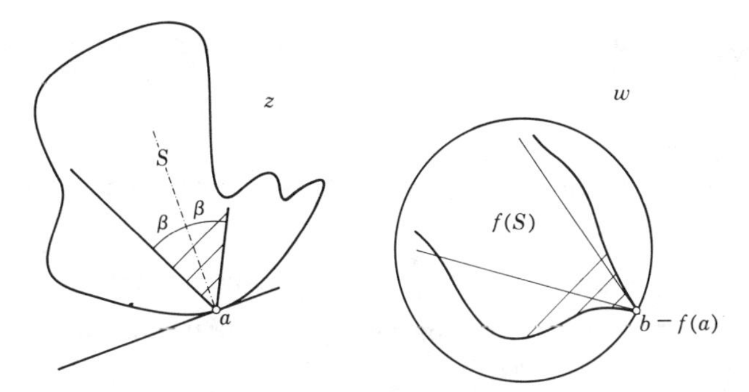

Fig. 16.3d. Angle-preserving map.

THEOREM 16.3h (Lindelöf–Carathéodory theorem). *Let D be a Jordan region whose boundary curve Γ has a tangent at the point $z=a$, and let f be a conformal map of D onto the unit disk E. Then the extension of f to the closure D' of D is angle preserving at $z=a$.*

Proof. Let $b:=f(a)$. Without loss of generality we may assume $b=1$. Then the function

$$w \mapsto \frac{g(w)-g(1)}{w-1},$$

where $g:=f^{[-1]}$, is analytic and $\neq 0$ in the simply connected region E. Hence it possesses an analytic logarithm, and consequently a single-valued and continuous argument

$$\psi(w) := \arg \frac{g(w)-g(1)}{w-1} = \operatorname{Im} \log \frac{g(w)-g(1)}{w-1}$$

which, by the Osgood-Carathéodory theorem, can be extended to a continuous function in $E'\backslash\{1\}$. The function ψ is bounded, because of every $\varepsilon>0$ it is bounded on $E'\backslash\{w\colon |w-1| \geqq \varepsilon\}$ as a continuous function on a compact set, and for sufficiently small ε it is bounded on the set $E' \cap \{w\colon 0<|w-1|<\varepsilon\}$ because $\arg[g(w)-g(1)]$ cannot change there by more than $\frac{3}{2}\pi$, say, in view of Γ having a tangent at $a=g(b)$.

Being the imaginary part of an analytic function, $\psi(w)$ is harmonic in E, and by the foregoing its boundary values $\psi(e^{i\theta})$ are continuous for $-\pi \leqq \theta < 0$ and for $0<\theta \leqq \pi$. The one-sided limits

$$\lambda^- := \lim_{\theta \uparrow 0} \psi(e^{i\theta}) \qquad \text{and} \qquad \lambda^+ := \lim_{\theta \downarrow 0} \psi(e^{i\theta})$$

both exist because, for instance, the limits

$$\lim_{\theta \uparrow 0} \arg[g(e^{i\theta})-g(1)] \qquad \text{and} \qquad \lim_{\theta \uparrow 0} \arg(e^{i\theta}-1)$$

both exist individually, the first on account of Γ having a tangent at $a=g(1)$. We moreover assert that

$$\lambda^{-}=\lambda^{+}=:\lambda.$$

This is so because when w passes from $e^{-i\varepsilon}$ to $e^{i\varepsilon}$ along a small semicircular arc of radius ε in E, both

$$\arg[g(w)-g(1)] \qquad \text{and} \qquad \arg(w-1)$$

decrease by an amount that approaches π as $\varepsilon\to 0$, the first argument again by virtue of Γ having a tangent at $g(1)$.

Thus the function

$$\psi_0(e^{i\theta}):=\begin{cases}\psi(e^{i\theta}), & \theta\not\equiv 0 \mod 2\pi\\ \lambda, & \theta\equiv 0 \mod 2\pi,\end{cases}$$

is continuous. By Poisson's integral it possesses a harmonic extension $\psi_0(w)$ into E which is continuous in E'. By Lemma 15.4c, the function $\psi(w)$, which is bounded and harmonic in E and which has the same boundary values as ψ_0 except possibly at one point, is identical with ψ_0. Thus for the function ψ,

$$\lim_{w\to 1}\psi(w)=\lambda$$

likewise exists. Setting $w=f(z)$, this means that the limit (16.3-7) exists. —

PROBLEMS

1. Let D be a Jordan region, and let Λ be a crosscut of D. Establish the following propositions, which were used repeatedly in the proof of Theorem 16.3a:
   (a) The set $D\backslash\{\Lambda\}$ has precisely two components, D_1 and D_2, say.
   (b) Let g be a topological map of D onto a Jordan region E, and let $\Gamma:=g(\Lambda)$ be a crosscut of E. If E_1 and E_2 are the two components of $E\backslash\{\Gamma\}$ defined in (a), then for a suitable numbering, $E_j=g(D_j)$, $j=1,2$. (Use the Jordan curve theorem and the definition of connectedness.)
2. Simplify the proof of Lemma 16.3e by using the monotonic convergence theorem of real variable theory.

NOTES

For Theorem 16.3a see Osgood and Taylor [1913] and Carathéodory [1913]. The above elementary proof follows lectures by S. Warschawski at UCLA in 1963. Hille [1962] has a proof making more use of real variable theory. A proof by Arsove [1968] is concise but less transparent. A source for Theorem 16.3h is Lindelöf [1916]. The literature on differentiability on the boundary is large; see Warschawski [1932] and many later papers. In some of these studies, the concept of extremal length (Ahlfors [1973], ch. 4) is particularly useful.

§ 16.4. BOUNDARY BEHAVIOR FOR PIECEWISE ANALYTIC JORDAN CURVES

Jordan curves that occur in practical mapping problems are not merely continuous or piecewise regular, but **piecewise analytic**, in the following sense. If $z = z(\tau)$, $\alpha \leqq \tau \leqq \beta$, is the parametric representation of a piecewise analytic Jordan curve Γ, then the parametric interval $[\alpha, \beta]$ has a finite subdivision $\alpha = \tau_0 < \tau_1 < \cdots < \tau_n = \beta$ such that the restriction of $z = z(\tau)$ to each closed subinterval $[\tau_{k-1}, \tau_k]$ agrees with a function $z_k = z_k(t)$ that is analytic in a region of the t plane containing the closed real interval $[\tau_{k-1}, \tau_k]$, and whose derivative is never zero. The points $\hat{z}_k := z_k(\tau_k)$ are called the **corners** of Γ.

The analytic functions z_k may be different from interval to interval, but since at each corner $\hat{z}_k$ the two derivatives $z_k'(\tau_k)$ and $z_{k+1}'(\tau_k)$ exist and are different from zero, the two subarcs emanating from $\hat{z}_k$ have well-defined tangents at $\hat{z}_k$ which form an interior angle

$$\arg \frac{z_{k+1}'(\tau_k)}{z_k'(\tau_k)} + \pi \quad \mathrm{mod}\, 2\pi.$$

(Here we set $z_n = z_0$.) Examples of piecewise analytic Jordan curves are polygons, or Jordan curves made up of a finite number of circular, elliptic, or parabolic arcs (see Fig. 16.4a).

For Jordan regions D bounded by piecewise analytic Jordan curves much stronger results on the behavior of the mapping function at the boundary are possible than for general Jordan regions. For simplicity of notation, we first consider these results for functions f that map D onto the upper half-plane U: $\mathrm{Im}\, w > 0$. Because the map from U to E is elementary, the reformulation of these results into results on the map from D to E is immediate. The results stated below are required for the rigorous treatment of the numerical mapping problem for regions bounded by piecewise analytic Jordan curves.

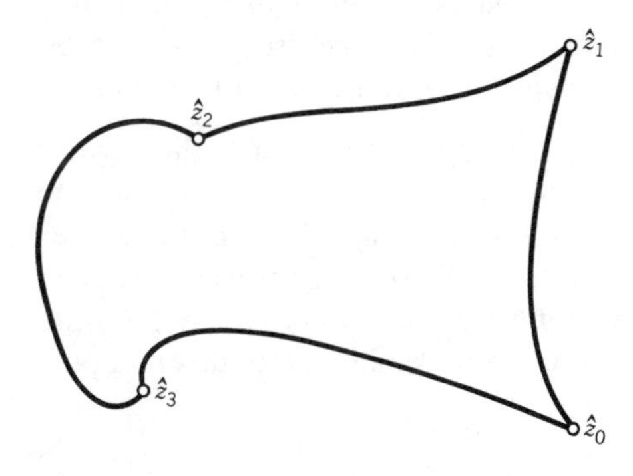

Fig. 16.4a. Piecewise analytic Jordan curve.

To begin with, let one of the analytic subarcs of Γ, say the arc Λ_k between the corners $\hat{z}_{k-1}$ and $\hat{z}_k$, be a straight line segment. The continuous extension of the map f (which by the Osgood–Carathéodory theorem certainly exists) then takes values on Λ_k which lie on the real axis. This is a situation to which the Schwarz reflection principle (Theorem 5.11b) applies. By assigning to points that lie symmetrically with respect to Λ_k values that lie symmetrically with regard to the real axis, the mapping function f is extended to a function that is analytic in a region which contains the interior points of Λ_k. Thus in this case, the extension of f to Λ_k is not merely continuous, but in fact analytic on the boundary.

Next let the arc Λ_k between $\hat{z}_{k-1}$ and $\hat{z}_k$ be analytic, but not necessarily a straight line segment. Let z^* be an interior point of Λ_k, and let the parametric representation of Λ_k in the neighborhood of z^* be the restriction of the analytic function

$$z = l(t) = z^* + \sum_{m=1}^{\infty} a_m (t - t^*)^m, \qquad |t - t^*| < \theta, \tag{16.4-1}$$

to real values of t. Because $l'(t) \neq 0$, $a_1 \neq 0$, and there exists $\eta > 0$ such that in the disk $|z - z^*| < \eta$, the inverse function $t = l^{[-1]}(z)$ exists. We may assume that $\eta \leqq \min(|z^* - z_{k-1}|, |z^* - z_k|)$.

Consider now the function $h := f \circ l$. If θ is small enough, h is defined, and analytic, in the intersection of the disk $|t - t^*| < \theta$ with either the upper or the lower half-plane, and it is continuous on the intersection of that disk with the closed half-plane. For real values of t it takes values on the real w axis. Again by the Schwarz reflection principle, h may be extended to a function that is analytic on the whole disk $|t - t^*| < \eta$. It follows that $f = l^{[-1]} \circ h$ can be extended to a function that is analytic on the disk $|z - z^*| < \eta$. Thus in this case also the Osgood–Carathéodory extension of f is not merely continuous, but in fact analytic, on the boundary. Mapping U to E we obtain:

THEOREM 16.4a. *The Osgood–Carathéodory extension of any function f mapping a Jordan region D whose boundary Γ is piecewise analytic to the unit disk is analytic at every boundary point of D that is not a corner of Γ.*

It remains to settle the behavior of the mapping function at the corners $\hat{z}_k$. Here of course the extended mapping function cannot be expected to remain analytic, because in general it is not even angle preserving. To see what can be expected, we consider the function f mapping the wedge-shaped region bounded by the rays $\arg z = 0$ and $\arg z = \alpha\pi$, where $0 < \alpha \leqq 2$, onto U (see Fig. 16.4b). If the points 0 and ∞ are to remain fixed, this mapping is

$$w = f(z) = \gamma z^{1/\alpha}, \qquad \gamma > 0, \tag{16.4-2}$$

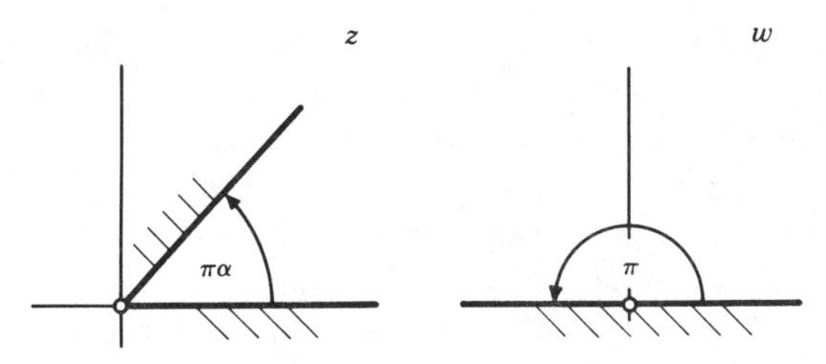

Fig. 16.4b. Mapping of wedge.

where, in forming $z^{1/\alpha}$, arg z is selected in the interval $(0, 2\pi)$. The mapping function f thus has the property that

$$z^{-1/\alpha}f(z) = \gamma,$$
$$z^{1-1/\alpha}f'(z) = \frac{\gamma}{\alpha}. \qquad (16.4\text{-}3)$$

Remarkably, the behavior of the mapping function at a corner of a piecewise analytic Jordan curve is completely analogous.

THEOREM 16.4b (Lichtenstein–Warschawski). *Let $z = 0$ be a corner of a piecewise analytic, positively oriented Jordan curve which enters 0 at the angle* arg $z = \pi\alpha$, $0 < \alpha \leqq 2$, *and which leaves 0 at the angle* arg $z = 0$. *If f maps the interior D of Γ onto the upper half-plane U such that $f(0) = 0$, then there exists $\gamma > 0$ such that*

$$\lim_{z\to 0} z^{-1/\alpha}f(z) = \gamma, \qquad (16.4\text{-}4a)$$

$$\lim_{z\to 0} z^{1-1/\alpha}f'(z) = \frac{\gamma}{\alpha}, \qquad (16.4\text{-}4b)$$

where in either limit the approach of z to 0 is only subjected to the condition that z remain in the closure D' of D (see Fig. 16.4c).

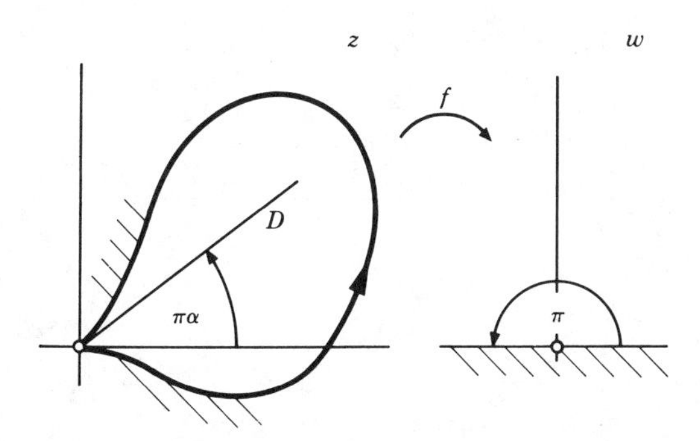

Fig. 16.4c. Theorem 16.4b.

In the limit (16.4-4b) it makes sense to talk about values of f' at points $z \in D'$, $z \neq 0$, because we already know from Theorem 16.4a that f can be extended analytically to points of the closure that are not corners of the boundary. At the point $z=0$ itself, f of course may become singular—as indeed it is in the simple case of a wedge—but then Theorem 16.4b tells us very precisely, say, how fast $|f'(z)|$ approaches ∞.

The proof of Theorem 16.4b is not as simple at it may appear at the outset. Our method of proof consists in reducing the situation of the theorem successively to simpler, more standardized situations.

LEMMA 16.4c. *Theorem* 16.4b *holds if there exists* $\rho > 0$ *such that the conclusion holds for the intersection* D_1 *of* D *with the disk* $|z| < \rho$ *(see Fig.* 16.4d).

Proof. If $\rho > 0$ is sufficiently small, D_1 is a Jordan region having precisely three corners $\hat{z}_0 = 0, \hat{z}_1, \hat{z}_2$. Let f_1 map D_1 onto U such that $f_1(0) = 0$. By Osgood-Carathéodory this may be extended to a map of the closure D_1' to U', which is continuous except at the pre-image of ∞. In particular, f_1 maps the two analytic arcs meeting at 0 onto a segment Σ_1 of the real axis having 0 as an interior point. The function $k := f \circ f_1^{[-1]}$ then maps U onto a subset $U_1 \subset U$ and takes Σ_1 into a real segment Σ which again has 0 as an interior point. By the reflection principle k can be extended to a function that is analytic in a neighborhood of 0. Because the extended function is still one-to-one, $k'(0) \neq 0$ and in fact, because the orientation is preserved, $k'(0) = \kappa > 0$. But now $f = k \circ f_1^{[-1]}$, hence if f_1 satisfies the relations (16.4-4) with γ replaced by γ_1, then

$$
\begin{aligned}
z^{-1/\alpha} f(z) &= z^{-1/\alpha} k(f_1(z)) \\
&= [z^{-1/\alpha} f_1(z)]\{[f_1(z)]^{-1} k(f_1(z))\} \\
&\rightarrow \gamma_1 \kappa
\end{aligned}
$$

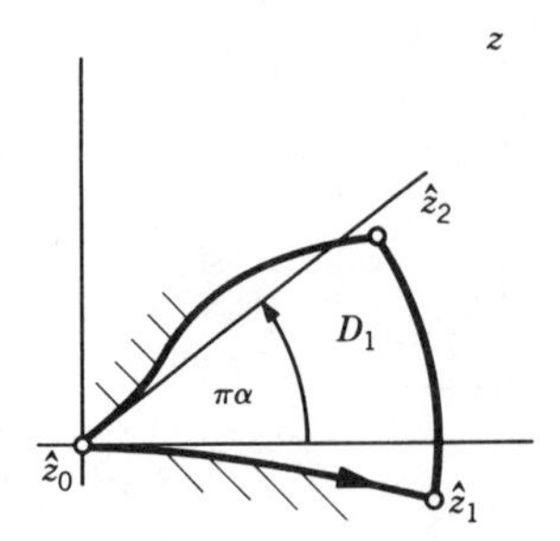

Fig. 16.4d. The region D_1.

$$z^{1-1/\alpha}f'(z) = z^{1-1/\alpha}k'(f_1(z)) \cdot f_1'(z)$$
$$\to \gamma_1\kappa/\alpha.$$

Thus the relations (16.4-4) are true for f with the constant $\gamma := \gamma_1\kappa$. —

In the following, D denotes a region such as the region D_1 defined in Lemma 16.4c.

LEMMA 16.4d. *Theorem* 16.4b *holds if for any function h mapping the region D onto the wedge W*: $0 < \arg w < \pi\alpha$ *such that* $h(0)=0$ *there exists* $\mu > 0$ *with the property that*

$$\lim_{z\to 0} z^{-1}h(z) = \mu, \tag{16.4-5a}$$

$$\lim_{z\to 0} h'(z) = \mu, \tag{16.4-5b}$$

where the approach of z to 0 *is unrestricted in D* (*see Fig.* 16.4e).

Proof. If f is the mapping function considered in Theorem 16.4b,

$$h(z) := [f(z)]^\alpha$$

realizes the mapping required in the lemma. Consequently, $f(z) = [h(z)]^{1/\alpha}$, and

$$z^{-1/\alpha}f(z) = [z^{-1}h(z)]^{1/\alpha} \to \gamma := \mu^{1/\alpha},$$

by (16.4-5a). Furthermore, using both relations (16.4-5),

$$z^{1-1/\alpha}f'(z) = \frac{1}{\alpha}[z^{-1}h(z)]^{1/\alpha-1} \cdot h'(z) \to \frac{1}{\alpha}\mu^{1/\alpha-1}\mu = \frac{\gamma}{\alpha}.$$

Thus (16.4-4) holds with $\gamma := \mu^{1/\alpha}$. —

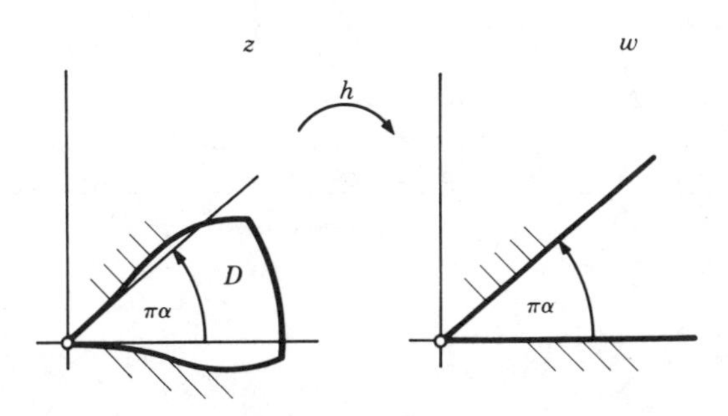

Fig. 16.4e. Lemma 16.4d.

LEMMA 16.4e. *Lemma* 16.4d *holds for angles* $\pi\alpha$ *such that* $0<\alpha\leqq 2$ *if it holds for angles* $\pi\alpha$ *such that* $0<\alpha\leqq 1$.

Proof. Let the region D at 0 have an angle $\pi\alpha$, where $1<\alpha\leqq 2$. Let D_1 be the image of D under the map $z\to z_1:=z^{1/2}$ (principal value). It is easy to see that D_1 has an analytic corner at 0 with angle $\pi\alpha/2$. Let h_1 be the map for D_1 as required in Lemma 16.4d. Then

$$h(z):=[h_1(z^{1/2})]^2$$

is a corresponding map for D. If (16.4-5) holds for h_1, with μ replaced by μ_1, then

$$z^{-1}h(z)=[z^{-1/2}h_1(z^{1/2})]^2\to\mu_1^2$$

and

$$h'(z)=[z^{-1/2}h_1(z^{1/2})]\cdot h_1'(z^{1/2})\to\mu_1\cdot\mu_1=\mu_1^2.$$

We thus see that the limits required in Lemma 16.4d hold for the constant $\mu:=\mu_1^2$. —

LEMMA 16.4f. *The limits* (16.4-5) *hold for unrestricted approach if they hold for nontangential approach, that is, if they hold when z is restricted to any sector S*: $\pi\alpha'\leqq\arg z\leqq\pi\beta$, *where* $0<\alpha'<\beta'<\alpha$ (*see Fig.* 16.4f).

Proof. Using Lemma 16.4e twice, we may suppose that $0<\alpha\leqq\frac{1}{2}$. Let Σ, the boundary arc emanating from 0 in the direction of the positive real axis,

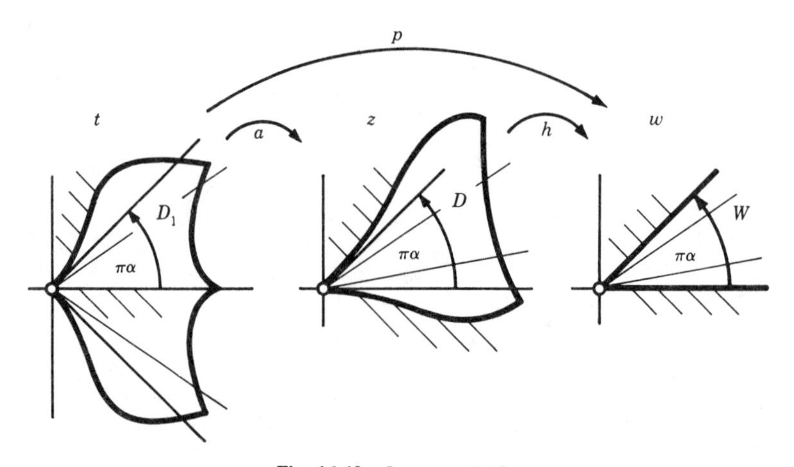

Fig. 16.4f. Lemma 16.4f.

have the parametric representation

$$z=a(\tau)=\sum_{m=1}^{\infty} a_m\tau^m, \qquad 0\leqq\tau\leqq\delta, \tag{16.4-6}$$

where $a_1>0$. By choosing ρ sufficiently small, we may assume that the whole arc from $\hat{z}_0=0$ to the corner $\hat{z}_1$ is represented by the series (16.4-6). This series for $|t|<\delta$ defines an analytic function $t\to z=a(t)$. Because $a_1\neq 0$, the inverse function $t=a^{[-1]}(z)$ exists for $|z|<\rho_1$. By choosing a smaller ρ if necessary we may assume $\rho\leqq\rho_1$. Consider the image of D under the map $a^{[-1]}$. This takes Σ into a segment Σ_1 of the real axis, the circular arc from $\hat{z}_1$ to $\hat{z}_2$ into some analytic arc, and the arc from $\hat{z}_2$ to 0 into another analytic arc. If ρ is sufficiently small, the image $D_1:=a^{[-1]}(D)$ lies in $\operatorname{Im} t>0$. The function $h\circ a$ maps D_1 onto the wedge $0<\arg w<\pi\alpha$, and Σ_1 on a part of the real axis. By the reflection principle, $h\circ a$ can be extended to an analytic function p mapping $D_1\cup\Sigma\cup\overline{D_1}$ onto the symmetric wedge $-\pi\alpha<\arg t<\pi\alpha$. Let now for each subwedge $-\pi\alpha'<\arg t<\pi\alpha'$, where $0<\alpha'<\alpha$, the relations

$$\lim_{t\to 0} t^{-1}p(t)=\nu,$$

$$\lim_{t\to 0} p'(t)=\nu$$

hold for some $\nu>0$. Then, since the map $a(t)$ is analytic at 0 and preserves angles, the function $h=p\circ a^{[-1]}$ satisfies (16.4-5), where $z\in D$ is restricted to the subset $\arg z\leqq\pi\alpha'$ of D. A similar treatment applied to the analytic arc from 0 to $\hat{z}_2$ yields the same result for an approach in any subset $\pi\beta'\leqq\arg z$ of D, where $0<\beta'<\alpha$. Choosing α' and β' so that the two subsets overlap, it follows that the limits obtained by the two approaches are identical. —

LEMMA 16.4g. *From the existence of the limit* (16.4-5a) *for nontangential approach there follows the existence of the limit* (16.4-5b) *for nontangential approach.*

Proof. Making both in the z plane and in the w plane the auxiliary transformations $z\mapsto z^{1/\alpha}$ and $w\mapsto w^{1/\alpha}$, we see by the Lindelöf–Carathéodory theorem (Theorem 16.3h) that the map h is angle preserving. Thus to a nontangential approach in the z plane there corresponds a nontangential approach in the w plane. In terms of the inverse function $g:=h^{[-1]}$ the hypothesis is that

$$\lim_{w\to 0}\frac{g(w)}{w}=\frac{1}{\mu} \tag{16.4-7}$$

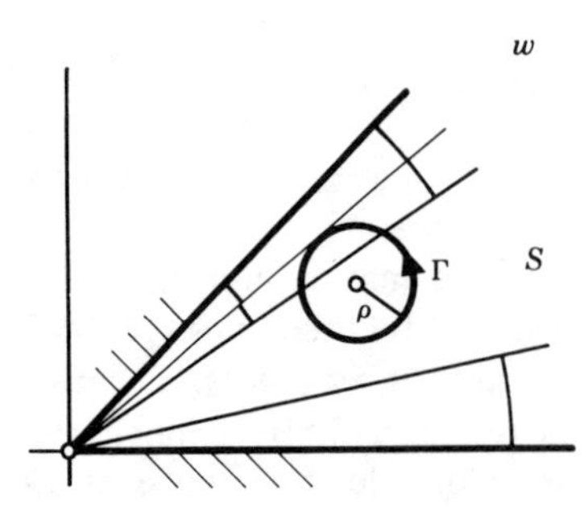

Fig. 16.4g. Proof of Lemma 16.4g.

exists, where g is analytic in $0 < \arg w < \pi\alpha$, and where the approach is restricted to a subangle S: $\pi\varepsilon \leqq \arg w \leqq \pi(\alpha - \varepsilon)$, where $2\varepsilon < \alpha$ (see Fig. 16.4g).

We represent the derivative of the function

$$k(w) := g(w) - \frac{w}{\mu}$$

by means of Cauchy's formula, applied to a circle Γ of radius ρ about w that touches one of the rays of the arguments $\pi\varepsilon/2$ and $\pi(\alpha - \varepsilon/2)$. This yields

$$k'(w) = \frac{1}{2\pi i} \int_\Gamma \frac{k(u)}{(u-w)^2}\, du = \frac{1}{2\pi i} \int_\Gamma \frac{k(u)}{u} \frac{u}{(u-w)^2}\, du.$$

Because $|u| \leqq |u - w| + |w|$ and $|u - w| = \rho$ for $u \in \Gamma$, we have

$$|k'(w)| \leqq \frac{1}{2\pi} \int_\Gamma \left| \frac{k(u)}{u} \right| \frac{\rho + |w|}{\rho^2} |du|,$$

and since $\rho \geqq |w| \sin(\varepsilon/2)$,

$$|k'(w)| \leqq \frac{2}{\sin(\varepsilon/2)} \max_{u \in \Gamma} \left| \frac{k(u)}{u} \right|.$$

By (16.4-7), $k(u)/u \to 0$ as $w \to 0$, and there follows

$$\lim_{w \to 0} k'(w) = 0$$

for nontangential approach. This is the same as $g'(w) \to 1/\mu$ or, returning to $h = g^{[-1]}$, $h'(z) \to \mu$. —

We are thus finally left with the problem of showing that the function h described in Lemma 16.4d satisfies

$$\lim_{z \to 0} z^{-1} h(z) = \mu \tag{16.4-8}$$

for nontangential approach.

Let Γ^* be a Jordan curve in the exterior of D which consists of two circular arcs intersecting at the angle $\pi\alpha$ at $z=0$ and at some other point, z^* say (see Fig. 16.4h).

The Moebius transformation

$$h^*: \quad z \mapsto \frac{z}{1-zz^{*-1}}$$

maps ext Γ^* onto the wedge W, and it satisfies

$$\lim_{z\to 0} \frac{h^*(z)}{z} = 1 \qquad (16.4\text{-}9)$$

for unrestricted approach. The function

$$z \to z_1 := [h^*(z)]^{1/\alpha}$$

maps ext Γ^* onto $\operatorname{Im} z_1 > 0$, D onto a subset of $\operatorname{Im} z_1 > 0$, and Γ onto a Jordan curve Γ_1 in $\operatorname{Im} z_1 \geqq 0$, which has $\operatorname{Im} z_1 = 0$ as a tangent at $z_1 = 0$. Let p map the region $D_1 := \operatorname{int} \Gamma_1$ onto $\operatorname{Im} w > 0$, $p(0) = 0$. Then

$$h(z) = [p(h^{*1/\alpha}(z))]^\alpha,$$

and in view of

$$z^{-1}h(z) = z^{-1}h^*(z)[h^{*-1/\alpha}(z)p(h^{*1/\alpha}(z))]^\alpha$$

and of (16.4-9) the desired relation (16.4-8) is proved if it is shown that

$$\lim_{z_1\to 0} \frac{p(z_1)}{z_1} = \mu \qquad (16.4\text{-}10)$$

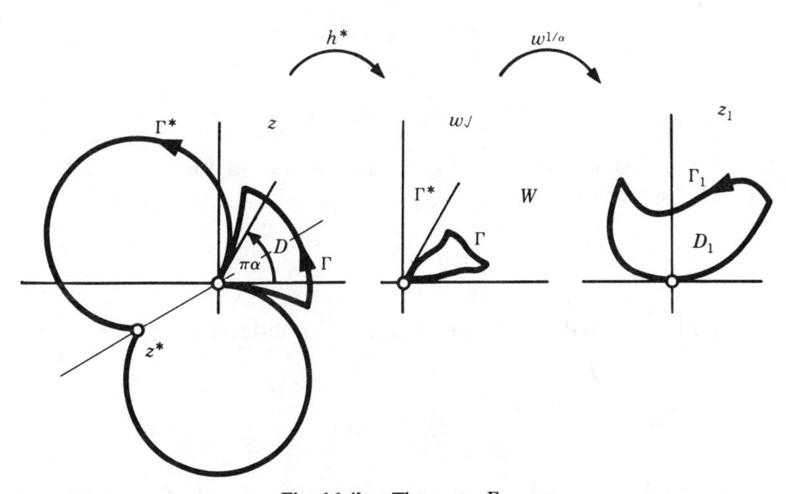

Fig. 16.4h. The curve Γ_1.

exists and is >0. The approach of z_1 to 0 is again nontangential, since by the Lindelöf-Carathéodory theorem the nontangential approach is preserved in all of the foregoing transformations.

Also by the Lindelöf-Carathéodory theorem, the nontangential approach is preserved when going from the z_1 plane to the w plane. In terms of the inverse function $p^{[-1]}$ it thus suffices to show that

$$\lim_{w\to 0} \frac{w}{p^{[-1]}(w)} = \mu > 0$$

exists for nontangential approach. Setting $w' := -1/w$, the nontangential approach to 0 corresponds to an approach to ∞ in a sector S: $\pi\delta < \arg w' < \pi(1-\delta)$, and on setting

$$q(w') := -\frac{1}{p^{[-1]}(-1/w')}$$

we are finally to show that

$$\lim_{\substack{w'\to\infty \\ w'\in S}} \frac{q(w')}{w'} = \mu > 0$$

exists. Writing z for w', the existence of a real limit $\mu \geqq 0$ is a consequence of the following theorem of independent interest:

THEOREM 16.4h (Landau–Valiron). *Let q be analytic in* $\operatorname{Im} z > 0$, *and let* $\operatorname{Im} q(z) > 0$ *for* $\operatorname{Im} z > 0$. *Then*

$$\mu := \lim_{\substack{z\to\infty \\ z\in S}} \frac{q(z)}{z} \tag{16.4-11}$$

exists, and $\mu \geqq 0$.

Proof. According to the Nevanlinna representation formula (12.10-5), q has the representation

$$q(z) = \alpha z + \beta + \int_{-\infty}^{\infty} \frac{1+\tau z}{\tau - z}\, d\chi(\tau),$$

where α and β are real, $\alpha \geqq 0$, and where χ is nondecreasing and bounded. There follows

$$\frac{q(z)}{z} = \alpha + \frac{\beta}{z} + \frac{1}{z}\int_{-\infty}^{\infty} \frac{1}{\tau - z}\, d\chi(\tau) + \int_{-\infty}^{\infty} \frac{\tau}{\tau - z}\, d\chi(\tau),$$

and since the first integral on the right is bounded, say, for $\operatorname{Im} z \geqq 1$, the theorem will be proved, with $\mu=\alpha$, if it is shown that

$$\int_{-\infty}^{\infty} \frac{\tau}{\tau-z}\, d\chi(\tau) \to 0, \qquad z\to\infty, \quad z\in S. \tag{16.4-12}$$

Let ε be given, and choose τ_1 so that

$$\left| \int_{|\tau|\geqq\tau_1} d\chi(\tau) \right| \leqq \frac{\varepsilon}{2}\sin\delta.$$

Then since $|\tau-z|\geqq\tau\sin\delta$ if $z\in S$,

$$\left| \int_{|\tau|\geqq\tau_1} \frac{\tau}{\tau-z}\, d\chi(\tau) \right| \leqq \frac{\varepsilon}{2}.$$

Next choose ρ such that $\rho>\tau_1$,

$$\frac{\tau_1}{\rho-\tau_1} \leqq \frac{\varepsilon}{2\nu},$$

where $\nu := \int_{-\infty}^{\infty} d\chi(\tau)$. Then since for $|z|>\rho$,

$$\left| \frac{\tau}{\tau-z} \right| \leqq \frac{|\tau|}{|z|-|\tau|} \leqq \frac{\tau_1}{\rho-\tau_1},$$

there follows

$$\left| \int_{|\tau|\leqq\tau_1} \frac{\tau}{\tau-z}\, d\chi(\tau) \right| \leqq \frac{\varepsilon}{2\nu}\cdot\nu = \frac{\varepsilon}{2}.$$

Hence for all $z\in S$ with $|z|\geqq\rho$,

$$\left| \int_{-\infty}^{\infty} \frac{\tau}{\tau-z}\, d\chi(\tau) \right| \leqq \frac{\varepsilon}{2}+\frac{\varepsilon}{2} = \varepsilon,$$

proving the assertion (16.4-12) and hence Theorem 16.4h. —

We now have established the existence of the crucial limit (16.4-10),

$$\mu = \lim_{z\to 0} \frac{p(z)}{z}, \tag{16.4-13}$$

where p maps the interior of Γ_1 onto U such that $p(0)=0$. We recall that Γ_1 has the real axis as a tangent at $z=0$, and that the positive imaginary axis points to the interior of Γ_1. To establish Theorem 16.4b, it remains to be shown that $\mu>0$. This is accomplished by comparing p with another mapping function p_i for which the limit is known to be positive (see Fig. 16.4i).

Let Γ_i be the circle that touches the real axis at $z=0$ and, with the exception of the point $z=0$, lies in int Γ_1. The pre-image of the half-line

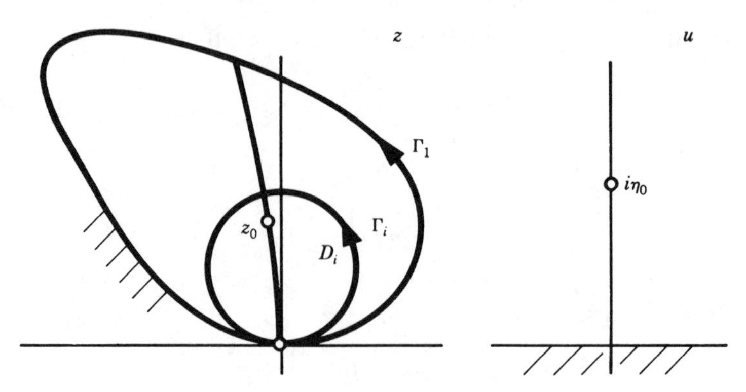

Fig. 16.4i. Showing that $\mu > 0$.

$u = i\eta$, $\eta \geqq 0$, under the map p by Theorem 16.3h (Lindelöf-Carathéodory) meets the real axis at a right angle. Thus for sufficiently small $\eta > 0$, the points $p^{[-1]}(i\eta)$ lie within Γ_i. Let z_0 be such a point, and let $i\eta := p(z_0)$. Let p_i be the function mapping the interior D_i of Γ_i onto U and satisfying

$$p_i(0) = 0, \qquad p_i(z_0) = i\eta. \tag{16.4-14}$$

The function p_i is a Moebius transformation that could be written down explicitly, but there will be no need to do so.

The Moebius transformation

$$w = \frac{u - i\eta}{u + i\eta}$$

maps U onto E. Denoting by $u(w)$ the inverse transformation, the function

$$q(w) := \frac{p(p_i^{[-1]}(u(w))) - i\eta}{p(p_i^{[-1]}(u(w))) + i\eta}$$

by (16.4-14) satisfies $q(0) = 0$. Moreover, since all values of p lie in U, $|q(w)| < 1$. Hence by the lemma of Schwarz (Theorem 5.10b),

$$|q(w)| \leqq |w|, \qquad |w| < 1,$$

which is to say,

$$\left|\frac{p(p_i^{[-1]}(u)) - i\eta}{p(p_i^{[-1]}(u)) + i\eta}\right| \leqq \left|\frac{u - i\eta}{u + i\eta}\right|, \qquad u \in U$$

or

$$\left|\frac{p(z) - i\eta}{p(z) + i\eta}\right| \leqq \left|\frac{p_i(z) - i\eta}{p_i(z) + i\eta}\right|, \qquad z \in D_i.$$

Letting $p := p(z)$, $p_i := p_i(z)$, this implies

$$|(p-i\eta)(p_i+i\eta)| \leqq |(p+i\eta)(p_i-i\eta)|$$

or

$$\left|p-p_i-i\frac{\eta^2+pp_i}{\eta}\right| \leqq \left|p-p_i+i\frac{\eta^2+pp_i}{\eta}\right|. \tag{16.4-15}$$

We let $z = iy$, where $y > 0$, and take into account that both $p(iy)$ and $p_i(iy)$ are $O(y)$ as $y \to 0$. If divided by iy, (16.4-15) shows that the point

$$\frac{p(iy)-p_i(iy)}{iy}$$

lies closer to a point

$$z^* := \frac{\eta^2+O(y^2)}{\eta y}$$

than to $-z^*$. Since z^* for $y \to 0$ approaches the real axis, this is possible only if

$$\lim_{y\to 0} \frac{p(iy)-p_i(iy)}{iy}$$

(which is already known to exist) is nonnegative. Hence

$$\mu = \lim_{y\to 0} \operatorname{Re} \frac{p(iy)}{iy} \geqq \lim_{y\to 0} \operatorname{Re} \frac{p_i(iy)}{iy} =: \mu_i.$$

But evidently $\mu_i = \operatorname{Re} p_i'(0) = p_i'(0) > 0$, because the derivative of the Moebius transformation p_i never vanishes and at $z = 0$ is positive because the image $p(\Gamma_i)$ is tangent to Γ_i at $z = 0$. We conclude that $\mu > 0$, and thus have concluded the proof of Theorem 16.4b. —

PROBLEM

1. Extend Theorem 16.4b to show that for $n = 1, 2, \ldots$,

$$\lim_{z\to 0} \{z^{n-1/\alpha} f^{(n)}(z)\} = \gamma \frac{1}{\alpha}\left(\frac{1}{\alpha}-1\right) \cdots \left(\frac{1}{\alpha}-n+1\right).$$

(See Warschawski [1955].)

NOTES

Our proof of Theorem 16.4b follows Warschawski [1955]. Lichtenstein [1911] had proved the result for irrational α. For Theorem 16.4h, see Landau and Valiron [1929] or Carathéodory

[1929], where the proofs are different. See also Carathéodory [1961], § 298, and Littlewood [1944], p. 148.

§ 16.5. CONFORMAL MAPPING AND POTENTIAL THEORY

With this section we return to our central problem of constructing mapping functions for specified regions. Unlike the osculation methods presented in § 16.2, the methods about to be discussed require special assumptions on the boundaries of the region to be mapped. In the present section we present some methods that directly exploit the connection between the mapping function and potential theory that was established in § 15.6. Here we assume that the boundary of the region to be mapped is a Jordan curve.

I. A Dirichlet Problem for the Interior Mapping Function

Let R be the *interior* of a Jordan curve Γ, let $a \in R$, and consider the problem of determining a function f mapping R onto $|w|<1$ such that $f(a)=0$. If γ is Green's function for R, then the representation (15.6-10) yields

$$\operatorname{Log}|f(z)| = -2\pi\gamma(z, a)$$

and thus

$$\operatorname{Log}\left|\frac{f(z)}{z-a}\right| = -2\pi\gamma(z, a) + \operatorname{Log}\frac{1}{|z-a|} = -2\pi u(z, a),$$

where u is the regular part of the fundamental solution γ, that is, the harmonic function satisfying the boundary condition (15.6-6). The function $f(z)(z-a)^{-1}$ is analytic and $\neq 0$ in the simply connected region R and hence possesses an analytic logarithm in R, whose real part is $-2\pi u(z, a)$. If v is a conjugate harmonic function of u, and if $h := u + iv$, then this analytic logarithm satisfies

$$\log\frac{f(z)}{z-a} = -2\pi h(z) + i\alpha,$$

where α is a real constant. There follows

$$f(z) = (z-a)\, e^{i\alpha}\, e^{-2\pi h(z)}.$$

The appearance of the factor $e^{i\alpha}$ is natural, because by our conditions f is determined only up to a constant factor of modulus 1. The appearance of the factor -2π in the above formula is due only to the normalization of Green's function. In streamlined form the foregoing result reads as follows:

THEOREM 16.5a. *Let R be a Jordan region, and let $a \in R$. Then the function f mapping R onto $|w|<1$ such that $f(a)=0, f'(a)>0$ may be constructed as* *follows*:

(i) *Solve the Dirichlet problem for R with the boundary condition*

$$u(z) = \operatorname{Log} \frac{1}{|z-a|}, \qquad z \in R. \tag{16.5-1}$$

(ii) *Determine the conjugate harmonic function v of u such that*

$$v(a) = 0. \tag{16.5-2}$$

(iii) *Set*

$$f(z) = (z-a)e^{u(z)+iv(z)}. \tag{16.5-3}$$

Thus in principle, any method for solving the Dirichlet problem can be used to construct the Riemann mapping function. After the solution of the Dirichlet problem (16.5-1) has been determined, the conjugate function v may be found from the formula

$$v(z) = \int_a^z (v_x \, dx + v_y \, dy),$$

which by virtue of the Cauchy–Riemann equations is the same as

$$v(z) = \int_a^z (-u_y \, dx + u_x \, dy), \tag{16.5-4}$$

the path of integration being irrelevant.

If a finite-difference method is used to construct u, finite-difference analogs of the Cauchy–Riemann equations may be used to approximate the relation (16.5-4). A computer program written by Kaiser [1975] does all of this automatically and also draws the images of the finite-difference net lines. In Fig. 16.5a we show the image of the unit square $R := \{(x, y)\colon -1 < x < 1, -1 < y < 1\}$, where $a = -\frac{1}{4} - \frac{1}{2}i$.

Also in the same program, Kaiser calculates and draws, by means of inverse interpolation, the inverse images of selected circles $|w| = \rho$ and of rays emanating from $w = 0$. (In electrostatic applications where a represents the location of a thin wire suspended in a hollow cylinder with cross section R whose boundary is at potential zero, these inverse images are, respectively, curves of constant potential and field lines.) In Fig. 16.5b we show these inverse images for the mapping given above.

In some cases the construction indicated in Theorem 16.5a may be carried through analytically.

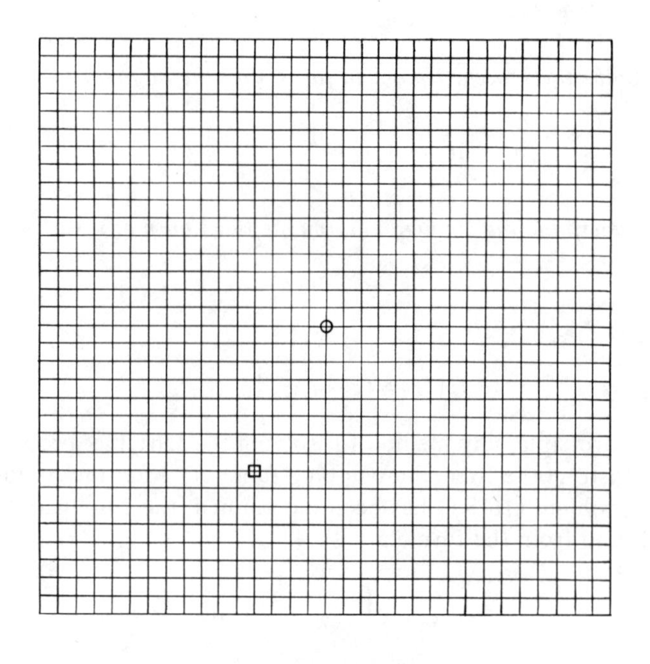

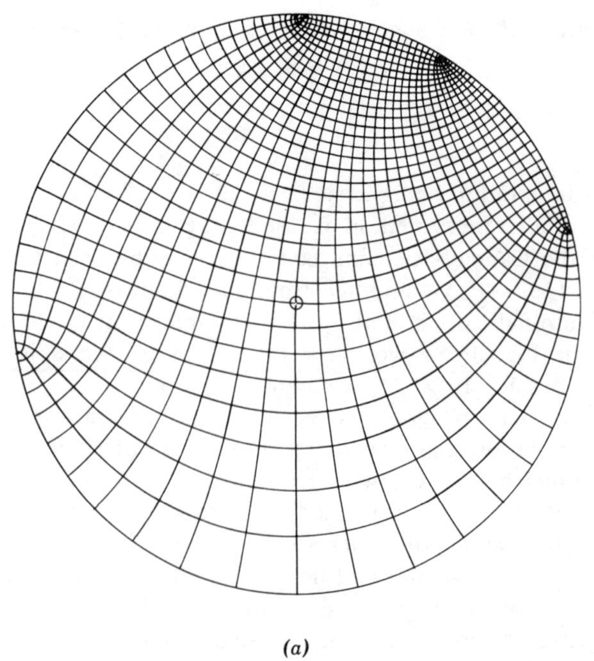

(a)

Fig. 16.5a. Mapping the unit square.

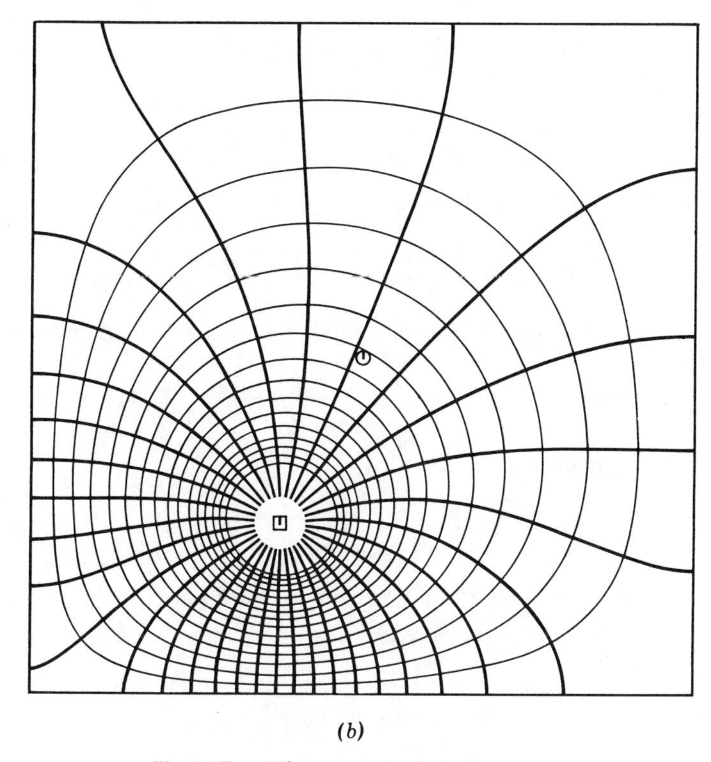

(*b*)

Fig. 16.5b. Wire suspended in hollow square.

EXAMPLE 1

Let $\rho > 1$, and let R be the region bounded by the ellipse

$$z(\tau) = \tfrac{1}{2}(\rho\, e^{i\tau} + \rho^{-1}\, e^{-i\tau}), \qquad 0 \leqq \tau \leqq 2\pi, \tag{16.5-5}$$

with semiaxes $\frac{1}{2}(\rho + \rho^{-1})$ and $\frac{1}{2}(\rho - \rho^{-1})$. We use the method of Theorem 16.5a to construct the Riemann mapping function for $a = 0$, which in Example 2 of § 5.11 was shown to be not elementary.

The Dirichlet problem (16.5-1) may be solved by the method sketched in Problem 3 of § 15.2. The first step is to find the Fourier series of the boundary function:

$$\begin{aligned} -\mathrm{Log}|z(\tau)| &= -\tfrac{1}{2}\,\mathrm{Log}\, z(\tau)\overline{z(\tau)} \\ &= -\tfrac{1}{2}\,\mathrm{Log}[\tfrac{1}{4}(\rho\, e^{i\tau} + \rho^{-1}\, e^{-i\tau})(\rho\, e^{-i\tau} + \rho^{-1}\, e^{i\tau})] \\ &= -\frac{1}{2}\,\mathrm{Log}\,\frac{\rho^2}{4} - \frac{1}{2}\,\mathrm{Log}(1 + \rho^{-2}\, e^{-2i\tau}) - \tfrac{1}{2}\,\mathrm{Log}(1 + \rho^{-2}\, e^{2i\tau}) \end{aligned}$$

$$= -\operatorname{Log}\frac{\rho}{2} + \operatorname{Re}\sum_{n=1}^{\infty}\frac{(-1)^n}{n}\rho^{-2n}e^{2in\tau}$$

$$= -\operatorname{Log}\frac{\rho}{2} + \sum_{n=1}^{\infty}\frac{(-1)^n}{n}\rho^{-2n}\cos 2n\tau.$$

It then follows that the solution of the Dirichlet problem is

$$u(z) = -\operatorname{Log}\frac{\rho}{2} + \sum_{n=1}^{\infty}\frac{(-1)^n}{n}\frac{1}{\rho^{4n}+1}\operatorname{Re} T_{2n}(z),$$

where $T_n(z) :=$ nth Chebyshev polynomial. Hence we immediately have

$$u(z) + iv(z) = -\operatorname{Log}\frac{\rho}{2} + \sum_{n=1}^{\infty}\frac{(-1)^n}{n}\frac{1}{\rho^{4n}+1}T_{2n}(z),$$

and the mapping function is given by

$$f(z) = \frac{2z}{\rho}\exp\left\{\sum_{n=1}^{\infty}\frac{(-1)^n}{n}\frac{1}{\rho^{4n}+1}T_{2n}(z)\right\}, \qquad (16.5\text{-}6)$$

where the series converges rapidly except for very flat ellipses. In view of $T_{2n}(0) = (-1)^n$, the formula yields

$$f'(0) = \frac{2}{\rho}\exp\left\{\sum_{n=1}^{\infty}\frac{2}{n(\rho^{4n}+1)}\right\}. \quad —$$

II. A Dirichlet Problem for the Exterior Mapping Function

Again let Γ be a Jordan curve, and let $\mathring{D}$ be the *exterior* of Γ. In Corollary 5.10c of the Riemann mapping theorem we have already stated the fact that there exists a unique function $\mathring{f}$ mapping $\mathring{D}$ onto $|w| > 1$ such that $\mathring{f}(\infty) = \infty$ and the Laurent series of $\mathring{f}$ at ∞ has the form

$$\mathring{f}(z) = \gamma^{-1}z + a_0 + a_1 z^{-1} + \cdots, \qquad \gamma > 0. \qquad (16.5\text{-}7)$$

This is the **exterior mapping function** of Γ, and γ is called the **capacity** of Γ. We now show that the exterior mapping function can likewise be constructed by solving a Dirichlet problem.

LEMMA 16.5b. *Let a be a point in the interior of* Γ. *Then there exists an analytic logarithm of* $\mathring{f}(z)/(z-a)$ *in* $\mathring{D}$. *That is, an analytic function* $l(z)$ *can be defined in* $\mathring{D}$ *whose value at every* $z \in \mathring{D}$ *is one of the values of* $\log[\mathring{f}(z)/(z-a)]$. *The function l is analytic even at* ∞, *and can be so defined that* $l(\infty)$ *is real.*

Proof. The result does not follow from Theorem 4.3f, because $\mathring{D}$ is not simply connected. However, if R is a region of arbitrary connectivity, and if g is analytic and $\neq 0$ in R, it is still true that

$$\log g(z) - \log g(z_0) = \int_{z_0}^{z} \frac{g'(z)}{g(z)}\,dz,$$

provided that the integral defines a single-valued function in R. In our case, $R = \mathring{D}$,

$$g(z) = \frac{\mathring{f}(z)}{z-a} \neq 0, \qquad z \in \mathring{D},$$

and the integral defines a single-valued function provided that

$$\int_{\Gamma_1} \frac{g'(z)}{g(z)}\,dz = 0, \tag{16.5-8}$$

where Γ_1 is any closed curve surrounding Γ. In particular, Γ_1 may be taken as a large circle. Since

$$\frac{g'(z)}{g(z)} = \frac{\mathring{f}'(z)}{\mathring{f}(z)} - \frac{1}{z-a} = O(z^{-2}), \qquad z \to \infty,$$

the Laurent series at ∞ of g'/g has no term in z^{-1}, and the integral (16.5-8) clearly vanishes. The same expansion shows that by defining

$$l(z) := \operatorname{Log} \gamma^{-1} + \int_{\infty}^{z} \frac{g'(z)}{g(z)}\,dz$$

we obtain a branch of $\log[f(z)/(z-a)]$ with the desired properties. —

Now let

$$l(z) = u(z) + iv(z).$$

The real part u is harmonic in $\mathring{D}$ and bounded. By Theorem 15.4d, u is thus determined uniquely as the solution of an exterior Dirichlet problem. In view of $|\mathring{f}(z)| = 1$, the values on the boundary Γ of $\mathring{D}$ are

$$u(z) = \operatorname{Log}\left|\frac{\mathring{f}(z)}{z-a}\right| = \operatorname{Log}\frac{1}{|z-a|}.$$

Theorem 16.5a thus has the following counterpart:

THEOREM 16.5c. *Let $\mathring{D}$ be the exterior of a Jordan curve Γ, and let a be a point in the interior of Γ. If u denotes the solution of the exterior Dirichlet problem,*

$$\Delta u = 0, \qquad z \in \mathring{D}, \tag{16.5-9a}$$

u bounded in $\mathring{D}$,

$$u(z) = \operatorname{Log}\frac{1}{|z-a|}, \qquad z \in \Gamma, \tag{16.5-9b}$$

then the conjugate harmonic function v of u exists, and if $v(\infty)=0$, the exterior mapping function of Γ is given by

$$\mathring{f}(z)=(z-a)\, e^{u(z)+iv(z)}. \tag{16.5-10}$$

PROBLEMS

1. Use the method of Theorem 16.5a to construct (analytically) the function f which maps the disk $|z|<1$ onto itself while sending the point $z=a$, $0<a<1$, to $w=0$. Compare the result with the well-known expression

$$f(z)=\frac{z-a}{1-az}.$$

2. *Mapping the square.* Use Theorem 16.5a to find the map f for the square

$$-1<\operatorname{Re} z<1, \qquad -1<\operatorname{Im} z<1$$

onto the unit disk, $f(0)=0, f'(0)>0$.

(a) Assuming the expansion

$$\operatorname{Log}|1+i\sin \pi y|^{-1}=a_0+\sum_{k\,\text{odd}} a_k \cos\frac{k\pi y}{2}, \qquad -1\leqq y\leqq 1,$$

show that the boundary value problem (i) is solved by

$$u(x,y)=a_0+\sum_{k\,\text{odd}}\frac{a_k}{\cosh(k\pi/2)}\left(\cos\frac{k\pi y}{2}\cosh\frac{k\pi x}{2}+\cos\frac{k\pi x}{2}\cosh\frac{k\pi y}{2}\right).$$

(b) Show that

$$u(z)+iv(z)=a_0+\sum_{k\,\text{odd}} a_k\left(\cosh\frac{k\pi}{2}\right)^{-1}\left(\cos\frac{k\pi z}{2}+\cosh\frac{k\pi z}{2}\right).$$

(c) To find the a_k, first show that for all real y,

$$\operatorname{Log}\left|1+i\alpha\sin\frac{\pi y}{2}\right|^{-1}=\operatorname{Log}\frac{1-b^2}{1+b^2}+b^2\cos\pi y+\frac{b^4}{2}\cos 2\pi y+\frac{b^6}{3}\cos 3\pi y+\cdots,$$

where

$$b:=\frac{\alpha}{\sqrt{1+\alpha^2}+1}.$$

Then expand each term $\cos m\pi y$ in a series of terms $\cos(k\pi y/2)$, where k runs through all odd integers.

3. Using the method of the preceding problem, find the mapping function for an arbitrary rectangle. (The Schwarz-Christoffel function furnishes the inverse map.)

NOTES

The connection between Dirichlet's problem and conformal mapping is basic in the theory of the latter and was used in early attempts to prove the Riemann mapping theorem. The connection is explored at length by Courant [1950]. The mapping function for the interior of the ellipse is calculated by means of elliptic functions by Ostrowski [1955] and in a different elementary manner by Szegö [1950].

§ 16.6. INTEGRAL EQUATIONS OF THE FIRST KIND

The Dirichlet methods discussed in the preceding section reduce the problem of finding the interior or exterior mapping function to the solution of an interior or exterior boundary value problem of the Dirichlet type, that is, to the determination of an unknown real function of *two* real variables. Although the solution of this boundary problem is basically a simple matter, it would seem even more promising to reduce the mapping problem to the determination of a real function of *one* real variable. The methods to be discussed next all are of this kind.

I. Symm's Integral Equations

Let Γ be a piecewise analytic Jordan curve, given by

$$z = z(\tau), \qquad 0 \leqq \tau \leqq \beta. \tag{16.6-1}$$

(It is sometimes assumed that the parameter τ is the arc length on Γ. This assumption, although convenient for theoretical work, is numerically objectionable because it introduces an additional source of error. We therefore explicitly admit arbitrary (piecewise analytic) parametric representations.) Let D be the interior and $\overset{\circ}{D}$ the exterior of Γ. We assume that $0 \in D$. Our concern is with both (a) the interior and (b) the exterior mapping function of Γ. As to the interior mapping function, we are looking for the mapping function f satisfying the normalizing conditions

$$f(0) = 0, \qquad f'(0) > 0. \tag{16.6-2}$$

(a) If $a = 0$, the Dirichlet problem (16.5-1) is

$$\Delta u(z) = 0, \qquad z \in D, \tag{16.6-3a}$$

$$u(z) = -\operatorname{Log}|z|, \qquad z \in \Gamma. \tag{16.6-3b}$$

We seek a solution of this problem in the form of a logarithmic single-layer potential,

$$u(z) = -\frac{1}{2\pi} \int_0^\beta \operatorname{Log}|z - z(\tau)| \mu(\tau)\, d\tau, \qquad z \in D. \tag{16.6-4}$$

Here μ is a real function that is to be determined. If u can be thus represented with a continuous μ, then in view of the continuity of the logarithmic potential the boundary condition (16.6-3b) yields, for $z = z(\sigma)$,

$$\frac{1}{2\pi}\int_0^\beta \operatorname{Log}|z(\sigma)-z(\tau)|\mu(\tau)\,d\tau = \operatorname{Log}|z(\sigma)|, \qquad 0 \leqq \sigma \leqq \beta. \quad (16.6\text{-}5)$$

This is called **Symm's interior equation** because it was first explored by Symm [1966] for purposes of numerical conformal mapping. The function $z(\tau)$ being given, Symm's equation is a linear integral equation of the first kind for the unknown function μ. If a solution has been found, then $u(z)$ may be calculated from (16.6-4) for any $z \in D$. The conjugate harmonic function v, which is also required in (16.5-3), is obtained by replacing the harmonic kernel $\operatorname{Log}|z - z(\tau)|$ in (16.6-4) by the conjugate harmonic function $\arg[z(\tau) - z]$,

$$v(z) = -\frac{1}{2\pi}\int_0^\beta \arg[z(\tau)-z]\mu(\tau)\,d\tau + \text{const.}$$

The value of the constant and the choice of the argument are not clear. However, from the condition $v(0) = 0$ we get

$$0 = -\frac{1}{2\pi}\int_0^\beta \arg[z(\tau)]\mu(\tau)\,d\tau + \text{const},$$

which on subtraction yields

$$v(z) = -\frac{1}{2\pi}\int_0^\beta \arg\left[1 - \frac{z}{z(\tau)}\right]\mu(\tau)\,d\tau. \qquad (16.6\text{-}6)$$

Here for each $\tau \in [0, \beta]$,

$$\alpha(z) := \arg\left[1 - \frac{z}{z(\tau)}\right]$$

denotes that value of the argument that is continuous in D and satisfies $\alpha(0) = 0$. If Γ is starlike with respect to 0, this is the *principal value* of the argument.

(b) In order to construct the solution u of the exterior Dirichlet problem (16.5-9),

$$\Delta u(z) = 0, \qquad z \in \mathring{D}, \qquad (16.6\text{-}7a)$$

$$u(z) = -\operatorname{Log}|z|, \qquad z \in \Gamma, \qquad u \text{ bounded in } \mathring{D}, \qquad (16.6\text{-}7b)$$

it does not make sense to use the representation (16.6-4), because for $|z|$

large this behaves like

$$-\frac{1}{2\pi}\operatorname{Log}|z|\int_0^\infty \mu(\tau)\,d\tau,$$

which remains bounded only if

$$\int_0^\beta \mu(\tau)\,d\tau=0.$$

One might try to use the representation

$$u(z)=\frac{1}{2\pi}\int_0^\beta \operatorname{Log}\left|1-\frac{z(\tau)}{z}\right|\mu(\tau)\,d\tau.$$

However, this always tends to 0 as $z\to\infty$. If $\mathring{f}(z)=\gamma^{-1}z+\ldots$, then $u(\infty)=-\operatorname{Log}\gamma$. The correct ansatz for $u(z)$ therefore must contain the unknown capacity γ as follows:

$$u(z)=-\operatorname{Log}\gamma+\frac{1}{2\pi}\int_0^\beta \operatorname{Log}\left|1-\frac{z(\tau)}{z}\right|\mu(\tau)\,d\tau,\qquad z\in\mathring{D}. \tag{16.6-8}$$

If μ is continuous, the boundary condition (16.6-7b) yields in view of the continuity of the logarithmic potential **Symm's exterior equation,**

$$\frac{1}{2\pi}\int_0^\beta \operatorname{Log}\left|1-\frac{z(\tau)}{z(\sigma)}\right|\mu(\tau)\,d\tau=\operatorname{Log}\gamma-\operatorname{Log}|z(\sigma)|,\qquad 0\leqq\sigma\leqq\beta. \tag{16.6-9}$$

If this equation is solved, u can be obtained from (16.6-8) and the conjugate harmonic function v from

$$v(z)=\frac{1}{2\pi}\int_0^\beta \arg\left[1-\frac{z(\tau)}{z}\right]\mu(\tau)\,d\tau. \tag{16.6-10}$$

Here for each $\tau\in[0,\beta]$,

$$\alpha(z):=\arg\left[1-\frac{z(\tau)}{z}\right]$$

denotes the value of the argument that is continuous in $\mathring{D}$ and satisfies $\alpha(\infty)=0$.

II. The Boundary Correspondence Function

By virtue of the Osgood-Carathéodory theorem (Theorem 16.3a), the interior mapping function can be extended under the hypotheses made earlier to a homeomorphism of the closure D' to the closure E'. Thus as

$z(\tau)$ moves along Γ, the image point $f(z(\tau))$ describes the unit circle. Thus by Theorem 4.6a,

$$\theta(\tau) := \arg f(z(\tau)), \qquad 0 \leqq \tau \leqq \beta,$$

may be defined as a continuous function. Any such continuous argument of $f(z(\tau))$ is called an **interior boundary correspondence function** for the map f. This correspondence refers to a particular parametric representation of Γ, and thus is not determined by the geometry of Γ alone.

Because $\theta(\tau)$ locally equals one of the branches of $\operatorname{Im}\log f(z(\tau))$, any boundary correspondence function for a piecewise analytic boundary curve Γ is not only continuous, but also analytic, except at the corners of Γ. The same holds for the derivative $\theta'(\tau)$. As to the behavior at the corners, the following is a direct consequence of the Warschawski–Lichtenstein theorem (Theorem 16.4b):

THEOREM 16.6a. *If τ_0 is a parameter value such that the piecewise analytic boundary curve Γ has a corner at $z(\tau_0)$ with interior angle $\alpha\pi$, where $0 < \alpha \leqq 2$, the one-sided limits of $(\tau-\tau_0)^{1-1/\alpha}\theta'(\tau)$ exist at τ_0 and are $\neq 0$. It follows that in a neighborhood of any such corner, $|\theta'(\tau)|^p$ is integrable for every exponent $p<2$. If the interior angle at the corner is $<2\pi$, then $[\theta'(\tau)]^2$ is integrable.*

If a boundary correspondence function is known, the values of both the mapping function f and the inverse mapping function $g = f^{[-1]}$ may be calculated by quadrature at arbitrary interior points of their domains of definition. If $z \in D$, then by Cauchy's integral formula,

$$w = f(z) = \frac{1}{2\pi i}\int_\Gamma \frac{f(t)}{t-z}\,dt,$$

which on introducing $t = z(\tau)$ becomes

$$w = \frac{1}{2\pi i}\int_0^\beta \frac{e^{i\theta(\tau)}}{z(\tau)-z} z'(\tau)\,d\tau. \tag{16.6-11}$$

If $w \in E$, then again by Cauchy's formula,

$$z = g(w) = \frac{1}{2\pi i}\int_{|u|=1} \frac{g(u)}{u-w}\,du = \frac{1}{2\pi}\int_0^{2\pi} \frac{g(e^{i\theta})\,e^{i\theta}}{e^{i\theta}-w}\,d\theta,$$

which on introducing $\theta = \theta(\tau)$ turns into

$$z = \frac{1}{2\pi}\int_0^\beta \frac{z(\tau)\theta'(\tau)}{1-e^{-i\theta(\tau)}w}\,d\tau. \tag{16.6-12}$$

A boundary correspondence function may be defined also for the exterior mapping function $\mathring{f}$. Under the hypotheses made, $\mathring{f}$ can be extended to a homeomorphism of the closure $\mathring{D}'$ to the closure $\mathring{E}'$. Any continuous argument

$$\mathring{\theta}(\tau) := \arg \mathring{f}(z(\tau)), \qquad 0 \leqq \tau \leqq \beta,$$

is called an **exterior boundary correspondence function**. Again $\mathring{\theta}(\tau)$ is not merely continuous but piecewise analytic, and Theorem 16.6a holds for $\mathring{\theta}'(\tau)$ *if the angles are measured from the interior of* $\mathring{D}$.

To express the exterior mapping function $\mathring{f}$ and its inverse $\mathring{g}$ in terms of $\mathring{\theta}$, we consider

$$f^*(z) = \frac{1}{\mathring{f}(1/z)},$$

which maps the interior of the curve

$$\Gamma^*: \quad z = z^*(\sigma) := \frac{1}{z(-\sigma)}, \qquad -\beta \leqq \sigma \leqq 0,$$

onto E and satisfies $f^*(0) = 0, f^{*\prime}(0) > 0$. Its boundary correspondence function is

$$\theta^*(\sigma) = \arg f^*(z^*(\sigma)) = -\arg \mathring{f}(z(-\sigma)) = -\mathring{\theta}(-\sigma).$$

In view of $\mathring{f}(z) = [f^*(z^{-1})]^{-1}$, the foregoing formulas yield, for $z \in \mathring{D}$,

$$w = \mathring{f}(z) = \left[\frac{z}{2\pi i} \int_0^\beta \frac{e^{-i\mathring{\theta}(\tau)}}{z - z(\tau)} \frac{z'(\tau)}{z(\tau)} \, d\tau \right]^{-1}. \tag{16.6-13}$$

For the inverse function we obtain in a similar manner

$$z = \mathring{g}(w) = \left[\frac{1}{2\pi} \int_0^\beta \frac{1}{z(\tau)} \frac{1}{1 - w^{-1} e^{i\mathring{\theta}(\tau)}} \mathring{\theta}'(\tau) \, d\tau \right]^{-1}. \tag{16.6-14}$$

III. Solution of Symm's Equations

In our discussion of Symm's integral equations, the questions of whether these equations possess solutions, and whether any existing solutions are unique, were left open. Here we shall not only answer these questions, but actually solve the equations in terms of the boundary correspondence functions defined above.

Let z be exterior to Γ, and let $g := f^{[-1]}$. Then the function of w,

$$w \to z - g(w),$$

is analytic in E and continuous and different from 0 on E'. It therefore has an analytic logarithm

$$h(w) := \log[z - g(w)], \tag{16.6-15}$$

which is again continuous on E'. In view of the extended form of Cauchy's formula (Theorem 4.7d),

$$\frac{1}{2\pi}\int_0^{2\pi} h(e^{i\theta})\, d\theta = h(0).$$

In view of $g(0)=0$ this is the same as

$$\frac{1}{2\pi}\int_0^{2\pi} \log[z - g(e^{i\theta})]\, d\theta = \log z,$$

where the branch of $\log z$ is determined by the analytic branch of $\log[z-g(w)]$ that has been selected. In the last integral, introduce the parameter τ by $\theta = \theta(\tau)$, where $\theta(\tau)$ is an interior boundary correspondence function. By virtue of $g(e^{i\theta(\tau)}) = z(\tau)$ there follows:

LEMMA 16.6b. *With the above notation, there holds for each z in the exterior of Γ*

$$\frac{1}{2\pi}\int_0^{\beta} \log[z - z(\tau)]\theta'(\tau)\, d\tau = \log z, \tag{16.6-16}$$

where the logarithms are, respectively, the values on $|w|=1$ and at $w=0$ of the function h defined by (16.6-15).

We shall use this lemma repeatedly. At this time we take real parts in (16.6-16). This yields

$$\frac{1}{2\pi}\int_0^{\beta} \operatorname{Log}|z - z(\tau)|\theta'(\tau)\, d\tau = \operatorname{Log}|z|,$$

now without ambiguity about the logarithms. Let $\sigma \in [0, \beta]$, and let $z \to z(\sigma)$. By Theorem 15.8b and Theorem 16.6a, the logarithmic potential on the left is continuous, and there follows

$$\frac{1}{2\pi}\int_0^{\beta} \operatorname{Log}|z(\sigma) - z(\tau)|\theta'(\tau)\, d\tau = \operatorname{Log}|z(\sigma)|. \tag{16.6-17}$$

A comparison with (16.6-5) shows:

THEOREM 16.6c. *Let Γ be a piecewise analytic Jordan curve, and let $\theta(\tau)$ be an interior boundary correspondence function of Γ. Then Symm's interior equation* (16.6-5) *in every space $L_p(0, \beta)$ where $p<2$ has the solution $\mu(\tau) = \theta'(\tau)$. If the interior angles at all corners are $<2\pi$, then this solution is in $L_2(0, \beta)$.*

A new representation of the mapping function now follows from Theorem 16.5a. From (16.5-3),

$$f(z) = z\exp\{u(z) + iv(z)\},$$

where

$$u(z) = -\frac{1}{2\pi}\int_0^\beta \operatorname{Log}|z - z(\tau)|\theta'(\tau)\,d\tau,$$

and where v is the conjugate harmonic function satisfying $v(0) = 0$,

$$v(z) = -\frac{1}{2\pi}\int_0^\beta \arg\left[1 - \frac{z}{z(\tau)}\right]\theta'(\tau)\,d\tau.$$

We evidently have

$$u(z) + iv(z) = -\frac{1}{2\pi}\int_0^\beta \log\left[1 - \frac{z}{z(\tau)}\right]\theta'(\tau)\,d\tau - \omega,$$

where for each $\tau \in [0, \beta]$ the log function is the analytic branch having the value 0 at $z = 0$, and

$$\omega := \frac{1}{2\pi}\int_0^\beta \operatorname{Log}|z(\tau)|\theta'(\tau)\,d\tau.$$

To evaluate ω, we let $\theta = \theta(\tau)$ be the new variable of integration and get

$$\begin{aligned}\omega &= \frac{1}{2\pi}\int_0^{2\pi} \operatorname{Log}|g(e^{i\theta})|\,d\theta \\ &= \operatorname{Re}\frac{1}{2\pi i}\int_{|w|=1} \frac{1}{w}\operatorname{Log}\frac{g(w)}{w}\,dw \\ &= \operatorname{Log} g'(0) = -\operatorname{Log} f'(0),\end{aligned}$$

by the mean value theorem for analytic functions. We thus find:

THEOREM 16.6d. *Under the hypotheses of the preceding theorem, the interior mapping function is*

$$f(z) = z\exp\left\{-\frac{1}{2\pi}\int_0^\beta \log\left[1 - \frac{z}{z(\tau)}\right]\theta'(\tau)\,d\tau\right\}f'(0), \qquad (16.6\text{-}18)$$

where

$$f'(0) = \exp\left\{-\frac{1}{2\pi}\int_0^\beta \operatorname{Log}|z(\tau)|\theta'(\tau)\,d\tau\right\}. \qquad (16.6\text{-}19)$$

The log in (16.6-18) is the principal value if Γ is starlike with respect to 0.

We next consider the *exterior* mapping problem. Let $\mathring{g} := \mathring{f}^{[-1]}$, and let z be in the *interior* of Γ. Then

$$\mathring{h}(w) := \log \frac{\mathring{g}(w) - z}{w} \tag{16.6-20}$$

may be defined as a function that is analytic in $\mathring{E}$ (and also at $w = \infty$) and continuous on $\mathring{E}'$, because

$$\frac{d}{dw} \log \frac{\mathring{g}(w) - z}{w} = O(w^{-2}).$$

If Γ_ρ denotes the circle $|w| = \rho$, $\rho > 1$, it thus follows that

$$\frac{1}{2\pi i} \int_{\Gamma_\rho} \frac{\mathring{h}(w)}{w}\, dw = \frac{1}{2\pi i} \int_{|w|=1} \frac{\mathring{h}(w)}{w}\, dw.$$

For $|w|$ large, $\mathring{g}(w) = w + O(1)$, hence

$$\mathring{h}(w) = \operatorname{Log} \gamma + O(w^{-1}).$$

Hence for $\rho \to \infty$ the limit of the first integral above equals $\operatorname{Log} \gamma$. Letting $w = e^{i\theta}$ in the second integral, we obtain

$$\frac{1}{2\pi} \int_0^{2\pi} \mathring{h}(e^{i\theta})\, d\theta = \operatorname{Log} \gamma$$

or

$$\frac{1}{2\pi} \int_0^{2\pi} \log\{e^{-i\theta}[\mathring{g}(e^{i\theta}) - z]\}\, d\theta = \operatorname{Log} \gamma.$$

Here we let $\theta = \mathring{\theta}(\tau)$, where $\mathring{\theta}(\tau)$ is an exterior boundary correspondence function. In view of $\mathring{g}(e^{i\mathring{\theta}(\tau)}) = z(\tau)$, there results:

LEMMA 16.6e. *For each z interior to Γ,*

$$\frac{1}{2\pi} \int_0^{\beta} \log\{e^{-i\mathring{\theta}(\tau)}[z(\tau) - z]\}\mathring{\theta}'(\tau)\, d\tau = \operatorname{Log} \gamma, \tag{16.6-21}$$

where the logarithm is the value at $w = e^{i\mathring{\theta}(\tau)}$ of the function (16.6-20).

Taking real parts and letting $z \to z(\sigma)$, we get, in view of the continuity of the logarithmic potential,

$$\frac{1}{2\pi} \int_0^{\beta} \operatorname{Log}|z(\tau) - z(\sigma)|\mathring{\theta}'(\tau)\, d\tau = \operatorname{Log} \gamma \quad \text{for all} \quad \sigma \in [0, \beta]. \tag{16.6-22}$$

There follows

$$\frac{1}{2\pi}\int_0^\beta \operatorname{Log}\left|1-\frac{z(\tau)}{z(\sigma)}\right| \mathring{\theta}'(\tau)\,d\tau = \operatorname{Log}\gamma - \operatorname{Log}|z(\sigma)|\frac{1}{2\pi}\int_0^\beta \mathring{\theta}'(\tau)\,d\tau,$$

which, in view of

$$\int_0^\beta \mathring{\theta}'(\tau)\,d\tau = \mathring{\theta}(\beta)-\mathring{\theta}(0)=2\pi,$$

yields

$$\frac{1}{2\pi}\int_0^\beta \operatorname{Log}\left|1-\frac{z(\tau)}{z(\sigma)}\right| \mathring{\theta}'(\tau)\,d\tau = \operatorname{Log}\gamma - \operatorname{Log}|z(\sigma)|, \qquad 0\leqq\sigma\leqq\beta. \tag{16.6-23}$$

A comparison with (16.6-9) yields:

THEOREM 16.6f. *If the boundary curve Γ is piecewise analytic, Symm's exterior integral equation (16.6-9) in every space $L_p(0,\beta)$ where $p<2$ has the solution $\mu(\tau)=\mathring{\theta}'(\tau)$, where $\mathring{\theta}$ is an exterior boundary correspondence function of Γ. If the angles at the corners of Γ (measured from the exterior of Γ) all are $<2\pi$, this solution is in $L_2(0,\beta)$.*

It follows from Theorem 16.5c and from (16.6-9) that the exterior mapping function is also represented by

$$\mathring{f}(z) = z\exp\{u(z)+iv(z)\},$$

where

$$u(z)=\frac{1}{2\pi}\int_0^\beta \operatorname{Log}\left|1-\frac{z(\tau)}{z}\right|\mathring{\theta}'(\tau)\,d\tau - \operatorname{Log}\gamma,$$

$$v(z)=\frac{1}{2\pi}\int_0^\beta \arg\left[1-\frac{z(\tau)}{z}\right]\mathring{\theta}'(\tau)\,d\tau.$$

Evidently,

$$u(z)+iv(z)=\frac{1}{2\pi}\int_0^\beta \log\left[1-\frac{z(\tau)}{z}\right]\mathring{\theta}'(\tau)\,d\tau - \operatorname{Log}\gamma,$$

where the logarithm is analytic in $\mathring{D}$ and 0 at $z=\infty$. We thus get:

THEOREM 16.6g. *Under the hypotheses of the preceding theorem the exterior mapping function is*

$$\mathring{f}(z)=\gamma^{-1}z\exp\left\{\frac{1}{2\pi}\int_0^\beta \log\left[1-\frac{z(\tau)}{z}\right]\mathring{\theta}'(\tau)\,d\tau\right\}, \tag{16.6-24}$$

where γ is the capacity of Γ.

The constant γ is expressed in terms of $\mathring{\theta}(\tau)$ by (16.6-22), but it remains a disadvantage of Symm's exterior equation that both $\mathring{\theta}$ and γ are unknown.

IV. Uniqueness of the solutions

We now have demonstrated that Symm's equations have solutions. But are the solutions that we have found the only ones?

The solutions that we have found are in $L_p(0, \beta)$ for every $p<2$. Moreover, on each closed subinterval of $[0, \beta]$ that does not contain a point corresponding to a corner of Γ they are analytic and therefore certainly uniformly Hölder continuous. It is in this sense that the term "solution of Symm's equation" will be understood.

Assume now that there are two solutions of Symm's interior equation (16.6-5), and let $\delta(\tau)$ be their difference. There follows

$$\frac{1}{2\pi}\int_0^\beta \operatorname{Log}|z(\sigma)-z(\tau)|\delta(\tau)\,d\tau=0, \qquad 0\leqq\sigma\leqq\beta. \tag{16.6-25}$$

For all $z\in\mathbb{C}$ let

$$u(z):=\frac{1}{2\pi}\int_0^\beta \operatorname{Log}|z-z(\tau)|\delta(\tau)\,d\tau. \tag{16.6-26}$$

The function u is harmonic for $z\in D$ and for $z\in\mathring{D}$, and it assumes the values 0 on Γ. There immediately follows

$$u(z)=0, \qquad z\in D.$$

To compute $u(z)$ for $z\in\mathring{D}$, observe that $u(\mathring{g}(w))$ is harmonic in $\mathring{E}$, zero on $|w|=1$, and $O(\operatorname{Log}|w|)$ for $w\to\infty$. Such a function is of the form $\kappa \operatorname{Log}|w|$, where the real constant κ remains undetermined. There follows

$$u(z)=\kappa \operatorname{Log}|\mathring{f}(z)|, \qquad z\in\mathring{D}. \tag{16.6-27}$$

We now compute, in two ways, the conjugate harmonic functions of u for $z\in D$ and for $z\in\mathring{D}$. For $z\in D$, $u=0$ has the conjugate harmonic function $v=0$. In any simply connected subregion $R\subset\mathring{D}$, u has the conjugate harmonic function $\kappa \arg\mathring{f}(z)$. On the other hand, the conjugate harmonic functions may be determined by replacing the kernel $\operatorname{Log}|z-z(\tau)|$ by the conjugate harmonic kernel $\arg[z-z(\tau)]$. Since conjugate functions are determined up to constants, this yields

$$0=\frac{1}{2\pi}\int_0^\beta \log[z-z(\tau)]\delta(\tau)+\text{const}, \qquad z\in D$$

$$\kappa \log\mathring{f}(z)=\frac{1}{2\pi}\int_0^\beta \log[z-z(\tau)]\delta(\tau)\,d\tau+\text{const}, \qquad z\in R\subset\mathring{D}.$$

We remove the constants as well as the ambiguities in the logarithms by differentiation. This yields

$$\frac{1}{2\pi}\int_0^\beta \frac{1}{z-z(\tau)}\,\delta(\tau)\,d\tau = \begin{cases} 0, & z\in D \\ \kappa\dfrac{\mathring{f}'(z)}{\mathring{f}(z)}, & z\in \mathring{D}. \end{cases} \tag{16.6-28}$$

Letting

$$m(t) := -\frac{i}{z'(\tau)}\,\delta(\tau), \qquad t = z(\tau), \tag{16.6-29}$$

the integral on the left becomes a Cauchy integral,

$$p(z) := \frac{1}{2\pi i}\int_\Gamma \frac{m(t)}{t-z}\,dt.$$

For points $z\in\Gamma$, it follows from (16.6-28) that

$$p^+(z) = 0,$$

and if z is not a corner of Γ,

$$p^-(z) = \kappa\frac{\mathring{f}'(z)}{\mathring{f}(z)}.$$

From the Sokhotskyi formulas (Theorem 14.1c) we have

$$p^+(z) - p^-(z) = m(z).$$

Letting $z = z(\tau)$, using (16.6-29) and considering the relations

$$\mathring{f}(z(\tau)) = e^{i\mathring{\theta}(\tau)},$$
$$\mathring{f}'(z(\tau))z'(\tau) = i\mathring{\theta}'(\theta)\,e^{i\mathring{\theta}(\tau)},$$

there follows

$$\delta(\tau) = \kappa\mathring{\theta}'(\tau) \tag{16.6-30}$$

for all τ that do not correspond to corners of Γ.

It remains to determine κ. We compute the asymptotic behavior of $u(z)$ as $z\to\infty$ in two ways. From (16.6-27) we find, using $\mathring{f}(z) = \gamma^{-1}z + a_0 + \ldots$,

$$u(z) = \kappa\ \mathrm{Log}|z| - \kappa\ \mathrm{Log}\ \gamma + O(z^{-1}).$$

On the other hand, by writing (16.6-26) in the form

$$u(z) = \frac{1}{2\pi}\int_0^\infty \left\{\mathrm{Log}|z| + \mathrm{Log}\left|1-\frac{z(\tau)}{z}\right|\right\}\delta(\tau)\,d\tau$$

we see that

$$u(z)=\left[\frac{1}{2\pi}\int_0^\beta \delta(\tau)\,d\tau\right]\mathrm{Log}|z|+O(z^{-1}).$$

Comparing the constant terms in these expressions, there follows

$$\kappa\ \mathrm{Log}\ \gamma=0. \tag{16.6-31}$$

Thus if $\gamma\neq 1$, $\kappa=0$ follows, and $\mu(\tau)=\theta'(\tau)$ is the only solution of Symm's interior equation. If $\gamma=1$, then (16.6-22) already shows that the solution is determined only up to arbitrary multiples of $\mathring{\theta}'(\tau)$. We now have seen that there are no further solutions. In summary we have proved:

THEOREM 16.6h. *Let Γ be a piecewise analytic Jordan curve, let γ be the capacity of Γ, and let $\theta(\tau)$ and $\mathring{\theta}(\tau)$ respectively be the interior and the exterior boundary correspondence functions of Γ. If $\gamma\neq 1$, then $\mu(\tau)=\theta'(\tau)$ is the only solution of Symm's interior equation* (16.6-5). *If $\gamma=1$, every real solution of the equation is of the form $\mu(\tau)=\theta'(\tau)+\kappa\mathring{\theta}'(\tau)$ where κ is real. In any case, $\mu(\tau)=\theta'(\tau)$ is the only solution that satisfies the normalizing condition*

$$\frac{1}{2\pi}\int_0^\beta \mu(\tau)\,d\tau=1. \tag{16.6-32}$$

In order to construct the exterior mapping function, it is not necessary to solve Symm's exterior equation (16.6-9), because we now know from (16.6-22) that its solution $\mu(\tau)=\mathring{\theta}'(\tau)$ also solves the simpler equation

$$\frac{1}{2\pi}\int_0^\beta \mathrm{Log}|z(\sigma)-z(\tau)|\mu(\tau)\,d\tau=\mathrm{Log}\ \gamma. \tag{16.6-33}$$

The difference of any two solutions again satisfies (16.6-25). We thus conclude from the foregoing development that the solution of (16.6-33) is unique if $\gamma\neq 1$, and is determined up to a constant factor if $\gamma=1$. We thus may state:

THEOREM 16.6i. *Under the hypotheses of the preceding theorem, $\mu(\tau)=\mathring{\theta}'(\tau)$ is the only solution of* (16.6-33) *satisfying the normalizing condition* (16.6-32).

The fact, somewhat surprising at first sight, that the uniqueness of Symm's equation is not invariant under homothetic transformations $z\to\lambda z$, is easily explained by noting that the equation itself is not invariant under such transformations. Crude estimates for the capacity γ are obtained as follows. Here the **inner radius** of Γ is the radius of the largest disk contained in the interior of Γ, and the **outer radius** is the radius of the smallest disk containing the interior of Γ.

THEOREM 16.6j. *Let ρ_i, ρ_a, and γ denote the inner radius, the outer radius, and the capacity of the Jordan curve Γ, respectively. Then*

$$\rho_i \leqq \gamma \leqq \rho_a. \tag{16.6-34}$$

Proof. Let

$$z = \mathring{g}(w) = \gamma w + c_0 + c_1 w^{-1} + \cdots$$

be the normalized function mapping the exterior of $|w| = 1$ onto the exterior $\mathring{D}$ of Γ. We may adjust c_0 so that the center of the largest disk contained in the interior is at 0. Now consider the function

$$h(w) := \operatorname{Log}\left|\frac{\mathring{g}(w)}{w}\right|,$$

which is harmonic in $|w| > 1$, continuous in $|w| \geqq 1$, and which tends to $\operatorname{Log} \gamma$ for $w \to \infty$. Its values on $|w| = 1$ satisfy $h(w) \geqq \operatorname{Log} \rho_i$. By the maximum principle, $\operatorname{Log} \gamma \geqq \operatorname{Log} \rho_i$, implying the left inequality (16.6-34). The right inequality is proved similarly. —

Thus in order to make sure that the capacity of Γ is not equal to 1, it suffices to replace Γ by a homothetic curve which either contains or is contained in a disk of radius 1.

V. The Numerical Solution of Symm's Equation

The numerical solution of Symm's equation at first appears to present some difficulties. The kernel has a logarithmic singularity, and there is no obvious way to obtain the solution by iteration. The principle of discretization has nevertheless been used with good success by Symm [1966, 1967] and by Hayes et al. [1972], where an accurate program using spline interpolation is described.

Here we describe the algorithmic aspects of a solution method proposed by Berrut [1976] which uses Fourier series. We assume the parametric interval to be $[0, 2\pi]$, so that Symm's interior equation is

$$\frac{1}{2\pi}\int_0^{2\pi} \operatorname{Log}|z(\sigma) - z(\tau)|\mu(\tau)\, d\tau = \operatorname{Log}|z(\sigma)|, \qquad 0 \leqq \sigma \leqq 2\pi. \tag{16.6-35}$$

The idea now is to expand the unknown function μ, the inhomogeneous term, and the kernel, all in Fourier series. This is appropriate because both $\operatorname{Log}|z(\sigma)|$ and the desired solution are functions of period 2π, and the normalizing condition (16.6-32) is easily taken into account by requiring the constant term in the Fourier series to be 2π. The difficulty will be encountered that the Fourier series for the kernel, Log $|z(\sigma)-z(\tau)|$, converges only slowly due to the logarithmic singularity at $\tau=\sigma$. However, it is easy to transform the equation into one with a smooth kernel by writing it in the form

$$\frac{1}{2\pi}\int_0^{2\pi} \operatorname{Log}|e^{i\sigma}-e^{i\tau}|\mu(\tau)\,d\tau+\frac{1}{2\pi}\int_0^{2\pi} \lambda(\sigma,\tau)\mu(\tau)\,d\tau=\operatorname{Log}|z(\sigma)|, \tag{16.6-36}$$

where the kernel

$$\lambda(\sigma,\tau):=\operatorname{Log}\left|\frac{z(\sigma)-z(\tau)}{e^{i\sigma}-e^{i\tau}}\right|=\sum_{m,n=-\infty}^{\infty} a_{mn}\,e^{im\tau}\,e^{in\sigma}$$

for sufficiently smooth $z(\tau)$ now is smooth. Assuming the solution in the form

$$\mu(\tau)=\sum_{n=-\infty}^{\infty} x_n\,e^{in\tau}, \qquad x_0=2\pi, \quad x_{-n}=\overline{x_n},$$

the second integral in (16.6-36) is easily calculated with the result

$$\frac{1}{2\pi}\int_0^{2\pi} \lambda(\sigma,\tau)\mu(\tau)\,d\tau=\sum_{n=-\infty}^{\infty}\left(\sum_{m=-\infty}^{\infty} a_{mn}x_{-m}\right)e^{in\sigma}.$$

As to the first integral, we use the formula, valid for any integer n,

$$\frac{1}{2\pi}\int_0^{2\pi} \operatorname{Log}|e^{i\sigma}-e^{i\tau}|\,e^{in\tau}\,d\tau=\begin{cases}0, & n=0\\ -\dfrac{1}{2|n|}\,e^{in\sigma}, & n\neq 0,\end{cases} \tag{16.6-37}$$

which is easily established (see Problem 2). We thus obtain

$$\frac{1}{2\pi}\int_0^{2\pi} \operatorname{Log}|e^{i\sigma}-e^{i\tau}|\mu(\tau)\,d\tau=-\frac{1}{2}\sum_{n\neq 0}\frac{1}{|n|}x_n\,e^{in\sigma}.$$

Letting

$$\operatorname{Log}|z(\sigma)|=\sum_{n=-\infty}^{\infty} b_n\,e^{in\sigma},$$

substituting in (16.6-36) and comparing Fourier coefficients, we thus obtain for the unknown Fourier coefficients x_n, $n\neq 0$, the infinite system of linear equations

$$-\frac{1}{2|n|}x_n+\sum_{m=-\infty}^{\infty} a_{mn}\overline{x_m}=b_n, \qquad n=\pm 1, \pm 2, \ldots, \tag{16.6-38}$$

which may be solved by iteration in an obvious manner.

EXAMPLE 1

The process described can be carried out analytically for the curve

$$\Gamma: \quad z(\tau) = e^{i\tau} + \varepsilon\, e^{-i\tau}, \qquad 0 \leqq \tau \leqq 2\pi, \tag{16.6-39}$$

where $0 < \varepsilon < 1$, which describes an ellipse with semiaxes $1 \pm \varepsilon$. Here we have

$$\frac{z(\sigma) - z(\tau)}{e^{i\sigma} - e^{i\tau}} = 1 - \varepsilon\, e^{-i(\sigma+\tau)}.$$

Hence

$$\lambda(\sigma, \tau) = \operatorname{Log}\left|\frac{z(\sigma) - z(\tau)}{e^{i\sigma} - e^{i\tau}}\right| = -\sum_{n=1}^{\infty} \frac{\varepsilon^n}{2n}\left[e^{in(\sigma+\tau)} + e^{-in(\sigma+\tau)}\right],$$

and we see that

$$a_{mn} = \begin{cases} 0, & m \neq n \\ -\dfrac{\varepsilon^n}{2n}, & m = n \neq 0. \end{cases}$$

In view of

$$\operatorname{Log}|z(\sigma)| = \operatorname{Log}|e^{i\sigma} + \varepsilon\, e^{-i\sigma}| = \sum_{n=1}^{\infty} (-1)^{n-1} \frac{\varepsilon^n}{2n}\left[e^{2in\sigma} + e^{-2in\sigma}\right],$$

the system (16.6-38) for $n = 1, 2, \ldots$ thus reads

$$-\frac{1}{2n} x_n - \frac{\varepsilon^n}{2n} \overline{x_n} = \begin{cases} 0, & n \text{ odd} \\ (-1)^{m-1} \dfrac{\varepsilon^m}{2m}, & n = 2m. \end{cases}$$

We immediately find the solution

$$x_n = \begin{cases} 0, & n \text{ odd} \\ 2\dfrac{(-1)^m \varepsilon^m}{1 + \varepsilon^{2m}}, & n \text{ even}, \quad n = 2m, \end{cases}$$

$n = 1, 2, \ldots$, and $x_{-n} = x_n$. Thus the derivative of the boundary correspondence function for the ellipse (16.6-39) is

$$\theta'(\tau) = 1 + 4 \sum_{m=1}^{\infty} (-1)^m \frac{\varepsilon^m}{1 + \varepsilon^{2m}} \cos 2m\tau,$$

and the boundary correspondence function itself is

$$\theta(\tau) = \tau + 2 \sum_{m=1}^{\infty} \frac{(-1)^m}{m} \frac{\varepsilon^m}{1 + \varepsilon^{2m}} \sin 2m\tau.$$

PROBLEMS

1. Integrals such as (16.6-11) that express the mapping function in terms of the boundary correspondence function may be numerically inconvenient because of the near singularity of the integrand when z is close to the boundary. Show that the singularity may be weakened by integration by parts, and obtain in this manner the representations

$$w = f(z) = -\frac{1}{2\pi}\int_0^\beta e^{i\theta(\tau)}\log\left[1 - \frac{z}{z(\tau)}\right]\theta'(\tau)\,d\tau$$

$$z = g(w) = -\frac{1}{2\pi i}\int_0^\beta z'(\tau)\,\mathrm{Log}[1 - e^{-i\theta(\tau)}w]\,d\tau$$

for the interior mapping functions. Also find similar expressions for the exterior mapping function and its inverse.

2. Establish (16.6-37) by considering the integral as the limit as $w \to e^{i\tau}$ of the function

$$l_n(w) = \frac{1}{2\pi}\int_0^{2\pi}\mathrm{Log}|w - e^{i\tau}|\,e^{in\tau}\,d\tau,$$

which is the $(-n)$th Fourier coefficient of

$$\lambda(\tau, w) := \mathrm{Log}|w - e^{i\tau}| = \mathrm{Re}\log(w - e^{i\tau}).$$

Compute these Fourier coefficients for $|w| > 1$ by expanding λ in powers of w^{-1}.

3. Let Γ be the ellipse considered in Example 1.
   (a) Show that the capacity of Γ equals 1.
   (b) Why is the solution of Symm's equation obtained by Fourier series nevertheless unique?
   (c) Determine the exterior boundary correspondence function $\mathring{\theta}$ of Γ.
   (d) Verify that Symm's exterior equation holds.

NOTES

Symm's equations were stated, but not solved, by Symm [1966, 1967]. Gaier [1976] gave a rigorous discussion, valid for arbitrary rectifiable curves Γ, in which he expressed the solutions in terms of the boundary correspondence functions and completely resolved the question of the uniqueness of the solutions. For properties of the capacity γ, there called the "outer conformal radius," see Problems IV 97–120 in Polya and Szegö [1925].

As to the numerical solution of Symm's equation, Hayes et al. [1972] advocate the use of spline functions. Christiansen [1981] discusses the condition of the linear systems which thus arise. Wendland [1980] and Hsiao, Kopp and Wendland [1980] discuss a class of integral equations of the first kind for solving elliptic boundary value problems which contains Symm's equation as a special case. They also have a formulation of Symm's equation which has a unique solution for all values of γ (see also Reichel [1985]). Arnold [1983], Arnold and Wendland [1983], and Wendland [1983] discuss the asymptotic convergence of their "boundary element methods" for solving these integral equations. Lamp, Schleicher, and Wendland [1984] further discuss the solution of one-dimensional integral equations. They obtain the counterintuitive result that for the numerical evaluation of the boundary integrals that occur, the trapezoidal rule furnishes less accurate results than certain composite Gaussian rules, which they are likewise able to evaluate using an FFT. Hoidn [1983] uses a regularization method due to Marti [1980] to solve Symm's equation.

The Fourier series method to solve Symm's equation was developed by Berrut [1976]. Reichel [1985] proposes a modification of the iteration procedure which in many cases gives faster convergence.

§ 16.7. INTEGRAL EQUATIONS OF THE SECOND KIND

Here certain integral equations of the second kind will be established for both the interior and the exterior boundary correspondence functions and their derivatives. Equations of the second kind enjoy the advantage that both their theoretical properties and their numerical solutions are, in general, better understood than for equations of the first kind. Throughout this section the boundary curve Γ of the region to be mapped is assumed to be at least piecewise analytic; in fact, some of the equations to be discussed hold only for curves Γ that have no corners.

I. Interior Boundary Correspondence Function

Our starting point is Lemma 16.6b. Let z_0 be exterior to Γ. For each $\tau \in [0, \beta]$, the function $\log[z - z(\tau)]$, the integrand in (16.6-16), can be defined locally as an analytic function of z near z_0. Because a continuous argument of $z_0 - z(\tau)$ can be defined for $\tau \in [0, \beta]$, the integrand is continuous also in τ. It then follows from Theorem 4.1a that the derivative of (16.6-16) with respect to z may be calculated by differentiating under the integral sign. Differentiating (16.6-16) we thus get

$$\frac{1}{2\pi}\int_0^\beta \frac{1}{z - z(\tau)}\theta'(\tau)\,d\tau = \frac{1}{z}. \tag{16.7-1}$$

This holds for z near z_0 and hence, since z_0 was arbitrary in the exterior of Γ, for *all* z in the exterior of Γ.

We let $z \to z(\tau)$, a point on Γ. Because the integral then becomes singular, an appeal to the theory of Cauchy integrals is required. The expression on the left of (16.7-1) is

$$p(z) := \frac{1}{2\pi i}\int_\Gamma \frac{m(t)}{t - z}\,dt,$$

where

$$m(t) := -i\frac{\theta'(\tau)}{z'(\tau)}, \qquad t = z(\tau).$$

By the Sokhotskyi formulas (Theorem 14.1c)

$$p^-(z(\sigma)) := \lim_{\substack{z \to z(\sigma) \\ z \in \operatorname{ext} \Gamma}} p(z)$$

exists at every $z(\sigma)$ which is not a corner of Γ, and

$$p^-(z(\sigma)) = \frac{1}{2\pi i} \operatorname{PV} \int_\Gamma \frac{m(t)}{t - z(\sigma)}\, dt - \frac{1}{2} m(z(\sigma)).$$

Undoing our abbreviations, we thus get from (16.7-1)

$$\frac{i}{2} \frac{\theta'(\sigma)}{z'(\sigma)} + \frac{1}{2\pi} \operatorname{PV} \int_0^\beta \frac{\theta'(\tau)}{z(\sigma) - z(\tau)}\, d\tau = \frac{1}{z(\sigma)}. \tag{16.7-2}$$

We multiply by $2z'(\sigma)$ and take the imaginary part. This yields

$$\theta'(\sigma) + \operatorname{PV} \int_0^\beta \frac{1}{\pi} \operatorname{Im}\left[\frac{z'(\sigma)}{z(\sigma) - z(\tau)}\right] \theta'(\tau)\, d\tau$$

$$= 2 \operatorname{Im} \frac{z'(\sigma)}{z(\sigma)} = 2 \frac{d}{d\sigma} \arg z(\sigma). \tag{16.7-3}$$

The function[1]

$$\nu(\sigma, \tau) := \frac{1}{\pi} \operatorname{Im} \frac{z'(\sigma)}{z(\sigma) - z(\tau)} \tag{16.7-4}$$

is related to the *Neumann kernel* $\kappa(s, t)$ introduced in § 15.9 as follows. If $s = z(\sigma)$, $t = z(\tau)$, then

$$\nu(\sigma, \tau) = |z'(\sigma)| \kappa(s, t). \tag{16.7-5}$$

The two functions have identical values at corresponding points if the parametric representation is such that $|z'(\sigma)| = 1$, that is, if the arc length is used as a parameter. For numerical work the advantage of not being tied to that special representation is obvious. In the present context we therefore prefer to use the more general definition (16.7-4) and call this function the **parametric Neumann kernel**.

It follows from Theorem 15.9a via (16.7-5) that $\nu(\sigma, \tau)$ is a continuous function of τ for all σ such that $z(\sigma)$ is not a corner of Γ. For such σ, the principal value symbol in (16.7-3) may thus be omitted, and we obtain:

[1] Not to be confused with the *Neumann function* (see § 15.7), which was denoted by the same symbol.

THEOREM 16.7a (Warschawski's integral equation). *For every $\sigma \in [0, \beta]$ such that $z(\sigma)$ is not a corner of Γ, there holds*

$$\theta'(\sigma) + \int_0^\beta \nu(\sigma, \tau)\theta'(\tau)\,d\tau = 2\frac{d}{d\sigma}[\arg z(\sigma)], \tag{16.7-6}$$

where ν is the parametric Neumann kernel defined by (16.7-4).

Our next goal is a similar equation for the integrated derivative $\theta(\sigma)$. Once again we start from (16.6-16). We now integrate by parts and obtain on the left

$$\frac{1}{2\pi}[\theta(\tau)\log(z - z(\tau))]_0^\beta + \frac{1}{2\pi}\int_0^\beta \frac{z'(\tau)}{z - z(\tau)}\theta(\tau)\,d\tau.$$

Since Γ has winding number 0 with respect to z, the logarithm on the left returns to its original value as τ increases from 0 to β. In view of $\theta(\beta) = \theta(0) + 2\pi$, the integrated part thus equals $\log(z - z(0))$. The integral may be written

$$\frac{1}{2\pi i}\int_\Gamma \frac{-i\theta(\tau)}{t - z}\,dt.$$

We again let $z \to z(\sigma)$. If $z(\sigma)$ is not a corner of Γ, the Sokhotskyi formulas are applicable and yield for the limit of the integral

$$\frac{i}{2}\theta(\sigma) - \frac{1}{2\pi}\,\mathrm{PV}\int_0^\beta \frac{z'(\tau)}{z(\tau) - z(\sigma)}\theta(\tau)\,d\tau.$$

Letting $z = z(\sigma)$ also on the right in (16.6-16) and taking imaginary parts, we obtain

$$\theta(\sigma) - \mathrm{PV}\int_0^\beta \mathrm{Im}\left[\frac{1}{\pi}\frac{z'(\tau)}{z(\tau) - z(\sigma)}\right]\theta(\tau)\,d\tau$$
$$= 2\{\arg z(\sigma) - \arg[z(\sigma) - z(0)]\}.$$

The kernel appearing here is the parametric Neumann kernel, with the arguments interchanged. This being continuous at $\tau = \sigma$, we may omit the instruction PV and obtain:

THEOREM 16.7b (Gerschgorin's integral equation). *For every $\sigma \in [0, \beta]$ such that $z(\sigma)$ is not a corner of Γ, there holds*

$$\theta(\sigma) - \int_0^\beta \nu(\tau, \sigma)\theta(\tau)\,d\tau = 2\{\arg z(\sigma) - \arg[z(\sigma) - z(0)]\}. \tag{16.7-7}$$

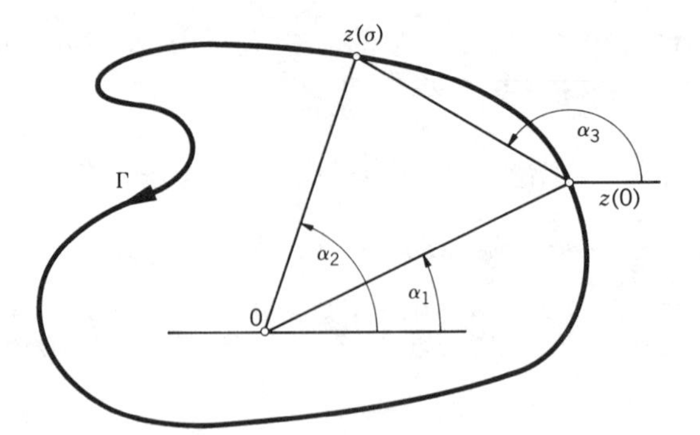

Fig. 16.7a. Choice of arguments in (16.7-7).

The two arguments on the right are determined as follows: $\arg z(\sigma)$ is obtained by continuous continuation from $\arg z(0)$, which may be selected arbitrarily. The value of $\arg[z(\sigma)-z(0)]$ then is obtained by continuous continuation of $\arg[z(\sigma)-0]=\arg z(\sigma)$, where z travels from 0 to $z(0)$ in the interior of Γ (see Fig. 16.7a, where $\alpha_1 := \arg z(0)$, $\alpha_2 := \arg z(\sigma)$, $\alpha_3 := \arg[z(\sigma)-z(0)]$.

II. Exterior Boundary Correspondence Function

Here our starting point is Lemma 16.6e. Let z_0 be in the interior of Γ. For each $\tau \in [0, \beta]$, $\log[e^{-i\mathring{\theta}(\tau)}(z(\tau)-z)]$ can be defined as an analytic function of z in the neighborhood of z_0. Moreover for each such z the integrand is continuous in τ. Thus the derivative of the integral with respect to z may be calculated by differentiating under the integral sign. There results

$$\frac{1}{2\pi}\int_0^\beta \frac{1}{z(\tau)-z}\mathring{\theta}'(\tau)\,d\tau = 0 \quad \text{for all} \quad z \in D.$$

In order to let $z \to z(\sigma)$ we write the integral in the form

$$\frac{1}{2\pi i}\int_\Gamma \frac{m(t)}{t-z}\,dt,$$

where

$$m(t) := i\frac{\mathring{\theta}'(\tau)}{z'(\tau)}, \qquad t = z(\tau).$$

The Sokhotskyi formulas then yield, since z now approaches Γ from the left,

$$0 = \frac{1}{2\pi i}\,\mathrm{PV}\int_\Gamma \frac{m(t)}{t-z(\sigma)}\,dt + \frac{1}{2}m(z(\sigma)),$$

which is the same as

$$i\frac{\mathring{\theta}'(\sigma)}{z'(\sigma)}+\frac{1}{\pi}\,\mathrm{PV}\int_0^\beta \frac{\mathring{\theta}'(\tau)}{z(\tau)-z(\sigma)}\,d\tau=0.$$

Multiplying by $z'(\sigma)$ and taking imaginary parts, there again appears the parametric Neumann kernel, and we obtain;

THEOREM 16.7c (Banin's integral equation). *For every σ such that $z(\sigma)$ is not a corner of Γ, there holds*

$$\mathring{\theta}'(\sigma)-\int_0^\beta \nu(\sigma,\tau)\mathring{\theta}'(\tau)\,d\tau=0. \tag{16.7-8}$$

In other words, $\mathring{\theta}'$ is a nonvanishing solution of a certain homogeneous integral equation of the second kind, that is, an eigenfunction.

We finally obtain an equation for the exterior boundary correspondence function itself by integrating (16.6-21) by parts. This yields

$$\frac{1}{2\pi}\{\log[e^{-i\mathring{\theta}(\tau)}(z(\tau)-z)]\mathring{\theta}(\tau)\}_0^\beta$$

$$-\frac{1}{2\pi}\int_0^\beta\left\{-i\mathring{\theta}'(\tau)+\frac{z'(\tau)}{z(\tau)-z}\right\}\mathring{\theta}(\tau)\,d\tau=\mathrm{Log}\,\gamma.$$

Since the logarithm returns to its initial value, the integrated term equals

$$\log[e^{-i\mathring{\theta}(0)}(z(0)-z)]=-i\mathring{\theta}(0)+\log(z(0)-z).$$

The integral of the first term under the integral may be evaluated:

$$\frac{i}{2\pi}\int_0^\beta \mathring{\theta}(\tau)\mathring{\theta}'(\tau)\,d\tau=\frac{i}{4\pi}[\mathring{\theta}(\tau)^2]_0^\beta=i\mathring{\theta}(0)+i\pi.$$

The limit of the second integral as $z\to z(\sigma)$ from the interior of Γ by Sokhotskyi thus equals

$$-\frac{i}{2}\mathring{\theta}(\sigma)-\frac{1}{2\pi}\,\mathrm{PV}\int_0^\beta \frac{z'(\tau)}{z(\tau)-z(\sigma)}\mathring{\theta}(\tau)\,d\tau.$$

Thus after taking imaginary parts, we get:

THEOREM 16.7d (Equation of Kantorovich–Krylov). *For every σ such that $z(\sigma)$ is not a corner of Γ, there holds*

$$\mathring{\theta}(\sigma)+\int_0^\beta \nu(\tau,\sigma)\mathring{\theta}(\tau)\,d\tau=2\arg[z(\sigma)-z(0)]. \tag{16.7-9}$$

III. Solvability of the Integral Equations

If Γ is an analytic curve without corners, the integral equations established above hold for all $\sigma \in [0, \beta]$, and the kernel $\nu(\sigma, \tau)$ by Theorem 15.9a is continuous for $(\sigma, \tau) \in [0, \beta] \times [0, \beta]$. The basic facts concerning such Fredholm integral equations were summarized in § 15.9.

We recall that $\lambda = 1$ is an eigenvalue of ν. We now see that $\mathring{\theta}'(\tau)$ is a corresponding eigenfunction of $\nu(\sigma, \tau)$, and we recall that $\omega(\sigma) = 1$ is a corresponding eigenfunction of the transposed kernel $\nu(\tau, \sigma)$. Since the multiplicity of $\lambda = 1$ is 1, there are no other (linearly independent) eigenfunctions. It now follows from the Fredholm theory (Theorem 15.9b) that the inhomogeneous equation (16.7-7) has a solution only if the inhomogeneous part is orthogonal to $\mathring{\theta}'(\sigma)$, that is, if

$$\int_0^\beta [\arg z(\sigma) - \arg(z(\sigma) - z(0))]\mathring{\theta}'(\sigma)\, d\sigma = 0. \tag{16.7-10}$$

Although (16.7-10) follows implicitly because we know that (16.7-7) in fact has the solution $\theta(\sigma)$, a direct verification may be of interest. Taking imaginary parts in (16.6-21), there follows for every z in the interior of Γ

$$\int_0^\beta [-\mathring{\theta}(\tau) - \arg(z(\tau) - z)]\mathring{\theta}'(\tau)\, d\tau = 0.$$

Subtracting this from the corresponding relation for $z = 0$, we obtain

$$\int_0^\beta [\arg z(\tau) - \arg(z(\tau) - z)]\mathring{\theta}'(\tau)\, d\tau = 0,$$

and (16.7-10) follows on letting $z \to z(0)$. The Fredholm theory also states that the solution of (16.7-7) is determined only up to an arbitrary multiple of the corresponding homogeneous equation, that is, up to a constant. This was to be expected, because the normalizing condition $f'(0) > 0$ did not enter into (16.7-7), and the boundary correspondence function thus *is* only determined up to an additive constant.

We also recall that $\lambda = -1$ is not an eigenvalue of ν. It follows that the solutions of (16.7-6) and (16.7-9), the integral equations for θ' and $\mathring{\theta}$, are uniquely determined.

IV. Berrut's Integral Equation

The Neumann kernel integral equations given in Sections I and II all have been known at least since 1955; only their common derivation from Lemma 16.6b may have any claim to novelty. By virtue of Theorem 15.9a this derivation requires Γ to have a continuously turning tangent. We now present an integral equation for θ' that was recently derived from Symm's equation by Berrut. Berrut's equation is a formalization of the solution method for Symm's equation given in § 16.6V. It has the advantage of holding whenever the derivative θ' of the boundary correspondence function θ belongs to $L_2(I)$, the space of square integrable functions on the interval $I := [0, \beta]$. By the results of § 16.4, $\theta' \in L_2(I)$ certainly holds for curves Γ that are piecewise analytic and have no cusps, that is, no corners with interior angle 2π. Such curves Γ will be called **admissible**.

Here it is convenient to use some of the abbreviating notations of functional analysis. If s is a function in $L_2(I \times I)$, we denote by Int s the integral operator mapping a function $x \in L_2(I)$ onto the function

$$y: \quad \sigma \to y(\sigma) := \int_0^\beta s(\sigma, \tau) x(\tau)\, d\tau.$$

For instance, if $z(\tau)$ is the parametric representation of Γ and

$$l(\sigma, \tau) := \frac{1}{2\pi} \operatorname{Log}|z(\sigma) - z(\tau)|,$$

then Symm's integral equation (16.6-5) by writing $L := \text{Int}\, l$ takes the simple form

$$Lx = y, \tag{16.7-11}$$

where $y(\sigma) := \operatorname{Log}|z(\sigma)|$. By Theorem 16.6c, it has the solution $x(\tau) := \theta'(\theta)$.

Following and slightly generalizing the approach taken in § 16.6V, we now split L into a regular part depending on Γ, and a singular part which is the same for all Γ. Without loss of generality, we may assume $I = [0, 2\pi]$. For theoretical reasons it is helpful to introduce a parameter $\rho > 0, \rho \neq 1$. For $(\sigma, \tau) \in I \times I$, we define

$$r(\sigma, \tau) := \frac{1}{2\pi} \operatorname{Log}\left|\frac{z(\sigma) - z(\tau)}{\rho(e^{i\sigma} - e^{i\tau})}\right|,$$

$$s(\sigma, \tau) := \frac{1}{2\pi} \operatorname{Log}|\rho(e^{i\sigma} - e^{i\tau})|$$

and set $R := \text{Int}\, r$, $S := \text{Int}\, s$. Equation (16.7-11) then is equivalent to

$$(S + R)x = y. \tag{16.7-12}$$

The point now is that S has a left inverse S_l^{-1}, which will be calculated presently. Multiplying (16.7-12) from the left by S_l^{-1}, we obtain

$$x + S_l^{-1} Rx = S_l^{-1} y. \tag{16.7-13}$$

We proceed to compute S_I^{-1} and $S_I^{-1}R$. If $x \in L_2(I)$ has the Fourier series

$$x(\tau) = \sum_{n=-\infty}^{\infty} x_n e^{in\tau}, \tag{16.7-14}$$

then it follows from (16.6-37) that

$$y(\tau) := (Sx)(\tau) = \sum_{n=-\infty}^{\infty} y_n e^{in\tau}, \tag{16.7-15}$$

where

$$y_n = \begin{cases} x_0 \operatorname{Log} \rho, & n = 0 \\ -\dfrac{1}{2|n|} x_n, & n \neq 0. \end{cases}$$

It follows that if y is in the range of S and is given by (16.7-15), then x is given by (16.7-14), where

$$x_n = \begin{cases} \dfrac{y_0}{\operatorname{Log} \rho}, & n = 0 \\ -2n(\operatorname{sgn} n) y_n & n \neq 0. \end{cases}$$

In view of $n(\operatorname{sgn} n)y_n = (-i \operatorname{sgn} n)iny_n$, the contribution of the terms with $n \neq 0$ to x may be written as $\mathcal{K}Dy$, where D denotes differentiation, and where $\mathcal{K}$ is the conjugation operator defined in § 13.5. Since $y_0 = My$, the linear mean of the function y, we thus have

$$S^{-1} = \frac{1}{\operatorname{Log} \rho} M - 2\mathcal{K}D,$$

and (16.7-13) yields

$$x = c + 2\mathcal{K}DRx - 2\mathcal{K}Dy, \tag{16.7-16}$$

where c is a constant function,

$$c = \frac{1}{\operatorname{Log} \rho}(My - MRx).$$

The constant c is readily identified by applying M to both sides of (16.7-16). In view $M\mathcal{K} = 0$ and $Mc = c$ we get

$$c = Mx.$$

For the solution $x = \theta'$ of Symm's equation it is known that $Mx = 1$, and there follows $c = 1$. It furthermore may be shown that the operator $\mathcal{K}D$ commutes with $R = \operatorname{Int} r$, that is, that conjugation and differentiation may be performed under the integral sign. We thus arrive at the following result, the complete proof of which is given by Berrut:

THEOREM 16.7e. *For an admissible Jordan curve Γ of capacity $\gamma \neq 1$, θ' is the only solution in $L_2(I)$ of the Fredholm integral equation of the second kind,*

$$\theta'(\sigma) = \frac{1}{\pi}\int_0^{2\pi} \mathscr{K}_\sigma\left[\frac{\partial}{\partial\sigma}\operatorname{Log}\left|\frac{z(\sigma)-z(\tau)}{e^{i\sigma}-e^{i\tau}}\right|\right]\theta'(\tau)\,d\tau + 1 - 2\mathscr{K}_\sigma \operatorname{Re}\frac{z'(\sigma)}{z(\sigma)}. \tag{16.7-17}$$

The subscript σ in $\mathscr{K}_\sigma$ indicates the variable with respect to which the conjugation is to be performed.

Concerning the numerical solution of (16.7-17), Berrut recommends to approximate directly θ', Dr, and Dy by trigonometric polynomials of degree M, and to solve the resulting (finite) system of linear equations. The convergence to the exact solution as $M \to \infty$ then is a consequence of a result of Schleiff [1968]. In numerical tests involving curves with corners, the performance of Berrut's method was vastly superior to that of Warschawski's equation (16.7-6), and superior also to the iteration method described in § 16.6V.

PROBLEMS

1. *Lichtenstein's integral equation.* Obtain an integral equation with the Neumann kernel for the periodic function

$$\delta(\tau) := \theta(\tau) - \arg z(\tau)$$

by applying Cauchy's formula to

$$h(z) := \log\frac{f(z)}{z},$$

letting $z \to z(\sigma)$, and taking imaginary parts. (See Gaier [1964], p. 7.)

2. Solve Berrut's equation analytically for the eccentric circle

$$\Gamma:\quad z = a + e^{i\tau}, \qquad 0 \leqq \tau \leqq 2\pi,$$

where $|a| < 1$. Show that $r(\sigma, \tau) = 0$, and hence that (16.7-17) reduces to

$$\theta' = 1 - 2\mathscr{K}\operatorname{Re}\frac{z'(\tau)}{z(\tau)}.$$

Conclude that for real a,

$$\theta'(\tau) = \frac{1-a^2}{1+a^2+2a\cos\tau}.$$

3. Solve Berrut's equation analytically for the ellipse

$$z(\tau) = e^{i\tau} + \varepsilon\, e^{-i\tau}, \qquad 0 < \varepsilon < 1.$$

Compare the result with the solution obtained in Example 1, § 16.6.

NOTES

Stiefel [1956] and Arbenz [1958] present a modification of Warschawski's equation where Γ is permitted to have corners. Schober [1967] likewise deals with corners. Numerous further integral equations of the second kind are given in Gaier's survey [1964]. For Berrut's work, see Berrut [1985a, 1985b].

§ 16.8. METHODS BASED ON CONJUGATE FUNCTIONS

I. The Inverse Boundary Correspondence Function

The methods discussed so far have aimed at finding the normalized function f mapping a given Jordan region D, $0 \in D$, onto the unit disk, or equivalently at finding the boundary correspondence function $\theta(\tau) = \arg f(z(\tau))$, where

$$\Gamma: \quad z = z(\tau), \qquad 0 \leqq \tau \leqq \beta, \tag{16.8-1}$$

is a parametric representation of the boundary of D. Contrary to this, the methods to be discussed presently have as their goal the determination of the inverse map $g = f^{[-1]}$, or equivalently, of the boundary correspondence of that map. To describe that boundary correspondence analytically, we continue the function $z(\tau)$ occurring in (16.8-1) to $\mathbb{R}$ as a continuous function of period β. Any function $\tau(\theta)$ on $\mathbb{R}$ such that

$$\tau(\theta) - \frac{\beta}{2\pi}\theta$$

is periodic with period 2π and such that

$$g(e^{i\theta}) = z(\tau(\theta)), \qquad \theta \in \mathbb{R},$$

is called an **inverse boundary correspondence function** of Γ. This function is called **normalized** if $0 \leqq \tau(0) < \beta$; all inverse boundary correspondence functions are obtained from the normalized function by adding integral multiples of β.

The normalized inverse boundary correspondence function may be characterized very simply:

THEOREM 16.8a. *Let the Jordan curve Γ be given by* (16.8-1), *and let τ be any continuous real function on $\mathbb{R}$ such that*

(i) $0 \leqq \tau(0) < \beta$.
(ii) $\tau(\theta) - \beta\theta/2\pi$ *is periodic with period* 2π.
(iii) *The values $z(\tau(\theta))$ for $\theta \in \mathbb{R}$ are the boundary values $g_1^+(e^{i\theta})$ of an analytic function g_1 on E: $|w| < 1$, such that*

$$g_1(0) = 0, \qquad g_1'(0) > 0.$$

Then τ is the normalized inverse boundary correspondence function of Γ, and $g_1 = g$, the normalized mapping from E to the interior of Γ.

Proof. As θ increases from 0 to 2π, the point $z = z(\tau(\theta))$ traces Γ once in the positive direction. Let z_0 be any point not on Γ. By the principle of the argument, the number of times g_1 assumes z_0 on E equals the number of times the image of the boundary of E under g_1 winds around z_0. This image being Γ, g_1 assumes every value in the interior of Γ precisely once, and it assumes no value in the exterior of Γ. Because g_1 also satisfies the proper normalization, $g_1 = g$. —

The general procedure is now this. With the desired mapping function g (unknown, but known to exist) we form an auxiliary function h which is analytic on E. By virtue of $g(e^{i\theta}) = z(\tau(\theta))$ both $\operatorname{Re} h$ and $\operatorname{Im} h$ can be related to $\tau(\theta)$. The fact that

$$\operatorname{Im} h(e^{i\theta}) = \mathscr{K} \operatorname{Re} h(e^{i\theta}) + \operatorname{Im} h(0)$$

then yields a functional equation for the inverse boundary correspondence function $\tau(\theta)$.

II. The Method of Theodorsen: The Interior Map

Here we assume Γ to be given in polar coordinates,

$$\Gamma: \quad z = z(\phi) = \rho(\phi)\, e^{i\phi}, \qquad 0 \leqq \phi \leqq 2\pi, \tag{16.8-2}$$

where ρ is a given positive continuous function of period 2π (see Fig. 16.8a). We thus have $\beta = 2\pi$, and we require D to be starlike with respect to 0.

The auxiliary function here is

$$h(w) := \log \frac{g(w)}{w}, \tag{16.8-3}$$

where the logarithm is analytic in E and is made unique by requiring that $h(0) = \log g'(0)$ is real.

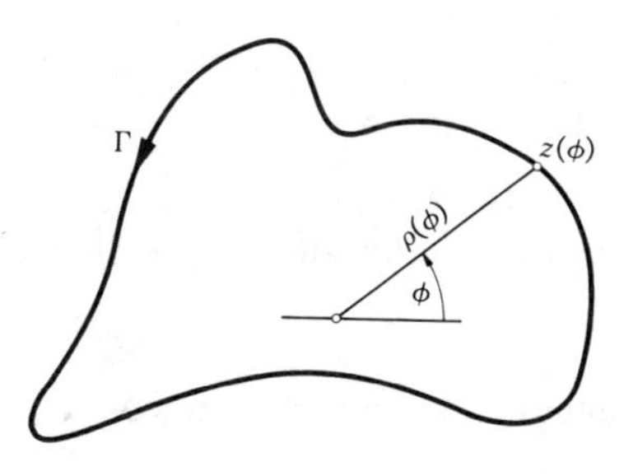

Fig. 16.8a. Data for Theodorsen's method.

The normalized inverse boundary correspondence function is now

$$\phi(\theta)=\arg g(e^{i\theta}),$$

with a continuous choice of the argument, $0\leqq\phi(0)<2\pi$. On $|w|=1$, (16.8-3) yields

$$\operatorname{Re} h(e^{i\theta})=\operatorname{Log}\rho(\phi(\theta)), \qquad \operatorname{Im} h(e^{i\theta})=\phi(\theta)-\theta,$$

and since $\operatorname{Im} h(0)=0$, our general principle yields the functional equation

$$\phi(\theta)-\theta=\mathscr{K}[\operatorname{Log}\rho(\phi(\theta))], \tag{16.8-4}$$

called **Theodorsen's integral equation**. The operator $\mathscr{K}$ here is to be understood as acting on and yielding functions of period 2π. It is explicitly defined by the principal value integral

$$\mathscr{K}\phi(\sigma)=\frac{1}{2\pi}\operatorname{PV}\int_0^{2\pi}\phi(\tau)\cot\frac{\sigma-\tau}{2}\,d\tau.$$

THEOREM 16.8b. *Let $\rho(\phi)$ be any continuous positive function of period 2π. Then Theodorsen's integral equation has exactly one continuous solution $\phi(\theta)$ such that $0\leqq\phi(0)<2\pi$, namely, the normalized inverse boundary correspondence function of the curve Γ defined by* (16.8-2).

Proof. We already have seen that the boundary correspondence function in question is a solution of (16.8-4). Conversely, let ϕ_1 be any continuous solution such that $0\leqq\phi_1(0)<2\pi$. The function $\phi_1(\theta)-\theta$, being a conjugate, clearly has period 2π. Thus ϕ_1 satisfies conditions (i) and (ii) of Theorem 16.8a. To show that (iii) is satisfied, let h_1 be the unique function analytic in E and continuous in E' such that

$$\operatorname{Re} h_1(e^{i\theta})=\operatorname{Log}\rho(\phi_1(\theta)), \qquad \operatorname{Im} h_1(e^{i\theta})=\phi_1(\theta)-\theta.$$

This function satisfies $\operatorname{Im} h_1(0)=0$. We assert that the values $z(\phi_1(\theta))$ are the boundary values of

$$g_1(w)=w\,e^{h_1(w)}.$$

In fact,

$$g_1(e^{i\theta})=e^{i\theta}\exp\{\operatorname{Log}\rho(\phi_1(\theta))+i[\phi_1(\theta)-\theta]\}=z(\phi_1(\theta)).$$

Evidently $g_1(0)=0$, $g_1'(0)>0$. Thus by Theorem 16.8a, $\phi_1=\phi$. —

Our next concern is with the construction of the unique solution. Letting

$$\delta(\theta):=\phi(\theta)-\theta,$$

where δ now is periodic with period 2π, Theodorsen's equation becomes

$$\delta(\theta)=\mathscr{K}[\operatorname{Log}\rho(\theta+\delta(\theta))], \tag{16.8-5}$$

and the following obvious iteration suggests itself. Choose $\delta_0(\theta)$ (usually, $\delta_0(\theta)=0$ is a natural choice), and construct the sequence $\{\delta_k(\theta)\}$ by

$$\delta_{k+1}(\theta)=\mathcal{K}[\mathrm{Log}\,\rho(\theta+\delta_k(\theta))], \qquad k=0,1,\ldots. \tag{16.8-6}$$

The numerical evaluation of the conjugation operator $\mathcal{K}$ proceeds as described in Theorem 13.5a. One computes, by means of an FFT, the Fourier coefficients of the balanced trigonometric polynomial that interpolates the given function (here: $\theta \mapsto \mathrm{Log}\,\rho(\theta+\delta_k(\theta))$) on a set of n equidistant sampling points. One applies the conjugation formulas (13.5-7) to these Fourier coefficients; and one evaluates the conjugated polynomial at the sampling points, again using an FFT. Concerning the errors committed in this process, see Theorem 13.6g and its corollary.

To discuss the convergence of the iteration (16.8-6), we assume $\rho(\theta)$ to be piecewise differentiable and set

$$\varepsilon := \sup_{0\le\phi\le 2\pi}\left|\frac{\rho'(\theta)}{\rho(\theta)}\right|. \tag{16.8-7}$$

To understand the geometric meaning of ε, we recall that the outward normal of Γ at the point $z(\phi)$ has the direction of $-iz'(\phi)$. From (16.8-2),

$$-iz'(\phi)=z(\phi)-i\rho'(\phi)\,e^{i\phi}.$$

The angle which the outward normal forms with the radius vector thus equals

$$\arg\left[-\frac{iz'(\phi)}{z(\phi)}\right]=\arg\left[1-i\frac{\rho'(\phi)}{\rho(\phi)}\right],$$

and (16.8-7) means that this angle does not exceed arctan ε.

THEOREM 16.8c. *Let $\rho(\phi)$ be piecewise differentiable, and let the quantity ε defined by* (16.8-7) *satisfy $\varepsilon<1$. Then the sequence $\{\delta_k(\theta)\}$ converges to $\delta(\theta)$ in the L_2 norm.*

Proof. Subtracting (16.8-5) from (16.8-6) we have

$$\delta_{k+1}(\theta)-\delta(\theta)=\mathcal{K}[\mathrm{Log}\,\rho(\theta+\delta_k(\theta))-\mathrm{Log}\,\rho(\theta+\delta(\theta))].$$

Taking norms and observing Theorem 14.2f we get

$$\|\delta_{k+1}-\delta\|\le\|\mathrm{Log}\,\rho(\theta+\delta_k)-\mathrm{Log}\,\rho(\theta+\delta)\|.$$

To estimate the difference on the right, we use the fact that for any two real numbers ϕ_1 and ϕ_2, using (16.8-7),

$$|\mathrm{Log}\,\rho(\phi_2)-\mathrm{Log}\,\rho(\phi_1)|=\left|\int_{\phi_1}^{\phi_2}\frac{\rho'(\theta)}{\rho(\theta)}\,d\theta\right|\le\varepsilon|\phi_2-\phi_1|.$$

On letting $\phi_1 := \theta + \delta_k(\theta)$, $\phi_2 := \theta + \delta(\theta)$, and integrating there follows

$$\|\text{Log}\,\rho(\theta+\delta_k) - \text{Log}\,\rho(\theta+\delta)\| \leqq \varepsilon \|\delta_k - \delta\|.$$

We thus have

$$\|\delta_{k+1} - \delta\| \leqq \varepsilon \|\delta_k - \delta\|,$$

and hence

$$\|\delta_k - \delta\| \leqq \varepsilon^k \|\delta_0 - \delta\|,$$

implying that $\|\delta_k - \delta\| \to 0$ since $\varepsilon < 1$. —

From the fact that $\delta_k \to \delta$ in norm it is possible, under the hypotheses of Theorem 16.8c, to establish pointwise convergence, that is, that $\delta_k(\theta) \to \delta(\theta)$ for every $\theta \in [0, 2\pi]$; see Warschawski [1945]. The hypothesis that $\varepsilon < 1$ is used again in an essential manner.

The condition that $\varepsilon < 1$ is known as the ε **condition**. It is readily demonstrated by means of numerical experiments that the iteration (16.8-6) may diverge for curves Γ that violate the ε condition. The fact that the ε condition is necessary for convergence appears to limit Theodorsen's method in its original form to curves Γ that are not too far from circular, which appears to be a serious constraint on the applicability of the method.

However, two remedies that have proved themselves useful in practical work can be recommended:

(a) *Preliminary Auxiliary Transformations.* Often it is possible to find an elementary function f_0 that transforms the region D to be mapped into a region D_0 that is nearly circular. Theodorsen himself, whose interest in conformal mapping was motivated by problems of flows around airfoils, used a Joukowski map (see § 5.1) to make the cross section of the airfoil nearly circular. Another preliminary map that recommends itself is to perform a few steps of an osculation algorithm (see § 16.2). Provided that $D \subset E$, this will automatically make D_0 as circular as desired, although there is no a priori guarantee that the boundary of any D_0 thus obtained will satisfy an ε condition. Preliminary transformations can also be used to remove corners of Γ, and thus to increase the accuracy of the numerical conjugation algorithm.

(b) *Underrelaxation.* This device, which is well known in numerical linear algebra, consists in replacing the iteration (16.8-6) by

$$\delta_{k+1}(\theta) = (1-\omega)\delta_k(\theta) + \omega\mathcal{K}[\text{Log}\,\rho(\theta+\delta_k(\theta))], \qquad k = 0, 1, \ldots. \tag{16.8-8}$$

Here ω is a relaxation factor, to be suitably chosen in the interval $0 < \omega < 1$. (In certain applications in linear algebra, one chooses $1 < \omega < 2$.) If $\rho(\theta)$ is suitably restricted (ρ even; $\text{sgn}\,\rho'(\phi) = \text{sgn}\,\phi$ for $-\pi < \phi < \pi$) Gutknecht

[1981] proved for a discretized version of (16.8-8) that given any $\varepsilon < \infty$, an $\omega > 0$ can be explicitly computed such that the iteration (16.8-8) converges at least locally. It is plausible that a similar result also holds when (16.8-8) is viewed as a process in L_2. Yet faster convergence is obtained by replacing (16.8-8) by the analog of the second-order Richardson iteration in numerical linear algebra,

$$\delta_{k+1}(\theta) = (1-\omega)\delta_{k-1}(\theta) + \omega\mathscr{K}[\operatorname{Log} \rho(\theta + \delta_k(\theta))], \qquad k = 0, 1, \ldots, \tag{16.8-9}$$

where $\delta_{-1}(\theta) = \delta_0(\theta) = 0$.

III. Theodorsen's Method for the Exterior Map

Theodorsen's method can be set up with equal ease for the construction of the *exterior* mapping function of a starlike Jordan curve Γ given in the form (16.8-2). A short discourse on exterior conjugate periodic functions is required.

Let ϕ be a real 2π-periodic function in Lip,

$$\phi(\theta) = \sum_{n=-\infty}^{\infty} a_n e^{in\theta}, \qquad a_{-n} = \overline{a_n}.$$

The solution u of Dirichlet's problem in $\mathring{E}$, bounded at ∞ and assuming the boundary values ϕ on $|w| = 1$, then is given by

$$u(w) = a_0 + \sum_{n=1}^{\infty} (a_n \overline{w^{-n}} + a_{-n} w^{-n}), \qquad |w| \geqq 1.$$

Although $\mathring{E}$ is not simply connected, it is easily seen that the condition of Theorem 15.1d for the existence of a conjugate harmonic function is satisfied. Since we may write

$$u(w) = a_0 + 2 \operatorname{Re} f(w),$$

where

$$f(w) = \sum_{n=1}^{\infty} \overline{a_n} w^{-n},$$

the harmonic conjugate v of u that vanishes at ∞ is given by

$$v(w) = 2 \operatorname{Im} f(w) = \frac{1}{i} \sum_{n=1}^{\infty} (\overline{a_n} w^{-n} - a_n \overline{w^{-n}}).$$

Its values for $w = e^{i\theta}$ are

$$\psi(\theta) = \sum_{n=-\infty}^{\infty} b_n e^{in\theta},$$

where

$$b_n = \begin{cases} ia_n, & n>0 \\ 0, & n=0 \\ -ia_n, & n<0. \end{cases}$$

The ψ thus found is called the **exterior periodic conjugate function** of ϕ. Denoting by $\mathcal{K}$ the conjugation operator as defined in Theorem 13.5a, we evidently have:

LEMMA 16.8d. *If the periodic conjugate function $\mathcal{K}\phi$ of a given periodic function ϕ exists, the exterior periodic conjugate function likewise exists and is given by $\psi = -\mathcal{K}\phi$,*

Returning to the mapping problem, let Γ be a Jordan curve as indicated, and let

$$\mathring{g}(w) = \gamma w + a_0 + a_1 w^{-1} + \cdots \tag{16.8-10}$$

be the normalized function mapping $\mathring{E}$ onto the exterior of Γ. As before we consider the auxiliary function

$$h(w) := \log \frac{\mathring{g}(w)}{w}, \qquad |w| \geqq 1.$$

This is analytic also at ∞ and may be normalized by requiring that $h(\infty) = \operatorname{Log} \gamma$ is real. If the exterior inverse boundary correspondence function $\mathring{\phi}$ of Γ is defined by

$$\mathring{g}(e^{i\theta}) = \rho(\mathring{\phi}(\theta))\, e^{i\mathring{\phi}(\theta)}, \qquad 0 \leqq \theta \leqq 2\pi,$$

we have

$$\operatorname{Re} h(e^{i\theta}) = \operatorname{Log}|\mathring{g}(e^{i\theta})| = \operatorname{Log} \rho(\mathring{\phi}(\theta))$$

$$\operatorname{Im} h(e^{i\theta}) = \arg \frac{\mathring{g}(e^{i\theta})}{e^{i\theta}} = \mathring{\phi}(\theta) - \theta.$$

For the periodic function

$$\mathring{\delta}(\theta) := \mathring{\phi}(\theta) - \theta,$$

Lemma 16.8d thus yields **Theodorsen's exterior equation** for the inverse boundary correspondence function,

$$\mathring{\delta}(\theta) = -\mathcal{K}[\operatorname{Log} \rho(\theta + \mathring{\delta}(\theta))]. \tag{16.8-11}$$

All statements made earlier on the uniqueness of continuous solutions and on the convergence of iterative methods for solving Theodorsen's equation are likewise applicable to (16.8-11).

IV. Timman's Method for the Exterior Map

Let Γ: $z = z(\sigma)$ be a piecewise analytic Jordan curve, where the function $z(\sigma)$ is a periodic function with period β. It is not assumed that Γ is starlike. However, in this special instance we do assume that the parameter σ is the *arc length* on Γ, measured from some point $z(0)$ on Γ, so that

$$|z'(\sigma)| = 1. \tag{16.8-12}$$

We furthermore assume that

$$\phi(\sigma) := \arg z'(\sigma) \tag{16.8-13}$$

can be defined as a function in Lip such that $\phi(\sigma) - 2\pi\sigma/\beta$ is periodic with period β. An explicit knowledge of the function $\phi(\sigma)$ is required for Timman's method. In many cases ϕ is easy to calculate. For instance, if Γ is composed of straight line segments and circular arcs, the function ϕ is piecewise linear.

In Timman's method we seek to determine the function $\mathring{g}$ which maps $|w| > 1$ onto the *exterior* of Γ such that $\mathring{g}(\infty) = \infty$ and such that $w = 1$ corresponds to a preassigned point $z(\sigma_0)$ on Γ. This uniquely determined function is of the form

$$\mathring{g}(w) = cw + a_0 + a_1 w^{-1} + \cdots, \tag{16.8-14}$$

where $c = \gamma e^{i\alpha}$, both γ and α being unknown. Our aim is to determine the exterior inverse boundary correspondence function $\mathring{\sigma}(\theta)$, which up to irrelevant multiples of β is determined by

$$\mathring{g}(e^{i\theta}) = z(\mathring{\sigma}(\theta)). \tag{16.8-15}$$

In view of $\mathring{g}(1) = z(\sigma_0)$, we have

$$\mathring{\sigma}(0) = \sigma_0. \tag{16.8-16}$$

We assume that Γ is such that $\mathring{g}'$ can be extended to a continuous function in $|w| \geqq 1$.

The auxiliary function for Timman's method is

$$h(w) := i \log \frac{\mathring{g}'(w)}{\mathring{g}'(\infty)}, \tag{16.8-17}$$

where the analytic logarithm is chosen such that $h(\infty) = 0$. It is then seen from (16.8-14) that

$$h(w) = O(w^{-2}), \qquad w \to \infty. \tag{16.8-18}$$

Differentiating (16.8-15), we find

$$ie^{i\theta}\mathring{g}'(e^{i\theta}) = z'(\mathring{\sigma}(\theta))\mathring{\sigma}'(\theta)$$

or

$$\mathring{g}'(e^{i\theta}) = -ie^{-i\theta}z'(\mathring{\sigma}(\theta))\mathring{\sigma}'(\theta).$$

Recalling the definition of ϕ and observing that $|z'(\sigma)| = 1$, we thus have

$$\text{Re } h(e^{i\theta}) = \theta + \frac{\pi}{2} + \alpha - \phi(\mathring{\sigma}(\theta)),$$

$$\text{Im } h(e^{i\theta}) = \text{Log } \mathring{\sigma}'(\theta) - \text{Log } \gamma.$$

By virtue of Lemma 16.8d there follows Timman's equation,

$$\text{Log } \mathring{\sigma}'(\theta) - \text{Log } \gamma = \mathscr{K}\left[\phi(\mathring{\sigma}(\theta)) - \theta - \frac{\pi}{2} - \alpha\right]. \qquad (16.8\text{-}19)$$

Timman's equation is a functional equation for the unknown function

$$\lambda(\theta) := \text{Log } \mathring{\sigma}'(\theta) - \text{Log } \gamma = \text{Im } h(e^{i\theta})$$

in the following sense. If $\lambda(\theta)$ is known, then evidently

$$\mathring{\sigma}'(\theta) = \gamma e^{\lambda(\theta)}.$$

Hence, using (16.8-16),

$$\mathring{\sigma}(\theta) = \sigma_0 + \gamma \int_0^\theta e^{\lambda(\theta)}\, d\theta.$$

Here the unknown quantity γ may be found from the fact that $\mathring{\sigma}(2\pi) - \mathring{\sigma}(0) = \beta$, hence

$$\gamma = \frac{\beta}{\int_0^{2\pi} e^{\lambda(\theta)}\, d\theta}.$$

Knowing $\sigma(\theta)$, we can determine α from the fact that the 2π-periodic function

$$\phi(\mathring{\sigma}(\theta)) - \theta - \frac{\pi}{2} - \alpha = -\text{Re } h(e^{i\theta})$$

has the mean value 0.

The above construction is easily turned into an iterative process for finding $\lambda(\theta)$ as follows. Let $\lambda_k(\theta)$ be an approximation to $\lambda(\theta)$, given in terms of its Fourier series (or approximating trigonometric polynomial),

$$\lambda_k(\theta) = \sum_{n=-\infty}^{\infty} l_n^{(k)}\, e^{in\theta} \qquad (l_{-n}^{(k)} = \overline{l_n^{(k)}}). \qquad (16.8\text{-}20)$$

Since (16.8-18) shows that in the Fourier series for the exact λ,

$$\lambda(\theta) = \sum_{n=-\infty}^{\infty} l_n\, e^{in\theta}, \qquad l_{-n} = \overline{l_n},$$

we have $l_0 = l_1 = 0$, the same may be assumed in (16.8-20). By means of a discrete Fourier transform we now compute the Fourier series

$$e^{\lambda_k(\theta)} = \sum_{n=-\infty}^{\infty} b_n^{(k)} e^{in\theta}.$$

Since

$$\int_0^{2\pi} e^{\lambda_k(\theta)}\, d\theta = 2\pi b_0^{(k)},$$

the approximate value

$$\gamma_k = \frac{\beta}{2\pi b_0^{(k)}} \tag{16.8-21}$$

is now found. By analytic integration we obtain

$$\mathring{\sigma}(\theta) - \sigma_0 = \frac{\beta}{2\pi}\theta + \frac{\beta}{2\pi i b_0^{(k)}} \sum_{n\neq 0} \frac{b_n^{(k)}}{n}(e^{in\theta} - 1)$$

or

$$\mathring{\sigma}_k(\theta) = \frac{\beta}{2\pi}\theta + \sum_{n=-\infty}^{\infty} c_n^{(k)} e^{in\theta},$$

say. By means of another Fourier analysis we now compute

$$\phi(\mathring{\sigma}_k(\theta)) - \theta = \sum_{n=-\infty}^{\infty} d_n^{(k)} e^{in\theta}. \tag{16.8-22}$$

Because it is known that the zeroth Fourier coefficient in

$$-\text{Re}\, h(e^{i\theta}) = \phi(\mathring{\sigma}(\theta)) - \theta - \frac{\pi}{2} - \alpha \tag{16.8-23}$$

is zero, we obtain the approximation

$$\alpha_k := d_0^{(k)} - \frac{\pi}{2} \tag{16.8-24}$$

for α. Because also the first Fourier coefficient of (16.8-23) is zero, we may artificially suppress it. Using the formulas (13.5-7), we now may compute

$$\lambda_{k+1}(\theta) = \mathscr{K}\left[\sum_{|n|\geqq 2} d_n^{(k)} e^{in\theta}\right], \tag{16.8-25}$$

and the iteration step is complete.

Little appears to be known about mathematical conditions under which the sequence of approximations $\{\lambda_k(\theta)\}$ thus constructed converges to $\lambda(\theta)$.

However, the method has performed well in actual numerical practice, where the convergence seems to be helped by the fact that certain low-order Fourier coefficients are a priori known to be zero. Also from the practical point of view, the method has the advantage of not being restricted to starlike regions.

If the mapping function $\mathring{g}_1$ satisfying $\mathring{g}_1'(\infty) > 0$ is desired, it may be found from the $\mathring{g}$ considered above by the formula

$$\mathring{g}_1(w) = \mathring{g}(e^{-i\alpha}w).$$

PROBLEMS

1. Let $0 < \varepsilon < 1$.

   (a) Show that the curve

$$\Gamma: \quad z(\tau) = e^{i\tau} + \varepsilon e^{-i\tau}, \qquad 0 \leqq \tau \leqq 2\pi,$$

   is an ellipse in normal position with semiaxes $1 \pm \varepsilon$.

   (b) Show that the representation (16.8-2) of Γ is

$$\rho(\phi) = \frac{1-\varepsilon}{\sqrt{1 - [4\varepsilon/(1+\varepsilon)^2](\cos\phi)^2}}.$$

   (c) Show that the standardized exterior mapping function for Γ is

$$\mathring{g}(w) = w + \varepsilon w^{-1}.$$

   (d) Show that the exterior inverse boundary correspondence function for Γ is

$$\mathring{\phi}(\theta) = \arctan\left[\frac{1-\varepsilon}{1+\varepsilon}\tan\theta\right].$$

   (e) Verify the relation (16.8-11).

2. *A perturbation method for Theodorsen's equation* (Yoshikawa [1960]). Let ρ depend on a parameter ε, $\rho = \rho(\phi, \varepsilon)$, such that $\rho(\phi, 0) = 1$, and assume that

$$\lambda(\phi, \varepsilon) := \operatorname{Log} \rho(\phi, \varepsilon) = \sum_{n=1}^{\infty} \lambda_n(\phi)\varepsilon^n,$$

   where the λ_n are trigonometric polynomials of degree not exceeding n.

   (a) Show that there exist trigonometric polynomials $\delta_n(\theta)$, $n = 1, 2, \ldots$, such that the power series in ε,

$$\delta(\theta) := \sum_{n=1}^{\infty} \delta_n(\theta)\varepsilon^n,$$

   satisfies Theodorsen's interior equation (16.8-5) at least formally.

   (b) Show how to construct $\delta_n(\theta)$ rationally from $\lambda_1(\phi), \lambda_2(\phi), \ldots, \lambda_n(\phi)$.

3. *Melentiev's method.* Under the hypotheses and with the notations of Theodorsen's method for the interior map, show that the use of the auxiliary

function

$$h(w) := \frac{g(w)}{w}$$

yields the functional equation for $\delta(\theta)$,

$$\delta(\theta) = \arctan \frac{\mathcal{K}[\rho(\theta+\delta(\theta)) \cos \delta(\theta)]}{\rho(\theta+\delta(\theta)) \cos \delta(\theta)}.$$

4. Let $0 \in \operatorname{int} \Gamma$. Develop analogs of Timman's method for the interior mapping function g
   (a) by considering

$$h(w) := \log \frac{g'(w)}{g'(0)}.$$

   (b) by applying Timman's method to the exterior mapping function for $1/\Gamma$,

$$\mathring{g}_1(w) := \frac{1}{g(1/w)}.$$

NOTES

The basic mathematical results on Theodorsen's method are due to Warschawski [1945, 1950], who also studies the pointwise convergence of the iteration sequence (16.8-6). Gaier [1964] in his thorough discussion of the method considers the effect of discretization on the basic integral equation (16.8-3). Later developments are summarized by Gaier [1983]. On the existence of solutions of the discretized equation (16.8-3) see Gutknecht [1977] and Hübner [1982]. On securing and speeding up convergence by underrelaxation and by similar devices, see Niethammer [1966] and Gutknecht [1981]. Hübner [1985] solves Theodorsen's equation by applying Newton's method in function space; the solution of a Riemann problem is required at each step. The use of FFTs for solving (16.8-3) was advocated by Lundwall-Skaar [1975], Henrici [1976], and Ives [1976], but already Jeltsch [1969] used the FFT (for $n = 64$ points) in a similar context. Extensive numerical experiments on Theodorsen's method, with special attention to the treatment of corners, are reported by Gutknecht [1983a].

Gutknecht [1985] provides a common theoretical basis for the various mapping techniques using conjugate functions. For Timman's method see Timman [1951], Woods [1961], and James [1971]. Halsey [1979, 1980a, b] reports extensive experiments in an industrial environment (Douglas). He compares Timman's method with Theodorsen's original method (without underrelaxation) and finds Timman's method superior. The mapping method used by Bauer et al. [1975] is virtually identical with Timann's method. This version of the method again appears to have passed the test of large-scale industrial use (Boeing).

§ 16.9. METHODS USING FUNCTION-THEORETIC BOUNDARY VALUE PROBLEMS

We continue to use the notation of the preceding section; the curve Γ is assumed in the form (16.8-1), where $z'(\tau) \in \text{Lip}$, $z'(\tau) \neq 0$. In the methods to be discussed here, the problem of finding the normalized inverse boundary correspondence function as characterized by Theorem 16.8a is attacked more directly than in the methods discussed in § 16.8. Suppose we possess an approximation $\tau_k(\theta)$ to the boundary correspondence function that satisfies conditions (i) and (ii) of that theorem. It then only remains to find a correction $\eta(\theta)$, with period 2π, such that the values

$$z(\tau_k(\theta)+\eta(\theta)) \tag{16.9-1}$$

are the boundary values at $e^{i\theta}$ of a function g analytic in E and satisfying the normalizing conditions

$$g(0)=0, \qquad g'(0)>0. \tag{16.9-2}$$

To find the required η exactly is as difficult as finding the exact boundary correspondence function. However, to find an approximation to the desired correction is easier. Suppose the required correction η is small. We then may require that the *linearized* form of (16.9-1) is the boundary function of an analytic function g with the required properties. That is, we may try to determine $\eta(\theta)$ such that

$$z(\tau_k(\theta))+z'(\tau_k(\theta))\eta(\theta)=g^+(e^{i\theta}), \tag{16.9-3}$$

where g is analytic in E and satisfies (16.9-2). If such a g exists, and if the required η is sufficiently small, the expression on the left of (16.9-3) traces a simple closed curve which, although not identical with Γ, is close to it. It then follows from Theorem 16.8a that g furnishes a conformal map of E onto a region which is close to R. Since the process described may be iterated, we may hope in this manner to obtain arbitrarily close approximations to the desired mapping function.

Two implementations of (16.9-3) will be discussed.

1. The Method of Fornberg

If a Hölder continuous function h is given on a Jordan curve Γ, then according to Theorem 14.3a, h can be continued continuously to a function that is analytic in the interior of Γ if and only if

$$\frac{1}{2\pi i}\int_\Gamma h(t)t^k\,dt=0, \qquad k=0,1,2,\ldots.$$

If Γ is the unit circle, this means the same as

$$\int_0^{2\pi} h(e^{i\theta})e^{ik\theta}\,d\theta=0, \qquad k=1,2,\ldots,$$

that is, the Fourier coefficients of *negative* index of $h(e^{i\theta})$ must vanish.

Applying this to the problem on hand and letting

$$z(s) := z(\tau_k(\theta)), \qquad t(s) := z'(\tau_k(\theta)), \tag{16.9-4}$$

where $s = e^{i\theta}$, we find that a function g, analytic in E and satisfying the limit relation (16.9-3), exists provided that

$$\int_0^{2\pi} \{z(e^{i\theta}) + \eta(\theta)t(e^{i\theta})\}\, e^{ik\theta}\, d\theta = 0 \tag{16.9-5}$$

holds for $k = 1, 2, \ldots$. Moreover, the first condition (16.9-2) requires the integral (16.9-5) to be zero also for $k = 0$. We momentarily disregard the second condition (16.9-2). However, since the required correction $\eta(\theta)$ must be real, the relations

$$\int_0^{2\pi} \{\overline{z(e^{i\theta})} + \eta(\theta)\overline{t(e^{i\theta})}\}\, e^{-ik\theta}\, d\theta = 0 \tag{16.9-6}$$

must likewise hold for $k = 0, 1, 2, \ldots$. We are thus faced with the curious problem of finding a 2π-periodic function $\eta(\theta)$ such that one-half of the Fourier coefficients of $\eta(\theta)t(e^{i\theta})$ and one-half of the Fourier coefficients of $\eta(\theta)[t(e^{i\theta})]^{-}$ have prescribed values. Here $t(e^{i\theta})$ is a given function that is never zero.

The solution method proposed by Fornberg [1980] proceeds by discretization. Let n be a (large) integer; we use discrete Fourier analysis in the space Π_{2n}. Thus let

$$w := \exp\left(\frac{2\pi i}{2n}\right),$$

$z_k := z(w^k)$, $t_k := t(w^k)$, $k \in \mathbb{Z}$, and denote by η_k a number intended to approximate $\eta(k\pi/n)$. If Fourier coefficients are approximated in the usual manner, condition (16.9-5) is then simulated by

$$\sum_{j=0}^{2n-1} (z_j + \eta_j t_j) w^{jk} = 0, \qquad k = 0, 1, 2, \ldots, n-1,$$

or by

$$\sum_{j=0}^{2n-1} t_j \eta_j w^{jk} = u_k, \qquad k = 0, 1, \ldots, n-1, \tag{16.9-7}$$

where

$$u_k := -\sum_{j=0}^{2n-1} z_j w^{jk}.$$

The system (16.9-7) comprises n equations for the $2n$ unknowns η_j. An additional set of n equations is obtained by conjugation, using $\overline{\eta_j} = \eta_j$.

To study the linear system, we introduce the vectors

$$\boldsymbol{\eta}_0 := \begin{pmatrix} \eta_0 \\ \eta_2 \\ \vdots \\ \eta_{2n-2} \end{pmatrix}, \qquad \boldsymbol{\eta}_1 := \begin{pmatrix} \eta_1 \\ \eta_3 \\ \vdots \\ \eta_{2n-1} \end{pmatrix}, \qquad \mathbf{u} := \begin{pmatrix} u_0 \\ u_1 \\ \vdots \\ u_{n-1} \end{pmatrix},$$

and the $n \times n$ matrices

$$\mathbf{F} := (w^{2ij})_{i,j=0}^{n-1}, \qquad \mathbf{W} := \operatorname{diag}(w^j)_{j=0}^{n-1},$$

$$\mathbf{T}_0 := \operatorname{diag}(t_{2j})_{j=0}^{n-1}, \qquad \mathbf{T}_1 := \operatorname{diag}(t_{2j+1})_{j=0}^{n-1}.$$

The system to be solved may then be written

$$\begin{aligned} \mathbf{F}\mathbf{T}_0\boldsymbol{\eta}_0 + \mathbf{W}\mathbf{F}\mathbf{T}_1\boldsymbol{\eta}_1 &= \mathbf{u}, \\ \mathbf{F}^*\mathbf{T}_0^*\boldsymbol{\eta}_0 + \mathbf{W}^*\mathbf{F}^*\mathbf{T}_1^*\boldsymbol{\eta}_1 &= \mathbf{u}^*, \end{aligned} \tag{16.9-8}$$

where * denotes complex conjugation. From this we readily find

$$\begin{aligned} \operatorname{Im}\{\mathbf{T}_1^{-1}\mathbf{F}^{-1}\mathbf{W}^{-1}\mathbf{F}\mathbf{T}_0\}\boldsymbol{\eta}_0 &= \operatorname{Im}\{\mathbf{T}_1^{-1}\mathbf{F}^{-1}\mathbf{W}^{-1}\mathbf{u}\}, \\ \operatorname{Im}\{\mathbf{T}_0^{-1}\mathbf{F}^{-1}\mathbf{W}\mathbf{F}\mathbf{T}_1\}\boldsymbol{\eta}_1 &= \operatorname{Im}\{\mathbf{T}_0^{-1}\mathbf{F}^{-1}\mathbf{u}\}. \end{aligned} \tag{16.9-9}$$

This represents two *real* systems, of order n each, for the real vectors $\boldsymbol{\eta}_0$ and $\boldsymbol{\eta}_1$. The matrices of these systems are the imaginary parts of products of the matrices $\mathbf{T}_i$, $\mathbf{W}$, and $\mathbf{F}$. Multiplication of a vector by the diagonal matrices $\mathbf{W}, \mathbf{W}^{-1}$, $\mathbf{T}_i, \mathbf{T}_i^{-1}$ of course requires just n scalar multiplications. Multiplication of a vector by $\mathbf{F}$ amounts, up to a scalar factor, to a discrete Fourier transform in the space Π_n, and thus may be performed in $O(n \operatorname{Log} n)$ operations. The two systems (16.9-9) thus have the form $\mathbf{C}\boldsymbol{\eta} = \mathbf{v}$, where the multiplication by $\mathbf{C}$ of any vector $\boldsymbol{\eta}$ is cheap. Iterative methods of solution are therefore indicated. Because of an observed clustering of the eigenvalues of $\mathbf{C}$, the conjugate gradient method gives excellent results.

Above, we have disregarded the conditions that the function g satisfying (16.9-3) should be such that $g'(0) > 0$. In place of this constraint we also may prescribe the location of one boundary point, for instance by arbitrarily setting $\eta_0 = 0$. If this is to be possible, the rank of the system (16.9-9) must be $<2n$. It is a weak point in Fornberg's theory that there is no transparent proof of this. However, in numerous computed examples Fornberg did find that the system (16.9-9), or a system closely related to it, is indeed uniquely solvable after setting $\eta_0 = 0$.

II. Wegmann's Method

Since $\eta(\theta)$ must be real, it follows from (16.9-3) that

$$z'(\tau_k(\theta))g^+(e^{i\theta}) - \overline{z'(\tau_k(\theta))}z(\tau_k(\theta))$$

is real. Wegmann's method (Wegmann [1978, 1984]) is based on the observation that this condition may be transformed into a Riemann problem for g, which together with the normalizing conditions (16.9-2) serves to determine g uniquely. Using the abbreviations (16.9-4), the foregoing condition becomes

$$\operatorname{Im}[\overline{t(s)}g^+(s) - \overline{t(s)}z(s)] = 0, \qquad |s| = 1,$$

or

$$\operatorname{Re}[\overline{it(s)}g^+(s)] = \operatorname{Re}[\overline{it(s)}z(s)]. \tag{16.9-10}$$

Since the expression on the right is known, this is precisely a Riemann problem for g in the sense of § 14.5. Since Γ already is the unit circle, no preliminary transformation is required for its solution. The associated Privalov problem is best expressed in terms of

$$\phi(s) := \arg t(s),$$

where the argument is continuous save at one point. The Privalov problem then becomes

$$g^+(s) = e^{2i\phi(s)}g^-(s) + b(s), \tag{16.9-11}$$

where

$$b(s) := z(s) - e^{2i\phi(s)}\overline{z(s)}. \tag{16.9-12}$$

Since $\phi(s)$ increases by 2π as s travels around the unit circle, the index of the Privalov problem is clearly 2.

We solve the Riemann problem by means of the formulas of § 14.5. We have

$$a_1(s) = \log[s^{-2}e^{2i\phi(s)}] = 2i(\phi(s) - \theta),$$

where $\theta := \arg s$. Hence $\psi(s) = 2\delta(s)$, where

$$\delta(s) := \phi(s) - \theta, \tag{16.9-13}$$

and

$$l(w) = \frac{1}{\pi}\int_{|s|=1} \frac{\delta(s)}{s-w}\,ds. \tag{16.9-14}$$

The canonical solution of the homogeneous Privalov problem thus is

$$g_0(w) := \begin{cases} e^{-i\hat{\delta}}e^{l(w)}, & |w| < 1 \\ w^{-2}e^{-i\hat{\delta}}e^{l(w)}, & |w| > 1. \end{cases}$$

Here $\hat{\delta}$ as usual is the mean value of $\delta(s)$. To construct a solution of the inhomogeneous problem (16.9-11) we require the values $g_0^+(s)$, which by (14.5-20) are

$$g_0^+(s) = e^{-\mathcal{H}\delta(s) + i\delta(s)}.$$

The function h defined by (14.5-21) becomes

$$h(s) = \frac{b(s)}{g_0^+(s)} = e^{\mathscr{K}\delta(s) - i\delta(s)} 2ie^{i\phi(s)} \operatorname{Im}[\overline{t(s)}z(s)]$$

or, on account of $\phi(s) - \delta(s) = \theta$, $e^{i\theta} = s$,

$$h(s) = 2i\lambda(s),$$

where

$$\lambda(s) := e^{\mathscr{K}\delta(s)} \operatorname{Im}[\overline{t(s)}z(s)] \tag{16.9-15}$$

is real. The function k defined by (14.5-22) thus is

$$k(w) = \frac{1}{\pi} \int_{|s|=1} \frac{s\lambda(s)}{s-w}\, ds = c + k_1(w),$$

where

$$c := \frac{1}{\pi} \int_{|s|=1} \lambda(s)\, ds, \qquad k_1(w) := \frac{1}{2\pi i} \int_{|s|=1} \frac{2i\lambda(s)}{s-w}\, ds. \tag{16.9-16}$$

Since $2i\lambda(s)$ is pure imaginary, we have from (14.2-25)

$$k^*(w) = \bar{c} + \frac{1}{w}[k_1(w) - 2i\hat{\lambda}]. \tag{16.9-17}$$

By Theorem 14.5b, the general solution of the Riemann problem (16.9-10) is

$$g(w) = \{p(w) + \tfrac{1}{2}[k(w) + w^2 k^*(w)]\} g_0(w),$$

where p is an arbitrary self-reciprocal polynomial of degree 2. Using (16.9-17), this becomes

$$g(w) = \{p_0(w) + wk_1(w) - iw\hat{\lambda}\} g_0(w),$$

where p_0 is another self-reciprocal polynomial.

The polynomial $p_0(w) = a + \beta w + \bar{a}w^2$ now must be chosen so that the normalizing conditions for g hold. To satisfy $g(0) = 0$, we must have $a = 0$; to satisfy $g'(0) > 0$, it is necessary that

$$(\beta + k_1(0) - i\hat{\lambda}) g_0(0)$$

be real. We have

$$k_1(0) = 2i\hat{\lambda}, \qquad g_0(0) = e^{i\hat{\delta}}$$

and therefore want

$$(\beta + i\hat{\lambda}) e^{i\hat{\delta}} = (\beta - i\hat{\lambda}) e^{-i\hat{\delta}}.$$

This equation is solvable for β if $\sin\hat{\delta} \neq 0$, and the unique solution then is

$$\beta = -\hat{\lambda}\cot\hat{\delta}. \tag{16.9-18}$$

The solution g has now been completely determined. The value

$$g'(0) = -\frac{\hat{\lambda}}{\sin\hat{\delta}} \tag{16.9-19}$$

thus obtained is real, though not necessarily positive.

THEOREM 16.9a. *Wegmann's Riemann problem* (16.9-10) *has a unique solution satisfying* $g(0)=0$, $\operatorname{Im} g'(0)=0$ *provided that* $\hat{\delta} \not\equiv 0 \bmod \pi$. *This solution is given by*

$$g(w) = w\left\{-\frac{\hat{\lambda}}{\sin\hat{\delta}} + e^{-i\hat{\delta}}k_1(w)\right\}e^{l(w)}, \tag{16.9-20}$$

where l *and* k_1 *are defined by* (16.9-14) *and* (16.9-16), *and where* $\hat{\delta}$ *and* $\hat{\lambda}$ *are the mean values of the functions* δ *and* λ *defined by* (16.9-13) *and* (16.9-15).

To proceed, we require the boundary values $g^+(s)$. From (14.2-24) we have

$$l^+(s) = -\mathcal{K}\delta(s) + i\hat{\delta} + i\delta(s),$$

$$k_1^+(s) = -\mathcal{K}\lambda(s) + i\hat{\lambda} + i\lambda(s).$$

Using $e^{i\delta(s)} = e^{i\phi(s)}s^{-1}$, we thus get:

COROLLARY 16.9b. *The boundary values* g^+ *of the function* g *determined in Theorem* 16.9a *on* $|s|=1$ *are*

$$g^+(s) = e^{i\phi(s)-\mathcal{K}\delta(s)}\{-\mathcal{K}\lambda(s) - \hat{\lambda}\cot\hat{\delta} + i\lambda(s)\}. \tag{16.9-21}$$

It is seen that $g^+(s)$ may be computed from the data by merely two applications of the operator $\mathcal{K}$, that is, by four discrete Fourier transforms.

The function g thus determined is not yet the desired mapping function for the interior of the curve Γ. Instead, it maps the unit disk onto the interior of a curve Γ_k with parametric representation

$$z_k(\theta) = z(\tau_k(\theta)) + z'(\tau_k(\theta))\eta(\theta),$$

where the real function $\eta(\theta)$ is found from the basic relation (16.9-3) as

$$\eta(\theta) = \frac{1}{z'(\tau_k(\theta))}[g^+(e^{i\theta}) - z(\tau_k(\theta))]$$

(see Fig. 16.9a). If Γ is smooth and η is small, Γ_k resembles Γ closely.

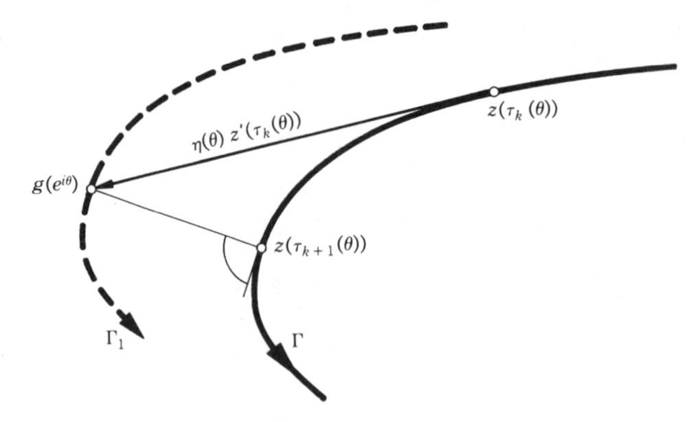

Fig. 16.9a. Correction in Wegmann's method.

To obtain the exact mapping function for the interior of Γ, the above process must be iterated. To this end, we extract from the above a new approximation $\tau_{k+1}(\theta)$ to the inverse boundary correspondence function $\tau(\theta)$. This may be done in two ways.

(a) In the spirit of the linearization used previously, we simply set

$$\tau_{k+1}(\theta) = \tau_k(\theta) + \eta(\theta). \tag{16.9-22}$$

(b) We project the point $g(e^{i\theta})$ perpendicularly onto Γ (see Fig. 16.9a). For each θ for which $\tau_{k+1}(\theta)$ is sought, this requires finding $\tau = \tau_{k+1}(\theta)$ as the solution of the transcendental equation

$$\operatorname{Re} \frac{z(\tau) - g(e^{i\theta})}{z'(\tau)} = 0. \tag{16.9-23}$$

At the time of this writing, it is not yet clear which of the two methods is numerically preferable. In either case, if $\tau_k(\theta)$ is at all reasonable, $\tau_{k+1}(\theta)$ may be expected to be a better approximation to the true boundary correspondence function, and since the terms that are being neglected are quadratic in η, the convergence of the process may ultimately be expected to be quadratic. That quadratic convergence in fact takes place under suitable assumptions on the smoothness of Γ is proved by Wegmann [1984].

How close the approximations of Wegmann's method are to the true mapping function may be estimated by the closeness of $\hat{\delta}$ and $\hat{\lambda}$ to their limiting values, which are easily calculated. If g is the true mapping function for the interior of Γ, and if $\tau(\theta)$ is the true inverse boundary correspondence function so that $z(\tau(\theta)) = g(e^{i\theta})$, we find from

$$z'(\tau(\theta))\tau'(\theta) = g'(e^{i\theta})ie^{i\theta}$$

that

$$\phi(s) = \arg z'(\tau(\theta)) = \arg g'(s) + \frac{\pi}{2} + \theta.$$

Hence

$$\delta(s) = \phi(s) - \theta = \text{Im} \log[ig'(s)],$$

and thus

$$\begin{aligned}\hat{\delta} &= \frac{1}{2\pi} \int_0^{2\pi} \text{Im} \log[ig'(e^{i\theta})]\, d\theta = \text{Im} \frac{1}{2\pi i} \int_{|s|=1} \log[ig'(s)\, \frac{ds}{s} \\ &= \text{Im} \log[ig'(0)]\end{aligned}$$

or

$$\hat{\delta} = \frac{\pi}{2}. \tag{16.9-24}$$

Similarly from

$$\lambda(s) = e^{\mathcal{H}\delta(s)}\, \text{Im}\{e^{-i\phi(s)} g(s)\} = -g'(0)\, \text{Re}\left[\frac{g(s)}{sg'(s)}\right]$$

there follows

$$\hat{\lambda} = -g'(0)\, \text{Re} \frac{1}{2\pi i} \int_{|s|=1} \frac{g(s)}{s^2 g'(s)}\, ds$$

or

$$\hat{\lambda} = -1. \tag{16.9-25}$$

Thus in the limiting case, if g is the exact mapping function, we find that the quantity $-\hat{\lambda} \cot \hat{\delta}$ appearing in (16.9-21) equals zero.

PROBLEMS

1. Let $\tau_k(\theta) = \tau(\theta)$, the exact inverse boundary correspondence function, and suppose that $\tau(\theta)$ has a continuous derivative. Show that the homogeneous system corresponding to (16.9-5),

$$\int_0^{2\pi} t(e^{i\theta})\eta(\theta) e^{ik\theta}\, d\theta = 0, \qquad k = 0, 1, 2, \ldots,$$

has the nontrivial solution $\eta(\theta) = \tau'(\theta)$. (The relations

$$\int_0^{2\pi} z(\tau(\theta + \delta)) e^{ik\theta}\, d\theta = 0, \qquad k = 0, 1, 2, \ldots,$$

hold for all $\delta \in \mathbb{R}$. Differentiate with respect to δ and set $\delta = 0$.)

2. *Wegmann's method for the exterior mapping problem*

(a) Using the notation of the main text, show how to obtain an improved approximation $\mathring{g}(w)$ to the exterior mapping (normalized by $\mathring{g}(\infty)=\infty$, $\mathring{g}'(\infty)$ real) by solving the Riemann problem

$$\operatorname{Re}[\overline{it(s)}\mathring{g}^-(s)] = \operatorname{Re}[\overline{it(s)}z(s)].$$

(b) Show that the unique solution satisfying the required conditions at ∞ is

$$\mathring{g}(w) = \left[e^{-i\hat{\delta}}k_1(w) - \frac{\hat{\lambda}}{\sin\hat{\delta}}\right]we^{l(w)},$$

where

$$\delta(s) := \theta - \phi(s),$$

$$l(w) := \frac{1}{\pi}\int_{|s|=1}\frac{\lambda(s)}{s-w}\,ds,$$

$$\lambda(s) := e^{\mathcal{K}\delta(s)}\operatorname{Im}[t(s)\overline{z(s)}],$$

$$k_1(w) := \frac{1}{\pi}\int_{|s|=1}\frac{\lambda(s)}{s-w}\,ds,$$

and where $\hat{\delta}$ and $\hat{\lambda}$ denote mean values.

(c) Show that $\mathring{g}$ on $|s|=1$ has the boundary values

$$\mathring{g}^-(s) = e^{i\phi(s)-\mathcal{K}\delta(s)}\{-\mathcal{K}\lambda(s) - \hat{\lambda}\cot\hat{\delta} - i\lambda(s)\}.$$

NOTES

For Fornberg's method, see Fornberg [1980]. In Fornberg [1981a] he applies his method in a hydrodynamical context. Widlund [1982] lends theoretical support to the fact, experimentally noted by Fornberg, that the eigenvalues of the matrix **C** cluster around 1. Golub and van Loan [1983] explain why the conjugate gradient method is suitable for such systems. For Wegmann's method, see Wegmann [1978]. In view of its quadratic convergence, thoroughly discussed in Wegmann [1984], and its $O(n \operatorname{Log} n)$ operations count per iteration step, Wegmann's method may well be the best mapping method in existence, but sufficient experimental documentation of the method is as yet lacking. Gutknecht [1985] provides a superstructure which contains these and many other methods as special cases.

§ 16.10. PARAMETER DETERMINATION IN SCHWARZ–CHRISTOFFEL MAPS

When discussing numerical methods of conformal mapping, we should not forget that a quasiexplicit formula for the mapping function exists for the case where the region D to be mapped is polygonal, that is, where D is a simply connected region the boundary Γ of which consists of a finite number of straight line segments, half-lines, or lines. Let the vertices of Γ be $z_0, z_1, \ldots, z_n$, numbered according to the orientation of Γ. We write $z_{n+1} := z_0$. It is permitted that some $z_i = \infty$. However, no two consecutive z_i shall be at infinity. (The case of a polygonal region the boundary of which contains an infinite straight line can nevertheless be handled by introducing artificial corners.) If z_k is finite, let the interior angle of Γ at z_k be $\alpha_k\pi$, where $0<\alpha_k\leqq 2$. If $\alpha_k=2$, this means that the boundary has a slit, see Fig. 5.12a; this is explicitly permitted. If $z_k=\infty$, let the interior angle of Γ at z_k (measured on the Riemann sphere) be $-\alpha_k\pi$, where $0\leqq -\alpha_k<2$. If $\alpha_k=0$, this means that the boundary contains two parallel half-lines extending to ∞; again this is explicitly permitted.

Now let a be an interior point of D. According to Theorem 5.12e, any function g mapping the unit disk E onto D and satisfying $g(0)=a$ is given by the **Schwarz–Christoffel** (S-C) **formula,**

$$z=g(w)=a+c\int_0^w \prod_{k=0}^{n}\left(1-\frac{w}{w_k}\right)^{\alpha_k-1} dw, \qquad (16.10\text{-}1)$$

where c is a suitable complex constant, and where the w_k are the pre-images (or **prevertices**) of the z_k, $|w_k|=1$. The map g may be made unique by requiring that $g'(0)=c$ be positive, or by prescribing the position of one prevertex w_k.

A formula similar to (16.10-1) also holds for functions mapping the upper half-plane onto D. In this case the pre-images w_k of the vertices z_k lie on the real axis.

If the parameters c and $w_0, w_1, \ldots, w_n$ are known, the Schwarz-Christoffel formula is as explicit a representation of the mapping function g as could be desired. Before the formula can be used, however, the correct values of these parameters must be determined. This is the **parameter problem** of the Schwarz-Christoffel map. For $n=1, 2$, or if D has a high degree of symmetry, it was shown in § 5.12 how to solve the parameter problem analytically. For polygons without symmetry, however, no analytical methods exist if $n+1$, the number of vertices of Γ, exceeds 3, and numerical methods must be resorted to. Although from a purely theoretical point of view the parameter problem may be viewed as trivial—after all, the solution is known to exist, and to be unique—, it turns out that by current standards of scientific computation neither the adequate formulation nor the solution of the parameter problem is completely trivial. We therefore briefly report on work by Trefethen [1980] which, at least for unbounded polygons, appears to be the first really effective solution of the parameter problem if the number of vertices does not exceed, say, 20.

I. The Equivalent System of Real Equations

The first decision that must be made in applying the Schwarz-Christoffel formula concerns the domain of g. Examples show that if the upper half-plane is chosen, then in many instances the orders of magnitude of the prevertices w_k differs so widely as to cause numerical difficulties. For this reason the map from the *unit disk* E to D is to be preferred. This map is normalized by the condition

$$w_0 = 1. \tag{16.10-2}$$

To formulate a nonlinear system for the remaining parameters $w_1, w_2, \ldots, w_{n-1}$ and c now at first sight seems to be an easy matter, because the relations

$$z_k - a = c \int_0^{w_k} \prod_{j=0}^{n} \left(1 - \frac{w}{w_j}\right)^{\alpha_j - 1} dw \tag{16.10-3}$$

are to be satisfied for $k = 0, 1, \ldots, n$. However, these relations cannot be used for vertices which are at ∞. (Even the simple examples of § 5.12 show that almost all bona fide applications of the S–C map require one or several corners at ∞.) To deal with unbounded polygons, the following assumptions are made:

(a) Each component of ∂D contains at least one finite vertex.
(b) One connected component of ∂D contains at least two finite vertices.

None of these assumptions involves any loss of generality, because if necessary artificial vertices z_k with $\alpha_k = 1$ can be introduced. Again without loss of generality, we may assume that the two finite vertices required in (b) are z_n and z_0.

We now eliminate c from our set of equations by using (16.10-3) for $k = 0$, yielding

$$c = \frac{z_0 - a}{\int_0^1 \prod_{j=0}^{n} (1 - w/w_j)^{\alpha_j - 1}\, dw}. \tag{16.10-4a}$$

This defines c as a function of $w_1, w_2, \ldots, w_n$. It remains to find n real equations for these n parameters. One equation is supplied by the case $k = n$ of (16.10-3),

$$z_n - a = c \int_0^{w_n} \prod_{j=0}^{n} \left(1 - \frac{w}{w_j}\right)^{\alpha_j - 1} dw, \tag{16.10-4b}$$

which amounts to two real equations. To obtain further usable equations, denote by $\Gamma_1, \Gamma_2, \ldots, \Gamma_m$ the distinct connected components of ∂D, numbered in counterclockwise order, and let z_{k_l} $(l = 1, 2, \ldots, m)$ be the last corner on Γ_l. We then impose for $l = 2, \ldots, m$ the $2m - 2$ additional real conditions

$$z_k - a = c \int_0^{w_{k_l}} \prod_{j=0}^{n} \left(1 - \frac{w}{w_j}\right)^{\alpha_j - 1} dw. \tag{16.10-4c}$$

Finally $n - 2m$ conditions of side length are imposed. For each pair (z_k, z_{k+1})

where both vertices are finite, there holds

$$|z_{k+1}-z_k|=\left|c\int_{w_k}^{w_{k+1}}\prod_{j=0}^{n}\left(1-\frac{w}{w_j}\right)^{\alpha_j-1}dw\right|. \tag{16.10-4d}$$

If D is unbounded, and if ∂D has m components, D has $n+1-m$ finite corners, and thus there exist $n+1-2m$ pairs (z_k, z_{k+1}) of consecutive finite vertices. We then require (16.10-4d) for the first $n-2m$ of these; the pair (z_n, z_0) has already been taken into account by (16.10-4a) and (16.10-4b). If D is bounded, there will be $n+1$ distinct pairs of consecutive finite vertices. We then require (16.10-4d) for any $n-1$ consecutive pairs (z_k, z_{k+1}). Precisely one vertex, say z_j, will then not occur explicitly in the system of equations (16.10-4). Unless the interior angle at the omitted vertex is 2π, the polygon will nevertheless close at the correct point. The case where $\alpha_j=2$ must be avoided in the selection of pairs (z_k, z_{k+1}).

We now have obtained a set of n real equations for the n real parameters θ_k, where

$$w_k=e^{i\theta_k}, \qquad k=1,2,\ldots,n.$$

For the w_k to be correctly ordered it is necessary that the θ_k satisfy the constraints

$$\theta_0:=0<\theta_1<\theta_2<\cdots<\theta_n<\theta_{n+1}:=2\pi.$$

To eliminate the constraints, we introduce the new variables

$$\lambda_k:=\operatorname{Log}\frac{\theta_{k+1}-\theta_k}{\theta_k-\theta_{k-1}}, \qquad k=1,2,\ldots,n,$$

which are real if and only if the constraints are satisfied. If the λ_k are known, the θ_k are found by solving the tridiagonal system of linear equations

$$\theta_{k+1}-\theta_k=e^{\lambda_k}(\theta_k-\theta_{k-1}), \qquad k=1,2,\ldots,n, \tag{16.10-5}$$

where $\theta_0:=0$, and where $\theta_{n+1}=2\pi$ is the sole nonhomogeneous term.

II. Solution of the Nonlinear System

In Section I we have obtained a system of nonliner equations of the form

$$\phi_k(\lambda_1,\lambda_2,\ldots,\lambda_n)=0, \qquad k=1,2,\ldots,n, \tag{16.10-6}$$

for the unconstrained real unknowns λ_i. The analytic computation of the partial derivatives $\partial\phi_k/\partial\lambda_j$, although possible in principle, is awkward in practice. Thus the system (16.10-6) is best solved by a method that does not use partial derivatives. Trefethen employs an algorithm due to Powell (routine NS01A, Powell [1968]) for this purpose.

A comment is required about the numerical evaluation of the functions ϕ_k, which contains the evaluation of the definite integrals in (16.10-4) as its nontrivial element.

The first decision to be made concerns the path of integration. Mathematically any path in the unit disk may be chosen, since the integrand is analytic there. Numerically it seems simplest and best to integrate along straight line segments. By a linear change of variables, all integrals to be evaluated may be assumed to have the form

$$\int_{-1}^{1} (1-\tau)^{\alpha}(1+\tau)^{\beta} f(\tau)\, d\tau, \tag{16.10-7}$$

where α and β are suitable real numbers > -1, and where f is complex valued and analytic on the *closed* interval of integration.

Since integration in closed form is out of the question, the next choice concerns the method of numerical integration. For integrals such as (16.10-7) a natural choice would seem to be Gauss-Jacobi quadrature. A **Gauss–Jacobi quadrature formula** (see Atkinson [1978], ch. 5) is a sum

$$\sum_{i=1}^{q} \omega_i f(\tau_i),$$

where the **weights** ω_i and the **nodes** τ_i are chosen in such a way that the formula computes the integral (16.10-7) exactly for $f(\tau)$ a polynomial of the highest possible degree $2q-1$. For any given α, β, and q, the required weights and nodes can be computed accurately by a program due to Golub and Welsch [1969].

Disappointingly, straightforward application of Gauss-Jacobi quadrature to the integrals in question was not found to yield the integrals with sufficient accuracy. The explanation lies in the fact that due to the irregular spacing of the singularities w_i on $|w|=1$, the function f in (16.10-7), although analytic on $[-1, 1]$, frequently has singularities *close* to the interval of integration. Fortunately the location of these singularities is known. Thus the following **compound Gauss–Jacobi quadrature** was found to be successful. The interval $[-1, 1]$ is split up into subintervals such that no singularity is closer to any subinterval than half the length of the subinterval. Over each subinterval, the integral is then evaluated by the appropriate Gauss-Jacobi rule. (If both endpoints of the subinterval are interior points of $[-1, 1]$, then the rule for $\alpha=\beta=0$ is used.) Trefethen found that by using this device he was able to reduce the integration error in typical situations from 10^{-2} to 10^{-7} at little additional computational expense.

III. Computation of the Mapping Function

Once the points $w_k = e^{i\theta_k}$ or, equivalently, the parameters λ_k have been computed, the mapping from the disk E to the region D is given by the explicit formula

$$z = g(w) = g(w^*) + c \int_{w^*}^{w} \prod_{j=0}^{n} \left(1 - \frac{w}{w_j}\right)^{\alpha_j - 1} dw. \qquad (16.10\text{-}8)$$

Here w^* may be taken as any point for which the image $z^* = g(w^*)$ is known. Trefethen chooses $w^* = 0$, implying $g(0) = a$, unless w is close to a prevertex w_k, in which case the recommended choice is $w^* = w_k$. In either case the integral in (16.10-8) is evaluated by a compound Gauss-Jacobi rule, as indicated earlier. This method has the advantage of producing accurate results also in the neighborhood of the vertices. In the solution of potential problems by conformal transplantation, the method thus will produce accurate results also near corners, an advantage that is not shared by finite-difference or finite-element methods.

In applications of conformal mapping to partial differential equations other than Laplace's equation, the derivative $g'(w)$ is often required. This is immediately available from

$$g'(w) = c \prod_{j=0}^{n} \left(1 - \frac{w}{w_j}\right)^{\alpha_j - 1}. \qquad (16.10\text{-}9)$$

Many applications also require the values of the mapping $f = g^{[-1]}$ (region to disk) at specified points $z \in D$. Since $w = f(z)$ is the unique solution of the equation for w,

$$g(w) = z, \qquad (16.10\text{-}10)$$

w may be found by solving (16.10-10). Since g is analytic, the complex version of Newton's method (see § 6.12) recommends itself; the required derivative g' is easily available from (16.10-9).

Another approach to the computation of $f(z)$ may be based on the fact that the function $w = f(z)$ satisfies the differential equation

$$w' = f'(z) = [g'(w)]^{-1},$$

which by (16.10-9) takes the simple form

$$w' = c \prod_{j=0}^{n} \left(1 - \frac{w}{w_j}\right)^{1 - \alpha_j}. \qquad (16.10\text{-}11)$$

Starting from any point z^* such that $w^* = f(z^*)$ is known, this differential equation may be integrated numerically along any path Γ in D. The point $w = f(z)$ will then trace the image path $f(\Gamma)$.

Trefethen recommends that one integrates the differential equation (written as a first-order system for two real functions u, v) by a robust numerical method in order to obtain good starting values for the computation of $f(z)$ by Newton's method as indicated above.

Finally if values of f' are required, they may be obtained easily from (16.10-10), where $w=f(z)$.

PROBLEM

1. Study numerically the Schwarz-Christoffel map of a rectangle with corners $\pm a \pm ib$. In particular, establish (numerically or analytically) the following curious fact. If $g(0)=0$, and if the prevertices are $w=\pm e^{\pm i\theta_1}$, then the angle θ_1 becomes very small even for moderately small values of the ratio b/a such as $b/a=\frac{1}{3}$ (This effect points to a kind of ill-conditioning of the S-C map, and of conformal mapping problems in general.)

NOTES

For an historical perspective on the Schwarz-Christoffel map, see Pfluger [1981]. Trefethen [1980] gives numerous examples (with figures) and applications of his method, additional hints for its implementation, as well as (in Trefethen [1982]) a user's guide for his program SCPACK. For earlier implementations of the map see Howe [1973], Vecheslavov and Kokoulin [1973], and additional references provided by Trefethen [1980]. The implementation of Reppe [1979] likewise uses the idea of compound quadrature and provides interesting applications.

Trefethen ([1984] and papers to appear) introduces the *generalized S-C parameter problem* for which some of the equations in the nonlinear system have a totally different form, and analytical computation of partial derivatives is impossible. This problem comes up, for example, in slit resistor design, where the slit length must be determined; in free-streamline flows, where some conditions in the physical plane must be matched; in S-C maps on Riemann surfaces, where conditions on the branch point are imposed; see also Elcrat [1982] and a forthcoming book by Trefethen.

The crowding phenomenon alluded to in Problem 1 points to a very basic difficulty in all of numerical conformal mapping. A uniform distribution of points in one plane may yield a highly nonuniform distribution of the image points in the other plane. (Grassmann [1979] called this the Geneva effect.) Berrut [1985b] and Kerzman and Trummer [1985] in their numerical experiments counteracted the crowding phenomenon by introducing special ad hoc parametrizations of the boundary curve. However, more research appears to be needed on how to deal with crowding.

§ 16.11. THE MODULUS OF A QUADRILATERAL

Let Γ be a positively oriented Jordan curve, and let three distinct points a_1, a_2, a_3 on Γ be given, arranged in the direction of increasing parameters. Let T be the interior of Γ. The system $(T; a_1, a_2, a_3)$ is called a **trilateral**.

Two trilaterals $(T; a_1, a_2, a_3)$ and $(S; b_1, b_2, b_3)$ are called **conformally equivalent** if there exists a conformal map from T onto S such that the continuous extension of f to the boundary of T (the existence of which is guaranteed by Theorem 16.3a) satisfies

$$f(a_i)=b_i, \qquad i=1,2,3. \tag{16.11-1}$$

The equivalent relation thus defined clearly satisfies the required properties of symmetry, reflexivity, and transitivity.

THEOREM 16.11a. *All trilaterals are conformally equivalent.*

For the *proof* it suffices to show that any given trilateral $(T; a_1, a_2, a_3)$ is conformally equivalent to some fixed trilateral, for instance to the unit disk E with the three distinguished boundary points $b_1=-i$, $b_2=1$, $b_3=i$. Let r be any conformal map of T to E. As shown by the principle of the argument, the points $r(a_i)$, $i=1,2,3$, then are oriented in the positive sense on the unit circle. Therefore there exists a Moebius transformation t mapping E onto E such that $t(r(a_i))=b_i$, $i=1,2,3$. The map $f:=t\circ r$ is the required map. —

A **quadrilateral** now obviously is a system $(Q; a_1, a_2, a_3, a_4)$, where Q is a Jordan region and where the a_i are four distinct points on the boundary of Q, arranged in the sense of increasing parameter values (see Fig. 16.11a).

We occasionally write $\mathbf{Q}=(Q, a, b, c, d)$. We note that two quadrilaterals obtained from each other by a permutation of the distinguished boundary points are *not* considered to be identical; thus $(Q; a, b, c, d)\neq(Q; b, c, d, a)$.

Two quadrilaterals $(Q_i; a_{i1}, a_{i2}, a_{i3}, a_{i4})$, $i=1,2$, are called **conformally equivalent** if there exists a conformal map f from Q_1 to Q_2 such that its Osgood–Carathéodory extension satisfies

$$f(a_{1j})=a_{2j}, \qquad j=1,2,3,4.$$

Obviously not all quadrilaterals are conformally equivalent, for this would mean that given two sets $(a_{i1}, a_{i2}, a_{i3}, a_{i4})$ of four points on the unit

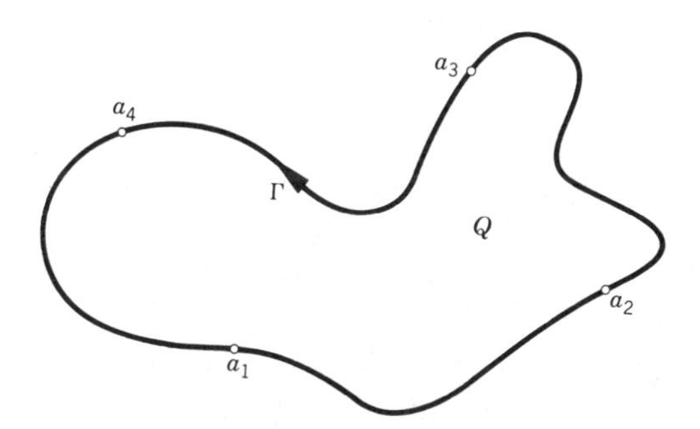

Fig. 16.11a. A quadrilateral.

circle, $i = 1, 2$, both oriented in the positive sense, there would exist a conformal map, and hence a Moebius transformation t, of the unit disk onto itself such that $t(a_{1j}) = a_{2j}$, $j = 1, 2, 3, 4$. Such a map in general does not exist, because any such Moebius transformation is determined already by *three* pairs of corresponding points.

The problem thus arises to describe the equivalence classes of quadrilaterals that *are* conformally equivalent. It turns out that this description is possible in terms of a single positive real number which can be assigned to any quadrilateral, and which is called its **modulus**.

Let $\mathbf{Q} = (Q; a, b, c, d)$ be a quadrilateral, and let h be the conformal map of Q onto the upper half-plane U such that

$$h(a) = 0, \qquad h(b) = 1, \qquad h(d) = \infty.$$

The map h is uniquely determined by these conditions, and therefore $\xi := h(c)$ is uniquely determined. Because the orientation of the boundary is preserved, $1 < \xi < \infty$. We now assert the unique existence of a conformal map t of U onto itself (which must be a Moebius transformation) such that

$$t(0) = 1, \qquad t(1) = \eta, \qquad t(\xi) = -\eta, \qquad t(\infty) = -1,$$

where η is a suitable real number, $\eta > 1$, which is uniquely determined by ξ. To see this, it suffices to show that the inverse Moebius transformation is uniquely determined. By the conditions on the values of t at 0, 1, and ∞, $t^{[-1]}$ must have the form

$$t^{[-1]}(w) = \frac{\eta+1}{\eta-1}\,\frac{w-1}{w+1}.$$

The fourth condition yields

$$t^{[-1]}(-\eta) = \left(\frac{\eta+1}{\eta-1}\right)^2 = \xi,$$

where $\xi > 1$ is given. Thus we find

$$\frac{\eta+1}{\eta-1} = \pm\sqrt{\xi}.$$

Since $t^{[-1]}$ maps U onto U, only the positive sign is possible, and we have

$$t^{[-1]}(w) = \sqrt{\xi}\,\frac{w-1}{w+1}.$$

We now recall a result from § 5.13 stating that for any given $\eta > 1$, there exists a unique number $\mu > 0$ such that U can be mapped onto the rectangular region R with the four corners μ, $\mu + i$, i, 0 such that the pre-images of these corners are the points $1, \eta, -\eta, -1$, in that order. The number μ may be expressed in terms of η by means of elliptic integrals, but this is not essential in the present context.

Composing the foregoing conformal maps, we see that there exists a unique real number $\mu > 0$ such that the given quadrilateral $\mathbf{Q}$ is conformally equivalent to the rectangular quadrilateral $\mathbf{R} = (R; \mu, \mu + i, i, 0)$. This number μ is called the **modulus** of the given quadrilateral, and will be denoted by $\mu(\mathbf{Q})$. We have proved the following:

THEOREM 16.11b. *Two quadrilaterals are conformally equivalent if and only if they have the same modulus* (*see* Fig. 16.11b).

If the boundary of Q is piecewise regular, the modulus of the quadrilateral $\mathbf{Q}$ has a simple physical interpretation. Let Q consist of a thin sheet of metal of specific resistance 1. Let the segments (a, b) and (c, d) of the boundary be kept at the potentials 1 and 0, respectively, and let the segments (b, c) and (d, a) be insulated. How much current passes through the sheet? Evidently, the total current is

$$I = \int_a^b \frac{\partial \phi}{\partial n}\, ds, \tag{16.11-2}$$

where $\partial/\partial n$ denotes differentiation in the direction of the *exterior* normal, and where ϕ is the solution of the mixed boundary value problem

$$\begin{aligned} \Delta\phi &= 0 \quad \text{in} \quad Q, \\ \phi &= 1 \quad \text{on} \quad (a, b), \\ \phi &= 0 \quad \text{on} \quad (c, d), \\ \frac{\partial \phi}{\partial n} &= 0 \quad \text{on} \quad (b, c) \quad \text{and} \quad (d, a). \end{aligned} \tag{16.11-3}$$

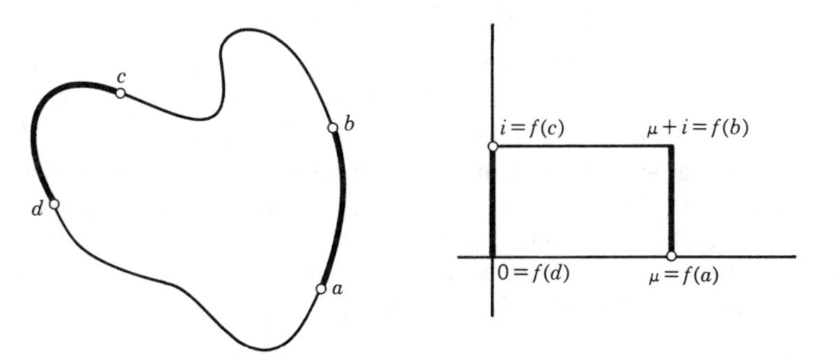

Fig. 16.11b. Modulus of a quadrilateral.

This problem may be solved by conformal transplantation. Let f map **Q** onto the conformally equivalent quadrilateral **R**. The transplanted potential $\psi(w) := \phi(f^{[-1]}(w))$ then satisfies the boundary conditions

$$\begin{aligned} \mu &= 1 \quad \text{on} \quad (\mu, \mu+i), \\ \psi &= 0 \quad \text{on} \quad (0, i), \\ \frac{\partial \psi}{\partial n} &= 0 \quad \text{on} \quad (0, \mu) \quad \text{and} \quad (i, i+\mu) \end{aligned} \tag{16.11-4}$$

and thus is given by

$$\psi(w) = \mu^{-1} \operatorname{Re} w.$$

The integral (16.11-2) is invariant under conformal transplantation (see (5.7-16)) and therefore may be calculated in the w plane. There it has the value

$$I = \mu^{-1}.$$

If the constant potential on (a, b) is U in place of 1, the total current is $\mu^{-1}U$. We thus have $U = \mu I$, and we see that μ is the resistance of the sheet Q between the "electrodes" (a, b) and (c, d) when the remaining parts of the boundary are insulated.

We state some simple properties of the modulus of a quadrilateral.

If the quadrilateral $\mathbf{Q} = (Q; a, b, c, d)$ is conformally equivalent to the rectangular quadrilateral $\mathbf{R} = (R; \mu, \mu+i, i, 0)$, then the quadrilateral $\mathbf{Q}_1 := (Q; c, d, a, b)$ is conformally equivalent to $(R; i, 0, \mu, \mu+i)$, which (by virtue of the map $z \mapsto (\mu+i) - z$) is in turn conformally equivalent to **R**. Thus the moduli of **Q** and $\mathbf{Q}_1$ are the same.

THEOREM 16.11.c. *For any quadrilateral there holds*

$$\mu(Q; c, d, a, b) = \mu(Q; a, b, c, d). \tag{16.11-5}$$

By virtue of Theorem 16.11c, the quadrilaterals $(Q; a, b, c, d)$ and $(Q; c, d, a, b)$ may be regarded as identical.

If $\mathbf{Q} = (Q; a, b, c, d)$ is a quadrilateral, the quadrilateral $(Q; b, c, d, a)$ is called the **reciprocal** of Q, and is denoted by $\mathbf{Q}^{-1}$. If **Q** is conformally equivalent to $\mathbf{R} = (R; \mu, \mu+i, i, 0)$, then $\mathbf{Q}^{-1}$ is conformally equivalent to $\mathbf{R} = (R; \mu+i, i, 0, \mu)$, which (by virtue of $z \mapsto i - i\mu^{-1}z$) is conformally equivalent to the rectangular quadrilateral $(R_1; \mu^{-1}, \mu^{-1}+i, i, 0)$, which has modulus μ^{-1}. Thus $\mathbf{Q}^{-1}$ has modulus μ^{-1}.

THEOREM 16.11d. *Reciprocal quadrilaterals have reciprocal moduli.*

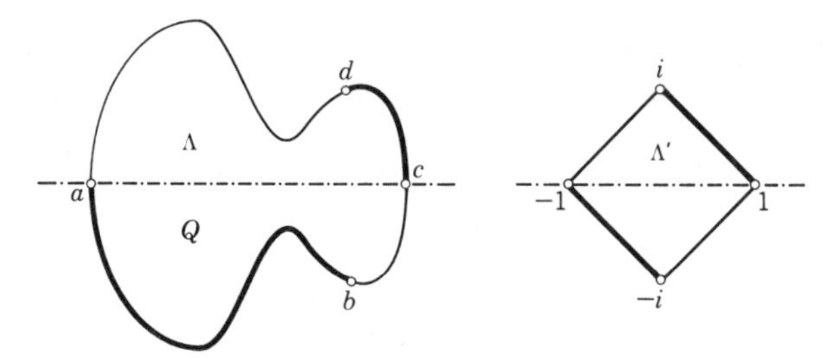

Fig. 16.11c. A symmetric quadrilateral and its conformal image.

The quadrilateral $\mathbf{Q} = (Q; a, b, c, d)$ is called **symmetric** if the region Q is symmetric with respect to the straight line Λ through a and c, and if the points b and d are symmetric with respect to Λ (see Fig. 16.11c).

THEOREM 16.11e. *Every symmetric quadrilateral has modulus 1.*

Proof. The line Λ cuts Q into two halves, T and T', say. Let T be the half of the boundary which contains d, and consider the trilateral $(T; a, c, d)$. Because all trilaterals are conformally equivalent, this trilateral is conformally equivalent to the triangular trilateral with the three corners $-1, 1, i$. By the Schwarz reflection principle, this conformal mapping may be continued to a map of $\mathbf{Q}$ to the square quadrilateral with the four corners $-1, -i, 1, i$, which obviously has modulus 1. —

EXAMPLE 1

Let $|w| = 1$, $\operatorname{Im} w > 0$. The rhombus-shaped quadrilateral $\mathbf{P} = (P; 1, 1+w, w, 0)$ is symmetric with respect to the line passing through 0 and $1+w$, and therefore has modulus 1 (see Fig. 16.11d). —

We next characterize the modulus of a quadrilateral by a variational property. This makes use of the *Dirichlet integral* of a piecewise differentiable function ϕ defined on a region Q. A function here is called piecewise differentiable in Q if it is continuous in Q and if Q can be subdivided into a finite number of nonintersecting subregions $Q_1, Q_2, \ldots, Q_m$ such that for each $i \in \{1, 2, \ldots, m\}$, ϕ can be extended to a function that is in $C^2(Q_i)$ as

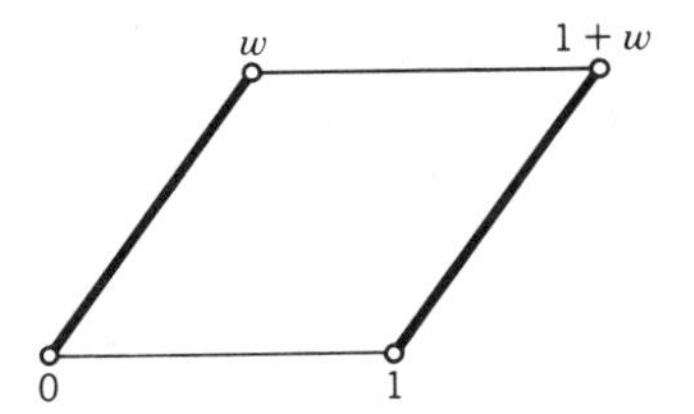

Fig. 16.11d. Rhombus-shaped quadrilateral.

well as in $C^1(Q_i')$, where Q_i' is the closure of Q_i. The **Dirichlet integral** of ϕ over Q is then defined to be the functional

$$D(\phi) := \iint_Q |\text{grad}\,\phi|^2 \boxed{dz}. \tag{16.11-6}$$

We require:

THEOREM 16.11f. *The Dirichlet integral is invariant under conformal transplantation.*

Proof. Let R denote the image of Q under a conformal map f, and let ψ be the conformal transplant of ϕ, $\psi(w) := \phi(f^{[-1]}(w))$. It is asserted that

$$D(\phi) = \iint_R |\text{grad}\,\psi|^2 \boxed{dw}.$$

This readily follows on introducing in (16.11-6) $w = f(z)$ as a new variable of integration. By the first transplantation theorem (Theorem 5.6a),

$$|\text{grad}\,\phi| = |\text{grad}\,\psi| \cdot |f'(z)|.$$

Moreover,

$$\boxed{dz} = \frac{1}{|f'(z)|^2} \boxed{dw}.$$

Thus the factor $|f'(z)|^2$ cancels, and the desired identity is established. —

THEOREM 16.11g. *Let $(Q; a, b, c, d)$ be any quadrilateral, and let μ be its modulus. Then*

$$\mu^{-1} = \min D(\phi), \tag{16.11-7}$$

where the minimum is taken with respect to all piecewise differentiable functions ϕ on Q that satisfy $\phi = 1$ on (a, b) and $\phi = 0$ on (c, d). The minimum is assumed for the solution ϕ_0 of the boundary value problem (16.11-3) *or, more precisely, for $\phi_0(z) = \mu^{-1}\,\text{Re}\,f(z)$, where f maps $(Q; a, b, c, d)$ onto the conformally equivalent quadrilateral $(R; \mu, \mu + i, i, 0)$.*

Proof. Let ϕ be any function meeting the conditions of the theorem. We may write $\phi = \phi_0 + \phi_1$, where ϕ_1 has the same smoothness as ϕ and satisfies $\phi_1 = 0$ on (a, b) as well as on (c, d). By Theorem 16.11f, $D(\phi_0 + \phi_1)$ may be computed in the conformally equivalent rectangular quadrilateral $(R; \mu, \mu + i, i, 0)$ with the transplanted functions $\psi_i = \phi_i \circ f^{[-1]}$. There it equals, using Green's formula,

$$D(\psi_0 + \psi_1) = D(\psi_0) + D(\psi_1) + 2 \iint_R \text{grad}\,\psi_0\,\text{grad}\,\psi_1 \boxed{dw}.$$

In view of $\psi_0(w) = \mu^{-1}\operatorname{Re} w$, there holds grad $\psi_0 = (\mu^{-1}, 0)$, hence on writing $w = u + iv$,

$$\iint_R \operatorname{grad}\psi_0 \operatorname{grad}\psi_1 \boxed{dw} = \mu^{-1}\int_0^1 dv \int_0^{\mu^{-1}} \frac{\partial \psi_1}{\partial u}\, du$$

$$= \mu^{-1}\int_0^1 [\psi_1(\mu, v) - \psi_1(0, v)]\, dv = 0,$$

on account of the boundary conditions satisfied by ψ_1. Since $D(\psi_1) \geqq 0$, with equality holding only for $\psi_1 = 0$, there follows

$$D(\psi_0 + \psi_1) \geqq D(\psi_0) = \mu^{-1}. \quad —$$

By letting ϕ range over appropriate subfamilies of the family of all functions ϕ satisfying the required boundary conditions, upper bounds for μ^{-1}, that is, lower bounds for μ, may be obtained. By applying the same minimum property to the modulus of the reciprocal quadrilateral, we can also obtain upper bounds for μ. If the boundary of Q is polygonal, the *method of finite elements* recommends itself. Proceeding as in § 15.12.II, we triangularize Q, making sure that the distinguished boundary points are nodes of the triangularization. We now consider only those functions ϕ that are, say, linear on each triangule of the triangularization. Each such ϕ is determined by its values at the nodes, and the Dirichlet integral (16.11-7) is a quadratic function in these finitely many values. The determination of its minimum is then an algebraic problem. In appropriate situations, Q may be subdivided into squares, and we may consider functions ϕ that are bilinear on each square of the subdivision.

As another application of Theorem 16.11g, we obtain certain monotonicity properties that can be useful for the estimation of moduli.

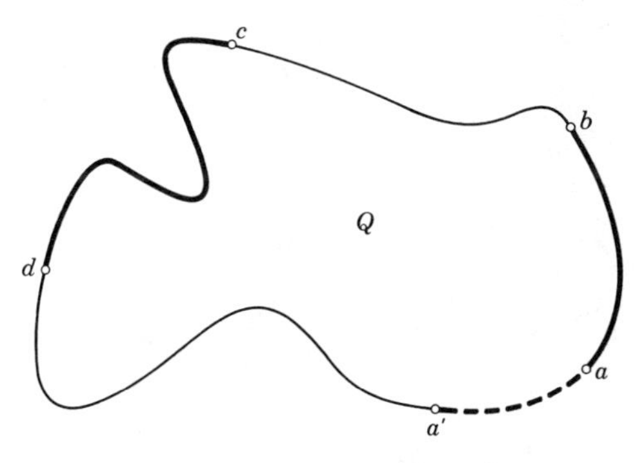

Fig. 16.11e. Two quadrilaterals.

To begin with, let $(Q; a, b, c, d)$ be a quadrilateral, and let $(Q; a', b, c, d)$ be quadrilateral obtained from the first by moving the point a along the boundary of Q in the direction of d (see Fig. 16.11e). Denoting the moduli of the two quadrilaterals by μ and μ', respectively, we have:

THEOREM 16.11h. $\mu > \mu'$.

Proof. In the minimum (16.11-7), every ϕ which satisfies the conditions for μ' also satisfies the conditions for μ, but not vice versa. In particular, the winner in the competition for μ may not compete in the competition for μ', because it does not have the value 1 on (a', a). Without any computation at all, $\mu^{-1} < \mu'^{-1}$ follows. —

Next consider two quadrilaterals $(Q; a, b, c, d)$ and $(Q'; a, b, c, d)$, which have the boundary segments (b, c) and (d, a) in common, but which are such that $Q \subset Q'$, the inclusion being proper. (For instance, Q' may protrude beyond Q along the boundary segment (a, b), see Fig. 16.11f.) Denoting the moduli of the two quadrilaterals again by μ and μ', we have:

THEOREM 16.11i. $\mu < \mu'$.

Proof. In competing for the minimum (16.11-7), every function which competes for the minimum for μ^{-1} may be extended (by defining it to be 0 or 1 in $(Q' \backslash Q)$ to a function competing for the minimum for μ'^{-1}. On the other hand, not every function competing for the minimum μ'^{-1} can be restricted to a function competing for the minimum μ^{-1}. The winner in the competition for μ'^{-1}, in particular, may not compete for μ^{-1}, because it does not have the value 1 on $Q' \backslash Q$. Thus the competition for μ'^{-1} is larger, and the minimum therefore smaller. —

We finally establish a property of superadditivity of the modulus of a quadrilateral. Let Q be a Jordan region with *six* distinguished boundary

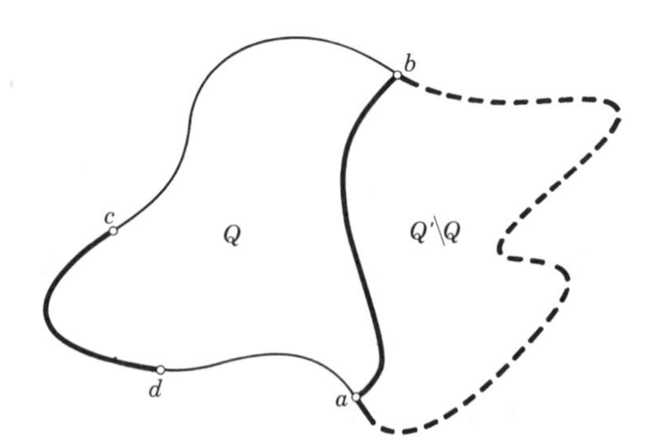

Fig. 16.11f. Enlarged quadrilateral.

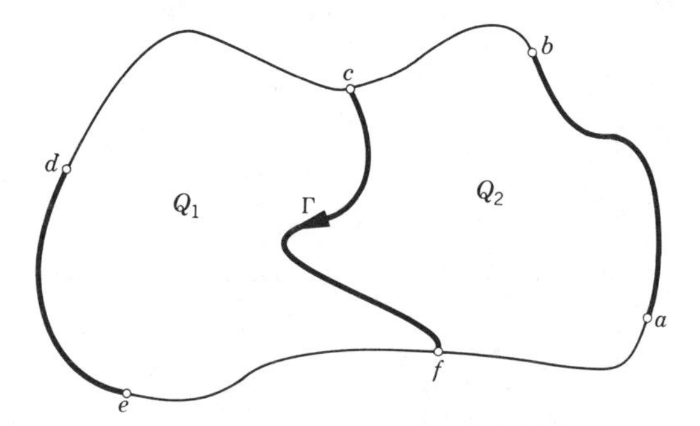

Fig. 16.11g. Union of two quadrilaterals.

points a, b, c, d, e, f. By means of a crosscut Γ from c to f, let Q be decomposed into two disjoint Jordan regions Q_1 and Q_2 whose boundaries have the crosscut Γ in common (see Fig. 16.11g). We shall compare the moduli

$$\mu := (Q; a, b, d, e),$$

$$\mu_1 := (Q_1; f, c, d, e),$$

$$\mu_2 := (Q_2; a, b, c, f).$$

THEOREM 16.11j. $\mu \geqq \mu_1 + \mu_2$.

Proof. Given any two functions ϕ_1 and ϕ_2 that may compete for μ_1^{-1} and μ_2^{-1}, respectively, we can construct a $\tilde{\phi}$ that may compete for μ^{-1} by defining

$$\tilde{\phi}(z) := \begin{cases} \alpha\phi_1(z), & z \in Q_1 \\ \alpha + (1-\alpha)\phi_2(z), & z \in Q_2. \end{cases}$$

Here α is an arbitrary real parameter. (Of course, not every ϕ competing for μ^{-1} may be thus represented.) It follows from Theorem 16.11g that

$$\begin{aligned} \mu^{-1} &= \min_{\phi} D(\phi) \leqq \min_{\tilde{\phi}} D(\tilde{\phi}) \\ &= \alpha^2 \min_{\phi_1} D(\phi_1) + (1-\alpha)^2 \min_{\phi_2} D(\phi_2) \\ &= \alpha^2 \cdot \mu_1^{-1} + (1-\alpha)^2 \cdot \mu_2^{-1}. \end{aligned}$$

In the last expression we take the minimum with respect to α. By calculus it readily follows that the minimum is achieved for

$$\alpha = \frac{\mu_1}{\mu_1 + \mu_2}, \qquad 1 - \alpha = \frac{\mu_2}{\mu_1 + \mu_2},$$

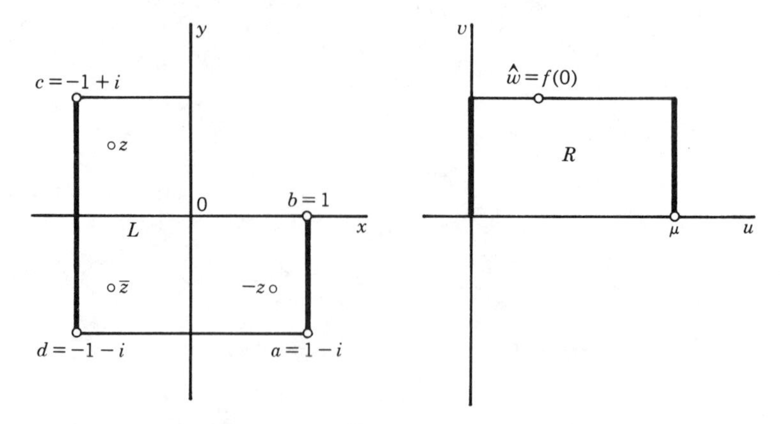

Fig. 16.11h. Modulus of an L-shaped quadrilateral.

and that it consequently has the value $(\mu_1+\mu_2)^{-1}$. There follows

$$\mu^{-1} \leqq (\mu_1+\mu_2)^{-1},$$

tantamount to the assertion. —

Equality is clearly achieved if Γ is such that the potential satisfying the boundary conditions (16.11-3) for Q is constant along Γ.

Following Hersch [1982], we use some of the foregoing results in order to determine the modulus of the L-shaped quadrilateral $\mathbf{L} :=$ $(L; 1-i, 1, -1+i, -1-i)$ shown in Fig. 16.11h.

THEOREM 16.11k. $\mu(\mathbf{L}) = \sqrt{3}$.

Proof. Let ϕ denote the solution of the boundary value problem (16.11-3). It is easily seen that the harmonic function

$$\phi(z)+\phi(\bar{z})+\phi(-z)$$

equals 1 on the whole boundary of L. Therefore it is identically equal to 1. Setting $z=0$, there follows

$$\phi(0)=\tfrac{1}{3}.$$

If f maps $\mathbf{L}$ onto the conformally equivalent rectangular quadrilateral $\mathbf{R}=(R; \mu, \mu+i, i, 0)$, we have $\phi(z)=\mu^{-1}\operatorname{Re} f(z)$, and if $\hat{w} := f(0)$, there follows

$$\operatorname{Re} \hat{w} = \tfrac{1}{3}\mu.$$

In addition to $\mathbf{L}$, we also consider the quadrilateral $\hat{\mathbf{L}} :=$ $(L; 1-i, 0, -1+i, -1-i)$, which is conformally equivalent to $\hat{\mathbf{R}} :=$ $(R; \mu, \hat{w}, i, 0)$ (see Fig. 16.11i).

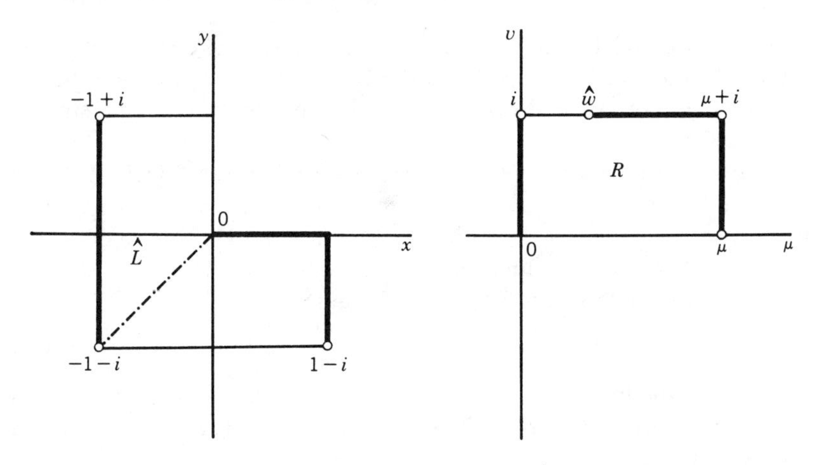

Fig. 16.11i. The modulus of $\hat{\mathbf{L}}$.

Since $\hat{\mathbf{L}}$ is symmetric (with respect to the line through 0 and $-1-i$), $\mu(\hat{\mathbf{L}})=1$ by Theorem 16.11e, hence also

$$\mu(\hat{\mathbf{R}})=1.$$

Augmenting $\hat{\mathbf{R}}$ by its reflection at the real axis, the modulus is halved, and augmenting the augmented quadrilateral once again by reflection at the imaginary axis, the modulus is doubled again. Denoting by T the rectangle with the four corners $\pm\mu\pm i$ and letting

$$\mathbf{T}:=(T;\ \bar{\hat{w}},\ \hat{w},\ -\bar{\hat{w}},\ -\hat{w})$$

(see Fig. 16.11j), we thus also have

$$\mu(\mathbf{T})=1.$$

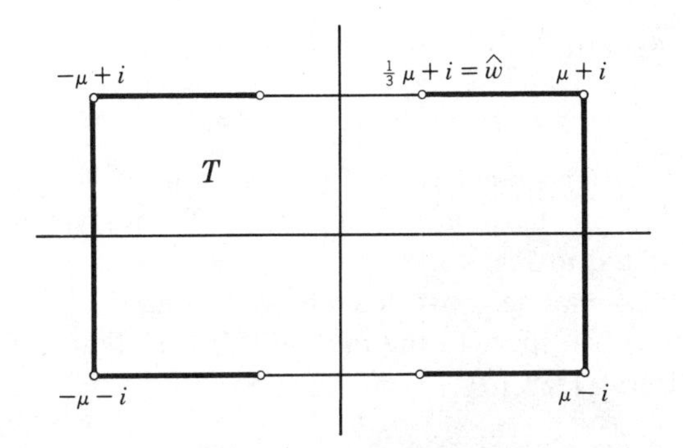

Fig. 16.11j. The quadrilateral **T**.

The modulus of **T** being equal to 1, there exists a conformal map g mapping T onto the square Q with the four corners $\pm 1 \pm i$ such that the points $\pm\frac{1}{3}\mu \pm i$ are mapped onto the four corners of Q. By symmetry, the map is such that $g(0)=0$.

We now recall the concept of *harmonic measure* (see § 15.5). Let $\omega(z)$ denote the harmonic measure of the straight line segment $[-\frac{1}{3}\mu + i, \frac{1}{3}\mu + i]$ with respect to T, that is, let ω be the function which is harmonic in T, equal to 1 on the straight line segment mentioned, and equal to 0 on the remaining parts of the boundary of T. We assert that

$$\omega(0) = \tfrac{1}{4}. \tag{16.11-8}$$

Indeed, the function $\omega \circ g^{[-1]}$, which is defined in the square Q, is the harmonic measure of the top side of the square with regard to the square. Thus the function

$$\omega \circ g^{[-1]}(w) + \omega \circ g^{[-1]}(\bar{w}) + \omega \circ g^{[-1]}(-w) + \omega \circ g^{[-1]}(-\bar{w})$$

is equal to 1 on the whole boundary of the square, and therefore identically 1. Setting $w=0$, we obtain

$$4\omega \circ g^{[-1]}(0) = 4\omega(0) = 1,$$

proving (16.11-8).

Consider next the harmonic measure of the top side of an arbitrary rectangle with respect to the rectangle. If the sides of the rectangle are α and β, where α is the length of the top side, then the value of the harmonic measure at the center of the rectangle depends only on α/β. We denote it by $\nu(\alpha/\beta)$. Adding the harmonic measure of the four sides of the rectangle, we obtain a function which is identically equal to 1. Evaluating it at the center, we obtain

$$2\nu\left(\frac{\alpha}{\beta}\right) + 2\nu\left(\frac{\beta}{\alpha}\right) = 1. \tag{16.11-9}$$

We next express the value $\omega(0) = \frac{1}{4}$ in terms of ν. Evidently for $z \in T$,

$$\omega(z) = \tfrac{1}{2}(\omega_1(z) + \omega_2(z)), \tag{16.11-10}$$

where ω_1 is the harmonic measure of the top side of T with respect to T, and where ω_2 is the harmonic function in T satisfying the boundary conditions indicated in Fig. 16.11k.

The function ω_2 may be constructed by reflecting the harmonic measure of the top side with respect to the middle third of T at the vertical sides of the middle third. Thus

$$\omega_2(0) = \nu(\tfrac{1}{3}\mu).$$

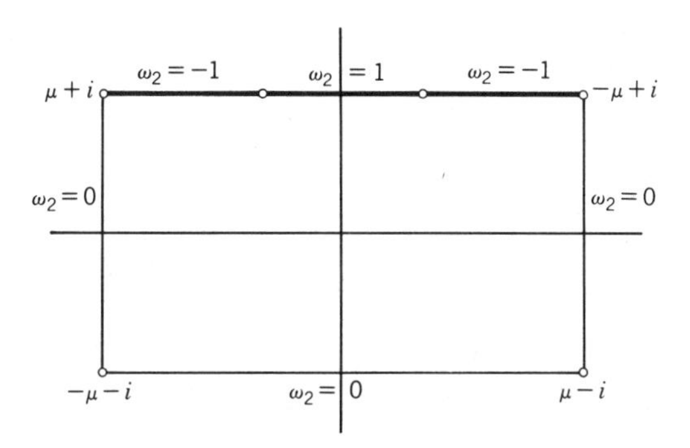

Fig. 16.11k. The function ω_2.

Evidently $\omega_1(0)=\nu(\mu)$. Thus from (16.11-10),

$$\nu(\mu)+\nu(\tfrac{1}{3}\mu)=2\omega(0)=\tfrac{1}{2}.$$

Together with (16.11-9), which for $\alpha/\beta=\mu$ yields

$$\nu(\mu)=\nu(\mu^{-1})=\tfrac{1}{2},$$

this implies

$$\nu(\mu^{-1})=\nu(\tfrac{1}{3}\mu).$$

Because ν is strictly monotonic, there follows

$$\mu^{-1}=\tfrac{1}{3}\mu \qquad \text{or} \qquad \mu=\sqrt{3}. \quad —$$

PROBLEMS

1. Let Γ denote the curve consisting of the upper half of the ellipse with foci ± 1 and major axis $\eta>1$, and of the straight line segment joining the vertices $\pm\eta$. Let

$$\mathbf{E}:=(E;-1,1,\eta,-\eta),$$

where E is the interior of Γ. Determine $\mu(\mathbf{E})$.

2. Estimate the modulus μ of the L-shaped quadrilateral of Theorem 16.11k by the method of finite elements, using a decomposition of L into three congruent squares and approximating ϕ on each square by the bilinear trial functions described in Example 1 of § 15.12. In this manner show that

$$\tfrac{9}{7}<\mu<\tfrac{7}{3}.$$

3. Let $\mathbf{Q}=(Q;a,b,c,d)$ be a quadrilateral, and let ω be the harmonic measure of (a,b) with respect to Q. The quantity

$$\gamma:=-\frac{1}{4\pi}\int_{(c,d)}\frac{\partial\omega}{\partial n}|dz|$$

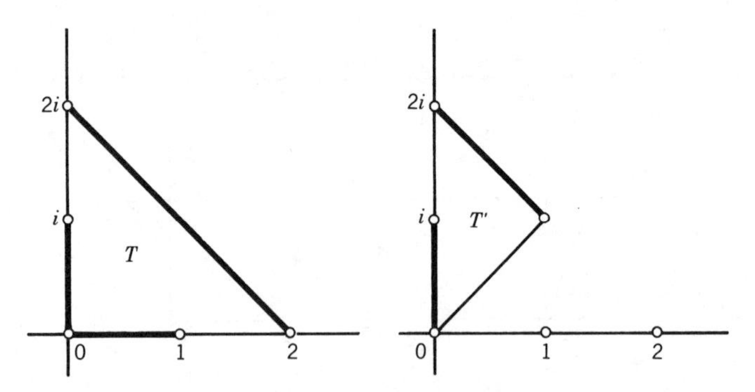

Fig. 16.11l. The quadrilateral **T**.

($n :=$ outward normal) is called the **capacitance** between (a, b) and (c, d); see Campbell [1975]. Show that γ may be expressed in terms of the modulus μ of **Q** as follows:

(a)
$$\gamma = \frac{2}{\pi^2} \sum_{n=1,3,5,\ldots} \frac{1}{n \sinh[n\pi\mu^{-1}]},$$

(b)
$$\gamma = -\frac{\operatorname{Log} k}{2\pi^2},$$

where

$$k := 4q^{1/2} \prod_{n=1}^{\infty} \left(\frac{1+q^{2n}}{1+q^{2n-1}} \right)^2, \qquad q := e^{-\pi\mu}.$$

(Use conformal transplantation; see Gaier [1979].)

4. Let T denote the triangle with the three corners 0, 2, $2i$, and let

$$\mathbf{T} := (T; 2, 2i, i, 1)$$

(see Fig. 16.11l). Show that $\mu(\mathbf{T}) = \frac{1}{2}$.

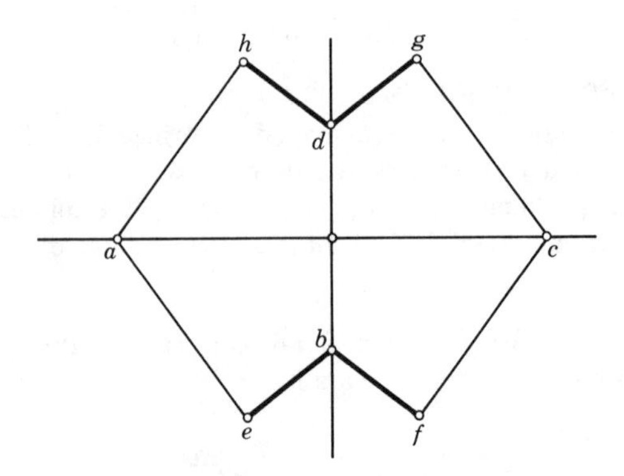

Fig. 16.11m. Octagon.

(By symmetry, μ equals one half of the modulus of the quadrilateral $\mathbf{T}'$; now apply Theorem 16.11e.)

5. Let a, b, c, d be the corners of a rhombus R with center at 0. Reflect each triangle $0ab$, $0bc$, $0cd$, $0da$ at its hypotenuse, thereby obtaining an irregular octagon Q with corners a, e, b, f, c, g, d, h (see Fig. 16.11m).

   (a) If $\mathbf{Q} = (Q; e, f, g, h)$, show that $\mu(\mathbf{Q}) = 1$.

   (b) Show that the value of the harmonic measure of each of the eight sides of Q at 0 is $\frac{1}{8}$.

NOTES

Trilaterals and quadrilaterals are special cases of a concept called configuration; see Ahlfors [1973], ch. 4. The equivalence classes of quadrilaterals could also be characterized by the cross ratio χ of the corners of the conformally equivalent circular quadrilateral; Trefethen [1984] gives the connection between χ and μ. Theorem 16.11e is stated by Hersch [1984], but Lienhard [1981] has a proof of an equivalent result based on considerations of heat flow. The minimum property of μ given in Theorem 16.11g is stated by Gaier [1971], who also has error estimates for its application to finite-element techniques. Theorem 16.11k as well as its elementary proof are due to Hersch [1982]. Hersch [1984] has many additional results of this type, all obtained by elementary methods. Some authors, such as Gaier [1971, 1978], denote by μ^{-1} what is here denoted by μ.

17

CONSTRUCTION OF CONFORMAL MAPS FOR MULTIPLY CONNECTED REGIONS

Here we discuss the construction of standardized mapping functions for regions of connectivity greater than one. Following the outline of the preceding chapter, we begin by discussing the existence of certain standardized maps. We next should turn to a discussion of the boundary behavior of these maps. However, for reasons to be explained, the corresponding discussion in Chapter 16 is adequate also for maps of multiply connected regions. For doubly connected regions, we then discuss the analogs of the methods of osculation, of linear integral equations of the first and second kinds, and of the methods of conjugate functions that were presented in the preceding chapter. There follows a discussion of the little-known possibility of extending the Schwarz–Christoffel map to polygonal regions of connectivity two.

For regions of connectivity greater than two we restrict our attention to Koebe's circular map, which appears to be particularly suited to applications in fluid dynamics. We first present an existence proof due to Schiffer and Hawley, which is based on variational methods, and then proceed to Koebe's own construction of the circular map as elaborated by Gaier.

The remarks in the introduction to Chapter 16 concerning applications of conformal maps to grid generation hold with even greater force for multiply connected regions since the regions that occur in applications most frequently are of this kind.

§ 17.1. EXISTENCE AND PROPERTIES OF MAPPING FUNCTIONS FOR MULTIPLY CONNECTED REGIONS

To begin with, let D be a *doubly* connected region in the complex plane. This means that the complement of D with respect to the extended complex plane consists of precisely two components, B and U say, exactly one of which is unbounded. We denote by U the unbounded component.

Again we wish to map D conformally onto a standard region. For topological reasons, any such standard region likewise must be doubly connected. The circular annulus naturally recommends itself as a standard region. However, it follows from the reflection principle (see Theorem 5.11c) that no single annulus can be conformally equivalent to all doubly connected regions. Rather, the ratio of the radii of the two circles bounding the annulus is a constant that is uniquely determined by D. The appropriate mapping theorem has already been stated (Theorem 5.10h), but not proved.

We need not consider the situation where one or both components B and U reduce to a single point, because in these cases the content of Theorem 5.10h is easily deduced from the Riemann mapping theorem. A doubly connected region such that neither component of its complement reduces to a single point will be called **nondegenerate**. For nondegenerate regions the mapping theorem in question reads as follows:

THEOREM 17.1a. *Let D be a nondegenerate doubly connected region. Then there exists a unique real number μ, $0<\mu<1$, such that there exists a one-to-one analytic function f that maps D onto the annulus A: $\mu<|w|<1$. If the outer boundaries correspond to each other, then f is determined up to a rotation of the annulus.*

Proof. Building on hard work done previously, we present a proof that is based on two-dimensional potential theory.

The first important fact is that without loss of generality the boundaries of D may be assumed to be smooth, even analytic. For let f_1 be a map that maps the complement of B onto the exterior of the unit disk in such a way that ∞ remains fixed. (This map exists by Corollary 5.10d; see Fig. 17.1a) The function f_1 maps D onto another doubly connected region D_1 whose complement has some unbounded component U_1 and the bounded component $B_1 = E'$, the closure of the unit disk. Next we apply a map f_2 that takes the complement of U_1 onto E. This will map D_1 onto a doubly connected region D_2 bounded on the outside by the unit circle and on the inside by the analytic curve $z=f_2(e^{i\tau})$, $0\leqq\tau\leqq 2\pi$, bounding the set $B_2 :=$ $f_1(B_1)$. If f maps D_2 onto an annulus, the map required by the theorem is $f\circ f_2\circ f_1$. Clearly, we may assume that $0\in B_2$.

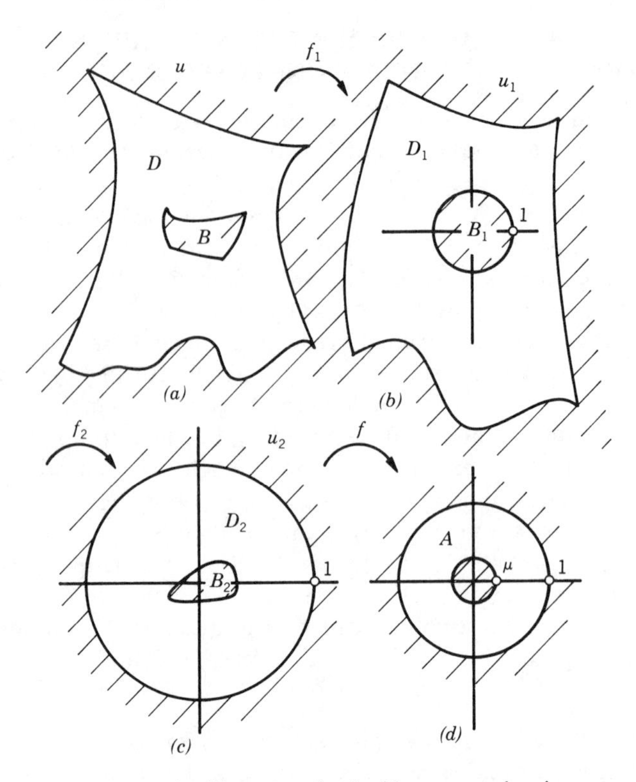

Fig. 17.1a. Auxiliary maps for doubly connected regions.

Let the nondegenerate region D therefore be bounded by two analytic, positively oriented Jordan curves Γ_0 (outer boundary) and Γ_1 (inner boundary) (see Fig. 17.1b). It is not required that Γ_0 be a circle. By the assumption just made, the winding numbers (see § 4.6) of both Γ_0 and Γ_1 with respect to 0 are

$$n(\Gamma_0, 0) = n(\Gamma_1, 0) = 1. \tag{17.1-1}$$

We first prove the assertions concerning uniqueness in Theorem 17.1a. The uniqueness proof at the same time will motivate the existence proof that follows.

Let f be a function that maps D onto *some* annulus A: $\mu < |w| < 1$, where $0 < \mu < 1$. Because the boundary curves Γ_0 and Γ_1 are analytic, it follows as in the proof of Theorem 16.4a that f may be extended analytically to the boundary of D. (The arguments used in the proof of Theorem 16.4a were local; the connectivity of D played no part in the proof.) We assume that f is such that Γ_0 is mapped onto $|w| = 1$, and Γ_1 onto $|w| = \mu$. Because

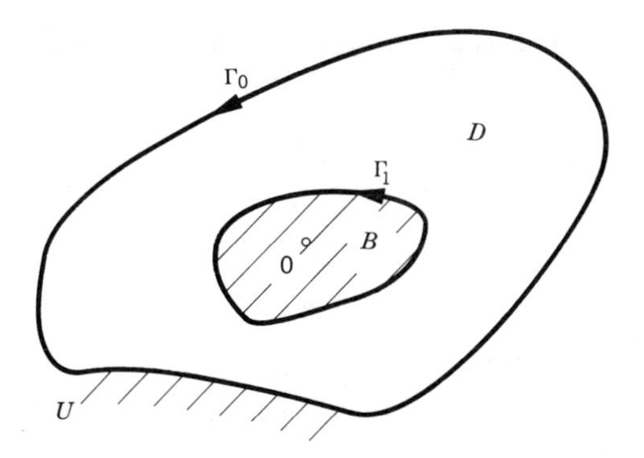

Fig. 17.1b. Notations used in proof of Theorem 17.1a.

f is analytic on the closure of D, we may integrate f, or any analytic function of f, along Γ_0 or Γ_1. For $w_0 \neq 0, 1$, we consider

$$d(w_0) := \frac{1}{2\pi i} \int_{\Gamma_0} \frac{f'(z)}{f(z) - w_0}\, dz - \frac{1}{2\pi i} \int_{\Gamma_1} \frac{f'(z)}{f(z) - w_0}\, dz. \qquad (17.1\text{-}2)$$

This equals a difference of winding numbers,

$$d(w_0) = n(f(\Gamma_0), w_0) - n(f(\Gamma_1), w_0).$$

Because the equation $f(z) = w_0$ has precisely one solution $z \in D$ for every $w_0 \in A$ and no solution $z \in D$ for $w_0 \notin A$, the general principle of the argument yields

$$d(w_0) = \begin{cases} 1, & \mu < |w_0| < 1 \\ 0, & |w_0| < \mu \quad \text{or} \quad |w_0| > 1. \end{cases}$$

The first integral in (17.1-2) is continuous and integer valued for $|w_0| \neq 1$, the second for $|w_0| \neq \mu$. Because both integrals are zero for $w_0 \to \infty$, the first integral has the value $+1$ for $|w_0| < 1$, and the second has the value $+1$ for $|w_0| < \mu$. For $w_0 = 0$ there follows, in particular, that

$$\frac{1}{2\pi i} \int_{\Gamma_j} \frac{f'(z)}{f(z)}\, dz = 1, \qquad j = 0, 1.$$

Thus the period of f'/f with respect to B is $2\pi i$. Because 0 is interior to Γ_1,

$$\frac{f'(z)}{f(z)} - \frac{1}{z}$$

has period 0, and hence by Theorem 15.1d possesses an indefinite integral

$u(z)+iv(z)$ in D, which is an analytic branch of

$$\log \frac{f(z)}{z}.$$

There follows

$$f(z)=z\,e^{u(z)+iv(z)}. \tag{17.1-3}$$

Noting that

$$u(z)=\operatorname{Log}\left|\frac{f(z)}{z}\right|$$

is uniquely characterized as the solution of the Dirichlet problem

$$\begin{aligned} \Delta u(z)&=0, \qquad z\in D,\\ u(z)&=-\operatorname{Log}|z|, \qquad z\in\Gamma_0,\\ u(z)&=\operatorname{Log}\mu-\operatorname{Log}|z|, \qquad z\in\Gamma_1, \end{aligned} \tag{17.1-4}$$

the conjugate function v of u is determined up to an additive real constant γ. This proves that f is determined up to a factor of modulus 1, that is, up to a rotation of A. The function f could be made unique by prescribing that

$$f(a)>0 \tag{17.1-5}$$

for some predetermined point a of D.

We turn to the proof of the existence of a mapping function f with the required properties. It would be tempting to base the proof directly on the Dirichlet problem (17.1-4), the existence of a solution of which we take for granted by Proposition 15.4b. This does not immediately work, because the problem (17.1-4) involves the constant $\lambda := \operatorname{Log}\mu$, which is unknown. However, for any μ the solution of (17.1-4) is

$$u=u_0+\lambda u_1,$$

where

$$u_0(z) := -\operatorname{Log}|z|,$$

and where u_1 is the harmonic measure (see § 15.5) of Γ_1 with respect to D. Because the Γ_j are analytic, u_0 may be continued harmonically beyond the Γ_j. We now determine λ by the condition that $u=u_0+\lambda u_1$ has a conjugate harmonic function v in D. By Theorem 15.1d this is so if and only if the conjugate period of u with respect to B is zero. Because the conjugate period of $\operatorname{Log}|z|$ equals 2π, our condition will be met precisely if

$$\lambda p = 2\pi,$$

where p is the conjugate period of u_1,

$$p := \int_{\Gamma_1} [-(u_1)_y\, dz + (u_1)_x\, dy]. \tag{17.1-6}$$

We assert that $p \neq 0$. For if $p = 0$, then u_1 has a conjugate function v_1 in D. The function $g_1 := u_1 + iv_1$ is analytic in D and, by reflection, can be extended to a function that is analytic on the closure of D. The images of Γ_0 and Γ_1 lie on $\text{Re } w = 1$ and $\text{Re } w = 0$, respectively. If w_0 is any point not on these lines, continuous arguments of $g_1(z) - w_0$ can be defined on both $g_1(\Gamma_0)$ and $g_1(\Gamma_1)$. Each of these arguments omits values congruent to $\pi/2 \bmod \pi$, and therefore lies in an interval of length $< \pi$. It follows that

$$[\arg(g_1(z) - w_0)]_{\Gamma_1 - \Gamma_0} = 0,$$

hence that the value w_0 is not taken by g_1. This implies that g_1 is constant, for if it were not, then by the theorem on the local mapping (Theorem 2.4f) g_1 would have to assume values off the lines $\text{Re } w = 0$ and $\text{Re } w = 1$. However, g_1 is not constant because $u_1 = \text{Re } g_1$ is not constant. This contradiction proves that $p \neq 0$.

We thus may let

$$\lambda := \frac{2\pi}{p}. \tag{17.1-7}$$

With this (and only with this) choice of λ,

$$u(z) = -\text{Log}|z| + \lambda u_1(z)$$

possesses a conjugate harmonic function v in D. Let

$$f(z) := z\, e^{u(z) + iv(z)}. \tag{17.1-8}$$

It remains to show that f has the required properties.

For $z \in \Gamma_0$,

$$|f(z)| = |z|\, e^{u(z)} = |z|\, e^{-\text{Log}|z|} = 1; \tag{17.1-9a}$$

for $z \in \Gamma_1$,

$$|f(z)| = |z|\, e^{u(z)} = |z|\, e^{\lambda - \text{Log}|z|} = e^{\lambda}. \tag{17.1-9b}$$

Thus f has constant modulus on each of the bounding curves; however, we do not know as yet that $e^{\lambda} < 1$. We calculate the number of times f takes a given value w_0. By the principle of the argument, this equals

$$d(w_0) := n(f(\Gamma_0), w_0) - n(f(\Gamma_1), w), \qquad |w_0| \neq 1, \quad |w_0| \neq e^{\lambda},$$

where

$$n(f(\Gamma_j), w_0) = \frac{1}{2\pi i} \int_{\Gamma_j} \frac{f'(z)}{f(z) - w_0}\, dz, \qquad j = 0, 1.$$

Because

$$\log \frac{f(z)}{z} = u(z) + iv(z)$$

is single valued in D, the integral of the derivative of this function along any closed curve is zero. Hence

$$\frac{1}{2\pi i} \int_{\Gamma_j} \frac{f'(z)}{f(z)} dz = \frac{1}{2\pi i} \int_{\Gamma_j} \frac{1}{z} dz$$

or, because both Γ_0 and Γ_1 have winding number 1 with respect to 0,

$$n(f(\Gamma_j), 0) = 1, \qquad j = 0, 1.$$

Both winding numbers $n(f(\Gamma_j), w_0)$ are zero for $|w_0|$ large. Because these winding numbers are constant for w_0 off the curves $f(\Gamma_j)$, and because they are zero for $|w_0|$ large, we have by (17.1-9)

$$n(f(\Gamma_0), w_0) = \begin{cases} 1, & |w_0| < 1 \\ 0, & |w_0| > 1 \end{cases}$$

$$n(f(\Gamma_1), w_0) = \begin{cases} 1, & |w_0| < e^{\lambda} \\ 0, & |w_0| > e^{\lambda}. \end{cases}$$

As a nonconstant function, f by the theorem on the local map takes values off the circles $|w_0| = 1$ and $|w_0| = e^{\lambda}$. Thus $d(w_0) > 0$ for some w_0. By the above, this is possible only if $e^{\lambda} < 1$. Then

$$d(w_0) = \begin{cases} 0, & |w_0| > 1 \\ 1, & 1 > |w_0| > e^{\lambda}, \\ 0, & |w_0| < e^{\lambda}, \end{cases}$$

that is, f takes in D precisely once every value in the annulus A: $\mu = e^{\lambda} < |w_0| < 1$, and it does not take any other values in D. —

The number $\mu^{-1} := e^{-\lambda} > 1$ is called the **modulus** of the doubly connected region D. As follows from the above proof, and as we have seen already in § 5.10, the modulus of a doubly connected region is uniquely determined by D. The quantity

$$\mu^* := \frac{1}{2\pi} \operatorname{Log} \mu^{-1},$$

called the **logarithmic modulus** of D, is closely related to the modulus of a quadrilateral; see Problems 1–5.

Theorem 17.1a completely solves the mapping problem for doubly connected regions. The standard regions in that case are circular annuli.

We sketch briefly what can be done by similar methods for regions D of arbitrary connectivity $k+1$, where $k>1$. Let the components of the complement of D be U (the unbounded component), $B_1, B_2, \ldots, B_k$. Without loss of generality it may be assumed that none of these components reduces to a single point. Using the Riemann mapping theorem for simply connected regions, we map the complement of U onto the unit disk E and label the images of the B_j again B_j. We then map the complement of B_1 onto the exterior of E, thereby transforming the boundary of B_1 into the unit circle and the boundary of U into some analytic Jordan curve. Always retaining the same notation, we then apply the same process to the complement of B_2, etc. Continuing in this fashion we arrive at a region D of connectivity $k+1$ whose boundary consists of $k+1$ *analytic* Jordan curves Γ_0 (the boundary of the unbounded component U of the complement of D) and $\Gamma_j, j=1,2,\ldots,k$. We assume that all Γ_j are positively oriented. Then we have

$$n(\Gamma_0, z)=\begin{cases}0, & z \text{ in the interior of } U,\\ 1, & z \text{ in the complement of } U,\end{cases} \tag{17.1-10a}$$

and for $j=1,2,\ldots,k$,

$$n(\Gamma_1, z)=\begin{cases}1, & z \text{ in the interior of } B_j\\ 0, & z \text{ in the complement of } B_j.\end{cases} \tag{17.1-10b}$$

With this notation, there holds:

THEOREM 17.1b. *Under the above hypotheses there exist k real numbers $\mu_j, j=1,2,\ldots,k$, such that $0<\mu_k<\mu_j<1, j=1,2,\ldots,k-1$, such that there is an analytic function f that maps D conformally onto the annulus $\mu_k<|w|<1$, cut along $k-1$ mutually disjoint arcs Λ_j located on the circles $|w|=\mu_j, j=1,2,\ldots,k-1$. The mapping function f can be extended analytically to the curves Γ_j bounding D. The images of Γ_0 and of Γ_k are the circles Λ_0: $|w|=1$ and Λ_k: $|w|=\Gamma_k$, respectively. The images of the curves Γ_j are the arcs $\Lambda_j, j=1,2,\ldots,k-1$, traversed twice. The function f is determined up to a factor of modulus* 1.

The notations used and the assertions of the theorem are illustrated in Fig. 17.1c for the case $k=3$.

The *proof* of Theorem 17.1b is analogous to that of the preceding theorem. We prove uniqueness first and thereby show how to construct f by means of solving a Dirichlet problem. Let f be a function with the properties described by the theorem. Then by the principle of the argument, the number of times f takes a given value not on any of the curves $f(\Gamma_j)$ equals

$$m(w_0)=n(f(\Gamma_0), w_0)-\sum_{j=1}^{k} n(f(\Gamma_j), w_0),$$

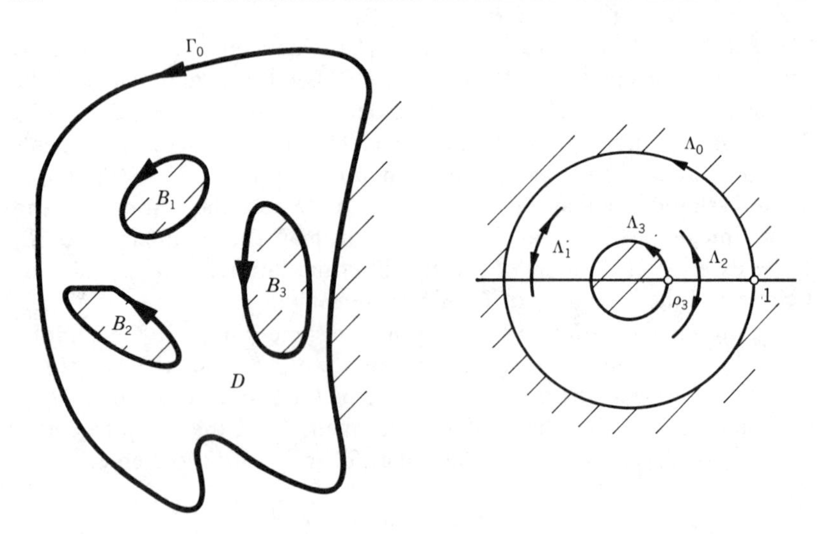

Fig. 17.1c. Mapping of a region of connectivity 4.

where

$$n(f(\Gamma_j), w_0) := \frac{1}{2\pi i} \int_{\Gamma_j} \frac{f'(z)}{f(z) - w_0} dz.$$

We know that all these winding numbers are zero for $|w_0| \to \infty$. Because they are constant on every component of the complement of $f(\Gamma_j)$, it also follows that

$$n(f(\Gamma_j), w_0) = 0, \qquad w_0 \notin \Lambda_j, \quad j = 1, 2, \ldots, k-1.$$

As in the proof of Theorem 17.1a we conclude that

$$n(f(\Gamma_0), 0) = n(f(\Gamma_k), 0) = 1.$$

Without loss of generality we may assume that $z = 0$ is in the interior of Γ_k. As in the foregoing proof we then may conclude that

$$f(z) = z\, e^{u(z) + iv(z)},$$

where $u + iv$ is analytic in D. Setting $\mu_0 := 1$, u is uniquely characterized as the solution of the Dirichlet problem

$$\begin{aligned} &\Delta u(z) = 0, \qquad z \in D \\ &u(z) = \operatorname{Log} \mu_j - \operatorname{Log}|z|, \qquad z \in \Gamma_j, \quad j = 0, 1, \ldots, k. \end{aligned} \tag{17.1-11}$$

It follows that the conjugate harmonic function v is determined up to a constant, showing that f is determined up to a rotation.

To construct f (and thereby show the existence of f) without knowing the values of the constants

$$\lambda_j := \operatorname{Log} \mu_j, \qquad j = 1, 2, \ldots, k,$$

we try to find u in the form

$$u(z) = \sum_{j=1}^{k} \lambda_j u_j(z) - \operatorname{Log}|z|, \qquad z \in D, \tag{17.1-12}$$

where the u_j are the harmonic measures of the Γ_j with respect to D. They satisfy the relation

$$u_0(z) + u_1(z) + \cdots + u_k(z) = 1, \qquad z \in D,$$

because the function on the left is harmonic in D and takes the value 1 on all boundary curves Γ_j. We now try to find values $\lambda_1, \ldots, \lambda_k$ such that u, as determined by (17.1-12), has a conjugate harmonic function in D. By Theorem 15.1d, this is the case if and only if the conjugate periods of u with respect to each Γ_j, $j > 0$, are zero,

$$\int_{\Gamma_j} (u_x\, dy - u_y\, dx) = 0, \qquad j = 1, \ldots, k. \tag{17.1-13}$$

We let

$$\alpha_{ij} := \int_{\Gamma_i} [(u_j)_x\, dy - (u_j)_y\, dx]$$

and use the fact that for $l(z) := \operatorname{Log}|z|$ by virtue of (17.1-10),

$$\int_{\Gamma_i} (l_x\, dy - l_y\, dx) = \begin{cases} 0, & i = 1, 2, \ldots, k-1 \\ 2\pi, & i = k. \end{cases}$$

The condition (17.1-13) thus is satisfied if and only if

$$\begin{aligned} \alpha_{11}\lambda_1 + \alpha_{12}\lambda_2 + \cdots + \alpha_{1k}\lambda_k &= 0, \\ \alpha_{21}\lambda_1 + \alpha_{22}\lambda_2 + \cdots + \alpha_{2k}\lambda_k &= 0, \\ &\vdots \\ \alpha_{k-1,1}\lambda_1 + \alpha_{k-1,2}\lambda_2 + \cdots + \alpha_{k-1,k}\lambda_k &= 0, \\ \alpha_{k1}\lambda_1 + \alpha_{k2}\lambda_2 + \cdots + \alpha_{kk}\lambda_k &= 2\pi. \end{aligned} \tag{17.1-14}$$

We wish to show that this system of k equations with k unknowns has a solution. If not, then the corresponding homogeneous system would have a nontrivial solution, which would mean that there exists a nontrivial linear combination

$$u^*(z) := \lambda_1 u_1(z) + \cdots + \lambda_k u_k(z)$$

that has zero periods with respect to all Γ_j and hence has a conjugate harmonic function v^*. As in the foregoing proof, the resulting analytic function $f^* = u^* + iv^*$ could be shown to be constant by means of the principle of the argument, which is a contradiction. Hence the system (17.1-14) does have a unique solution $\lambda_1, \lambda_2, \ldots, \lambda_k$, and the function (17.1-12) formed with these λ_j possesses a conjugate harmonic function v in D. The function

$$f(z) := z\, e^{u(z)+iv(z)} \tag{17.1-15}$$

then is the required mapping function, as can be shown by means of the principle of the argument, much as in the proof of Theorem 17.1a. For details see Ahlfors [1966], pp. 247–249. —

It is instructive to interpret Theorem 17.1b in terms of plane electrostatics as discussed in § 5.7. Consider a dielectric medium bounded by $k+1$ conducting cylindrical surfaces Σ_j with generators perpendicular to the (x, y) plane, and let the intersection of the Σ_j with that plane be denoted by Γ_j, $j = 0, 1, \ldots, k$. We assume that the Γ_j can be represented analytically as piecewise analytic Jordan curves. The numbering is chosen so that Γ_0 contains the remaining Γ_j in its interior, as in Fig. 17.1c. Let D be the cross section of the medium, that is, the region bounded by the Γ_j. As above, f is the function that maps D onto the canonical region described in Theorem 17.1b.

If $\psi(w) := \operatorname{Log}|w|$, then ψ may be described electrostatically as a potential that assumes the value 0 on $|w| = 1$, the value $\operatorname{Log} \mu_k$ on $|w| = \mu_k$, and the constant values $\operatorname{Log} \mu_j$ on the circular slits Λ_j, $j = 1, 2, \ldots, k-1$. If the dielectric constant is assumed to be 1, the charge per unit length sitting on the cylindrical conductors Σ_j^* intersecting the w plane in the Λ_j is

$$Q_j = -\int_{\Lambda_j} \frac{\partial \psi}{\partial n} |dw|$$

($n :=$ interior normal), which in view of (5.7-16) may be written

$$Q_j = -\operatorname{Im} \int_{\Gamma_j} \overline{\operatorname{grd} \psi(w)}\, dw, \qquad j = 1, 2, \ldots, k.$$

Because in the absence of sources the total charge is zero,

$$Q_0 = -Q_1 - Q_2 - \cdots - Q_k. \tag{17.1-16}$$

In view of

$$\operatorname{grd} \psi(w) = \frac{1}{\bar{w}},$$

we find

$$Q_j = -\text{Im} \int_{\Lambda_j} \frac{1}{w}\, dw, \qquad j = 1, 2, \ldots, k.$$

For $j = k$ we get

$$Q_k = -2\pi,$$

whereas for the slits $\Lambda_j, j = 1, 2, \ldots, k-1$, we obtain

$$Q_j = 0, \qquad j = 1, 2, \ldots, k-1,$$

because the integration must be performed on both sides of the slits. For Λ_0 there results

$$Q_0 = +2\pi.$$

All of the above remains unchanged by conformal transplantation. Thus the function

$$\phi(z) := \psi(f(z)) = \text{Log}|f(z)|$$

solves the problem of finding the potential in D, if the surface Σ_0 is kept at zero potential, a charge of -2π is placed onto Σ_k, and if the total charge on each of the surfaces $\Sigma_1, \ldots, \Sigma_{k-1}$ is zero. The numbers $\lambda_j = \text{Log}\, \mu_j$ are the values which ϕ automatically assumes on the conductors $\Sigma_j, j = 1, 2, \ldots, k$. The fact that the λ_j are uniquely determined, which we have established as a mathematical result, is intuitively plausible on the basis of the electrostatic analogy.

The more standard problem to find a potential ϕ that assumes prescribed values σ_j on *all* conductors $\Sigma_j, j = 0, 1, \ldots, k$, is solved in terms of the harmonic measures u_j by means of the formula

$$\phi(z) = \sum_{j=0}^{k} \sigma_j u_j(z).$$

In this case it is the charges sitting on each conductor that are determined uniquely. In terms of the coefficients α_{ij} introduced earlier they are given by

$$Q_i = \sum_{j=1}^{k} (\sigma_0 - \sigma_j)\alpha_{ij}, \qquad i = 1, 2, \ldots, k,$$

whereas Q_0 is determined by (17.1-16).

PROBLEMS

1. Let D denote the region bounded by the ellipse with foci ± 1 and major semiaxis $\eta > 1$, with the straight line segment joining the foci removed. Use the Joukowski

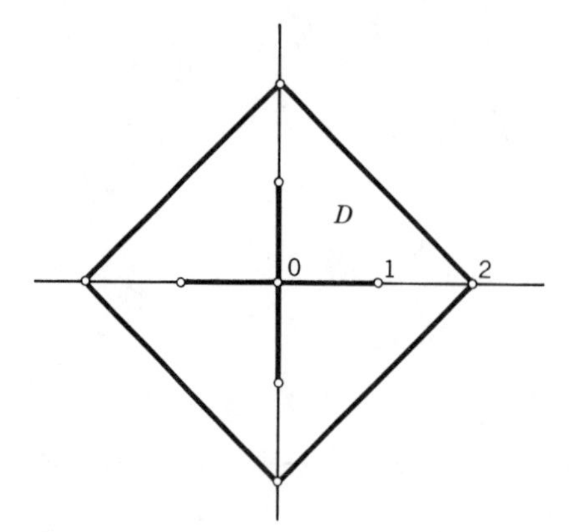

Fig. 17.1d. Doubly connected region.

map to show that the logarithmic modulus of D equals

$$\mu^* = \frac{1}{2\pi} \operatorname{Log}(\eta + \sqrt{\eta^2 - 1}).$$

2. Show that the logarithmic modulus μ^* of a doubly connected region R satisfies

$$\mu^{*-1} = \min_{\phi} D(\phi),$$

where $D(\phi)$ denotes the Dirichlet integral, and where the minimum is taken with respect to all piecewise differentiable functions that equal 0 on Γ_0 and 1 on Γ_1. Also show that the minimum is assumed for the harmonic measure of Γ_1 with respect to R. (Use conformal invariance.)

3. Let $\Gamma_0, \Gamma_1, \Gamma_2$ be three Jordan curves, each lying in the interior of the preceding one. Let the doubly connected region D_1 be bounded by Γ_0 and Γ_1, let D_2 be bounded by Γ_1 and Γ_2, and let D be bounded by Γ_0 and Γ_2. Denoting the corresponding moduli by ν_1, ν_2, and ν, prove that

$$\nu \geqq \nu_1 \nu_2.$$

4. The doubly connected region D shown in Fig. 17.1d is mapped conformally onto the annulus $1 < |w| < \rho$. Show that $\rho = e^{\pi/4}$.

5. Estimate the modulus of the annular region shown in Fig. 17.1e, using the variational property established in Problem 2 and working with finite elements. Show that

$$\tfrac{9}{74} < \mu^* < \tfrac{99}{784}$$

and hence that $2.147 < \mu^{-1} < 2.221$.

NOTES

Gaier [1978] gives a survey of the theory and the construction of conformal maps of multiply connected regions. The computation of the modulus of a doubly connected region is treated

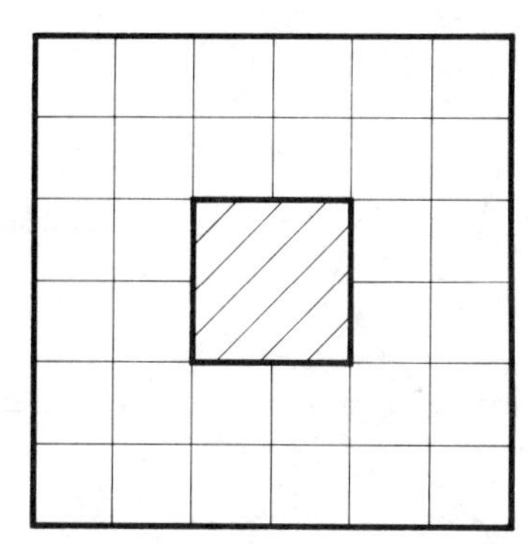

Fig. 17.1e. Finite elements for doubly connected region.

by Gaier [1964, 1974a], Opfer [1967, 1968, 1969], and Hersch [1984]. Problem 4 is due to Hersch (oral communication). For a direct implementation of the mapping method implied in Theorem 17.1b, see Ellacott [1978] and Mayo [1985], who uses a fast Poisson solver for the solution of the Dirichlet problems involved.

§ 17.2. DOUBLY CONNECTED REGIONS: OSCULATION METHODS

The osculation methods introduced in § 16.2 for the mapping of simply connected regions can also be used to map regions that are doubly connected. The same osculation families, such as the Koebe family or Grassmann's extended Koebe families, may be used.

Let $D = D_0$ be the given doubly connected region that is to be mapped onto an annulus. We assume that neither of the two components of the complement of D reduces to a point. We then may assume without loss of generality that D is contained in the unit disk E, and that $z = 0$ is an interior point in the bounded component of the complement of D.

Given an osculation family $\mathscr{F}$, a sequence of doubly connected regions $\{D_n\}$ will now be constructed which tends to an annulus A. All these regions will touch the unit circle from the inside, and $z = 0$ will be a point in the bounded component of the complement of every D_n. The exterior boundary of D_n is denoted by Γ_n, the interior boundary by Λ_n. Moreover we set

$$\begin{aligned} \alpha_n &:= \min_{z\in\Gamma_n} |z|, \\ \beta_n &:= \max_{z\in\Lambda_n} |z|, \\ \rho_n &:= \min_{z\in\Lambda_n} |z|, \end{aligned} \qquad (17.2\text{-}1)$$

$n = 0, 1, 2, \ldots$ (see Fig. 17.2a).

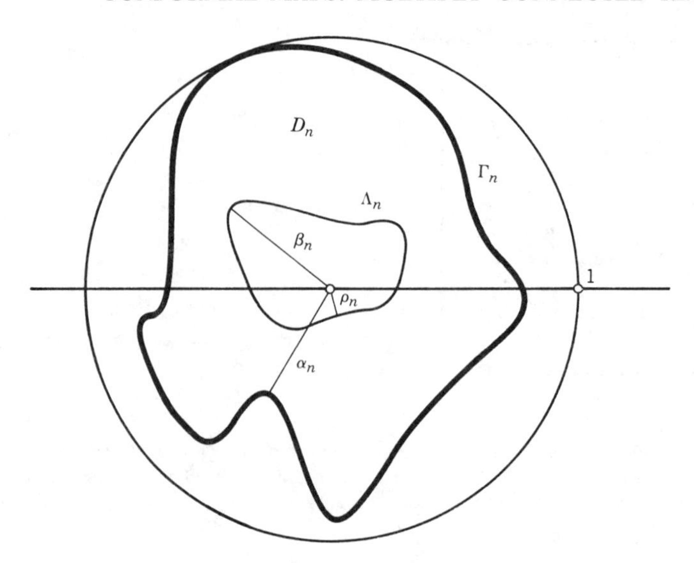

Fig. 17.2a. Notation for osculation method.

The interior of Γ_n, or more precisely the union of D_n with the bounded component of its complement, will be denoted by D_n^*.

Given D_n, the region D_{n+1} is constructed in two steps.

(a) We select $h \in \mathscr{F}(D_n^*)$ and compute $D'_n = h(D_n)$. This is a doubly connected region with exterior boundary Γ'_n such that

$$\alpha'_n := \min_{z \in \Gamma'_n} |z| \geqq \psi_0(\alpha_n, \alpha_n) > \alpha_n,$$

where ψ_0 is the function defined in Theorem 16.2h. Since all points interior to Γ_n are pushed outward, Γ'_n likewise touches $|z| = 1$, and $z = 0$ is an exterior point of D'_n. Moreover, denoting the inner boundary of D'_n by Λ'_n and letting

$$\rho'_n := \min_{z \in \Lambda'_n} |z|,$$

we have, if z'_n is a point on Λ'_n such that $|z'_n| = \rho'_n$, $z'_n = h(z_n)$,

$$\rho'_n = |z'_n| = |h(z_n)| \geqq \psi(\alpha_n, |z_n|) \geqq \psi_0(\alpha_n, \rho_n) > \rho_n$$

and thus in particular

$$\rho'_n > \rho_n. \tag{17.2-2}$$

Here we have used the properties of ψ_0 sketched after (16.2-24).

(b) We now set

$$D_{n+1} := \frac{\rho'_n}{D'_n}.$$

This is a doubly connected region with outer and inner boundaries

$$\Gamma_{n+1}=\frac{\rho'_n}{\Lambda'_n}, \qquad \Lambda_{n+1}=\frac{\rho'_n}{\Gamma'_n}.$$

If, in addition to the notation already introduced,

$$\beta'_n := \max_{z\in\Lambda'_n} |z|,$$

we thus have

$$\alpha_{n+1}=\frac{\rho'_n}{\beta'_n}, \qquad \beta_{n+1}=\frac{\rho'_n}{\alpha'_n}, \tag{17.2-3}$$

and, since Γ'_n touches $|z|=1$,

$$\rho_{n+1}=\frac{\rho'_n}{1}=\rho'_n. \tag{17.2-4}$$

THEOREM 17.2a. *The regions D_n tend to an annular region A: $\rho<|z|<1$.*

Proof. The inner radii by virtue of (17.2-2) form an increasing sequence. Since $\rho_n<1$ for all n,

$$\rho := \lim_{n\to\infty} \rho_n \tag{17.2-5}$$

exists. Because of the invariance of the modulus under a conformal map, $\rho<1$. We assert that

$$\lim_{n\to\infty} \alpha_n = 1. \tag{17.2-6}$$

For a proof by contradiction, suppose $\alpha := \liminf \alpha_n < 1$. We again use the function ψ_0 discussed in Theorem 16.2h. Let η be defined by $\psi_0(\alpha, \rho)=\rho+2\eta$; we know that $\eta>0$. By the continuity of ψ_0 there exists $\delta>0$ such that $\psi_0(\alpha_n, \rho_n)>\rho+\eta$ if $|\alpha_n-\alpha|<\delta$, $|\rho_n-\rho|<\delta$. The second condition is satisfied for all sufficiently large n, the first for infinitely many n. For such n it follows in view of

$$\rho_{n+1}=\rho'_n \geqq \psi_0(\alpha_n, \rho_n)$$

that

$$\rho_{n+1}>\rho+\eta,$$

contradicting the fact that the increasing sequence $\{\rho_n\}$ has the limit ρ. It thus follows that $\lim \alpha_n = 1$, and it only remains to be shown that

$$\lim_{n\to\infty} \beta_n = \rho.$$

In view of

$$1 \geqq \alpha'_n = \psi_0(\alpha_n, \alpha_n) > \alpha_n$$

there follows from (17.2-3), using $\rho'_n = \rho_{n+1}$,

$$\rho_{n+1} \leqq \beta_{n+1} \leqq \frac{\rho_{n+1}}{\alpha_n},$$

and the desired result follows by letting $n \to \infty$ in view of $\rho_n \to \rho$ and $\alpha_n \to 1$. —

Denoting by h_n the combined mapping function of steps (a) and (b), we set

$$f_n := h_n \circ h_{n-1} \circ \cdots \circ h_0.$$

It is of course not true that the sequence $\{f_n\}$ converges, since at each step the inner and the outer boundaries are interchanged. It is not even true that the sequence $\{f_{2n}\}$ converges. For this to be the case the maps f_n have to be normalized, for instance by prescribing that for a certain point $z_0 = D_0$ such that $\arg z_0 = 0$, all images should satisfy

$$\arg f_n(z_0) = 0.$$

Under this condition it can indeed by shown that the sequence $\{f_{2n}\}$ converges, locally uniformly in D_0, to a function f mapping D_0 onto the annulus A. We omit the proof.

Numerical experiments show, as in the case of simply connected regions, that the osculation algorithm converges well at the beginning. Further on, however, when the regions D_n are already nearly circular, the convergence becomes very slow.

PROBLEM

1. Komatu's method (Komatu [1949]) is similar to the osculation method described above, but assumes that the mapping problem for simply connected regions is trivial. Using the notation of the main text, one step of the iteration is described as follows:

   (a) Map D_n^* onto the full unit disk E (so that $\alpha'_n = 1$), and let D'_n be the image of D_n under this map.

   (b) $$D_{n+1} := \frac{\rho'_n}{D_n}.$$

   Obtain the converge of Komatu's method as a special case of Theorem 17.2a. (Let $\mathcal{F}$ be the Riemann family.)

NOTES

For the Koebe family, the iteration method described above was proposed by Graeser [1930] in a doctoral dissertation written under Koebe. Hoidn [1982] studies the method for general osculation families and supports his conclusions with numerous experiments. For related work see Cremer [1930] and Albrecht [1955].

§ 17.3. DOUBLY CONNECTED REGIONS: LINEAR INTEGRAL EQUATIONS

In § 17.1 the mapping problem for regions of arbitrary finite connectivity was reduced to the problem of solving a Dirichlet problem plus that of determining a conjugate harmonic function. For simply connected regions we have seen in § 16.5 that the reduction to a Dirichlet problem is feasible also for the numerical construction of the mapping function. The same has been shown by Opfer [1967, 1969] for doubly connected regions, and except for the increasing computational complexity there is no reason why the method should not work also for regions of connectivity >2.

Here we examine how the linear integral equations for the boundary correspondence function that were set up in § 16.6 and § 16.7 generalize to doubly connected regions. Let the outer and inner boundary curves be given in parametric representation as follows:

$$\Gamma_0: \quad z = z_0(\tau), \qquad 0 \leqq \tau \leqq \beta_0,$$

$$\Gamma_1: \quad z = z_1(\tau), \qquad 0 \leqq \tau \leqq \beta_1.$$

If f is a function mapping the region D bounded by Γ_0 and Γ_1 onto the annulus A: $\mu < |w| < 1$ so that the inner and the outer boundaries correspond to each other, the **boundary correspondence functions** θ_0 (outer boundary) and θ_1 (inner boundary) are continuous functions satisfying

$$f(z_0(\tau)) = e^{i\theta_0(\tau)}, \qquad 0 \leqq \tau \leqq \beta_0,$$

$$f(z_1(\tau)) = \mu\, e^{i\theta_1(\tau)}, \qquad 0 \leqq \tau \leqq \beta_1,$$

where the expressions on the left are to be understood as the continuous extensions of the mapping function to the boundary. If these boundary correspondence functions are known, both $f(z)$ and $f^{[-1]}(w)$ are calculated easily for arbitrary $z \in D$ and $w \in A$, for instance by Cauchy's integral formula. In the following formulas the strong singularity of Cauchy's integral for $|w|$ near 1 or μ has been weakened by integration by parts:

$$f(z)=\frac{1}{2\pi i}\int_0^{\beta_0} e^{i\theta_0(\tau)}\frac{z_0'(\tau)}{z_0(\tau)}\,d\tau$$

$$-\frac{1}{2\pi}\int_0^{\beta_0} e^{i\theta_0(\tau)}\log\left(1-\frac{z}{z_0(\tau)}\right)\theta_0'(\tau)\,d\tau$$

$$+\frac{1}{2\pi}\int_0^{\beta_0} e^{i\theta_1(\tau)}\log\left(1-\frac{z_1(\tau)}{z}\right)\theta_1'(\tau)\,d\tau. \qquad (17.3\text{-}1)$$

Here the logarithms are the continuous continuations inside Γ_0 from $z=0$, and outside Γ_1 from $z=\infty$, respectively; they are principal values if both curves are starlike. For the inverse mapping function we get

$$z=g(w)=\frac{1}{2\pi}\int_0^{\beta_0} z_0(\tau)\theta_0'(\tau)\,d\tau$$

$$-\frac{1}{2\pi i}\int_0^{\beta_0} z_0'(\tau)\,\mathrm{Log}\left(1-e^{-i\theta_0(\tau)}w\right)d\tau$$

$$-\frac{1}{2\pi i\mu}\int_0^{\beta_1} z_1'(\tau)\,\mathrm{Log}\left(1-\frac{e^{i\theta_1(\tau)}}{w}\right)d\tau. \qquad (17.3\text{-}2)$$

The logarithms here are always principal values.

We now represent the mapping function f in terms of the functions $\theta_i(\tau)$ in yet a different manner. Let D be such that the mapping function f from D to A may be differentiably extended to the boundary of D. Let $g:=f^{[-1]}$ be the inverse mapping function. If z is in the exterior of the outer boundary Γ_0, Γ_0 has winding number zero with respect to z, and

$$h_0(w):=\log(z-g(w))$$

can be defined as an analytic function in A. There follows

$$\frac{1}{2\pi i}\int_{|w|=1}\frac{h_0(w)}{w}\,dw=\frac{1}{2\pi i}\int_{|w|=\mu}\frac{h_0(w)}{w}\,dw$$

or

$$\frac{1}{2\pi}\int_0^{2\pi}\log(z-g(e^{i\theta}))\,d\theta=\frac{1}{2\pi}\int_0^{2\pi}\log(z-g(\mu\,e^{i\theta}))\,d\theta. \qquad (17.3\text{-}3)$$

Substituting $\theta=\theta_0(\tau)$ and $\theta=\theta_1(\tau)$ in these integrals, we get

$$\frac{1}{2\pi}\int_0^{\beta_0}\log(z-z_0(\tau))\theta_0'(\tau)\,d\tau=\frac{1}{2\pi}\int_0^{\beta_1}\log(z-z_1(\tau))\theta_1'(\tau)\,d\tau$$

and on taking real parts and letting $z \to z_0(\sigma)$,

$$\frac{1}{2\pi}\int_0^{\beta_0} \operatorname{Log}|z_0(\sigma)-z_0(\tau)|\theta_0'(\tau)\,d\tau - \frac{1}{2\pi}\int_0^{\beta_1} \operatorname{Log}|z_0(\sigma)-z_1(\tau)|\theta_1'(\tau)\,d\tau = 0. \tag{17.3-4}$$

Let now z be in the interior of Γ_1. Then both Γ_0 and Γ_1 have winding number $+1$ with respect to z, the same as the winding number of $f(\Gamma_0)$ and $f(\Gamma_1)$ with respect to 0. It follows that

$$h_1(w) := \log\frac{g(w)-z}{w} \tag{17.3-5}$$

may be defined as an analytic function in A. Then

$$\frac{1}{2\pi i}\int_{|w|=1} \frac{h_1(w)}{w}\,dw = \frac{1}{2\pi i}\int_{|w|=\mu} \frac{h_1(w)}{w}\,dw$$

or

$$\frac{1}{2\pi}\int_0^{2\pi} \log\frac{g(e^{i\theta})-z}{e^{i\theta}}\,d\theta = \frac{1}{2\pi}\int_0^{2\pi} \log\frac{g(\mu\, e^{i\theta})-z}{e^{i\theta}}\,d\theta$$

or, making the same substitutions as above,

$$\frac{1}{2\pi}\int_0^{\beta_0} \log\frac{z_0(\tau)-z}{e^{i\theta_0(\tau)}}\,\theta_0'(\tau)\,d\tau - \frac{1}{2\pi}\int_0^{\beta_1} \log\frac{z_1(\tau)-z}{e^{i\theta_1(\tau)}}\,\theta_1'(\tau)\,d\tau = \operatorname{Log}\frac{1}{\mu}. \tag{17.3-6}$$

Taking real parts and letting $z \to z_1(\sigma)$, an arbitrary point on Γ_1, we now obtain

$$\frac{1}{2\pi}\int_0^{\beta_0} \operatorname{Log}|z_0(\tau)-z_1(\sigma)|\theta_0'(\tau)\,d\tau - \frac{1}{2\pi}\int_0^{\beta_1} \operatorname{Log}|z_1(\tau)-z_1(\sigma)|\theta_1'(\tau)\,d\tau = \operatorname{Log}\frac{1}{\mu}. \tag{17.3-7}$$

We have obtained:

THEOREM 17.3a. *If the boundary curves Γ_0 and Γ_1 are such that the boundary correspondence functions θ_i have continuous derivatives, then the functions θ_0' and θ_1' satisfy the system of linear integral equations of the first* *kind,*

$$\frac{1}{2\pi}\int_0^{\beta_0} \operatorname{Log}|z_0(\sigma)-z_0(\tau)|\theta_0'(\tau)\,d\tau - \frac{1}{2\pi}\int_0^{\beta_1} \operatorname{Log}|z_0(\sigma)-z_1(\tau)|\theta_1'(\tau)\,d\tau = 0,$$

$$\frac{1}{2\pi}\int_0^{\beta_0} \operatorname{Log}|z_1(\sigma)-z_0(\tau)|\theta_0'(\tau)\,d\tau - \frac{1}{2\pi}\int_0^{\beta_0} \operatorname{Log}|z_1(\sigma)-z_1(\tau)|\theta_1'(\tau)\,d\tau = \operatorname{Log}\frac{1}{\mu}. \tag{17.3-8}$$

Bypassing the question of uniqueness of the solution of this system, we show how the mapping function f may be constructed from *any* solution of the above system. Thus let ξ_0 and ξ_1 be two functions satisfying (17.3-8) (in place of θ_0' and θ_1') satisfying the conditions

$$\frac{1}{2\pi}\int_0^{\beta_0} \xi_0(\tau)\,d\tau = \frac{1}{2\pi}\int_0^{\beta_1} \xi_1(\tau)\,d\tau = 1. \tag{17.3-9}$$

For z in the closure of D we define

$$u(z) := -\frac{1}{2\pi}\int_0^{\beta_0} \operatorname{Log}\left|1-\frac{z}{z_0(\tau)}\right|\xi_0(\tau)\,d\tau + \frac{1}{2\pi}\int_0^{\beta_1} \operatorname{Log}\left|1-\frac{z_1(\tau)}{z}\right|\xi_1(\tau)\,d\tau.$$

The function u is harmonic in D, continuous on the closure of D, and on the boundary curves assumes the values

$$u(z) = -\frac{1}{2\pi}\int_0^{\beta_0} \operatorname{Log}|z-z_0(\tau)|\xi_0(\tau)\,d\tau + \gamma + \frac{1}{2\pi}\int_0^{\beta_1} \operatorname{Log}|z-z_1(\tau)|\xi_1(\tau)\,d\tau - \operatorname{Log}|z|,$$

where

$$\gamma := \frac{1}{2\pi}\int_0^{\beta_0} \operatorname{Log}|z_0(\tau)|\xi_0(\tau)\,d\tau. \tag{17.3-10}$$

By (17.3-8) we thus have

$$u(z) = \gamma - \operatorname{Log}|z|, \qquad z \in \Gamma_0,$$
$$u(z) = \gamma + \operatorname{Log}\mu - \operatorname{Log}|z|, \qquad z \in \Gamma_1.$$

We conclude that $u-\gamma$ is the solution of the Dirichlet problem (17.1-4). An appropriate conjugate harmonic function is easily constructed, and we have

$$u(z)+iv(z)=-\frac{1}{2\pi}\int_0^{\beta_0}\log\left(1-\frac{z}{z_0(\tau)}\right)\xi_0(\tau)\,d\tau$$
$$+\frac{1}{2\pi}\int_0^{\beta_1}\log\left(1-\frac{z_1(\tau)}{z}\right)\xi_1(\tau)\,d\tau,$$

where the logarithms are defined by analytic continuation from $z=0$ and $z=\infty$, respectively. (They are principal values if Γ_0 and Γ_1 are starlike.)

The above holds for any pair of continuous solutions (ξ_0, ξ_1) of (17.3-8). In particular it holds for $\xi_0=\theta_0'$, $\xi_1=\theta_1'$. Substituting these solutions and using the resulting expression for $u+iv$ in the representation (17.1-8) for the mapping function, we obtain the following new representation for f:

$$w=f(z)=z\exp\{-\gamma+i\alpha\}\exp\left\{-\frac{1}{2\pi}\int_0^{\beta_0}\log\left(1-\frac{z}{z_0(\tau)}\right)\theta_0'(\tau)\,d\tau\right.$$
$$\left.+\frac{1}{2\pi}\int_0^{\beta_1}\log\left(1-\frac{z_1(\tau)}{z}\right)\theta_1'(\tau)\,d\tau\right\}. \tag{17.3-11}$$

Here γ is given by (17.3-10), and α is an undetermined constant that effects a rotation of the annulus A.

As in the case of a simply connected region, we can deduce from the above a system of integral equations of the second kind for θ_0', θ_1'. Differentiating (17.3-3) with respect to z, we get

$$\frac{1}{2\pi}\int_0^{\beta_0}\frac{1}{z-z_0(\tau)}\theta_0'(\tau)\,d\tau-\frac{1}{2\pi}\int_0^{\beta_1}\frac{1}{z-z_1(\tau)}\theta_1'(\tau)\,d\tau=0. \tag{17.3-12}$$

We wish to let $z\to z_0(\sigma)$, which in the second integral is harmless and in the first requires an appeal to Sokhotskyi's theorem. Defining

$$\begin{aligned}\nu_{0,0}(\tau,\sigma)&:=\frac{1}{\pi}\operatorname{Im}\frac{z_0'(\sigma)}{z_0(\sigma)-z_0(\tau)},\\ \nu_{0,1}(\tau,\sigma)&:=\frac{1}{\pi}\operatorname{Im}\frac{z_0'(\sigma)}{z_0(\sigma)-z_1(\tau)},\end{aligned} \tag{17.3-13}$$

the result is

$$\theta_0'(\sigma)+\mathrm{PV}\int_0^{\beta_0}\nu_{0,0}(\tau,\sigma)\theta_0'(\tau)\,d\tau-\int_0^{\beta_1}\nu_{0,1}(\tau,\sigma)\theta_1'(\tau)\,d\tau=0.$$

Differentiating (17.3-6) we similarly get

$$\frac{1}{2\pi}\int_0^{\beta_0}\frac{1}{z_0(\tau)-z}\theta_0'(\tau)\,d\tau-\frac{1}{2\pi}\int_0^{\beta_1}\frac{1}{z_1(\tau)-z}\theta_1'(\tau)\,d\tau=0. \qquad (17.3\text{-}14)$$

We let z approach a point $z_1(\sigma)$ on Γ_1 from the interior of Γ_1, which in the second integral requires an application of Sokhotskyi's theorem. Defining

$$\begin{aligned}\nu_{1,0}(\tau,\sigma)&=\frac{1}{\pi}\operatorname{Im}\frac{z_1'(\sigma)}{z_1(\sigma)-z_0(\tau)},\\ \nu_{1,1}(\tau,\sigma)&=\frac{1}{\pi}\operatorname{Im}\frac{z_1'(\sigma)}{z_1(\sigma)-z_1(\tau)},\end{aligned} \qquad (17.3\text{-}15)$$

there results

$$\theta_1'(\sigma)+\int_0^{\beta_0}\nu_{1,0}(\tau,\sigma)\theta_0'(\tau)\,d\tau-\mathrm{PV}\int_0^{\beta_1}\nu_{1,1}(\tau,\sigma)\theta_1'(\tau)\,d\tau=0.$$

If the curves Γ_0 and Γ_1 are such that $z_i''(\tau)$ is continuous, $i=1, 2$, then the kernels $\nu_{0,0}$ and $\nu_{1,1}$ are continuous by Theorem 15.9a. Then all integrals may be taken as ordinary integrals, and we get:

THEOREM 17.3b (Warschawski's equation for doubly connected regions). *If the boundary curves are such that z_i'' is continuous and the boundary correspondence functions θ_i have continuous derivatives, then the functions θ_0' and θ_1' satisfy the system of integral equations*

$$\begin{aligned}\theta_0'(\sigma)+\int_0^{\beta_0}\nu_{0,0}(\tau,\sigma)\theta_0'(\tau)\,d\tau-\int_0^{\beta_1}\nu_{0,1}(\tau,\sigma)\theta_1'(\tau)\,d\tau=0,\\ \theta_1'(\sigma)+\int_0^{\beta_0}\nu_{1,0}(\tau,\sigma)\theta_0'(\tau)\,d\tau-\int_0^{\beta_1}\nu_{1,1}(\tau,\sigma)\theta_1'(\tau)\,d\tau=0,\end{aligned} \qquad (17.3\text{-}16)$$

where the kernels $\nu_{i,j}$ are defined by (17.3-13) *and* (17.3-15).

We finally obtain a pair of integral equations for the functions θ_0 and θ_1 themselves. Integrating (17.3-4) by parts, we get

$$\begin{aligned}&\frac{1}{2\pi}[\log(z-z_0(\tau))\theta_0(\tau)]_0^{\beta_0}+\frac{1}{2\pi}\int_0^{\beta_0}\frac{z_0'(\tau)}{z-z_0(\tau)}\theta_0(\tau)\,d\tau\\ &\quad-\frac{1}{2\pi}[\log(z-z_1(\tau))\theta_1(\tau)]_0^{\beta_1}-\frac{1}{2\pi}\int_0^{\beta_1}\frac{z_1'(\tau)}{z-z_1(\tau)}\theta_1(\tau)\,d\tau=0,\end{aligned}$$

which is to say

$$\log\frac{z-z_0(0)}{z-z_1(0)}-\frac{1}{2\pi i}\int_{\Gamma_0}\frac{i\theta_0(\tau)}{t-z}\,dt-\frac{1}{2\pi}\int_0^{\beta_1}\frac{z_1'(\tau)}{z-z_1(\tau)}\theta_1(\tau)\,d\tau=0.$$

Letting $z \to z_0(\sigma)$ by virtue of the Sokhotskyi formulas yields

$$2\log\frac{z_0(\sigma)-z_0(0)}{z_0(\sigma)-z_1(0)}+i\theta_0(\sigma)-\frac{1}{\pi}\,\mathrm{PV}\int_0^{\beta_0}\frac{z_0'(\tau)}{z_0(\tau)-z_0(\sigma)}\theta_0(\tau)\,d\tau$$

$$+\frac{1}{\pi}\int_0^{\beta_1}\frac{z_1'(\tau)}{z_1(\tau)-z_0(\sigma)}\theta_1(\tau)\,d\tau=0.$$

On taking the imaginary part and using the abbreviations (17.3-13) and (17.3-15) we get

$$\theta_0(\sigma)-\int_0^{\beta_0}\nu_{0,0}(\sigma,\tau)\theta_0(\tau)\,d\tau+\int_0^{\beta_1}\nu_{1,0}(\sigma,\tau)\theta_1(\tau)\,d\tau=2\arg\frac{z_0(\sigma)-z_1(0)}{z_0(\sigma)-z_0(0)}.$$

Another equation is obtained by integrating (17.3-6) by parts. If z is interior to Γ_1, we find

$$\frac{1}{2\pi}\left[\theta_0(\tau)\log\frac{z_0(\tau)-z}{e^{i\theta_0(\tau)}}\right]_0^{\beta_0}-\frac{1}{2\pi}\int_0^{\beta_0}\left[\frac{z_0'(\tau)}{z_0(\tau)-z}-i\theta_0'(\tau)\right]\theta_0(\tau)\,d\tau$$

$$-\frac{1}{2\pi}\left[\theta_1(\tau)\log\frac{z_1(\tau)-z}{e^{i\theta_1(\tau)}}\right]_0^{\beta_1}$$

$$+\frac{1}{2\pi}\int_0^{\beta_1}\left[\frac{z_1'(\tau)}{z_1(\tau)-z}-i\theta_1'(\tau)\right]\theta_1(\tau)\,d\tau=\mathrm{Log}\,\frac{1}{\mu}.$$

Here the integrated parts are

$$\log\frac{z_j(0)-z}{e^{i\theta_j(0)}},\qquad j=0,1.$$

Furthermore

$$\frac{1}{2\pi}\int_0^{\beta_j}\theta_j(\tau)\theta_j'(\tau)\,d\tau=\frac{1}{4\pi}[\theta_j(\tau)^2]_0^{\beta_j}$$

$$=\frac{1}{4\pi}[\theta_j(\beta_j)+\theta_j(0)][\theta_j(\beta_j)-\theta_j(0)]$$

$$=\frac{1}{4\pi}2\theta_j(0)\cdot 2\pi=\theta_j(0).$$

Thus we get

$$\log\frac{z_0(0)-z}{z_1(0)-z}-\frac{1}{2\pi}\int_0^{\beta_0}\frac{z_0'(\tau)}{z_0(\tau)-z}\theta_0(\tau)\,d\tau+\frac{1}{2\pi}\int_0^{\beta_1}\frac{z_1'(\tau)}{z_1(\tau)-z}\theta_1(\tau)\,d\tau=\mathrm{Log}\,\frac{1}{\mu}.$$

Again we let $z \to z_1(\sigma)$ from the interior, using the Sokhotskyi formulas. On taking imaginary parts there results

$$\theta_1(\sigma)-\int_0^{\beta_0}\nu_{0,1}(\sigma,\tau)\theta_0(\tau)\,d\tau+\int_0^{\beta_1}\nu_{1,1}(\sigma,\tau)\theta_1(\tau)\,d\tau=2\arg\frac{z_1(\sigma)-z_1(0)}{z_1(\sigma)-z_0(0)}.$$

We have found:

THEOREM 17.3c (Gerschgorin's equation for doubly connected regions). *Under the hypotheses of Theorem* 17.3b *the functions* θ_0 *and* θ_1 *satisfy the system of integral equations*

$$\begin{aligned}\theta_0(\sigma)-\int_0^{\beta_0}\nu_{0,0}(\rho,\tau)\theta_0(\tau)\,d\tau+\int_0^{\beta_1}\nu_{1,0}(\sigma,\tau)\theta_1(\tau)\,d\tau\\ =2\arg\frac{z_0(\sigma)-z_1(0)}{z_0(\sigma)-z_0(0)},\\ \theta_1(\sigma)-\int_0^{\beta_0}\nu_{0,1}(\sigma,\tau)\theta_0(\tau)\,d\tau+\int_0^{\beta_1}\nu_{1,1}(\sigma,\tau)\theta_1(\tau)\,d\tau\\ =2\arg\frac{z_1(\sigma)-z_1(0)}{z_1(\sigma)-z_0(0)}.\end{aligned}\tag{17.3-17}$$

The integral equations (17.3-16) and (17.3-17) do not involve the modulus μ^{-1} of the given doubly connected region. If the functions θ_0, θ_1 and/or their derivatives are known, the modulus may be determined from (17.3-6), which holds for arbitrary z interior to Γ_1. Choosing $z=0$ and taking real parts, we find the simple formula

$$\operatorname{Log}\frac{1}{\mu}=\frac{1}{2\pi}\int_0^{\beta_0}\operatorname{Log}|z_0(\tau)|\theta_0'(\tau)\,d\tau-\frac{1}{2\pi}\int_0^{\beta_1}\operatorname{Log}|z_1(\tau)|\theta_1'(\tau)\,d\tau.$$

By an integration by parts, using

$$\frac{d}{d\tau}\operatorname{Log}|z_j(\tau)|=\operatorname{Re}\frac{z_j'(\tau)}{z_j(\tau)},$$

this yields

$$\operatorname{Log}\frac{1}{\mu}=\operatorname{Log}\left|\frac{z_0(0)}{z_1(0)}\right|-\int_0^{\beta_0}\operatorname{Re}\frac{z_0'(\tau)}{z_0(\tau)}\theta_0(\tau)\,d\tau+\int_0^{\beta_1}\operatorname{Re}\frac{z_1'(\tau)}{z_1(\tau)}\theta_1(\tau)\,d\tau,\tag{17.3-18}$$

a result which is obvious if D itself is a circular annulus.

PROBLEMS

1. Compute the kernels $\nu_{i,j}(\tau, \sigma)$, $i, j = 0, 1$, for the case where D is the annulus $\mu < |z| < 1$, and verify Gerschgorin's and Warschawski's integral equations in this trivial situation.
2. Let $\alpha, \rho \in (0, 1)$, $\alpha + \rho < 1$. Compute the inner and the outer boundary correspondence functions for the region bounded by the two circles

$$\Gamma_0: \quad z_0(\tau) = e^{i\tau}, \qquad 0 \leqq \tau \leqq 2\pi;$$

$$\Gamma_1: \quad z_1(\tau) = \alpha + \rho\, e^{i\tau}, \qquad 0 \leqq \tau \leqq 2\pi.$$

   Compute the modulus of this doubly connected region and verify Symm's equation.
3. *Potential flow through a periodic channel.* Let S be a simply connected, striplike region with period 2π, that is, let $z \in S$ if and only if all points $z + 2\pi k$, $(k \in \mathbb{Z}) \in S$, and let there exist $\eta < \infty$ such that S is contained in $|\text{Im } z| < \eta$ (see Fig. 17.3a). We wish to determine the potential flow through S (see § 5.8) and to this end have to map S onto a strip S_1 bounded by two straight lines parallel to the real w axis:

$$S_1: \quad \alpha < \text{Im } w < \beta.$$

   (a) Show that this map may be obtained in the following manner:

   (i) Map S onto a doubly connected region D by $z \mapsto z_1 := e^{iz}$. This maps all points $z + 2\pi k$ onto the *same* point z_1.
   (ii) Map D onto an annulus A: $\mu < |z_2| < 1$.
   (iii) Map A onto a strip by $z_2 \mapsto w = -i \log z_2$.

   (b) Which mapping is to be found in step (ii) if S is the meanderlike channel shown in Fig. 17.3b?

   (c) Carry out the required computations analytically in the case where D is an eccentric annulus. Determine the velocity vector at the boundary of S.

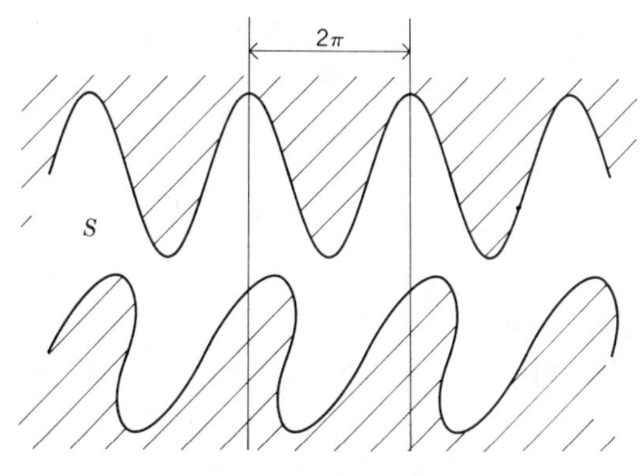

Fig. 17.3a. Periodic channel.

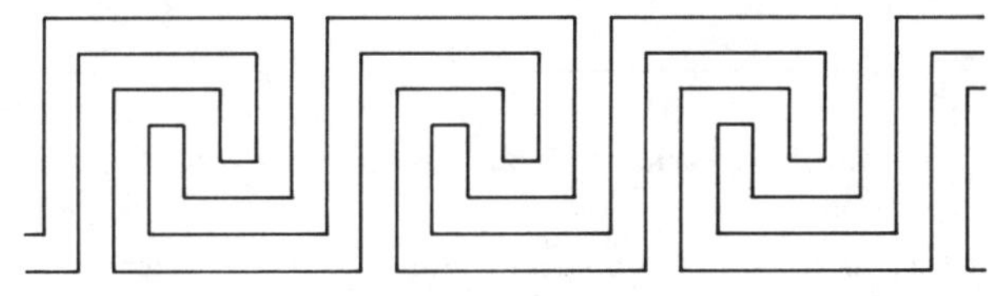

Fig. 17.3b. Meander.

NOTES

For a different derivation of Gerschgorin's integral equation (17.3-17) see Gaier [1964], p. 191. For Symm's integral equation for doubly connected regions, see Symm [1969], and for the fact that θ_0' and θ_1' is a solution pair, see Gaier [1981]. Hoidn [1976] has experimented numerically with the integral equations given here and found that the linear systems of equations that result from discretization may be ill-conditioned if D is not nearly circular.

§ 17.4. DOUBLY CONNECTED REGIONS: METHODS BASED ON CONJUGATE FUNCTIONS

The simple relation that exists between the real and the imaginary parts of the boundary values of a function that is analytic in the unit disk was made use of in § 16.8 to obtain various numerical mapping methods including the Theodorsen method. The same idea can be exploited to map doubly connected regions. The methods thus obtained furnish the boundary correspondence of the map from the standard region, that is, from the annulus, to the given region.

I. Conjugate Periodic Functions for an Annulus

Let ϕ_0 and ϕ_1 be two real 2π-periodic functions, given by their Fourier series

$$\phi_j(\theta) = \sum_{n=-\infty}^{\infty} a_{j,n}\, e^{in\theta}, \qquad j=0,1, \qquad a_{j,-n} = \overline{a_{j,n}}, \tag{17.4-1}$$

which we assume to be absolutely convergent:

$$\sum_{n=-\infty}^{\infty} |a_{j,n}| < \infty, \qquad j=0,1. \tag{17.4-2}$$

Let $0<\mu<1$, and let A denote the annulus $\mu<|w|<1$. We wish to carry out the following program:

(a) To solve the Dirichlet problem for the annulus A with the boundary conditions

$$\begin{aligned} u(e^{i\theta}) &= \phi_0(\theta), \\ u(\mu\, e^{i\theta}) &= \phi_1(\theta). \end{aligned} \tag{17.4-3}$$

(b) To construct, if possible, a conjugate harmonic function v of u.
(c) To extend v continuously to the boundary of A, and to express the boundary values of v directly in terms of the functions ϕ_0 and ϕ_1, or of their Fourier coefficients.

Step (a) was considered in § 15.2. Expressed in terms of Fourier series and letting $w = r e^{i\theta}$, the solution of the Dirichlet problem (17.4-3) is

$$u(w) = a_{0,0} + (a_{1,0} - a_{0,0}) \frac{\operatorname{Log} r}{\operatorname{Log} \mu}$$
$$+ \sum_{\substack{n=-\infty \\ n \neq 0}}^{\infty} \frac{1}{1-\mu^{2n}} \left\{ a_{0,n} \left(r^n - \frac{\mu^{2n}}{r^n} \right) + a_{1,0} \left(\frac{\mu^n}{r^n} - \mu^n r^n \right) \right\} e^{in\theta}. \quad (17.4\text{-}4)$$

Concerning step (b), we refer to § 15.1. There it was found that a conjugate harmonic function of u exists if and only if the conjugate period of u with respect to the bounded component of the complement of A is zero, which means that

$$\int_\Gamma (u_x \, dy - u_y \, dx) = 0$$

along any curve surrounding $|w| \leqq \mu$. Only the Log term contributes to the integral, and the contribution is

$$(a_{0,0} - a_{1,0}) \frac{2\pi}{\operatorname{Log} \mu}.$$

Therefore $a_{0,0} = a_{1,0}$ or

$$\int_0^{2\pi} \phi_0(\theta) \, d\theta = \int_0^{2\pi} \phi_1(\theta) \, d\theta \quad (17.4\text{-}5)$$

is a necessary and sufficient condition for the existence of a conjugate harmonic function. If this condition is met, one easily verifies that

$$u(w) = \operatorname{Re} h(w),$$

where h is an analytic function in A, which is defined by its Laurent series,

$$h(w) := a_{0,0} + 2 \sum_{\substack{n=-\infty \\ n \neq 0}}^{\infty} \frac{a_{0,n} - \mu^n a_{1,n}}{1-\mu^{2n}} w^n.$$

It follows that a conjugate function v of u is given by

$$v(w) = \frac{1}{2i} [h(w) - \overline{h(w)}],$$

that is, by

$$v(w)=\frac{1}{i}\sum_{\substack{n=-\infty\\ n\neq 0}}^{\infty}\frac{1}{1-\mu^{2n}}\left\{a_{0,n}\left(w^n+\frac{\mu^{2n}}{\bar{w}^n}\right)-a_{1,n}\left(\mu^n w^n+\frac{\mu^n}{\bar{w}^n}\right)\right\}.$$

Step (c) of our program is now trivial. We see that the boundary values of v are

$$\begin{aligned}\psi_0(\theta)&:=v(e^{i\theta})=-i\sum_{\substack{n=-\infty\\ n\neq 0}}^{\infty}\frac{1}{1-\mu^{2n}}[(1+\mu^{2n})a_{0,n}-2\mu^n a_{1,n}]\,e^{in\theta}\\ \psi_1(\theta)&:=v(\mu\,e^{i\theta})=-i\sum_{\substack{n=-\infty\\ n\neq 0}}^{\infty}\frac{1}{1-\mu^{2n}}[2\mu^n a_{0,n}-(1+\mu^{2n})a_{1,n}]\,e^{in\theta}.\end{aligned} \tag{17.4-6}$$

These Fourier series are absolutely convergent on account of (17.4-2).

For compact notation, let Π here denote the space of 2π- periodic real functions ϕ that are represented by an absolutely convergent Fourier series,

$$\phi(\theta)=\sum_{n=-\infty}^{\infty}a_n\,e^{in\theta},\qquad a_{-n}=\overline{a_n}.$$

For $0<\mu<1$, let $\mathcal{H}_\mu$ denote the operator $\Pi\to\Pi$ associating with ϕ the function

$$\mathcal{H}_\mu\phi(\theta)=\sum_{n=-\infty}^{\infty}b_n\,e^{in\theta}, \tag{17.4-7a}$$

where

$$b_n:=\begin{cases}0, & n=0\\ -i\dfrac{1+\mu^{2n}}{1-\mu^{2n}}a_n, & n\neq 0.\end{cases} \tag{17.4-7b}$$

Let $\mathcal{G}_\mu$: $\Pi\to\Pi$ be similarly defined by

$$\mathcal{G}_\mu\phi(\theta):=\sum_{n=-\infty}^{\infty}c_n\,e^{in\theta}, \tag{17.4-8a}$$

where

$$c_n:=\begin{cases}0, & n=0\\ -i\dfrac{2\mu^n}{1-\mu^{2n}}a_n, & n\neq 0.\end{cases} \tag{17.4-8b}$$

The formulas (17.4-6) are then expressed by

$$\begin{aligned}\psi_0&=\mathcal{H}_\mu\phi_0-\mathcal{G}_\mu\phi_1,\\ \psi_1&=\mathcal{G}_\mu\phi_0-\mathcal{H}_\mu\phi_1.\end{aligned} \tag{17.4-9}$$

The operator $\mathcal{H}_\mu$ is closely related to the conjugation operator $\mathcal{K}$ defined in § 13.5 and further studied in § 14.2. Clearly,

$$b_n = \begin{cases} -ia_n - i\dfrac{2\mu^{2n}}{1-\mu^{2n}}a_n, & n>0 \\ ia_n + i\dfrac{2\mu^{-2n}}{1-\mu^{-2n}}a_n, & n<0. \end{cases}$$

Thus for $\phi \in \Pi$ there follows

$$\mathcal{H}_\mu\phi = \mathcal{K}\phi + \mathcal{K}_\mu\phi, \qquad (17.4\text{-}10)$$

where

$$\mathcal{K}_\mu\phi(\theta) := -i\sum_{n=1}^{\infty}\frac{2\mu^{2n}}{1-\mu^{2n}}(a_n\, e^{in\theta} - a_{-n}\, e^{-in\theta}).$$

An integral representation for $\mathcal{K}_\mu$ is obtained by introducing

$$a_n = \frac{1}{2\pi}\int_{-\pi}^{\pi}\phi(\tau)\, e^{-in\tau}\, d\tau$$

and interchanging summation and integration. This yields

$$\mathcal{K}_\mu\phi(\theta) = \frac{1}{\pi}\int_{-\pi}^{\pi}\phi(\tau)\kappa_\mu(\theta-\tau)\, d\tau, \qquad (17.4\text{-}11)$$

where

$$\kappa_\mu(\sigma) := \sum_{n=1}^{\infty}\frac{2\mu^{2n}}{1-\mu^{2n}}\sin n\sigma.$$

In a similar manner one obtains

$$\mathcal{G}_\mu\phi(\theta) = \frac{1}{\pi}\int_{-\pi}^{\pi}\phi(\tau)\gamma_\mu(\theta-\tau)\, d\tau, \qquad (17.4\text{-}12)$$

where

$$\gamma_\mu := \sum_{n=1}^{\infty}\frac{2\mu^{n}}{1-\mu^{2n}}\sin n\sigma.$$

Relations (17.4-10)-(17.4-12) extend the domain of definition of the operators $\mathcal{H}_\mu$ and $\mathcal{G}_\mu$ from Π to $L_2(0, 2\pi)$. In fact, because κ_μ and γ_μ are analytic periodic functions, the functions $\mathcal{K}_\mu\phi$ and $\mathcal{G}_\mu\phi$ are analytic for every $\phi \in L_2(0, 2\pi)$. Statements (a), (b), and (c) made in § 14.2 concerning the operator $\mathcal{K}$ thus also hold for $\mathcal{H}_\mu$ and $\mathcal{G}_\mu$. For the explicit evaluation of $\mathcal{H}_\mu$ and $\mathcal{G}_\mu$, (17.4-7) and (17.4-8) should be used together with an FFT to compute the a_n. From the Fourier expansions it is evident that Theorem 14.2f can be replaced by

$$\|\mathcal{K}_\mu \phi\| \leqq \frac{2\mu^2}{1-\mu^2} \|\phi\|,$$

$$\|\mathcal{G}_\mu \phi\| \leqq \frac{2\mu}{1-\mu^2} \|\phi\|.$$

II. The Theodorsen–Garrick equations

Let D be a doubly connected region bounded on the outside by a Jordan curve Γ_0 and on the inside by a Jordan curve Γ_1. We assume that both Γ_0 and Γ_1 are piecewise analytic, that the origin lies in the interior of Γ_1, and that both Γ_0 and Γ_1 are starlike with respect to 0. The curves Γ_j thus possess parametric representations

$$z = \rho_j(\phi)\, e^{i\phi}, \qquad 0 \leqq \phi \leqq 2\pi, \tag{17.4-13}$$

where the functions ρ_j are positive and piecewise analytic.

Let the annulus A: $\mu < |w| < 1$ be conformally equivalent to D, and let g be a map from A to D. We know that g possesses a continuous extension from the closure A' of A to the closure D' of D. We thus may define the outer and the inner (inverse) boundary correspondence functions ϕ_0 and ϕ_1 as two continuous functions satisfying

$$\phi_0(\theta) = \arg g(e^{i\theta}),$$

$$\phi_1(\theta) = \arg g(\mu\, e^{i\theta}), \qquad 0 \leqq \theta \leqq 2\pi.$$

For $w \in A'$, let

$$h(w) := \log \frac{g(w)}{w}.$$

We know from § 17.1 that h can be defined as an analytic function in A. It can be extended continuously to A' with the boundary values

$$h(e^{i\theta}) = \operatorname{Log} \rho_0(\phi_0(\theta)) + i[\phi_0(\theta) - \theta]$$

and, if the undetermined additive multiple of 2π in ϕ_1 is chosen appropriately,

$$h(\mu\, e^{i\theta}) = \operatorname{Log} \rho_1(\phi_1(\theta)) - \operatorname{Log} \mu + i[\phi_1(\theta) - \theta], \qquad 0 \leqq \theta \leqq 2\pi.$$

By the general form of Cauchy's theorem, the integral

$$\frac{1}{2\pi i} \int_{\Gamma_\rho} \frac{h(w)}{w}\, dw$$

has the same value along all circles Γ_ρ: $w=\rho\, e^{i\theta}, 0\leqq\theta\leqq 2\pi$, for $\mu\leqq\rho\leqq 1$. Thus in particular

$$\frac{1}{2\pi}\int_0^{2\pi} h(e^{i\theta})\,d\theta=\frac{1}{2\pi}\int_0^{2\pi} h(\mu\, e^{i\theta})\,d\theta,$$

which on separating the real and imaginary parts yields

$$\operatorname{Log}\mu=\frac{1}{2\pi}\int_0^{2\pi}\operatorname{Log}\frac{\rho_1(\phi_1(\theta))}{\rho_0(\phi_0(\theta))}\,d0 \tag{17.4-14}$$

and

$$\int_0^{2\pi}[\phi_0(\theta)-\theta]\,d\theta=\int_0^{2\pi}[\phi_1(\theta)-\theta]\,d\theta. \tag{17.4-15}$$

Obviously, the function g is not uniquely determined by D, because together with g also the function g_γ: $w\to g_\gamma(w):=g(e^{i\gamma}w)$ has the required mapping properties for any real γ. We now show that it is possible to normalize g uniquely by choosing γ such that both integrals (17.4-15) vanish. Let γ be any real number, and let γ_0 be congruent to γ modulo 2π, $0\leqq\gamma_0<2\pi$. If g has the boundary correspondence functions (17.4-13), then for arbitrary integers k, g_γ has the boundary correspondence functions

$$\phi_{\gamma,j}(\theta)=\begin{cases}\phi_j(\theta+\gamma_0)+2\pi k, & 0\leqq\theta\leqq 2\pi-\gamma_0\\ \phi_j(\theta+\gamma_0-2\pi)+2\pi(k+1), & 2\pi-\gamma_0\leqq\theta\leqq 2\pi,\end{cases}$$

$j=0, 1$. Let

$$\eta:=\int_0^{2\pi}\phi_0(\theta)\,d\theta=\int_0^{2\pi}\phi_1(\theta)\,d\theta.$$

Then

$$\int_0^{2\pi}\phi_{\gamma,j}(\theta)\,d\theta=\eta+4\pi^2k+2\pi\gamma_0.$$

For this to equal

$$\int_0^{2\pi}\theta\,d\theta=2\pi^2,$$

it is necessary and sufficient that

$$\gamma_0+2\pi k=\frac{2\pi^2-\eta}{2\pi},$$

a relation that uniquely determines both γ_0 and k.

We assume that γ and the boundary correspondence functions have been chosen in the above manner and omit the subscript γ. Because now both functions

$$\beta_0(\theta) := \operatorname{Im} h(e^{i\theta}) = \phi_0(\theta) - \theta,$$

$$\beta_1(\theta) := \operatorname{Im} h(\mu\, e^{i\theta}) = \phi_1(\theta) - \theta$$

have mean value zero, the fomulas (17.4-9) are applicable. Also, the constant $-\operatorname{Log} \mu$ may be dropped from $h(\mu\, e^{i\theta})$ because $\mathscr{H}_\mu$ and $\mathscr{G}_\mu$ applied to constant functions yield zero. Altogether this yields:

THEOREM 17.4a (Theodorsen–Garrick integral equations). *Let D be a doubly connected region with modulus μ^{-1}, bounded by piecewise analytic Jordan curves Γ_0, Γ_1, both starlike with respect to 0 and given in the form* (17.4-13), *and let the function g mapping the annulus A: $\mu < |w| < 1$ onto D be normalized such that it has boundary correspondence functions ϕ_0 and ϕ_1 satisfying*

$$\int_0^{2\pi} [\phi_0(\theta) - \theta]\, d\theta = \int_0^{2\pi} [\phi_1(\theta) - \theta]\, d\theta = 0. \tag{17.4-16}$$

Then there holds the system of equations

$$\begin{aligned} \phi_0 - \theta &= \mathscr{H}_\mu \operatorname{Log} \rho_0(\phi_0) - \mathscr{G}_\mu \operatorname{Log} \rho_1(\phi_1), \\ \phi_1 - \theta &= \mathscr{G}_\mu \operatorname{Log} \rho_0(\phi_0) - \mathscr{H}_\mu \operatorname{Log} \rho_1(\phi_1), \end{aligned} \tag{17.4-17}$$

where $\mathscr{H}_\mu$ and $\mathscr{G}_\mu$ are defined by (17.4-10)–(17.4-12).

It follows as in the case of simply connected regions by an application of the principle of the argument that (ϕ_0, ϕ_1) is the *only* pair of continuous functions that satisfies the above equations.

In the numerical solution of system (17.4-17) there now is the difficulty that the number μ is unknown. However, this number can be determined from ϕ_0 and ϕ_1 via (17.4-14). This suggests the following iterative procedure for computing a sequence $\{\phi_0^{(m)}, \phi_1^{(m)}\}$ converging to the exact solution (ϕ_0, ϕ_1) of (17.4-17):

(a) Choose $\phi_0^{(0)}, \phi_1^{(0)}$ satisfying (17.4-16), for instance,

$$\phi_0^{(0)}(\theta) = \theta, \qquad \phi_1^{(0)}(\theta) = \theta.$$

(b) If the approximations $\phi_0^{(m)}, \phi_1^{(m)}$ are known, compute

$$\mu = \mu^{(m)} := \exp\left\{\frac{1}{2\pi}\int_0^{2\pi} \operatorname{Log} \frac{\rho_1(\phi_1^{(m)}(\theta))}{\rho_0(\phi_0^{(m)}(\theta))}\, d\theta\right\} \tag{17.4-18a}$$

(this is the geometric mean of the function $\rho_1(\phi_1)/\rho_0(\phi_0)$),

$$\phi_0^{(m+1)} - \theta = \mathcal{H}_\mu \operatorname{Log} \rho_0(\phi_0^{(m)}) - \mathcal{G}_\mu \operatorname{Log} \rho_1(\phi_1^{(m)}), \quad (17.4\text{-}18\text{b})$$

$$\phi_1^{(m+1)} - \theta = \mathcal{G}_\mu \operatorname{Log} \rho_0(\phi_0^{(m)}) - \mathcal{H}_\mu \operatorname{Log} \rho_1(\phi_1^{(m)}). \quad (17.4\text{-}18\text{c})$$

The convergence of the iteration as $m \to \infty$ may be studied as in § 16.8. Again the convergence proof requires that the functions $\rho_j(\phi)$ satisfy an ε condition with sufficiently small ε ($\varepsilon < 1$ is not sufficient here). If such an ε condition is violated, it may be possible, as it was in the case of the simple Theodorsen method, to secure convergence by underrelaxation.

The numerical evaluation of the operators $\mathcal{H}_\mu$ and $\mathcal{G}_\mu$ is done by means of Fourier series. At each step, numerical values of the Fourier coefficients $a_{jk}^{(m)}$ in the expansions

$$\operatorname{Log} \rho_j(\phi_j^{(m)}(\theta)) = \sum_{k=-\infty}^{\infty} a_{jk}^{(m)} e^{ik\theta}, \qquad j = 0, 1,$$

are required. Normally these values will be obtained by means of an FFT. In special situations, for example, when the derivatives ρ_j' are discontinuous, other methods may be more advantageous. It will be noted that the integrals

$$\frac{1}{2\pi} \int_0^{2\pi} \operatorname{Log} \rho_j(\phi_j^{(m)}(\theta))\, d\theta, \qquad j = 0, 1,$$

that are required to compute $\mu^{(m)}$, just equal $a_{j0}^{(m)}$, and thus need not be calculated separately. In fact we have

$$\mu^{(m)} = \exp(a_{10}^{(m)} - a_{00}^{(m)}). \quad (17.4\text{-}19)$$

PROBLEMS

1. Discuss the convergence of the iteration (17.4-18) under the simplifying assumption that the modulus μ is known exactly, $\mu^{(m)} = \mu$. In particular if

$$\phi_j(\theta) = \theta + \delta_j(\theta),$$

$$\boldsymbol{\delta} := \begin{pmatrix} \delta_0 \\ \delta_1 \end{pmatrix},$$

$$|\boldsymbol{\delta}| := \|\delta_0\| + \|\delta_1\|,$$

show that

$$|\boldsymbol{\delta}^{(m+1)} - \boldsymbol{\delta}| \leqq \frac{1+\mu}{1-\mu} \varepsilon |\boldsymbol{\delta}^{(m)} - \boldsymbol{\delta}|, \qquad m = 0, 1, 2, \ldots.$$

Conclude that the process converges if

$$\varepsilon < \frac{1-\mu}{1+\mu}.$$

2. Formulate Timman's method for doubly connected regions.

NOTES

The Theodorsen–Garrick method is due to Garrick [1936], an aeronautical engineer. (In those days, people were flying biplanes, hence the interest in the doubly connected case!). The convergence of the iteration (17.4-8) is discussed by Schauer [1963]; see Gaier [1964], pp. 194–207. Hammerschick [1966] discusses the discretization of Garrick's equation. For the implementation of the operators $\mathscr{H}_\mu$ and $\mathscr{G}_\mu$ by means of Fourier series and the FFT, see Ives [1976] or Henrici [1979a]; successful applications are shown by Goedbloed [1981]. Halsey [1979] gives an extension of Timman's method to doubly connected regions. Fornberg's method (see § 16.9) is similarly extended by Fornberg [1982], and Wegmann's method by Wegmann [1985], with impressive numerical results.

§ 17.5. THE SCHWARZ–CHRISTOFFEL MAP FOR DOUBLY CONNECTED POLYGONAL REGIONS

The Schwarz–Christoffel formula (see § 5.12 and § 16.10) affords an explicit representation of the conformal map g of the unit disk E onto an arbitrary *simply connected* region whose boundary is *polygonal* in the sense that it consists of a finite number of straight line segments. Here we wish to construct an analogous formula for the map of a nondegenerate annulus A onto a *doubly connected polygonal region D.*

Let the outer and the inner polygonal boundary curves of D be denoted by Γ_0 and Γ_1, respectively. For the time being, we assume that the corners of the boundary curves all are finite. Let the m vertices of Γ_0 be $z_{01}, z_{02}, \ldots, z_{0m}$, and let the n vertices of Γ_1 be $z_{11}, z_{12}, \ldots, z_{1n}$. The angle of Γ_j at z_{jk}, measured from the interior of D, is denoted by $\pi\alpha_{jk}$, where $0 < \alpha_{jk} \leqq 2$. The relations

$$\sum_{k=1}^{m} \alpha_{0k} = m-2, \qquad \sum_{k=1}^{n} \alpha_{1k} = n+2 \tag{17.5-1}$$

then hold (see Fig. 17.5a).

Let A: $\mu < |w| < 1$ be an annulus that is conformally equivalent to D, and let g be a map from A to D that takes $|w| = 1$ into the outer boundary of D. The pre-image or **prevertex** of z_{jk} is denoted by $w_{jk}, j = 0, 1; k = 1, 2, \ldots$. Thus all $|w_{0k}| = 1$, and all $|w_{1k}| = \mu$.

To find the desired representation of g, we recall one derivation of the Schwarz-Christoffel formula for simply connected polygonal regions D. Thus let us assume that the inner boundary Γ_1 is absent. The Schwarz reflection principle permits us to conclude that near each prevertex w_{0k} of a corner of the outer boundary,

$$g(w) - z_{0k} = \left(1 - \frac{w}{w_{0k}}\right)^{\alpha_{0k}} h_k(w),$$

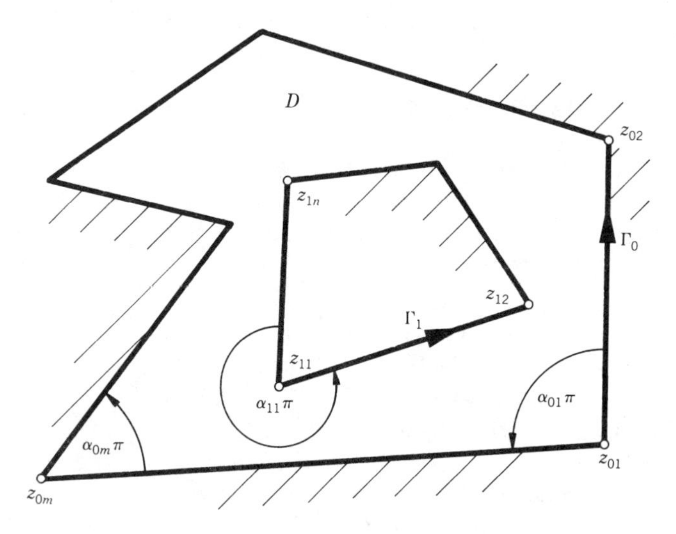

Fig. 17.5a. Doubly connected polygonal region.

where h_k is analytic at w_{0k}, $h(w_{0k}) \neq 0$. Thus the function

$$h(w) := g'(w) \prod_{k=1}^{m} \left(1 - \frac{w}{w_{0k}}\right)^{1-\alpha_{0k}}$$

is analytic and $\neq 0$ in the closed unit disk. We wish to show that h is constant and to this end compute arg $h(w)$ for $|w| = 1$. If w is between two successive w_{0k}, $w = e^{i\phi}$, the point $z(\phi) := g(e^{i\phi})$ moves along a straight line. Therefore the argument of

$$z'(\phi) = i\, e^{i\phi} g'(e^{i\phi})$$

is constant, and there follows

$$\arg g'(e^{i\phi}) = \text{const} - \phi.$$

Furthermore,

$$1 - \frac{w}{w_{0k}} = 1 - e^{i(\phi - \phi_{0k})} = 2ie^{i(\phi - \phi_{0k})/2} \sin[(\phi_{0k} - \phi)/2],$$

and thus

$$\arg \prod_{k=1}^{m} \left(1 - \frac{w}{w_{0k}}\right)^{1-\alpha_{0k}} = \text{const} + \frac{\phi}{2} \sum_{k=1}^{m} (1 - \alpha_{0k}).$$

On account of (17.5-1) there follows

$$\arg h(w) = \text{const} - \phi + \frac{\phi}{2} \sum_{k=1}^{m} (1 - \alpha_{0k}) = \text{const}.$$

Since h is analytic and $\neq 0$ on $|w| \leq 1$, arg h has the same constant value on the whole circle $|w| = 1$. The principle of the argument now permits us to conclude that h itself is a constant, and on integration we find the Schwarz-Christoffel formula in the form

$$g(w) = c \int_{w^*}^{w} \prod_{k=1}^{m} \left(1 - \frac{w}{w_{0k}}\right)^{\alpha_{0k}-1} dw + g(w^*). \tag{17.5-2}$$

If g maps $|w| > \mu$ onto the exterior of Γ_1 such that $g(\infty) = \infty$, essentially the same argument yields the formula

$$g(w) = c \int_{w^*}^{w} \prod_{k=1}^{n} \left(1 - \frac{w_{1k}}{w}\right)^{\alpha_{1k}-1} dw + g(w^*); \tag{17.5-3}$$

see Problems 6-9, § 5.12.

To deal with the case where D is a doubly connected polygonal region, we require an auxiliary function θ depending on a real parameter $\mu \in (0, 1)$ and defined for all $w \neq 0, \infty$ by

$$\theta(w) := \prod_{d} (1 - \mu^d w)(1 - \mu^d w^{-1}). \tag{17.5-4}$$

Here d runs through all *odd* positive integers, $d = 1, 3, 5, \ldots$. This function is related to the θ functions in the theory of elliptic functions. The following self-contained presentation is restricted to the facts that are required in the present context.

Because the product (17.5-4) converges absolutely on every compact subset of $0 < |w| < \infty$, θ is analytic on that set, with simple zeros at the points $w = \mu^{\pm 1}, \mu^{\pm 3}, \mu^{\pm 5}, \ldots$. We note that for all real ϕ,

$$\theta(e^{i\phi}) = \prod_{d} (1 - 2\mu^d \cos \phi + \mu^{2d}) > 0.$$

Thus on $|w| = 1$, $\theta(w)$ has the continuous argument

$$\arg \theta(e^{i\phi}) = 0. \tag{17.5-5}$$

We assert that an analytic logarithm of $\theta(w)$ can be defined in the annulus R: $\mu < |w| < \mu^{-1}$. Since $\theta(w) \neq 0$ in R, this by Theorem 15.1d will be the case if

$$\int_{|w|=1} d(\arg \theta(w)) = 0.$$

This however is an immediate consequence of (17.5-5).

We note the relations

$$\theta(\mu^{-1}w) = \prod_d (1-\mu^{d-1}w)(1-\mu^{d+1}w^{-1})$$

$$= (1-w)\prod_v (1-\mu^v w)(1-\mu^v w^{-1}) \tag{17.5-6a}$$

where v runs through the *even* positive integers, and similarly

$$\theta(\mu w) = \prod_d (1-\mu^{d+1}w)(1-\mu^{d-1}w^{-1})$$

$$= (1-w^{-1})\prod_v (1-\mu^v w)(1-\mu^v w^{-1}). \tag{17.5-6b}$$

As a consequence, the functions $\theta(\mu^{\pm 1}w)$ have on $|w|=1$, $w \neq 1$, the continuous arguments

$$\arg\theta(\mu^{-1}e^{i\phi}) = \arg(1-e^{i\phi}) + \arg(\text{positive}) = \frac{\phi}{2} - \frac{\pi}{2}, \tag{17.5-7a}$$

$$\arg\theta(\mu e^{i\phi}) = \arg(1-e^{-i\phi}) + \arg(\text{positive}) = -\frac{\phi}{2} + \frac{\pi}{2}, \tag{17.5-7b}$$

where $0 < \phi < 2\pi$.

After these preparations, let g be a map from A to the doubly connected polygonal region D described initially. It follows as in the simply connected case that at each prevertex w_{0k},

$$g(w) - z_{0k} = \left(1 - \frac{w}{w_{0k}}\right)^{\alpha_{0k}} h_{0k}(w),$$

where h_{0k} is analytic and $\neq 0$ in the closure of A, and at each prevertex w_{1k},

$$g(w) - z_{1k} = \left(1 - \frac{w_{1k}}{w}\right)^{\alpha_{1k}} h_{1k}(w),$$

where h_{1k} has similar properties. Thus we conclude as in the simply connected case that

$$h_0(w) := g'(w) \prod_{k=1}^{m} \left(1 - \frac{w}{w_{0k}}\right)^{1-\alpha_{0k}} \prod_{k=1}^{n} \left(1 - \frac{w_{1k}}{w}\right)^{1-\alpha_{1k}}$$

is analytic and $\neq 0$ in the closure of A. We next would try to show that h_0 is a constant. This does not work, however, because $\arg h_0$ is not constant on the boundary of A. However, the method works if the terms

$$1 - \frac{w}{w_{0k}} \quad \text{and} \quad 1 - \frac{w_{1k}}{w}$$

are replaced by the functions

$$\theta\left(\frac{w}{\mu w_{0k}}\right) \quad \text{and} \quad \theta\left(\frac{\mu w}{w_{1k}}\right),$$

respectively, which by (17.5-6) likewise have simple zeros at the prevertices and otherwise are analytic and $\neq 0$ on the closure of A. Moreover, since for $w \in A$ the arguments $\mu^{-1} w_{0k}^{-1} w$ and $\mu w_{1k}^{-1} w$ lie in the annulus R described earlier, analytic logarithms, and hence analytic fractional powers, can be defined for both functions. Thus let

$$h(w) := g'(w) \prod_{k=1}^{m} \left[\theta\left(\frac{w}{\mu w_{0k}}\right)\right]^{1-\alpha_{0k}} \prod_{k=1}^{n} \left[\theta\left(\frac{\mu w}{w_{1k}}\right)\right]^{1-\alpha_{1k}}.$$

Near each prevertex $h(w)$ differs from $h_0(w)$ only by an analytic nonzero factor. It follows that h again is analytic and $\neq 0$ in the closure of A.

We wish to show that h is constant and to this end compute $\arg h(w)$ on the boundary of A. If $w = e^{i\phi}$, and if ϕ varies between two consecutive values of ϕ_{0k}, the fact that $g(e^{i\phi})$ travels along a straight line implies as in the simply connected case

$$\arg g'(e^{i\phi}) = \text{const} - \phi.$$

Thus from (17.5-7a) and (17.5-5) there follows

$$\arg h(e^{i\phi}) = \text{const} - \phi + \frac{\phi}{2} \sum_{k=1}^{m} (1 - \alpha_{0k}),$$

which is constant on account of (17.5-1). Because $\arg h(e^{i\phi})$ is continuous for all ϕ, the constant is the same on all arcs between consecutive w_{0k}. Similarly if $w = \mu e^{i\phi}$, where ϕ varies between any two consecutive ϕ_{1k},

$$\arg g'(\mu e^{i\phi}) = \text{const} - \phi,$$

and thus from (17.5-5) and (17.5-7b),

$$\arg h(\mu e^{i\phi}) = \text{const} - \phi - \frac{\phi}{2} \sum_{k=1}^{n} (1 - \alpha_{1k}),$$

which by the second relation (17.5-1) is again constant. As above, the constant is the same for all ϕ.

To conclude that $h(w) = \text{const}$, we now invoke the principle of the argument, which states that the number of times h assumes a value z^* in A equals the difference of the winding numbers with respect to z^* of the closed curves Λ_0: $z = h(e^{i\phi})$ and Λ_1: $z = h(\mu e^{i\phi})$, $0 \leqq \phi \leqq 2\pi$. By the foregoing results about $\arg h$, each curve Λ_j lies entirely on a ray $\arg z = \lambda_j$.

If h were not constant, it would have to assume values z^* that lie on neither ray; however, the winding number of each Λ_j with respect to any such z^* is obviously zero. Thus h is constant, and on integration we find:

THEOREM 17.5a. *Let D be a doubly connected polygonal region with the finite vertices $z_{01}, \ldots, z_{0m}$ on the exterior boundary polygon and vertices $z_{11}, \ldots, z_{1n}$ on the interior boundary polygon, and let the interior angle at z_{jk} be $\pi\alpha_{jk}$, where $0 < \alpha_{jk} \leqq 2$. Any map g from the conformally equivalent annulus A: $\mu < |w| < 1$ to D for any w^* in the closure of A satisfies*

$$g(w) - g(w^*) = c \int_{w^*}^{w} \prod_{k=1}^{m} \left\{\theta\left(\frac{w}{\mu w_{0k}}\right)\right\}^{\alpha_{0k}-1} \prod_{k=1}^{n} \left\{\theta\left(\frac{\mu w}{w_{1k}}\right)\right\}^{\alpha_{1k}-1} dw. \qquad (17.5\text{-}8)$$

Here c is a complex constant, w_{0k} and w_{1k} are the preimages of z_{0k} and z_{1k}, respectively, and the function θ is defined by (17.5-4).

We mention without proof that Theorem 17.5a also holds if one or several vertices z_{0k} are at ∞, provided that the corresponding angles are measured as indicated in Theorem 5.12e.

In practical applications of formula (17.5-8) there is the difficulty, familiar already from the classical Schwarz-Christoffel map, that for a given polygonal region D the location of the points w_{jk} is a priori unknown. (From the general theory of maps of doubly connected regions, only one of these w_{jk} may be selected arbitrarily.) The difficulty is confounded by the fact that the integrand in (17.5-8) is more complicated than in the classical formula. For the simply connected case we have described in § 16.10 a number of devices, due to Trefethen [1980], that jointly afford a fully practical numerical solution of the parameter problem. There is no discernible reason why these devices should not also work in the doubly connected case, and in fact Daeppen [1984] has constructed a computer program effecting the doubly connected Schwarz-Christoffel map along these lines. Figures 17.5b and 17.5c show a map obtained by Daeppen's program. The region being mapped is bounded on the inside by a polygon with four vertices, and on the outside by a nonconvex polygon with eight vertices. There are no symmetries, and the angles are no obvious rational multiples of π. The inverse modulus of the annular image (Fig. 17.5c) has been determined as $\mu = 0.38094061$. Also shown are the inverse images of some lines arg $w = $ const and $|w| = $ const; in the electrostatic interpretation these are, respectively, the field lines and the potential lines if the inner and outer boundaries of the region are kept at the potentials 1 and 0, say.

In some cases, the parameter problem is alleviated by considerations of symmetry.

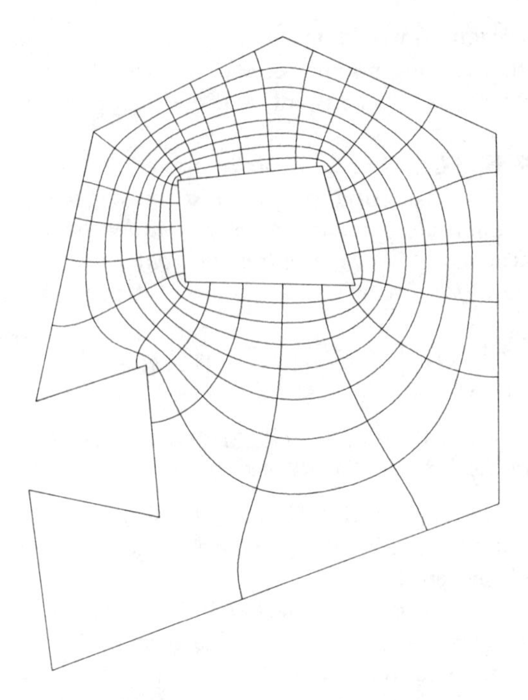

Fig. 17.5b. Doubly connected polygonal region.

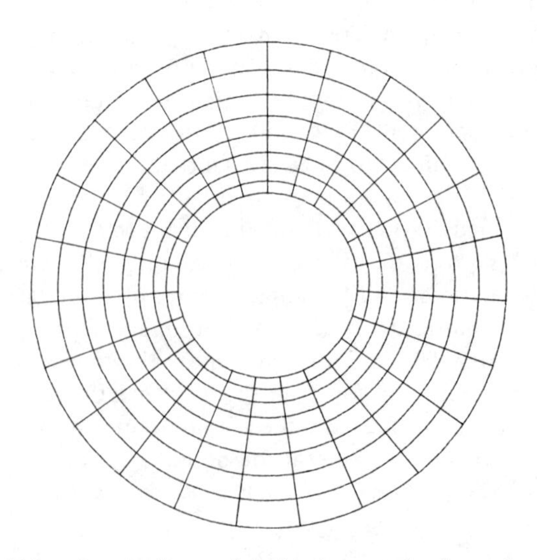

Fig. 17.5c. Annular image of doubly connected polygonal region.

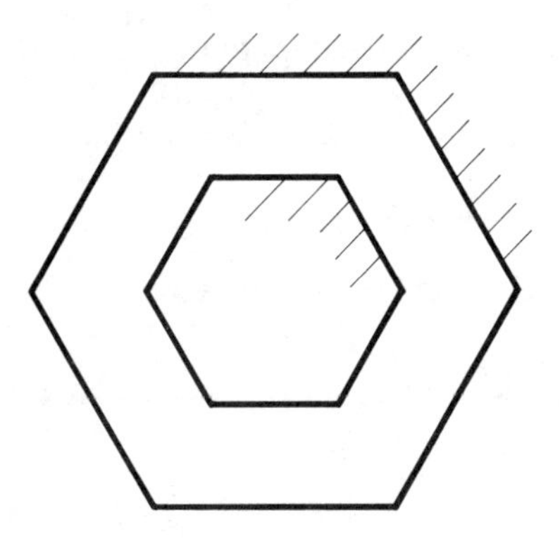

Fig. 17.5d. Region bounded by concentric 6-gons.

EXAMPLE 1

Let D be a region bounded by two regular n-gons, centered at 0 and positioned such that the vertices lie on common rays emanating from 0 (see Fig. 17.5d for $n=6$). Here all $\alpha_{0k}=1-2/n$ and all $\alpha_{1k}=1+2/n$. By symmetry, the prevertices are

$$w_{0k}=w_n^k \qquad \text{and} \qquad w_{1k}=\mu w_n^k, \qquad k=1,2,\ldots,n,$$

where $w_n := \exp(2\pi i/n)$ and μ is the modulus of D. Denoting the functions θ by $\theta(w, \mu)$ in order to indicate their dependence on μ, the products in (17.5-8) simplify on account of the identities

$$\prod_{k=1}^{n} \theta\left(\frac{w}{\mu w_n^k}, \mu\right) = \theta\left(\frac{w^n}{\mu^n}, \mu^n\right),$$

$$\prod_{k=1}^{n} \theta\left(\frac{w}{w_n^k}, \mu\right) = \theta(w^n, \mu^n),$$

which are consequences of the fact that

$$\left(1-\rho\frac{w}{w_n}\right)\left(1-\rho\frac{w}{w_n^2}\right)\cdots\left(1-\rho\frac{w}{w_n^n}\right)=1-\rho^n w^n.$$

Thus the mapping function for D satisfies

$$g(w)-g(w^*)=c\int_{w^*}^{w}\left[\frac{\theta(w^n, \mu^n)}{\theta(\mu^{-n}w^n, \mu^n)}\right]^{2/n} dw,$$

and μ is the only parameter that remains to be determined numerically in this case. —

Any numerical work with the formula (17.5-8) requires the efficient evaluation of the function θ. This is facilitated by some classical formulas which are mentioned in volume II, but which we here adapt to the notation on hand.

In terms of the function $g(z, t)$ defined by (8.2-7), we obviously have

$$\theta(w, \mu) = g(\mu, -w).$$

By Jacobi's triple product identity (Theorem 8.2b) there follows

$$\theta(w, \mu) = \frac{1}{q(\mu^2)} \sum_{k=-\infty}^{\infty} \mu^{k^2}(-w)^k, \tag{17.5-9}$$

where

$$q(\mu^2) := \prod_{k=1}^{\infty} (1 - \mu^{2k}).$$

For values of μ not very close to 1, the series (17.5-9) converges rapidly. For instance, if $\mu = \frac{1}{2}$, the terms where $|k| \leqq 3$ suffice to evaluate the series with a truncation error $< 10^{-16}$ for any w such that $\mu \leqq |w| \leqq 1$. When using the series in formula (17.5-8), the factors $q(\mu^2)$, being independent of w, may be pulled before the integral sign. In fact, since the total exponent of $q(\mu^2)$ is

$$\sum_{k=1}^{m} (\alpha_{0k} - 1) + \sum_{k=1}^{n} (\alpha_{1k} - 1) = 0,$$

these factors may be omitted entirely, and the function θ may be replaced by $\sigma(-w, \mu)$, where

$$\sigma(w, \mu) = \sum_{k=-\infty}^{\infty} \mu^{k^2} w^k = 1 + \sum_{k=1}^{\infty} \mu^{k^2}(w^k + w^{-k}). \tag{17.5-10}$$

For values of μ close to 1 the convergence of the series (17.5-10) still may be slow. In this case, yet another of Jacobi's identities may be used which earlier was given as (10.6-23). In the present notation it reads as follows. For $0 < \mu < 1$, $w \neq 0$, let $\lambda > 0$ and $t \in \mathbb{C}$ be such that

$$\mu = e^{-\pi\lambda}, \qquad w = e^{it}.$$

Set $u := e^{t\lambda^{-1}}$, $\nu := e^{-\pi\lambda^{-1}}$. Then

$$\sigma(w, \mu) = \frac{1}{\sqrt{\lambda}} e^{-t^2/4\pi\lambda} \sigma(u, \nu). \tag{17.5-11}$$

Since $\operatorname{Log} \mu \cdot \operatorname{Log} \nu = \pi^2$, the function σ thus may be evaluated by means of a series (17.5-10) where $\mu \leqq e^{-\pi} = 0.043214$. Only the terms where $|k| \leqq 3$ are required to achieve a truncation error $< 10^{-16}$.

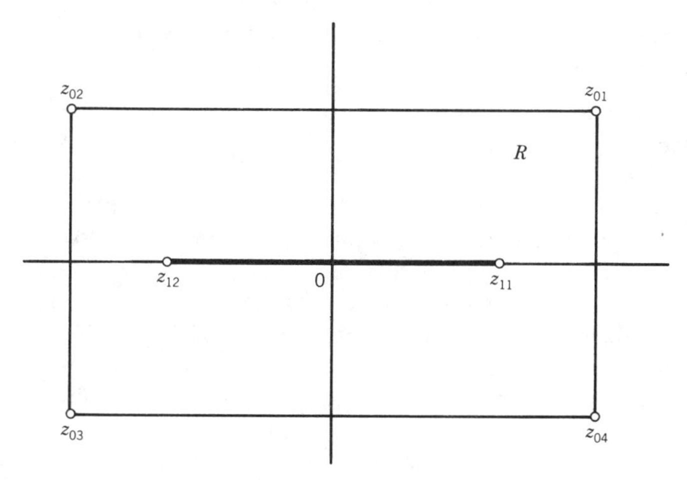

Fig. 17.5e. Rectangle with slit.

PROBLEMS

1. Let R be a rectangle with a slit located on one axis of symmetry and symmetric with respect to the other (see Fig. 17.5e). We have $z_{11} = -z_{12} = \overline{z_{11}}$, $z_{01} = -z_{03} = \overline{z_{04}} = -\overline{z_{02}}$. Assuming

$$w_{11} = \mu, \qquad w_{12} = -\mu,$$

and

$$w_{01} = e^{i\beta}, \qquad w_{02} = -e^{-i\beta}, \qquad w_{03} = -e^{i\beta}, \qquad w_{04} = e^{-i\beta},$$

where $0 < \beta < \pi/2$, show that the map from A to R satisfies

$$g(w) - g(w^*) = c \int_{w^*}^{w} \frac{\theta(w^2, \mu^2)}{[\theta(\mu^{-2}w^2e^{2i\beta}, \mu^2)\theta(\mu^{-2}w^2e^{-2i\beta}, \mu^2)]^{1/2}}\, dw.$$

2. Show directly from the expansion (17.5-9) that

$$\theta(\mu^{2n+1}, \mu) = 0$$

for all integers n.

3. If μ is sufficiently small, the series (17.5-10) converges so rapidly that in numerical computation it may be replaced by 1. Working in d-digit floating arithmetic, how close to 1 must μ be in order that (17.5-11) may be replaced by

$$\sigma(w, \mu) \approx \frac{1}{\sqrt{\lambda}} e^{-t^2/4\pi\lambda}\,?$$

NOTES

Theorem 17.5a is given with a different proof by Komatu [1945]; see also Koppenfels and Stallmann [1959].

§ 17.6. ARBITRARY CONNECTIVITY: EXISTENCE OF THE CIRCULAR MAP

Last in this chapter we study standardized conformal maps of regions R of connectivity $n \geqq 2$. One such map was already presented in § 17.1, where it was shown that such a region can always be mapped onto a circular annulus cut along $n-2$ circular arcs. Here now we show that such a region R can always be mapped onto the complement (with respect to the extended complex plane) of n closed circular disks (possibly points). Such regions will be called **circular of connectivity** n. (In this terminology, the circular regions considered in § 5.3 are circular of connectivity 1.) Without loss of generality, we may assume $\infty \in R$. The mapping function may then be normalized by the condition that its Laurent series at ∞ begins thus:

$$f(z) = z + O\left(\frac{1}{z}\right). \tag{17.6-1}$$

Our prime object is the proof of:

THEOREM 17.6a. *Let R be a region of connectivity $n \geqq 2$ in the extended complex plane such that $\infty \in R$. Then there exist a unique circular region C of connectivity n and a unique one-to-one analytic function f in R satisfying* (17.6-1), *such that $f(R) = C$.*

The function f of Theorem 17.6a is called the **circular map** of R, and the region C is the **normalized circular image** of R.

The following *proof* of Theorem 17.6a is due to Schiffer and Hawley [1962]. It combines ideas from potential theory and from the calculus of variations. A general outline of the proof is as follows. Assuming that a circular map exists, we derive a minimum property which the map necessarily satisfies. We then prove the unique existence of a function having the required minimum property, and we show that it defines a circular map.

Without loss of generality, the region R may be assumed to have the following properties:

(i) No component of the complement of R reduces to a point (otherwise, add it to R and omit its image from the image of R).

(ii) The boundary of R consists of n analytic Jordan curves Γ_i, $i = 1, 2, \ldots, n$ (see § 17.1). We assume each Γ_i to be parametrized by $z = z_i(\tau)$, where τ is the arc length on Γ_i, and the orientation is such that R lies to the left of Γ_i. If λ_i denotes the length of Γ_i, the parametric interval for Γ_i is chosen to be $[\tau_i, \tau_{i+1}]$, where

$$\tau_1 := 0, \qquad \tau_{i+1} := \tau_i + \lambda_i, \qquad i = 1, 2, \ldots, n. \tag{17.6-2}$$

We also set $\Gamma := \Gamma_1 + \Gamma_2 + \cdots + \Gamma_n$, $\lambda := \lambda_1 + \lambda_2 + \cdots + \lambda_n$, and we call $z(\tau)$ the union of all functions $z_i(\tau)$.

Let now f be a circular map of R. We do not yet require the normalization (17.6-1). By Theorem 16.4a, f may be extended to a function that is analytic on the closure of R. By hypothesis, the images of the Γ_i, that is, the curves

$$\Lambda_i: \quad w = w_i(\tau) := f(z_i(\tau)), \qquad \tau_i \leqq \tau \leqq \tau_{i+1}, \tag{17.6-3}$$

are circles. Note that, in general, τ is not the arc length on Λ_i.

We recall the intuitive (however rigorous) definition of curvature as the *speed by which the tangent turns* if the point of tangency runs along the curve with speed 1. Thus on Γ the curvature at a point $t := z(\tau)$ is

$$\begin{aligned}\kappa(t) &= \frac{d}{d\tau} \arg z'(\tau) = \frac{d}{d\tau} \operatorname{Im} \log z'(\tau)\\ &= \operatorname{Im} \frac{d}{d\tau} \log z'(\tau)\\ &= \operatorname{Im} \frac{z''(\tau)}{z'(\tau)}.\end{aligned}$$

Since $|z'(\tau)| = 1$, $\operatorname{Re} \log z'(\tau) = \operatorname{Log}|z'(\tau)| = 0$, therefore $\operatorname{Re}(d/d\tau) \log z'(\tau) = 0$, and we in fact have

$$\kappa(t) = \frac{1}{i} \frac{z''(\tau)}{z'(\tau)}. \tag{17.6-4}$$

If $\kappa^*(s)$ denotes the curvature of Λ at the point $s = w(\tau)$ we similarly have, denoting the arc length on Λ by σ and writing $\tau = \tau(\sigma)$,

$$\begin{aligned}\kappa^*(s) &= \frac{d}{d\sigma} \arg w'(\tau(\sigma))\\ &= \frac{d}{d\sigma} \arg\{f'(z(\tau)) z'(\tau)\}\\ &= \frac{d}{d\tau}\{\arg f'(z(\tau)) + \arg z'(\tau)\} \frac{d\tau}{d\sigma}.\end{aligned}$$

On account of $d\sigma/d\tau = |f'(z(\tau))|$, there follows

$$|f'(z(\tau))| \kappa^*(s) = \frac{d}{d\tau} \arg f'(z(\tau)) + \kappa(t).$$

By the Cauchy–Riemann equations (see the proof of Observation 15.1e), the derivative of the imaginary part of an analytic function in the direction of the tangent equals the derivative of the real part in the direction of the normal. Thus

$$|f'(t)| \kappa^*(s) = \frac{d}{dn} \operatorname{Log}|f'(t)| + \kappa(t) \tag{17.6-5}$$

at all points $t \in \Gamma$. Because the curves Λ_i are circles, $\kappa^*(s) = \text{const}$ on each Λ_i, and because the tangent vector turns by -2π in time $2\pi\rho_i$, where ρ_i is the radius of the circle Λ_i, the value of the constant is $-1/\rho_i$. We thus find from (17.6-5)

$$\frac{\partial}{\partial n} \operatorname{Log}|f'(t)| = -\frac{1}{\rho_i}|f'(t)| - \kappa(t), \qquad t \in \Gamma_i, \quad i = 1, 2, \ldots, n. \quad (17.6\text{-}6)$$

We now assume f to be normalized by the condition

$$\int_\Gamma \operatorname{Log}|f'(t)||dt| = 0, \quad (17.6\text{-}7)$$

which can always be achieved by multiplying f by a suitable nonzero constant. In this case (17.6-6) and (17.6-7) express the fact that the function

$$u(z) := \operatorname{Log}|f'(z)|$$

is the normalized solution of a Neumann problem for R with boundary data

$$\frac{\partial}{\partial n} u(t) = -\frac{1}{\rho_i} e^{u(t)} - \kappa(t), \qquad t \in \Gamma_i.$$

Denoting by $\nu(z, t)$ the Neumann function for R, we thus have by the representation formula (15.7-16)

$$u(z) = -\sum_{i=1}^{n} \int_{\Gamma_i} \nu(z, t)\left[\frac{1}{\rho_i} e^{u(t)} + \kappa(t)\right]|dt|. \quad (17.6\text{-}8)$$

In (17.6-8) the function

$$k(z) := -\int_\Gamma \nu(z, t)\kappa(t)\,|dt| \quad (17.6\text{-}9)$$

may, in principle, be regarded as known. By the theory of the Neumann problem (see § 15.7) it is the solution satisfying

$$\int_\Gamma k(z)\,|dz| = 0$$

of the Neumann problem for R with boundary data

$$\frac{\partial k}{\partial n}(t) = -\kappa(t) + \alpha, \qquad t \in \partial R$$

where the constant α is determined by

$$\int_\Gamma (-\kappa(t) + \alpha)\,|dt| = 0.$$

By the geometrical interpretation of curvature,

$$\int_{\Gamma_i} \kappa(t)\,|dt| = -2\pi, \qquad i = 1, 2, \ldots, n.$$

There follows $\alpha = -2\pi n/\lambda$, and the boundary condition satisfied by k is

$$\frac{\partial k}{\partial n}(t) = -\kappa(t) - \frac{2\pi n}{\lambda}, \qquad t \in \partial R. \tag{17.6-10}$$

Using the abbreviation (17.6-9), (17.6-8) may be written

$$u(z) = k(z) - \sum_{i=1}^{n} \frac{1}{\rho_i} \int_{\Gamma_i} \nu(z, t)\, e^{u(t)}\,|dt|, \tag{17.6-11}$$

and we have achieved the first step of our proof, namely, to establish a functional relation which the desired mapping function, or a function closely related to it, necessarily satisfies.

Our task now is to show that there exists a function u harmonic in $R \cup \partial R$, and satisfying both the condition

$$\int_{\Gamma} u(t)\,|dt| = 0$$

corresponding to (17.6-7) and the nonlinear integral equation (17.6-11). We use a variational method. Let Ω denote the space of all real functions v on R such that:

(a) v is harmonic in R and continuous on $R \cup \partial R$.

(b) $$\iint_R |\text{grad } v(z)|^2 \boxed{dz} < \infty.$$

(c) $$\int_{\Gamma} v(t)\,|dt| = 0.$$

Clearly, Ω is a linear space, and thus, since it contains at least the function $v = 0$, it is not empty. On Ω we consider the functional

$$\phi(v) := \frac{1}{2} \iint_R |\text{grad}(v - k)|^2 \boxed{dz} + 2\pi \sum_{i=1}^{n} \text{Log} \int_{\Gamma_i} e^{v(t)}\,|dt|, \tag{17.6-12}$$

where k is defined by (17.6-9). The connection between ϕ and (17.6-11) is as follows. *Suppose* there exists a $u \in \Omega$ minimizing ϕ. Then for any $v \in \Omega$ and any $\varepsilon \in \mathbb{R}$,

$$\phi(u + \varepsilon v) \geqq \phi(u).$$

We now use Euler's classical device in the calculus of variations to consider, for fixed $v \in \Omega$, $v \neq 0$, the function of ε,

$$\theta(\varepsilon) := \phi(u+\varepsilon v) = \frac{1}{2}\iint_R |\text{grad}(u+\varepsilon v - k)|^2 \boxed{dz}$$

$$+2\pi \sum_{i=1}^{n} \text{Log} \int_{\Gamma_i} e^{u+\varepsilon v}\,|dt|. \qquad (17.6\text{-}13)$$

Clearly, θ is differentiable as many times as required. Since θ has a minimum at $\varepsilon = 0$, there must hold

$$\theta'(0) = 0, \qquad \forall v \in \Omega. \qquad (17.6\text{-}14)$$

Expanding the integrands in powers of ε and defining

$$\rho_i := \frac{1}{2\pi}\int_{\Gamma_i} e^{u(t)}\,|dt|, \qquad (17.6\text{-}15)$$

we find by straightforward computation

$$\theta'(0) = \iint_R \text{grad}(u-k)\text{ grad } v \boxed{dz} + \sum_{i=1}^{n} \frac{1}{\rho_i}\int_{\Gamma_i} e^{u(t)} v(t)\,|dt|.$$

It follows that the minimizing function u is such that for every $v \in \Omega$,

$$\iint_R \text{grad}(u-k)\text{ grad } v \boxed{dz} + \sum_{i=1}^{n} \frac{1}{\rho_i}\int_{\Gamma_i} e^{u(t)} v(t)\,|dt| = 0.$$

For $z \in R$, the function $v(t) := \kappa(z, t)$, where κ here denotes the (real) Bergman kernel function of R, is in Ω. With this choice of v we get by the reproducing property of the Bergman kernel function (Theorem 15.7e), noting that $\kappa = \nu$ for $t \in \partial R$,

$$u(z) - k(z) + \sum_{i=1}^{n} \frac{1}{\rho_i}\int_{\Gamma_i} e^{u(t)} \nu(z, t)\,|dt| = 0.$$

For $z \to \partial R$ this reduces to (17.6-11). Note, however, that the constants ρ_i are now defined in terms of u by (17.6-15).

We thus have shown that any u minimizing the functional ϕ satisfies (17.6-11), with suitably chosen constants ρ_i. Euler's technique in this case also allows us to establish the uniqueness of u. Again by straightforward calculation we find for any $\varepsilon \in \mathbb{R}$,

$$\theta''(\varepsilon) = \iint_R |\text{grad } v|^2 \boxed{dz} + \sum_{i=1}^{n} \frac{1}{\rho_i(\varepsilon)}\int_{\Gamma_i} e^{u+\varepsilon v} v^2\,|dt|$$

$$-\frac{1}{2\pi}\sum_{i=1}^{n} \frac{1}{[\rho_i(\varepsilon)]^2}\left\{\int_{\Gamma} e^{u+\varepsilon v} v\,|dt|\right\}^2,$$

where

$$2\pi\rho_i(\varepsilon) := \int_{\Gamma_i} e^{u(t)+\varepsilon v(t)}\,|dt|,$$

so that the ρ_i used earlier equals $\rho_i = \rho_i(0)$. By the Schwarz inequality,

$$2\pi\rho_i(\varepsilon)\int_{\Gamma_i} e^{u+\varepsilon v} v^2\,|dt| \geqq \left\{\int_{\Gamma_i} e^{u+\varepsilon v} v\,|dt|\right\}^2,$$

and since any nonzero $v \in \Omega$ is not constant, there follows

$$\theta''(\varepsilon) > 0, \qquad \varepsilon \in \mathbb{R},$$

for any such v. Assuming the existence of two distinct u, say u_1 and u_2, minimizing ϕ on setting $u := u_1$, $v := u_2 - u_1$ now immediately leads to a contradiction inasmuch as the derivative of the function $\theta'(\varepsilon)$, which vanishes for $\varepsilon = 0$ and for $\varepsilon = 1$, would have to vanish at a point $\varepsilon \in (0, 1)$.

All of the preceding is based on the assumption that a $u \in \Omega$ minimizing ϕ exists. To *prove* the existence of such a u we must show, first of all, that the functional ϕ is bounded from below. By a limiting case of the inequality of the arithmetic and the geometric means,

$$\operatorname{Log}\frac{1}{\lambda_i}\int_{\Gamma_i} e^{v(t)}\,|dt| \geqq \frac{1}{\lambda_i}\int_{\Gamma_i} v(t)\,|dt|,$$

therefore

$$\sum_{i=1}^{n} \operatorname{Log}\int_{\Gamma_i} e^{v(t)}\,|dt| \geqq \sum_{i=1}^{n}\frac{1}{\lambda_i}\int_{\Gamma_i} v(t)\,|dt| + \sum_{i=1}^{n}\operatorname{Log}\lambda_i.$$

The difficulty now is to show that the first term on the right cannot become arbitrarily *negative*. We use the Schwarz inequality to estimate its square from *above*,

$$\begin{aligned}\left\{\sum_{i=1}^{n}\frac{1}{\lambda_i}\int_{\Gamma_i} v(t)\,|dt|\right\}^2 &\leqq \sum_{i=1}^{n}\frac{1}{\lambda_i^2}\sum_{j=1}^{n}\left\{\int_{\Gamma_j} v(t)\,|dt|\right\}^2\\ &\leqq \sum_{i=1}^{n}\frac{1}{\lambda_i^2}\sum_{j=1}^{n}\int_{\Gamma_j}|dt|\int_{\Gamma_j} v^2(t)\,|dt|\\ &\leqq \sum_{i=1}^{n}\frac{1}{\lambda_i^2}\sum_{j=1}^{n}\lambda_j\cdot\int_{\Gamma} v^2(t)\,|dt|.\end{aligned}$$

At this point we use the fact, not proven here, that there exists a constant $\beta > 0$, depending only on R, such that for any function $v \in \Omega$,

$$\int_{\Gamma} v^2(t)\,|dt| \leqq \beta \iint_R |\operatorname{grad} v|^2\,\boxed{dz}.$$

(β^{-1} is the lowest nontrivial *Steklov eigenvalue* of R.) It thus follows that for any $v \in \Omega$,

$$2\pi \sum_{i=1}^{n} \operatorname{Log} \int_{\Gamma_i} e^{v(t)} |dt| \geqq -\alpha_1 \left[\iint_R |\text{grad } v|^2 \boxed{dz} \right]^{1/2} + \alpha_2,$$

where α_1, α_2 are positive numbers depending only on R. There further follows, denoting by $\alpha_3, \alpha_4, \ldots$ similar constants,

$$\begin{aligned} \phi(v) &\geqq \frac{1}{2} \iint_R |\text{grad } v|^2 \boxed{dz} - \alpha_3 \left| \iint_R |\text{grad } v|^2 \boxed{dz} \right|^{1/2} + \alpha_4 \\ &= \frac{1}{2} \left\{ \left[\iint_R |\text{grad } v|^2 \boxed{dz} \right]^{1/2} - \alpha_3 \right\}^2 + \alpha_5 \\ &\geqq \alpha_5. \end{aligned}$$

Having established that ϕ is bounded from below, we conclude that

$$\mu := \inf_{v \in \Omega} \phi(v)$$

is finite. By the definition of an infimum it follows that there exists a sequence $\{u_k\}$, $u_k \in \Omega$, such that

$$\lim_{k \to \infty} \phi(u_k) = \mu.$$

We now would have to show that the sequence $\{u_k\}$, or at least a subsequence of it, converges, in a suitable topology, to a limit element u, and that $u \in \Omega$. We omit this part of the Schiffer–Hawley proof, which requires somewhat technical arguments based on enlarging the space Ω.

Accepting the existence of a unique function $u \in \Omega$ which minimizes ϕ and therefore satisfies the integral equation (17.6-11), we now shall use u to construct the mapping function f of Theorem 17.6a.

To begin with, we show that the harmonic function u possesses a conjugate harmonic function v in R. By Observation 15.1e, this will be the case if

$$\eta_i := \int_{\Gamma_i} \frac{\partial u}{\partial n}(t) \, |dt| = 0, \qquad i = 1, 2, \ldots, n. \tag{17.6-16}$$

By the definition of k, and by the defining property of the Neumann function, (17.6-11) states that u is a solution of the Neumann problem with boundary data

$$\frac{\partial u}{\partial n}(t) = -\kappa(t) - \frac{1}{\rho_i} e^{u(t)}, \qquad t \in \partial R,$$

where ρ_i is defined by (17.6-15). Therefore

$$\int_{\Gamma_i} \frac{\partial u}{\partial n}(t)\,|dt| = -\int_{\Gamma_i} \kappa(t)\,|dt| - \frac{1}{\rho_i}\int_{\Gamma_i} e^{u(t)}\,|dt|$$
$$= -(-2\pi) - 2\pi = 0, \qquad i = 1, 2, \ldots, n,$$

as desired. Thus v exists such that $u + iv$ is analytic in R. Let

$$g(z) := e^{u(z)+iv(z)},$$

which again is analytic in R. We claim that g has an indefinite integral in R. For this to be the case it is necessary and sufficient that

$$\int_{\Gamma_i} g(z)\,dz = 0, \qquad i = 1, 2, \ldots, n. \tag{17.6-17}$$

We recall the parametrization of Γ_i by $z = z(\tau)$, where τ is an arc length, $\tau_i \leqq \tau \leqq \tau_{i+1}$, and we let $z_{i0} := z(\tau_i)$. Then by the Cauchy–Riemann equations

$$\frac{dv(z(\tau))}{d\tau} = \frac{\partial u}{\partial n}(z(\tau)),$$

that is,

$$\frac{dv(z(\tau))}{d\tau} = -\kappa(z(\tau)) - \frac{1}{\rho_i}\,e^{u(z(\tau))},$$

that is, using (17.6-4),

$$v(z(\tau)) - v(z_{i0}) = -\frac{1}{i}\log\frac{z'(\tau)}{z'(\tau_i)} - \frac{1}{\rho_i}\int_{\tau_i}^{\tau} e^{u(z(\tau))}\,d\tau.$$

Letting

$$\alpha := v(z_{i0}) + \frac{1}{i}\log z'(\tau_i) \in \mathbb{R},$$

we have

$$\begin{aligned} g(z)\,dz &= e^{u(z(\tau))+iv(z(\tau))}z'(\tau)\,d\tau \\ &= e^{u(z(\tau))}\,e^{i\alpha}z'(\tau)\exp\left[-\frac{i}{\rho_i}\int_{\tau_i}^{\tau} e^{u(z(\tau))}\,d\tau - \log z'(\tau)\right]d\tau \\ &= i\rho_i e^{i\alpha}\frac{d}{d\tau}\left\{\exp\left[-\frac{i}{\rho_i}\int_{\tau_i}^{\tau} e^{u(z(\tau))}\,d\tau\right]\right\}d\tau, \end{aligned} \tag{17.6-18}$$

and on using (17.6-15), the desired relation (17.6-17) follows. We therefore may select $z_0 \in R$ and define

$$f(z) := f(z_0) + \int_{z_0}^{z} g(z)\,dz,$$

where $f(z_0)$ and the path of integration in R are chosen arbitrarily. By the maximum principle $u(\infty)$ is finite, and therefore $g(\infty) \neq 0$. By selecting the additive constant in $v(z)$ appropriately, we can in fact achieve $g(\infty) > 0$. There follows $f(\infty) = \infty$. We next trace the image of the curves Γ_i under the map f. By (17.6-18) we find for $\tau_i \leqq \tau \leqq \tau_{i+1}$,

$$f(z(\tau)) - f(z_{i0}) = i\rho_i\, e^{i\alpha}\left\{\exp\left[-\frac{i}{\rho_i}\int_{\tau_i}^{\tau} e^{u(z(\tau))}\, d\tau\right] - 1\right\}.$$

Because the function

$$\tau \to \frac{1}{\rho_i}\int_{\tau_i}^{\tau} e^{u(z(\tau))}\, d\tau$$

increases monotonically from 0 to 2π as τ increases from τ_i to τ_{i+1}, we see that the point $w = f(z(\tau))$ traces a circle, of radius ρ_i, in the negative sense. Calling these circles Λ_i, the principle of the argument may now be invoked to show that if $w \in \mathbb{C}$ is any point with respect to which all Λ_i have winding number zero, there exists precisely one $z \in R$ such that $f(z) = w$. That is, the function f maps R conformally and one-to-one onto the complement C of the n disks bounded by the Λ_i.

We thus have established the existence of a map of R onto a circular region of connectivity n. The normalization (17.6-1) is achieved by finally applying a linear transformation $z \to \alpha z + b$, where $\alpha > 0$ and $b \in \mathbb{C}$. The uniqueness of the final map f follows from the uniqueness of the map subjected to the side condition (17.6-7). —

It should be noted that the foregoing proof not only establishes the existence of the circular map, but also suggests a method for actually constructing it. One would have to approximate Ω by a finite-dimensional space spanned by suitable linearly independent harmonic functions $v_1, \ldots, v_N \in \Omega$, and to minimize

$$\psi(\omega_1, \ldots, \omega_N) := \phi(\omega_1 v_1 + \omega_2 v_2 + \cdots + \omega_N v_N)$$

as a function of the finitely many variables $\omega_1, \ldots, \omega_N$. However, more numerical experience with this method is required to demonstrate its effectiveness.

NOTES

The circular map is just one of many standardized mapping functions for multiply connected regions that have been considered. In addition to § 17.1, see Courant [1950] and Bergman [1950]. The proof of Theorem 17.6a is taken from Schiffer and Hawley [1962]. The Steklov problem used in proving the boundedness of ϕ is discussed in Bergmann and Schiffer [1953]. Some aspects of the numerical method described at the end are discussed by Löffler [1983].

§ 17.7. ARBITRARY CONNECTIVITY: ITERATIVE CONSTRUCTION OF THE CIRCULAR MAP

Let $\hat{\mathbb{C}}$ denote the extended complex plane, and let $R \subset \hat{\mathbb{C}}$ be a region of connectivity $n \geqq 2, \infty \in R$. We denote the components of the complement of R by $K_1, K_2, \ldots, K_n$. It could be assumed without loss of generality that the K_i are bounded by smooth, or even analytic, Jordan curves, but there is no need to make this assumption in the present section. By Theorem 17.6a there exists a unique function f which is normalized at ∞ by

$$f(z) = z + O(z^{-1}) \tag{17.7-1}$$

and which maps R onto a circular region of connectivity n. Here we present an effective method for constructing this **circular map**. This method was devised by Koebe as early as 1910. Much later (see Halsey [1979]) it was rediscovered and implemented by aircraft engineers, who found it to be highly effective.

In what follows, any function f mapping a neighborhood of ∞ and satisfying (17.7-1) will be called **normalized at** ∞. The building blocks of Koebe's algorithm are mappings, normalized at ∞, of simply connected regions $S \subset \hat{\mathbb{C}}$, $\infty \in S$, onto the complement of a disk. It is a direct consequence of the Riemann mapping theorem that such a map is uniquely determined. Any of the methods discussed in Chapter 16 may, in principle, be used to construct it.

Koebe's algorithm is as follows. Let

$$C_0 := R, \qquad D_{0,i} := K_i, \qquad i = 1, 2, \ldots, n.$$

Having obtained a region C_{k-1} of connectivity n such that $\infty \in C_{k-1}$, whose complement has the components $D_{k-1,i}$, $i = 1, 2, \ldots, n$, let $j \equiv k \bmod n$, $1 \leqq j \leqq n$, and let h_k be the map, normalized at ∞, of $\hat{\mathbb{C}} \backslash D_{k-1,j}$ onto the exterior of a *disk*. Let $D_{k,j}$ be that disk, and let

$$C_k := h_k(C_{k-1}).$$

C_k then is a region of connectivity n, the complement of which has the components

$$D_{k,j} \qquad \text{and} \qquad D_{k,i} := h_k(D_{k-1,i}), \qquad i \neq j.$$

Thus Koebe's algorithm runs through the components $D_{k,i}$, $i = 1, 2, \ldots, n$, cyclically, always making the one for which $k \equiv i \bmod n$ into a disk; see Fig. 17.7a for the first three steps of the iteration, where $n = 3$.

Naturally the $D_{k,i}$ do not remain disks for $k > i$ (they will become disks again for $k = i + mn$, $m = 1, 2, \ldots$), but the contention is that they become

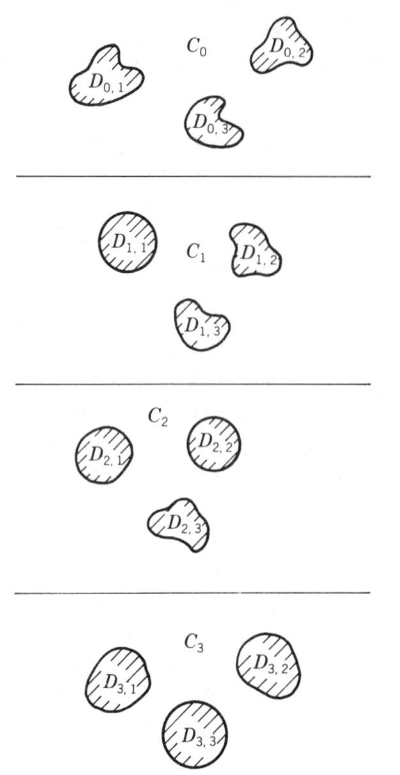

Fig. 17.7a. Koebe's algorithm.

more and more disklike, and that the restrictions of

$$f_k := h_k \circ h_{k-1} \circ \cdots \circ h_1$$

to R tend to the circular map f, normalized at ∞, of $C_0 = R$.

An equivalent assertion is obtained as follows. Let $C := f(R)$ be the normalized circular image of R. Then C is also the normalized circular image of each region C_k, because

$$C = f \circ f_k^{[-1]}(C_k),$$

and inverses as well as compositions of normalized maps are normalized. The above assertion is thus equivalent to saying that the maps $f \circ f_k^{[-1]}$, or equivalently, their inverses

$$g_k := f_k \circ f^{[-1]}, \tag{17.7-2}$$

which are maps, normalized at ∞, of C onto C_k, tend to the identity. The following is a quantitative version of this assertion:

THEOREM 17.7a. *There exist constants $\gamma > 0$ and μ, $0 < \mu < 1$, such that for $k = 1, 2, \ldots$ and for all $w \in C$,*

$$|g_k(w) - w| \leqq \gamma\mu^{4[k/n]}. \tag{17.7-3}$$

Here $[k/n]$ denotes the greatest integer not exceeding k/n. Explicit values for γ and μ will emerge from the proof.

The idea of the proof is this. We first show that the regions C_k, $k = 1, 2, \ldots$, increasingly enjoy the property of being capable of reflection at their boundaries. This fact is then used to obtain the estimate of the theorem.

To begin with, let us recall the precise notion of symmetry of two regions S, S' with respect to a closed curve Γ. It is clear what this means when Γ is a circle. There is a one-to-one correspondence between the points $z \in S$ and $z' \in S'$ such that z and z' are symmetric to Γ in the sense of § 5.4. If Γ is not a circle, S and S' are said to be symmetric with respect to Γ if there exists a region R containing S, S', and Γ, and a conformal map f of R such that $f(\Gamma)$ is a circle and $f(S)$ and $f(S')$ are symmetric with respect to $f(\Gamma)$. (The curve thus is necessarily analytic.) We express the fact that S and S' are symmetric with respect to Γ by the symbol

$$S \mid S' \quad (\Gamma).$$

Let now $C_0 = R$, C_1, $C_2, \ldots$ be the sequence of regions generated by the Koebe algorithm, and let the (set-theoretical) boundaries of the components $D_{k,i}$, $i = 1, 2, \ldots, n$, of $\hat{\mathbb{C}} \backslash C_k$ be denoted by $\Gamma_{k,i}$. Because $\Gamma_{1,1}$ is a circle, C_1 can be reflected at $\Gamma_{1,1}$. We denote the reflected region by C_1^1. Since the maps h_k for $k = 2, \ldots, n$ are defined on $C_k \cup D_{k,1}$, the regions

$$C_k^1 := h_k \circ \cdots \circ h_2(C_1^1), \qquad k = 2, \ldots, n,$$

satisfy

$$C_k^1 \mid C_k \quad (\Gamma_{k,1}). \tag{17.7-4}$$

The map h_{n+1} is not defined on $D_{n,1}$. However, it takes the boundary $\Gamma_{n,1}$ of $D_{n,1}$ into a circle and thus by the symmetry principle can be extended to a map which takes C_k^1 into a region C_{n+1}^1 which satisfies (17.7-4) for $k = n+1$. Continuing as above, we obtain a sequence of regions $\{C_k^1\}$ satisfying (17.7-4) for all $k \geqq 1$.

Similarly, because $\Gamma_{2,2}$ is a circle, we can define C_2^2 as the symmetric image of C_2 with respect to $\Gamma_{2,2}$. The regions

$$C_k^2 := h_k \circ h_{k-1} \circ \cdots \circ h_3(C_2^2)$$

(where every nth h_k has to be extended by the reflection principle) then satisfy

$$C_k^2 \mid C_k \quad (\Gamma_{k,2}), \qquad k \geqq 2.$$

In the same fashion, we define for $i=3,\ldots,n$ and for $k\geqq i$ regions C^i_k such that

$$C^i_k \,|\, C_k \quad (\Gamma_{k,i}), \qquad k=i. \tag{17.7-5}$$

After the first cycle of the Koebe algorithm has been performed, the regions C^i_k all are defined for $i=1,2,\ldots,n$. Since $\Gamma_{n+1,1}$ is again a circle, we now may define C^{i1}_{n+1} as the image of C^i_{n+1} under reflection at $\Gamma_{n+1,1}$. Of course, $C^{11}_{n+1}=C_{n+1}$, but the remaining regions C^{i1}_{n+1} are new. Applying the maps h_k, where $k=n+2$, extended if necessary, we obtain in this manner a sequence of regions C^{j1}_k such that for $i=1,2,\ldots,n$ and for $k\geqq n+1$,

$$C^{i1}_k \,|\, C^i_k \quad (\Gamma_{k,1}).$$

In a similar manner, we obtain for $k>n+j$ regions C^{ij}_k satisfying

$$C^{ij}_k \,|\, C^i_k \quad (\Gamma_{k,j}).$$

After m cycles of the algorithm, having defined the regions

$$C^{i_1 i_2 \cdots i_m}_k, \tag{$*$}$$

where each i_h runs from 1 to n, we define regions

$$C^{i_1 i_2 \cdots i_{m+1}}_k$$

satisfying

$$C^{i_1 i_2 \cdots i_{m+1}}_k \,|\, C^{i_1 \cdots i_m} \quad (\Gamma_{k,m+1})$$

for all combinations of the superscripts and for $k>(m+1)n$.

Along with the regions C_k, we transform their boundaries. The n boundary components of the region ($*$) are denoted by

$$\Gamma^{i_1 i_2 \cdots i_m}_{k,j}, \qquad j=1,2,\ldots,n.$$

The notation is defined inductively so that

$$\Gamma^{i_1 i_2 \cdots i_m}_{k,j} \,|\, \Gamma^{i_1 i_2 \cdots i_{m-1}}_{k,j} \quad (\Gamma_{k,i_m}).$$

LEMMA 17.7b. *For every $m>0$, for all possible combinations of indices $i_1 \cdots i_m$, and for $k>(m+1)n$ there holds*

$$C^{j i_1 \cdots i_m}_k \,|\, C^{i_1 \cdots i_m}_k \quad (\Gamma^{i_1 \cdots i_m}_{k,j}).$$

Proof. Let us determine the region X which satisfies

$$X \,|\, C^{i_1 \cdots i_m}_k \quad (\Gamma^{i_1 \cdots i_m}_{k,j}).$$

Reflection at Γ_{k,i_m} yields

$$X^{i_m} \,|\, C^{i_1 \cdots i_{m-1}}_k \quad (\Gamma^{i_1 \cdots i_{m-1}}_{k,j}),$$

and so on, until finally

$$X^{i_m i_{m-1} \cdots i_1} \,|\, C_k \quad (\Gamma_{k,j}),$$

which by (17.7-5) implies $X = C_k^j$. Successively reflecting at $\Gamma_{k,i_1}, \ldots, \Gamma_{k,i_m}$ yields

$$X = C^{j i_1 i_2 \cdots i_m},$$

as desired. —

By Lemma 17.7b, not only can the regions C_k for $k > mn$ be reflected at each of their boundary components, but the reflected regions have again the same property, and so on, until m levels of reflection have been obtained. We briefly say that C_k for $k > mn$ is m **times totally reflectable**. By virtue of their construction, for every $m = 1, 2, \ldots$, all regions

$$C_k^{i_1 i_2 \cdots i_h},$$

where $h \leqq m$ and where no two consecutive i_j are identical, are disjoint for every $k > mn$.

By means of the map $g_k^{[-1)} = f \circ f_k^{[-1]}$, the foregoing configurations are mapped into the w plane. Since

$$C = g_k^{[-1]}(C_k)$$

does not depend on k, the same is true of all reflections of C_k. We write

$$C^{i_1 \cdots i_h} := g_k^{[-1]}(C_k^{i_1 \cdots i_h}),$$

and similarly

$$\Gamma_j^{i_1 \cdots i_h} := g_k^{[-1]}(\Gamma_{k,j}^{i_1 \cdots i_h}).$$

The latter curves all are circles; the circles $\Gamma_j, j = 1, 2, \ldots, n$, are the boundary components of the circular image C of R. These circles are disjoint, as are all their mirror images.

In the w plane we now perform the following construction. We enlarge the n circles Γ_j concentrically by one and the same factor ρ until a first value of ρ is obtained where at least two of the enlarged circles touch. Let this critical value of ρ be denoted by μ^{-1}, and let the inflated circles be $\tilde{\Gamma}_j, j = 1, 2, \ldots, n$. We subject the circles $\tilde{\Gamma}_j$ to the same process of reflection as the circles Γ_j and the region C. We denote by

$$\tilde{\Gamma}_j^{i_1 i_2 \cdots i_h}$$

the circle obtained by successively reflecting $\tilde{\Gamma}_j$ at the circles $\Gamma_{i_1}, \Gamma_{i_2}, \ldots, \Gamma_{i_h}$.

The circles $\tilde{\Gamma}_j$ and $\tilde{\Gamma}_j^j$ evidently form the boundaries of an annulus A_j of modulus μ^{-2}, which is symmetric about Γ_j. The annuli

$$A_j^{i_1 i_2 \cdots i_h}$$

obtained by successively reflecting A_j at the circles $\Gamma_{i_1}, \Gamma_{i_2}, \ldots, \Gamma_{i_h}$ likewise all have the modulus μ^{-2}. If $i_1 \neq j$ and if no two consecutive i's are identical, all these annuli are disjoint. The above annulus is symmetric with regard to $\Gamma_j^{i_1 \cdots i_h}$; it is bounded on the outside by $\tilde{\Gamma}_j^{i_1 \cdots i_h}$ and on the inside by $\tilde{\Gamma}_j^{j i_1 \cdots i_h}$.

Finally, the annuli $A_j^{i_1 \cdots i_h}$ and their boundaries are transplanted into the z plane by the maps g_k, $k > nh$. We set

$$A_{k,j}^{i_1 \cdots i_h} := g_k(A_j^{i_1 \cdots i_h}),$$

$$\tilde{\Gamma}_{k,j}^{i_1 \cdots i_h} := g_k(\tilde{\Gamma}_j^{i_1 \cdots i_h}).$$

The $A_{k,j}^{i_1 \cdots i_h}$ are doubly connected regions which, because the modulus is a conformal invariant, again have modulus μ^{-2}.

We now can begin to estimate the function $g_k(w) - w$, which by the foregoing construction is analytic in W_m, the complement with respect to $\hat{\mathbb{C}}$ of the union of all disks $D_j^{i_1 \cdots i_m}$ bounded by the circles $\Gamma_j^{i_1 \cdots i_m}$, the union being taken with respect to all systems $(i) := i_1 \cdots i_m$ such that no two consecutive i_h are the same, and for each such system with respect to all $j \neq i_1$. Denoting by Γ_ρ a (large) circle containing all the disks D_j, we have by Cauchy's theorem for $w \in C_0$,

$$g_k(w) - w = \frac{1}{2\pi i} \int_{\Gamma_\rho} \frac{g_k(s) - s}{s - w}\, ds - \sum_{(i),j} \frac{1}{2\pi i} \int_{\Gamma_j^{(i)}} \frac{g_k(s) - s}{s - w}\, ds.$$

Here the first integral vanishes for $\rho \to \infty$ because $g(w) - w = O(w^{-1})$. In the remaining integrals, since w is outside all circles $\Gamma_j^{(i)}$, the terms

$$\frac{1}{2\pi i} \int_{\Gamma_j^{(i)}} \frac{s}{s - w}\, ds$$

may be omitted. On the other hand, one may choose arbitrary complex numbers $c_j^{(i)}$ and subtract terms

$$\frac{1}{2\pi i} \int_{\Gamma_j^{(i)}} \frac{c_j^{(i)}}{s - w}\, ds,$$

which are likewise zero. We thus have

$$g_k(w) - w = -\sum_{(i),j} \frac{1}{2\pi i} \int_{\Gamma_j^{(i)}} \frac{g_k(s) - c_j^{(i)}}{s - w}\, ds.$$

We estimate the integrals in an elementary manner. Letting

$$\delta := \min_{i \neq j} \operatorname{dist}(\Gamma_i, D_j^i),$$

we have $\delta > 0$ and, since $\Gamma_j^{(i)} \subset D_j^{i_1}$, $|s - w| > \delta$. Further, let

$$\delta_{k,j}^{(i)} := \operatorname{diam}(\Gamma_{k,j}^{(i)}).$$

The curve $\Gamma_{k,j}^{(i)} = g_k(\Gamma_j^{(i)})$ is then contained in a circle of radius $\delta_{k,j}^{(i)}$ (actually, a somewhat smaller circle would suffice). Choosing for $c_j^{(i)}$ the center of that circle, we have for $s \in \Gamma_j^{(i)}$

$$|g_k(s) - c_j^{(i)}| \leqq \delta_{k,j}^{(i)}.$$

The length of the path of integration is $\pi\delta_j^{(i)}$, where

$$\delta_j^{(i)} := \operatorname{diam} \Gamma_j^{(i)}.$$

There follows

$$\begin{aligned} |g_k(w) - w| &\leqq \sum_{(i)} \sum_j \frac{1}{\pi\delta} \delta_{k,j}^{(i)} \delta_j^{(i)} \\ &\leqq \frac{1}{2\pi\delta} \sum_{(i)} \sum_j \{[\delta_{k,j}^{(i)}]^2 + [\delta_j^{(i)}]^2\}, \end{aligned} \tag{17.7-6}$$

where we have used the inequality of the arithmetic and the geometric mean. To proceed, we require two results which use the notation $\alpha(\Gamma)$ for the area enclosed by a Jordan curve Γ.

LEMMA 17.7c. *Let B be a doubly connected region with finite modulus $\mu^{-1} > 1$, and let B be bounded on the outside by a Jordan curve Γ_0 and on the inside by a Jordan curve Γ_1. Then*

(a) $$\alpha(\Gamma_1) \leqq \mu^2 \alpha(\Gamma_0).$$

(b) $$[\operatorname{diam} \Gamma_1]^2 \leqq \frac{\pi}{2 \operatorname{Log} \mu^{-1}} \alpha(\Gamma_0).$$

Proof. Let g map the annulus $1 < |w| < \mu^{-1}$ onto B,

$$g(w) = \sum_{n=-\infty}^{\infty} a_n w^n.$$

Then

$$\alpha(\Gamma_1) = \pi \sum_{n=-\infty}^{\infty} n|a_n|^2$$

$$\alpha(\Gamma_0) = \pi \sum_{n=-\infty}^{\infty} n|a_n|^2 \mu^{-2n}.$$

Hence

$$\alpha(\Gamma_0) - \mu^2 \alpha(\Gamma_1) = \sum_{n=-\infty}^{\infty} n|a_n|^2 (\mu^{-2n} - \mu^{-2}) \geqq 0,$$

proving (a). As to (b), diam Γ_1 is bounded by the diameter of the larger set $g(1<|w|<\rho)$ for any $\rho\in(1,\mu^{-1})$, and this in turn is bounded by one-half of the length of the outer boundary $g(|w|=\rho)$. Thus

$$2\operatorname{diam}\Gamma_1 \leqq \int_{|w|=\rho} |g'(w)||dw| = \int_0^{2\pi} |g(\rho e^{i\theta})|\rho\, d\theta.$$

By the Schwarz inequality,

$$[2\operatorname{diam}\Gamma_1]^2 \leqq \int_0^{2\pi} |g(\rho e^{i\theta})|^2\rho\, d\theta \cdot \int_0^{2\pi} \rho\, d\theta$$

$$= 2\pi\rho \int_0^{2\pi} |g(\rho e^{i\theta})|^2 \rho\, d\theta$$

or

$$\frac{2}{\pi\rho}[\operatorname{diam}\Gamma_1]^2 \leqq \int_0^{2\pi} |g(\rho e^{i\theta})|^2\rho\, d\theta.$$

Integrating between the limits 1 and μ^{-1} yields (b). —

Returning to (17.7-6), we apply Lemma 17.7c a first time to the annular regions bounded by $\tilde{\Gamma}_{k,j}^{(i)}$ and $\Gamma_{k,j}^{(i)}$. Using (b), we have

$$[\delta_{k,j}^{(i)}]^2 \leqq \frac{\pi}{2\operatorname{Log}\mu^{-1}}\alpha(\tilde{\Gamma}_{k,j}^{(i)}),$$

while trivially

$$[\delta_j^{(i)}]^2 = \frac{4}{\pi}\alpha(\Gamma_j^{(i)}) = \frac{4}{\pi}\mu^2\alpha(\tilde{\Gamma}_j^{(i)}).$$

The circles $\tilde{\Gamma}_j^{(i)} = \tilde{\Gamma}_j^{i_1\cdots i_m}$ where $j\neq i_1$ all are contained in the region $C^{i_1\cdots i_m}$, which by Lemma 17.7b is bounded on the outside by $\Gamma_{i_1}^{i_2\cdots i_m}$. Hence they are also contained in the disk bounded by the circle $\tilde{\Gamma}_{i_1}^{(i)}$. Thus

$$\sum_{(i),j}\alpha(\tilde{\Gamma}_j^{(i)}) \leqq \sum_{(i)}\alpha(\tilde{\Gamma}_{i_1}^{(i)}),$$

and similarly,

$$\sum_{(i),j}\alpha(\tilde{\Gamma}_{k,j}^{(i)}) \leqq \sum_{(i)}\alpha(\tilde{\Gamma}_{k,i_1}^{(i)}).$$

Applying (a) to the annular regions $A_{i_1}^{i_2\cdots i_m}$ bounded by $\tilde{\Gamma}_{i_1}^{i_1 i_2\cdots i_m}$ and by $\tilde{\Gamma}_{i_1}^{i_2\cdots i_m}$, which all have modulus μ^{-2}, we get

$$\sum_{(i)}\alpha(\tilde{\Gamma}_{i_1}^{i_1\cdots i_m}) \leqq \mu^4 \sum_{(i)}\alpha(\tilde{\Gamma}_{i_1}^{i_2\cdots i_m})$$

and similarly for their images under g_k,

$$\sum_{(i)} \alpha(\tilde{\Gamma}_{k,i_1}^{i_1\cdots i_m}) \leqq \mu^4 \sum_{(i)} \alpha(\tilde{\Gamma}_{k,i_1}^{i_2\cdots i_m}).$$

The circles $\tilde{\Gamma}_{i_1}^{i_2\cdots i_m}$ are contained in $\tilde{\Gamma}_{i_2}^{i_2\cdots i_m}$. Hence

$$\sum_{i_1\cdots i_m} \alpha(\tilde{\Gamma}_{i_1}^{i_2\cdots i_m}) \leqq \sum_{i_2\cdots i_m} \alpha(\tilde{\Gamma}_{i_2}^{i_2\cdots i_m})$$

$$\leqq \mu^4 \sum_{i_2\cdots i_m} \alpha(\tilde{\Gamma}_{i_2}^{i_3\cdots i_m}).$$

Continuing in this fashion, we find

$$\sum_{(i),j} \alpha(\tilde{\Gamma}_j^{(i)}) \leqq \mu^4 \sum_{i_1\cdots i_m} \alpha(\tilde{\Gamma}_{i_1}^{i_2\cdots i_m}) \leqq \cdots \leqq \mu^{4m} \sum_{i_m} \alpha(\tilde{\Gamma}_{i_m}).$$

The last sum is μ^{-2} times the sum of the areas of all disks D_{i_m}, which we denote by γ_1. Putting it all together, we thus get

$$\sum_{(i),j} [\delta_j^{(i)}]^2 \leqq \frac{4}{\pi} \mu^{4m} \gamma_1.$$

Similarly, by working in the z plane in place of the w plane,

$$\sum_{(i),j} [\delta_{k,j}^{(i)}]^2 \leqq \frac{\pi}{2 \operatorname{Log} \mu^{-1}} \mu^{4m-2} \gamma_2,$$

where γ_2 is the sum of the areas enclosed by the curves Γ_{k,i_m}. If the disks D_i all are contained in a circle of radius ρ about 0, the functions $g_k(w) = w + \cdots$ are univalent in $|w| > \rho$ and the image of C therefore contains the set $|w| > 4\rho$ by the Koebe $\frac{1}{4}$ theorem (Theorem 19.1i). There follows

$$\gamma_2 \leqq 16\pi\rho^2.$$

Putting it all together, we get for $k > mn$,

$$|g_k(w) - w| \leqq \frac{1}{2\pi\delta}\left(\frac{\pi}{2\mu^2 \operatorname{Log} \mu^{-1}} 16\pi\rho^2 + \frac{4}{\pi}\pi\rho^2\right)\mu^{4m}, \qquad (17.7\text{-}7)$$

proving Theorem 17.7a with explicitly defined values of γ and μ. —

PROBLEMS

1. In a typical case where $n = 3$, make a sketch of the regions $C_0, C_1, \ldots, C_9$ and of their possible reflections as well as of their boundaries, and verify the statement of Lemma 17.7b.
2. Show that Koebe's algorithm for $n = 2$ is identical with Komatu's method (see Problem 1, § 17.2), and thus give a new convergence proof for that method.

NOTES

Koebe [1910] describes his iteration method in a few lines, and hints at the geometric rate of convergence. The actual convergence proof was given by Gaier [1959b]. The proof given above paraphrases Gaier's proof in a more elaborate notation. As to Lemma 17.7c, (a) is due to Carleman [1918] and (b) to Gaier [1959a], who has a more precise estimate. The iteration described in this chapter also underlies the methods of Fornberg [1982] and Wegmann [1985] for the doubly connected case. For other methods to obtain the circular map, see Koebe [1920], Gaier [1964], and the brief survey in Gaier [1978].

18

POLYNOMIAL EXPANSIONS AND CONFORMAL MAPS

The disk D: $|z|<1$ has the following property. There exists a standardized sequence $\{p_n\}$ of analytic functions (namely, $p_n(z) := z^n$) such that every function f that is analytic in D can be expanded in a series $f = \sum a_n p_n$ that converges uniformly on every compact subset of D. In this chapter we are concerned with constructing similar sequences of functions for arbitrary regions R, usually assumed to be simply connected. Two essentially different methods of construction will be presented: (a) the method of Faber polynomials, which requires the availability of a certain mapping function; (b) the method of orthogonalization, which may be used, at least in principle, to *construct* the Riemann mapping function for simply connected regions R. In the second case there are two essentially different methods of orthogonalization, associated with the names of Bergman and Szegö.

§ 18.1. FORMAL LAURENT SERIES AT INFINITY; FABER POLYNOMIALS

Faber polynomials are useful tools in conformal mapping and in approximation theory. To establish some of their formal properties, we need to reformulate certain results on formal Laurent series that were established in Chapter 1.

I. Formal Laurent Series at ∞

In § 1.8 we considered formal Laurent series (fLs)

$$L = \sum_{k=(-\infty)}^{\infty} a_k x^k, \tag{18.1-1}$$

where the coefficients are taken from some field $\mathscr{F}$ of characteristic 0, and where the parentheses serve as a reminder that only finitely many a_k with $k<0$ are different from 0. It was shown that with the obvious definitions of addition and multiplication these formal Laurent series form a *field* $\mathscr{L}$.

As far as the field $\mathscr{L}$ is concerned, it clearly does not matter whether the indeterminate in (18.1-1) is denoted by x or by x^{-1}. Thus we could as well have considered formal Laurent series of the form

$$L = \sum_{k=-\infty}^{(\infty)} a_k x^k, \tag{18.1-2}$$

where now only finitely many a_k with positive k are different from zero. We call (18.1-2) a **formal Laurent series at** ∞ and denote the field of these fLs by $\mathscr{L}_\infty$. If $\mathscr{F} = \mathbb{C}$, and if the series (18.1-2) is convergent, it represents a function which either is analytic at ∞ or has a pole there.

An obvious isomorphism exists between the fields $\mathscr{L}$ and $\mathscr{L}_\infty$ by virtue of the substitution $x \mapsto x^{-1}$. This isomorphism is not preserved if we consider differentiation. For this reason, and to achieve a somewhat greater generality than in § 1.8, we develop the theory of $\mathscr{L}_\infty$ briefly but independently from § 1.8.

The **derivative** of the fLs (18.1-2) is again defined by

$$L' := \sum_k k a_k x^{k-1}.$$

With this formal definition, the usual rules for manipulating derivatives again are established easily. If $L \in \mathscr{L}_\infty$, we still call the coefficient of x^{-1} the **residue** of L, and we denote it by res L. As in § 1.8, there holds:

LEMMA 18.1a. *The residue of a formal Laurent series at ∞ is zero if and only if L is a derivative, that is, if there exists a series $M \in \mathscr{L}_\infty$ such that $M' = L$.*

II. Composition

For any nonzero series

$$M = \sum_{k=-\infty}^{(\infty)} b_k x^k$$

in $\mathscr{L}_\infty$, the powers M^n can be formed for all integers n, positive, negative, or zero. We again shall use the notation

$$M^n = \sum_{k=-\infty}^{(\infty)} b_k^{(n)} x^k. \tag{18.1-3}$$

Here we study circumstances under which we can form the composition of a series

$$L = \sum_{k=-\infty}^{(\infty)} a_k x^k \tag{18.1-4}$$

with M, in the sense that substituting M for x in (18.1-4) and collecting coefficients of equal powers of x require algebraic operations only. One such situation was discussed in § 1.6. There the series L was a formal power series, and the series M was a nonunit in the integral domain of formal power series. In the present notation this means that L does not contain powers x^k where $k>0$, and M does not contain powers x^k where $k \geqq 0$.

However, this is not the only case in which a composition can be formed. All that is needed in order to define

$$L \circ M = \sum c_k x^k$$

formally is that all sums

$$c_k = \sum_n a_n b_k^{(n)} \tag{18.1-5}$$

contain only finitely many nonzero terms. Now if

$$M = \sum_{k=-\infty}^{m} b_k x^k, \tag{18.1-6}$$

where $m>0$ and $b_m \neq 0$, then the highest power appearing in M^n is x^{nm}, and this is true irrespectively of the sign of n. There follows

$$b_k^{(n)} = 0, \qquad k > nm.$$

Thus in the sums (18.1-5) there is no contribution for large n because $a_n = 0$ for such n, and if $m>0$, there is no contribution for $n < m^{-1}k$ because $b_k^{(n)} = 0$ for such n. We conclude that the composition $L \circ M$ can be formed according to (18.1-5) for any $L \in \mathscr{L}_\infty$ if M has the form (18.1-6), where $m>0$ and $b_m \neq 0$. Such M will be called **singular** formal Laurent series.

It is easy to see that for a singular series L and a nonsingular series M the composition $L \circ M$ can, in general, not be formed.

It may be shown, either directly or by means of a matrix isomorphism similar to that considered in § 1.7, that if the composition of several series in L_∞ can be formed at all, it is associative. Thus for instance, if the series M and N are singular,

$$L \circ (M \circ N) = (L \circ M) \circ N. \tag{18.1-7}$$

III. The Group of Almost Regular Formal Laurent Series

A series $F \in \mathscr{L}_\infty$ is called **almost regular** (a.r.) if it has the special form

$$F = \sum_{k=-\infty}^{1} a_k x^k,$$

where $a_1 \neq 0$. The composition of two a.r. fLs clearly is a.r.. In particular, composing any a.r. fLs F with the a.r. series

$$X := x$$

yields F. We have already noted that composition is associative. By the method undetermined coefficients it is easily seen that given any a.r. series F, there exists a unique a.r. series G such that $G \circ F = X$. We write $G = F^{[-1]}$ and call this the **reversion** of the series F. Denoting the totality of all almost regular fLs by $\mathscr{L}_\infty^{(1)}$, we have verified:

THEOREM 18.1b. *The series in $\mathscr{L}_\infty^{(1)}$ form a group under the operation of composition.*

Let now L be any series in $\mathscr{L}_\infty$, and let $F \in \mathscr{L}_\infty^{(1)}$, $G := F^{[-1]}$. We seek a formula of the Lagrange–Bürmann type (see § 1.9) for the coefficients c_k in

$$L \circ G = \sum_{k=-\infty}^{(\infty)} c_k x^k. \tag{18.1-8}$$

Composing (18.1-8) from the right with F, we have

$$L \circ G \circ F = L = \sum_{k=-\infty}^{(\infty)} c_k F^k.$$

The infinite sum on the right is algebraically meaningful, because each fixed power x^m occurs only in finitely many terms. Differentiation yields

$$L' = \sum_{k=-\infty}^{(\infty)} k c_k F^{k-1} F'$$

and, if n is any integer,

$$F^n L' = \sum_{k=-\infty}^{(\infty)} k c_k F^{n+k-1} F'.$$

In this identity we take the residue. Only finitely many terms on the right can contribute to the residue, because if $k < -n$, the highest power in $F^{n+k-1}F'$ has an exponent < -1. We thus have

$$\text{res}\,(F^n L') = \text{res} \sum_{k=-n}^{(\infty)} k c_k F^{n+k-1} F'.$$

However, because for $n+k \neq 0$,

$$F^{n+k-1}F' = \frac{1}{n+k}(F^{n+k})',$$

the terms where $n+k \neq 0$ by virtue of Lemma 18.1a do not contribute to the residue. By explicit computation we see that

$$\operatorname{res}(F^{-1}F') = 1.$$

Thus

$$kc_k = \operatorname{res}(F^{-k}L'),$$

and if $k \neq 0$,

$$c_k = \frac{1}{k}\operatorname{res}(F^{-k}L'). \tag{18.1-9}$$

An alternative version of this formula, which also holds for $k=0$, is obtained as follows. For any two series L, $M \in \mathscr{L}_\infty$ there holds

$$\operatorname{res}(L'M) = -\operatorname{res}(LM')$$

because $L'M + LM' = (LM)'$ is a derivative and therefore has zero residue. Hence

$$\operatorname{res}(F^{-k}L') = -\operatorname{res}((-k)F^{-k-1}F'L),$$

and there follows

$$c_k = \operatorname{res}(LF^{-k-1}F'), \qquad k \neq 0. \tag{18.1-10}$$

A separate verification shows that this also holds for $k=0$. We thus have proved the following version of the Lagrange–Bürmann theorem.

THEOREM 18.1c. *Let $L \in \mathscr{L}_\infty$, $F \in \mathscr{L}_\infty^{(1)}$, $G := F^{[-1]}$. Then*

$$L \circ G = \sum_{k=-\infty}^{(\infty)} \operatorname{res}(LF^{-k-1}F')x^k. \tag{18.1-11}$$

A striking special case is obtained for

$$L = X^n,$$

where n is any nonzero integer. On the left of (18.1-11) the coefficient of x^k then equals $b_k^{(n)}$, where

$$G^n = \sum_{k=-\infty}^{n} b_k^{(n)}x^k. \tag{18.1-12a}$$

Using the form (18.1-9) of c_k, the same coefficient on the right is

$$\frac{1}{k}\operatorname{res}(F^{-k}nX^{n-1}) = \frac{n}{k}a_{-n}^{(-k)},$$

where

$$F^n = \sum_{k=-\infty}^{n} a_k^{(n)} x^k. \tag{18.1-12b}$$

If $n \neq 0$ and $k \neq 0$, we thus obtain the following generalized form of the Schur–Jabotinsky theorem (Theorem 1.9a):

$$b_k^{(n)} = \frac{n}{k} a_{-n}^{(-k)}. \tag{18.1-13}$$

IV. Faber Polynomials

Let

$$G = \sum_{k=-\infty}^{1} b_k x^k$$

be in $\mathscr{L}^{(1)}$, $b_1 \neq 0$. If u is another indeterminate, the series

$$G - u = b_1 x + (b_0 - u) + \sum_{k=-\infty}^{-1} b_k x^k$$

may be viewed as an a.r. formal Laurent series over the field $\mathscr{L}_\infty(u)$, the set of formal Laurent series over $\mathscr{F}$ with indeterminate u. The series $xG'(G-u)^{-1}$ thus will likewise have coefficients in $\mathscr{L}_\infty(u)$. It is regular and may be written

$$\frac{xG'}{G-u} = \sum_{k=0}^{\infty} p_n(u) x^n, \tag{18.1-14}$$

where the $p_n(u) \in \mathscr{L}_\infty(u)$. Multiplying by $(G-u)x^{-1}$ we obtain the identity

$$\begin{aligned} &xb_1 - 1 \cdot b_{-1}x_{-1} - 2 \cdot b_{-2}x^{-2} - \cdots \\ &\quad = (p_0 + p_1 x^{-1} + p_2 x^{-2} + \cdots)[b_1 x + (b_0 - u) + b_{-1}x^{-1} + \cdots] \end{aligned}$$

which on comparing coefficients yields

$$\begin{aligned} p_0 &= 1, \\ p_1 &= \frac{1}{b_1}(u - b_0), \\ p_n &= \frac{1}{b_1}\{(u-b_0)p_{n-1} - b_{-1}p_{n-2} - \cdots - b_{-n+2}p_1 - nb_{-n+1}\}, \qquad n = 2, 3, \ldots, \end{aligned} \tag{18.1-15}$$

showing that p_n is a polynomial in u of precise degree n,

$$p_n(u) = \frac{1}{b_1^n}\{u^n - nb_0 u^{n-1} + \cdots\}. \tag{18.1-16}$$

The polynomials p_n are called the **Faber polynomials** associated with the series G.

We shall obtain an explicit representation for the Faber polynomials. By definition,

$$p_n(u) = \operatorname{res}\left(\frac{G'}{G-u}X^n\right)$$

or because $G' = (G-u)'$,

$$p_n(u) = \operatorname{res}\left(\frac{(G-u)'}{G-u}X^n\right).$$

By the Lagrange-Bürmann theorem (Theorem 18.1c), the residue equals the coefficient of x^0 in the series S^n where

$$S := (G-u)^{[-1]}.$$

To identify S, we compose

$$S^{[-1]} = G-u$$

from the right with S, which yields

$$X = G \circ S - u$$

or

$$G \circ S = X - u.$$

Composing from the left with $F = G^{[-1]}$ yields

$$S = F \circ (X+u).$$

Thus the coefficient of x^0 in S^n equals the coefficient of x^0 in $F^n \circ (X+u)$, which in the notation (18.1-12) is easily calculated. There follows:

THEOREM 18.1d. *Let G be an almost regular series in $\mathscr{L}_\infty$, $F := G^{[-1]}$, and let*

$$F^n = \sum_{k=-\infty}^{n} a_k^{(n)} x^k, \qquad n = 0, 1, 2, \ldots.$$

Then $p_n(u)$, the nth *Faber polynomial associated with G, is given by*

$$p_n(u) = \sum_{k=0}^{n} a_k^{(n)} u^k. \tag{18.1-17}$$

The sum (18.1-17) may be called the polynomial part of $F^n(u)$. It expresses the Faber polynomials associated with a series G in terms of the coefficients of the reversion F of G.

V. Faber Functions

Let $G \in \mathscr{L}_{\infty}^{(1)}$, and let $\{p_n\}$ be the associated sequence of Faber polynomials. Clearly, p_n may be regarded as a (singular) formal Laurent series. The **Faber functions** associated with G are formally defined as the formal Laurent series

$$H_n := p_n \circ G, \qquad n = 1, 2, \ldots. \tag{18.1-18}$$

By Theorem 18.1d,

$$p_n = F^n - R_n,$$

where

$$R_n := \sum_{k=-\infty}^{-1} a_k^{(n)} x^k$$

is regular, $R_n = O(x^{-1})$. There follows

$$H_n = F^n \circ G - R_n \circ G,$$

and in view of $F^n \circ G = (F \circ G)^n = X^n$,

$$H_n = x^n + \sum_{k=-\infty}^{-1} h_{nk} x^k. \tag{18.1-19}$$

We establish a symmetry relation for the **Faber coefficients**

$$c_{nk} := \frac{1}{n} h_{nk}, \tag{18.1-20}$$

where $n > 0$ and $k < 0$. By Theorem 18.1d and by the definition of composition, using the notation (18.1-12),

$$c_{nk} = \frac{1}{n} \sum_{m=0}^{n} a_m^{(n)} b_k^{(m)}.$$

Using (18.1-13),

$$c_{nk} = \frac{1}{nk} \sum_{m=0}^{n} m a_m^{(n)} a_{-m}^{(-k)}. \tag{18.1-21}$$

Now

$$\sum_{m=k}^{n} m a_m^{(n)} a_{-m}^{(-k)} = \operatorname{res}[(F^n)' F^{-k}] = 0,$$

because

$$(F^n)' F^{-k} = n F^{n-k-1} F' = \frac{n}{n-k} (F^{n-k})'$$

is a derivative. There follows

$$c_{nk} = -\frac{1}{nk} \sum_{m=k}^{-1} m a_m^{(n)} a_{-m}^{(-k)} = \frac{1}{nk} \sum_{m=1}^{-k} m a_m^{(-k)} a_{-m}^{(n)}.$$

In view of (18.1-21) we thus have:

THEOREM 18.1e. *The Faber coefficients c_{nk} defined by* (18.1-20) *satisfy*

$$c_{nk} = c_{-k,-n} \tag{18.1-22}$$

for all integers $n > 0$, $k < 0$.

VI. The Grunsky Series

Let L denote the logarithmic series introduced in § 1.7,

$$L(= \mathrm{Log}(1+x)) := x - \tfrac{1}{2}x^2 + \tfrac{1}{3}x^3 - \cdots.$$

If M is a formal Laurent series at ∞ with highest power x^m, where $m < 0$, the composition of L with M is possible and yields a nonsingular fLs with constant terms zero which satisfies the formal differential equation

$$(L \circ M)' = \frac{M'}{1+M}. \tag{18.1-23}$$

Let G be an almost regular fLs with highest coefficient 1,

$$G = x + \sum_{k=-\infty}^{0} b_k x^k.$$

We may consider G as a series over the coefficient field $\mathscr{L}_\infty(y)$ of formal Laurent series at ∞ in some other indeterminate y. The elements of that coefficient field will be denoted by $A(y)$, $B(y)$, ... In that sense,

$$\frac{G - G(y)}{x - y} = 1 + S$$

is a well-defined series over that coefficient field, and

$$S = \sum_{k=1}^{\infty} P_k(y) x^{-k},$$

where the $P_k(y)$ are certain fLs at ∞ in y. We therefore may form

$$L \circ S = \sum_{n=1}^{\infty} Q_n(y) x^{-n} \tag{18.1-24}$$

(which, if $\mathscr{F} = \mathbb{C}$ and G has a positive radius of convergence for complex

x and y of sufficiently large modulus, represents

$$\log \frac{G(x)-G(y)}{x-y}\Bigg).$$

Our goal is to identify the coefficients $Q_n(y)$. By differentiation we find, using (18.1-23),

$$(L \circ S)' = \frac{S'}{1+S} = \frac{G'}{G-G(y)} - \frac{1}{x-y}.$$

Expanding in powers of x^{-1} yields, using the definition of the Faber polynomials,

$$(L \circ S)' = \sum_{n=0}^{\infty} p_n(G(y))x^{-n-1} - \sum_{n=0}^{\infty} y^n x^{-n-1}$$

$$= \sum_{n=0}^{\infty} [p_n(G(y)) - y^n]x^{-n-1}.$$

By (18.1-19) and (18.1-20) the coefficients are

$$H_n(y) - y^n = n \sum_{k=-\infty}^{-1} c_{nk} y^k.$$

There follows

$$(L \circ S)' = \sum_{n=1}^{\infty} n \sum_{k=-\infty}^{-1} c_{nk} y^k x^{-n-1}$$

and on (formal) integration

$$L \circ S = -\sum_{n=1}^{\infty} \left(\sum_{k=1}^{\infty} c_{n,-k} y^{-k} \right) x^{-n}.$$

We thus have proved:

THEOREM 18.1f. *The coefficients in the expansion* (18.1-24) *are*

$$Q_n(y) = -\sum_{k=1}^{\infty} c_{n,-k} y^{-k}, \qquad (18.1\text{-}25)$$

where the $c_{n,-k}$ *are the Faber coefficients associated with* G.

The expansion (18.1-24) thus may be written

$$\log \frac{G(x)-G(y)}{x-y} = -\sum_{k,n=1}^{\infty} c_{n,-k} x^{-n} y^{-k}. \qquad (18.1\text{-}26)$$

This is called the **Grunsky series** associated with G. Its significance in the theory of univalent functions will become apparent in § 19.3.

PROBLEMS

1. In § 1.7, a matrix isomorphism was set up for the composition of almost units in the integral domain of formal power series. Set up a similar matrix isomorphism for the composition of almost regular formal Laurent series at ∞.
2. Verify relation (18.1-13) for the special case where

$$G = x - x^{-1}.$$

3. Show that the Faber polynomials associated with the series

$$G = \frac{1}{2}\left(x + \frac{\rho^2}{x}\right)$$

are $p_n(u) = \rho^n T_n(u/\rho)$, $n = 0, 1, 2, \ldots$, where T_n is the nth Chebyshev polynomial.

4. What are the Faber polynomials associated with $G := x - a$?

NOTES

The seminal paper on Faber polynomials is Faber [1903]. The formal properties given here are dealt with in the analytic context by Schur [1945], Schiffer [1948], and Curtiss [1971]. The formal treatment given here appears to be new.

§ 18.2. FABER EXPANSIONS

Let K be a simply connected, compact set in the complex plane, and let h be analytic on K (that is, analytic on an open set containing K). Is it possible to approximate h by polynomials uniformly on K, and if so, is there a systematic way to construct these approximating polynomials? These questions, which properly belong to complex approximation theory, are dealt with here because of the neat answers that are possible in terms of Faber polynomials.

I. A Naive Approach

Let $\tilde{D}$ be a simply connected region such that $K \subset \tilde{D}$ and h is analytic on $\tilde{D}$, and let f map $\tilde{D}$ onto E: $|w| < 1$. Then $\tilde{h} := h \circ f^{[-1]}$ is analytic on E and thus can be expressed as a power series,

$$\tilde{h}(w) = \sum_{m=0}^{\infty} a_m w^m, \qquad w \in E,$$

which converges uniformly on every compact subset of E. Since $K \subset \tilde{D}$ is compact, $f(K)$ is such a subset and thus, given $\varepsilon > 0$, there exists n such that

$$\left| \tilde{h}(w) - \sum_{m=0}^{n} a_m w^m \right| < \varepsilon, \qquad w \in f(K).$$

Setting $w = f(z)$, this in view of $\tilde{h}(f(z)) = h(z)$ becomes

$$\left| h(z) - \sum_{m=0}^{n} a_m [f(z)]^m \right| < \varepsilon, \qquad z \in K.$$

We thus have obtained an approximation to h. However, the approximations are polynomials in $f(z)$, and not in z. Thus in order to obtain approximating polynomials $p(z)$ by this method, it would be necessary to approximate the mapping function f, which may be as difficult a task as the original problem of approximating h. The fact that f depends on $\tilde{D}$, and thus on the region of analyticity of the function to be approximated, is a further flaw of the method.

II. The Faber Series

The disadvantages just mentioned are avoided in a method due to Faber. Let $\mathring{D}$ denote the complement of K, and let f map $\mathring{D}$ onto $|w| > 1$. The Laurent series of f has the form

$$f(z) = a_1 z + a_0 + a_{-1} z^{-1} + a_{-2} z^{-2} + \cdots, \tag{18.2-1}$$

where $a_1 \neq 0$. The inverse map $g = f^{[-1]}$ is represented by a similar Laurent series,

$$g(w) = b_1 w + b_0 + b_{-1} w^{-1} + \cdots, \tag{18.2-2}$$

where $b_1 = a_1^{-1}$. For $\rho > 1$, let Γ_ρ denote the curve $z = g(\rho e^{i\tau})$, $0 \leqq \tau \leqq 2\pi$. If ρ is sufficiently close to 1, h is analytic on and in the interior of Γ_ρ, and Cauchy's formula by Theorem 4.7d for $z \in K$ yields

$$h(z) = \frac{1}{2\pi i} \int_{\Gamma_\rho} \frac{h(t)}{t - z}\, dt.$$

Here we substitute $t = g(w)$ where $|w| = \rho$. This yields

$$h(z) = \frac{1}{2\pi i} \int_{|w|=\rho} \frac{h(g(w))g'(w)}{g(w) - z}\, dw. \tag{18.2-3}$$

For each $z \in K$, the function

$$w \mapsto \frac{g'(w)}{g(w) - z}$$

is analytic on $|w|>1$. Its Laurent series at ∞ is given by

$$\frac{g'(w)}{g(w)-z}=\sum_{n=0}^{\infty} p_n(z)w^{-n-1}, \tag{18.2-4}$$

where the $p_n(z)$ are the Faber polynomials associated with the series (18.2-2), considered as a formal Laurent series at ∞. For every $\sigma>1$ there holds the Cauchy coefficient estimate,

$$|p_n(z)|\leqq\frac{\mu(\sigma)}{\delta(\sigma)}\sigma^{n+1}, \qquad z\in K, \tag{18.2-5}$$

where $\delta(\sigma)$ is the distance of Γ_σ from K, and where

$$\mu(\sigma):=\sup_{|w|=\sigma}|g'(w)|.$$

For every $\rho>1$, choosing $1<\sigma<\rho$, the series (18.2-4) thus converges uniformly for $z\in K$ and for $|w|\geqq\rho$.

Substituting (18.2-4) into (18.2-3) and integrating term by term, we obtain:

THEOREM 18.2a. *Let h be analytic on the simply connected, compact set K, let $\{p_n(z)\}$ be the sequence of Faber polynomials of the series* (18.2-2), *where g maps $|w|>1$ onto the complement of K, and let*

$$c_n:=\frac{1}{2\pi i}\int_{|w|=\rho} h(g(w))w^{-n-1}\,dw, \qquad n=0,1,2,\ldots, \tag{18.2-6}$$

where $\rho>1$ is such that h is continuous on and analytic inside the image of $|w|=\rho$ under g. Then the representation

$$h(z)=\sum_{n=0}^{\infty} c_n p_n(z) \tag{18.2-7}$$

holds uniformly for $z\in K$.

The series (18.2-7) is called the **Faber series** of h with respect to K, and the coefficients c_n are its **Faber coefficients.**

III. Numerical Evaluation of the Faber Coefficients

If the boundary of K is a Jordan curve Γ, then the map g may be extended continuously to $|w|=1$, and the Faber coefficients (18.2-6) may be calculated by

$$c_n=\frac{1}{2\pi i}\int_{|w|=1} h(g(w))w^{-n-1}\,dw.$$

Here we set $w=e^{i\theta}$. If Γ is represented parametrically by $z=z(\tau)$, and if the outer inverse boundary correspondence function of the map g as in § 16.8 is defined by $z(\tau(\theta))=g(e^{i\theta})$, we obtain

$$c_n=\frac{1}{2\pi}\int_0^{2\pi} h(z(\tau(\theta)))\, e^{-in\theta}\, d\theta. \tag{18.2-8}$$

The Faber coefficients thus equal the Fourier coefficients of positive index of the function $h\circ z\circ \tau$, and thus may be calculated by FFT methods solely on the basis of the boundary correspondence function of the required map g. All coefficients are obtained simultaneously. See Ellacott [1983a] for practical experience concerning this method.

IV. Numerical Evaluation of the Faber Polynomials

We discuss two methods, both suggested by Ellacott [1983].

(a) If G is the Laurent series representing g, $F:=G^{[-1]}$, and

$$F^n=\sum_{k=-\infty}^{n} a_k^{(n)}x^k,$$

then by Theorem 18.1d,

$$p_n(z)=\sum_{k=0}^{n} a_k^{(n)}z^k, \qquad n=0,1,2,\ldots.$$

Because the $a_k^{(n)}$ are the Laurent coefficients of $[f(z)]^n$ where $f:=g^{[-1]}$, we thus have

$$a_k^{(n)}=\frac{1}{2\pi i}\int_\Gamma [f(z)]^n z^{-k-1}\, dz$$

for any Γ having winding number $+1$ with respect to K. Selecting for Γ a circle $z=\rho\, e^{i\tau}$ enclosing K, we get

$$a_k^{(n)}=\frac{1}{2\pi}\int_0^{2\pi}\left\{\frac{f(\rho\, e^{i\tau})}{\rho}\right\}^n e^{-ik\tau}\, d\tau.$$

Thus for fixed n, the coefficients $a_k^{(n)}$ are the Fourier coefficients of $[\rho^{-1}f(\rho\, e^{i\tau})]^n$, and may be computed simultaneously by an FFT. Ellacott [1983b] offers advice on how to choose ρ in this computation.

(b) If the Laurent series for g is explicitly available, the recurrence relation (18.1-15) may be used to generate the sequence of Faber polynomials. In numerous examples tried by Ellacott, this process appears to be numerically stable. The method is particularly appropriate in the case where the boundary curve Γ is a polygon, for then the power series for g is easily obtained from the Schwarz–Christoffel formula after the accessory parameters have been determined. Note that the formula for the *exterior* mapping is required.

V. The Faber Approximation Is Near Best

Under the hypotheses of Theorem 18.2a, let the Faber series (18.2-7) be truncated after the nth term, that is, consider

$$h_n(z) := \sum_{k=0}^{n} c_k p_k(z). \tag{18.2-9}$$

Since p_k is a polynomial of degree k, h_n is clearly a polynomial of degree $\leqq n$. By Theorem 18.2a, the polynomials h_n approximate the function h on K in the sense that they converge to h, uniformly for $z \in K$, as $n \to \infty$. Expressed differently, if we define the norm of a function k that is continuous on K by

$$\|k\| := \sup_{z \in K} |k(z)|,$$

then there holds

$$\|h - h_n\| \to 0 \qquad \text{as} \qquad n \to \infty.$$

Here we are now concerned with the quality of the approximation of h by h_n when n is fixed. In particular, we shall address the following question: how much worse can the approximation of h by h_n be than the approximation of h by *any* polynomial of degree n? To formalize this, let, for the given h and the given set K,

$$\rho_n = \rho_n(h, K) := \inf_{q \in \Pi_n} \|h - q\|, \tag{18.2-10}$$

where Π_n denotes the set of all polynomials of degree n. (It will be proved in a later chapter on approximation theory that the infimum is assumed, and that there exists a unique polynomial $q^* \in \Pi_n$ such that $\rho_n = \|h - q^*\|$. However, this result is not required here.)

THEOREM 18.2b. *Let the boundary Γ of the simply connected, compact set K be a Jordan curve. Then for each $n = 0, 1, 2, \ldots$ there exists a constant $\lambda_n > 0$ with the following property. If h is any analytic function on K, and if h_n is the truncated Faber series* (18.2-9) *of h, then there holds*

$$\|h - h_n\| \leqq (1 + \lambda_n)\, \rho_n(h, K). \tag{18.2-11}$$

As it stands, the theorem merely states that the approximation of h by h_n cannot be arbitrarily *worse* than the best polynomial approximation of corresponding degree. However, under weak additional hypotheses on Γ it will be shown that the numerical values of λ_n are rather small. It is in this sense, then, that the approximation of h by h_n is "near best."

Proof. Let $\varepsilon > 0$, and let $q \in \Pi_n$ be such that $\|h - q\| \leqq \rho_n + \varepsilon$. Since each p_k has precise degree k, q may be represented as a linear combination of the p_k,

$$q(z) = \sum_{k=0}^{n} c_k^* p_k(z), \tag{18.2-12}$$

and it is easily shown that this in fact is the Faber series for q,

$$c_k^* = \frac{1}{2\pi i} \int_{|w|=\rho} q(g(w)) w^{-k-1}\, dw, \qquad k = 0, 1, \ldots, n. \tag{18.2-13}$$

From

$$h(z) - h_n(z) = h(z) - q(z) + q(z) - h_n(z)$$

it is now obvious that

$$\|h - h_n\| \leqq \|h - q\| + \|q - h_n\|.$$

The first term on the right is bounded by $\rho_n + \varepsilon$. To estimate the second term, we use (18.2-6) and (18.2-13) to get

$$\begin{aligned} q(z) - h_n(z) &= \sum_{k=0}^{n} (c_k^* - c_k) p_k(z) \\ &= \sum_{k=0}^{n} \frac{1}{2\pi i} \int_{|w|=\rho} \{q(g(w)) - h(g(w))\} w^{-k-1}\, dw \cdot p_k(z) \\ &= \frac{1}{2\pi i} \int_{|w|=\rho} \{q(g(w)) - h(g(w))\} \sum_{k=0}^{n} p_k(z) w^{-k-1}\, dw. \end{aligned}$$

In view of the hypothesis on K, the map g may be extended continuously to $|w| \geqq 1$, and the integral may be computed on $|w| = 1$. There the expression in braces is bounded by $\rho_n + \varepsilon$. We therefore find

$$\|q - h_n\| \leqq (\rho_n + \varepsilon)\lambda_n,$$

where

$$\lambda_n := \sup_{z \in K} \frac{1}{2\pi} \int_0^{2\pi} \left| \sum_{k=0}^{n} p_k(z)\, e^{-ik\tau} \right| d\tau. \tag{18.2-14}$$

We have $\lambda_n < \infty$ because the integrand depends continuously on z, and the result follows on letting $\varepsilon \to 0$. —

We wish to estimate the constants λ_n and to this end require a new integral representation for the $p_k(z)$. The following simple lemma is needed:

LEMMA 18.2c. *Let the function $s(w)$ be analytic for $|w|>1$ and continuous for $|w|\geqq 1$ except for finitely many integrable singularities on $|w|=1$, and let*

$$s(w)=\sum_{n=0}^{\infty} s_n w^{-n}, \qquad |w|>1.$$

Then for $n=1, 2, \ldots$,

$$s_n=\frac{1}{\pi}\int_0^{2\pi} e^{in\theta} \operatorname{Re} s(e^{i\theta})\, d\theta. \tag{18.2-15}$$

Proof. By the formula for the coefficients of the Laurent series at ∞ of the function s,

$$\frac{1}{2\pi}\int_0^{2\pi} e^{in\theta} s(e^{i\theta})\, d\theta = s_n, \qquad n=0, 1, 2, \ldots, \tag{18.2-16}$$

and

$$\frac{1}{2\pi}\int_0^{2\pi} e^{-in\theta} s(e^{i\theta})\, d\theta = 0, \qquad n=1, 2, \ldots.$$

The second relation on conjugation yields

$$\frac{1}{2\pi}\int_0^{2\pi} e^{in\theta}\overline{s(e^{i\theta})}\, d\theta = 0$$

and on adding this to (18.2-16) we find

$$\frac{1}{2\pi}\int_0^{2\pi} e^{in\theta}[s(e^{i\theta})+\overline{s(e^{i\theta})}]\, d\theta = s_n,$$

the desired result. —

We now assume the boundary Γ of K to be piecewise analytic, with no angle at a corner (measured from ext K) being equal to zero. Then by § 16.4 the mapping function g is analytic on $|w|=1$, except at the pre-images of the corners of Γ, where $g'(w)$ has integrable singularities. If z is any point in the interior of K, we may thus apply Lemma 18.2c to

$$s(w):=\frac{wg'(w)}{g(w)-z}$$

and in view of (18.2-4) obtain

$$p_n(z)=\frac{1}{\pi}\int_0^{2\pi} e^{in\theta} \operatorname{Re}\left\{\frac{e^{i\theta}g'(e^{i\theta})}{g(e^{i\theta})-z}\right\} d\theta, \qquad n=1, 2, \ldots.$$

Here we use

$$\operatorname{Re}\left\{\frac{e^{i\theta}g'(e^{i\theta})}{g(e^{i\theta})-z}\right\} = \operatorname{Im}\left\{\frac{ie^{i\theta}g'(e^{i\theta})}{g(e^{i\theta})-z}\right\} = \operatorname{Im}\frac{d}{d\theta}\{\log[g(e^{i\theta})-z]\}$$

$$= \frac{d}{d\theta}\arg\{g(e^{i\theta})-z\}.$$

Setting

$$\nu(\theta, z) := \frac{d}{d\theta}\arg\{g(e^{i\theta})-z\}, \tag{18.2-17}$$

and noting that

$$\int_0^{2\pi} \nu(\theta, z)\, d\theta = 2\pi,$$

we thus obtain in view of $p_0(z) = 1$:

LEMMA 18.2d. *Let the boundary of K be a piecewise analytic Jordan curve, and let ν be defined by* (18.2-17). *Then for each z in the interior of K,*

$$p_n(z) = \begin{cases} \dfrac{1}{2\pi}\displaystyle\int_0^{2\pi} \nu(\theta, z)\, d\theta, & n = 0 \\[2ex] \dfrac{1}{\pi}\displaystyle\int_0^{2\pi} e^{in\theta}\nu(\theta, z)\, d\theta, & n > 0. \end{cases} \tag{18.2-18}$$

The sum required in the estimation of λ_n may now be simply expressed thus:

$$\sum_{k=0}^{n} p_k(z)\, e^{-ik\tau} = \frac{1}{2\pi}\int_0^{2\pi} \left\{1 + 2\sum_{k=1}^{n} e^{ik(\theta-\tau)}\right\} \nu(\theta, z)\, d\theta.$$

There follows:

$$\left|\sum_{k=0}^{n} p_k(z)\, e^{-ik\tau}\right| \leqq \frac{1}{2\pi}\int_0^{2\pi} |\nu(\theta, z)| \left|1 + 2\sum_{k=1}^{n} e^{ik(\theta-\tau)}\right| d\theta$$

and on interchanging integrations,

$$\frac{1}{\pi}\int_0^{2\pi} \left|\sum_{k=0}^{n} p_k(z)\, e^{-ik\tau}\right| d\tau$$

$$\leqq \frac{1}{2\pi}\int_0^{2\pi} |\nu(\theta, z)| \left\{\frac{1}{2\pi}\int_0^{2\pi} \left|1 + 2\sum_{k=1}^{n} e^{ik(\theta-\tau)}\right| d\tau\right\} d\theta.$$

The inner integral being independent of θ, we thus have:

LEMMA 18.2e. *If the boundary of K is a piecewise analytic Jordan curve, then the quantities λ_n occurring in Theorem 18.2b satisfy*

$$\lambda_n \leqq \omega(1+\sigma_n), \qquad n=1,2,\ldots, \tag{18.2-19}$$

where

$$\omega := \sup_{z\in K} \frac{1}{2\pi}\int_0^{2\pi} |\nu(\theta, z)|\, d\theta \tag{18.2-20}$$

is independent of n, and

$$\sigma_n := \frac{1}{\pi}\int_0^{2\pi} \left|\sum_{k=1}^{n} e^{ik\tau}\right| d\tau, \qquad n=1,2,\ldots, \tag{18.2-21}$$

is independent of K.

Some remarks concerning the two factors of the product estimating λ_n follow.

(a) The quantity ω may be shown to equal $(2\pi)^{-1}$ times the **total rotation** of the boundary curve Γ, that is,

$$\omega = \frac{1}{2\pi}\int_0^{2\pi} |\arg z'(\theta)|\, d\theta, \tag{18.2-22}$$

where $z(\theta) := g(e^{i\theta})$, $0 \leqq \theta \leqq 2\pi$. Yet a simpler statement is possible if K is assumed to be *convex*. Then for each z in the interior of K, $\arg\{g(e^{i\theta}) - z\}$ may be defined as a continuous increasing function for $0 \leqq \theta \leqq 2\pi$, and its total increase equals 2π, independently of z. Thus we then have

$$\int_0^{2\pi} |\nu(\theta, z)|\, d\theta = \int_0^{2\pi} \nu(\theta, z)\, d\theta = 2\pi,$$

and we find

$$\omega = 1. \tag{18.2-23}$$

(b) For an appraisal of the potential usefulness of Theorem 18.2b it is important to know numerical values as well as the correct asymptotic behavior as $n \to \infty$ of the quantities σ_n. We carry through the required computations for n even, $n = 2m$; the final results for odd values of n are identical.

Using the fact that a complex number does not change its modulus when multiplied by a number of modulus 1, we have

$$\sigma_n = \sigma_{2m} = \frac{1}{\pi}\int_0^{2\pi} \left|\sum_{k=0}^{2m-1} e^{i(-m+k+\frac{1}{2})\tau}\right| d\tau.$$

The sum of exponentials is real, and by summing geometrically is found to equal

$$\frac{\sin m\tau}{\sin(\tau/2)}.$$

Thus it is positive for small $\tau > 0$ and changes its sign at each of the points $\tau_j := j\pi/m$, $j = 1, 2, \ldots, 2m-1$. Setting

$$w_k := \exp\left\{\frac{(2k+1)i\pi}{2m}\right\},$$

we thus have

$$\begin{aligned}\sigma_{2m} &= \frac{1}{\pi}\sum_{j=0}^{2m-1}(-1)^j \int_{\tau_j}^{\tau_{j+1}} 2\,\mathrm{Re}\sum_{k=0}^{m-1} e^{i(k+\frac{1}{2})\tau}\,d\tau \\ &= \frac{4}{\pi}\sum_{j=0}^{2m-1}(-1)^j\,\mathrm{Re}\sum_{k=0}^{m-1}\frac{1}{(2k+1)i}[w_k^{j+1} - w_k^j]\end{aligned}$$

and on interchanging summations, using $w_k^{2m} = -1$,

$$\begin{aligned}\sigma_{2m} &= \frac{4}{\pi}\sum_{k=0}^{m-1}\frac{1}{2k+1}\,\mathrm{Im}(w_k - 1)\sum_{j=0}^{2m-1}(-w_k)^j \\ &= \frac{8}{\pi}\sum_{k=0}^{m-1}\frac{1}{2k+1}\,\mathrm{Im}\,\frac{w_k - 1}{w_k + 1}\end{aligned}$$

or finally

$$\sigma_{2m} = \frac{8}{\pi}\sum_{k=0}^{m-1}\frac{1}{2k+1}\tan\frac{(2k+1)\pi}{4m}. \tag{18.2-24}$$

We write $\sigma_{2m} = \sigma'_{2m} + \sigma''_{2m}$, where

$$\sigma'_{2m} := \frac{4}{\pi}\sum_{k=0}^{m-1}\left\{\frac{4m}{(2k+1)\pi}\tan\frac{(2k+1)\pi}{4m} - \frac{8m}{(2m-2k-1)\pi^2}\right\}\frac{\pi}{2m}$$

is a midpoint sum approximating the integral

$$I := \frac{4}{\pi}\int_0^{\pi/2}\left[\frac{1}{x}\tan x - \frac{4}{\pi(\pi - 2x)}\right]dx$$

and thus by standard results on numerical integration satisfies

$$\sigma'_{2m} = I + O(m^{-2}), \tag{18.2-25}$$

whereas

$$\sigma''_{2m} := \frac{16}{\pi^2}\sum_{k=0}^{m-1}\frac{1}{2m-2k-1} = \frac{16}{\pi^2}\sum_{k=0}^{m-1}\frac{1}{2k+1}.$$

Letting, for $n = 0, 1, 2, \ldots$,

$$h_n := 1 + \frac{1}{2} + \cdots + \frac{1}{n},$$

we have

$$\sum_{k=1}^{m-1} \frac{1}{2k+1} = h_{2m} - \frac{1}{2} h_m$$

and hence, using (11.11-16),

$$\sigma''_{2m} = \frac{8}{\pi^2} \operatorname{Log} 2m + \frac{8}{\pi^2} (\operatorname{Log} 2 + \gamma) + O(m^{-2}),$$

where $\gamma := 0.557216\ldots$ is Euler's constant. By numerical integration, $I = 0.949145\ldots$. For even n we thus have

$$1 + \sigma_n = \frac{8}{\pi^2} \operatorname{Log} n + 2.978861 \cdots + O(n^{-2}). \qquad (18.2\text{-}26)$$

We have thus obtained our final result:

THEOREM 18.2f. *If K is a convex set with a piecewise analytic boundary, then the estimate of Theorem* 18.2b *holds with* $\lambda_n = 1 + \sigma_n$, *where* σ_n *is defined by* (18.2-21), *may be calculated from* (18.2-24), *and satisfies the asymptotic relation* (18.2-26).

Some numerical values of $\lambda_n = 1 + \sigma_n$ are given in the following table:

n	$1+\alpha_n$ (exact)	$1+\sigma_n$ (asympt. formula)
2	3.546479	3.540705
4	4.104034	4.102549
8	4.664768	4.664393
16	5.226332	5.226237
32	5.788106	5.788081
64	6.349932	6.349925
128	6.911772	6.911769

Since $1 + \sigma_n \geqq 9$ only for $n \geqq 1682$, it is seen that for any $n < 1682$ at most 1 decimal digit is lost when the polynomial q_n^* of best approximation is replaced by the Faber series of degree n. This result is not without interest even in the case where K is a circular disk, when the Faber series reduces to Taylor's series. However, in the case of a disk approximations that are much closer to the best approximation can be obtained by means of Trefethen's Carathéodory–Fejer approximation, which is described in a later chapter.

PROBLEMS

1. If $s(w)=\sum_{n=0}^{\infty} a_n w^n$ is analytic for $|w|<1$ and continuous for $|w|\leqq 1$, Im $s(0)=0$, show that

$$a_n=\frac{\varepsilon_n}{2\pi}\int_0^{2\pi} e^{-in\theta}\,\mathrm{Re}\,s(e^{i\theta})\,d\theta,$$

where $\varepsilon_0=1$, $\varepsilon_1=\varepsilon_2=\cdots=2$. As a corollary, obtain the Schwarz formula (Theorem 14.2b).

2. Let $\{p_n(z)\}$ be the sequence of Faber polynomials associated with the simply connected compact set K. The **Faber transform** T maps the polynomial $p(w):=\sum_{k=0}^{n} a_k w^k$ onto the polynomial

$$(Tp)(z):=\sum_{k=0}^{n} a_k p_k(z).$$

(a) Show that T is one-to-one.

(b) If the boundary of K is a piecewise analytic Jordan curve Γ, show that for $z\in \mathrm{int}\,K$,

$$(Tp)(z)=\frac{1}{2\pi i}\int_{|w|=1}\frac{p(w)g'(w)}{g(w)-z}\,dw=\frac{1}{2\pi i}\int_{\Gamma}\frac{p(f(t))}{t-z}\,dt.$$

(c) Show that the inverse map T^{-1} is given by

$$(T^{-1}q)(v)=\frac{1}{2\pi i}\int_{|w|=1}\frac{q(g(w))}{w-v}\,dw.$$

3. (Continuation) The *norm* of the Faber operator T is defined by

$$\|T\|:=\sup_{p\neq 0}\left\{\frac{\sup_{z\in K}|Tp(z)|}{\sup_{|w|=1}|p(w)|}\right\}.$$

Show that under the hypotheses and with the notation of Lemma 18.2e,

$$\|T\|\leqq 1+2\omega.$$

4. By the foregoing problem, T is a bounded operator. Because the polynomials are dense in the Banach space $A(D)$ of functions analytic in $|w|<1$ and continuous in $|w|\leqq 1$, endowed with the supremum norm, T may be extended to $A(D)$. Show that for any $h\in A(D)$,

$$(Th)(z)=\frac{1}{2\pi i}\int_{\Gamma}\frac{h(f(t))}{t-z}\,dt,\qquad z\in \mathrm{int}\,K,$$

where Γ is the boundary of K.

5. Prove that the Faber transform of a rational function is rational. More precisely, let r be a rational function with poles of the respective orders m_k at the points w_k, $k = 1, 2, \ldots$, $|w_k| > 1$. Show that Tr is a rational function with poles of orders $\leqq m_k$ at the points $z_k := g(w_k)$.

NOTES

On the Faber series, see Faber [1903]. Ellacott [1983b] has experimental results on the methods proposed in Sections III and IV. With a proper interpretation of $\nu(\theta, z)$, the integral representation (18.2-18) may be shown to hold for arbitrary sets K whose boundary curve has bounded rotation; see Pommerenke [1964] or Gaier [1980], p. 49. The simple proof given here is based on a personal communication by S. W. Ellacott. Theorem 18.4b is essentially due to Kövari and Pommerenke [1967]. For a discussion of the constants σ_n see Geddes and Mason [1975]. On the Faber transform dealt with in Problems 2–5 see Gaier [1980]; for Problem 5 in particular, see Ganelius [1982], p. 24, or Ellacott [1983b].

§ 18.3. THE SPACE OF SQUARE INTEGRABLE FUNCTIONS OVER A REGION, THE BERGMAN KERNEL FUNCTION, AND CONFORMAL MAPPING

I. The space $L_2(R)$

Let $R \subset \mathbb{C}$ be a region of arbitrary connectivity, $R \neq \mathbb{C}$. We denote by $L_2(R)$ the space of functions that are analytic in R and for which

$$\iint_R |f(z)|^2 \, \boxed{dz} < \infty. \tag{18.3-1}$$

Our first task is to give a precise meaning to the integral appearing in (18.3-1). No problem arises if the integral is taken in the sense of Lebesgue. If the integral is to be defined constructively, however, then Riemann's notion of an integral must be used. For two-dimensional integrals, this is defined only for functions that are continuous on compact sets whose boundary is at least piecewise smooth. Thus if f does not possess a continuous extension to the closure of R, or if the boundary of R is not piecewise smooth, an integral such as (18.3-1) must be defined by a limit process, as follows. Let $\{R_n\}$ be a sequence of compact sets in R such that $R_n \subset R_{n+1}$, $n = 0, 1, 2, \ldots$, and such that every compact set $K \subset R$ is ultimately contained in some R_n (and hence in all subsequent R_n). (If R is simply connected, R_n may be taken as the image of $|w| \leqq 1 - 2^{-n}$ under a conformal map of $|w| < 1$ onto R.) Now if ϕ is any continuous, nonnegative function on R, the integral

$$\lambda_n := \iint_{R_n} \phi(z) \, \boxed{dz}$$

exists for $n = 0, 1, 2, \ldots$. The sequence $\{\lambda_n\}$ is obviously nondecreasing, hence either it tends to $+\infty$, or it has a finite limit

$$\lambda := \lim_{n\to\infty} \lambda_n.$$

It is easy to see that the occurrence of a finite or infinite limit, or the value of the limit if it is finite, does not depend on the particular sequence $\{R_n\}$ provided it satisfies the stipulated conditions. Hence we may define

$$\iint_R |f(z)|^2 \boxed{dz} := \lim_{n\to\infty} \iint_{R_n} |f(z)|^2 \boxed{dz}.$$

We now show that $L_2(R)$, with the obvious definitions of addition and scalar multiplication, is a linear space (see § 2.1). It is clear that the axioms (2.1-1) for scalar multiplication are satisfied. It remains to be shown that if $f, g \in L_2(R)$, then also $f+g \in L_2(R)$. Clearly $f+g$ is analytic in R. To show that the integral exists, we observe that for any $z \in R$,

$$|f(z)+g(z)|^2 \leqq |f(z)|^2 + 2|f(z)||g(z)| + |g(z)|^2.$$

Hence for $n = 0, 1, 2, \ldots$,

$$\iint_{R_n} |f(z)+g(z)|^2 \boxed{dz} \leqq \iint_{R_n} |f(z)|^2 \boxed{dz} + 2\iint_{R_n} |f(z)||g(z)| \boxed{dz} + \iint_{R_n} |g(z)|^2 \boxed{dz}.$$

By the Schwarz inequality, the second integral on the right is not greater than

$$\left\{\iint_{R_n} |f(z)|^2 \boxed{dz} \iint_{R_n} |g(z)|^2 \boxed{dz}\right\}^{1/2}.$$

Letting $n \to \infty$, the limits of all integrals on the right are finite by hypothesis, and we get

$$\iint_R |f(z)+g(z)|^2 \boxed{dz} \leqq \left\{\left[\iint_R |f(z)|^2 \boxed{dz}\right]^{1/2} + \left[\iint_R |g(z)|^2 \boxed{dz}\right]^{1/2}\right\}^2, \qquad (18.3\text{-}2)$$

showing that $f+g \in L_2(R)$.

We next define the *norm* of an element $f \in L_2(R)$ by

$$\|f\| := \left[\iint_R |f(z)|^2 \boxed{dz}\right]^{1/2}. \tag{18.3-3}$$

With this definition of a norm, $L_2(R)$ becomes a *normed linear space.* The first three of the axioms (2.1-2) of a norm are trivially satisfied, and (18.3-2) expresses the fact that the triangle inequality, $\|f+g\| \leqq \|f\| + \|g\|$, holds for arbitrary $f, g \in L_2(R)$.

We penetrate into analysis more deeply by showing that $L_2(R)$ is complete, that is, that it is a *Banach space.* As a preliminary result we require:

LEMMA 18.3a. *Let $z_0 \in R$, and let $f \in L_2(R)$. Then*

$$|f(z_0)| \leqq \frac{1}{\sqrt{\pi\delta}} \|f\|, \tag{18.3-4}$$

where δ is the distance of z_0 from the boundary of R.

Proof. Let $0 < \sigma < \delta$, and let D_σ: $|z - z_0| \leqq \sigma$. Because D_σ is a compact set contained in R,

$$\|f\|^2 \geqq \iint_{D_\sigma} |f(z)|^2 \boxed{dz} = \int_0^\sigma \rho\, d\rho \int_0^{2\pi} |f(z_0 + \rho\, e^{i\theta})|^2\, d\theta. \tag{18.3-5}$$

We have

$$\int_0^{2\pi} |f(z_0 + \rho\, e^{i\theta})|^2\, d\theta \geqq \left| \int_0^{2\pi} [f(z_0 + \rho\, e^{i\theta})]^2\, d\theta \right|,$$

and using the mean value property for analytic functions, the last expression equals $2\pi|f(z_0)|^2$, independently of ρ. Thus

$$\int_0^\sigma \rho\, d\rho \int_0^{2\pi} |f(z_0 + \rho\, e^{i\theta})|^2\, d\theta \geqq 2\pi|f(z_0)|^2 \int_0^\sigma \rho\, d\rho = \pi\sigma^2 |f(z_0)|^2,$$

and together with (18.3-5) this is seen to imply (18.3-4). —

We now can show:

THEOREM 18.3b. *In the metric defined by the norm* (18.3-3), *$L_2(R)$ is complete.*

Proof. It is to be shown that if $\{f_n\}$ is a Cauchy sequence in $L_2(R)$, that is, if, given any $\varepsilon > 0$, there exists $k = k(\varepsilon)$ such that

$$\|f_n - f_m\| < \varepsilon \qquad \text{whenever} \qquad m, n > k, \tag{18.3-6}$$

then an element $f \in L_2(R)$ can be found such that

$$\|f_n - f\| \to 0 \qquad \text{as} \qquad n \to \infty. \tag{18.3-7}$$

Let $\{f_n\}$ be a Cauchy sequence, and let K be a compact subset of R. If δ denotes the distance of K from the boundary of R, then $\delta > 0$, and Lemma 18.3a shows that

$$|f_n(z) - f_m(z)| \leqq \frac{1}{\sqrt{\pi}\,\delta} \|f_n - f_m\|.$$

It follows that the sequence $\{f_n\}$ converges uniformly on K, and hence that it converges locally uniformly on R. By Theorem 3.4b its limit function, which we call f, is analytic on R.

We next show that $f \in L_2(R)$. Because $\{f_n\}$ is a Cauchy sequence, the sequence of real numbers $\|f_n\|$ is bounded, say, by μ. Therefore, for any compact set $K \subset R$ and for any n,

$$\iint_K |f_n(z)|^2 \boxed{dz} \leqq \iint_R |f_n(z)|^2 \boxed{dz} \leqq \mu^2,$$

and because $f_n \to f$ uniformly on K,

$$\iint_K |f(z)|^2 \boxed{dz} \leqq \mu^2.$$

Because $K \subset R$ was arbitrary, it follows that

$$\iint_R |f(z)|^2 \boxed{dz} \leqq \mu^2.$$

We finally establish (18.3-7). We know that for any $\varepsilon > 0$ there exists $k = k(\varepsilon)$ such that

$$\iint_K |f_n(z) - f_m(z)|^2 \boxed{dz} < \frac{\varepsilon^2}{16}, \qquad n, m > k,$$

where K is any compact subset of R. Since $f_m \to f$ uniformly on any such K, there follows

$$\iint_K |f_n(z) - f(z)|^2 \boxed{dz} < \frac{\varepsilon^2}{4}, \qquad n > k.$$

Because this is true for any compact $K \subset R$, it is also true that

$$\|f_n - f\| < \varepsilon, \qquad n > k,$$

which proves (18.3-7). —

Going one step further, we shall now show that an inner product can be defined in $L_2(R)$ in such a way that it becomes a *Hilbert space.* In general, a **Hilbert space** is a Banach space $\mathscr{B}$ in which on the set of ordered pairs $\{\mathbf{x}, \mathbf{y}\}$ of elements $\mathbf{x}, \mathbf{y} \in \mathscr{B}$ a complex-valued function, called the **inner product** of $\mathbf{x}$ and $\mathbf{y}$ and ordinarily denoted by $(\mathbf{x}, \mathbf{y})$, is defined which satisfies the following axioms for arbitrary $\mathbf{x}, \mathbf{y}, \mathbf{z}, \cdots \in \mathscr{B}$ and arbitrary complex numbers a:

$$(\mathbf{x}, \mathbf{x}) \geqq 0, \qquad [(\mathbf{x}, \mathbf{x})]^{1/2} = \|\mathbf{x}\|, \tag{18.3-8a}$$

$$(\mathbf{x}+\mathbf{y}, \mathbf{z}) = (\mathbf{x}, \mathbf{z}) + (\mathbf{y}, \mathbf{z}), \tag{18.3-8b}$$

$$(a\mathbf{x}, \mathbf{y}) = a(\mathbf{x}, \mathbf{y}), \tag{18.3-8c}$$

$$(\mathbf{y}, \mathbf{x}) = \overline{(\mathbf{x}, \mathbf{y})}. \tag{18.3-8d}$$

From these axioms there follows the *Schwarz inequality*:

$$|(\mathbf{x}, \mathbf{y})| \leqq \|\mathbf{x}\| \, \|\mathbf{y}\|. \tag{18.3-9}$$

The following examples of Hilbert spaces are undoubtedly familiar:

EXAMPLE 1

$\mathbb{C}^n$, the complex Euclidean space of dimension n, becomes a Hilbert space if the inner product of the n-tuples

$$\mathbf{x} = \begin{pmatrix} x_1 \\ x_2 \\ \vdots \\ x_n \end{pmatrix} \quad \text{and} \quad \mathbf{y} = \begin{pmatrix} y_1 \\ y_2 \\ \vdots \\ y_n \end{pmatrix}$$

is defined by

$$(\mathbf{x}, \mathbf{y}) := \sum_{i=1}^{n} x_i \overline{y_i}.$$

EXAMPLE 2

The Banach space of infinite sequences $x = \{x_i\}_{i=0}^{\infty}$ satisfying

$$\sum_{i=0}^{\infty} |x_i|^2 < \infty$$

becomes a Hilbert space by defining the inner product

$$(x, y) := \sum_{i=0}^{\infty} x_i \overline{y_i}.$$

EXAMPLE 3

If $[\alpha, \beta]$ is a finite interval, the complex-valued measurable functions f defined on $[\alpha, \beta]$ that are square integrable form a Banach space with the norm

$$\|f\| := \left[\int_\alpha^\beta |f(\tau)|^2 \, d\tau \right]^{1/2}.$$

It becomes a Hilbert space by defining the inner product

$$(f, g) := \int_\alpha^\beta f(\tau)\overline{g(\tau)} \, d\tau. \quad —$$

In the case $\mathscr{B} = L_2(R)$ on hand, Example 3 suggests that we define the inner product of two elements f and g in $L_2(R)$ by

$$(f, g) := \iint_R f(z)\overline{g(z)} \, \boxed{dz}. \tag{18.3-10}$$

However, the integral as yet has no meaning because the only integrals over R that have been defined are those of nonnegative real functions. To define the integral, we use the algebraic identity

$$f\bar{g} = \frac{1}{2}|f+g|^2 + \frac{i}{2}|f+ig|^2 - \frac{1+i}{2}(|f|^2 + |g|^2).$$

The integral of each term on the right is defined, hence we may put

$$\iint_R f(z)\overline{g(z)} \, \boxed{dz} := \frac{1}{2}\|f+g\|^2 + \frac{i}{2}\|f+ig\|^2 - \frac{1+i}{2}(\|f\|^2 + \|g\|^2).$$

With this definition of the integral in (18.3-10) it is easy to verify that the inner product (f, g) indeed satisfies the axioms (18.3-8).

THEOREM 18.3c. *With the definition* (18.3-10) *of the inner product,* $L_2(R)$ *is a Hilbert space.*

II. Orthonormal Systems in Hilbert Space

We require some notions and computational facts that hold in any infinite-dimensional Hilbert space. In any linear space $\mathscr{L}$, the elements $\mathbf{x}_1, \mathbf{x}_2, \ldots, \mathbf{x}_n$ are called **linearly independent** if the relation

$$c_1\mathbf{x}_1 + c_2\mathbf{x}_2 + \cdots + c_n\mathbf{x}_n = \mathbf{0} \tag{18.3-11}$$

is possible only if all $c_i = 0$. If (18.3-11) is possible without all c_i being 0, then the elements $\mathbf{x}_1, \ldots, \mathbf{x}_n$ are called **linearly dependent.** If there exist integers n such that any $n+1$ elements in $\mathscr{L}$ are linearly dependent, then the smallest such n is called the **dimension** of $\mathscr{L}$. If no such n exists, $\mathscr{L}$ is called **infinite-dimensional.**

In a Hilbert space $\mathscr{H}$, two elements $\mathbf{x}$ and $\mathbf{y}$ are called **orthogonal** if

$$(\mathbf{x}, \mathbf{y}) = 0.$$

THEOREM 18.3d. *Given a sequence $\{\mathbf{x}_n\}_{n=0}^{\infty}$ of linearly independent elements of an infinite-dimensional Hilbert space $\mathscr{H}$, there exists a sequence $\{y_n\}_{n=0}^{\infty}$ of elements of $\mathscr{H}$ such that for $n = 0, 1, \ldots$,*

(i) $\mathbf{y}_n$ is a linear combination of $\mathbf{x}_0, \mathbf{x}_1, \ldots, \mathbf{x}_n$,
(ii) $\|\mathbf{y}_n\| = 1$.
(iii) $(\mathbf{y}_n, \mathbf{y}_m) = 0$ if $m \neq n$.

The *proof* is by an explicit construction known as the **Gram–Schmidt process.** Because $\mathbf{x}_0$ is linearly independent, $\mathbf{x}_0 \neq 0$, hence $\|\mathbf{x}_0\| \neq 0$. Condition (i) for $n = 0$ now requires $\mathbf{y}_0 = c_0\mathbf{x}_0$, and (ii) may be satisfied by taking $c_0 = \|\mathbf{x}_0\|^{-1}$. (Any number of modulus $\|\mathbf{x}_0\|^{-1}$ would do, thus the construction is not unique.) Suppose $\mathbf{y}_0, \mathbf{y}_1, \ldots, \mathbf{y}_n$ have been determined such that (i), (ii), and (iii) hold. We next find

$$\mathbf{y}_{n+1}^* = \mathbf{x}_{n+1} - \sum_{k=0}^{n} c_k \mathbf{y}_k$$

such that $(\mathbf{y}_{n+1}^*, \mathbf{y}_m) = 0$ for $m = 0, 1, \ldots, n$. Using (ii) and (iii), this requires

$$c_m = (\mathbf{x}_{n+1}, \mathbf{y}_m), \qquad m = 0, 1, \ldots, n.$$

We have $\mathbf{y}_{n+1}^* \neq \mathbf{0}$, because otherwise $\mathbf{x}_0, \mathbf{x}_1, \ldots, \mathbf{x}_n$ would be linearly dependent. We thus can satisfy (ii) for $n+1$ by forming

$$\mathbf{y}_{n+1} := \frac{y_{n+1}^*}{\|\mathbf{y}_{n+1}^*\|}. \quad —$$

A sequence $\{\mathbf{y}_n\}$ having the properties (ii) and (iii) is called an **orthonormal sequence.** Although the Gram–Schmidt process is theoretically straightforward, the actual numerical construction of an orthonormal sequence from a given sequence $\{\mathbf{x}_n\}$ is a nontrivial problem of numerical linear algebra, especially if the elements $\mathbf{x}_i$ are nearly linearly dependent; see Chapter 6 of Golub and van Loan [1983].

The elements of any orthonormal sequence $\{\mathbf{y}_m\}$ are linearly independent, for if

$$\sum_{k=0}^{n} c_k \mathbf{y}_k = \mathbf{0},$$

then scalar multiplication by $\mathbf{y}_m$ yields

$$\left(\sum_{k=0}^{n} c_k \mathbf{y}_k, \mathbf{y}_m\right) = \sum_{k=0}^{n} c_k(\mathbf{y}_k, \mathbf{y}_m) = c_m = 0, \qquad m = 0, 1, \ldots, n.$$

Now let $\mathbf{x}$ be an arbitrary element of $\mathcal{H}$, and let $\{\mathbf{y}_n\}$ be an orthonormal sequence. Then the numbers

$$c_k := (\mathbf{x}, \mathbf{y}_k), \qquad k = 0, 1, 2, \ldots \tag{18.3-12}$$

are called the **Fourier coefficients** of $\mathbf{x}$ with respect to the sequence $\{\mathbf{y}_n\}$. With these coefficients one may form the (formal) series

$$\mathbf{s} := \sum_{k=0}^{\infty} c_k \mathbf{y}_k, \tag{18.3-13}$$

called the **Fourier series** of $\mathbf{x}$ in the system $\{\mathbf{y}_n\}$. Judging from elementary special cases, one expects some sort of relationship between $\mathbf{s}$ and $\mathbf{x}$. While it is not always true that $\mathbf{s} = \mathbf{x}$, the following may be shown:

THEOREM 18.3e. *The Fourier series* (18.3-13) *of an element* $\mathbf{x} \in \mathcal{H}$ *with respect to the system* $\{\mathbf{y}_n\}$ *has the following properties*:

(i) *It converges, and its sum* $\mathbf{s}$ *satisfies*

$$(\mathbf{x} - \mathbf{s}, \mathbf{y}_m) = 0, \qquad m = 0, 1, 2, \ldots.$$

(ii)
$$\sum_{k=0}^{\infty} |c_k|^2 \leqq \|\mathbf{x}\|^2. \tag{18.3-14}$$

(iii) *For* $n = 0, 1, 2, \ldots$ *and for arbitrary complex numbers* $a_0, a_1, \ldots$ *there holds*

$$\left\|\mathbf{x} - \sum_{k=0}^{n} c_k \mathbf{y}_k\right\| \leqq \left\|\mathbf{x} - \sum_{k=0}^{n} a_k \mathbf{y}_k\right\|,$$

that is, the nth partial sum of the Fourier series gives the best approximations to $\mathbf{x}$ *in the set of all linear combinations of* $\mathbf{y}_0, \mathbf{y}_1, \ldots, \mathbf{y}_n$.

Proof. We begin by proving (iii). Using the fact that the system $\{\mathbf{y}_n\}$ is orthonormal,

$$\left\|\mathbf{x} - \sum_{k=0}^{n} a_k \mathbf{y}_k\right\|^2 = \left(\mathbf{x} - \sum_{k=0}^{n} a_k \mathbf{y}_k, \mathbf{x} - \sum_{k=0}^{n} a_k \mathbf{y}_k\right)$$

$$= \|\mathbf{x}\|^2 - \sum_{k=0}^{n} (a_k \overline{c_k} + \overline{a_k} c_k - |a_k|^2),$$

where c_k is given by (18.3-12). If $a_k = c_k$,

$$\left\|\mathbf{x} - \sum_{k=0}^{n} c_k \mathbf{y}_k\right\|^2 = \|\mathbf{x}\|^2 - \sum_{k=0}^{n} |c_k|^2, \tag{18.3-15}$$

and subtraction yields

$$\left\|\mathbf{x} - \sum_{k=0}^{n} a_k \mathbf{y}_k\right\|^2 - \left\|\mathbf{x} - \sum_{k=0}^{n} c_k \mathbf{y}_k\right\|^2 = \sum_{k=0}^{n} (|a_k|^2 - a_k\overline{c_k} - \overline{a_k}c_k + |c_k|^2)$$

$$= \sum_{k=0}^{n} (a_k - c_k)(\overline{a_k} - \overline{c_k}) \geqq 0,$$

proving (iii). From (18.3-15) we infer that

$$\sum_{k=0}^{n} |c_k|^2 \leqq \|\mathbf{x}\|^2$$

for every n, which establishes (ii). Thus given $\varepsilon > 0$, there exists $k > 0$ such that for $m > k$ and $n > m$,

$$\sum_{l=m}^{n} |c_l|^2 < \varepsilon.$$

The expression on the left equals

$$\left\|\sum_{l=m}^{n} c_l \mathbf{y}_l\right\|^2,$$

showing that the partial sums of the series $\mathbf{s}$ form a Cauchy sequence. Because $\mathscr{H}$ is complete, $\mathbf{s}$ thus exists as an element of $\mathscr{H}$. Finally if $n > m$,

$$\left(\sum_{l=0}^{n} c_l \mathbf{y}_l, \mathbf{y}_m\right) = c_m.$$

Because the scalar product is continuous in each of its arguments, on letting $n \to \infty$ there follows $(\mathbf{s}, \mathbf{y}_m) = c_m$, proving (i). —

Relation (18.3-14) is called **Bessel's inequality.**

Fourier series are of interest mainly if they can be used to *represent* an arbitrary element $\mathbf{x} \in \mathscr{H}$. For this to be the case, the series $\mathbf{s}$ must not only converge; it also must be equal to $\mathbf{x}$. An orthonormal system $\{\mathbf{y}_k\}$ such that

$$\mathbf{x} = \sum_{k=0}^{\infty} (\mathbf{x}, \mathbf{y}_k)\mathbf{y}_k \tag{18.3-16}$$

for all $\mathbf{x} \in \mathscr{H}$ is called **complete.** The following theorem states four different conditions for an orthonormal system to be complete.

THEOREM 18.3f. *Let $\{\mathbf{y}_n\}$ be an orthonormal system in the Hilbert space $\mathcal{H}$. Then the following statements are equivalent*:

(i) *The relation* (18.3-16) *holds for all* $\mathbf{x} \in \mathcal{H}$.
(ii) *Linear combinations of finitely many* $\mathbf{y}_n$ *are dense in* $\mathcal{H}$.
(iii) *For each* $\mathbf{x} \in \mathcal{H}$ *there holds* **Parseval's relation,**

$$\sum_{k=0}^{\infty} |(\mathbf{x}, \mathbf{y}_k)|^2 = \|\mathbf{x}\|^2. \tag{18.3-17}$$

(iv) If $(\mathbf{x}, \mathbf{y}_n)$ *for all* n, *then* $\mathbf{x} = \mathbf{0}$.

Proof. We show (i)⇒(ii)⇒(iii)⇒(iv)⇒(i). Thus suppose (i) holds. To prove (ii), we have to show that, given any $\varepsilon > 0$, then there exist finitely many constants $a_0, a_1, \ldots, a_n$ such that

$$\left\| \mathbf{x} - \sum_{k=0}^{n} a_k \mathbf{y}_k \right\| < \varepsilon.$$

Clearly, a partial sum of sufficiently high order of the Fourier series for $\mathbf{x}$ will do the trick. If

$$\left\| \mathbf{x} - \sum_{k=0}^{n} c_k \mathbf{y}_k \right\| < \varepsilon,$$

then by computing the norm of the element on the left we see that

$$\|\mathbf{x}\|^2 - \sum_{k=0}^{n} |c_k|^2 < \varepsilon^2.$$

Since ε was arbitrary, this proves (iii). If Parseval's relation holds and $c_k = (\mathbf{x}, \mathbf{y}_k) = 0$ for all k, then $\|\mathbf{x}\| = 0$ and thus $\mathbf{x} = \mathbf{0}$, proving (iv). To show that this implies (i), we recall from Theorem 18.3e that the Fourier series converges in any case, and that its sum $\mathbf{s}$ satisfies

$$(\mathbf{x} - \mathbf{s}, \mathbf{y}_n) = 0 \quad \text{for all} \quad n.$$

By (iv) this implies $\mathbf{x} - \mathbf{s} = \mathbf{0}$, or $\mathbf{x} = \mathbf{s}$. —

EXAMPLE **4**

Let $L_2(-\pi, \pi)$ be the space of Lebesgue integrable functions on $[-\pi, \pi]$ (more precisely: of the equivalence classes of such functions differing at most on sets of measure zero). If endowed with the inner product

$$(f, g) := \int_{-\pi}^{\pi} f(\tau)\overline{g(\tau)}\, d\tau,$$

$L_2(-\pi, \pi)$ becomes a Hilbert space $\mathcal{H}$. It can be shown that the functions

$$g_0(\tau) := \frac{1}{\sqrt{2\pi}}, \qquad g_{2k-1}(\tau) := \frac{1}{\sqrt{\pi}} \cos k\tau, \qquad g_{2k}(\tau) := \frac{1}{\sqrt{\pi}} \sin k\tau,$$

$k = 1, 2, \ldots$, form a complete orthonormal system in $\mathcal{H}$. Thus every $f \in \mathcal{H}$ can be expanded in a Fourier series $\sum a_k q_k(\tau)$ in the sense that

$$\left\| f - \sum_{k=0}^{n} a_k g_k \right\| = \left\{ \int_{-\pi}^{\pi} \left| f(\tau) - \sum_{k=0}^{n} a_k g_k(\tau) \right|^2 d\tau \right\}^{1/2} \to 0$$

as $n \to \infty$. This result does not mean that the relation

$$f(\tau) = \sum_{k=0}^{n} a_k g_k(\tau) \tag{18.3-18}$$

holds for all τ in $[-\pi, \pi]$. In fact it is known that there exist continuous $f \in \mathcal{H}$ such that (18.3-18) does not hold for certain τ. (On the other hand, a deep result due to L. Carleson states that for every $f \in \mathcal{H}$, (18.3-18) holds for almost all τ.)

III. Orthonormal Systems in $L_2(R)$

Here we apply the results of Section II to the space $L_2(R)$ studied in Section I. We begin by discussing a result that holds for any complete orthonormal system; various ways to construct such systems are discussed subsequently.

Let $\{f_n\}$ be an orthonormal system in $L_2(R)$, that is, let the functions f_n satisfy

$$\iint_R f_n(z)\overline{f_m(z)}\, \boxed{dz} = \begin{cases} 1, & n = m \\ 0, & n \neq m. \end{cases}$$

We also assume that the system $\{f_n\}$ is complete. Given any $f \in L_2(R)$, the series

$$\sum_{k=0}^{\infty} c_k f_k(z),$$

where

$$c_k := \iint_R f(z)\overline{f_k(z)}\, \boxed{dz}, \qquad k = 0, 1, 2, \ldots,$$

then by Theorem 18.3f converges to f in norm, that is,

$$\lim_{n \to \infty} \left\| f - \sum_{k=0}^{n} c_k f_k \right\| = 0. \tag{18.3-19}$$

Unlike the situation considered in Example 4, in the space $L_2(R)$ convergence in norm implies pointwise convergence.

THEOREM 18.3g. *Let $\{f_n\}$ be a complete orthonormal system in $L_2(R)$. Then for any $f \in L_2(R)$ the Fourier series of f in the system $\{f_n\}$ converges pointwise to f, uniformly on every compact subset of R.*

Proof. Let $K \subset R$ be compact, and let $\delta > 0$ be the distance of K from the boundary of R. By Lemma 18.3a we have for $z \in K$ and $n = 0, 1, 2, \ldots$,

$$\left| f(z) - \sum_{k=0}^{n} c_k f_k(z) \right| \leqq \frac{1}{\sqrt{\pi}\delta} \left\| f - \sum_{k=0}^{n} c_k f_k \right\|.$$

Since the expression on the right tends to 0 for $n \to \infty$, so does the expression on the left, establishing uniform convergence. —

We now concern ourselves with the construction of orthonormal systems. To begin with, let R be the unit disk, $R = D$: $|z| < 1$. Here it is not difficult to guess that the functions $1, z, z^2, \ldots$ after normalization will form an orthonormal system. Indeed, if n and m are nonnegative integers, then on writing $z = \rho e^{i\theta}$ we have

$$\iint_D z^n \overline{z^m} \, \boxed{dz} = \int_0^1 d\rho \int_0^{2\pi} \rho^{n+m} e^{i(n-m)\theta} \rho \, d\theta$$

$$= \begin{cases} \dfrac{\pi}{n+1}, & n = m \\ 0, & n \neq m, \end{cases}$$

showing that the functions

$$g_n(z) := \sqrt{\frac{n+1}{\pi}} z^n, \qquad n = 0, 1, 2, \ldots, \tag{18.3-20}$$

form an orthonormal system for D.

THEOREM 18.3h. *The system* (18.3-20) *is complete in $L_2(D)$.*

Proof. We use one of the criteria of Theorem 18.3f. Let $f \in L_2(D)$. Because f is analytic in D, it can be represented by a power series,

$$f(z) = \sum_{k=0}^{\infty} a_k z^k,$$

which converges uniformly for $|z| \leqq \rho$ for any $\rho < 1$. Integrating term by term we thus get

$$\iint_{|z|\leq\rho} f(z)\overline{g_n(z)}\,\boxed{dz} = \sqrt{\frac{\pi}{n+1}}\,a_n\rho^{2n+2},$$

which on letting $\rho\to 1$ yields

$$(f, g_n) = \iint_D f(z)\overline{g_n(z)}\,\boxed{dz} = \sqrt{\frac{\pi}{n+1}}\,a_n.$$

Thus if $(f, g_n) = 0$ for $n = 0, 1, 2, \ldots$, then $a_n = 0$ for all n, and it follows that f is the function 0. The system $\{g_n\}$ thus satisfies criterion (iv) of Theorem 18.3f, implying completeness. —

For functions in $L_2(D)$, the Fourier series in the system (18.3-20) is the Taylor series, which thus converges not only pointwise but also in norm.

Now let R be an arbitrary simply connected region, $R \neq \mathbb{C}$, and let $f \in L_2(R)$. We evaluate the integral defining the norm of f,

$$\|f\|^2 = \iint_R |f(z)|^2\,\boxed{dz},$$

by a transformation of variables. Let z_0 be a point in R, and let h be the function mapping R onto the unit disk D such that $h(z_0) = 0$, $h'(z_0) > 0$. If $w = h(z)$,

$$\iint_R |f(z)|^2\,\boxed{dz} = \iint_D |f(h^{[-1]}(w))|^2|h^{[-1]\prime}(w)|^2\,|\,dw\,|,$$

showing that the function $g := (f\circ h^{[-1]})h^{[-1]\prime} = (fh'^{-1})\circ h^{[-1]}$ is in $L_2(D)$. By Theorem 18.3g, the Taylor series of g, written in the form

$$g(w) = \sum_{k=0}^{\infty} a_k g_k(w),$$

where g_k is defined by (18.3-20), converges to g in norm as well as pointwise, which means that

$$\iint_D \left| g(w) - \sum_{k=0}^{n} a_k g_k(w)\right|^2 |\,dw\,| \to 0, \qquad n\to\infty.$$

Letting $w = h(z)$, we see that the integral equals

$$\iint_R \left| f(z)[h'(z)]^{-1} - \sum_{k=0}^{n} a_k g_k(h(z)) \right|^2 |h'(z)|^2 \boxed{dz}$$

$$= \iint_R \left| f(z) - \sum_{k=0}^{n} a_k g_k(h(z)) h'(z) \right|^2 \boxed{dz}.$$

The fact that this tends to 0 as $n \to \infty$ means that f can be represented, in the norm of $L_2(R)$, in terms of the functions

$$f_k(z) := g_k(h(z))h'(z) = \sqrt{\frac{k+1}{\pi}}[h(z)]^k h'(z). \tag{18.3-21}$$

The transformation of variables used above shows that the system $\{f_k\}$ is orthonormal. By the definition of completeness we have:

THEOREM 18.3i. *The system of functions* (18.3-21) *forms a complete orthogonal set in* $L_2(R)$ *for any simply connected region* $R \neq \mathbb{C}$.

In addition to exhibiting a complete system, Theorem 18.3i has the important implication that for $R \neq \mathbb{C}$ the space $L_2(R)$ is never empty. As to furnishing a complete system, the theorem is not always practical, because to construct the system (18.3-21) one first would have to find the mapping function h. The question may be raised whether a complete orthonormal system could not be built up by applying the Gram–Schmidt process to a simple system of functions such as

$$1, z, z^2, \ldots. \tag{18.3-22}$$

Clearly these functions are linearly independent in any nonempty region R. However, they do not always belong to $L_2(R)$, because the integrals

$$\iint_R |z^n|^2 \boxed{dz}$$

need not be finite for unbounded regions R. If R is bounded, then all powers z^n belong to $L_2(R)$, and the process of orthogonalization can be carried out. However, the resulting system need not be complete. To see this, let R be obtained from the unit disk D by removing from it the straight line segment $\frac{1}{2} \leq x < 1$. The functions (18.3-20) then still form an orthonormal system in R, because the cut does not affect the value of the inner product.

Let now g be an arbitrary function in R, and let

$$\sum_{k=0}^{\infty} a_k g_k(z) = \sum_{k=0}^{\infty} a_k \sqrt{\frac{k+1}{\pi}}\, z^k$$

be its Fourier series in the system $\{g_k\}$. If the system $\{g_k\}$ is complete, then by Theorem 18.3g the series converges to $g(z)$ pointwise for $z \in R$, uniformly on compact subsets. Because the series is also a power series, it converges for $|z| < 1$, uniformly on compact subsets, which means that g can be continued analytically into D. Clearly this is not true for every $g \in L_2(R)$. Hence the system (18.3-22) is not complete in $L_2(R)$.

The above considerations show that the following result is the best that can be hoped for:

THEOREM 18.3j. *Let R be the interior of a Jordan curve Γ. Then the polynomials in z form a complete system in $L_2(R)$.*

Proof. Let $R' := R \cup \Gamma$, the closure of R. We are to show that, given $f \in L_2(R)$ and $\varepsilon > 0$, there exists a polynmial p such that $\|f - p\| < \varepsilon$. To achieve this end, we construct a function f^* which is analytic on the compact set R' and satisfies $\|f - f^*\| < \varepsilon/2$. By means of the Faber series (Theorem 18.2a) we then may find a polynomial p such that

$$|f^*(z) - p(z)| < \frac{\varepsilon}{2\sqrt{\alpha}}, \qquad z \in R',$$

where α denotes the area of R'. We then have

$$\begin{aligned} \|f^* - p\| &\leqq \sup|f^*(z) - p(z)| \left[\iint_{R'} \boxed{dz}\right]^{1/2} \\ &< \frac{\varepsilon}{2}, \end{aligned}$$

and by the triangle inequality there follows

$$\|f - p\| \leqq \|f - f^*\| + \|f^* - p\| < \varepsilon.$$

To find f^* we use an auxiliary conformal map. Let $\{\Gamma_n\}$ be a sequence of Jordan curves that approximates Γ from the interior, let z_0 be a point in R, and let h_n be the function mapping the interior of Γ_n onto R such that $h_n(z_0) = z_0$, $h_n'(z_0) > 0$ for all large n. It follows from the Carathéodory convergence theorem (Theorem 19.4a) that for any compact set $K \subset R$

$$h_n(z) \to z, \qquad h_n'(z) \to 1 \tag{18.3-23}$$

uniformly for $z \in K$. We now set

$$f_n := (f \circ h_n) h_n', \qquad n = 0, 1, 2, \ldots.$$

All f_n are analytic on R'. Thus if we can find n such that $\|f - f_n\| < \varepsilon/2$, then $f^* = f_n$ will serve the required purpose.

For any compact $K \subset R$ there holds

$$\|f - f_n\|^2 = \iint_{R\setminus K} |f(z) - f_n(z)|^2 \, |dz| + \iint_K |f(z) - f_n(z)|^2 \boxed{dz}. \tag{18.3-24}$$

We wish this to be $< \varepsilon^2/4$ and shall choose K and n such that each of the integrals on the right is $< \varepsilon^2/8$. Using $|a - b|^2 \leqq 2|a|^2 + 2|b|^2$ we have

$$\iint_{R\setminus K} |f(z) - f_n(z)|^2 \boxed{dz} \leqq 2 \iint_{R\setminus K} |f(z)|^2 \boxed{dz} + 2 \iint_{R\setminus K} |f_n(z)|^2 \boxed{dz}.$$

We now choose K such that

$$\iint_{R\setminus K} |f(z)|^2 \boxed{dz} < \varepsilon^2/48. \tag{18.3-25}$$

Such a choice is possible by the very definition of the integral over R. Concerning the second integral on the right we have, using the definition of f_n and the transformation of variables $w = h_n(z)$,

$$\iint_{R\setminus K} |f_n(z)|^2 \boxed{dz} = \iint_{h_n(R\setminus K)} |f(z)|^2 \boxed{dz}.$$

In view of $h_n(R\setminus K) = h_n(R) \setminus h_n(K) \subset R \setminus h_n(K)$, the last integral is less than

$$\iint_{R\setminus h_n(K)} |f(z)|^2 \boxed{dz}.$$

In view of (18.3-23), $h_n(K) \to K$ as $n \to \infty$. Thus the foregoing integral approaches the integral on the left of (18.3-25) and thus by choosing n sufficiently large can be made $< \varepsilon^2/24$.

The first integral on the right of (18.3-24) is now $< \varepsilon^2/8$. By (18.3-23), $f_n \to f$ uniformly on K. By choosing n still larger, if necessary, the second integral can thus also be made $< \varepsilon^2/8$. —

As a consequence of Theorem 18.3j, if the Gram–Schmidt algorithm is used to orthonormalize the functions $1, z, z^2, \ldots$ in the interior of a Jordan curve, a complete orthonormal system will result. We add some remarks concerning the numerical implementation of the Gram–Schmidt algorithm.

This algorithm, as described in the proof of Theorem 18.3d, requires the computation of some $\frac{1}{2}n^2$ scalar products to compute the first n functions of the orthonormal system. If, as it would seem to be the case in $L_2(R)$, each scalar product requires the evaluation of a double integral, the amount of work could be considerable. However, if R is a smoothly bounded Jordan region, the scalar products can be reduced to simple integrals.

LEMMA 18.3k. *Let R be the interior of a piecewise smooth, positively oriented Jordan curve Γ, and let f and g be analytic on the closure of R. Then*

$$\iint_R f(z)\overline{g'(z)}\,\boxed{dz} = \frac{1}{2i}\int_\Gamma f(z)\overline{g(z)}\,dz. \tag{18.3-26}$$

Proof. This is a corollary of the complex form of Green's formula (15.10-4),

$$\iint_R \frac{\partial h}{\partial \bar{z}}(z)\,\boxed{dz} = \frac{1}{2i}\int_\Gamma h(z)\,dz,$$

that holds for any real-differentiable function h. To obtain (18.3-26), set

$$h(z) := f(z)\overline{g(z)}. \quad —$$

As a result of Lemma 18.3k we have, for instance,

$$\iint_R z^n\overline{z^m}\,\boxed{dz} = \frac{1}{2i(m+1)}\int_\Gamma z^n\overline{z^{m+1}}\,dz.$$

As the Gram–Schmidt process is extremely sensitive numerically to errors in the scalar products, these line integrals should be evaluated with the highest precision possible. If Γ is piecewise analytic, the line integrals should be evaluated separately for each analytic subarc. After introducing the parametric representation $z = z(\tau)$, the integrand is analytic on the closed interval of integration, and the integral may be evaluated by an algorithm which takes advantage of this fact, such as the Romberg algorithm explained in § 11.12. Care must also be taken of the fact that due to the oscillatory nature of the integrand the value of the integral may be much smaller than the absolute value of the integrand, which in numerical integration sometimes makes it difficult to select proper convergence criteria.

In the following example, Lemma 18.3k is used to verify analytically that a certain system of polynomials is orthogonal.

EXAMPLE **5.**

Let $\rho > 1$, and let R be the interior of the ellipse

$$\Gamma: \quad z = \tfrac{1}{2}(\rho\, e^{i\tau} + \rho^{-1}\, e^{-i\tau}), \qquad 0 \leqq \tau \leqq 2\pi.$$

We assert that the polynomials obtained by orthogonalizing the system $1, z, z^2, \ldots$ are

$$g_n(z) := \gamma_n T'_{n+1}(z), \qquad n = 0, 1, 2, \ldots, \tag{18.3-27}$$

where the T_{n+1} are the ubiquitous Chebyshev polynomials, and the γ_n are suitable positive factors. Indeed by Lemma 18.3k,

$$\frac{1}{\gamma_n \gamma_m} \iint_R g_m(z)\overline{g_n(z)}\, \boxed{dz} = \frac{1}{2i} \int_\Gamma T'_{m+1}(z)\overline{T_{n+1}(z)}\, dz.$$

To evaluate the integral on the right, we use the fact that Γ is the image of $|w| = \rho$ under the Joukowski map

$$w \to z = \frac{1}{2}\left(w + \frac{1}{w}\right),$$

and that for $z = z(w)$,

$$T_{n+1}(z) = \tfrac{1}{2}(w^{n+1} + w^{-n-1}),$$

$$T'_{m+1}(z) = \frac{m+1}{2}(w^m - w^{-m-2})\left(\frac{dz}{dw}\right)^{-1}.$$

Thus

$$\frac{1}{2i} \int_\Gamma T'_{m+1}(z)\overline{T_{n+1}(z)}\, dz = \frac{n+1}{8i} \int_{|w|=\rho} (w^{m+1} - w^{-m-1})(\overline{w^{n+1}} + \overline{w^{-n-1}})\, \frac{dw}{w}.$$

Using $w\bar{w} = \rho^2$, the last integral can be written as the integral of an analytic function with an isolated singularity at $w = 0$. The residue is evaluated immediately, and we find

$$\frac{1}{2i} \int_\Gamma T'_{m+1}(z)\overline{T_{n+1}(z)}\, \boxed{dz} = \begin{cases} 0, & n \neq m \\ \dfrac{(n+1)\pi}{4}(\rho^{2n+2} - \rho^{-2n-2}), & n = m. \end{cases}$$

It follows that for the elliptic region R the polynomials g_n defined by (18.3-27) form an orthonormal system if

$$\gamma_n := 2[(n+1)\pi(\rho^{2n+2}-\rho^{-2n-2})]^{-1/2}. \quad —$$

Complete orthonormal systems can also be constructed for the case $R: |z|>\rho$ (see Problem 4), and for R a nondegenerate annulus, $R: \mu<|z|<1$ (see Problem 3). By means of appropriate mapping functions we thus can find complete orthonormal systems for the exterior of a simply connected compact set, or for any doubly connected region. If R is a region of connectivity $m>2$, which is bounded on the outside by a Jordan curve Γ_0 and on the inside by Jordan curves $\Gamma_1, \ldots, \Gamma_{m-1}$, a complete orthonormal system may be constructed in a purely numerical manner by selecting points a_i in the interior of each curve Γ_i, $i=1, 2, \ldots, m-1$, and by applying the Gram–Schmidt algorithm to the system of functions

$$f_{mk}(z) := z^k,$$

$$f_{mk+j}(z) := (z-a_j)^{-k-1}, \qquad j=1, \ldots, m-1,$$

where $k=0, 1, 2, \ldots$.

IV. The Bergman Kernel Function

Let $R \neq \mathbb{C}$ be a region of arbitrary connectivity, and let $\{f_m\}$ be a complete orthonormal system in $L_2(R)$. This system will be used to construct the Bergman kernel function, a function defined on $R \times R$ which is akin in importance to Green's and Neumann's functions, to which it is in fact related. If R is simply connected, there are connections with the Riemann mapping function for R which can be exploited constructively.

We first show that the series

$$\sum_{k=0}^{\infty} |f_k(t)|^2 \tag{18.3-28}$$

converges for $t \in R$, uniformly on every compact $K \subset R$. For a fixed $t \in R$, let

$$h_n(z) := \sum_{k=0}^{n} f_k(z)\overline{f_k(t)}.$$

This function is in $L_2(R)$ and has the norm

$$\|h_n\| = \left\{\sum_{k=0}^{n} |f_k(t)|^2\right\}^{1/2}.$$

Therefore, by Lemma 18.3a,

$$|h_n(z)| \leq \frac{1}{\sqrt{\pi}\,\delta(z)} \|h_n\|,$$

where $\delta(z)$ is the distance of z from the boundary of R. In particular for $z=t$,

$$|h_n(t)| = \sum_{k=0}^{n} |f_k(t)|^2 = \|h_n\|^2 \leqq \frac{1}{\sqrt{\pi}\,\delta(t)} \|h_n\|,$$

which on cancellation yields $\sqrt{\pi}\,\delta(t)\|h_n\| \leqq 1$ or

$$|h_n(t)| \leqq \frac{1}{\pi\delta^2(t)}.$$

Because the bound on the right is independent of n, this implies the convergence of the series (18.3-28) for every fixed $t \in R$.

Because of the convergence of (18.3-28) for each fixed $t \in R$, the series

$$k(z,t) := \sum_{m=0}^{\infty} f_m(z)\overline{f_m(t)} \tag{18.3-29}$$

converges as a function of z in the norm of $L_2(R)$ and also pointwise, uniformly for z in every compact subset K of R. This result is in marked contrast with the Hilbert space considered in Example 4, where the corresponding series formed with the orthonormal system converges neither in norm nor pointwise.

The function k has another remarkable property. We write $k(z,t) = k_t(z)$ if k is regarded as a function of z depending on the parameter t.

THEOREM 18.3l. *For any $f \in L_2(R)$ and any $t \in R$,*

$$(f, k_t) = f(t). \tag{18.3-30}$$

Proof. Let

$$f(z) = \sum_{k=0}^{\infty} c_k f_k(z)$$

be the Fourier representation of f in the system $\{f_k\}$. Then

$$(f, k_t) = \iint_R \sum_{k=0}^{\infty} c_k f_k(z) \sum_{m=0}^{\infty} \overline{f_k(z)} f_k(t) \boxed{dz},$$

which on account of the orthogonality of the system $\{f_k\}$ collapses into

$$\sum_{k=0}^{\infty} c_k f_k(t) = f(t). \quad \text{—}$$

The property expressed by (18.3-30) is called the **reproducing property** of k, and a function k with this property is called a **reproducing kernel**. It is quite exceptional for a Hilbert space to possess a reproducing kernel; for an axiomatic study of such spaces see Aronszajn [1950].

On the face of it it would seem that the function k depends on the choice of the orthonormal system $\{f_k\}$. That all functions k obtained from the various possible choices of the systems $\{f_k\}$ should have the reproducing property will arouse some suspicion. In fact there holds:

THEOREM 18.3m. *The function k is uniquely characterized by the reproducing property* (18.3-30).

Proof. Let k and k^* be two different reproducing kernels. Expression (18.3-30) then implies that

$$(f, k_t - k_t^*) = 0 \quad \text{for all} \quad f \in L_2(R).$$

Thus, in particular, this relation would have to hold for $f := k_t - k_t^*$, implying $\|k_t - k_t^*\| = 0$ or $k_t^* = k_t$ for all $t \in R$. —

The function k, which is thus independent of the choice of the particular orthonormal system $\{f_k\}$ used to form the infinite series (18.3-29), is called the (complex) **Bergman kernel function** of the region R.

Because k does not depend on the particular orthonormal system $\{f_k\}$, we may choose a system which is particularly convenient for the derivation of further properties of k. If R is simply connected, one such system is given by Theorem 18.3i. Let the mapping function h_t be chosen such that

$$h_t(t) = 0, \qquad h_t'(t) > 0.$$

For this choice, all terms in the series (18.3-29) except the zeroth term vanish, and we have:

THEOREM 18.3n. *Let R be a simply connected region, $R \neq \mathbb{C}$, let $t \in R$, and let h_t be the Riemann mapping function for R normalized at t. Then the Bergman kernel function for the region R is given by the formula*

$$k(z, t) = \frac{1}{\pi} h_t'(z) h_t'(t). \tag{18.3-31}$$

This result provides us with yet another method for constructing the Riemann mapping function for smoothly bounded Jordan regions R. One constructs an orthonormal system $\{f_n\}$ for R, for instance by orthonormalizing $1, z, z^2, \ldots$ by the Gram-Schmidt process. Formula (18.3-29) then yields $k(z, t)$ to arbitrary accuracy. Having chosen the normalizing point t, one next computes $h_t'(t)$ from

$$h_t'(t) = \sqrt{\pi k(t, t)}.$$

One then finds $h_t(z)$ from

$$h_t(z)=\frac{\pi}{h_t'(t)}\int_t^z k(s,t)\,ds.$$

For details of the numerical procedure, see Gaier [1964].

We present two examples where the above process can be carried out analytically.

EXAMPLE 6

If R is the unit disk, $R=D\colon |z|<1$, an orthonormal system was given in Theorem 18.3h. There follows

$$k(z,t)=\frac{1}{\pi}\sum_{k=1}^{\infty}(k+1)z^k\overline{t^k}=\frac{1}{\pi}\frac{1}{(1-z\bar{t})^2}.$$

If t is any point in D, we thus have

$$h_t'(t)=\sqrt{\pi k(t,t)}=\frac{1}{1-|t|^2},$$

and the mapping function is, as is well known,

$$\begin{aligned}h_t(z)&=(1-|t|^2)\int_t^z\frac{1}{(1-s\bar{t})^2}\,ds\\&=\frac{1-|t|^2}{\bar{t}}\left\{\frac{1}{1-z\bar{t}}-\frac{1}{1-t\bar{t}}\right\}\\&=\frac{z-t}{1-z\bar{t}}.\end{aligned}$$

EXAMPLE 7

Once again, let R be the elliptical region considered in Example 5. From the complete system constructed there we get

$$k(z,t)=\frac{4}{\pi}\sum_{k=0}^{\infty}\frac{T_{k+1}'(z)\overline{T_{k+1}'(t)}}{(k+1)(\rho^{2k+2}-\rho^{-2k-2})}.$$

Selecting $t=0$ and observing that $T_{2m}'(0)=0$, $T_{2m+1}'(0)=(-1)^m(2m+1)$, we obtain

$$h'(0)=\left\{4\sum_{m=0}^{\infty}\frac{2m+1}{\rho^{4m+2}-\rho^{-4m-2}}\right\}^{1/2}$$

and

$$h(z)=\frac{4}{h'(0)}\sum_{m=0}^{\infty}\frac{(-1)^m T_{2m+1}(z)}{\rho^{4m+2}-\rho^{-4m-2}}.$$

PROBLEMS

1. Let D be a simply connected region, $a \in D$, and let f be the Riemann mapping function for D such that $f(a)=0$, $f'(a)>0$. Show that the Bergmann kernel function for D is

$$k(z, t)=\frac{1}{\pi}\frac{\overline{f'(t)}f'(z)}{[1-\overline{f(t)}f(z)]^2}.$$

2. Let $R \subset \mathbb{C}$ be a bounded region of arbitrary connectivity, and let the system of boundary curves Γ of R be sufficiently smooth. Let $\gamma(z, t)$ denote Green's function for R (see § 15.6).

(a) Using the known nature of the singularity of γ, show that

$$h(z, t):=-4\frac{\partial^2\gamma}{\partial z\,\partial \bar{t}}(z, t)$$

(the derivatives understood in the sense of the Wirtinger calculus) is continuous for $z \in R$, $t \in R$.

(b) Applying Green's formula (15.10-4) to the region R with a small disk around the point z removed, show that for any f analytic in R there holds

$$\iint_R f(t)\overline{h(t, z)}\,\boxed{dt}=f(z).$$

(c) Using Theorem 18.3m, conclude that

$$k(z, t)=h(z, t), \qquad (z, t) \in R \times R.$$

3. Let $0<\mu<1$.

(a) Show that an orthonormal system for the annulus A: $\mu<|z|<1$ is given by $f_n(z)=\omega_n z^n$, $n \in \mathbb{Z}$, where

$$\omega_n:=\begin{cases}\left[\dfrac{n+1}{\pi(1-\mu^{2n+2})}\right]^{1/2}, & n \neq -1\\[2ex] \left[2\pi \operatorname{Log}\dfrac{1}{\mu}\right]^{-1/2}, & n=-1.\end{cases}$$

(b) Show that the foregoing system is complete in A.

(c) Show that the Bergmann kernel function for A is

$$k(z, t)=\frac{1}{\pi}\frac{1}{z\bar{t}}\left\{\frac{1}{2\operatorname{Log}\mu^{-1}}+\sum_{n=1}^{\infty}\frac{n}{\mu^{-n}-\mu^{n}}\left[\left(\frac{z\bar{t}}{\mu}\right)^n+\left(\frac{\mu}{z\bar{t}}\right)^n\right]\right\}.$$

(d) Obtain the same result also from the representation (15.6-13) of Green's function for A via Problem 2.

(e) Using (15.6-18), show that the kernel function for the annulus A is also given by

$$k(z, t)=\frac{1}{\pi}\frac{1}{z\bar{t}}\left\{\frac{1}{\operatorname{Log}\mu^{-2}}-\frac{\Sigma_1\Sigma_3-\Sigma_2^2}{\Sigma_1^2}\right\},$$

where the Σ_i denote the rapidly converging sums,

$$\Sigma_1 := 1 + \sum_{k=1}^{\infty} (-1)^k \mu^{k^2} \left[\left(\frac{z\bar{t}}{\mu} \right)^k + \left(\frac{\mu}{z\bar{t}} \right)^k \right],$$

$$\Sigma_2 := \sum_{k=1}^{\infty} (-1)^k \mu^{k^2} k \left[\left(\frac{z\bar{t}}{\mu} \right)^k - \left(\frac{\mu}{z\bar{t}} \right)^k \right\rangle,$$

$$\Sigma_3 := \sum_{k=1}^{\infty} (-1)^k \mu^{k^2} k^2 \left[\left(\frac{z\bar{t}}{\mu} \right)^k + \left(\frac{\mu}{z\bar{t}} \right)^k \right].$$

4. Show that the functions

$$f_n(z) := \sqrt{\frac{n}{\pi}}\, z^{-n-1}, \qquad n = 1, 2, \ldots,$$

form a complete orthonormal system for the region $R: |z| > 1$. Show, consequently, that the kernel function for R is

$$k(z, t) = \frac{1}{\pi} \frac{1}{(z\bar{t} - 1)^2}.$$

5. Let R be the exterior of a Jordan curve Γ, and let $\mathring{f}$ be the mapping function for R onto $|w| > 1$, normalized such that

$$\mathring{f}(\infty) = \infty, \qquad \mathring{f}'(\infty) > 0.$$

show that the kernel function for R is

$$k(z, t) = \frac{1}{\pi} \frac{\mathring{f}'(z)\overline{\mathring{f}'(t)}}{(\mathring{f}(z)\overline{\mathring{f}(t)} - 1)^2}.$$

6. If $\mathring{f}$ is defined as in the preceding problem, $\gamma := [\mathring{f}'(\infty)]^{-1}$ is the capacity of Γ (see § 16.5)

   (a) Show that in terms of the kernel function for R, γ is given by

$$\gamma = \frac{1}{\pi} \lim_{\substack{z \to \infty \\ \bar{t} \to \infty}} [z^2 \overline{t^2} k(z, t)]^{1/2},$$

the square root having its principal value.

   (b) If $\{f_k\}_{k=1}^{\infty}$ is a complete orthonormal system for R, and

$$f_k(z) = \frac{a_k}{z^k} + O(z^{-k-1}), \qquad k = 1, 2, \ldots,$$

show that the capacity of Γ is also given by

$$\gamma = \left[\pi \sum_{k=1}^{\infty} |a_k|^2 \right]^{-1}.$$

7. If Γ is a Jordan curve with capacity γ, and if $k(z, t)$ is the kernel function for the exterior of Γ, show that the exterior mapping function $\mathring{f}$ for Γ may be

calculated from

$$k(z) := \lim_{t \leftarrow \infty} \overline{t^2} k(z, t)$$

by means of the formula

$$\mathring{f}(z) = \left[\frac{\pi}{\gamma} \int_z^\infty k(t) \, dt \right]^{-1}.$$

NOTES

Bergmann [1922] and Bochner [1922] wrote the seminal papers on $L_2(R)$ and on expansions of analytic functions into series of orthogonal functions. Bergman [1950] gives an authoritative account, but Bergman and Schiffer [1953] is more readable. Elementary presentations are in Nehari [1952], Hille [1962], and P. J. Davis [1963]. The reproducing property of the kernel functions is treated axiomatically by Aronszajn [1950]; see also Meschkowski [1962]. Gaier [1964] devotes chapter III to the construction of mapping functions by methods of orthogonalization; see also Bergman and Chalmers [1967]. The practical performance of some of these methods is subject to difficulties if a high accuracy is desired. Levin et al. [1978] recommend the use of singular functions in the orthogonalization process to improve the performance; see also Gaier [1980]. Further experiments in this direction are reported by Papamichael and Warby [1983], Papamichael, Warby, and Hough [1984], and Papamichael and Kokkinos [1984]. A comparison of various mapping methods based on kernel functions is given by Ellacott [1981].

§ 18.4. ORTHOGONALIZATION ON THE BOUNDARY; THE SZEGÖ KERNEL FUNCTION

For any bounded region $R \subset \mathbb{C}$, the Bergman kernel function may be obtained by means of the formula $k(z, t) = \sum p_n(z)\overline{p_n(t)}$ from any complete orthonormal system $\{p_n\}$ of functions in $L_2(R)$. If R is a Jordan region, such a system is constructed by applying the Gram–Schmidt process to a system of functions, such as $1, z, z^2, \ldots$, that is known to be complete in $L_2(R)$.

A similar construction is possible if orthogonalization over R is replaced by orthogonalization over the *boundary* of R. There again results a kernel function, called the Szegö kernel function, that is related to the Riemann mapping functions for R.

I. The Space $A_2(\Gamma)$

To avoid technicalities, we assume that the boundary Γ of R is an analytic Jordan curve. We denote by $A_2(\Gamma)$ the space of functions g that are analytic in R and continuous in $R \cup \Gamma$. If endowed with the inner product

$$(g, h) := \int_\Gamma g(t)\overline{h(t)} \, |dt| \qquad (18.4\text{-}1)$$

and the resulting norm

$$\|g\| := \left(\int_\Gamma |g(t)|^2\,|dt|\right)^{1/2},$$

$A_2(\Gamma)$ becomes an inner product space (see § 15.12). However, as shown in Problem 1, $A_2(\Gamma)$ is not complete, hence it is not a Hilbert space. Several interesting facts nevertheless hold for $A_2(\Gamma)$.

LEMMA 18.4a. *Let $h \in A_2(\Gamma)$, let $z \in R$, and let $\delta > 0$ be the distance from z to Γ. Then*

$$|h(z)| \leqq \frac{1}{\sqrt{2\pi\delta}}\|h\|. \tag{18.4-2}$$

Proof. By Cauchy's formula,

$$[h(z)]^2 = \frac{1}{2\pi i}\int_\Gamma \frac{[h(t)]^2}{t-z}\,dt,$$

hence

$$|h(z)|^2 \leqq \frac{1}{2\pi\delta}\int_\Gamma |h(t)|^2\,|dt|. \quad —$$

There immediately follows:

COROLLARY 18.4b *Let the functions h_n, $n = 0, 1, 2, \ldots$, and h be in $A_2(\Gamma)$, and let*

$$\|h_n - h\| \to 0, \qquad n \to \infty. \tag{18.4-3}$$

Then the sequence $\{h_n\}$ converges to h locally uniformly in R.

Indeed, if K is any compact subset of R and if δ is the distance of K from Γ, then by Lemma 18.4a,

$$\sup_{z \in K}|h_n(z) - h(z)| \leqq \frac{1}{\sqrt{2\pi\delta}}\|h_n - h\| \to 0,$$

establishing uniform convergence on K. —

It is easy to show by means of examples that even if condition (18.4-3) holds, the convergence of the sequence $\{h_n\}$ to h need not be uniform in $R \cup \Gamma$.

Let a be any point of R. We denote by f_a the Riemann mapping function for R with normalizing point a, that is, the unique function mapping R onto $|w| < 1$ such that

$$f_a(a) = 0, \qquad f_a'(a) > 0.$$

We recall that, due to Γ being analytic, f_a can be extended to a function that is analytic on $R \cup \Gamma$, and that $f_a'(z) \neq 0$ for $z \in R \cup \Gamma$. Thus f_a' has an analytic square root in $R \cup \Gamma$; we denote by $r_a := (f_a')^{1/2}$ the square root that is positive at a.

LEMMA 18.4c. *Let $h \in A_2(\Gamma)$, and let $a \in R$. Then*

$$(h, r_a) = 2\pi \frac{h(a)}{r_a(a)}. \tag{18.4-4}$$

Proof. Let $g_a := f_a^{[-1]}$. Making the substitution $t = g_a(u)$, we have

$$\begin{aligned}(h, r_a) &= \int_\Gamma h(t)\overline{r_a(t)}\,|dt| \\ &= \int_\Gamma \frac{h(t)}{r_a(t)}|f_a'(t)|\,|dt| \\ &= \int_\Theta \frac{h(g_a(u))}{r_a(g_a(u))}|du|,\end{aligned}$$

where Θ denotes the unit circle. If $|u| = 1$, then $|du| = du/iu$, and the last integral by residues equals

$$\frac{1}{i}\int_\Theta \frac{h(g_a(u))}{r_a(g_a(u))}\frac{du}{u} = 2\pi \frac{h(g_a(0))}{r_a(g_a(0))},$$

which by virtue of $g_a(0) = a$ equals (18.4-4). —

II. Orthogonalization on the Boundary

We apply the Gram–Schmidt process to the system of functions $1, z, z^2, \ldots$ to obtain a sequence of polynomials $\{p_n\}$, where p_n has the precise degree n, and where

$$(p_n, p_m) = \begin{cases} 1, & n = m \\ 0, & n \neq m. \end{cases} \tag{18.4-5}$$

(All inner products are now taken in the sense (18.4-1).) The polynomials p_n are determined by (18.4-5) only up to a constant factor of modulus 1. We make p_n unique by stipulating that the coefficient of z^n (which cannot be zero) is positive.

The **Szegö kernel function** of Γ will for $z \in R$ and $s \in R$ be defined by

$$k(z, s) = \sum_{m=0}^{\infty} \overline{p_m(z)}\,p_m(s), \tag{18.4-6}$$

but first it must be shown that this definition is meaningful.

To this end we consider the partial sums

$$\hat{k}(z, s) := \sum_{m=0}^{n} \overline{p_m(z)} p_m(s), \qquad n = 0, 1, 2, \ldots,$$

which are defined for arbitrary $z, s \in \mathbb{C}$. (Our notation does not indicate the dependence of $\hat{k}$ on n.) If for fixed z, $\hat{k}(z, s)$ is to be regarded as a function of the second variable, we denote it by $\hat{k}_z$. The orthogonality of the p_m then immediately implies

$$\|\hat{k}_z\|^2 = \hat{k}(z, z) = \hat{k}_z(z). \tag{18.4-7}$$

Let K be a compact subset of R, and let $\delta > 0$ be the distance of K to the boundary of R. Applying Lemma 18.4a to $f := \hat{k}_z$ and using (18.4-7), there follows

$$[\hat{k}_z(z)]^2 \leqq \frac{1}{2\pi\delta} \hat{k}_z(z), \qquad z \in K,$$

hence

$$\hat{k}_z(z) = \hat{k}(z, z) \leqq \frac{1}{2\pi\delta}, \qquad z \in K.$$

This shows that the partial sums of the series of nonnegative terms,

$$k(z, z) = \sum_{m=0}^{\infty} |p_m(z)|^2 \tag{18.4-8}$$

are uniformly bounded for $z \in K$. Since the terms are continuous functions, Dini's theorem now implies that the series converges uniformly on K. The Schwarz inequality now shows that the series $k(z, s)$ converges uniformly for $z \in K$, $s \in K$, and thus that the definition (18.4-6) is meaningful.

We shall express $k(z, s)$ in terms of f_z, but first we have to establish certain extremal properties of the polynomials p_m.

Let $\hat{\Pi}$ denote the class of all polynomials of degree $\leqq n$, and let h be any function in $A_2(\Gamma)$. For which $p \in \hat{\Pi}$ does $\|h - p\|$ become smallest?

Any $p \in \hat{\Pi}$ can be expressed in the form

$$p(z) = \sum_{m=0}^{n} a_m p_m(z),$$

where the a_m are suitable complex numbers, and any function of this form is in $\hat{\Pi}$. Thus we are asking for the minimum of

$$\left\| h - \sum_{m=0}^{n} a_m p_m \right\|$$

as a function of the a_m. We have

$$\left\| h - \sum_{m=0}^{n} a_m p_m \right\|^2 = \left(h - \sum_{m=0}^{n} a_m p_m, h - \sum_{m=0}^{n} a_m p_m \right)$$

$$= \|h\|^2 - 2 \operatorname{Re} \sum_{m=0}^{n} b_m \overline{a_m} + \sum_{m=0}^{n} |a_m|^2$$

where

$$b_m := (h, p_m), \qquad m = 0, 1, \ldots, n. \tag{18.4-9}$$

The last expression may be written

$$\|h\|^2 - \sum_{m=0}^{n} |b_m|^2 + \sum_{m=0}^{n} |b_m - a_m|^2,$$

and this clearly becomes smallest for $a_m = b_m$, $m = 0, 1, \ldots, n$. We thus find:

THEOREM 18.4d. *Let $h \in A_2(\Gamma)$. Then the best approximation in the norm of $A_2(\Gamma)$ of h by a polynomial p of degree $\leqq n$ is given by*

$$\hat{p}(z) := \sum_{m=0}^{n} (h, p_m) p_m(z), \tag{18.4-10}$$

which may also be written as

$$\hat{p}(z) = (h, \hat{k}_z). \tag{18.4-11}$$

Several applications follow. First, consider the special case where $h(z) = z^{n+1}$. We are thus asking for the polynomial $p \in \hat{\Pi}$ that realizes the minimum of $\|z^{n+1} - p\|$. By Theorem 18.4d, the solution is

$$p(z) = (t^{n+1}, \hat{k}_z).$$

The polynomial

$$q(z) := z^{n+1} - p(z) = z^{n+1} - \sum_{m=0}^{n} (t^{n+1}, p_m) p_m(z)$$

has the property that for $m = 0, 1, \ldots, n$,

$$(q, p_m) = (t^{n+1}, p_m) - (t^{n+1}, p_m) = 0.$$

Thus q, being orthogonal to $p_0, p_1, \ldots, p_n$, is proportional to p_{n+1}. We thus have proved:

COROLLARY 18.4e. *For $n = 1, 2, \ldots$, the monic polynomial q_n of degree n that minimizes $\|q_n\|$ is given by*

$$q(z) = \kappa_n^{-1} p_n(z),$$

where κ_n is the leading coefficient of p_n.

If h is analytic on $R \cup \Gamma$, then by Theorem 18.2a there exists a sequence of polynomials $\{q_n\}$, $\deg q_n \leqq n$, such that

$$\sup_{z \in R \cup \Gamma} |h(z) - q_n(z)| \to 0, \qquad n \to \infty.$$

By a basic theorem in complex approximation theory (Mergelyan's theorem, to be proved in a later chapter), this fact holds for any $h \in A_2(\Gamma)$. Then certainly also $\|h - q_n\| \to 0$. In view of the minimum property of the polynomial $\hat{p}$ defined in Theorem 18.4d, $\|h - \hat{p}\| \leqq \|h - q_n\|$, and there follows $\|h - \hat{p}\| \to 0$. Applying Corollary 18.4b, we obtain

COROLLARY 18.4f. *For any $h \in A_2(\Gamma)$, the series*

$$\sum_{m=0}^{\infty} (h, p_m) p_m(z) \tag{18.4-12}$$

converges to $h(z)$, locally uniformly on R.

As noted earlier, the nth partial sum of the series (18.4-12) may be expressed in the form $(h, \hat{k}_z)$. By the Schwarz inequality,

$$(h, \hat{k}_z)^2 \leqq \|h\|^2 \|\hat{k}_z\|^2, \qquad z \in R.$$

Letting $n \to \infty$, the expression on the left by Corollary 18.4f tends to $[h(z)]^2$, and by Theorem 18.4d the second member on the right tends to $k(z, z)$. There follows:

COROLLARY 18.4g. *For any $h \in A_2(\Gamma)$, $h \neq 0$, and for any $z \in R$ there holds*

$$k(z, z) \geqq \frac{[h(z)]^2}{\|h\|^2}.$$

An interesting special case, which will be required later, is obtained for $h = r_a$, where $r_a = \sqrt{f'_a}$ as defined in Lemma 18.4c. Here

$$\|h\|^2 = \|r_a\|^2 = \int_\Gamma |f'_a(t)|\,|dt| = \int_{f(\Gamma)} |du|,$$

where $u = f(t)$. Since $f(\Gamma)$ is the unit circle, the value of the last integral is 2π, and there follows:

LEMMA 18.4h. *For any $a \in R$, if f_a is the Riemann map of R normalized at a,*

$$k(a, a) \geqq \frac{1}{2\pi} f'_a(a). \tag{18.4-13}$$

We shall see presently that equality holds in (18.4-13).

III. Connection with the Riemann Map

Continuing to denote by r_a the function $[f'_a]^{1/2}$, $r_a(a)>0$, introduced in Lemma 18.4c, we prove:

THEOREM 18.4i. *For $(a, z)\in R\times R$ there holds*

$$k_a(z)=\frac{1}{2\pi}r_a(a)r_a(z). \tag{18.4-14}$$

Proof. For fixed $a\in R$, let

$$\gamma:=\frac{1}{2\pi}r_a(a)>0.$$

The expression on the right of (18.4-14) then equals $\gamma r_a(z)$. In order to show that this is the sum of the series $k(a, z)$ for all $z\in R$, it suffices, by Corollary 18.4b, to show that

$$\delta_n:=\|\hat{k}_a-r_a\|^2\to 0, \qquad n\to\infty.$$

We have

$$\delta_n=\|\hat{k}_a\|^2-2\gamma\,\mathrm{Re}(\hat{k}_a, r_a)+\gamma^2\|r_a\|^2.$$

The first term on the right by (18.4-7) equals $\hat{k}_a(a)$. The second term by Lemma 18.4c is calculated to be

$$2\gamma\,\mathrm{Re}(\hat{k}_a, r_a)=2\gamma\,\mathrm{Re}\,2\pi\frac{\hat{k}_a(a)}{r_a(a)}=2\hat{k}_a(a).$$

The third term is

$$\gamma^2\|r_a\|^2=\gamma^2\int_\Gamma|f'_a(t)||dt|=2\pi\gamma^2=\frac{f'_a(a)}{2\pi}.$$

We thus have

$$\delta_n=\frac{1}{2\pi}f'_a(a)-\hat{k}_a(a).$$

As $n\to\infty$, $\hat{k}_a(a)\to k(a, a)$ which by Lemma 18.4h is at least $(2\pi)^{-1}f'_a(a)$. Since $\delta_n\geqq 0$, there follows $\delta_n\to 0$. —

Setting $z=a$ in Theorem 18.4i yields:

COROLLARY 18.4j. *For all $a\in R$,*

$$k(a, a)=\frac{1}{2\pi}f'_a(a). \tag{18.4-15}$$

By means of Theorem 18.4i and its corollary, the function $f = f_a$ mapping R onto $|w| < 1$ such that $f(a) = 0$, $f'(a) > 0$ may, in principle, be calculated via the polynomials $p_n(z)$, as follows. From the polynomials we get

$$k(a, z) = \sum_{m=0}^{\infty} \overline{p_m(a)} p_m(z).$$

Then by (18.4-14) and (18.4-15),

$$f'(z) = 2\pi \frac{[k(a, z)]^2}{k(a, a)}, \tag{18.4-16}$$

and thus finally

$$f(z) = \frac{2\pi}{k(a, a)} \int_a^z [k(a, s)]^2 \, ds.$$

IV. The Integral Equation of Kerzman and Trummer

The Gram–Schmidt orthogonalization of the sequence of monomials $1, z, z^2, \ldots$ that is required to construct the Szegö kernel via its defining relation (18.4-6) is a numerically difficult and, if high accuracy is desired, frequently ill-conditioned process. It is therefore a matter of great interest that for fixed $a \in R$ and for $t \in \Gamma$ the function $k_a(t) := k(a, t)$ may be obtained as the solution of a certain Fredholm integral equation of the second kind that was discovered by Kerzman and Trummer.

If $z = z(\tau)$ is the parametric representation of Γ, and if $t = z(\tau)$, we denote by

$$q(t) := \frac{z'(\tau)}{|z'(\tau)|} \tag{18.4-17}$$

the unit tangent vector in the direction of increasing parameters at the point t. If f is any Riemann map of R and $g := f^{[-1]}$, then Γ is also described by

$$z(\theta) = g(e^{i\theta}), \qquad 0 \leqq \theta \leqq 2\pi.$$

Hence

$$z'(\theta) = i\, e^{i\theta} g'(e^{i\theta}).$$

Using $f(t) = e^{i\theta}$, a tangent vector at the point $t \in \Gamma$ thus is

$$if(t) \frac{1}{f'(t)}.$$

Writing $r := \sqrt{f'}$ and using $|f(t)| = 1$, $r\bar{r} = |f'|$, the unit tangent vector thus is given by

$$q(t) = if(t)\frac{\overline{r(t)}}{r(t)}, \tag{18.4-18}$$

a formula that will be used repeatedly.

The kernel of the Kerzman–Trummer integral equation for $t, s \in \Gamma$, $t \neq s$, is

$$b(s, t) := \frac{1}{2\pi i}\left\{\frac{q(t)}{t-s} - \frac{\overline{q(s)}}{\bar{t}-\bar{s}}\right\}. \tag{18.4-19}$$

It will be noted that the first term in view of $q(t)\,|dt| = dt$ just represents the Cauchy kernel. The second term has the effect of making the kernel $b(s, t)$ **skew hermitian**, that is, of making it satisfy

$$b(t, s) = -\overline{b(s, t)}. \tag{18.4-20}$$

To study the behavior of the kernel for $t \to s$, we consider the special parametrization of Γ where the parameter τ is the arc length. Also without loss of generality we may assume $s = 0$ and $q(0) = 1$. We then may write

$$z'(\tau) = e^{i\phi(\tau)},$$

where $\phi(\tau)$ is analytic,

$$\phi(\tau) = \alpha\tau + \beta\tau^2 + \cdots,$$

and we then have

$$t = z(\tau) = \int_0^\tau e^{i\phi(\tau)}\,d\tau = \tau + \tfrac{1}{2}i\alpha\tau^2 + \cdots$$

$$q(t) = z'(\tau) = 1 + i\alpha\tau + \cdots.$$

There follows

$$\begin{aligned}
2\pi i b(0, t) &= \frac{q(t)\bar{t} - t}{t\bar{t}} \\
&= \frac{(1 + i\alpha\tau + \cdots)(\tau - i\alpha\tau^2/2 + \cdots) - (\tau + i\alpha\tau^2/2 + \cdots)}{\tau^2(1 + O(\tau^2))} \\
&= O(\tau).
\end{aligned}$$

It follows that be defining $b(0, 0) := 0$ the function $b(0, z(\tau))$ becomes analytic at $\tau = 0$. This result clearly holds for any $s \in \Gamma$, and for any regular analytic parametrization of Γ. Thus we have:

LEMMA 18.4k. *If the curve* Γ: $z = z(\tau)$, $\alpha \leqq \tau \leqq \beta$, *is analytic, then the function* $(\tau, \sigma) \to b(z(\tau), z(\sigma))$ *is analytic on* $[\alpha, b] \times [\alpha, \beta]$.

Now let a be a fixed point $\in R$, let the function $z \mapsto k_a(z)$ be defined by (18.4-14) on the closure of R, and let, for $s \in \Gamma$,

$$g_a(s) := \left[\frac{1}{2\pi i}\frac{q(s)}{s-a}\right]^{-}. \tag{18.4-21}$$

We then have:

THEOREM 18.4l. *For $s \in \Gamma$, the function $s \mapsto k_a(s)$ satisfies the Fredholm integral equation of the second kind,*

$$k_a(s) - \int_\Gamma b(s,t)k_a(t)\,|dt| = g_a(s), \tag{18.4-22}$$

and it is the only continuous solution of this integral equation.

Proof. Evidently by (18.4-18),

$$\int_\Gamma b(s,t)k_a(t)\,|dt| = \mathrm{PV}\frac{1}{2\pi i}\int_\Gamma \frac{k_a(t)q(t)}{t-s}|dt|$$
$$- \mathrm{PV}\frac{1}{2\pi i}\int_\Gamma \frac{k_a(t)\overline{q(s)}}{\bar t - \bar s}|dt|.$$

We evaluate the two integrals on the right separately. In the first integral, we use $q(t)\,|dt| = dt$ and obtain by a straightforward application of Sokhotskyi's formula, since $r_a(z)$ is analytic on the closure of R,

$$\frac{r_a(a)}{2\pi}\mathrm{PV}\frac{1}{2\pi i}\int_\Gamma \frac{r_a(t)}{t-s}\,dt = \frac{r_a(a)}{2\pi}\cdot\frac{1}{2}r_a(s) = \frac{1}{2}k_a(s).$$

In the second integral we use (18.4-18) to obtain

$$\overline{r_a(t)} = -i\frac{q(t)r_a(t)}{f_a(t)}.$$

Hence

$$r_a(t) = i\frac{\overline{q(t)}}{\overline{f_a(t)}}\overline{r_a(t)}.$$

Thus in view of $\overline{q(t)}\,|dt| = \overline{dt}$ the second integral becomes

$$\frac{ir_a(a)\overline{q(s)}}{2\pi}\mathrm{PV}\frac{1}{2\pi i}\int_\Gamma \frac{\overline{r_a(t)}}{\overline{f_a(t)}(\bar t - \bar s)}\overline{dt}$$
$$= -\frac{ir_a(a)\overline{q(s)}}{2\pi}\left[\mathrm{PV}\frac{1}{2\pi i}\int_\Gamma \frac{r_a(t)}{f_a(t)(t-s)}\,dt\right]^{-}.$$

By the calculus of residues (using $f'_a = r_a^2$) and by Sokhotskyi's formulas the expression in brackets equals

$$\frac{1}{r_a(a)(a-s)}+\frac{1}{2}\frac{r_a(s)}{f_a(s)},$$

and on using (18.4-18) once more there results

$$-\frac{i}{2\pi}\frac{\overline{q(s)}}{\bar{a}-\bar{s}}-\frac{1}{2}k_a(s).$$

Thus in summary,

$$\int_\Gamma b(s,t)k_a(t)\,|dt| = k_a(s)-\frac{1}{2\pi i}\frac{\overline{q(s)}}{\bar{a}-\bar{s}},$$

which is equivalent to (18.4-22).

The uniqueness of the solution k_a follows from general facts of functional analysis, as follows. By (18.4-20) and by Lemma 18.4k, the kernel $b(s, t)$ is continuous and skew hermitian on $\Gamma\times\Gamma$. Hence the associated integral operator is a compact operator from $L_2(\Gamma)$ to $L_2(\Gamma)$ and has a purely imaginary spectrum. Consequently, (18.4-22) has a unique solution even in the class $L_2(\Gamma)$. —

By virtue of Theorem 18.4l, the function

$$k_a(s)=\frac{1}{2\pi}r_a(a)r_a(s)$$

($r_a(s):=[f_a(s)]^{1/2}$) may be constructed as the solution of a Fredholm integral equation of the second kind, the kernel of which is smooth and easily calculated. This construction may be performed via one of the usual discretization methods. Since $r_a(z(\tau))$ is a periodic function of the parameter τ, the Fourier series method used by Berrut in his solution of Symm's equation is likewise feasible here.

The Kerzman-Trummer equation furnishes the function k_a. By virtue of the relation

$$f'_a(t)=\frac{2\pi}{k_a(a)}[k_a(t)]^2$$

this determines the derivative of the mapping function up to the positive factor $\alpha := k_a(a)$. It is important to note, however, that the boundary correspondence function $\theta(\tau)$ corresponding to any representation $z = z(\tau)$, $0 \leqq \tau \leqq \beta$, of Γ may be calculated without knowing α. By differentiating the defining relation

$$f(z(\tau)) = e^{i\theta(\tau)}$$

we obtain

$$f'(z(\tau))z'(\tau) = i\theta'(\tau)\, e^{i\theta(\tau)}, \tag{18.4-23}$$

hence

$$\theta(\tau) = \arg[-if'(z(\tau))z'(\tau)]$$

or

$$\theta(\tau) = \arg\{-i[k_a(z(\tau))]^2 z'(\tau)\}. \tag{18.4-24}$$

If $\theta'(\tau)$ is desired, we obtain it by comparing absolute values in (18.4-23),

$$\theta'(\tau) = \frac{2\pi}{\alpha}|k_a(z(\tau))|^2|z'(\tau)|.$$

Here α now is determined from the condition that

$$2\pi = \int_0^\beta \theta'(\tau)\, d\tau = \frac{2\pi}{\alpha}\int_0^\beta |k_a(z(\tau))|^2|z'(\tau)|\, d\tau = \frac{2\pi}{\alpha}\int_\Gamma |k_a(t)|^2\, |dt|,$$

yielding

$$\alpha = \int_\Gamma |k_a(t)|^2\, |dt|. \tag{18.4-25}$$

V. Connection with the Exterior Mapping Function

So far the orthogonal polynomials $p_n(z)$ have been related, via the Szegö kernel function, to the interior mapping functions f_a of Γ. Rather surprisingly the p_n are also linked to the normalized exterior mapping function of Γ, that is, to the function mapping the exterior of Γ onto $|w| > 1$ and normalized at ∞ by

$$\mathring{f}(z) = \gamma^{-1}z + a_0 + O(z^{-1}),$$

where $\gamma > 0$ is the capacity of Γ. This link exists through the asymptotic behavior of the p_n as $n \to \infty$.

It is easy to guess what this asymptotic behavior might be. Since Γ is assumed to be analytic, $\mathring{f}$ is analytic on $\mathring{R} \cup \Gamma$, where $\mathring{R}$ denotes the exterior of Γ. Moreover, $\mathring{f}'$ is different from 0 and analytic in $\mathring{R} \cup \Gamma \cup \{\infty\}$. Thus

$$\mathring{r}(z) := [\mathring{f}'(z)]^{1/2}$$

may be defined as an analytic function; the square root is chosen such that $\mathring{r}(\infty) > 0$. The functions

$$h_n(z) := \frac{1}{\sqrt{2\pi}}[\mathring{f}(z)]^n \mathring{r}(z), \qquad n = 0, 1, 2, \ldots, \tag{18.4-26}$$

now share with the p_n the following properties. They are analytic in $\mathring{R} \cup \Gamma$; they have a pole of order n at ∞; they form an orthonormal system in $A_2(\Gamma)$. (This follows from the fact that on setting $u = \mathring{f}(z)$,

$$(h_n, h_m) = \frac{1}{2\pi} \int_{|u|=1} u^n \bar{u}^m \, |du| = \delta_{nm}.)$$

Postulating that all systems of such functions are asymptotically the same, we are led to conjecture that

$$p_n(z) \sim h_n(z), \qquad n \to \infty,$$

or, more precisely, that

$$\frac{p_n(z)}{h_n(z)} \to 1, \qquad z \in \mathring{R}. \tag{18.4-27}$$

In fact we shall prove:

THEOREM 18.4m. *The limit relation* (18.4-27) *holds, uniformly in any closed subset* $K \subset \mathring{R}$.

Proof. The functions $p_n h_n^{-1}$ are clearly analytic at ∞. We begin the proof by showing that their values at ∞ tend to 1. Denoting the leading coefficient of p_n by κ_n, we have

$$\frac{p_n(z)}{h_n(z)} = (2\pi)^{1/2} \gamma^{n+1/2} \kappa_n + O(z^{-1}),$$

and we therefore are to prove:

LEMMA 18.4n. *As* $n \to \infty$,

$$(2\pi)^{1/2} \gamma^{n+1/2} \kappa_n \to 1. \tag{18.4-28}$$

Proof. The method consists in squeezing the expression on the left between two limits that approach 1. First, since $|\mathring{f}(z)| = 1$ for $z \in \Gamma$,

$$\begin{aligned} 1 &= \| p_n \|^2 \\ &= \int_\Gamma |\mathring{f}(z)|^{-2n-1} |p_n(z)|^2 \, |dz| \\ &= \frac{1}{2\pi} \int_\Gamma \left| \frac{p_n(z)}{h_n(z)} \right|^2 \left| \frac{\mathring{f}'(z)}{\mathring{f}(z)} \right| |dz| \\ &\geqq \left| \frac{1}{2\pi i} \int_\Gamma \left[\frac{p_n(z)}{h_n(z)} \right]^2 \frac{\mathring{f}'(z)}{\mathring{f}(z)} \, dz \right|. \end{aligned}$$

Because

$$\frac{\mathring{f}'(z)}{\mathring{f}(z)}=\frac{1}{z}+O(z^{-2}),$$

the last expression by Laurent's theorem equals the value of $[p_n(z)/h_n(z)]^2$ at ∞, that is, $2\pi\gamma^{2n+1}\kappa_n^2$. There follows

$$1\geqq(2\pi)^{1/2}\gamma^{n+1/2}\kappa_n. \tag{18.4-29}$$

To obtain an estimate from below we use Corollary 18.4e, which asserts that among all monic polynomials of degree n, $\kappa_n^{-1}p_n(z)$ has the smallest norm. We seek a comparison polynomial that resembles p_n. The function h_n cannot be used because it is not a polynomial. However, we can use the polynomial part of the expansion of h_n at ∞ which we call $\hat{h}_n$. To make the leading coefficient 1, we multiply by $(2\pi)^{1/2}\gamma^{n+1/2}$. This yields

$$\begin{aligned}\kappa_n^{-1}&=\|\kappa_n^{-1}p_n\|\\&\leqq(2\pi)^{1/2}\gamma^{n+1/2}\|\hat{h}_n\|\\&\leqq(2\pi)^{1/2}\gamma^{n+1/2}(\|h_n\|+\|\hat{h}_n-h_n\|).\end{aligned}$$

Here

$$\|h_n\|^2=\frac{1}{2\pi}\int_\Gamma|\mathring{f}(z)|^{2n}|\mathring{f}'(z)|\,|dz|=\frac{1}{2\pi}\int_{|u|=1}|u|^{2n}\,|du|=1.$$

We thus need to estimate $\|h_n-\hat{h}_n\|$. Let Γ_ρ denote the image of $|w|=\rho$ under the map $z=\mathring{g}(w)$, where $\mathring{g}:=\mathring{f}^{[-1]}$. Since $\mathring{f}$ is analytic on Γ, $\mathring{g}$ is defined for $|w|\geqq\rho$ where $\rho<1$, ρ sufficiently close to 1. By Cauchy's formula for annular regions we have for $z\in\Gamma$, if Γ is contined in $|z|<\mu$,

$$\hat{h}_n(z)-h_n(z)=\frac{1}{2\pi i}\int_{|t|=\mu}\frac{\hat{h}_n(t)-h_n(t)}{t-z}\,dt-\frac{1}{2\pi i}\int_{\Gamma_\rho}\frac{\hat{h}_n(t)-h_n(t)}{t-z}\,dt.$$

The first integral vanishes for $\mu\to\infty$. In the second integral the term $\hat{h}_n$ may be omitted since it is analytic inside Γ_ρ. Thus

$$\hat{h}_n(z)-h_n(z)=\frac{1}{2\pi i}\int_{\Gamma_\rho}\frac{h_n(t)}{z-t}\,dt.$$

If $\delta:=\operatorname{dist}(\Gamma_\rho,\Gamma)$,

$$\mu:=\sup_{t\in\Gamma_\rho}(2\pi)^{-1/2}|\mathring{r}(t)|,$$

we thus in view of $|\mathring{f}(t)|=\rho$ have the estimate

$$|\hat{h}_n(z)-h_n(z)|\leqq\frac{\mu}{2\pi\delta}\rho^n,\qquad z\in\Gamma,$$

implying

$$\|\hat{h}_n - h_n\| \leqq \mu_1 \rho^n,$$

where μ_1 is a suitable constant. Thus there follows

$$\kappa_n^{-1}(2\pi)^{-1/2}\gamma^{-n-1/2} \leqq 1 + \mu_1 \rho^n,$$

which together with (18.4-29) proves Lemma 18.4n in the stronger form

$$1 - \mu_2 \rho^n \leqq (2\pi)^{1/2}\gamma^{n+1/2}\kappa_n \leqq 1,$$

where $\mu_2 > 0$ is another constant. —

Returning to the proof of Theorem 18.4m, we establish the equivalent assertion that

$$\frac{p_n(z)}{[\mathring{f}(z)]^n} - \frac{1}{\sqrt{2\pi}}\mathring{r}(z) \to 0 \qquad \text{uniformly for} \qquad z \in K. \qquad (18.4\text{-}30)$$

As a consequence of Lemma 18.4n, the value at ∞ of the function on the left,

$$\varepsilon_n := \kappa_n \gamma^n - (2\pi)^{-1/2}\gamma^{-1/2} \to 0.$$

Let $\mathring{A}_2(\Gamma)$ denote the space of functions that are analytic in $\mathring{R} \cup \{\infty\}$ ($\mathring{R} :=$ exterior of Γ), zero at infinity, and continuous on $\mathring{R} \cup \Gamma$. If endowed with the inner product (18.4-1), $\mathring{A}_2(\Gamma)$ again is an inner product space, and since for functions $g \in \mathring{A}_2(\Gamma)$ the Cauchy formula representation

$$g(z) = \frac{1}{2\pi i}\int_\Gamma \frac{g(t)}{z-t}\,dt, \qquad z \in \mathring{R},$$

holds, the analogs of Lemma 18.4a and of Corollary 18.4b hold for $\mathring{A}_2(\Gamma)$.

The functions $\mathring{f}^{-n}p_n - (2\pi)^{-1/2}\mathring{r} - \varepsilon_n$ belong to $\mathring{A}_2(\Gamma)$, and to establish (18.4-30) it suffices by Corollary 18.4b to show that

$$\|\mathring{f}^{-n}p_n - (2\pi)^{-1/2}\mathring{r} - \varepsilon_n\| \to 0,$$

or equivalently that

$$\|\mathring{f}^{-n}p_n - (2\pi)^{-1/2}\mathring{r}\| \to 0.$$

We have

$$\|\mathring{f}^{-n}p_n - (2\pi)^{-1/2}\mathring{r}\|^2 = \|\mathring{f}^{-n}p_n\|^2 + \frac{1}{2\pi}\|\mathring{r}\|^2 - 2(2\pi)^{-1/2}\operatorname{Re}(\mathring{f}^{-n}p_n, \mathring{r}).$$

Here, considering that $|\mathring{f}(z)| = 1$ on Γ,

$$\|\mathring{f}^{-n}p_n\|^2 = \int_\Gamma |\mathring{f}(z)|^{-2n}|p_n(z)|^2\,|dz| = 1.$$

Likewise we see by putting $u := \mathring{f}(z)$,

$$\frac{1}{2\pi}\|\mathring{r}\|^2 = \frac{1}{2\pi}\int_\Gamma |\mathring{f}'(z)|\,|dz| = \frac{1}{2\pi}\int_{|u|=1} |du| = 1.$$

The remaining term

$$2(2\pi)^{-1/2}\operatorname{Re}(\mathring{f}^{-n}p_n, \mathring{r}) = 2(2\pi)^{-1/2}\operatorname{Re}\int_\Gamma [\mathring{f}(z)]^{-n}p_n(z)\overline{\mathring{r}(z)}\,|dz|$$

by virtue of (18.4-18) (which holds if f, r are replaced by $\mathring{f}$, $\mathring{r}$, respectively) equals

$$2(2\pi)^{1/2}\operatorname{Re}\frac{1}{2\pi i}\int_\Gamma [\mathring{f}(z)]^{-n-1}p_n(z)\mathring{r}(z)q(z)\,|dz|.$$

Since $q(z)|dz| = dz$, the term following Re equals the coefficient of z^{-1} in the expansion at ∞ of

$$[\mathring{f}(z)]^{-n-1}p_n(z)\mathring{r}(z),$$

that is, $\gamma^{n+1/2}\kappa_n$. Thus we find

$$\|\mathring{f}^{-n}p_n - (2\pi)^{-1/2}\mathring{r}\|^2 = 2 - 2(2\pi)^{1/2}\gamma^{n+1/2}\kappa_n,$$

which tends to 0 by Lemma 18.4n. —

From Theorem 18.4m there immediately follows:

COROLLARY 18.4o. *If Γ is an analytic Jordan curve, and if $\{p_n\}$ is the sequence of orthogonal polynomials associated with Γ, the normalized mapping function $\mathring{f}$ for the exterior of Γ at every point z exterior to Γ is given by*

$$\mathring{f}(z) = \lim_{n\to\infty}\frac{p_{n+1}(z)}{p_n(z)}. \tag{18.4-31}$$

Our proof shows that the limit relation (18.4-31) holds uniformly in every closed subset of the exterior $\mathring{R}$ of Γ. Using stronger methods it may actually be shown that the relation holds uniformly in $\mathring{R} \cup \Gamma$. Thus by constructing the sequence $\{p_n\}$, it is possible in principle to find not only the interior but also the exterior mapping function, the latter as the limit of the rational functions

$$\frac{p_{n+1}(z)}{p_n(z)}.$$

However, as was pointed out earlier, the construction of a long sequence of orthogonal functions by means of the Gram–Schmidt process may run into numerical difficulties, and the author knows of no nontrivial example where an accurate determination of $\mathring{f}$ via the limit relation (18.4-31) was actually carried out.

PROBLEMS

1. Show that the space $A_2(\Gamma)$ is *not complete* (and hence is not a Hilbert space) by studying the special case where Γ is the unit circle,

$$h_n(z) := z + \frac{1}{2}z^2 + \cdots + \frac{1}{n}z^n, \qquad n = 1, 2, \ldots.$$

(The sequence $\{h_n\}$ is a Cauchy sequence, but is limit function does not belong to $A_2(\Gamma)$.)

2. Let ϕ_δ denote the *odd* 2π-periodic function such that

$$\phi_\delta(\tau) := \begin{cases} \dfrac{\tau}{\delta}, & 0 \leqq \tau \leqq \delta \\ 2 - \dfrac{\tau}{\delta}, & \delta \leqq \tau \leqq 2\delta \\ 0, & 2\delta \leqq \tau \leqq \pi. \end{cases}$$

Let h_n denote the function $\in A_2(\Gamma))$ ($\Gamma :=$ unit circle) such that

$$\operatorname{Re} h_n(e^{i\tau}) = \phi_\delta(\tau), \qquad \delta := \frac{1}{n}.$$

Show that the functions h_n satisfy the hypotheses of Corollary 18.4b where $h = 0$, but that

$$\sup_{|z|=1} |h_n(z) - h(z)| \geqq 1.$$

Conclude that the sequence $\{h_n\}$ need not converge uniformly to h on the set $R \cup \Gamma$.

3. The only case where the orthogonal polynomials p_n are known explicitly appears to be the circle $\Gamma\colon |z - a| = \rho$. Carry through the construction, and verify the statements concerning the Szegö kernel function.

4. Show that the kernel $b(s, t)$ of the Kerzman–Trummer integral equation vanishes if and only if Γ is a circle.

NOTES

Orthogonal polynomials with respect to a curve Γ were introduced by Szegö [1921]; see also Szegö [1959], ch. 16; Bergman [1950], ch. 7; and Walsh [1935], ch. 6. Equation (18.4-21) is derived by Kerzman and Trummer [1985] as the concrete realization of an operator equation given by Kerzman and Stein [1978]. Kerzman and Trummer [1985] also discuss the numerical solution, and provide relations (18.4-24) and (18.4-25), which are basic for numerical work. Taking advantage of the fact that the kernel is skew symmetric, Müller [1984] uses an acceleration device due to Niethammer and Varga [1983] for the efficient iterative solution of the Kerzman–Trummer equation. For the numerical solution of integral equations in general, see Atkinson [1976].

19

UNIVALENT FUNCTIONS

Conformal mapping functions may also be described as analytic functions that are one-to-one. Such functions are also called **univalent**. Thus if f is univalent, then $f(z_1)=f(z_2)$ implies $z_1=z_2$. In this chapter we examine properties of certain standardized classes of univalent functions. The theory of univalent functions does not aim at the construction of specific mapping functions. However, at least one special result (Theorem 19.1j) has been found useful in the study of certain numerical mapping techniques and will, in a later chapter on approximation theory, play an important role in the proof of Mergelyan's theorem. Our overview of univalent function theory culminates in the presentation of de Branges' recently discovered proof of the Bieberbach conjecture.

NOTES

Duren [1983] gives an excellent and readable account of the theory of univalent functions, which does full justice also to its historical aspects. Pommerenke [1975] gives another comprehensive account. Ahlfors [1973] most elegantly treats some special topics.

§ 19.1 ELEMENTARY THEORY OF UNIVALENT FUNCTIONS

Two classes of univalent functions are considered throughout this chapter:

(a) The class Σ_0 of functions f that are analytic and univalent in $D\colon |z|<1$, normalized so that the Taylor expansion at 0 has the form

$$f(z)=z+\sum_{n=2}^{\infty} a_n z^n. \tag{19.1-1}$$

(b) The class Σ_∞ of functions g that are analytic and univalent in $\mathring{D}$: $|z|>1$, with a simple pole at ∞, normalized such that the Laurent series at ∞ has the form

$$g(z)=z+\sum_{n=0}^{\infty} b_{-n}z^{-n}. \tag{19.1-2}$$

We first discuss some simple transformations that map the classes Σ_0 and Σ_∞ into each other and into themselves.

LEMMA 19.1a. *If $f\in\Sigma_0$, then the function*

$$g(z):=\frac{1}{f(1/z)}+w_0 \tag{19.1-3}$$

belongs to Σ_∞ for any choice of the constant w_0.

Proof. Because f assumes the value 0 only at $z=0$, g is analytic in $\mathring{D}$. If $g(z_1)=g(z_2)$, then $f(z_1^{-1})=f(z_2^{-1})$ and hence $z_1=z_2$. Computation shows that the expansion of g at ∞ has the required form (19.1-2). —

LEMMA 19.1b. *Let $g\in\Sigma_\infty$, and let w_1 not be in $g(\mathring{D})$. Then the function*

$$f(z):=\frac{1}{g(1/z)-w_1} \tag{19.1-4}$$

is in Σ_0.

Proof. Because g does not assume w_1, f is analytic in D. The fact that f is univalent follows as before. Computation shows that the expansion of f at 0 has the form (19.1-1). —

As an example, consider the Joukowski map,

$$j:\quad z\mapsto w=z+\frac{1}{z}$$

(normalized so as to satisfy (19.1-2)), which is in Σ_∞. The image of $\mathring{D}$ is the complement of the straight line segment $[-2, 2]$. By considering

$$g(z)=j(z)-2=z-2+\frac{1}{z}=\frac{(z-1)^2}{z},$$

which likewise is in Σ_∞, the straight line segment is shifted to $[-4, 0]$. Thus 0 is not in $g(\mathring{D})$. Lemma 19.1b now implies that the function

$$k(z):=\frac{1}{g(1/z)}=\frac{z}{(1-z)^2} \tag{19.1-5}$$

is in Σ_0. The function k is known as the **Koebe map**. The image of D under the Koebe map is the complex plane, cut along the negative real axis from $-\infty$ to $-\frac{1}{4}$.

We now discuss some transformations that map Σ_0 into Σ_0. Let $f \in \Sigma_0$. Then $z^{-1}f(z)$ is analytic and $\neq 0$ in D, because the only z for which $f(z)=0$ is $z=0$. By Theorem 4.3f we can define

$$s(z) := [z^{-1}f(z)]^{1/2}$$

as an analytic function in D satisfying $s(0)=1$.

LEMMA 19.1c. *If $f \in \Sigma_0$, then the function*

$$h(z) := [f(z^2)]^{1/2} = zs(z^2) \tag{19.1-6}$$

is in Σ_0; moreover, h is odd.

Proof. It is clear that h is an odd analytic function in D; it remains to be shown that h is univalent. Let $h(z_1)=h(z_2)$. This implies $f(z_1^2)=f(z_2^2)$ and, because f is univalent, $z_1^2=z_2^2$. There follows $s(z_1^2)=s(z_2^2)$ and, since $h(z)=zs(z^2)$, $z_1=z_2$. —

The passage from f to h defined by (19.1-6) will be called a (quadratic) Faber transformation. In a like manner we could define a Faber transformation of arbitrary integer order m by

$$h(z) := [f(z^m)]^{1/m}.$$

We have seen that the Faber transform of any $f \in \Sigma_0$ is an odd function. Is every odd $h \in \Sigma_0$ the Faber transform of some $f \in \Sigma_0$? Let $h \in \Sigma_0$ be odd, and put $h(z)=zq(z^2)$, where q is analytic, $q(0)=1$. Inversion of (19.1-6) yields

$$f(z) = [h(\sqrt{z})]^2 = zq(z).$$

It is clear that f is analytic in D and has the correct behavior at 0. To see whether f is univalent, assume $f(z_1)=f(z_2)$. Then $h(\sqrt{z_1})=h(\sqrt{z_2})$ and (because h is univalent and odd) $\sqrt{z_1}=\pm\sqrt{z_2}$. There follows $z_1=z_2$. We thus have:

LEMMA 19.1d. *Every odd $h \in \Sigma_0$ is the Faber transform of some $f \in \Sigma_0$.* —

LEMMA 19.1e. *If $f \in \Sigma_0$ and w is in the complement of $f(D)$, then the function*

$$h(z) := \frac{f(z)}{1-w^{-1}f(z)} \tag{19.1-7}$$

is in Σ_0.

Proof. Because $f(z) \neq w$ for $z \in D$, h is analytic in D. Because the Moebius transformation

$$u \mapsto \frac{u}{1 - w^{-1}u}$$

is one-to-one, h is univalent. It is obvious that $h(z) = z + \cdots$. Thus $h \in \Sigma_0$. —

LEMMA 19.1f. *Let $f \in \Sigma_0$, and let $|t| < 1$. Then the function*

$$h(z) := \frac{1}{(1 - |t|^2)f'(t)} \left\{ f\left(\frac{z+t}{1+\bar{t}z}\right) - f(t) \right\}$$

is in Σ_0.

Proof. The Moebius transformation

$$z \mapsto \frac{z+t}{1+\bar{t}z}$$

is a one-to-one map of D onto itself. Therefore

$$z \mapsto f\left(\frac{z+t}{1+\bar{t}z}\right)$$

is a univalent function on D. Subtracting $f(t)$ yields the value zero at 0, and dividing by $(1 - |t|^2)f'(t)$ makes the derivative equal to 1. Thus $h \in \Sigma_0$. —

By the foregoing lemmas, results which are proved for one of the classes Σ_0 and Σ_∞ can be reinterpreted as results for the other. The following result, called the **area theorem**, concerns the class Σ_∞.

THEOREM 19.1g. *For any function $g \in \Sigma_\infty$ given by* (19.1-2),

$$\sum_{n=1}^{\infty} n|b_{-n}|^2 \leqq 1. \tag{19.1-8}$$

Proof. For $\rho > 1$, let Γ_ρ denote the curve

$$w = g(\rho e^{i\tau}), \qquad 0 \leqq \tau \leqq 2\pi.$$

Because g is univalent, Γ_ρ is a positively oriented Jordan curve. By elementary calculus (see also Problem 1, § 4.2)

$$\alpha(\rho) := \frac{1}{2i} \int_{\Gamma_\rho} \bar{w}\, dw = \frac{1}{2} \int_{\Gamma_\rho} (v\, du - u\, dv)$$

is the area enclosed by Γ_ρ, hence

$$\alpha(\rho) \geqq 0. \tag{19.1-9}$$

Explicit calculation yields on the other hand

$$\alpha(\rho)=\frac{1}{2}\int_0^{2\pi}\left(\rho e^{-i\tau}+\sum_{n=0}^{\infty}\overline{b_{-n}}\rho^{-n}e^{-in\tau}\right)$$
$$\cdot\left(1-\sum_{n=1}^{\infty}nb_{-n}\rho^{-n-1}e^{-i(n+1)\tau}\right)\rho e^{i\tau}\,d\tau.$$

All the cross terms cancel, and we get

$$\alpha(\rho)=\pi\left(\rho^2-\sum_{n=1}^{\infty}n|b_{-n}|^2\rho^{-2n}\right).$$

The result follows by (19.1-9) on letting $\rho\to 1$. —

As a simple consequence of the area theorem,

$$|b_{-n}|\leqq\frac{1}{\sqrt{n}},\quad n=1,2,\ldots,$$

and in particular,

$$|b_{-1}|\leqq 1. \tag{19.1-10}$$

One function for which the equality sign holds is the Joukowski map, $z\mapsto z+z^{-1}$. As suggested by the above proof, the curve Γ_1 here encloses zero area.

The remainder of this section presents some applications of (19.1-10) by means of the foregoing lemmas.

THEOREM 19.1h. *Let $f\in\Sigma_0$ be given by* (19.1-1). *Then* $|a_2|\leqq 2$.

Proof. The direct passage from Σ_0 to Σ_∞ via Lemma 19.1a does not yield the desired result. We first apply a quadratic Faber transformation. Choosing the square root as indicated before Lemma 19.1c, we have

$$[f(z^2)]^{1/2}=z(1+\tfrac{1}{2}a_2z^2+\cdots),$$

hence

$$g(z):=[f(z^2)]^{-1/2}=z(1-\tfrac{1}{2}a_2z^{-2}+\cdots).$$

Here the coefficient of z^{-1} is $-\frac{1}{2}a_2$, and (19.1-10) yields $|a_2|\leqq 2$. —

The bound of Theorem 19.1h is attained by the Koebe function,

$$k(z)=\frac{z}{(1-z)^2}=z+2z^2+3z^3+\cdots.$$

Moreover, if $|a_2|=2$ for some $f\in\Sigma_0$, then the coefficient b_{-1} in the corresponding function g satisfies $|b_{-1}|=1$. This means that all remaining $b_{-n}=0$, with the consequence that $g(z)=z+e^{i\alpha}z^{-1}$. This in turn means that

$$f(z)=\frac{1}{g(1/\sqrt{z})^2}=\frac{z}{(1+e^{i\alpha}z)^2}$$

is a rotation of the Koebe function.

Bieberbach conjectured as early as 1916 that for any $f\in\Sigma_0$, $f(z)=z+\sum a_n z^n$, there holds $|a_n|\leqq n$ for all n. Many partial results concerning this **Bieberbach conjecture** were obtained in the intervening years, until the conjecture was finally proved early in 1984 by L. de Branges. An elementary account of de Branges' proof is given in § 19.6. Here we mention some consequences of $|a_2|\leqq 2$.

THEOREM 19.1i (Koebe $\frac{1}{4}$ theorem). *Let $f\in\Sigma_0$. Then $f(D)$ contains the disk $|w|<\frac{1}{4}$.*

Proof. Let w be in the complement of $f(D)$. By Lemma 19.1e, the function

$$h(z):=\frac{f(z)}{1-w^{-1}f(z)}$$

is in Σ_0. The Taylor series of h at 0 is

$$\begin{aligned}h(z)&=(z+a_2z^2+\cdots)\left(1+\frac{1}{w}z+\cdots\right)\\&=z+\left(a_2+\frac{1}{w}\right)z^2+\cdots.\end{aligned}$$

By the preceding theorem, $|a_2+1/w|\leqq 2$. But also $|a_2|\leqq 2$, hence $|w^{-1}|\leqq 4$ or $|w|\geqq\frac{1}{4}$. —

The statement of Theorem 19.1i is again sharp, because for the Koebe map, $w=-\frac{1}{4}$ is in the complement of $k(D)$.

THEOREM 19.1j. *Let $g\in\Sigma_\infty$, and let C be the complement of $g(\mathring{D})$. Then the diameter of C is at most* 4.

Proof. Let w_1, w_2 be any two points of C. By Lemma 19.1b, the function

$$f(z):=\frac{1}{g(1/z)-w_1}$$

is in Σ_0. Because g does not assume w_2, f does not assume the value $1/(w_2-w_1)$. By Theorem 19.1i there follows $|w_2-w_1|^{-1}\geqq\frac{1}{4}$, or $|w_1-w_2|\leqq 4$. —

Once again, the statement of Theorem 19.1j is sharp, because for the Joukowski map $z\mapsto z+z^{-1}$, C is the straight line segment [−2, 2], and thus it has the precise diameter 4.

There follow some results which regulate the growth of functions $f \in \Sigma$, and of their derivatives. Let $f \in \Sigma$, let $t \in D$ be fixed, and let h be associated with f in the manner of Lemma 19.1f. Setting

$$h(z) = z + a_2^*(t) z^2 + \cdots,$$

computation yields

$$a_2^*(t) = \frac{1}{2}\left\{(1-|t|^2)\frac{f''(t)}{f'(t)} - 2\bar{t}\right\}.$$

Since $h \in \Sigma_0$, Theorem 19.1h implies $|a_2^*(t)| \leqq 2$. Writing z for t and letting $\rho := |z|$, this may be written

$$\left|\frac{zf''(z)}{f'(z)} - \frac{2\rho^2}{1-\rho^2}\right| \leqq \frac{4\rho}{1-\rho^2}, \qquad 0 < |z| = \rho < 1.$$

Thus in particular,

$$\frac{2\rho^2 - 4\rho}{1-\rho^2} \leqq \operatorname{Re}\frac{zf''(z)}{f'(z)} \leqq \frac{2\rho^2 + 4\rho}{1-\rho^2}.$$

Because $f'(z) \neq 0$ and $f'(0) = 1$, $\log f'(z)$ can be chosen as an analytic function in D that vanishes at 0. Its derivative with respect to z is $f''(z)/f'(z)$, and if $z = \rho\, e^{i\theta}$, its partial derivative with respect to ρ is

$$\frac{\partial}{\partial\rho}\log f'(z) = \frac{z}{|z|}\frac{f''(z)}{f'(z)},$$

which implies

$$\rho\frac{\partial}{\partial\rho}\operatorname{Re}\log f'(z) = \operatorname{Re}\frac{zf''(z)}{f'(z)}.$$

We thus have

$$\frac{2\rho - 4}{1-\rho^2} \leqq \frac{\partial}{\partial\rho}\operatorname{Log}|f'(\rho\, e^{i\theta})| \leqq \frac{2\rho + 4}{1-\rho^2}.$$

Keeping θ fixed, we integrate this inequality between 0 and ρ. Calculus yields

$$\operatorname{Log}\frac{1-\rho}{(1+\rho)^3} \leqq \operatorname{Log}|f'(\rho\, e^{i\theta})| \leqq \operatorname{Log}\frac{1+\rho}{(1-\rho)^3}.$$

Exponentiating we obtain:

THEOREM 19.1k (Distortion theorem). *For any $f \in \Sigma$, if $|z| = \rho$,*

$$\frac{1-\rho}{(1+\rho)^3} \leqq |f'(z)| \leqq \frac{1+\rho}{(1-\rho)^3}, \qquad 0 \leqq \rho < 1. \tag{19.1-11}$$

The bounds (19.1-11) are again sharp; they are attained by the Koebe function.

By integrating the inequalities of the distortion theorem, we now obtain corresponding inequalities for $f(z)$. Let $z=\rho e^{i\theta}$ be fixed, $0<\rho<1$. Since $f(0)=0$,

$$f(z)=\int_0^\rho f'(\sigma e^{i\theta})\, e^{i\theta}\, d\sigma,$$

and by the distortion theorem there follows

$$|f(z)|\leqq\int_0^\rho |f'(\sigma e^{i\theta})|\, d\sigma=\int_0^\rho \frac{1+\sigma}{(1-\sigma)^3}\, d\sigma=\frac{\rho}{(1-\rho)^2}.$$

This yields the upper bound of:

THEOREM 19.1l (Growth theorem). *For any $f\in\Sigma_0$, if $|z|=\rho$,*

$$\frac{\rho}{(1+\rho)^2}\leqq|f(z)|\leqq\frac{\rho}{(1-\rho)^2},\qquad 0\leqq\rho<1. \tag{19.1-12}$$

To obtain the lower bound in (19.1-12), we need not be concerned with z for which $|f(z)|\geqq\frac{1}{4}$, because $\rho(1+\rho)^{-2}<\frac{1}{4}$ for $0\leqq\rho<1$. If $|f(z)|<\frac{1}{4}$, then by the Koebe $\frac{1}{4}$ theorem the straight line segment from 0 to $w:=f(z)$ lies entirely within $f(D)$. Let Γ be the pre-image of this segment. Then Γ is a simple arc from 0 to z, along which $\arg f'(z)\, dz=\arg dw=\text{const}$. Then

$$|f(z)|=\left|\int_\Gamma f'(z)\, dz\right|=\int_\Gamma |f'(z)||dz|.$$

Using the distortion theorem and the fact that $|dz|\geqq d|z|$, the last expression is bounded by

$$\int_0^\rho \frac{1-\sigma}{(1+\sigma)^3}\, d\sigma=\frac{\rho}{(1+\rho)^2},$$

which completes the proof of the growth theorem. —

Clearly, the bounds of the growth theorem are again sharp; they are attained by the Koebe function for positive and negative values of z, respectively.

The growth theorem allows us to perform a modest step in the direction of proving the Bieberbach conjecture. Let $f\in\Sigma_0$, $f(z)=z+\sum a_n z^n$. Then if $n\geqq 2$, we have by the Cauchy formula

$$a_n=\frac{1}{2\pi i}\int_{|z|=\rho}\frac{f(z)}{z^{n+1}}\, dz,$$

and hence

$$|a_n| \leqq \frac{1}{\rho^{n-1}(1-\rho)^2}, \qquad 0<\rho<1.$$

The minimum of the expression on the right is attained for

$$\rho = \frac{n-1}{n+1},$$

and it has the value

$$\left(\frac{n+1}{2}\right)^2 \left(1+\frac{2}{n-1}\right)^{n-1} \leqq \frac{e^2}{4}(n+1)^2.$$

This does not even have the correct order of magnitude as $n \to \infty$, but still it proves:

THEOREM 19.1m. *The suprema*

$$\alpha_n := \sup_{f \in \Sigma_0} |a_n|, \qquad n = 2, 3, \ldots,$$

are all finite.

PROBLEMS

1. Show that the Faber transform of the Koebe function is

$$h(z) = [k(z^2)]^{1/2} = \frac{z}{1-z^2} = z + z^3 + z^5 + \cdots.$$

(It thus was conjectured by Littlewood that if h is any *odd* function in Σ_0, $h(z) = z + b_3 z^3 + b_5 z^5 + \cdots$, then $|b_{2k+1}| \leqq 1$. This was shown to be false by Fekete and Szegö.)

2. If $f \in \Sigma_0$, then the function $z \mapsto e^{-i\alpha} f(e^{i\alpha} z) \in \Sigma_0$ for any real α. Hence prove: the Bieberbach conjecture is true if for any $f \in \Sigma_0, f(z) = z + \sum a_n z^n$, there holds

$$\operatorname{Re} a_n \leqq n, \qquad n = 2, 3, \ldots.$$

3. Let $g \in \Sigma_\infty$, and let C be defined as in Theorem 19.1j. Show that the diameter δ of C is at least 2. (For every $\rho > 1$, the diameter δ_ρ of the curve Γ_ρ: $w = g(\rho e^{i\tau}), 0 \leqq \tau \leqq 2\pi$, is at least $|g(z) - g(-z)|$, where $z = \rho e^{i\tau}$. Now use $z^{-1}\{g(z) - g(-z)\} = 2 + O(z^{-2})$ and let $\rho \to 1$.)

4. Let

$$g(w) = \int_0^w \prod_{k=0}^{n} \left(1 - \frac{w}{w_k}\right)^{\alpha_k - 1} dw$$

be the Schwarz-Christoffel map from the unit disk to the interior of a polygon, where the notation is chosen as in § 16.10. Show that the parameters $\beta_k := \alpha_k - 1$ satisfy the constraint

$$\left| \sum_{k=0}^{n} \beta_k w_k \right| \leqq 4.$$

(If $g(w) = w + a_2 w^2 + \ldots$, compute a_2 and use Theorem 19.1h.) Obtain further constraints of this type by assuming the Bieberbach conjecture.

5. Let $g \in \Sigma_\infty$ be the Schwarz-Christoffel map from $\mathring{D}$ to the exterior of a polygon P with the vertices $z_1, \ldots, z_n$ and the exterior angles $\pi\alpha_k$, $0 < \alpha_k \leqq 2$, $k = 1, 2, \ldots, n$, so that

$$g(w) = \int_0^w \prod_{k=0}^{n} \left(1 - \frac{w}{w_k}\right)^{\alpha_k - 1} dw$$

where the w_k are the prevertices of P. Show that the parameters $\beta_k := \alpha_k - 1$ and w_k satisfy the constraints

(a) $$\sum_{k=1}^{n} \beta_k w_k = 0;$$

(b) $$\left| \sum_{k=1}^{n} \beta_k w_k^2 \right| \leqq 2;$$

(c) $$\left| \sum_{k=1}^{n} \beta_k w_k^3 \right| \leqq 3\sqrt{2}.$$

6. Let $f(z) = z + \sum a_n z^n \in \Sigma_0$. Show that

$$|a_3 - a_2^2| \leqq 1.$$

(Apply Lemma 19.1a and use the area theorem.)

7. Let $g \in \Sigma_\infty$. Show that for $|z| > 1$,

$$|g'(z)| \leqq \frac{1}{1 - |z|^{-2}}.$$

(Apply Cauchy's inequality to

$$g'(z) = 1 - \sum_{n=1}^{\infty} \frac{nb_{-n}}{z^{n+1}}$$

and use Theorem 19.1g.)

NOTES

Chapter 1 of Duren [1983] has all results presented here, and more. See also Polya and Szegö [1925], vol. II, problems 136–163. Concerning Problem 3, Pfluger [1976] proves that $\delta = 2$ is attained if and only if $g(z) = z$. For Problems 4 and 5 see Pfaltzgraff [1985], where further inequalities of this type are given.

§ 19.2. PROPOSITIONS THAT IMPLY THE BIEBERBACH CONJECTURE

Let $f \in \Sigma_0$,

$$f(z) = a_1 z + a_2 z^2 + \cdots,$$

$a_1 = 1$, and let h be the Faber transform of f,

$$h(z) = b_1 z + b_3 z^3 + b_5 z^5 + \cdots,$$

$b_1 := 1$. Then $f(z) = [h(\sqrt{z})]^2$ or, written out in terms of power series,

$$a_1 z + a_2 z^2 + \cdots = z(b_1 + b_3 z + b_5 z^2 + \cdots)^2.$$

Comparing coefficients of z^n, we find

$$a_n = b_1 b_{2n-1} + b_3 b_{2n-3} + \cdots + b_{2n-1} b_1, \qquad n = 1, 2, \ldots.$$

Thus by the Schwarz inequality,

$$|a_n|^2 \leqq (|b_1|^2 + |b_3|^2 + \cdots + |b_{2n-1}|^2)^2. \tag{19.2-1}$$

Thus we clearly have:

THEOREM 19.2a. *Bieberbach's conjecture is true if for each odd $h \in \Sigma_0$, $h(z) = b_1 z + b_3 z^3 + \cdots$ $(b_1 := 1)$ and for $n = 0, 1, 2, \ldots$ there holds*

$$|b_1|^2 + |b_3|^2 + \cdots + |b_{2n+1}|^2 \leqq n + 1. \tag{19.2-2}$$

The conjecture that the premise of Theorem 19.2a is true is called **Robertson's conjecture**.

To continue, we require a simple inequality on the coefficients of certain formal power series.

LEMMA 19.2b (Lebedev–Milin inequality). *Let*

$$P := p_1 x + p_2 x^2 + \cdots$$

be a nonunit in the integral domain $\mathcal{P}$ of formal power series over $\mathbb{C}$, and let

$$Q := E \circ P = q_0 + q_1 x + q_2 x^2 + \ldots, \tag{19.2-3}$$

where E is the exponential series. Then for $n = 0, 1, 2, \ldots$,

$$|q_0|^2 + |q_1|^2 + \cdots + |q_n|^2 \leqq (n+1) \exp\left\{\frac{1}{n+1} \sum_{k=1}^{n} (n+1-k)\left[k|p_k|^2 - \frac{1}{k}\right]\right\}. \tag{19.2-4}$$

Proof. Formal differentiation of (19.1-3) yields

$$q_1+2q_2x+3q_3x^2+\cdots=(q_0+q_1x+q_2x^2+\cdots)(p_1+2p_2x+3p_3x^2+\cdots),$$

and on comparing coefficients,

$$nq_n=\sum_{k=0}^{n-1}(n-k)p_{n-k}q_k, \qquad n=1,2,\ldots.$$

By the Schwarz inequality there follows

$$n^2|q_n|^2\leqq\sum_{k=1}^{n}k^2|p_k|^2\sum_{k=0}^{n-1}|q_k|^2.$$

For $n=1,2,\ldots,$ *let*

$$\pi_n:=\sum_{k=1}^{n}k^2|p_k|^2, \qquad \gamma_n:=\sum_{k=0}^{n}|q_k|^2.$$

Then the above may be written

$$\gamma_n-\gamma_{n-1}\leqq\frac{1}{n^2}\pi_n\gamma_{n-1}.$$

Using $1+x\leqq e^x$, this yields

$$\begin{aligned}
\gamma_n&\leqq\left(1+\frac{1}{n^2}\pi_n\right)\gamma_{n-1}\\
&=\frac{n+1}{n}\left(\frac{n}{n+1}+\frac{\pi_n}{n(n+1)}\right)\gamma_{n-1}\\
&=\frac{n+1}{n}\left(1+\frac{\pi_n-n}{n(n+1)}\right)\gamma_{n-1}\\
&\leqq\frac{n+1}{n}\exp\left\{\frac{\pi_n-n}{n(n+1)}\right\}\gamma_{n-1}.
\end{aligned}$$

Estimating $\gamma_{n-1},\gamma_{n-2},\ldots$ similarly, there follows by $\gamma_0=1$,

$$\begin{aligned}
\gamma_n&\leqq(n+1)\exp\left\{\sum_{k=1}^{n}\frac{\pi_k-k}{k(k+1)}\right\}\\
&=(n+1)\exp\left\{\sum_{k=1}^{n}\frac{\pi_k}{k(k+1)}+1-\sum_{k=1}^{n+1}\frac{1}{k}\right\}.
\end{aligned}$$

Since

$$\frac{1}{k(k+1)}=\frac{1}{k}-\frac{1}{k+1},$$

we obtain by summation by parts

$$\sum_{k=1}^{n} \pi_k \frac{1}{k(k+1)} = \sum_{k=1}^{n} \pi_k \left(\frac{1}{k} - \frac{1}{k+1}\right)$$

$$= \sum_{k=1}^{n} \frac{1}{k}(\pi_k - \pi_{k-1}) - \frac{\pi_n}{n+1}$$

$$= \sum_{k=1}^{n} k|p_k|^2 - \frac{1}{n+1} \sum_{k=1}^{n} k^2|p_k|^2.$$

Thus

$$\gamma_n \leqq (n+1) \exp\left\{\sum_{k=1}^{n} \left(1 - \frac{k}{n+1}\right) k|p_k|^2 + 1 - \sum_{k=1}^{n+1} \frac{1}{k}\right\}$$

$$= (n+1) \exp\left\{\frac{1}{n+1} \sum_{k=1}^{n} (n+1-k)\left[k|p_k|^2 - \frac{1}{k}\right]\right\},$$

as desired. —

Let now h be any odd function in Σ_0,

$$h(z) = z + b_3 z^3 + b_5 z^5 + \ldots,$$

and let $f(z) = [h(\sqrt{z})]^2$ be its inverse Faber transform which by Lemma 19.1d exists in Σ_0. Using the Lebedev–Milin inequality, we convert Robertson's inequality (19.2-2) into an inequality for the coefficients c_k in

$$\log \frac{f(z)}{z} = \sum_{k=1}^{\infty} c_k z^k. \tag{19.2-5}$$

We have

$$\log \frac{f(z)}{z} = \log \frac{[h(\sqrt{z})]^2}{z} = 2 \log(1 + b_3 z + b_5 z^2 + \cdots),$$

hence

$$1 + b_3 z + b_5 z^2 + \cdots = \exp\left\{\frac{1}{2} \sum_{k=1}^{\infty} c_k z^k\right\},$$

and on using Lemma 19.2b where $q_k = b_{2k+1}$, $p_k = \frac{1}{2}c_k$ there follows

$$|b_1|^2 + |b_3|^2 + \cdots + |b_{2n+1}|^2$$

$$\leqq (n+1) \exp\left\{\frac{1}{4(n+1)} \sum_{k=1}^{n} (n+1-k)\left[k|c_k|^2 - \frac{4}{k}\right]\right\}, \qquad n = 0, 1, 2, \ldots.$$

The exponential is $\leqq 1$ if the exponent is $\leqq 0$. Thus in conjunction with Theorem 19.2a there follows:

THEOREM 19.2c. *The Bieberbach conjecture is true if for each $f \in \Sigma_0$ the coefficients c_k defined by* (19.2-5) *satisfy, for $n = 1, 2, \ldots$,*

$$\sum_{k=1}^{n} (n+1-k)\left[k|c_k|^2 - \frac{4}{k}\right] \leqq 0. \qquad (19.2\text{-}6)$$

It was conjectured by I. M. Milin [1971] that the inequality (19.2-6) in fact holds. This was proved by de Branges [1985] early in 1984 by means of a surprising application of a lemma due to Askey and Gasper [1976]. A simplified version of this proof is presented in § 19.6. The material in § 19.4 and § 19.5 is preparatory to this proof.

PROBLEMS

1. Show that in the Lebedev–Milin inequality (19.2-4), equality holds for some n if and only if

$$kp_k = q^k, \qquad k = 1, 2, \ldots, n,$$

for some complex q such that $|q| = 1$.

2. Under the hypotheses of Lemma 19.2b, show that for $n = 1, 2, \ldots$,

$$|q_n|^2 \leqq \exp\left\{\sum_{k=1}^{n}\left(k|p_k|^2 - \frac{1}{k}\right)\right\},$$

with equality only for $kp_k = q^k$, $|q| = 1$.

NOTES

See Duren [1983], § 2.11 and § 5.1.

§ 19.3. FABER POLYNOMIALS AND UNIVALENT FUNCTIONS

Here we prove some analytical consequences of the formal identities derived in § 18.1. The results are of historical importance because they provided proofs of the Bieberbach conjecture for the fourth and sixth coefficients. The essential tool is the following generalization of the area theorem (Theorem 19.1g):

THEOREM 19.3a. *Let p be a polynomial of degree $m \geqq 1$, let $g \in \Sigma_\infty$, and let*

$$p(g(z)) = \sum_{k=-\infty}^{m} b_k z^k. \qquad (19.3\text{-}1)$$

Then

$$\sum_{k=1}^{m} k|b_k|^2 \geqq \sum_{k=1}^{\infty} k|b_{-k}|^2. \qquad (19.3\text{-}2)$$

Proof. As in the proof of the area theorem, let, for $\rho>1$, Γ_ρ be the image of Θ_ρ: $|z|=\rho$ under g, and let R_ρ denote the interior of Γ_ρ. Let

$$\alpha(\rho):=\iint_{R_\rho}|p'(w)|^2\,\boxed{dw}.$$

Since

$$|p'(w)|^2=p'(w)\overline{p'(w)}=\frac{\partial}{\partial\bar w}[p'(w)\overline{p(w)}],$$

there follows by the complex form (15.10-4) of Green's identity

$$\alpha(\rho)=\frac{1}{2i}\int_{\Gamma_\rho}p'(w)\overline{p(w)}\,dw$$

or, on introducing $w=g(z)$,

$$\alpha(\rho)=\frac{1}{2i}\int_{\Theta_\rho}\overline{p(g(z))}p'(g(z))g'(z)\,dz.$$

Introducing (19.3-1), this becomes

$$\begin{aligned}\alpha(\rho)&=\frac{1}{2}\int_0^{2\pi}\left(\sum_{k=-\infty}^{m}\overline{b_k}\rho^k e^{-ik\tau}\right)\left(\sum_{k=-\infty}^{m}kb_k\rho^k e^{ik\tau}\right)d\tau\\&=\pi\sum_{k=-\infty}^{m}k|b_k|^2\rho^{2k}\\&=\pi\sum_{k=1}^{m}k|b_k|^2\rho^{2k}-\pi\sum_{k=1}^{\infty}k|b_{-k}|^2\rho^{-2k}.\end{aligned}$$

The result now follows on letting $\rho\to1$ in view of $\alpha(\rho)\geqq0$. —

Let now g be analytic in $\mathring{D}$, with the Laurent series

$$g(z)=z+\sum_{k=-\infty}^{0}b_kz^k \tag{19.3-3}$$

at ∞. We may regard this series as an almost regular series in $\mathscr{L}_\infty$, where $b_1=1$, and compute for it the Faber polynomials $p_n(u)$ and the Faber functions $h_n(z):=p_n(g(z))$, which possess the Laurent series

$$h_n(z)=z^n+n\sum_{k=-\infty}^{-1}c_{nk}z^k, \tag{19.3-4}$$

now convergent for $|z|>1$, the coefficients of which satisfy $c_{n,k}=c_{-k,-n}$ and may be computed rationally from the b_k. Moreover, the series (18.1-25) converge for $y=w$, where $|w|>1$, and the Grunsky series

$$l(z):=-\sum_{n=1}^{\infty}\left(\sum_{k=1}^{\infty}c_{n,-k}w^{-k}\right)z^{-n} \tag{19.3-5}$$

converges for $|z|$ sufficiently large and represents the branch of

$$\log \frac{g(z)-g(w)}{z-w}$$

that tends to zero for $z \to \infty$.

A fundamental problem in the theory of univalent functions consists in finding necessary and sufficient conditions for a function of type (19.3-3) to be univalent. The following condition is especially striking:

THEOREM 19.3b (Grunsky inequality). *The function g belongs to* Σ_∞ *if and only if for* $m = 1, 2, \ldots$ *and for arbitrary complex numbers* $x_1, x_2, \ldots,$

$$\sum_{n=1}^{\infty} n \left| \sum_{k=1}^{m} c_{n,-k} x_k \right|^2 \leqq \sum_{k=1}^{m} \frac{1}{k} |x_k|^2. \qquad (19.3\text{-}6)$$

Proof. (a) Let (19.3-6) hold as indicated. Setting all $x_j = 0$ with the exception of $x_k = 1$, there results

$$\sum_{n=1}^{\infty} n |c_{n,-k}|^2 \leqq \frac{1}{k}.$$

Thus certainly

$$|c_{n,-k}| \leqq 1.$$

For any $w \in \mathring{D}$ there follows

$$\left| \sum_{k=1}^{\infty} c_{n,-k} w^{-k} \right| \leqq \frac{1}{1-|w|}.$$

Thus the Grunsky series converges for $|z| > 1$, which means that

$$g(z) \neq g(w), \qquad z \neq w,$$

which means that g is univalent.

(b) Now let $g \in \Sigma_\infty$. For any choice of the complex numbers $x_1, x_2, \ldots, x_m$, the polynomial

$$p(u) := \sum_{n=1}^{m} \frac{1}{n} x_n p_n(u)$$

has degree $\leqq m$. By (18.1-19),

$$\begin{aligned} p(g(z)) &= \sum_{n=1}^{m} \frac{1}{n} x_n h_n(z) \\ &= \sum_{n=1}^{m} \frac{1}{n} x_n z^n + \sum_{k=1}^{\infty} \left(\sum_{n=1}^{m} c_{n,-k} x_n \right) z^{-k}. \end{aligned}$$

Applying Theorem 19.3a, (19.3-6) follows immediately. —

PROBLEM

1. Prove that $|a_4| \leqq 4$, as follows:

   (a) Let $f \in \Sigma_0, f(z) = z + \sum a_n z^n$, and assume w.l.o.g. that $a_4 > 0$ (see Problem 2, § 19.1).

   (b) Let $g(z) := [f(z^2)]^{-1/2} = z + \sum_{n=1}^{\infty} b_{-n} z^{-n}$, and express b_{-1}, b_{-3}, b_{-5} explicitly in terms of the a_n.

   (c) Express the coefficients $c_{1,-1}, c_{1,-3}, c_{3,-3}$ in terms of the b_n, and then in terms of the a_n. Thus show, for instance, that

$$c_{3,-3} = -\tfrac{13}{14} a_2^3 + a_2 a_3 - \tfrac{1}{2} a_4.$$

   (d) Use the inequality (19.3-6) for $m = 3$, $x_1 = x$, $x_2 = 0$, $x_3 = 1$, to obtain

$$|a_4 + 4b_3(a_2 - x) - \tfrac{5}{12} a_2^3 + a_2 x^2| \leqq 2|x|^2 + \tfrac{2}{3}.$$

   (e) Use the area theorem (Theorem 19.1g) to yield

$$|b_1|^2 + 3|b_3|^2 \leqq 1, \quad 4|b_3| \leqq \frac{2}{\sqrt{3}} [4 - |a_2|^2]^{1/2}$$

and conclude that

$$a_4 \leqq \frac{2}{\sqrt{3}} \sqrt{4 - |a_2|^2} (a_2 - x) + \operatorname{Re}\left\{ \frac{5}{12} a_2^3 + a_2 x^2 \right\} + 2|x|^2 + \frac{2}{3}.$$

From here onward, the proof is a matter of clever application of calculus, using $|a_2| \leqq 2$. It is convenient to set $a_2 = 2\rho\, e^{i\theta}$ and to choose $x = 2\rho\, e^{-i\theta/2} \cos(3\theta/2)$. (See Duren [1983], p. 132.)

NOTES

Theorem 19.3b is due to Pommerenke [1964], but a similar result was already found by Grunsky [1939]; see also Ahlfors [1973]. FitzGerald [1972] uses inequalities of this type to prove coefficient estimates for univalent functions that come close to proving the Bieberbach conjecture.

§ 19.4. SLIT MAPPINGS

Let D be the unit disk. A function $f \in \Sigma_0$ is called a **slit mapping** if $f(D)$ equals the complex plane minus a set of Jordan arcs. (Because $f(D)$ is simply connected, these arcs necessarily tend to infinity.) The function f is called a **single-slit mapping** if the complement of $f(D)$ is a single Jordan arc. Here we shall prove that, with respect to locally uniform convergence in D, the single-slit mappings are dense in Σ_0.

A special concept of convergence of a sequence of regions is required. Let $\{R_n\}$, $n = 1, 2, \ldots$, be a sequence of simply connected regions in $\mathbb{C}$ such that $R_n \neq \mathbb{C}$ and $0 \in R_n$ for all n. We also assume that the origin is an interior point of the intersection of all R_n, or equivalently, that some disk $|z| < \rho$

lies in R_n for all n. Let $\mathcal{K}$ be the family of regions K with the following properties: $0 \in K$, and each compact subset $\hat{K}$ of K belongs to almost all (that is, all but finitely many) R_n. $\mathcal{K}$ is not empty, because it contains the disk $|z| < \rho$. Let

$$R := \bigcup_{\mathcal{K}} K. \tag{19.4-1}$$

We assert that R itself belongs to $\mathcal{K}$. Obviously, $0 \in R$. Let $\hat{R}$ be a compact subset of R. The disks E such that their closure E' belongs to some $K \in \mathcal{K}$ form an open covering of $\hat{R}$. By the Heine–Borel theorem, there exists a finite subcovering of disks $E_1, E_2, \ldots, E_m$. The closure E_i' of each E_i belongs to some $K \in \mathcal{K}$ and therefore is contained in almost all R_n. The same is therefore true of their union, and thus of $\hat{R}$.

The region R defined by (19.4-1) is called the **kernel** of the sequence $\{R_n\}$. The sequence $\{R_n\}$ is said to **converge to its kernel in the sense of Carathéodory** if every subsequence of $\{R_n\}$ has the same kernel as the whole sequence.

THEOREM 19.4a (Carathéodory convergence theorem). *Let $\{R_n\}$ be a sequence of simply connected regions such that $0 \in R_n$, $R_n \neq \mathbb{C}$ for all n. Let 0 be an interior point of the intersection of all R_n, and let the kernel R of $\{R_n\}$ satisfy $R \neq \mathbb{C}$. Let f_n map the unit disk D conformally onto R_n such that $f_n(0) = 0, f'_n(0) > 0$. Then $f_n \to f$ uniformly on each compact subset of D if and only if $\{R_n\}$ converges to R in the sense of Carathéodory. In the case of convergence, R is a simply connected region, and the inverse maps $g_n := f_n^{[-1]}$ converge to $g := f^{[-1]}$ uniformly on each compact subset of R.*

Proof. (a) Let $f_n \to f$ uniformly on compact subsets of D. By the theorem of Hurwitz (Corollary 4.10f), f is either constant or univalent.

(a_1) Let f be constant. Since $f_n(0) = 0$ for all $n, f(z) = 0$ for $z \in D$. Let the disk $|w| < \rho$, $\rho > 0$, belong to all R_n. Then the functions $g_n(w)$ are defined for $|w| < \rho$ and have the property that $|g_n(w)| < 1$, $g_n(0) = 0$. By the lemma of Schwarz, $|g_n(w)| < \rho^{-1}|w|$, and thus $g'_n(0) < \rho^{-1}$. This implies $f'_n(0) > \rho$ for all n, which makes it impossible for the sequence $\{f_n\}$ to converge to the zero function. Thus:

(a_2) f is univalent. Let $S := f(D)$. We must show that $S = R$, and that $\{R_n\}$ converges to S in the sense of Carathéodory.

We first show that $S \subset R$. Let $\hat{S}$ be a compact subset of S, and surround $\hat{S}$ by a regular Jordan curve Γ in $S \backslash \hat{S}$. Let $\delta > 0$ be the distance from $\hat{S}$ to Γ, and let $\Gamma_1 := f^{[-1]}(\Gamma)$. We shall prove that $\hat{S} \subset R_n$ for all sufficiently large n. Let $w_0 \in \hat{S}$ be fixed. We have, on the one hand, $|f(z) - w_0| \geqq \delta$ for $z \in \Gamma_1$, and on the other, by the uniform convergence of the sequence $\{f_n\}$ on $\Gamma_1, |f_n(z) - f(z)| < \delta$ for all $z \in \Gamma_1$ and for $n > n_0$, say. Rouché's theorem

(Theorem 4.10b) now in view of

$$|f_n(z)-f(z)|<|f(z)-w_0|, \qquad z\in\Gamma,$$

implies that $f_n(z)-w_0=f(z)-w_0+[f_n(z)-f(z)]$ has the same number of zeros inside Γ_1 as $f(z)-w_0$, namely, one. This shows that $w_0\in R_n$ for all $n>n_0$, where n_0 depends on $\hat{S}$ but not on w_0. In other words, $\hat{S}\subset R_n$ for all $n>n_0$. By virtue of the definition of the kernel R, this means that $S\subset R$.

The inverse functions $g_n:=f_n^{[-1]}$ are defined on $\hat{S}$ for all $n>n_0$ and are uniformly bounded there, $|g_n(w)|\leqq 1$. By Montel's theorem (Theorem 12.8a), the functions g_n form a normal family. Thus there exists a subsequence $\{g_{k_n}\}$ which converges, locally uniformly on S, to a function g analytic on S with $g(0)=0$ and $g'(0)\geqq 0$. In fact,

$$0<\frac{1}{f'(0)}=\lim_{n\to\infty}\frac{1}{f'_n(0)}=\lim_{n\to\infty}g'_{n_k}(0)=g'(0).$$

Thus g is univalent by the theorem of Hurwitz.

We next show that $g=f^{[-1]}$. Fix $z_0\in D$ and let $w_0:=f(z_0)$. Choose $\varepsilon>0$ so that the circle Θ: $|z-z_0|=\varepsilon$ lies in D, and let $\Gamma:=f(\Theta)$. Let δ be the distance of w_0 from Γ. Then $|f_n(z)-w_0|\geqq\delta$ for $z\in\Theta$, while $|f_n(z)-f(z)<\delta$ on Θ for all $n>n_1$. As above it follows by Rouché's theorem that for each $n>n_1$ there exists precisely one z_n inside Θ such that $f_n(z_n)=w_0$. Thus $|z_n-z_0|<\varepsilon$ and $z_n=g_n(w_0)$. Therefore, if n is so large that $|g_{k_n}(w_0)-g(w_0)|<\varepsilon$,

$$|g(w_0)-z_0|\leqq|g(w_0)-g_{k_n}(w_0)|+|z_{k_n}-z_0|<2\varepsilon.$$

Since $\varepsilon>0$ may be taken arbitrarily small, there follows $g(w_0)=z_0$. Since $z_0\in D$ was arbitrary, this shows that $g=f^{[-1]}$.

We now know that every convergent subsequence of $\{g_n\}$ converges, uniformly on compact subsets $\hat{S}$ of S, to one and the same function, namely, to $f^{[-1]}$. Since the family $\{g_n\}$ is uniformly bounded, a further application of Montel's theorem shows that the whole sequence $\{g_n\}$ converges to $f^{[-1]}$. In fact, let R be the kernel of $\{R_n\}$, and let $\hat{R}$ be a compact subset of R. Almost all g_n are defined on $\hat{R}$, and since $S\subset R$, the sequence $\{g_n\}$ satisfies the hypotheses of Vitali's theorem (Theorem 12.8d). It follows that $\{g_n\}$ converges (to $f^{[-1]}$) uniformly on $\hat{R}$. Since $g_n(\hat{R})\subset D$, we have $\hat{R}\subset S$, which is possible only if $R=S$.

The desired conclusion that $R_n\to R$ in the sense of Carathéodory is now clear. For if we had started with any subsequence $\{R_{n_k}\}$, the above argument would again show that its kernel is $S=f(D)$. Thus all subsequences of $\{R_n\}$ have the same kernel, as desired.

(b) Suppose now that $R_n\to R\neq\mathbb{C}$ in the sense of Carathéodory. We assert that the sequence $\{f'_n(0)\}$ is bounded. Indeed if $f'_{n_k}(0)>k$ for some increasing sequence $\{n_k\}$, the Koebe $\frac{1}{4}$ theorem would show that $f_{n_k}(D)$ contains the disk $|w|<\frac{1}{4}k$, and it would follow that the sequence $\{R_{n_k}\}$ has the kernel $\mathbb{C}$. This contradiction shows that there exists $\gamma\in\mathbb{R}$ such that $f'_n(0)<\gamma$ for all n. By the growth theorem (Theorem 19.11),

$$|f_n(z)|\leqq f'_n(0)\frac{|z|}{(1-|z|)^2}, \qquad z\in D,$$

which shows that the sequence $\{f_n\}$ is uniformly bounded on each compact subset of D. By Montel's theorem, the family $\{f_n\}$ is normal. Thus there exists a subsequence $\{f_{n_k}\}$ which converges to an analytic function f, uniformly on compact subsets $\hat{D}$ of D. By part (a) of the proof, f maps D onto R. It again follows by Montel that the whole sequence $\{f_n\}$ converges to f, uniformly on compact subsets $\hat{D}$ of D. —

Carathéodory's convergence theorem now enables us to prove:

THEOREM 19.4b. *Let $f\in\Sigma_0$. Then there exists a sequence of single-slit mappings $f_n\in\Sigma_0$ such that $f_n\to f$ uniformly on each compact subset of D.*

Proof. It suffices to produce, for any $\varepsilon>0$ and any $\rho<1$, a single-slit mapping g such that

$$|f(z)-g(z)|<\varepsilon, \qquad |z|\leqq\rho.$$

As a first step in the construction, let

$$f_\sigma(z):=\frac{1}{\sigma}f(\sigma z), \qquad 0<\sigma<1.$$

Evidently, $f_\sigma\in\Sigma_0$, and if $\sigma\to 1$, then $f_\sigma\to f$ uniformly on $|z|\leqq\rho$. We choose $\sigma<1$ so that

$$|f(z)-f_\sigma(z)|<\tfrac{1}{2}\varepsilon, \qquad |z|\leqq\rho,$$

and approximate f_σ in place of f. This has the advantage that the boundary of the set $R:=f_\sigma(D)$, that is, the curve

$$w=f_\sigma(e^{i\theta})=\frac{1}{\sigma}f(\sigma e^{i\theta}), \qquad 0\leqq\theta\leqq 2\pi,$$

is an analytic Jordan curve. Let $w_0:=f_\sigma(1)$, $w_n:=f_\sigma(e^{2\pi i(1-1/n)})$, and let Γ_n denote the Jordan arc consisting (a) of an arc running from ∞ to w_0 in the exterior of R; (b) of the arc

$$z=f_\sigma(e^{i\theta}), \qquad 0\leqq\theta\leqq 2\pi\left(1-\frac{1}{n}\right),$$

which runs from w_0 to w_n along the boundary of R. We denote by R_n the complement of Γ_n. Let g_n map D conformally onto R_n such that $g_n(0)=0$, $g_n'(0)>0$. It is geometrically clear that R is the kernel of $\{R_n\}$, and that $R_n \to R$ in the sense of Carathéodory. Therefore in view of the Carathéodory convergence theorem, $g_n \to f_\sigma$ uniformly on compact subsets of D. By the Weierstrass convergence theorem (Theorem 3.4b), or equivalently by Cauchy's formula for the derivative, this implies

$$g_n'(0) \to f_\sigma'(0) = 1.$$

Hence the functions

$$h_n(z) := \frac{g_n(z)}{g_n'(0)}$$

are single-slit mappings in Σ_0 that converge to f_σ uniformly on compact subsets of D. —

The approximation theorem thus proved has an important consequence. Let ϕ be a real-valued functional defined on Σ which is continuous in the sense that $\phi(f_n) \to \phi(f)$ if $f_n \to f$ uniformly on compact subsets of D. Consider the problem of finding the number

$$\sigma := \sup_{f \in \Sigma_0} \phi(f).$$

If Σ_0' is the set of single-slit mappings in Σ_0, then by Theorem 19.4b,

$$\sigma = \sup_{f \in \Sigma_0'} \phi(f),$$

that is, it suffices to look for the supremum in the much smaller set Σ_0'.

In particular the above is true for the functionals

$$\phi_n(f) := \sum_{k=1}^{n} (n+1-k)\left[k|c_k|^2 - \frac{4}{k}\right], \qquad n = 1, 2, \ldots, \tag{19.4-2}$$

that occur in Milin's conjecture. It will be remembered that Bieberbach's conjecture is true if for every $f \in \Sigma_0$,

$$\phi_n(f) \leqq 0$$

holds for $n = 1, 2, \ldots$. We thus conclude:

COROLLARY 19.4c. *Bieberbach's conjecture is true if*

$$\phi_n(f) \leqq 0, \qquad n = 1, 2, \ldots,$$

holds for every single-slit mapping $f \in \Sigma_0$.

PROBLEM

1. Let $\{R_n\}$ be a sequence of regions, and let

$$R^* := \bigcap_{n=1}^{\infty} \bigcup_{k=n}^{\infty} R_k,$$

$$R_* := \bigcup_{n=1}^{\infty} \bigcap_{k=n}^{\infty} R_k.$$

(a) Show that R^* consists of the points belonging to infinitely many of the R_k.

(b) Show that R_* consists of the points belonging to all but finitely many R_k.

(c) The sequence $\{R_n\}$ is said to **converge topologically** to R if $R = R^* = R_*$. Show that if the sequence is *increasing*, that is, if $R_n \subset R_{n+1}$ for all n, then it converges to $R := \cup R_n$ topologically as well as in the sense of Carathéodory.

(d) Let R_n be the complement with respect to $\mathbb{C}$ of the set of points

$$z = \begin{cases} e^{2\pi i\tau}, & \frac{1}{n} \leqq \tau \leqq 1 \\ \tau, & 1 \leqq \tau < \infty. \end{cases}$$

Show that the sequence $\{R_n\}$ converges topologically to $D \cup \mathring{D}$, but that its limit in the sense of Carathéodory is merely D. Thus conclude that if the sequence $\{R_n\}$ is not increasing, its topological limit need not agree with its limit in the sense of Carathéodory.

NOTES

See Duren [1983], § 3.1 and § 3.2.

§ 19.5. LOEWNER'S DIFFERENTIAL EQUATION

To prove the Bieberbach conjecture, it now suffices to establish Milin's inequalities (19.2-6) for functions $f \in \Sigma_0$ that are single-slit mappings. Let f be such a mapping, and let Γ be the Jordan arc which is the complement of $R := f(D)$. We assume the parametric representation of Γ in the form

$$w = s(\tau), \qquad 0 \leqq \tau < \beta.$$

The point $w_0 := s(0)$ is the initial point of the slit; the function s is continuous and satisfies $\lim_{\tau \to \beta} |s(\tau)| = \infty$.

The basic idea of de Branges' proof of Milin's inequalities is simple. It consists in linking f through a one-parameter family of functions f_τ to a case where the inequality is obvious, namely, to the identical map. Such a one-parameter family is easily constructed. For $0 \leqq \tau < \beta$, denote by Γ_τ the portion of Γ lying between $s(\tau)$ and ∞, and let R_τ be the complement of

Γ_τ. Then $R_0 = R$, and $R_\sigma \subset R_\tau$ for $\sigma < \tau$. Let

$$f_\tau(z) = \alpha(\tau)\{z + a_2(\tau)z^2 + \cdots\}$$

be the conformal mapping of D onto R_τ for which $f_\tau(0) = 0$ and $f'_\tau(0) = \alpha(\tau) > 0$, so that $f_0 = f$. For any $\tau \in [0, \beta)$, the Carathéodory convergence theorem shows that the functions $f_\sigma(z)$ converge to $f_\tau(z)$ as $\sigma \to \tau$, uniformly on compact subsets of D. Therefore the coefficients α and a_k, $k = 2, 3, \ldots$, are continuous functions of τ. Moreover, $\alpha(\tau)$ is strictly increasing because if $\tau > \sigma$, the function $f_\tau^{[-1]} \circ f_\sigma$ satisfies the hypotheses of the Schwarz lemma and is not the identity; hence its derivative at 0, which equals $\alpha(\sigma)\alpha^{-1}(\tau)$, is <1. It follows that the parametric representation of Γ may be rechosen such that $\alpha(\tau) = e^\tau$. We call this the **standard parametrization** of Γ.

In the standard parametrization of Γ, the endpoint β of the parameter interval must be ∞. For let $\mu > 0$ be arbitrary. Then Γ_τ lies outside $|w| = \mu$ for τ sufficiently close to β. The function

$$z \mapsto f_\tau^{[-1]}(\mu z)$$

then again satisfies the hypotheses of the Schwarz lemma, and therefore its derivative at 0 satisfies

$$\mu e^{-\tau} < 1.$$

Since μ may be chosen arbitrarily large, it follows that $e^\tau \to \infty$ as $\tau \to \beta$. Thus $\beta = \infty$. The family f_τ now has the form

$$f_\tau(z) = e^\tau \left\{ z + \sum_{n=2}^{\infty} a_n(\tau) z^n \right\}.$$

The a_n still are continuous functions.

Consider now the functions $g_\tau := f_\tau^{[-1]} \circ f$, which map D conformally onto D minus an arc extending in from the boundary. We have

$$g_\tau(z) = e^{-\tau} \left\{ z + \sum_{n=2}^{\infty} b_n(\tau) z^n \right\},$$

where (for example, by the Lagrange–Bürmann formula) b_n is a polynomial in $a_2, \ldots, a_n$, and therefore continuous. Obviously, g_0 is the identical mapping. Moreover, there holds:

THEOREM 19.5a. *Let $f \in \Sigma_0$ be a single-slit mapping with the omitted arc Γ, given in its standard representation $w = s(\tau)$, $0 \leqq \tau < \infty$. Let f_τ be defined as above. Then the functions $g_\tau := f_\tau^{[-1]} \circ f$ satisfy*

$$\lim_{\tau \to \infty} e^\tau g(z) = f(z) \tag{19.5-1}$$

for $z \in D$, the convergence being uniform on each compact subset of D. Moreover there exists a continuous complex-valued function $q(\tau)$ with $|q(\tau)|=1$ such that for each $z \in D$,

$$\frac{\partial g_\tau}{\partial \tau} = -g_\tau \frac{q(\tau)+g_\tau}{q(\tau)-g_\tau}. \tag{19.5-2}$$

The differential equation (19.5-2) is known as **Loewner's differential equation**. It describes a family of curves (or a "flow"), depending on the parameter z, which continuously link the points $g_0(z)=z$ to an infinitesimal image of $f(z)$.

Proof. If z ranges over a compact subset of D, $w=f(z)$ ranges over a compact subset of $\mathbb{C}$. Therefore (19.5-1) is proved if it is shown that

$$\lim_{\tau\to\infty} e^\tau f_\tau^{[-1]}(w) = w$$

uniformly on all compact subsets of $\mathbb{C}$. By the growth theorem (Theorem 19.11),

$$\frac{e^\tau |z|}{(1+|z|)^2} \leq |f_\tau(z)| \leq \frac{e^\tau |z|}{(1-|z|)^2}.$$

Setting $z=f_\tau^{[-1]}(w)$ and rearranging these inequalities, there follows

$$[1-|f_\tau^{[-1]}(w)|]^2 \leq e^\tau \left| \frac{f_\tau^{[-1]}(w)}{w} \right| \leq [1+|f_\tau^{[-1]}(w)|]^2. \tag{19.5-3}$$

In particular, $|f_\tau^{[-1]}(w)| \leq 4|w|\, e^{-\tau}$, and we see that $f_\tau^{[-1]}(w) \to 0$ as $\tau\to\infty$, uniformly on compact sets. Looking at (19.5-3) again, we now see that

$$e^\tau \left| \frac{f_\tau^{[-1]}(w)}{w} \right| \to 1.$$

Thus the functions

$$w \mapsto h_\tau(w) := e^\tau \frac{f_\tau^{[-1]}(w)}{w}$$

form a normal family, and there exists a sequence $\{\tau(k)\}$ such that the corresponding sequence $h_{\tau(k)}$ tends to an analytic limit function h. Since $|h(w)|=1$ and $h(0)=1$, $h(w)$ is the constant 1. Since the limit function is independent of the subsequence selected, it follows as before that $h_\tau(w)\to 1$ as $\tau\to\infty$, uniformly on compact sets. This proves (19.5-1).

To establish Loewner's differential equation, let, for $0 \leq \sigma < \tau < \infty$,

$$h_{\sigma,\tau} := f_\tau^{[-1]} \circ f_\sigma.$$

This again maps D onto D minus a Jordan arc $\Lambda_{\sigma,\tau}$ extending from the boundary. Let $\Theta_{\sigma,\tau}$ be the arc on $|z|=1$ which corresponds to the two sides of $\Lambda_{\sigma,\tau}$, and let $q(\tau):=f_\tau^{[-1]}(s(\tau))$ be the point on $|z|=1$ which by f_τ is mapped onto the tip of Γ_τ. Then $q(\sigma)$ is an interior point of $\Theta_{\sigma,\tau}$, while $q(\tau)$ is the point where $\Lambda_{\sigma,\tau}$ meets the unit circle. It may be shown that the function $q(\tau)$ is continuous. We postpone the proof and turn to the derivation of Loewner's equation.

We have $h_{\sigma,\tau}(z)=e^{\sigma-\tau}z+\ldots$. Therefore we may select a branch of the logarithm such that

$$l(z)=l_{\sigma,\tau}(z)=\log\frac{h_{\sigma,\tau}(z)}{z}$$

is analytic in D and satisfies $l(0)=\sigma-\tau$. Moreover l is continuous on the closure of D. In view of the mapping properties just mentioned, $\operatorname{Re} l(z)=0$ everywhere on the circle with the exception of the arc $\Theta_{\sigma,\tau}$ where $\operatorname{Re} l(z)\leqq 0$. Therefore by the Schwarz formula (Theorem 14.2b),

$$l(z)=\frac{1}{2\pi}\int_\alpha^\beta \operatorname{Re}[l(e^{i\theta})]\frac{e^{i\theta}+z}{e^{i\theta}-z}\,d\theta, \tag{19.5-4}$$

where $e^{i\alpha}$ and $e^{i\beta}$ are the endpoints of $\Theta_{\sigma,\tau}$. In particular for $z=0$,

$$\sigma-\tau=l(0)=\frac{1}{2\pi}\int_\alpha^\beta \operatorname{Re}[l(e^{i\theta})]\,d\theta.$$

Since $h_{\sigma,\tau}\circ g_\sigma=f_\tau^{[-1]}\circ f_\sigma\circ f_\sigma^{[-1]}\circ f=f_\tau^{[-1]}\circ f=g_\tau$, substituting $g_\sigma(z)$ for z in (19.5-4) yields

$$\log\frac{g_\tau(z)}{g_\sigma(z)}=\frac{1}{2\pi}\int_\alpha^\infty \operatorname{Re}[l(e^{i\theta})]\frac{e^{i\theta}+g_\sigma(z)}{e^{i\theta}-g_\sigma(z)}\,d\theta. \tag{19.5-5}$$

The mean-value theorem of the integral calculus, applied separately to the real and imaginary parts of the integral in (19.5-5), gives

$$\log\frac{g_\tau(z)}{g_\sigma(z)}=\frac{1}{2\pi}\left\{\operatorname{Re}\left[\frac{e^{i\gamma}+g_\sigma(z)}{e^{i\gamma}-q_\sigma(z)}\right]+i\operatorname{Im}\left[\frac{e^{i\delta}+g_\sigma(z)}{e^{i\delta}-g_\sigma(z)}\right]\right\}\int_\alpha^\beta \operatorname{Re}[l(e^{i\theta})]\,d\theta,$$

where the points $e^{i\gamma}$ and $e^{i\delta}$ both lie on $\Theta_{\sigma,\tau}$. We now divide by

$$\sigma-\tau=\frac{1}{2\pi}\int_\alpha^\beta \operatorname{Re}[l(e^{i\theta})]\,d\theta$$

(see (19.5-4)) and let $\tau\to\sigma$. The arc $\Theta_{\sigma,\tau}$ then shrinks to $q(\sigma)$, and by the definition of the derivative we get

$$\frac{\partial}{\partial\sigma}\log g_\sigma(z)=-\frac{q(\sigma)+g_\sigma(z)}{q(\sigma)-g_\sigma(z)}. \tag{19.5-6}$$

Here the derivative $\partial/\partial\sigma$ is calculated as a right derivative. However, by the same reasoning the left derivative likewise exists and has the same value. The differential equation (19.5-6), which thus has been established, obviously is equivalent to (19.5-2).

We finally prove that the function $q(\tau)$ is continuous. Let Θ denote the unit circle, and let again $0 \leqq \sigma < \tau < \infty$. Since $h = h_{\sigma,\tau}$ maps $\Theta \backslash \Theta_{\sigma,\tau}$ onto a circle, h by the Schwarz reflection principle may be extended to a function that maps $\mathbb{C} \backslash \Theta_{\sigma,\tau}$ onto $\mathbb{C} \backslash \{\Lambda_{\sigma,\tau} \cup \Lambda^*_{\sigma,\tau}\}$, where $\Lambda^*_{\sigma,\tau}$ is the reflection of $\Lambda_{\sigma,\tau}$ at Θ. By the Koebe $\frac{1}{4}$ theorem $\Lambda_{\sigma,\tau}$ lies outside the disk $|u| < \frac{1}{4} e^{\sigma-\tau}$, and therefore $\Lambda^*_{\sigma,\tau}$ lies inside the disk $|u| < 4\, e^{\tau-\sigma}$. Furthermore, by the reflection principle,

$$\lim_{z\to\infty} \frac{h(z)}{z} = \lim_{z\to 0} \frac{z}{h(z)} = e^{\tau-\sigma}.$$

In particular, $z^{-1}h$ is bounded at ∞. Thus by the principle of the maximum,

$$|h(z)| < 4\, e^{\tau-\sigma}|z|$$

throughout the complement of $\Theta_{\sigma,\tau}$.

We now recall that $\Theta_{\sigma,\tau}$ contracts to $q(\sigma)$ as $\tau \downarrow \sigma$. By an application of Montel's theorem, the functions $z^{-1}h_{\sigma,\tau}$ then converge, uniformly on compact sets, to a function $k(z)$ analytic and bounded on the extended complex plane punctured at $q(\sigma)$. Since $k(0) = 1$, Liouville's theorem implies that $k(z) = 1$. It follows that $h_{\sigma,\tau}(z) \to z$ as $\tau \downarrow \sigma$ uniformly on any compact set not containing $q(\sigma)$.

Let now $\sigma \geqq 0$ be fixed. Given $\varepsilon > 0$, choose $\delta > 0$ so small that the arc $\Theta_{\sigma,\tau}$ lies inside the circle Δ: $|z - q(\sigma)| < \varepsilon$ for all τ such that $\sigma < \tau < \sigma + \delta$. Let Δ_1 be the image of Δ under the map $h_{\sigma,\tau}$ as extended above. Then Δ_1 is a Jordan curve which contains $\Lambda_{\sigma,\tau} \cup \Lambda^*_{\sigma,\tau}$ in its interior. The point $q(\tau)$ in particular is surrounded by Δ_1. Because $h_{\sigma,\tau}(z) \to z$ uniformly on Δ as $\tau \to \sigma$, the diameter of Δ_1 is less than 3ε for all τ sufficiently close to σ. Thus for any point $z_0 \in \Delta$

$$|q(\sigma) - q(\tau)| \leqq |q(\sigma) - z_0| + |z_0 - h_{\sigma,\tau}(z_0)| + |h_{\sigma,\tau}(z_0) - q(\tau)|.$$

The first two terms on the right are $< \varepsilon$ each, and the third term is $< 3\varepsilon$ because, as was seen above, $\Lambda_{\sigma,\tau}$ meets Θ at $q(\tau)$. Thus if $\tau > \sigma$ is sufficiently close to σ,

$$|q(\sigma) - q(\tau)| < 5\varepsilon,$$

proving that at σ, q is continuous from the right. An analogous argument proves continuity from the left. Thus the function q is continuous, proving Theorem 19.5a. —

To apply Loewner's equation we need to convert Theorem 19.5a into statements on f_τ.

COROLLARY 19.5b. *Let $f \in \Sigma_0$ be a single-slit mapping with the omitted arc Γ given in its standard parametrization $w = s(\tau)$, $0 \leqq \tau < \infty$. Let the functions*

$$f_\tau(z) = e^\tau \left(z + \sum_{n=2}^{\infty} a_n(\tau) z^n \right), \qquad z \in D, \quad 0 \leqq \tau < \infty,$$

be defined as above. Then

(i)
$$\lim_{\tau \to \infty} f_\tau(e^{-\tau} z) = z \tag{19.5-7}$$

uniformly on compact subsets of $\mathbb{C}$;

(ii) *for each $z \in D$, $\partial f_\tau / \partial \tau$ exists as a continuous function of τ, and with the continuous function $q(\tau)$ defined earlier there holds*

$$\frac{\partial f_\tau}{\partial \tau} = \frac{q(\tau) + z}{q(\tau) - z} z \frac{\partial f_\tau}{\partial z}, \qquad 0 \leqq \tau < \infty; \tag{19.5-8}$$

(iii) *the coefficients a_n are differentiable functions of τ, and there holds*

$$\frac{\partial}{\partial \tau} \{ e^{-\tau} f_\tau(z, \tau) \} = \sum_{n=2}^{\infty} a_n'(\tau) z^n, \tag{19.5-9}$$

where the series converges locally uniformly on D.

Proof. (i) Let $K \subset \mathbb{C}$ be a compact set, and let $K \subset D_\rho$, where D_ρ is the disk $|z| < \rho$. We already know that the functions $w \mapsto e^\tau f_\tau^{[-1]}(w)$ approach the identical map for $\tau \to \infty$, uniformly on compact sets. Thus for every $\varepsilon > 0$ there exists τ_0 such that for all $\tau > \tau_0$,

$$|e^\tau f_\tau^{[-1]}(w) - w| < \varepsilon, \qquad w \in D_{2\rho}. \tag{19.5-10}$$

By the theorem of Rouché, if $2\varepsilon < \rho$, K for $\tau > \tau_0$ is contained in the image of $D_{2\rho}$ under $w \mapsto e^\tau f_\tau^{[-1]}(w)$. Thus if $z \in K$, then $w = f_\tau(e^{-\tau} z) \in D_{2\rho}$, and from (19.5-10) there follows

$$|z - f_\tau(e^{-\tau} z)| < \varepsilon$$

for all $\tau > \tau_0$.

(ii) From $g_\tau = f_\tau^{[-1]} \circ f$ we have

$$f_\tau \circ g_\tau = f.$$

Since $\partial g_\tau / \partial \tau$ exists and is $\neq 0$, $\partial f_\tau / \partial \tau$ exists, and by differentiating with respect to τ we get, considering that f does not depend on τ,

$$\frac{\partial f_\tau}{\partial \tau} \circ g_\tau + f'_\tau \circ g_\tau \cdot \frac{\partial g_\tau}{\partial \tau} = 0,$$

hence

$$\frac{\partial g_\tau}{\partial \tau} = -\frac{1}{f'_\tau \circ g_\tau} \cdot \frac{\partial f_\tau}{\partial \tau} \circ g_\tau.$$

Loewner's equation thus may be written

$$\frac{\partial f_\tau}{\partial \tau} \circ g_\tau = g_\tau \frac{q(\tau) + g_\tau}{q(\tau) - g_\tau} f'_\tau \circ g_\tau,$$

and (19.5-8) results by composing the above with $g_\tau^{[-1]}$.

(iii) We see from (19.5-8) that $\partial f_\tau / \partial \tau$ is continuous on $D \times [0, \infty)$. By Cauchy's formula, if Θ is a suitable circle about the origin,

$$a_n(\tau) = \frac{1}{2\pi i} \int_\Theta e^{-\tau} f_\tau(t) t^{-n-1}\, dt,$$

and thus, by differentiating under the integral sign,

$$a'_n(\tau) = \frac{1}{2\pi i} \int_\Theta \frac{\partial}{\partial \tau} \{e^{-\tau} f_\tau(t)\} t^{-n-1}\, dt.$$

Let $|z| \leqq \rho < 1$, and let Θ have the radius $\frac{1}{2}(1+\rho)$. Then for $n = 2, 3, \ldots,$ since $a_0(\tau) = 0$ and $a_1(\tau) = 1$,

$$\sum_{k=2}^{n} a'_k(\tau) z^k = \frac{1}{2\pi i} \int_\Theta \frac{\partial}{\partial \tau} \{e^{-\tau} f_\tau(t)\} \frac{1 - (z/t)^{n+1}}{t - z}\, dt.$$

For $n \to \infty$ the integrand tends to

$$\frac{\partial}{\partial \tau} \{e^{-\tau} f_\tau(t)\} \frac{1}{t - z},$$

uniformly for $t \in \Theta$. Thus by Cauchy's formula, the integral tends to $\partial / \partial \tau \{e^{-\tau} f_\tau(z)\}$, as desired. —

PROBLEMS

1. If f denotes the Koebe function, show that

$$f_\tau(z) = \frac{e^\tau z}{(1-z)^2},$$

and verify that the statements of Corollary 19.5b hold for a suitable (constant) q. Also construct the functions g_τ, and verify Theorem 19.5a.

2. Let $f \in \Sigma_0$, $f(z) = z + \Sigma\, a_n z^n$, and let g_τ be associated with f as in Theorem 19.5a.
   (a) Show that

$$g_\tau(z) = e^{-\tau}\left\{z + \sum_{n=2}^{\infty} b_n(\tau) z^n\right\}$$

where $b_n(0) = 0$ and $\lim_{\tau\to\infty} b_n(\tau) = a_n$, $n = 2, 3, \ldots$.
   (b) By substituting g_τ into Loewner's (original) differential equation, expanding in powers of z, and comparing coefficients, show that

$$b_2'(\tau) = -2e^{-\tau}\bar{q}(\tau),$$
$$b_3'(\tau) = -2e^{-2\tau}[\bar{q}(\tau)]^2 - 4\, e^{-\tau}\bar{q}(\tau) b_2(\tau). \qquad (*)$$

   (c) Show that $|a_2| \leqq 2$ by noting that

$$a_2 = \int_0^\infty b_2'(\tau)\, d\tau = -2\int_0^\infty e^{-\tau}\bar{q}(\tau)\, d\tau,$$

where $|q(\tau)| = 1$.
   (d) Show that $|a_3| \leqq 3$ by writing (*) in the form

$$a_3 = \int_0^\infty b_3'(\tau)\, d\tau = -2\int_0^\infty e^{-2\tau}[\bar{q}(\tau)]^2\, d\tau + 4\left\{\int_0^\infty e^{-\tau}\bar{q}(\tau)\, d\tau\right\}^2.$$

Now set $q(\tau) = e^{i\theta(\tau)}$ and use the Schwarz inequality.

NOTES

See Duren [1983], § 3.3 and Ahlfors [1973], ch. 6.

§ 19.6. THE DIFFERENTIAL INEQUALITY OF DE BRANGES

To prove Milin's conjecture for single-slit mappings in Σ_0, let f be such a mapping, and denote by f_τ the family defined in the preceding section. Milin's conjecture concerns the coefficients c_k in the expansion

$$\log\frac{f(z)}{z} = \sum_{k=1}^{\infty} c_k z^k, \qquad |z| < 1.$$

We write for $|z| < 1$

$$l_\tau(z) = \log\frac{f_\tau(z)}{e^\tau z} = \sum_{k=1}^{\infty} c_k(\tau) z^k, \qquad (19.6\text{-}1)$$

so that $c_k(0) = c_k$ by virtue of $f_0 = f$. Dividing Loewner's equation (19.5-8) by f_τ and noting that

$$\frac{\partial}{\partial\tau} l_\tau(z) = \frac{(\partial/\partial\tau)f_\tau(z)}{f_\tau(z)} - 1,$$

$$l'_\tau(z) = \frac{f'_\tau(z)}{f_\tau(z)} - \frac{1}{z},$$

the equation may be written

$$1 + \frac{\partial}{\partial\tau} l_\tau(z) = \frac{q(\tau)+z}{q(\tau)-z}(1 + zl'_\tau(z)).$$

Using the series (19.6-1) as well as the trivial expansion

$$\frac{q(\tau)+z}{q(\tau)-z} = 1 + 2\sum_{k=1}^{\infty} \overline{q(\tau)}^k z^k,$$

there results

$$1 + \sum_{k=1}^{\infty} c'_k(\tau) z^k = \left(1 + 2\sum_{k=1}^{\infty} \overline{q(\tau)}^k z^k\right)\left(1 + \sum_{k=1}^{\infty} kc_k(\tau) z^k\right). \tag{19.6-2}$$

Comparing coefficients we have, for $k = 1, 2, \ldots,$

$$c'_k(\tau) = 2\overline{q(\tau)}^k + 2\sum_{m=1}^{k-1} mc_m(\tau)\overline{q(\tau)}^{k-m} + kc_k(\tau),$$

which, on noting that $\bar{q} = q^{-1}$ and introducing $s_0(\tau) := 0$,

$$s_k(\tau) := \sum_{m=1}^{k} mc_m(\tau) q(\tau)^m, \qquad k = 1, 2, \ldots, \tag{19.6-3}$$

simplifies to

$$c'_k(\tau) = q(\tau)^{-k}[s_k(\tau) + s_{k-1}(\tau) + 2]. \tag{19.6-4}$$

Milin's conjecture is that

$$\sum_{k=1}^{n} \left(k|c_k(0)|^2 - \frac{4}{k}\right)(n+1-k) \leqq 0.$$

We try to prove this by introducing, for each fixed $n = 1, 2, \ldots,$ real-differentiable functions $\nu_k(\tau)$, $k = 1, 2, \ldots, n$, satisfying

$$\nu_k(0) = n + 1 - k, \tag{A}$$

$$\lim_{\tau\to\infty} \nu_k(\tau) = 0, \qquad k = 1, 2, \ldots, n. \tag{B}$$

Setting

$$\phi(\tau) := \sum_{k=1}^{n} \left(k|c_k(\tau)|^2 - \frac{4}{k} \right) \nu_k(\tau),$$

we then have (in view of the uniform boundedness of the coefficient c_k over Σ_0, which follows from Theorem 19.1m)

$$\lim_{\tau\to\infty} \phi(\tau) = 0,$$

and Milin's inequality will be proved if the $\nu_k(\tau)$ can be chosen such that

$$\phi'(\tau) \geqq 0, \qquad \tau \geqq 0. \tag{19.6-5}$$

We compute ϕ', using the fact that

$$\begin{aligned}\frac{\partial}{\partial\tau}|c_k(\tau)|^2 &= \frac{\partial}{\partial\tau}[c_k(\tau)\overline{c_k(\tau)}] = c_k'(\tau)\overline{c_k(\tau)} + \bar{c}'(\tau)c_k(\tau) \\ &= 2 \operatorname{Re}[c_k'(\tau)\overline{c_k(\tau)}].\end{aligned}$$

For simplicity we drop the variable τ. (Thus c_k now denotes $c_k(\tau)$, and no longer $c_k(0)$.) Since

$$k\overline{c_k} = (\overline{s_k} - \overline{s_{k-1}})q^k,$$

by the definition of s_k, we obtain by (19.6-4)

$$\phi' = \sum_{k=1}^{n} 2 \operatorname{Re}\{(\overline{s_k} - \overline{s_{k-1}})(s_k + s_{k-1} + 2)\}\nu_k + \sum_{k=1}^{n} (|s_k - s_{k-1}|^2 - 4)\frac{\nu_k'}{k}.$$

We use the fact that

$$\operatorname{Re}\{(s_k - s_{k-1})(s_k + s_{k-1} + 2)\} = |s_k|^2 - |s_{k-1}|^2 + 2 \operatorname{Re} s_k - 2 \operatorname{Re} s_{k-1}.$$

A summation by parts therefore yields

$$\phi' = \sum_{k=1}^{n} (2|s_k|^2 + 4 \operatorname{Re} s_k)(\nu_k - \nu_{k+1}) + \sum_{k=1}^{n} (|s_k - s_{k-1}|^2 - 4)\frac{\nu_k'}{k},$$

where $\nu_{n+1} := 0$. We now subject the functions ν_k to the further condition

$$\frac{\nu_k'}{k} + \frac{\nu_{k+1}'}{k+1} = \nu_{k+1} - \nu_k, \; k = 1, \ldots, n. \tag{C}$$

For such ν_k we have

$$\begin{aligned}\phi' &= -\sum_{k=1}^{n} (2|s_k|^2 + 4 \operatorname{Re} s_k)\left(\frac{\nu_k'}{k} + \frac{\nu_{k+1}'}{k+1}\right) + \sum_{k=1}^{n} (|s_k - s_{k-1}|^2 - 4)\frac{\nu_k'}{k} \\ &= -\sum_{k=1}^{n} \{2|s_k|^2 + 2|s_{k-1}|^2 + 4 \operatorname{Re} s_k + 4 \operatorname{Re} s_{k-1} + 4 - |s_k - s_{k-1}|^2\}\frac{\nu_k'}{k}.\end{aligned}$$

This simplifies to

$$\phi' = -\sum_{k=1}^{n} |s_k + s_{k-1} + 2|^2 \frac{\nu_k'}{k},$$

and we see that $\phi'(\tau) \geqq 0$ holds if the functions ν_k satisfy one last condition:

$$\nu_k'(\tau) \leqq 0, \qquad \tau \geqq 0, \quad k = 1, 2, \ldots, n. \tag{D}$$

The relations (C) may be regarded as a system of ordinary differential equations for the system of functions $(\nu_1, \nu_2, \ldots, \nu_n)$. The initial conditions required to fix the solution are given by (A). We thus see:

THEOREM 19.6a. *The Bieberbach conjecture is true if, for $n = 1, 2, \ldots$, the system of functions $(\nu_1, \nu_2, \ldots, \nu_n)$ defined by* (A) *and* (C) *satisfies* (B) *and* (D).

At the time of this writing, the only way to verify (B) and (D) appears to be to solve the system explicitly, and to manipulate the solution. The apparatus for solving the system was introduced in § 9.3. We define the vectors

$$\boldsymbol{\nu} := \begin{pmatrix} \nu_1 \\ \nu_2 \\ \vdots \\ \nu_n \end{pmatrix}, \qquad \mathbf{c} := \begin{pmatrix} n \\ n-1 \\ \vdots \\ 1 \end{pmatrix}$$

and the matrices

$$\mathbf{D} := \begin{pmatrix} 1 & & & & 0 \\ & 2 & & & \\ & & 3 & & \\ & & & \ddots & \\ 0 & & & & n \end{pmatrix}, \qquad \mathbf{Z} := \begin{pmatrix} 0 & 1 & & & 0 \\ & 0 & 1 & & \\ & & \ddots & \ddots & \\ & & & 0 & 1 \\ 0 & & & & 0 \end{pmatrix}.$$

Denoting by $\mathbf{I}$ the unit matrix, the system (C) then is given by

$$(\mathbf{I} + \mathbf{Z})\mathbf{D}^{-1}\boldsymbol{\nu}' = -(\mathbf{I} - \mathbf{Z})\boldsymbol{\nu},$$

that is, by

$$\boldsymbol{\nu}' = \mathbf{A}\boldsymbol{\nu}, \tag{19.6-6}$$

where

$$\mathbf{A} := -\mathbf{D}(\mathbf{I} + \mathbf{Z})^{-1}(\mathbf{I} - \mathbf{Z}). \tag{19.6-7}$$

By Example 1, § 9.3, the solution satisfying the initial condition $\boldsymbol{\nu}(0) = \mathbf{c}$ is

$$\boldsymbol{\nu}(\tau) = e^{\mathbf{A}\tau}\mathbf{c}.$$

The matrix $\mathbf{A}$ is a product of upper triangular matrices and thus is itself upper triangular. Its main diagonal equals the diagonal of $-\mathbf{D}$, hence the eigenvalues of $\mathbf{A}$ are $-1, -2, \ldots, -n$. It is thus already clear that $\boldsymbol{\nu}(\tau) \to \mathbf{0}$ as $\tau \to \infty$, showing that (B) holds. To verify (D) we require an explicit representation of $\boldsymbol{\nu}$. We recall from § 2.6 that if $\mathbf{T}$ is such that

$$\mathbf{T}^{-1}\mathbf{A}\mathbf{T} = -\mathbf{D}, \tag{19.6-8}$$

then

$$e^{\mathbf{A}\tau} = \mathbf{T}\, e^{-\mathbf{D}\tau}\mathbf{T}^{-1}.$$

Here $e^{-\mathbf{D}\tau}$ is the diagonal matrix with kth element $e^{-k\tau}$.

To find $\mathbf{T}$, we note that (19.6-8) implies

$$\mathbf{A}\mathbf{T} = -\mathbf{T}\mathbf{D}.$$

Looking at the mth column of this equation and denoting by $\mathbf{t}_m$ the mth column of $\mathbf{T}$, this yields

$$\mathbf{A}\mathbf{t}_m = -m\mathbf{t}_m$$

or in view of (19.6-7),

$$-\mathbf{D}(\mathbf{I}+\mathbf{Z})^{-1}(\mathbf{I}-\mathbf{Z})\mathbf{t}_m = -m\mathbf{t}_m,$$

$$(\mathbf{I}-\mathbf{Z})\mathbf{t}_m = m(\mathbf{I}+\mathbf{Z})\mathbf{D}^{-1}\mathbf{t}_m.$$

Letting

$$\mathbf{t}_m = \begin{pmatrix} t_{1m} \\ t_{2m} \\ \vdots \\ t_{nm} \end{pmatrix},$$

the kth row of the above relations states that

$$t_{km} - t_{k+1,m} = \frac{m}{k}\, t_{km} + \frac{m}{k+1}\, t_{k+1,m}.$$

Since $\mathbf{A}$ is upper triangular, $\mathbf{T}$ is upper triangular, thus $t_{km} = 0$ for $k > m$. We arbitrarily set $t_{mm} = 1$. The above relation yields

$$\left(1 - \frac{m}{k}\right) t_{km} = \left(1 + \frac{m}{k+1}\right) t_{k+1,m},$$

and we immediately find

$$t_{km} = (-1)^{m-k} \frac{(m+k+1)_{m-k}}{(m-k)!}\frac{k}{m}, \qquad k = 1, 2, \ldots, m. \tag{19.6-9}$$

Instead of inverting $\mathbf{T}=(t_{km})$, the matrix $\mathbf{T}^{-1}=\mathbf{S}=(s_{km})$ may be found directly from the fact that

$$\mathbf{SA}=-\mathbf{DS},$$

and that $\mathbf{S}$ again is an upper triangular matrix with main diagonal $\mathbf{I}$. Denoting the kth row of $\mathbf{S}$ by $\mathbf{s}_k^T$, we have

$$\mathbf{s}_k^T\mathbf{A}=-k\mathbf{s}_k^T$$

or in view of (19.6-7), because the matrices $\mathbf{I}-\mathbf{Z}$ and $(\mathbf{I}+\mathbf{Z})^{-1}$ commute,

$$-\mathbf{s}_k^T\mathbf{D}(\mathbf{I}-\mathbf{Z})(\mathbf{I}+\mathbf{Z})^{-1}=-k\mathbf{s}_k^T,$$

$$-\mathbf{s}_k^T\mathbf{D}(\mathbf{I}-\mathbf{Z})=-k\mathbf{s}_k^T(\mathbf{I}+\mathbf{Z}),$$

or

$$(k+m-1)s_{k,m-1}=(m-k)s_{km}, \qquad m=1,2,\ldots,n.$$

This is solved by $s_{km}=0$, $m<k$; $s_{kk}=1$, and

$$s_{km}=\frac{(2k)_{m-k}}{(m-k)!}, \qquad m\geqq k. \tag{19.6-10}$$

The tools are now on hand to compute the solution

$$\boldsymbol{\nu}(\tau)=\mathbf{T}\,e^{-\mathbf{D}\tau}\mathbf{T}^{-1}\mathbf{c}.$$

We first find the vector $\mathbf{v}:=\mathbf{T}^{-1}\mathbf{c}=\mathbf{Sc}$, the kth component of which is

$$v_k=\sum_{m=1}^{n}s_{km}c_m=\sum_{m=k}^{n}\frac{(2k)_{m-k}}{(m-k)!}(n+1-m)$$

$$=\sum_{p=0}^{n-k}\frac{(2k)_p}{p!}(n+1-k-p).$$

In the notation of hypergeometric series, we have

$$v_k=(n-k+1)\,{}_2F_1(-n+k,2k;-n+k-1;1)$$

and applying Vandermonde's formula (1.4-11), there follows

$$v_k=\frac{(2k+2)_{n-k}}{(n-k)!}, \qquad k=1,2,\ldots,n.$$

From the fact that

$$\boldsymbol{\nu}(\tau)=\mathbf{T}\,e^{-\mathbf{D}\tau}\mathbf{v}$$

we now immediately find the desired representation

$$\nu_k(\tau)=\sum_{m=1}^{n}t_{km}v_m\,e^{-m\tau}.$$

The lower limit of summation may be replaced by k, and on using (19.6-9) and setting $m = k+p$ we get

$$\nu_k(\tau) = k \sum_{p=0}^{n-k} (-1)^p \frac{(2k+p+1)_p}{p!(k+p)} \frac{(2k+2p+2)_{n-k-p}}{(n-k-p)!} e^{-k\tau-p\tau},$$
$$k = 1, 2, \ldots, n. \quad (19.6\text{-}11)$$

Evidently,

$$\nu_k'(\tau) = -k\, e^{-k\tau} \sum_{p=0}^{n-k} (-1)^p \frac{(2k+p+1)_p (2k+2p+2)_{n-k-p}}{p!(n-k-p)!} e^{-p\tau}.$$

By virtue of the identities,

$$(2k+1+p)_p(2k+2p+2)_{n-k-p} = \frac{(2k+1)_{2p}}{(2k+1)_p} \frac{(2k+2)_{n-k+p}}{(2k+2)_{2p}}$$
$$= \frac{(k+\frac{1}{2})_p}{(2k+1)_p} \frac{(2k+2)_{n-k}(n+k+2)_p}{(k+\frac{3}{2})_p},$$

$$\frac{1}{(n-k-p)!} = (-1)^p \frac{(-n+k)_p}{(n-k)!},$$

the sum on the right may be expressed as a hypergeometric series of type ${}_3F_2$, and we have

$$\nu_k'(\tau) = -k \frac{(2k+2)_{n-k}}{(n-k)!} e^{-k\tau} {}_3F_2\left[\begin{matrix} -n+k,\ n+k+2,\ k+\frac{1}{2};\ e^{-\tau} \\ k+\frac{3}{2},\ 2k+1 \end{matrix}\right].$$

The factor of the ${}_3F_2$ is negative. Letting $m := n-k$, we thus have in view of Theorem 19.6a:

THEOREM 19.6b. *The Bieberbach conjecture is true if*

$$ {}_3F_2\left[\begin{matrix} -m,\ m+2k+2,\ k+\frac{1}{2};\ x \\ k+\frac{3}{2},\ 2k+1 \end{matrix}\right] \geqq 0, \qquad 0 \leqq x \leqq 1, \qquad (19.6\text{-}12)$$

holds for all $m = 0, 1, 2, \ldots$ *and for all* $k = 1, 2, \ldots$.

The truth of the Bieberbach conjecture is thus finally implied by:

THEOREM 19.6c (Askey–Gasper lemma). *The inequalities* (19.6-12) *hold for the required values of* m *and* k *and for all real* x.

Proof. We are faced with the task of showing that a certain set of polynomials, which are given in explicit hypergeometric form and which have rational coefficients, have nonnegative values on [0, 1]. For small values of

m this could be accomplished by careful numerical computation, or more elegantly, by forming Sturmian chains. But of course we are interested in the general result.

One way of showing that a quantity is positive is to represent it as the square of some other real quantity. There in fact exists one famous case where a hypergeometric series of type ${}_3F_2$ can be represented as a square of an ordinary hypergeometric series. This is **Clausen's formula,**

$$ {}_3F_2\left[\begin{matrix} 2a, 2b, a+b; x \\ 2a+2b, a+b+\frac{1}{2} \end{matrix}\right] = \left\{ {}_2F_1\left[\begin{matrix} a, b; x \\ a+b+\frac{1}{2} \end{matrix}\right]\right\}^2, \tag{19.6-13} $$

which holds for arbitrary a and b such that the series on the left is defined. Clausen's formula may be proved, for instance, by showing that the functions on either side satisfy the same Fuchsian differential equation.

The series on the left of (19.6-12) just fails to be of the type to which Clausen's formula applies. We therefore settle for the next best possibility, which is to represent our ${}_3F_2$ as a sum of ${}_3F_2$'s to which Clausen's formula applies.

It helps to pull the discussion down to the level of functions of type ${}_2F_1$ about which more is known. This we do by means of the integral operator

$$ (Ih)(x) := \frac{\Gamma(2k+1)}{[\Gamma(k+\frac{1}{2})]^2} \int_0^1 [\tau(1-\tau)]^{k-1/2} h(\tau x)\, d\tau, $$

where Γ denotes the gamma function. By Euler's beta integral (8.7-2),

$$ \frac{\Gamma(2k+1)}{[\Gamma(k+\frac{1}{2})]^2} \int_0^1 [\tau(1-\tau)]^{k-1/2} (\tau x)^p\, d\tau = \frac{(k+\frac{1}{2})_p}{(2k+1)_p} x^p. $$

Thus our function

$$ f(x) := {}_3F_2\left[\begin{matrix} -m, m+2k+2, k+\frac{1}{2}; x \\ k+\frac{3}{2}, 2k+1 \end{matrix}\right] $$

is given by $f = Ig$, where

$$ g(x) := {}_2F_1\left[\begin{matrix} -m, m+2k+2; x \\ k+\frac{3}{2} \end{matrix}\right]. $$

Now if

$$ h(x) = {}_2F_1\left[\begin{matrix} \alpha, \beta; x \\ \gamma \end{matrix}\right] $$

is any hypergeometric function with real parameters, then

$$ Ih(x) = {}_3F_2\left[\begin{matrix} \alpha, \beta, k+\frac{1}{2}; x \\ \gamma, 2k+1 \end{matrix}\right] $$

is a series to which Clausen's formula applies, and which therefore is nonnegative for all real x, provided that

$$\alpha+\beta=2k+1,$$
$$\gamma=k+1.$$

We therefore endeavor to represent the polynomial g as a linear combination of the polynomials

$$h_p^k(x):={}_2F_1\left[\begin{matrix}-p,p+2k+1;x\\k+1\end{matrix}\right],\qquad p=0,1,\ldots,m,$$

and all hinges now on the question whether the coefficients in this combination will turn out positive.

The calculation of these coefficients is facilitated by the fact that the sequence of polynomials $\{h_p^k\}_{p=0}^{\infty}$ is *orthogonal* with respect to the weight function

$$\omega(x):=[x(1-x)]^k,\qquad 0\leqq x\leqq 1.$$

This could be shown by relating the h_p^k to certain ultraspherical polynomials. For a self-contained proof, it suffices to establish the relations

$$\int_0^1[x(1-x)]^k h_p^k(x)x^q\,dx=0$$

for $q<p$. If q is any nonnegative integer, then we have by the definition of h_p^k, again using Euler's beta integral,

$$\begin{aligned}\int_0^1&[x(1-x)]^k h_p^k(x)x^q\,dx\\&=\sum_{l=0}^{p}\frac{(-p)_l(p+2k+1)_l}{l!(k+1)_l}\int_0^1(1-x)^k x^{k+q+l}\,dx\\&=\sum_{l=0}^{p}\frac{(-p)_l(p+2k+1)_l}{l!(k+1)_l}\frac{\Gamma(k+1)\Gamma(k+1+q+l)}{\Gamma(2k+2+q+l)}\\&=\frac{\Gamma(k+1)\Gamma(k+1+q)}{\Gamma(2k+2)\Gamma(2k+2+q)}\sum_{l=0}^{p}\frac{(-p)_l(p+2k+1)_l(k+1+q)_l}{l!(k+1)_l(2k+2+q)_l}.\end{aligned}$$

The sum by the Saalschütz formula (1.5-11) equals

$$\begin{aligned}{}_3F_2\left[\begin{matrix}-p,p+2k+1,q+k+1;1\\k+1,q+2k+2\end{matrix}\right]&=\frac{(-p-k)_p(-q)_p}{(k+1)_p(-2k-1-p-q)_p}\\&=\frac{(-q)_p}{(2k+2+q)_p}.\end{aligned}$$

Since $(-q)_p=0$ for $q<p$, we thus have

$$\int_0^1 [x(1-x)]^k h_p^k(x) x^q\, dx = \begin{cases} 0, & q<p \\ (-1)^p \dfrac{[\Gamma(k+1)]^2}{\Gamma(2k+2)} \dfrac{p!(k+1)_p}{(2k+2)_{2p}}, & q=p \\ \dfrac{[\Gamma(k+1)]^2}{\Gamma(2k+2)} \dfrac{(k+1)_q(-q)_p}{(2k+2)_{p+q}}, & q>p. \end{cases} \tag{19.6-14}$$

The highest coefficient in h_p^k being

$$\frac{(-p)_p(p+2k+1)_p}{p!(k+1)_p} = (-1)^p \frac{(p+2k+1)_p}{(k+1)_p},$$

there also follows

$$\begin{aligned} \alpha_p &:= \int_0^1 [x(1-x)]^k \{h_p^k(x)\}^2\, dx \\ &= \frac{[\Gamma(k+1)]^2}{\Gamma(2k+1)} \frac{(p+2k+1)_p}{(k+1)_p} \frac{p!(k+1)_p}{(2k+2)_{2p}} \\ &= \frac{[\Gamma(k+1)]^2}{\Gamma(2k+1)} \frac{1}{(2k+1+2p)(2k+1)_p}. \end{aligned}$$

Of course $\alpha_p>0$.

Since h_p^k is a polynomial of precise degree p, it is clear that an expansion

$$g(x) = \sum_{q=0}^{m} \beta_q h_q^k(x) \tag{19.6-15}$$

holds with uniquely determined β_q. To compute β_q, we multiply on either side by $[x(1-x)]^k h_p^k(x)$ and integrate. By virtue of the orthogonality of the h_q^k, the result on the right clearly is $\alpha_p\beta_p$. Thus there follows

$$\beta_p = \alpha_p^{-1} \int_0^1 [x(1-x)]^k g(x) h_p^k(x)\, dx,$$

and the question now is whether these integrals are nonnegative. By the definition of g, using (19.6-14), we have

$$\begin{aligned} &\int_0^1 [x(1-x)]^k g(x) h_p^k(x)\, dx \\ &= \sum_{q=0}^{m} \frac{(-m)_q(m+2k+2)_q}{q!(k+\frac{3}{2})_q} \int_0^1 [x(1-x)]^k x^q h_p^k(x)\, dx \\ &= \frac{[\Gamma(k+1)]^2}{\Gamma(2k+2)} \sum_{q=p}^{m} \frac{(-m)_q(m+2k+2)_q}{q!(k+\frac{3}{2})_q} \frac{(k+1)_q(-q)_p}{(2k+2)_{p+q}}. \end{aligned}$$

which is

$$\frac{(a+c+d-f)_n}{n!}\,{}_3F_2\left[\begin{matrix}-n, f-c, f-d; 1\\ f, 1+f-a-c-d-n\end{matrix}\right].$$

We thus find that for $n=0, 1, 2, \ldots$,

$${}_3F_2\left[\begin{matrix}-n, c, d; 1\\ f, 1-a-n\end{matrix}\right]=\frac{(a+c+d-f)_n}{n!}\,{}_3F_2\left[\begin{matrix}-n, f-c, f-d; 1\\ f, 1+f-a-c-d-n\end{matrix}\right]. \qquad (19.6\text{-}16)$$

For $a=1-2c-n$, $d=2b+n$, $f=b+\frac{1}{2}$, this yields

$${}_3F_2\left[\begin{matrix}-n, 2b+n, c; 1\\ b+\frac{1}{2}, 2c\end{matrix}\right]=\frac{(b-c+\frac{1}{2})_n}{(1-2c-n)_n}\,{}_3F_2\left[\begin{matrix}-n, b-c+\frac{1}{2}, \frac{1}{2}-b-n; 1\\ b+\frac{1}{2}, c-b+\frac{1}{2}-n\end{matrix}\right].$$

The series on the right has the structure required by Dixon's theorem. It thus is zero if n is odd, and for $n=2m$ the expression on the right equals

$$\frac{(b-c+\frac{1}{2})_{2m}}{(1-2c-2m)_{2m}}\,\frac{(2m)!}{m!}\,\frac{(1-c-2m)_m}{(b-c+\frac{1}{2}+m)_m(\frac{1}{2}-b-m)_m}.$$

The formula of Lemma 19.6d results after trivial simplification. —

By virtue of the lemma, $s=0$ if $m-p$ is odd, and

$$s=\left[\frac{(\frac{1}{2})_q}{(k+p+\frac{3}{2})_q}\right]^2>0$$

if $m-p=2q$, $q=0, 1, 2, \ldots$. The sums s thus are nonnegative, which completes the proof of the Askey-Gasper lemma. Thus we can state with confidence:

THEOREM 19.6e. *The Bieberbach conjecture is true. That is, if*

$$f(z)=z+\sum_{n=2}^{\infty} a_n z^n$$

is any function in Σ_0, *then* $|a_n|\leqq n$, $n=2, 3, \ldots$.

It is possible to show, by a careful examination of the argument, that if $|a_n|=n$ for any n, then f is the Koebe function or one of its rotations,

$$f(z)=\frac{z}{(1-e^{i\alpha}z)^2}.$$

PROBLEM

1. *Clausen's formula.* If u satisfies the differential equation

$$u''+mu'+nu=0,$$

Letting $q=p+l$, the sum becomes

$$(-1)^p \frac{(-m)_p(m+2k+2)_p(k+1)_p}{(k+\frac{3}{2})_p(2k+2)_{2p}} \sum_{l=0}^{m-p} \frac{(-m+p)_l(m+2k+2+p)_l}{l!(k+\frac{3}{2}+p)_l(2k+2-$$

Since $(-1)^p(-m)_p>0$, the factor of the last sum is clearly po sum itself equals

$$s:={}_3F_2\left[\begin{matrix}-m+p,\, m+2k+2+2p,\, k+1+p;\, 1\\ k+\frac{3}{2}+p,\, 2k+2+2p\end{matrix}\right],$$

and this terminating hypergeometric sum may be summed as the s

$$n=m-p\geqq 0, \qquad b=c=k+p+1$$

of the following lemma:

LEMMA 19.6d. *For $n=0, 1, 2, \ldots$ and for complex b, c such t $b+\frac{1}{2}$ nor c is zero or a negative integer, there holds*

$${}_3F_2\left[\begin{matrix}-n,\, 2b+n,\, c;\, 1\\ b+\frac{1}{2},\, 2c\end{matrix}\right]=\begin{cases}\dfrac{(\frac{1}{2})_m(b-c+\frac{1}{2})_m}{(b+\frac{1}{2})_m(c+\frac{1}{2})_m}, & n \text{ even},\\ 0, & n \text{ odd}.\end{cases}$$

Proof. One of the known cases where a terminating ${}_3F_2$ with unit can be summed is given by Dixon's formula (Problem 10, § 1.6, special case of Theorem 8.6d). We therefore endeavor to transfor of the lemma into one to which Dixon's formula is applicable. Tl by a general transformation rule for terminating ${}_3F_2$'s, which is in itself. Let a, d, e, f be complex numbers such that $f\neq 0, -1, -2$ the coefficient of x^n in the expansion of

$$(1-x)^{-a}{}_2F_1\left[\begin{matrix}c,\, d;\, x\\ f\end{matrix}\right]$$

is given by

$$\frac{(a)_n}{n!}\,{}_3F_2\left[\begin{matrix}-n,\, c,\, d;\, 1\\ f,\, 1-a-n\end{matrix}\right],$$

as is readily seen by forming the Cauchy product. However, by E identity (9.9-21),

$${}_2F_1\left[\begin{matrix}c,\, d;\, x\\ f\end{matrix}\right]=(1-x)^{f-c-d}{}_2F_1\left[\begin{matrix}f-c,\, f-d;\, x\\ f\end{matrix}\right].$$

Thus the coefficient in question also equals the coefficient of x^n

$$(1-x)^{f-a-c-d}{}_2F_1\left[\begin{matrix}f-c,\, f-d;\, x\\ f\end{matrix}\right],$$

then $v := u^2$ satisfies

$$v''' + 3mv'' + (2m^2 + m' + 4n)v' + (4mn + 2n')v = 0.$$

Thus if

$$u = {}_2F_1\left[\begin{matrix} a, b; z \\ a+b+\frac{1}{2} \end{matrix}\right],$$

show that $v := u^2$ satisfies the same differential equation as

$$v = {}_3F_2\left[\begin{matrix} 2a, 2b, a+b; z \\ 2a+2b, a+b+\frac{1}{2} \end{matrix}\right]$$

(see § 1.5). Show that this differential equation has a singular point of the first kind at $z = 0$, with precisely one characteristic exponent equal to 0. Conclude that the differential equation has precisely one solution that is analytic at 0 and assumes the value 1 there.

NOTES

See de Branges [1985] for his own account of his proof. Our presentation of the proof is inspired by FitzGerald and Pommerenke [1985]. For Theorem 19.6c see Askey and Gasper [1976]; our self-contained version of the proof is somewhat different from theirs. Lemma 19.6d is due to Watson [1925].

BIBLIOGRAPHY

Chapters or sections to which a reference is relevant are given at the end of each reference.

Abramowitz, M., and I. A. Stegun [1965]. *Handbook of mathematical functions.* Dover, New York. § 13.4.

Achieser, N. L., and I. M. Glasmann [1960]. *Theorie der linearen Operatoren im Hilbert-Raum.* Akademie-Verlag, Berlin. § 14.11.

Ahlfors, L. [1966]. *Complex analysis*, 2nd ed. McGraw-Hill, New York, §§ 15.2, 15.3, 15.4, 16.0, 16.1, 17.1.

—— [1973]. *Conformal invariants. Topics in geometric function theory.* McGraw-Hill, New York, §§ 15.5, 15.7, 16.3, 16.11, 19.1, 19.3.

Aho, A. V., J. E. Hopcroft, and J. D. Ullmann [1974]. *The design and analysis of computer algorithms.* Addison-Wesley, Reading, MA. §§ 13.1, 13.7.

——, K. Steiglitz, and J. D. Ullmann [1975]. Evaluating polynomials at fixed sets of points. *SIAM J. Comput.* **4**, 533–539. § 13.9.

Albrecht, R. [1952]. Zum Schmiegungsverfahren der konformen Abbildung. *Z. angew. Math. Mech.* **32**, 316–318. § 16.2.

—— [1955]. Iterationsverfahren zur konformen Abbildung eines Ringgebietes auf einen konzentrischen Kreisring. *Sitz.-Ber. Bayer. Akad. Wiss., Math.-Naturw. Kl. 1954.* pp. 169–178. § 17.2.

Anderson, B. D. O., K. L. Hitz, and N. D. Diem [1974]. Recursive algorithm for spectral factorization. *IEEE Trans. Circuits Syst.*, **CAS-21**, 742–750. § 13.4.

Arbenz, K. [1958]. *Integralgleichungen für einige Randwertprobleme für Gebiete mit Ecken.* Diss., ETH Zürich. § 16.7.

Arnold, D. N. [1983]. A spline-trigonometric Galerkin method and an exponentially convergent boundary integral method. *Math. of Comp.* **41**, 383–397. § 16.6.

—— and W. L. Wendland [1983]. On the asymptotic convergence of collocation methods. *Math. of Comp.* **41**, 349–381. § 16.6.

Aronszajn, N. [1950]. Theory of reproducing kernels. *Trans. Amer. Math. Soc.* **68**, 337–404. § 18.3.

Arsove, M. G. [1968]. The Osgood–Taylor–Carathéodory theorem. *Proc. Amer. Math. Soc.* **19**, 38–44. § 16.3.

Askey, R., and G. Gasper [1976]. Positive Jacobi polynomial sums, II. *Amer. J. Math.* **98**, 709–737. § 19.6.

Atkinson, K. E. [1976]. *A survey of numerical methods for the solution of Fredholm integral equations of the second kind.* SIAM, Philadelphia, PA. §§ 16.7, 18.4.

—— [1978]. *An introduction to numerical analysis.* Wiley, New York. § 16.10.

Auslander, L., and R. Tolmieri [1979]. Is computing with the finite Fourier transform pure or applied mathematics? *Bull. Amer. Math. Soc.* (*New Ser.*) **1**, 847–897. § 13.1.

Bauer, F. L. [1955]. Ein direktes Iterationsverfahren zur Hurwitz-Zerlegung eines Polynoms. *Arch. Elektr. Übertr.* **9**, 285–290. § 13.4.

Bauer, F., P. Garabedian, D. Korn, and A. Jameson [1975]. *Supercritical wing sections* (*Lecture notes in economics and mathematical systems*, vol. 108). Springer, New York, § 14.2, 16.8.

Beckenbach, E. F., ed. [1952]. *Construction and applications of conformal maps* (*Applied mathematics ser.* 18). National Bureau of Standards. § 16.0.

Bedrosian, A. [1963]. A product theorem for Hilbert transforms. *Proc. IEEE* **51**, 868–869. § 14.12.

Bellmann, R., and K. L. Cooke [1963]. *Differential-difference equations.* Academic, New York. § 14.10.

Bergman, S. [1922]. Über die Entwicklung der harmonischen Funktionen der Ebene und des Raumes in Orthogonalfunktionen. *Math. Ann.* **86**, 238–271. § 18.3.

—— [1950]. *The kernel function and conformal mapping* (*Mathematical surveys, V*). Amer. Math. Soc., New York. §§ 17.6, 18.3, 18.4.

—— and B. Chalmers [1967]. A procedure for the conformal mapping of triply-connected domains. *Math. of Comp.* **21**, 527–542. § 18.3.

—— and M. Schiffer [1953]. *Kernel functions and elliptic differential equations in mathematical physics.* Academic, New York. §§ 15.6, 15.7, 17.6, 18.3.

Berrut, J.-P. [1976]. *Numerische Lösung der Symmschen Integralgleichung durch Fourier-Methoden.* Diploma thesis, ETH Zürich. § 16.6.

——[1984]. Baryzentrische Formeln zur trigonometrischen Interpolation, I. II. *Z. angew. Math. Physik* **35**, 91–105, 193–205. § 13.6.

—— [1985a]. A Fredholm integral equation of the second kind for conformal mapping. *J. Appl. Comp. Math.*, § 16.7.

—— [1985b]. *Ueber Integralgleichungen und Fourier-Methoden zur numerischen konformen Abbildung.* Diss. No. 7754, Swiss Federal Institute of Technology, Zurich. §§ 16.7, 16.10.

Bieberbach, L. [1934]. *Lehrbuch der Funktionentheorie*, vol. 2. Teubner, Leipzig. §§ 16.1, 19.1.

—— [1953]. *Conformal mapping* (Translated from the 4th German edition by F. Steinhardt). Chelsea, New York. § 16.0.

Blackman, R. B., and J. W. Tukey [1959]. *The measurement of power spectra.* Dover, New York. § 13.7.

Bloomfield, P. [1976]. *Fourier analysis of time series.* Wiley, New York. § 13.7.

Blue, J. L. [1978]. Boundary integral solutions of Laplace's equation. *Bell. Sys. Tech. J.* **57**, 2797–2822. § 15.9.

Bluestein, L. I. [1970]. A linear filtering approach to the computation of discrete Fourier transform. *IEEE Trans. Audio Electroacoust.* **AU-18**, 451–455. § 13.7.

Blumenfeld, J., and W. Mayer [1914]. Über Poincarésche Fundamentalfunktionen. *Sitz.-Ber. Wien. Akad. Wiss. IIa* **123**, 2011–2047. § 15.9.

Boas, R. P. [1972]. Summation formulas and band-limited signals. *Tôhoku Math. J.* **24**, 121–125. § 13.3.

Boatwright, J. [1978]. Detailed spectral analysis of two small New York State earthquakes. *Bull. Seis. Soc. Amer.* **68**, 1117–1131. § 13.7.

Bochner, S. [1922]. Über orthogonale Systeme analytischer Funktionen. *Math. Z.* **14**, 180–207. § 18.3.

Brandt, A. [1977]. Multi-level adaptive solutions to boundary-value problems. *Math. Comp.* **31**, 339–390. § 15.12.

Brass, H. [1982]. Zur numerischen Berechnung konjugierter Funktionen, in *Numerical methods of approximation theory*, vol. 6, L. Collatz, G. Meinardus, and H. Werner, eds., Birkhäuser, Basel, pp. 43–66. § 13.6.

Brent, R. P. [1976]. Fast multiple-precision evaluation of elementary functions. *J. ACM* **23**, 242–251. § 13.9.

——, F. G. Gustavson, and D. Y. Y. Yun [1980]. Fast solution of Toeplitz systems of equations and computation of Padé approximants. *J. Algorithms* **1**, 259–295. § 13.9.

—— and H. T. Kung [1976]. $O((n \log n)^{3/2})$ algorithms for composition and reversion of power series, in *Analytic computational complexity* (Proc. Symp., Carnegie-Mellon Univ., Pittsburgh, PA). Academic, New York, pp. 217–225. § 13.9.

—— and H. T. Kung [1978]. Fast algorithms for manipulating formal power series. *J. ACM* **25**, 581–595. § 13.9.

—— and J. F. Traub [1980]. On the complexity of composition and generalized compposition of power series. *SIAM J. Comput.* **9**, 54–66. § 13.9.

Bühlmann, H. [1984]. Numerical evaluation of the compound Poisson distribution: recursion or Fast Fourier Transform? *Scand. Actural J.* **1984**, 116–126. § 13.9.

Bunemann, O. [1969]. A compact non-iterative Poisson solver. Rep. SUIPR-294, Inst. for Plasma Research, Stanford Univ., Stanford, CA. § 15.13.

Burniston, E. E., and C. E. Siewert [1972]. Exact analytical solutions basic to a class of two-body orbits. *Celest. Mech.* **7**, 225–235. § 14.10.

—— [1973a]. Further results concerning exact analytical solutions basic to two-body orbits. *Celest. Mech.* **10**, 5–15. § 14.10.

—— [1973b]. The use of Riemann problems in solving a class of transcendental equations. *Proc. Camb. Phil. Soc.* **73**, 111–118. § 14.10.

—— [1978]. On the solution of certain algebraic equations. *J. Comp. Appl. Math.* **4**, 37–39. § 14.10.

Butkovskyi, A. G. [1982]. *Green's functions and transfer functions handbook.* Ellis Horwood, Chichester. § 15.6.

Buzbee, B. L., and F. W. Dorr [1974]. The direct solution of the biharmonic equation on rectangular regions and the Poisson equation on irregular regions. *SIAM J. Numer. Anal.* **11**, 753–763. § 15.13.

——, F. W. Dorr, J. A. George, and G. H. Golub [1971]. The direct solution of the discrete Poisson equation. *SIAM J. Numer. Anal.* **8**, 722–736. § 15.13.

——, G. H. Golub, and C. W. Nielson [1970]. On direct methods for solving Poisson's equation. *SIAM J. Numer. Anal.* **7**, 627–656. § 15.13.

Calderon, A. P. [1977]. On the Cauchy integral on Lipschitz curves and related operators. *Proc. Nat. Acad. Sci.* **4**, 1324–1327. § 14.1.

Campbell, J. B. [1975]. Finite difference techniques for ring capacitors. *J. Eng. Math.* **9**, 21–28. § 16.11.

Carasso, A. S., J. G. Sanderson, and J. M. Hyman [1978]. Digital removal of random media image degradations by solving the diffusion equation backwards in time. *SIAM J. Numer. Anal.* **15**, 344–367. § 13.10.

Carathéodory, C. [1913]. Über die Begrenzung einfach zusammenhängender Gebiete. *Math. Ann.* **73**, 305–370. § 16.3.

—— [1929]. Über die Winkelderivierte von beschränkten analytischen Funktionen. *Sitz.-Ber. Preuss. Akad. Wiss. 1929.* pp. 39–54. § 16.4.

—— [1932]. *Conformal representation* (*Cambridge tracts in mathematics and mathematical physics* 28). Cambridge Univ. Press, London. § 16.0.

—— [1961]. *Funktionentheorie*, vol. 2. Birkhäuser, Basel. §§ 16.3, 16.4.

Carleman, T. [1918]. Über ein Minimalproblem der mathematischen Physik. *Math. Z.* **1**, 208–212. § 17.7.

—— [1922]. Über die Abelsche Integralgleichung mit konstanten Integrationsgrenzen. *Math. Z.* **15**, 111–120. § 14.9.

Carrier, G. F., M. Krook, and C. E. Pearson [1966]. *Functions of a complex variable, theory and technique.* McGraw-Hill, New York. §§ 14.1, 14.9.

Chapman, C. H. [1978]. Body waves in seismology, in *Modern Problems in Electric Wave Propagation* (J. Miklowitz and J. D. Achenbach, eds.) Wiley, New York, pp. 477–498. § 13.7.

Cheng, H. K., and N. Rott [1954]. Generalizations of the inversion formula of thin airfoil theory. *J. Rat. Mech. Anal.* **3**, 357–382. § 14.9.

Chew, G. F. [1961]. *S-Matrix theory of strong interactions.* (*Frontiers of Physics*, a lecture note and reprint series). Benjamin, New York. § 14.11.

Christiansen, S. [1981]. Condition number of matrices derived from two classes of integral equations. *Math. Meth. in Appl. Sci.* **3**, 364–392. § 16.6.

Ciarlet, P. G. [1978]. *The finite element method for elliptic problems.* North-Holland, New York. § 15.12.

Clements, D. J., and B. D. O. Anderson [1976]. Polynomial factorization via the Riccati equation. *SIAM J. Appl. Math.* **31**, 179–205. § 13.4.

Coifman, R. R., A. McIntosh, and Y. Meyer [1982]. L'intégrale de Cauchy définit un opérateur borné sur L^2 pour les courbes Lipschitziennes. *Ann. Math.* **116**, 361–388. § 14.1.

Collatz, L. [1960]. *The numerical treatment of differential equations* (*Grundlehren*, vol. 60). Springer, Berlin. § 15.12.

Concus, P., and G. H. Golub [1973]. Use of fast direct methods for efficient numerical solution of nonseparable elliptic equations. *SIAM J. Numer. Anal.* **10**, 1103–1120. § 15.13.

Conway, J. B. [1973]. *Functions of a complex variable.* Springer, New York. § 16.1.

Cooley, J. W. R., A. W. Lewis, and P. D. Welch [1967a]. Application of the Fast Fourier Transform to the computation of Fourier integrals, Fourier series, and convolution integrals. *IEEE Trans. Audio Electroacoust.* **AU-15**, 79–84. § 13.7.

——, ——, and —— [1967b]. The Fast Fourier transform and its applications. Research paper RC-1743, IBM Research. §§ 13.1, 13.7.

——, ——, and —— [1967c]. Historical notes on the Fast Fourier Transform. *Proc. IEEE* **55**, 1674–1677. § 13.1.

——, ——, and —— [1970a]. The Fast Fourier Transform algorithm: Programming considerations in the calculation of sine, cosine, and Laplace transforms. *J. Sound Vib.* **12**, 315–337. § 13.1.

Cooley, J. W. R., A. W. Lewis, and P. D. Welch [1970b]. The application of the fast Fourier transform to the estimation of spectra and cross-spectra. *J. Sound Vib.* **12**, 339–352. § 13.7.

—— and J. W. Tukey [1965]. An algorithm for the machine calculation of complex Fourier series. *Math. Comp.* **19**, 297–301. § 13.1.

Courant, R. [1943]. Variational methods for the solution of problems of equilibrium and vibrations. *Bull. Amer. Math. Soc.* **49**, 1–23. § 15.12.

—— [1950]. *Dirichlet's principle, conformal mapping, and minimal surfaces.* Interscience, New York. §§ 16.5, 17.6.

—— and D. Hilbert [1951, 1963]. *Methods of mathematical physics,* 2 vols. Wiley-Interscience, New York. §§ 15.6, 15.9, 15.12.

Cremer, H. [1930]. Ein Existenzsatz der Kreisringabbildung zweifach zusammenhängender schlichter Bereiche. *Sitz.-Ber. sächs. Akad. Wiss. Leipzig, Math.-Phys. Kl. 82.* pp. 190–192. § 17.2.

Curtiss, J. [1971]. Faber polynomials and the Faber series. *Amer. Math. Month.* **78**, 577–596. §§ 18.1, 18.2.

Daeppen, H. [1982]. *Die schnelle Fouriertransformation für beliebiges n.* Dipoma thesis, ETH Zurich. § 13.7.

—— [1984]. *An implementation of the Schwarz–Christoffel map for doubly connected regions.* Unpubl. § 17.5.

Dällenbach, W. [1921]. Verschärftes rechnerisches Verfahren der harmonischen Analyse. *Arch. Elektrotech.* **10**, 277–281. § 13.2.

Davis, P. J. [1959]. On the numerical integration of periodic analytic functions, in *On numerical approximation,* R. E. Langer, ed. Univ. of Wisconsin Press, Madison, pp. 45–59. § 13.2.

—— [1963]. *Interpolation and approximation.* Blaisdell, New York. § 18.5.

Davis, R. T. [1979]. Numerical methods for coordinate generation based on Schwarz–Christoffel transformations, in *4th AIAA Computational Fluid Dynamics Conf. proc.* (Williamsburg, VA). §§ 16.0, 16.9.

de Boor, C. [1980]. FFT as nested multiplication with a twist. *SIAM J. Sci. Stat. Comput.* **1**, 173–178. § 13.1.

de Branges, L. [1985]. A proof of the Bieberbach conjecture. *Acta Math.* **154**, 137–152.

Demchenko, B. [1931]. Sur la formula de H. Villat resolvant le problème de Dirichlet dans un anneau circulaire. *J. math. pures appli.* **10**, 201–211. § 15.2.

Dorr, F. W. [1970]. The direct solution of the discrete Poisson equation on a rectangle. *SIAM Rev.* **12**, 248–263. § 15.13.

Duren, P. L. [1983]. *Univalent functions* (*Grundlehren,* vol. 259). Springer, Berlin. §§ 19.1–19.5.

Elcrat, A. R. [1982]. Separated flow past a plate with spoiler. *SIAM J. Math. Anal.* **13**, 632–639. § 16.10.

Ellacott, S. W. [1978]. *On the approximate conformal mapping of multiply connected domains.* Dept. of Math., Brighton Polytechnic, Brighton. § 17.7.

—— [1981]. On the convergence of some approximate methods of conformal mapping. *IMA J. Numer. Anal.* **1**, 185–192. § 18.3.

—— [1983a]. Computation of Faber series with application to numerical polynomial approximation in the complex plane. *Math. Comp.* **40**, 575–587. § 18.2.

—— [1983b]. On the Faber transform and efficient numerical rational approximation. *SIAM J. Numer. Anal.* **20**, 989–1000. § 18.2.

Elliott, D., and D. F. Paget [1975]. On the convergence of a quadrature rule for evaluating certain principal value integrals. *Numer. Math.* **23**, 311–319; **25**, 287–289. § 14.6.

Engeli, M., Th. Ginsburg, H. Rutishauser, and E. Stiefel [1959]. *Refined iterative methods for computation of the solution and the eigenvalues of self-adjoint boundary value problems* (*Mitt. Inst. Angew. Math.* 8), Birkhäuser, Basel. § 15.12.

Erdélyi, A., ed. [1954]. *Tables of integral transforms*, vol. 2. MacGraw-Hill, New York. § 14.11.

Faber, G. [1903]. Über polynomische Entwicklungen. *Math. Ann.* **57**, 389–408. §§ 18.1, 18.2.

Fiduccia, Ch. M. [1972]. Polynomial evaluation via the division algorithm—the fast Fourier transform revisited, in *Proc. 4th Ann. ACM Symp. on Theory of Computing.* pp. 88–93. § 13.1.

Filon, L. N. G. [1929]. On a quadrature formula for trigonometric integrals. *Proc. Roy. Soc. Edinburgh* **49**, 38–47. § 13.2.

Fischer, D., G. H. Golub, O. Hald, C. Leiva, and O. Widlund [1974]. On Fourier-Toeplitz methods for separable elliptic problems. *Math. Comp.* **28**, 349–368. § 15.13.

Fitzgerald, C. H. [1972]. Quadratic inequalities and coefficient estimates for schlicht functions. *Arch. Rat. Mech. Anal.* **46**, 356–368. § 19.3.

—— and Ch. Pommerenke [1985]. The de Branges theorem on univalent functions. *Trans. Amer. Math. Soc.* **290**, 683–690. § 19.6.

Fornberg, B. [1980]. A numerical method for conformal mapping. *SIAM J. Sci. Stat. Comput.* **1**, 386–400. § 16.9.

—— [1981a]. Numerical computation of nonlinear waves, in *Nonlinear phenomena in physics and biology*, R. H. Evans, B. L. Jones, R. M. Miura, and S. S. Rangnekar, eds. Plenum, New York, pp. 159–184. §§ 16.0, 16.9.

—— [1981b]. A vector implementation of the Fast Fourier Transform algorithm. *Math. of Comp.* **36**, 189–191. § 13.1.

—— [1981c]. Algorithm 579: CPSC: Complex power series coefficients. *ACM Trans. Math. Software* **7**, 542–547. § 13.4.

—— [1981d]. Numerical differentiation of analytic functions. *ACM Trans. Math. Software* **7**, 512–526. § 13.4.

—— [1982]. *A numerical method for conformal mapping of doubly connected regions.* Rept., Dept. Appl. Math., California Inst. of Tech., Pasadena. §§ 17.4, 17.7.

Forsythe, G. E., and W. Wasow [1960]. *Finite difference methods for partial differential equations.* Wiley, New York. § 15.12.

Frautschi, S. C. [1963]. *Regge poles and S-matrix theory* (*Frontiers in physics*, a lecture note and reprint series). Benjamin, New York. § 14.11.

Fuchs, K., and G. Müller [1971]. Computation of synthetic seismograms with the reflectivity method and comparison with observations. Geophys. J. Roy. Astr. Soc. **23**, 417–433. § 13.7.

Gaier, D. [1959a]. Über ein Extremalproblem der konformen Abbildung. *Math. Z.* **71**, 83–88. § 17.7.

—— [1959b]. Untersuchung zur Durchführung der konformen Abbildung mehrfach zusammenhängender Gebiete. *Arch. Rat. Mech. Anal.* **3**, 149–178. § 17.7.

—— [1964]. *Konstruktive Methoden der konformen Abbildung* (*Springer tracts in natural philosophy*, vol. 3). Springer, Berlin. §§ 16.0, 16.7, 16.8, 17.1, 17.3, 17.4, 17.7, 18.3.

—— [1971]. Ermittlung des konformen Moduls von Vierecken mit Differenzenmethoden. *Numer. Math.* **19**, 179–194. § 16.11.

—— [1974a]. Determination of conformal modules of ring domains and quadrilaterals, in *Functional analysis and applications* (*Lecture notes in mathematics* 399). Springer, pp. 180–188. §§ 16.11, 17.1.

—— [1974b]. Ableitungsfreie Abschätzungen bei trigonometrischer Interpolation und Konjugiertenbestimmung. *Computing* **12**, 145–148. § 13.6.

Gaier, D. [1976]. Integralgleichungen erster Art und konforme Abbildung. *Math. Z.* **147**, 113-129. §§ 15.8, 16.6.

—— [1978]. Konforme Abbildung mehrfach zusammenhängender Gebiete. *Jahresber. Dt. Math.-Ver.* **81**, 25-44. §§ 17.1, 17.7.

—— [1979]. Capacitance and the conformal modulus of quadrilaterals. *J. Math. Anal. Appl.* **70**, 236-239. § 16.11.

—— [1980]. *Vorlesungen über Approximation in Komplexen.* Birkhäuser, Basel. §§ 18.2, 18.3.

—— [1981]. Das logarithmische Potential und die konforme Abbildung mehrfach zusammenhängender Gebiete, in *E. B. Christoffel*, P. L. Butzer and F. Féher, eds. Birkhäuser, Basel, pp. 290-303. § 17.3.

—— [1983]. Numerical methods in conformal mapping, in *Computational aspects of complex analysis*, H. Werner et al., eds. Reidel, Dordrecht, pp. 51-78. § 16.8.

Gander, W., and A. Mazzario [1972]. Numerische Prozeduren I (in memorian Heinz Rutishauser), *Ber. Fachgruppe Comput. Wiss.*, vol. 4, ETH Zürich. § 13.1.

Ganelius, T. [1982]. Degree of rational approximation, in *Lectures on approximation and value distribution* (*Séminaire de mathmatiques supérieures* 79). Les presses de l'université de Montréal. § 18.2.

Gårding, L. [1980]. The Dirichlet problem. *Math. Intelligencer* **2**, 43-53. § 15.0.

Garrick, I. E. [1936]. Potential flow about arbitrary biplane wing sections. NACA Rep. 542. § 17.4.

Gauss, C. F. [1866]. Theoria interpolationis methodo novo tractata, in *Collected works*, vol. 3. pp. 265-327. § 13.1.

Gautschi, W. [1972]. Attenuation factors in practical Fourier analysis. *Numer. Math.* **18**, 373-400. § 13.2.

—— [1981]. A survey of Gauss-Christoffel quadrature formulae, in *E. B. Christoffel*, P. L. Butzer, and F. Féher, eds. Birkhäuser, Basel, pp. 72-134. § 14.6.

Geddes, K. O., and J. C. Mason [1975]. Polynomial approximation by projections on the unit circle. *SIAM J. Numer. Anal.* **12**, 111-120. § 18.3.

Gehring, F. W., W. K. Hayman, and A. Hinkkanen [1982]. Analytic functions satisfying Hölder conditions on the boundary. *J. Approx. Theory* **36**, 243-249. §§ 14.3, 14.5.

Geiger, P. [1981]. Nullstellenbestimmung bei Polynomen und allgemeinen analytischen Funktionen als Anwendung der schnellen Fouriertransformation. Diss. ETH 6759, ADAG Zurich. § 13.4.

Gerhold, D. [1979]. Numerische Lösung eines Kavitationsproblems. *Z. angew. Math. Phys.* **30**, 491-502. § 14.5.

Glassman, J. A. [1970]. A generalization of the fast Fourier transform. *IEEE Trans. Comput.* **C-19**, 105-116. § 13.1.

Goedbloed, J. P. [1981]. Conformal mapping methods in two-dimensional magneto-hydrodynamics. Comput. Phys. Comm. **24**, 311-321. §§ 16.0, 17.4.

Goldstine, H. H. [1977]. *A history of numerical analysis from the 16th century through the 19th century.* Springer, New York. §§ 13.1, 13.6.

Golub, G. H., and C. F. van Loan [1983]. *Matrix computations.* Johns Hopkins Univ. Press, Baltimore MD. §§ 15.12, 16.9, 18.3.

—— and J. H. Welsch [1969]. Calculation of Gaussian quadrature rules. *Math. Comp.* **23**, 221-230. § 16.10.

Goluzin, G. M. [1969]. *Geometric theory of functions of a complex variable.* (*Trans. Math. monographs*, vol. 26). Amer. Math. Soc., Providence, RI. §§ 16.0, 16.4.

Gonzalez, R. C., and P. Wintz [1977]. *Digital image processing.* Addison-Wesley, Reading, MA. § 13.10.

Good, I. J. [1958]. The interaction algorithm and practical Fourier analysis. *J. Roy. Stat. Soc. Ser. B* **20**, 361–372; Addendum, **22**, 372–375. § 13.1.

—— [1969]. Polynomial algebra: An application of the Fast Fourier Transform. *Nature* **222**, 1302. § 13.1.

Graeser, E. [1930]. *Über die konforme Abbildung des allgemeinen zweifach zusammenhängenden schlichten Bereichs auf die Fläche eines Kreisrings.* Diss., Leipzig. § 17.2.

Grassmann, E. [1979]. Numerical experiments with a method of successive approximation for conformal mapping. *Z. angew. Math. Phys.* **30**, 873–884. §§ 16.2, 16.10.

Grunsky, H. [1939]. Koeffizientenbedingungen für schlicht abbildende meromorphe Funktionen. *Math. Z.* **45**, 29–61. § 19.3.

Gutknecht, M. H. [1977]. Existence of a solution of the discrete Theodorsen equation for conformal mappings. *Math. Comp.* **31**, 478–480. § 16.8.

—— [1979]. Fast algorithms for the conjugate periodic function. *Computing* **22**, 79–91. § 13.6.

—— [1981]. Solving Theodorsen's integral equation for conformal maps with the Fast Fourier Transform and various iterative methods. *Numer. Math.* **36**, 405–429. § 16.8.

—— [1983a]. Numerical experiments on solving Theodorsen's integral equation for conformal maps with the Fast Fourier Transform and various iterative methods. *SIAM J. Sci. Stat. Comput.* **4**, 1–30. § 16.8.

—— [1983b]. On the computation of the conjugate trigonometric rational function and on a related splitting problem. *SIAM J. Numer. Anal.* **20**, 1198–1205. §§ 13.5, 13.6.

—— [1985]. Numerical conformaly mapping based on function conjugation. *J. Comp. Appl. Math.*, to appear. §§ 16.8, 16.9.

Hackbusch, W., and U. Trottenberg, eds. [1982]. *Multi-grid methods, proceedings* (Köln-Porz, 1981) (*Lecture notes in mathematics* 960). Springer, Berlin. § 15.12.

Halsey, N. D. [1979]. Potential flow analysis of multielement airfoils using conformal mapping. Paper 79-0271, 17th Aerospace Sciences Meet., New Orleans, LA. §§ 16.8, 17.4, 17.7.

—— [1980a]. *Comparison of the convergence characteristics of two conformal mapping methods.* Preprint, Douglas Aircraft Company, Long Beach, CA. § 16.8.

—— [1980b]. Conformal-mapping analysis of multielement airfoils with boundary layer corrections. Paper AIAA-80-0069, 18th Aerospace Sciences Meet., Pasadena, CA. § 16.8.

Hammerschick, J. [1966]. Über die diskrete Form der Integralgleichungen von Garrick zur konformen Abbildung von Ringgebieten. *Mitt. Math. Sem. Giessen* **70**, Giessen. § 17.4.

Hamming, R. [1962]. *Numerical methods for scientists and engineers.* McGraw-Hill, New York. § 13.3.

Hardy, G. H., and W. W. Rogosinski [1944]. *Fourier series.* Cambridge Univ. Press, Cambridge, § 14.2.

—— and E. M. Wright [1954]. *An introduction to the theory of numbers,* 3rd ed. Oxford Univ. Press, Clarendon. § 13.1.

Harris, D. B., J. H. McClellan, D. S. K. Chan, and H. W. Schuessler [1977]. Vector radix Fast Fourier Transform, in *Proc. IEEE Int. Conf. Acoustics, Speech, Signal Process.* pp. 548–551. § 13.10.

Havin, V. P., S. V. Hruscëv, and N. K. Nikol'skii [1984]. *Linear and complex analysis problem book* (*Lecture notes in mathematics* 1043). Springer, Berlin. § 14.1.

Hayes, J. K., D. K. Kahaner, and R. Keller [1972]. An improved method for numerical conformal mapping. *Math. Comp.* **26**, 327–334. § 16.6.

Heinhold, J. [1947]. Zur Praxis der konformen Abbildung. *Ber. Math.-Tagung Tübingen 1946.* pp. 75–77. § 16.2.

—— [1948]. Ein Schmiegungsverfahren der konformen Abbildung. *Sitz-Ber. Bayer. Akad. Wiss., Math.-Naturw. Kl. 1948.* pp. 203–222. § 16.2.

Heins, A. E. [1956]. The scope and limitations of the method of Wiener and Hopf. *Comm. Pure Appl. Math.* **9**, 447–466. § 14.9.

—— and R. C. McCamy [1957, 1958]. A function-theoretic solution of certain integral equations I, II. *Quart. J. Math. Oxford* **9**, 132–143; **10**, 280–293. § 14.9.

Hellinger, E. [1935]. Hilberts Arbeiten über Integralgleichungen und unendliche Gleichungssysteme, in *David Hilberts gesammelte Abhandlungen*, vol. 3. Springer, Berlin, pp. 94–145. § 14.2.

Henrici, P. [1976]. Einige Anwendungen der schnellen Fouriertransformation. *Int. Ser. Numer. Math.* **32**, 111–124. §§ 13.6, 14.2, 16.8, 17.5.

—— [1979a]. Fast Fourier methods in computational complex analysis. *SIAM Rev.* **21**, 481–527. §§ 13.4, 13.7, 17.4.

—— [1979b]. Barycentric formulas for interpolating trigonometric polynomials and their conjugates. *Numer. Math.* **33**, 225–234. § 13.6.

—— [1980]. A model for the propagation of rounding error in floating arithmetic, in *Interval mathematics 1980*, K. L. E. Nickel, ed. New York, pp. 49–73. § 13.1.

—— [1982a]. *Essentials of numerical analysis.* Wiley, New York. § 13.1.

—— [1982b]. Pointwise error estimates for trigonometric interpolation and conjugation. *J. Comp. Appl. Math.* **8**, 131–132. § 13.6.

—— [1983]. A general theory of osculation algorithms for conformal mapping. *J. Lin. Alg. Appl.* **52/53**, 361–382. § 16.2.

—— [1985]. Poisson's equation in a hypercube: Discrete Fourier methods, eigenfunction expansions, Padé approximation to eigenvalues, in *Studies in numerical analysis*, G. H. Golub, ed. Math. Assoc. America, Washington, DC, pp. 371–411. § 15.13.

Hermite, Ch. [1885]. Sur une identité trigonométrique. *Nouv. Ann. Math. Sér. 3* **4**, 57–59, in *Oeuvres*, vol. 4 (1917). pp. 206–208. § 13.6.

Hersch, J. [1982]. Représentation conforme et symétries: une detérmination élémentaire du module d'un quadrilatère en forme de *L. El. Math.* **37**, 1–5. § 16.11.

—— [1984]. On harmonic measures, conformal moduli, and some elementary symmetry methods. *J. d'Analyse* **42**, 211–228. § 16.11.

——, A. Pfluger, and A. Schopf [1956]. Über ein simultanes Differenzenverfahren zur Abschätzung der Torsionssteifigkeit und der Kapazität nach beiden Seiten. *Z. angew. Math. Phys.* **7**, 89–113. § 15.12.

Hilbert, D. [1905]. Grundzüge einer allgemeinen Theorie der linearen Integralgleichungen, 3. Note. *Nachr. Ges. Wiss. Göttingen*, 307–338. *Grundzüge*, chap. 10. § 14.4.

—— [1912]. *Grundzüge einer allgemeinen Theorie der linearen Integralgleichungen.* Teubner, Leipzig. § 14.2.

Hildebrand, F. [1956]. *Introduction to numerical analysis.* McGraw-Hill, New York. § 13.8.

Hille, E. [1962]. *Analytic function theory*, vol. 2. Ginn, Boston. §§ 16.1, 18.5.

Hockney, R. W. [1965]. A fast direct solution of Poisson's equation using Fourier analysis. *J. ACM* **12**, 95–113. § 15.13.

—— [1970]. The potential calculation and some applications. *Meth. Comp. Phys.* **9**, 135–211. § 15.13.

—— [1972a]. Plasma, gravitational, and vortex simulation, in *Computing as a language of physics.* Int. Atomic Energy Agency, Vienna, pp. 95–107. § 15.13.

—— [1972b]. The solution of Poisson's equation, in *Computing as a language of physics.* Int. Atomic Energy Agency, Vienna, pp. 119–127. § 15.13.

—— and T. R. Brown [1975]. A lambda transition in classical electron film. *J. Phys. C: Solid State Phys.* **8**, 1813–1822. § 15.13.

—— and D. R. K. Brownrigg [1974]. Effect of population II stars and three-dimensional motion on spiral structure. *Mon. Not. R. Astr. Soc.* **167**, 351–357. § 15.13.

—— and S. P. Goel [1975]. Phase transitions of two-dimensional potassium chloride. *Chem. Phys. Lett.* **35**, 500–507. § 15.13.

——, R. A. Warriner, and M. Reiser [1974]. Two-dimensional particle models in semiconductor-device analysis. *Electron. Lett.* **10**, No. 23. § 15.13.

Hoidn, H.-P. [1976]. *Numerische konforme Abbildung zweifach zusammenhängender Gebiete durch Integralgleichungen mit Neumannschmem Kern.* Diploma thesis, ETH Zürich. § 17.3.

—— [1982]. Osculation methods for the conformal mapping of doubly connected regions. *Z. angew. Math. Phys.* **33**, 640–651. § 17.2.

—— [1983]. Die Kollokationsmethode angewandt auf die Symmsche Integralgleichung. Diss. ETH Nr. 7365. ADAG, Zürich. § 16.6.

Houstis, E. N., and T. S. Papatheodorou [1977]. Comparison of fast direct methods for elliptic problems, in *Advances in computer methods for partial differential equations*, vol. 2, R. Vichnevetskyi, ed. IMACS (AICA). § 15.13.

—— and —— [1979]. High order fast elliptic equation solver. *ACM Trans. Math. Software* **5**, 431–441. § 15.13.

Howe, D. [1973]. The application of numerical methods to the conformal transformation of polygonal boundaries. *J. Inst. Math. Appl.* **12**, 125–136. § 16.10.

Hsiao, G. C., P. Kopp, and W. L. Wendland [1980]. A Galerkin collocation method for some integral equations of the first kind. *Computing* **25**, 89–130. § 16.6.

Hübner, O. [1964a]. Das Ausschöpfungsverfahren zur konformen Abbildung eines einfach zusammenhängenden Gebietes auf die obere Halbebene. *Z. angew. Math. Mech.* **44**, T34–T36. § 16.2.

—— [1964b]. Funktionentheoretische Iterationsverahren zur konformen Abbildung ein- und mehrfach zusammenhängender Gebiete. *Mitt. Math. Sem. Giessen.* **63**. § 16.2.

—— [1982]. Über die Anzahl der Lösungen der diskreten Theodorsen-Gleichung. *Numer. Math.* **39**, 195–204. § 16.8.

—— [1985]. The Newton method for solving the Theodorsen integral equation, to appear. § 16.8.

Hurwitz, A. [1929]. *Vorlesungen über allgemeine Funktionentheorie und elliptische Funktionen.* 3rd ed., Herausgegeben und ergänzt durch einen Abschnitt über geometrische Funktionentheorie von R. Courant. Springer, Berlin. § 16.1.

Ives, D. C. [1976]. A modern look at conformal mapping including multiply connected regions. *AIAA J.* **14**, 1006–1011. §§ 14.2, 16.8, 17.4, 17.5.

—— [1982]. Conformal grid generation. *Appl. Math. Comp.* **10/11**, 107–135. § 16.0.

Jackson, J. D. [1975]. *Classical electrodynamics*, 2nd ed. Wiley, New York. § 14.11.

James, R. M. [1971]. *A new look at two-dimensional incompressible airfoil theory.* Rep. MDC-J0918/01, Douglas Aircraft Company. § 16.8.

Jänich, K. [1977]. *Einführung in die Funktionentheorie.* Springer, Berlin. §§ 15.10, 16.1.

Jaswon, M. A., and G. T. Symm [1977]. *Integral equation methods in potential theory and elastostatics.* Academic, London. § 15.9.

Jeltsch, R. [1969]. Numerische konforme Abbildung mit Hilfe der Formel von Cisotti. Diploma thesis, ETH Zürich. § 16.8.

John, F. [1978]. *Partial differential equations*, 3rd ed. Springer, New York. § 15.11.

Joyner, W. B., and A. T. F. Chen [1975]. Calculation of nonlinear ground response in earthquakes. Bull. Seismol. Soc. Amer. **65**, 1315–1336. § 13.7.

——, R. E. Warrick, and A. A. Oliver, III [1976]. Analysis of seismograms from a downhole array in sediments near San Francisco Bay. *Bull. Seismol. Soc. Amer.* **66**, 937–958. § 13.7.

Julia, G. [1931]. *Leçons sur la représentation conforme des aires simplement connexes.* Gauthiers-Villars, Paris. § 16.0, 16.2.

Kahaner, D. K. [1970]. Matrix description of the Fast Fourier Transform. *IEEE Trans. Audio Electroacoust.* **AU-18**, 442–450. § 13.1.

—— [1978]. The Fast Fourier Transform by polynomial evaluation. *Z. angew. Math. Phys.* **29**, 387–394. § 13.1.

Kaiser, A. [1975]. *Bestimmung konformer Abbildungsfunktionen durch Lösung eines Randwertproblems.* Diploma thesis, ETH Zürich. § 16.5.

Källén, G. [1964]. *Elementary particle physics.* Addison-Wesley, Reading, MA. § 14.11.

Karatsuba, A., and Y. Ofman [1962]. Multiplication of multidigit numbers on automata. *Dokl. Akad. Nauk SSSR* **145**, 293–294. § 13.8.

Katznelson, Y. [1968]. *An introduction to harmonic analysis.* Wiley, New York. §§ 13.3, 14.2.

Kerzman, N., and E. M. Stein [1978]. The Cauchy kernel, the Szegö kernel, and the Riemann mapping function. *Math. Ann.* **236**, 85–93. § 18.4.

—— and M. Trummer [1985]. Numerical conformal mapping via the Szegö kernel. *J. Comp. Appl. Math.*, to appear. §§ 16.10, 18.4.

Koebe, P. [1910]. Über die konforme Abbildung mehrfach-zusammenhängender Bereiche. *Jahresber. Dt. Math. Ver.* **19**, 339–348. § 17.7.

—— [1912]. Über eine neue Methode der konformen Abbildung und Uniformisierung. *Nachr. Kgl. Ges. Wiss. Göttingen, Math.-Phys. Kl. 1912.* pp. 844–848. § 16.2.

—— [1920]. Abhandlungen zur Theorie der konformen Abbildung VI. Abbildung mehrfach zusammenhängender Bereiche auf Kreisbereiche. *Math. Z.* **7**, 235–301. § 17.7.

Komatu, Y. [1945]. Darstellung der in einem Kreisringe analytischen Funktionen nebst den Anwendungen auf konforme Abbildung über Polygonalringgebiete. *Japan J. Math.* 203-215. § 17.5.

—— [1949]. Ein alternierendes Approximationsverfahren für konforme Abbildung von einem Ringgebiet auf einen Kreisring. *Proc. Japan Acad.* **21**, 146–158. § 17.2.

Koppenfels, W., and F. Stallmann [1959]. *Praxis der konformen Abbildung.* Springer, Berlin. § 17.5.

Kövari, T., and Ch. Pommerenke [1967]. On Faber polynomials and Faber expansions. *Math. Z.* **99**, 193–206. § 18.2.

Kral, J. [1980]. *Integral operators in potential theory* (*Lecture notes in mathematics*, 823). Springer, Berlin, §§ 15.7, 15.8.

Kress, R., and E. Martensen [1970]. Anwendung der Rechteckregel auf die relle Hilberttransformation mit unendlichem Intervall. *Z. angew. Math. Mech.* **50**, T61–T64. § 14.11.

Ku, C. C., W. M. Telford, and H. S. Lim [1971]. The use of linear filtering in gravity problems. *Geophysics* **36**, 1174–1203. § 13.7.

Kung, H. T. [1974]. On computing reciprocals of power series. *Numer. Math.* **22**, 341-348. § 13.9.

Lamp, U., K.-T. Schleicher, and W. L. Wendland [1984]. The fast Fourier transform and the numerical solution of one-dimensional integral equations. *Numer. Math.*, to appear. § 16.6.

Landau, E., and G. Valiron [1929]. A deduction from Schwarz's lemma. *J. London Math. Soc.* **4**, 162-163. § 16.4.

Laura, P. A. A. [1975]. A survey of modern applications of conformal mapping. *Rev. Unión Matemat. Argent.* **27**, 167-179. § 16.0.

Laurie, D. P. [1980]. Efficient implementation of Wilson's algorithm for factorizing a self-reciprocal polynomial. *BIT* **20**, 257-259. § 13.4.

Lavrentiev, M. A., and B. V. Shabat [1967]. *Methoden der komplexen Funktionentheorie.* VEB, Berlin, §§ 14.1, 14.4, 14.5, 15.3, 15.4, 15.7.

Leuthold, P. E. [1974]. Die Bedeutung der Hilberttransformation in der Nachrichtentechnik. *Scientia Elec.* **20**, 127-157. § 14.12.

Levin, D., N. Papamichael, and A. Sideridis [1978]. The Bergman kernel method for the numerical conformal mapping of simply connected regions. *J. Inst. Math. Appl.* **22**, 171-178. § 18.3.

Levinson, N. [1949]. The Wiener RMS error criterion in filter design and prediction, in N. Wiener, *Extrapolation, interpolation, and smoothing stationary time series with engineering applications.* Wiley, New York, app. B. § 13.4.

Lichtenstein, L. [1911]. Über die konforme Abbildung ebener analytischer Gebiete mit Ecken. *J. f. Math.* **140**, 100-119. § 16.4.

Lienhard, J. H. [1981]. Heat conduction through "Yin-Yang" bodies. *J. Heat Transfer* **103**, 299-300. § 16.11.

Lindelöf, E. [1916]. Sur la représentation conforme d'une aire simplement connexe sur l'aire d'un cercle, in *Compl. Rend 4^a Congr. de Math. Scand. à Stockholm.* pp. 55-90. § 16.3.

Littlewood, J. E. [1944]. *Lectures on the theory of functions.* Oxford Univ. Press, Oxford. § 16.4.

Löffler, H. [1983]. *Konstruktion von Green-Funktionen für mehrfach zusammenhängende Gebiete im* $\mathbf{R}^2$. Diploma thesis, TH Darmstadt. § 17.6.

Longman, I. M. [1958]. On the numerical evaluation of Cauchy principal value of integrals. *Math. Comp.* **12**, 205-207. § 14.6.

Lundwall-Skaar, C. [1975]. *Konforme Abbildung mit Fast Fourier Transformation.* Diploma thesis, ETH Zürich. §§ 16.8, 17.5.

Lynch, R. E., J. R. Rice, and D. H. Thomas [1964]. Direct solution of partial difference equations by tensor product methods. *Numer. Math.* **6**, 185-199. § 15.13.

Lyness, J. N., and C. Moler [1967]. Numerical differentiation of analytic functions. *SIAM J. Numer. Anal.* **4**, 202-210. § 13.4.

—— and G. Sandee [1971]. ENTCAF and ENTCRE: Evaluation of normalized Taylor coefficients of an analytic function. Algorithm 413. *Comm. ACM* **14**, 669-675. § 13.4.

Marti, J. [1978]. An algorithm recursively computing the exact Fourier coefficients of B-splines with nonequidistant knots. *Z. angew. Math. Phys.* **29**, 301-305. § 13.2.

—— [1980]. On the convergence of an algorithm for computing minimum-norm solutions of ill-posed problems. *Math. Comp.* **34**, 521-527. § 16.6.

Mayo, A. [1985]. Rapid method for the conformal mapping of multiply connected regions in the plane. *J. Comp. Appl. Math.*, to appear. § 17.7.

McClellan, J. H., and T. W. Parks [1972]. Eigenvalue and eigenvector decomposition of the discrete Fourier transform. *IEEE Trans. Audio Electroacoust.* **AU-20**, 66-74. § 13.1.

Meinardus, G. [1978]. Schnelle Fourier-Transformation (*International Ser. Num. Math.* vol. 42) pp. 192-203. Birkhäuser, Basel. § 13.1.

Meister, E. [1983]. *Randwertaufgaben der Funktionentheorie.* Teubner, Stuttgart. § 14.0.

Menikoff, R., and C. Zemach [1980]. Methods for numerical conformal mapping. *J. Comput. Phys.* **36**, 366-410. §§ 16.0, 17.3.

Merz, G. [1983]. Fast Fourier transform algorithms with applications, in *Computational aspects of complex analysis*, H. Werner et al., eds. Reidel, Dordrecht, pp. 249-278. § 13.1.

Meschkowski, H. [1962]. *Hilbertsche Räume mit Kernfunktion* (*Grundlehren*, vol. 113). Springer, Berlin. § 18.4.

Milin, I. M. [1971]. *Univalent functions and orthonormal systems.* Izdat. Nauka, Moscow. § 19.2.

Milne, W. E. [1953]. *The numerical solution of differential equations.* Wiley, New York. § 15.12.

Morrey, C. B. [1966]. *Multiple integrals in the calculus of variations.* (*Grundlehren*, vol. 130). Springer, Berlin. § 14.11.

Müller, J. [1984]. Das Verfahren von Kerzman und Trummer zur konformen Abbildung einfach zusammenhängender Gebiete. Diploma thesis, ETH Zürich. § 18.4.

Muschelishvili, N. I. [1965]. *Singuläre Integralgleichungen.* Akademie-Verlag, Berlin. This is a German translation of the second Russian edition. There exists an English translation of the first Russian edition: *Singular integral equations.* Noordhoff, Groningen, 1953. ch. 14.

NASA [1980]. *Numerical grid generation techniques.* Publ. 2166. § 16.0.

Nehari, Z. [1952]. *Conformal mapping.* Dover, New York. chap. 16. §§ 18.4, 19.1.

Neumann, C. [1877]. *Untersuchungen über das logarithmische und Newtonsche Potential.* Teubner, Leipzig. § 15.9.

Nevanlinna, R. [1936]. *Eindeutige analytische Funktionen* (*Grundlehren*, vol. 162). Springer, Berlin. § 15.5.

Nickel, K. [1951]. Lösung eines Integralgleichungssystems aus der Tragflügeltheorie. *Math. Z.* **54**, 81-96. § 14.9.

—— [1953]. Lösung von zwei verwandten Integralgleichungssystemen. *Math. Z.* **58**, 49-62. § 14.9.

Niethammer, W. [1966]. Iterationsverfahren bei der konformen Abbildung. *Computing* **1**, 146-153. § 16.8.

—— and R. S. Varga [1983]. The analysis of k-step iterative methods for linear systems from summability theory. *Numer. Math.* **41**, 177-206. § 18.4.

Noble, B. [1958]. *Methods based on the Wiener-Hopf technique.* Pergamon, New York. § 14.9.

Olver, F. W. J. [1983]. Error analysis of complex arithmetic, in *Computational aspects of complex analysis*, H. Werner et al., eds. Reidel, Dordrecht, pp. 279-292. § 13.1.

Opfer, G. [1967]. *Untere, beliebig verbesserbare Schranken für den Modul eines zweifach zusammenhängenden Gebietes mit Hilfe von Differenzenverfahren.* Diss., Univ. Hamburg. §§ 17.1, 17.3.

—— [1968]. Angenäherte Bestimmung des Moduls eines zweifach zusammenhängenden Gebietes mit Differenzenverfahren. *Z. angew. Math. Mech.* **48**, T96-T97. § 17.1.

—— [1969]. Die Bestimmung des Moduls zweifach zusammenhängender Gebiete mit Hilfe von Differenzenverfahren. *Arch. Rat. Mech. Anal.* **32**, 281-297. §§ 17.1, 17.3

—— [1974a]. Konforme Abbildungen und ihre numerische Behandlung, in *Überblicke Mathematik*, vol. 7, G. Laugwitz, ed. pp. 34-113. chap. 16.

——[1974b]. Eine Bemerkung zur Gewinnung diskret-harmonischer Funktionen mit funktionentheoretischen Mitteln (*Lecture notes in mathematics* 395). pp. 215-222. § 17.1.

Oppenheim, A. V., and R. W. Schafer [1975]. *Digital signal processing.* Prentice-Hall, Englewood Cliffs, NJ, chap. 13. § 14.11.

Osgood, W. F., and E. H. Taylor [1913]. Conformal transformations and the boundaries of their regions of definition. *Trans. Amer. Math. Soc.* **14**, 277-298. § 16.3.

Ostrowski, A. [1929]. Mathematische Miszellen XV: Zur konformen Abbildung einfach zusammenhängender Gebiete. *Jahresber. Dt. Math.-Ver.* **29**, 168-182. § 16.2.

—— [1955]. Conformal mapping of a special ellipse on the unit circle. *Appl. Math. Ser.* **42**, 1-2. § 16.5.

Paget, D. F., and D. Elliott [1972]. An algorithm for the numerical evaluation of certain Cauchy principal value integrals. *Numer. Math.* **19**, 373-385. § 14.6.

Papamichael, N., and D. M. Hough [1984]. The determination of the poles of the mapping function and their use in numerical conformal mapping. *J. Comp. Appl. Math.* **9**, 155-160. § 18.3.

—— and C. A. Kokkinos [1984]. The use of singular functions for the approximate conformal mapping of doubly-connected domains. *SIAM J. Sci. Stat. Comput.*, to appear. § 18.3.

—— and M. K. Warby [1983]. Pole type singularities and the numerical conformal mapping of doubly-connected domains. TR/02/83, Dept. of Math. and Stat., Brunel Univ. § 18.3.

Peschl, E. [1932]. Über die Krümmung der Niveaukurven bei der konformen Abbildung einfachzusammenhängender Gebiete auf das Innere eines Kreises. Eine Verallgemeinerung eines Satzes von E. Study. *Math. Ann.* **106**, 574-594. § 15.10.

Peters, A. S. [1969]. Some integral equations related to Abel's equation and the Hilbert transform. *Comm. Pure Appl. Math.* **22**, 539-560. § 14.9.

Petrovsky, I. G. [1954]. *Lectures on partial differential equations.* Interscience, New York. § 15.11.

Pfaltzgraff, J. A. [1985]. Univalence constraints on the Schwarz-Christoffel parameters. SIAM J. Math. Anal., to appear. § 19.1.

Pfluger, A. [1976]. On a uniqueness theorem in conformal mapping. *Mich. Math. J.* **23**, 363-365. § 19.1.

—— [1981]. Die Bedeutung der Arbeiten Christoffels für die Funktionentheorie, in *E. B. Christoffel*, P. L. Butzer and L. Féher, eds. Birkhäuser, Basel, pp. 244-252. § 16.10.

Pickering, W. M. [1977]. Some comments on the solution of Poisson's equation using Bickley's formula and Fast Fourier Transforms. *J. Inst. Math. Appl.* **19**, 337-338. § 15.13.

Pines, D. [1963]. *Elementary excitations in solids.* Benjamin, New York. § 14.11.

Plutchok, R., and P. Broome [1969]. Modeling of seismic signals from large underwater explosions to predict the spectra and covariance functions. *Bull. Seismol. Soc. Amer.* **59**, 1137-1147. § 13.7.

Polya, G., and G. Szegö [1925]. *Aufgaben und Lehrsätze der Analysis*, 2 vols. Springer, Berlin §§ 16.6, 19.1

Pommerenke, Ch. [1964]. Über die Faberschen Polynome schlichter Funktionen. *Math. Z.* **85**, 197-208. § 18.2.

—— [1975]. *Univalent functions.* Vandenhoeck & Ruprecht, Göttingen. § 18.1.

Pompeiu, M. D. [1912]. Sur une classe de fonctions d'une variable complexe. *Rendiconti Circ. Mat. Palermo* **33**, 108-113. § 15.10.

—— [1913]. Sur une classe de fonctions d'une variable complexe et sur certaines équations intégrales. *Rendiconti Circ. Mat. Palermo* **35**, 277-281. § 15.10.

Powell, M. J. D. [1968]. *A Fortran subroutine for solving systems of non-linear algebraic equations.* Tech. Rep. AERE-R- 5947, Harwell, England. § 16.10.

Price, T. E. [1984]. Pointwise error estimates for interpolation. Preprint, Kent State Univ. OH. § 13.6.

Proskurowski, W., and O. Widlund [1976]. On the numerical solution of Helmholtz's equation by the capacitance matrix method. *Math. Comp.* **30**, 433–468. § 15.13.

Rabiner, L. R., and B. Gold [1975]. *Theory and applications of digital signal processing.* Prentice-Hall, Englewood Cliffs, NJ. chap. 13.

Rabinowitz, P. [1978]. The numerical evaluation of Cauchy principal value integrals, in *Proc. Symp. Numerical Mathematics*, G. Joubert, ed. Univ. of Natal, Durban, pp. 53–82. § 14.6.

Ramos, G. U. [1971]. Roundoff error analysis of the Fast Fourier Transform. *Math. Comp.* **25**, 757–768. § 13.8.

Reichel, L. [1985]. A fast method for solving certain integral equations. *J. Comp. Appl. Math.*, to appear. § 16.6.

Reppe, K. [1979]. Berechnung von Magnetfeldern mit Hilfe der konformen Abbildung durch numerische Integration der Abbildungsfunktion von Schwarz-Christoffel. *Siemens Forsch-Entwicklungsber.* **8**, 190–195. § 16.10.

Richter, G. R. [1977]. Numerical solution of Laplace's equation as an integral equation of the first kind. *Math. Comput. in Simulation* **19**, 192–197. §§ 15.9, 16.6.

Riemann, B. [1953]. *Collected works.* Dover, New York. § 14.5.

Rissanen, J. [1973]. Algorithms for triangular decomposition of block Hankel and Toeplitz matrices with application to factoring positive matrix polynomials. *Math. Comp.* **27**, 147–154. § 13.4.

Ritzmann, P. M. [1984]. Ein numerischer Algorithmus zur Komposition von Potenzreihen und Komplexitätsschranken für die Nullstellenbestimmung bei Polynomen. Diss., Univ. Zurich. § 13.9.

Rosenfeld, A., and A. C. Kak [1976]. *Digital picture processing.* Academic, New York. § 13.10.

Rosser, J. B. [1974]. *Fourier series in the computer age.* MRC Rep. 1401 AD, Math. Res. Center, Madison, WI. § 15.13.

Runge, C., and H. König [1924]. *Vorlesungen über numerisches Rechnen.* Springer, Berlin. § 13.1.

Salzer, H. E. [1948]. Coefficients for facilitating trigonometric interpolation. *J. Math. Phys.* **27**, 274–278. § 13.6.

Schauer, U. [1963]. *Über das Verfahren von Theodorsen für einfach und zweifach zusammenhängende Gebiete.* Studienarbeit, TH Stuttgart. § 17.4.

Schiffer, M. [1948]. Faber polynomials in the theory of univalent functions. *Bull. Amer. Math. Soc.* **54**, 503–517. § 18.1.

—— and N. S. Hawley [1962]. Connections and conformal mapping. *Acta Math.* **107**, 175–274. § 17.6.

Schleiff, M. [1968]. Über Näherungsverfahren zur Lösung einer singulären linearen Integralgeleichung. *Z. angew. Math. Mech.* **48**, 477–483. § 16.7.

Schmidt, H., and G. P. Meyer [1977]. Zur Existenz und analytischen Darstellung von Nullstellenfolgen bei Exponentialsummen. *Bayr. Akad. Wiss., Math.-Naturw. Kl 1976.* pp. 125–133. § 14.10.

Schober, G. [1967]. A constructive method for the conformal mapping of domains with corners. *J. Math. Mech.* **16**, 1095–1116. § 16.7.

Schönhage, A., and V. Strassen [1971]. Schnelle Multiplikation grosser Zahlen. *Computing* **7**, 281–292. § 13.8.

Schröder, J., U. Trottenberg, and K. Witsch [1978]. On fast Poisson solvers and applications *Lecture notes in mathematics* 631, pp. 153-187. § 15.13.

Schur, I. [1945]. On Faber polynomials. *Amer. J. Math.* **67**, 33-41. § 18.1.

Schwarz, H. A. [1890]. Zur Integration der partiellen Differentiagleichung . . . *J. reine angew. Math.* **74**, 218-253, in *Gesammelte Abhandlungen*, vol. 2. Springer, Berlin, pp. 175-210. § 14.2.

Schwarz, H.-R. [1977]. Elementare Darstellung der schnellen Fouriertransformation. *Computing* **19**, 107-116. § 13.1.

—— [1978]. The Fast Fourier Transform for general order. *Computing* **19**, 341-350. § 13.1.

—— [1984]. *Methode der finiten Elemente*, 2nd ed. Teubner, Stuttgart. § 15.12.

Seewald, W. [1985]. *Die effiziente Lösung der Helmholtzgleichung auf dem Rechteck und der Kreisscheibe.* Diss. No. 7716, Swiss Federal Institute of Technology, Zurich.

Segel, L. A. [1961a]. Application of conformal mapping to boundary perturbation problems for the membrane equation. *Arch. Rat. Mech. Anal.* **8**, 228-237. § 16.0.

—— [1961b]. Application of conformal mapping to viscous flow between moving circular cylinders. *Quart. Appl. Math.* **18**, 335-353. § 16.0.

Seidel, W. [1952]. Bibliography of numerical methods in conformal mapping. (*Applied mathematics series* 18), pp. 269-280. Natl. Bureau Standards. § 16.0.

Sieveking, M. [1972]. An algorithm for the division of powerseries. *Computing* **10**, 153-156. § 13.9.

Siewert, C. E., and E. E. Burniston [1972a]. An exact analytical solution of Kepler's equation. *Celest. Mech.* **6**, 294-304. § 14.10.

—— and —— [1972b]. An exact closed-form result for the discrete eigenvalue in studies of polarized flight. *Astrophys. J.* **173**, 405-406. § 14.10.

—— and —— [1974]. Solutions of the equation . . . . *J. Math. Anal. Appl.* **46**, 329-337. § 14.10.

—— and C. J. Essig [1973]. An exact solution of a molecular field equation in the theory of ferromagnetism. *Z. angew. Math. Phys.* **24**, 281-286. § 14.10.

—— and J. S. Phelps [1979]. On solutions of a transcendental equation basic to the theory of vibrating plates. *SIAM J. Math. Anal.* **10**, 105-111. § 14.10.

Singleton, R. C. [1968]. Algorithm 339. An Algol procedure for the Fast Fourier transform with arbitrary factors. *Comm. ACM* **11**, 776. § 13.1.

Sköllermo, G. [1975]. A Fourier method for the numerical solution of Poisson's equation. *Math. Comp.* **29**, 697-711. § 15.13.

Smith, J. O. III [1983]. *Techniques of digital filter design and system identification with application to the violin.* Diss., Stanford Univ. § 13.4.

Söhngen, H. [1939]. Die Lösung der Integralgleichung . . . in der Tragflügeltheorie. *Math. Z.* **45**, 245-264. § 14.9.

Steger, J. L., and D. S. Chaussee [1980]. Generation of body-fitted coordinates using hyperbolic partial differential equations. *SIAM J. Sci. Stat. Comput.* **1**, 431-437. § 16.0.

Stiefel, E. [1956]. On solving Fredholm integral equations. Applications to conformal mapping and variational problems of potential theory. *SIAM J. Appl. Math.* **4**, 63-85. § 16.7.

Strang, G., and G. I. Fix [1973]. *An analysis of the finite element method.* Prentice-Hall, Englewood Cliffs, NJ. § 15.12.

Swarztrauber, P. N. [1977]. The methods of cyclic reduction, Fourier analysis and the FACR algorithm for the discrete solution of Poisson's equation on a rectangle. *SIAM Rev.* **19**, 490-501. § 15.13.

Swarztrauber, P. N., and R. A. Sweet [1973]. The direct solution of the discrete Poisson equation on a disk. *SIAM J. Numer. Anal.* **10**, 900–907. § 15.13.

Sweet, R. A. [1977]. A cyclic reduction algorithm for solving block tridiagonal systems of arbitrary dimension *SIAM J. Numer. Anal.* **14**, 706–720. § 15.13.

Symm, G. T. [1966]. An integral equation method in conformal mapping. *Numer. Math.* **9**, 250–258. § 16.6.

—— [1967]. Numerical mapping of exterior domains. *Numer. Math.* **10**, 437–445. § 16.6.

—— [1969]. Conformal mapping of doubly-connected domains. *Numer. Math.* **13**, 448–457. § 17.3.

Szegö, G. [1921]. Über orthogonale Polynome, die zu einer gegebenen Kurve der komplexen Ebene gehören. *Math. Z.* **9**, 218–270. § 18.4.

—— [1950]. Conformal mapping of the interior of an ellipse onto a circle. *Amer. Math. Month.* **57**, 474–478. § 16.5.

—— [1959]. *Orthogonal polynomials.* Amer. Math. Soc. Coll. Publ., New York. § 14.6.

Talbot, A. [1979]. The accurate numerical inversion of Laplace transforms. *J. Inst. Math. Appl.* **23**, 97–120. § 13.4.

Teichmüller, O. [1939]. Extremale quasikonforme Abbildungen und quadratische Differentiale. *Abh. Preuss. Akad. Wiss., Math.-Naturw. Kl.* **22**, 1–197. § 15.6.

Temperton, C. [1983]. Self-sorting mixed-radix Fast Fourier Transforms. *J. Comp. Phys.* **52**, 1–23. § 13.10.

Thacker, W. C. [1980]. A brief review of techniques for generating irregular computational grids. *Int. J. Numer. Meth. in Eng.* **15**, 1335–1341. § 16.0.

Theilheimer, F. [1969]. A matrix version of the Fast Fourier Transform. *IEEE Trans. Audio Electroacoust.* **17**, 158–161. § 13.1.

Thompson, J. F., ed. [1982]. Numerical grid generation. Special double vol., *Appl. Math. Comp.* **10/11**. North-Holland, Amsterdam. § 16.0.

Timman, R. [1951]. *The direct and inverse problem of airfoil theory. A method to obtain numerical solutions.* Nat. Luchtv. Labor, Amsterdam, Rep. F 16.

Titchmarsh, E. C. [1937]. *Introduction to the theory of Fourier integrals.* Oxford Univ. Press, Oxford. § 14.11.

—— [1939]. *The theory of functions*, 2nd ed. Univ. Press, Oxford. § 16.1.

Todd, J., ed. [1955]. Experiments in the computation of conformal maps. *Appl. Math. Ser.* **42**. § 16.0.

Trefethen, L. N. [1980]. Numerical computation of the Schwarz-Christoffel transformation. *SIAM J. Sci. Stat. Comput.* **1**, 82–102. § 16.10.

—— [1982]. *SCPACK version 2 user's guide.* Internal Rep. 24, Inst. Computer Appl. in Sci. and Eng., NASA Langley Res. Center, Hampton, VA. § 16.10.

—— [1984]. Analysis and design of polygonal resistors by conformal mapping. *Z. angew. Math. Phys.* **35**, 692–704. §§ 16.10, 16.11.

Tsuji, M. [1959]. *Potential theory in modern function theory.* Maruzen, Tokyo. ch. 15.

Tuel, W. G., Jr. [1968]. Computer algorithm for spectral factorization of rational matrices. *IBM J. Res. Develop.* **12**, 163–170. § 13.4.

Tukey, J. W. [1967]. An introduction to the calculation of numerical spectrum analysis, in *The spectral analysis of time series*, B. Harris, ed. Wiley, New York. § 13.7.

Uhrich, M. L. [1969]. Fast Fourier Transforms without sorting. *IEEE Trans. Audio Electroacoust.* **17**, 170–172. § 13.1.

Unbehauen, R. [1972]. Determination of the transfer function of a digital filter from the real part of the frequency response. *Arch. Elect. Übertr.* **26**, 551–557. § 14.12.

Varga, R. S. [1962]. *Matrix iterative analysis.* Prentice-Hall, Englewood Cliffs, NJ. § 15.12.

Vecheslavov, V. V., and V. I. Kokoulin [1974]. Determination of the parameters of the conformal mapping of simply connected polygonal regions. USSR Comput. Math. and Math. Phys. **13**, 57–65. § 16.10.

Vekua, I. N. [1962]. *Generalized analytic functions.* Pergamon, Oxford. § 15.10.

Villat, H. [1912]. Le problème de Dirichlet dans une aire annulaire. *Rendiconti Circ. Mat. Palermo* **33**, 134–175. § 15.2.

Vostry, Z. [1975]. New algorithm for polynomial spectral factorization with quadratic convergence I. *Kybernetika* **11**, 415–422. § 13.4.

Walsh, J. L. [1935]. *Interpolation and approximation by rational functions in the complex plane.* Amer. Math. Soc., Providence, RI. §§ 18.3, 18.4.

Warschawski, S. [1932]. Über das Randverhalten der Ableitung der Abbildungsfunktion bei konformer Abbildung. *Math. Z.* **35**, 322–456. § 16.3.

—— [1945]. On Theodorsen's method of conformal mapping of nearly circular regions. *Quart. J. Appl. Math.* **3**, 12–28. § 16.8.

—— [1950]. On conformal mapping of nearly circular regions. *Proc. Amer. Math. Soc.* **1**, 562–574. § 16.8.

—— [1955]. On a problem of L. Lichtenstein. *Pac. J. Math.* **5**, 835–839. § 16.4.

Watson, G. N. [1925]. A note on generalized hypergeometric series. *Proc. London Math. Soc.* (2) **23**, 13–15. § 19.6.

Weber, H. [1870]. Über die Integration der partiellen Differentialgleichung .... *Math. Ann.* **1**, 1–36. § 15.12.

Weeks, W. T. [1966]. Numerical inversion of the Laplace transform. *J. Assoc. Comp. Mach.* **13**, 419–429. § 13.4.

Wegmann, R. [1978]. Ein Iterationsverfahren zur konformen Abbildung. *Numer. Math.* **30**, 453–466. § 16.9.

—— [1984]. Convergence proofs and error estimates for an iterative method for conformal mapping. *Numer. Math.* **44**, 435–461. § 16.9.

—— [1985]. An iterative method for the conformal mapping of doubly connected regions. *J. Appl. Comp. Math.*, to appear. § 17.7.

Weissinger, J. [1941]. Ein Satz über Fourierreihen und seine Anwendung in der Tragflügeltheorie. *Math. Z.* **47**, 16–33. § 14.9.

Wendland, W. L. [1980]. On Galerkin collocation methods for integral equations of elliptic boundary value problems, in *Numerical treatment of integral equations*, J. Albrecht and L. Collatz, eds. Birkhäuser, Basel, pp. 244–275. § 16.6.

—— [1983]. Boundary element methods and their asymptotic convergence, in *Theoretical acoustics and numerical techniques*, P. Filippi, ed. Springer, Wien, pp. 135–216. § 16.6.

Wermer, J. [1974]. *Potential theory* (*Lecture notes in mathematics, vol. 408*). Springer, Berlin, Ch. 15.

Whiteside, D. T., ed. [1968]. *The mathematical papers of Isaac Newton, vol.* 2. Cambridge Univ. Press, Cambridge. § 13.9.

Widlund, O. [1982]. *On a numerical method for conformal mapping due to Fornberg.* Courant Institute of Math. Sci., New York. § 16.9.

Wiener, N., and E. Hopf [1931]. Über eine Klasse singulärer Integralgleichungen. *Sitz.-Ber. Preuss. Akad. Wiss., Phys.-Math. Kl.* **30–32**, 696–706. § 14.9.

Wilson, G. [1969]. Factorization of the covariance generating function of a pure moving average process. *SIAM J. Numer. Anal.* **6**, 1–7. § 13.4.

—— [1972]. The factorization of matricial spectral densities. *SIAM J. Appl. Math.* **23**, 420–426. § 13.4.

Wing, O. [1967]. An efficient method for the numerical inversion of Laplace transforms. *Computing* **2**, 11–17. § 13.4.

Winograd, S. [1976]. On computing the discrete Fourier transform. *Proc. Nat. Acad. Sci. US* **73**, 1005–1006. § 13.7.

—— [1978]. On computing the discrete Fourier transform. *Math. Comp.* **32**, 175–199. § 13.7.

Wirtinger, W. [1927]. Zur formalen Theorie der Funktionen von mehrenen komplexen Veränderlichen. *Math. Ann.* **97**, 357–375. § 15.10.

Woods, L. C. [1961]. *The theory of supersonic plane flow.* Cambridge Univ. Press, Cambridge. § 16.8.

Wright, E. M. [1961]. Stability criteria and the roots of a transcendental equation. *J. Soc. Ind. Appl. Math.* **9**, 136–148. § 14.10.

Yoshikawa, H. [1960]. On the conformal mapping of nearly circular domains. *J. Math. Soc. Japan* **12**, 174–186. § 16.8.

Young, D. M. [1971]. *Iterative solution of large linear systems.* Academic, New York. § 15.12.

Zygmund, A. [1959]. *Trigonometric series*, vols. 1 and 2, 2nd ed. Cambridge Univ. Press, Cambridge. § 13.5.

INDEX

Principal references (referring to definitions) are printed in **boldface.**